EVALUATION GUIDE FOR REFRESHER MATHEMATICS

REFRESHER MATHEMATICS is a comprehensive introductory or remedial textbook in general mathematics with emphasis on arithmetic and consumer applications.

Specially designed features help students identify areas of weakness and then direct them to specific practice exercises to achieve mastery. Other features prepare students for applying their math skills to important real-life experiences.

To review examples of these features please turn to the following pages.

(pp. 10–13, 15)   INVENTORY TESTS diagnose general difficulties and weaknesses and are keyed to the *Exercises* to provide individual assignments.

(pp. 43, 128, 244)   DIAGNOSTIC TESTS pinpoint math weaknesses and are keyed to specific examples within each *Exercise*.

(pp. 87, 339, 543)   RELATED PRACTICE EXAMPLES and PRACTICE PROBLEMS provide ample opportunity for skill mastery. Problems are graded by difficulty.

(pp. 154, 180, 241)   PRACTICAL APPLICATIONS help students to explore how math is a part of their everyday world.

(pp. 16–17, 38, 56, 107)   PROBLEM SOLVING helps students develop their problem-solving techniques.

(pp. 109, 145, 463)   REFRESH YOUR SKILLS and UNIT REVIEW reinforce and maintain key math skills learned throughout the text.

(pp. 70, 189, 233)   ESTIMATING ANSWERS help students develop logical thinking strategies and important mental arithmetic skills.

(pp. 71, 190, 234)   CHECK BY CALCULATOR develops the skilled and accurate use of hand held calculators.

(pp. 125, 249, 330)   COMPETENCY CHECK TESTS give students the experience of taking multiple choice tests.

(pp. 126, 236, 279)   KEYED ACHIEVEMENT TESTS check mastery of each major part of the text and are keyed to the *Exercises*.

(pp. 76, 198, 603)   ANNOTATIONS in the Teacher's Edition include exercise objectives, new vocabulary, assignment guides, answers, as well as references to supplementary tests, activities, related resources, and extended answers.

(pp. T40, T41, T68)   THE TEACHER'S GUIDE section includes a pacing chart, teaching suggestions, activities, computer activities, 32 supplementary tests, a student performance record, and extended answers.

SUPPLEMENTARY MATERIALS

THE TESTING PROGRAM and PERFORMANCE RECORD provides 32 masters for supplementary tests and a student progress and achievement chart.

PRACTICAL APPLICATIONS IN MATHEMATICS is a workbook to accompany *Stein's Refresher Mathematics*.

Stein's
REFRESHER MATHEMATICS

Eighth Edition

Edwin I. Stein

Allyn and Bacon, Inc.

Boston   Rockleigh, N.J.   Atlanta   Dallas   San Jose
London   Sydney   Toronto

Editor: Andrew P. Mastronardi
Senior Designer: L. Christopher Valente
Buyers: Martha Ballentine, Roger Powers
Technical Art: Graphics Etcetera
Contributor: Marilyn S. Lieberman, Meadowbrook School, Meadowbrook, PA

ISBN 0-205-07778-1

Library of Congress Catalog Card Number 82-70072

Printed in the United States of America

3 4 5 6 7 8 9   90 89 88 87 86 85 84

CONTENTS

Easy-to-use diagnostic tools.

Inventory Tests

Invaluable help in assessing abilities, these tests can be used to place the class or prescribe for weaknesses of individual students. Because each question is keyed to a specific lesson, it's easy to direct students to appropriate assignments. In fact, it's so foolproof, students can work independently.

Nine inventory tests are provided in the book—five for Part I (four Basic Skills and one Problem Solving), and one each for Measurement, Geometry, Algebra, and Graphs, Statistics, and Probability.

This is an example of an inventory test for Part I, Basic Skills in Arithmetic and Problem Solving. Note how the red numbers clearly indicate where to go to learn the skill being tested in the question.

This is the inventory test for the section on Graphs, Statistics, and Probability.

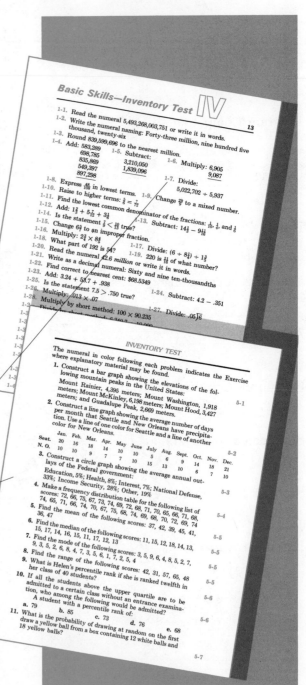

Lessons structured for success!

Step-by-step learning

From the *Aim, Procedure,* and *Sample Solutions* students gain a clear understanding of how to solve a specific type of math problem. Where helpful, *Preliminary Examples* provide practice in the skill being taught.

Diagnostic Test

These tests pinpoint any areas where the skill may be weak. Each question in the test is keyed to a set of related exercises for more practice.

Related Practice Examples

These exercises are carefully graded by difficulty and are keyed to the questions in the *Diagnostic Test.*

Note the wide range of examples presented in the Sample Solution section.

If a student misses question number 4 on the Diagnostic Test, he or she can be assigned Set 4 in the Related Practice Examples for more practice. The problems increase in difficulty from #1 to #10.

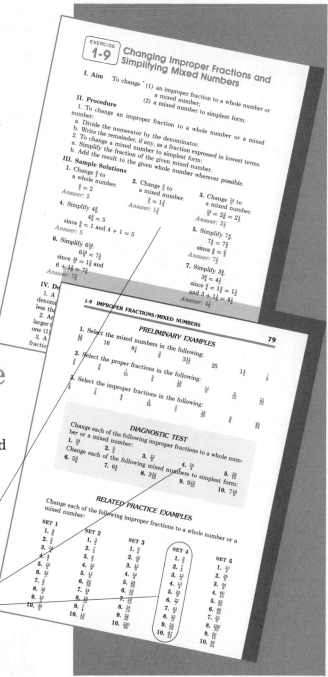

Builds problem-solving skills!

Problem-Solving Strategies

New sections on the method of attacking word problems have been included to help students master techniques of problem-solving.

No more fear and confusion! Students learn an easy, step-by-step approach that takes the worry out of problem-solving!

The comprehensive, general problem-solving strategy is located near the front of the book and can be easily referred to by the student for help at any time.

Two more specific strategies are located in sections where they are most applicable. One helps students attack geometry problems and another shows them how to solve problems using the equation method.

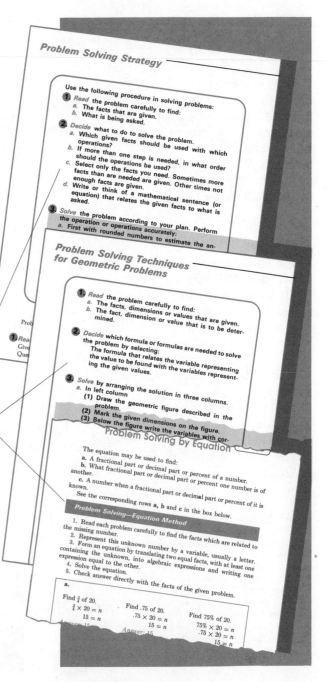

Problem Solving Strategy

Use the following procedure in solving problems:

1. **Read** the problem carefully to find:
   a. The facts that are given.
   b. What is being asked.

2. **Decide** what to do to solve the problem.
   a. Which given facts should be used with which operations?
   b. If more than one step is needed, in what order should the operations be used?
   c. Select only the facts you need. Sometimes more facts than are needed are given. Other times not enough facts are given.
   d. Write or think of a mathematical sentence (or equation) that relates the given facts to what is asked.

3. **Solve** the problem according to your plan. Perform the operation or operations accurately:
   a. First with rounded numbers to estimate the an-

Problem Solving Techniques for Geometric Problems

1. **Read** the problem carefully to find:
   a. The facts, dimensions or values that are given.
   b. The fact, dimension or value that is to be determined.

2. **Decide** which formula or formulas are needed to solve the problem by selecting:
   The formula that relates the variable representing the value to be found with the variables representing the given values.

3. **Solve** by arranging the solution in three columns.
   a. In left column
   (1) Draw the geometric figure described in the problem.
   (2) Mark the given dimensions on the figure.
   (3) Below the figure write the variables with cor-

Problem Solving by Equation

The equation may be used to find:
a. A fractional part or decimal part or percent of a number.
b. What fractional part or decimal part or percent one number is of another.
c. A number when a fractional part or decimal part or percent of it is known.
See the corresponding rows a, b and c in the box below.

Problem Solving—Equation Method

1. Read each problem carefully to find the facts which are related to the missing number.
2. Represent this unknown number by a variable, usually a letter.
3. Form an equation by translating two equal facts, with at least one containing the unknown, into algebraic expressions and writing one expression equal to the other.
4. Solve the equation.
5. Check answer directly with the facts of the given problem.

a.

Find $\frac{3}{4}$ of 20.
$\frac{3}{4} \times 20 = n$
$15 = n$
Answer: 15

Find .75 of 20.
$.75 \times 20 = n$
$15 = n$
Answer: 15

Find 75% of 20.
$75\% \times 20 = n$
$.75 \times 20 = n$
$15 = n$

Problem-Solving Practice

Sections within the lesson give students plenty of chances to practice and develop crucial problem-solving techniques. Most of the problems in these sections ask the student to apply basic math skills to practical, everyday situations.

Students get a chance to use their math skills in a practical way and practice problem-solving at the same time.

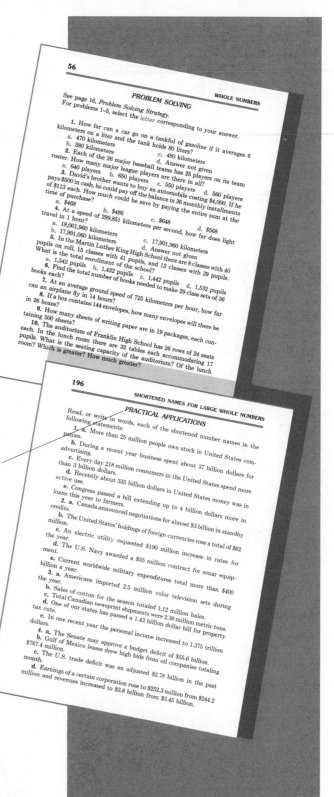

56

PROBLEM SOLVING

WHOLE NUMBERS

See page 16, *Problem Solving Strategy.*
For problems 1–5, select the letter corresponding to your answer.

1. How far can a car go on a tankful of gasoline if it averages 6 kilometers on a liter and the tank holds 80 liters?
  a. 470 kilometers
  b. 380 kilometers
  c. 480 kilometers
  d. Answer not given

2. Each of the 26 major baseball teams has 25 players on its team roster. How many major league players are there in all?
  a. 640 players
  b. 650 players
  c. 550 players
  d. 560 players

3. David's brother wants to buy an automobile costing $4,000. If he pays $500 in cash, he could pay off the balance in 36 monthly installments of $113 each. How much could he save by paying the entire sum at the time of purchase?
  a. $468
  b. $486
  c. $648
  d. $568

4. At a speed of 299,851 kilometers per second, how far does light travel in 1 hour?
  a. 18,001,960 kilometers
  b. 17,991,060 kilometers
  c. 17,901,960 kilometers
  d. Answer not given

5. In the Martin Luther King High School there are 8 classes with 40 pupils on roll, 15 classes with 41 pupils, and 13 classes with 39 pupils. What is the total enrollment of the school?
  a. 1,542 pupils
  b. 1,432 pupils
  c. 1,442 pupils
  d. 1,532 pupils

6. Find the total number of books needed to make 29 class sets of 36 books each?

7. At an average ground speed of 725 kilometers per hour, how far can an airplane fly in 14 hours?

8. If a box contains 144 envelopes, how many envelopes will there be in 26 boxes?

9. How many sheets of writing paper are in 19 packages, each containing 500 sheets?

10. The auditorium of Franklin High School has 26 rows of 24 seats each. In the lunch room there are 32 tables each accommodating 17 pupils. What is the seating capacity of the auditorium? Of the lunch room? Which is greater? How much greater?

196

SHORTENED NAMES FOR LARGE WHOLE NUMBERS

PRACTICAL APPLICATIONS

Read, or write in words, each of the shortened number names in the following statements:

1. a. More than 25 million people own stock in United States companies.
  b. During a recent year business spent about 37 billion dollars for advertising.
  c. Every day 218 million consumers in the United States spend more than 3 billion dollars.
  d. Recently about 335 billion dollars in United States money was in active use.
  e. Congress passed a bill extending up to 4 billion dollars more in loans this year to farmers.

2. a. Canada announced negotiations for almost $3 billion in standby credits.
  b. The United States' holdings of foreign currencies rose a total of $62 million.
  c. An electric utility requested $190 million increase in rates for the year.
  d. The U.S. Navy awarded a $55 million contract for sonar equipment.
  e. Current worldwide military expenditures total more than $400 billion a year.

3. a. Americans imported 2.5 million color television sets during the year.
  b. Sales of cotton for the season totaled 1.12 million bales.
  c. Total Canadian newsprint shipments were 2.38 million metric tons.
  d. One of our states has passed a 1.43 billion dollar bill for property tax cuts.
  e. In one recent year the personal income increased to 1.375 trillion dollars.

4. a. The Senate may approve a budget deficit of $55.6 billion.
  b. Gulf of Mexico leases drew high bids from oil companies totaling $767.4 million.
  c. The U.S. trade deficit was an adjusted $2.78 billion in the past month.
  d. Earnings of a certain corporation rose to $252.3 million from $244.2 million and revenues increased to $3.8 billion from $3.45 billion.

Strong, built-in program of review and testing.

The most comprehensive system of review and testing you'll find in any text!

Refresh Your Skills

Scattered throughout the text these sections reinforce and maintain key arithmetic skills.

Reviews

Each part of the book is followed by a comprehensive review keyed for easy reference to lessons for reteaching.

In addition, reviews for each Unit in the Parts on Basic Skills and Geometry are provided.

Handy references make it easy for students to go back and relearn specific skills.

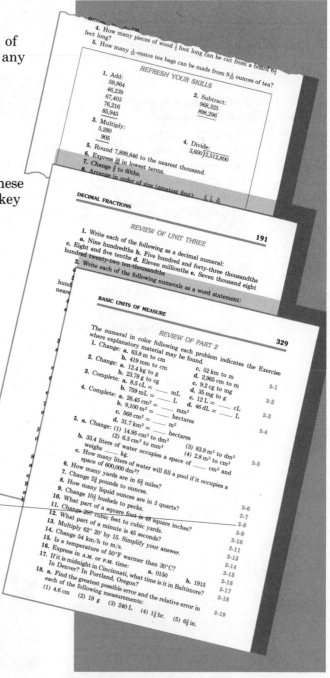

4. How many pieces of wood $\frac{3}{8}$ foot long can be cut from a board $6\frac{3}{4}$ feet long?
5. How many $\frac{1}{10}$-ounce tea bags can be made from $9\frac{3}{10}$ ounces of tea?

REFRESH YOUR SKILLS

1. Add:
59,864
46,239
67,403
76,216
85,945

2. Subtract:
968,325
898,296

3. Multiply:
5,280
905

4. Divide:
$3,600\overline{)2,512,800}$

5. Round 7,899,846 to the nearest thousand.
6. Express $\frac{54}{dm}$ in lowest terms.
7. Change $\frac{3}{5}$ to 60ths.
8. Arrange in order of size (greatest first): $\frac{3}{5}, \frac{7}{8}, \frac{8}{9}$

DECIMAL FRACTIONS

REVIEW OF UNIT THREE   191

1. Write each of the following as a decimal numeral:
a. Nine hundredths b. Five hundred and forty-three thousandths
c. Eight and five tenths d. Eleven millionths e. Seven thousand eight hundred twenty-two ten-thousandths
2. Write each of the following numerals as a word statement:

BASIC UNITS OF MEASURE

REVIEW OF PART 2   329

The numeral in color following each problem indicates the Exercise where explanatory material may be found.
1. Change: **a.** 63.8 m to cm   **c.** 52 km to m
   b. 419 m to cm   **d.** 2,965 cm to m   2-1
2. Change: **a.** 12.4 kg to g   **c.** 9.2 cg to mg
   b. 23.78 g to cg   **d.** 35 mg to g   2-2
3. Complete: **a.** 8.5 cL = _____ mL   **c.** 12 L = _____ cL
   b. 759 mL = _____ L   **d.** 46 dL = _____ L   2-3
4. Complete: **a.** 26.45 cm² = _____ mm²
   b. 9,100 m² = _____ mm²
   c. 568 cm² = _____ hectares
   d. 31.7 km² = _____ m²   2-4
5. **a.** Change: (1) 14.95 cm³ to dm³   (3) 83.9 m³ to dm³
   (2) 6.3 cm³ to mm³   (4) 2.8 m³ to cm³   2-5
   b. 33.4 liters of water occupies a space of _____ cm³ and weighs _____ kg.
   c. How many liters of water will fill a pool if it occupies a space of 600,000 dm³?
6. How many yards are in $6\frac{2}{3}$ miles?
7. Change $2\frac{3}{8}$ pounds to ounces.   2-6
8. How many liquid ounces are in 3 quarts?   2-7
9. Change $10\frac{1}{2}$ bushels to pecks.
10. What part of a square foot is 48 square inches?   2-8
11. Change 297 cubic feet to cubic yards?   2-9
12. What part of a minute is 45 seconds?
13. Multiply 62° 20′ by 15. Simplify your answer.   2-10
14. Change 54 km/h to m/s.   2-11
15. Is a temperature of 50°F warmer than 30°C?   2-13
16. Express in A.M. or P.M. time: 2-14
17. If it is midnight in Cincinnati, what time is it in Baltimore?
    In Denver? In Portland, Oregon?   **a.** 0150   **b.** 1915 2-15
18. **a.** Find the greatest possible error and the relative error in 2-16
    each of the following measurements: 2-17
    (1) 4.6 cm   (2) 19 g   (3) 240 L   2-18
    (4) $1\frac{3}{8}$ hr.   (5) $6\frac{3}{8}$ in.   2-19

T6

Competency Check Tests

These tests give students valuable practice taking multiple choice tests. They are keyed to lessons for easy reference.

Keyed to lessons!

Keyed Achievement Tests

These tests check mastery of each major part of the text. They are keyed to lessons, making it a snap to refer students back to relearn skills.

Keyed to lessons!

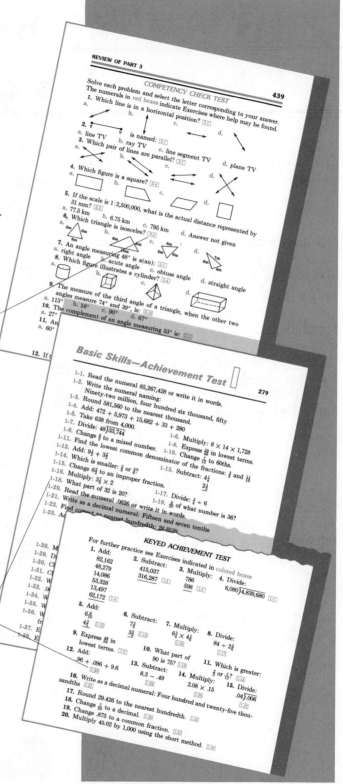

Lively learning features!

Estimating Answers

These sections help students develop the logical thinking strategies and important mental arithmetic skills they will need in everyday life.

Check by Calculator

Use these sections to develop the skilled and accurate use of hand-held calculators!

Unit Openers

Units in the Basic Skills portion of the text open with background information relevant to the lessons that follow.

Note: The vital modern skill of understanding shortened names for large whole numbers is covered on pages 195–205.

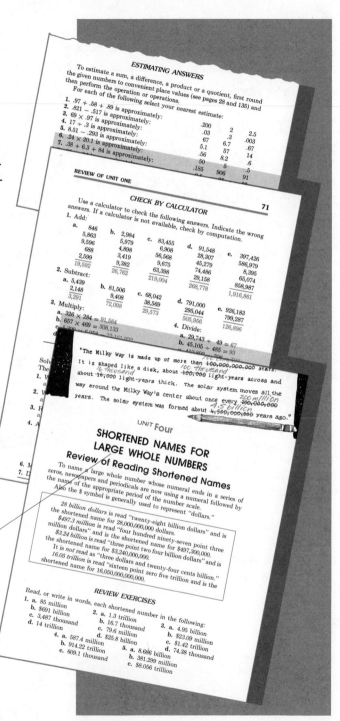

ESTIMATING ANSWERS

To estimate a sum, a difference, a product or a quotient, first round the given numbers to convenient place values (see pages 28 and 135) and then perform the operation or operations.

For each of the following select your nearest estimate:

1. .97 + .58 + .89 is approximately:
2. .821 − .517 is approximately:
3. 69 × .97 is approximately:
4. 17 ÷ .3 is approximately:
5. 8.51 − .293 is approximately:
6. .24 × 20.1 is approximately:
7. .38 + 6.3 + 84 is approximately:

.200	2	2.5
.03	.3	.003
67	6.7	.67
5.1	57	14
.56	8.2	.6
50	5	.5
.185	906	91

REVIEW OF UNIT ONE

CHECK BY CALCULATOR 71

Use a calculator to check the following answers. Indicate the wrong answers. If a calculator is not available, check by computation.

1. Add:

a. 846	b. 2,984	c. 83,455	d. 91,548	e. 397,426
5,863	5,979	6,908	28,307	586,979
9,596	4,898	56,568	45,279	8,395
688	3,419	9,675	74,486	65,074
2,599	9,382	63,398	29,158	858,987
19,592	26,762	219,004	268,778	1,916,861

2. Subtract:

a. 5,439	b. 81,506	c. 68,042	d. 791,000	e. 926,183
2,148	9,408	38,569	295,044	799,287
3,291	72,098	29,573	505,956	126,896

3. Multiply:
a. 326 × 284 = 91,584
b. 657 × 469 = 306,133

4. Divide:
a. 29,743 ÷ 49 = 67
b. 45,105 ÷ 485 = 93

"The Milky Way is made up of more than ~~100,000,000,000~~ stars. *200 million* It is shaped like a disk, about ~~100,000~~ light-years across and *100 thousand* about ~~16,000~~ light-years thick. The solar system moves all the *16 thousand* way around the Milky Way's center about once every ~~200,000,000~~ *200,000,000* years. The solar system was formed about ~~4,500,000,000~~ years ago." *4.5 billion*

UNIT Four
SHORTENED NAMES FOR LARGE WHOLE NUMBERS
Review of Reading Shortened Names

To name a large whole number whose numeral ends in a series of zeros, newspapers and periodicals are now using a numeral followed by the name of the appropriate period of the number scale.

Also the $ symbol is generally used to represent "dollars."

28 billion dollars is read "twenty-eight billion dollars" and is the shortened name for 28,000,000,000 dollars.

$497.3 million is read "four hundred ninety-seven point three million dollars" and is the shortened name for $497,300,000.

$3.24 billion is read "three point two four billion dollars" and is the shortened name for $3,240,000,000. It is *not* read as "three dollars and twenty-four cents billion."

16.05 trillion is read "sixteen point zero five trillion and is the shortened name for 16,050,000,000,000.

REVIEW EXERCISES

Read, or write in words, each shortened number in the following:

1. a. 85 million
   b. $691 billion
   c. 3,487 thousand
   d. 14 trillion

2. a. 1.3 trillion
   b. 16.7 thousand
   c. 79.6 million
   d. $25.8 billion

3. a. 4.95 billion
   b. $23.09 million
   c. $1.42 trillion
   d. 74.38 thousand

4. a. 587.4 million
   b. 914.22 trillion
   c. 609.1 thousand

5. a. 8.686 billion
   b. 381.299 million
   c. $8.056 trillion

Helpful, easy-to-use teacher materials.

Annotated Teacher's Edition

Teacher's Guide

The Teacher's Guide Section includes a handy pacing chart, teaching suggestions, activities, computer activities, 32 supplementary tests, a student performance record, and extended answers.

Annotated Pages

Annotations include:
1. Answers
2. Assignment guides
3. New vocabulary
4. Lesson objectives
5. References to:
   Supplementary Tests
   Activities
   Computer Activities
   Related Resources
   Extended Answers

Supplementary Materials

Testing Program and Performance Record

This convenient package includes masters for 32 supplementary tests plus a student progress and achievement chart.

Practical Applications in Mathematics

This workbook provides 224 pages of additional practice for reinforcing many of the lessons. Answer book available.

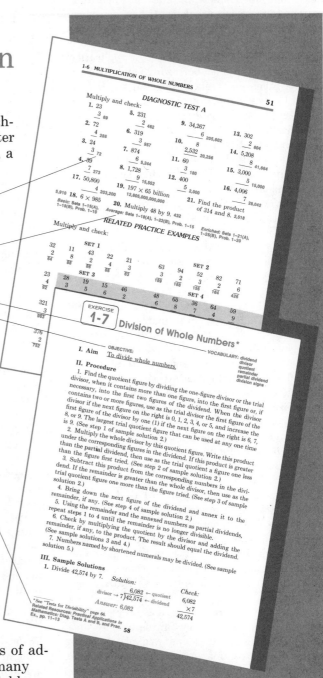

OVERVIEW OF TEXT

Refresher Mathematics is designed to be used in arithmetic classes of any grade level for the development and maintenance of proficiency in arithmetic and problem solving skills and their applications to everyday problems. It also contains a comprehensive treatment of the metric system, informal geometry, introductory algebra, statistics, probability, and basic consumer applications of earning, spending, borrowing, and investing money. Inventory Tests throughout the text are keyed to these subject areas.

Refresher Mathematics is ideal as a refresher or remedial textbook in general mathematics with emphasis on arithmetic and consumer applications. It may also be used as a basal textbook featuring individualized assignments according to each student's need.

Refresher Mathematics is divided into six parts. Each part contains several exercises which are arranged under unit topics.

In Part 1, "Basic Skills in Arithmetic and Problem Solving," each exercise gives (I) the aim, (II) the procedure to be followed, (III) sample solutions completely worked out, and wherever necessary, (IV) the definitions of terms used. Following this developmental material come the diagnostic tests which are keyed to the immediately succeeding practice examples.

The examples in the diagnostic tests and the related practice examples are closely graded in difficulty. Since one example is not perceptibly more difficult than the preceding one, the student is led, step by step, to think clearly and to develop a full understanding of a mathematical concept. The final examples in each exercise provide an automatic end test. The final keyed Achievement Tests serve as mastery tests to end the section.

The exercises and problems in Part 1 are organized so that each student finds his or her own deficiences in arithmetic by a simple testing procedure and is directed to specific assignments based on his or her needs. Thus students do not waste time drilling on problems they can solve, and individual achievement is accelerated.

Included in Part 1 are four Basic Skills—Inventory Tests in arithmetic and one Problem Solving—Inventory Test. They can be used to diagnose general difficulties and weaknesses before beginning the section. Also, a Problem Solving Strategy is provided to help students develop their problem-solving techniques.

In Part 2 basic units of measure are studied. The Inventory Test directs students to the topics in which they need practice. Both metric and customary units are discussed in this section.

Part 3 introduces informal geometry and geometric measurement. The Inventory Test directs students to topics in which they need practice. A Problem Solving Strategy for solving Geometric problems is provided. In the exercises the practice examples are followed by problems which show the value and use of the related skills and principles. The applications are both functional and interesting. They may be used to introduce topics and to motivate instruction.

In Part 4 the number system is extended to include positive and negative numbers. The language of algebra, the solutions of equations and inequalities in one variable, and ratio and proportion are studied. Part 5 presents some basic concepts used in statistics. Topics include the bar, line, and circle graphs; averages; and the study of probability. Inventory Tests in Parts 4 and 5 help to direct students to the practice they require.

Part 6 is a comprehensive section on everyday consumer problems. This part includes such practical topics as take-home pay, withholding tax, income tax, sales tax, property tax, unit pricing, buying at a sale, making change, installment buying, annual percentage rates, cost of home and automobile ownership, household expenses, life and fire insurance, banking, simple and compound interest, loans, mutual funds, money rates, and other investments, discounts, commission, and profit and loss, reading road maps and timetables.

Maintenance tests, called "Refresh Your Skills," Competency Check Tests, Estimation, Check by Calculator, and Reviews are provided throughout the book. The Competency Check Tests give students the experience of taking multiple-choice tests as they may see in state or district competency exams. However, they are not intended to take place of state or district exams.

The flexible organization of *Refresher Mathematics* allows for individual differences and at the same time provides maximum and minimum materials for group methods of instruction. Practice material for class, home, and optional assignments is included.

How to Use the Text for Initial Teaching

To use *Refresher Mathematics* as a basal textbook in general mathematics, emphasize Parts 1, 2, and 6. Topics from Parts 3, 4, and 5 can be discussed as desired or as class interests dictate.

Part 1, "Basic Skills in Arithmetic and Problem Solving," provides complete coverage of computational skills, including separate units on whole numbers, common fractions, decimal fractions, percents, and squares and square roots. Each of these units is divided into a number of exercises which are the basic subdivisions of the book. Each exercise begins with its aim (its objective), procedure (rules for computation), and sample solutions (worked-out examples). These are followed by a Diagnostic Test on the content of the exercise and then extensive sets of problems grouped under the heading "Related Practice Examples." Problem Solving sections throughout Part 1, keyed to the Problem Solving Strategy, provide students with everyday applications of computational skills.

Part 2, "Basic Units of Measure," provides coverage of length, weight or mass, area, and volume, including separate units on the metric system, the customary system, other basic measures, and measurement. The format of Part 2 is similar to that of Part 1, and Part 2 may be used in the same way that Part 1 is used.

Part 6, "Everyday Applications—Problem Solving," provides an extensive selection of practical topics and problems related to them, including separate topics on income, take-home pay, income tax, spending money, managing money, and applications in business and travel.

The Inventory Tests, Diagnostic Tests, and Acievement Tests may be used for evaluation, and the Individualized Pupil Performance Record may be used as a record of students' daily work.

How to Use the Text for Reteaching

To use *Refresher Mathematics* as a remedial textbook in general mathematics, you may wish to present Parts 1 and 2 (arithmetic skills and basic units of measure) as a review. The many Diagnostic Tests, Inventory Tests, and Achievement Tests allow students to proceed through this material at their own pace. Parts 3 through 6 can then be treated as new material; the Inventory Tests to these parts provide students with guidance about the topics that should be reviewed, the topics that are new to them and should be studied in detail, and the topics in which they are proficient and need no further review.

Inventory Tests, Diagnostic Tests, and Achievement Tests may be used to individualize remedial instruction in the following way:

1. The student works Inventory Tests I, II, III, and IV. A student who misses the same problem number on two or more tests should turn to the exercise with that number. For example a student who misses problem 12 on Inventory Tests I and III has trouble adding fractions and mixed numbers and should turn to Exercise 1-12, which is devoted to that type of computation.

2. After turning to the exercise that deals with a particular type of computation, the student should study the procedures and sample solutions, and then work the Diagnostic Test. Problems missed on the Diagnostic Test should pinpoint the exact areas of weakness the student still has. Continuing the example in item 1, a student who misses problem 6 on the Diagnostic Test in Exercise 1-12 has difficulty adding simple fractions with different denominators. The student should then work Set 6 in the Related Practice Examples (because it is the problem set with the corresponding number) for further practice.

3. After working through each area of difficulty, the student may be assigned one of the Achievement Tests to determine if mastery has been achieved.

The Individualized Pupil Performance Record on pp. T68 and T69 may be used by students to keep records of their work.

Note the section of tests included in the teacher's guide on pp. T41–T67. These tests include: additional Inventory Tests for both basic skills and problem solving; Unit Tests and an Achievement Test for Part One; end-of-section tests for Parts Two through Six; an alternate Problem Solving Competency Check Test and Two Competency Check Tests which provide practice for state and district competency exams. Both the student performance record and the tests are also available as Duplicating Masters in *Stein's REFRESHER MATHEMATICS Testing Program and Performance Record.*

The pacing chart is provided to demonstrate one possible use of Refresher Mathematics through a school year. Individual needs of teachers and students may require alternate year plans.

PACING CHART

Day	Exercise or Review and Test	Enriched	Average	Basic
1	1-1	1-25, 1-5	1-22, 1-4	1-18, 1-2
2	1-2	1-11, 1-5, 1-4	1-10, 1-4, 1-3	1-8, 1-2, 1-2
3	1-3	1(a)-8(a), 1(b)-8(b)	1(a)-8(a), 1(b)-6(b)	1(a)-8(a)
4	1-4	1-26, 1-10, 1-20	1-22, 1-9, 1-15	1-19, 1-7, 1-10
5	1-5	1-34, 1-19	1-30, 1-15	1-25, 1-10
6	1-6	1-21(A), 1-25(B), 1-20	1-18(A), 1-22(B), 1-15	1-15(A), 1-18(B), 1-10
7	1-7	1-26(A), 1-33, 1-2, 1-5, 1-12, 1-20	1-23(A), 1-30, 1-2, 1-4, 1-10, 1-15	1-19(A), 1-24, 1, 1-2, 1-8, 1-12
8	Unit Review Unit Test			
9	1-8	1-7, 1-5	1-6, 1-3	1-5, 1-2
10	1-9	1-10, 1-5	1-9, 1-4	1-7, 1-2
11	1-10	1-5, 1-6	1-4(a), 1-4	1-3, 1-3
12	1-11	1-4, 1-5	1-3, 1-4	1-2, 1-2
13	1-12	1-27, 1-15	1-24, 1-12	1-20, 1-9
14	1-13	1-24, 1-15	1-21, 1-12	1-17, 1-9
15	1-14	1-8, 1-2	1-7, 1-2	1-5, 1
16	1-15	1-4(c), 1-5, 1-5	1-4(b), 1-4, 1-4	1-3, 1-2, 1-2
17	1-16	1-38(j), 1-20	1-33, 38(a)-38(h), 1-15	1-27, 38(a)-38(d), 1-10
18	1-17	1-31(f), 1-16	1-30(c), 1-11	1-27, 1-8
19	1-18	1-10, 1-14, 1-8	1-9, 1-10, 1-7	1-7, 1-7, 1-5
20	1-19	1(a)-2(e), 1-2, 1-7	1(a)-2(c), 1-2, 1-6	1(a)-2(b), 1, 1-4
21	Unit Review Unit Test			
22	1-20	1-17	1-15	1-13
23	1-21	1-17	1-15	1-13
24	1-22	1-8(b), 1-3	1-8(a), 1-2	1-7(a), 1
25	1-23	1-25, 1-10, 1-5	1-22(b), 1-8, 1-4	1-20, 1-5, 1-2
26	1-24	1-32, 1-12	1-29, 1-10	1-25, 1-6
27	1-25	1-8, 1-3	1-7, 1-2	1-5, 1
28	1-26	1-32, 1-10	1-29, 1-8	1-25, 1-5
29	1-27	1-14(A), 1-11(c)(B), 1-11(b)(C), 1-10(c)(D), 1-12	1-12(c)(A), 1-11, (a)(B), 1-10(c)(C), 1-10(a)(D), 1-9	1-10(A), 1-9(B), 1-10(a)(C), 1-8(D), 1-5
30	1-28	1-18, 1-5	1-16, 1-4	1-13, 1-2
31	1-29	1-17, 1-6	1-15, 1-5	1-13, 1-3
32	1-30	1-18, 1-5, 1-5	1-16, 1-4, 1-4	1-13, 1-2, 1-2
33	1-31	1-14, 1-6	1-12, 1-4	1-9, 1-2
34	1-32	1-14	1-12	1-9
35	1-33	1-16	1-13	1-10
36	Unit Review Unit Test			
37	1-34	1-20, 1-3	1-16, 1-2	1-13, 1
38	1-35	1-23, 1-15	1-20, 1-13	1-16, 1-10
39	Unit Review Unit Test			
40	1-36	1-4, 1-2	1-3, 1	1-2
41	1-37	1-17, 1-3	1-15, 1-2	1-13, 1

Day	Exercise or Review and Test	Enriched	Average	Basic
42	1-38	1–16, 1–5	1–14, 1–4	1–12, 1–2
43	1-39	1–4, 1–6	1–3, 1–4	1–3, 1–3
44	1-40	1–14, 1–5	1–12, 1–4	1–9, 1–2
45	1-41	1–16(b), 1–14	1–15(a), 1–12	1–12, 1–8
46	1-42	1(a)–16, 1–2, 1–10	1(a)–13, 1–2, 1–8	1(a)–11, 1, 1–5
47	1-43	1–15, 1–5	1–12, 1–4	1–10, 1–2
48	Unit Review Unit Test			
49	1-44	1–6, 1–4	1–5, 1–3	1–4, 1
50	1-45	1–15, 1–5	1–12, 1–4	1–9, 1–2
51	Unit Review Unit Test			
52	1-46	1–5	1–4	1–2
53	1-47	1–3	1–2	1
54	1-48	1–10	1–9	1–7
55	1-49	1–3	1–2	1
56	1-50	1–4	1–3	1–2
57	1-51	1–3	1–2	1
58	1-52	1–10	1–9	1–7
59	1-53	1–10	1–9	1–7
60	1-54	1–10	1–9	1–7
61	1-55	1–5	1–4	1–2
62	1-56	1–12	1–10	1–8
63	1-57	1–5	1–4	1–2
64	1-58	1–4	1–3	1–2
65	1-59	1–6	1–5	1–3
66	1-60	1–3	1–2	1
67	1-61	1–25	1–22	1–18
68	1-62	1–15	1–12	1–9
69	Unit Review Unit Test			
70	2-1	1–25	1–22	1–18
71	2-2	1–22	1–19	1–15
72	2-3	1–20	1–16	1–13
73	2-4	1–10	1–9	1–7
74	2-5	1–15	1–12	1–9
75	Unit Review Unit Test			
76	2-6	1–15	1–12	1–9
77	2-7	1–15	1–12	1–9
78	2-8	1–12	1–10	1–8
79	2-9	1–9	1–7	1–5
80	2-10	1–12	1–10	1–8
81	2-11	1–12	1–10	1–8
82	2-12	1–13	1–12	1–9
83	Unit Review Unit Test			
84	2-13	1–25	1–22	1–18
85	2-14	1–12	1–10	1–8
86	2-15	1–10	1–9	1–7
87	2-16	1–5	1–4	1–2
88	2-17	1–5	1–4	1–2
89	2-18	1–6	1–5	1–3
90	Unit Review Unit Test			

Day	Exercise or Review and Test	Enriched	Average	Basic
91	2-19	1-10	1-9	1-7
92	2-20	1-30	1-27	1-24
93	2-21	1-4	1-3	1-2
94	Unit Review Unit Test			
95	Midterm Test			
96	3-1	1-27	1-24	1-20
97	3-2	1-7	1-6	1-4
98	3-3	1-30	1-26	1-23
99	3-4	1-11	1-10	1-8
100	3-5	1-10, 1-4, 1-6	1-9, 1-3, 1-5	1-7, 1-2, 1-3
101	3-6	1-16	1-13	1-10
102	Unit Review Unit Test			
103	3-7	1-10, 1-10	1-9, 1-9	1-7, 1-7
104	3-8	1-5, 1-3	1-4, 1-2	1-2, 1
105	3-9	1-10	1-9	1-7
106	3-10	1-7, 1-4	1-6, 1-3	1-4, 1-2
107	3-11	1-5, 1-4, 1-5	1-4, 1-3, 1-4	1-3, 1-2, 1-3
108	3-12	1-9	1-8	1-6
109	3-13	1-7	1-6	1-4
110	Unit Review Unit Test			
111	3-30	1-10, 1-5, 1-3, 1-3, 1-3, 1-15, 1-3, 1-2, 1-4, 1-3	1-9, 1-4, 1-2, 1-2, 1-2, 1-12, 1-2, 1-2, 1-3, 1-2	1-7, 1-2, 1, 1, 1, 1-10, 1, 1, 1-2, 1
112	3-31	1-7, 1-4, 1-10, 1-3, 1-3, 1-3	1-6, 1-3, 1-9, 1-2, 1-2, 1-2	1-4, 1-2, 1-7, 1, 1, 1
113	Unit Review Unit Test			
114	4-1	1-5	1-4	1-3
115	4-2	1-5	1-4	1-3
116	4-3	1-7	1-6	1-4
117	4-4	1-4	1-3	1-2
118	4-5	1-15	1-12	1-10
119	4-6	1-10	1-9	1-7
120	4-7	1-6	1-5	1-4
121	4-8	1-6	1-5	1-4d
122	Unit Review Unit Test			
123	4-9, 4-10	1-5, 1-5	1-4, 1-4	1-2, 1-2
124	4-11	1-5	1-4	1-2
125	4-12	1-5	1-4	1-2
126	Unit Review Unit Test			
127	4-13	1-2	1-2	1
128	4-14	1-4	1-3	1-2
129	4-15	1-2	1-2	1
130	4-16	1-2	1-2	1
131	4-17	1-12	1-10	1-8
132	4-18	1-20	1-17	1-15
133	Unit Review Unit Test			
134	4-27	1-15	1-12	1-10
135	4-28	1-17	1-14	1-12

PACING CHART (cont.)

Day	Exercise or Review and Test	Enriched	Average	Basic
136	Unit Review Unit Test			
137	6-1	1–40	1–5, 7–16, 19–20, 22–28, 31–32, 34–36, 38–39	1–3, 7–14, 19, 22–26 31, 34–35, 38
138	6-2	1–4	1–3	1–2
139	6-3	1–2	1–2	1
140	6-4	1–5	1–4	1–3
141	6-5	1–5	1–4	1–3
142	6-6	1–5	1–4	1–3
143	Unit Review Unit Test			
144	6-7	1–6	1–5	1–3
145	6-8	1–20	1–17	1–13
146	6-9	1–10	1–9	1–7
147	6-10	1–2	1–2	1
148	6-11	1–3	1–2	1
149	6-12	1–8	1–7	1–5
150	6-13	1–7	1–6	1–5
151	6-14	1–12	1–10	1–8
152	6-15	1–6	1–5	1–3
153	6-16	1–8	1–7	1–5
154	6-17	1–15	1–12	1–10
155	6-18	1–5	1–4	1–3
156	Unit Review Unit Test			
157	6-19	1–7	1–6	1–5
158	6-20	1–20	1–16	1–13
159	6-21	1–6, 1–5, 1–6	1–5, 1–4, 1–5	1–4, 1–3, 1–4
160	6-22	1–6	1–5	1–4
161	6-23	1–15	1–12	1–10
162	Unit Review Unit Test			
163	6-24	1–15	1–12	1–10
164	6-25	1–15	1–12	1–10
165	6-26	1–20	1–16	1–13
166	6-27	1–2	1–2	1
167	6-28	1–15	1–4, 6–9, 11–14	1–3, 6–8, 11–13
168	Unit Review Unit Test			
169	Year-end Review			
170	Year-end Test			

PART 1 BASIC SKILLS IN ARITHMETIC and PROBLEM SOLVING

The two main objectives of Part I are to help students master fundamental computational skills and develop problem solving techniques. Coverage in this part includes the four basic operations (addition, subtraction, multiplication, and division) using whole numbers, common fractions, and decimal fractions. Additional topics include short names for large whole numbers; percentages; squares and square roots; exponents, powers of ten, and scientific notation; and factors and related topics. Throughout Part I problem solving sections, keyed to the Problem Solving Strategy (pp. 16–17), allow students to apply computational skills to everyday word problems. An introduction to Part I familiarizes students with concepts of counting, names for numbers, and number systems, and with the properties of the four basic operations. Four Basic Skills—Inventory Tests and one Problem Solving—Inventory Test in Part I can be used to diagnose general difficulties and weaknesses before beginning the section. Part I closes with four Basic Skills—Achievement Tests that can be used as mastery tests for the section.

UNIT ONE WHOLE NUMBERS

The main objective of the exercises in Unit One is to help students master the four basic operations of addition, subtraction, multiplication, and division when working with whole numbers. Students will learn how to read and write whole numbers using proper punctuation or spacing, and how to round whole numbers. In addition, the students will be introduced to the use of shortened names for large numbers.

Teaching Suggestions

1-1 (p. 19) Encourage students to read numerals aloud, emphasizing numerals in hundreds, thousands, and millions. Also have students practice writing the names of numerals in words. Additional practical examples can be found in the various world almanacs that are published yearly.

1-2 (p. 23) Encourage students to practice writing numerals, emphasizing numerals in hundreds, thousands, and millions. Also have students practice reading their answers. Additional practical examples of current interest to students can be found in news magazines and local newspapers.

1-3 (p. 28) If students have difficulty with this exercise, have them draw a horizontal line between the number to be rounded and the number to the right of it. This technique may help them to see how to proceed. For additional practice use some of the practical applications in Exercise 1-1.

1-4 (p. 30) The preliminary examples give all of the basic addition combinations; students should be familiar with these and confident in their use before proceeding to the Diagnostic Test and Practice Examples. Have students practice addition with units only, with tens only, and with hundreds only.

1-5 (p. 41) The preliminary examples give all of the basic subtraction combinations; students should be familiar with these and confident in their use before proceeding to the Diagnostic Test and Practice Examples. Have students practice subtraction with units only, with tens only, and with hundreds only. Be sure students check their work.

1-6 (p. 49) The preliminary examples give all of the basic multiplication combinations; students should be familiar with these and confident in their use before proceeding to the Diagnostic Test and Practice Examples. Diagnostic Test A shows multiplication with one-digit multipliers; Diagnostic Test B shows more complex multiplication problems. Be sure students check their work.

1-7 (p. 58) The preliminary examples give all of the basic division combinations; students should be familiar with these and confident in their use before proceeding to the Diagnostic Test and Practice Examples. Diagnostic Test A shows division with one-digit divisors; Diagnostic Test B shows more complex division problems. Be sure students check their work.

The test for divisibility shows how to determine whether a number is divisible by 2, 3, 4, 5, 6, 7, 8, 9, or 10. The concept of averages is explored in the Problem Solving section. (Note that in addition to the usual provision for review and testing, students are introduced to the idea of estimation in "Estimating Answers" at the end of this unit (optional). "Check by Calculator" (optional) provides practice in using the calculator.)

UNIT TWO   COMMON FRACTIONS

The main objective of the exercises in Unit Two is to help students work with common fractions. The student will learn how to name fractions and add, subtract, multiply, and divide fractions.

Teaching Suggestions

1-8 (p. 76)   Students will use division and factoring in this exercise.

1-9 (p. 78)   The preliminary examples help students identify the various types of fractions. The Diagnostic Test and Practice Examples involve rewriting fractions. Division and addition will be used in this exercise.

1-10 (p. 81)   Students will use division and addition in this exercise. Practice in reading a ruler is provided in the Miscellaneous Examples.

1-11 (p. 83)   Point out to students that a group of fractions can have many common denominators, but only the smallest of these is the lowest common denominator. Division and multiplication will be used in this exercise.

1-12 (p. 86)   Point out to students that if fractions have unlike denominators, it is easiest to change the denominators to the lowest common denominator in order to add the fractions. The Problem Solving section shows some everyday uses of addition of fractions.

1-13 (p. 92)   The procedures here are very similar to those in the previous exercise. Students may need additional practice in those problems that involve regrouping (taking one from the whole number in the minuend and changing it to a fraction). The Preliminary Examples provide practice in regrouping. The Problem Solving section gives some everyday uses of fractions.

1-14 (p. 98)   Point out to students that they do not need to find the lowest common denominator for these problems. Students will use the ideas of less than and greater than to compare the fractions. The Miscellaneous Examples show how to round a mixed number to the nearest whole number.

1-15 (p. 100)   Multiplication and addition are needed for this exercise. Again the Practical Applications show some everyday uses of fractions.

1-16 (p. 102)   Students should be reminded to change mixed numbers to improper fractions before proceeding. The idea of greatest common factor may need some additional explanation. The ideas of fractional part and reduction are also demonstrated in this exercise. The Problem Solving section shows many different uses of fractions.

1-17 (p. 110)   Again students should be reminded to change mixed numbers to improper fractions before proceeding. Students do a problem in this exercise by finding the reciprocal of the divisor and then multiplying by the dividend. Point out that they should simplify complex fractions before proceeding. The Problem Solving section shows examples of everyday uses of fractions.

1-18 (p. 117)   Both objectives of this exercise involve making a fraction by using the two given numbers. Then the fraction is simplified. A ratio is a comparison of two numbers by division.

1-19 (p. 121)   This exercise involves division by fractions or multiplication by reciprocals. Two methods of solving sample problems are given.

UNIT THREE   DECIMAL FRACTIONS

The main objective of the exercises in Unit Three is to help students work with decimal fractions. The student will learn how to read and name decimal fractions and how to round, add, subtract, multiply, and divide decimal fractions. In addition, the students will apply their knowledge of shortened names for numbers when reading decimal fractions.

Teaching Suggestions

1-20 (p. 128)   Two ways are suggested for reading decimals: 1) using the name that applies to the place value of the last figure (tenths, hundredths, and so on) and 2) using the word "point" for the decimal point. Numerous practical examples can be found in world almanacs, news magazines, and local newspapers.

1-21 (p. 130)   Decimals written as words are given in two ways: 1) with the place names of the last figure (tenths, hundredths, and so on) and 2) with the word "point." Encourage students to read their answers aloud. Numerous practical examples can be found in world almanacs, news magazines, and local newspapers.

1-22 (p. 135) The actual rounding procedures here are similar to those used in Exercise 1-3, "Rounding Whole Numbers." Also, students are told the three different ways in which calculators round decimals (see Sample Solutions 14 and 15).

1-23 (p. 138) Encourage students to place decimal points one under the other. Addition is then accomplished as it is for whole numbers. (See Exercise 1-4, "Addition of Whole Numbers.") The "Concept" relates addition of decimals to addition of common fractions. The Problem Solving section and the section entitled "Micrometer Caliper Readings" show some everyday uses of decimal fractions.

1-24 (p. 146) Again encourage students to place decimal points one under the other. Subtraction is then accomplished as it is for whole numbers. (See Exercise 1-5, "Subtraction of Whole Numbers.") Answers can be checked by addition. The "Concept" relates subtraction of decimals to subtraction of common fractions. The Problem Solving section shows some everyday uses of decimal fractions.

1-25 (p. 153) Students use the ideas of greater than and less than to compare decimals. Point out to students that mixed decimals or whole numbers are greater than decimal fractions. The Practical Applications give some examples from the field of quality control.

1-26 (p. 155) Point out to students that multiplication of decimals is the same as multiplication of whole numbers; the only difference occurs in locating the decimal point in the answer. The number of decimal places in the answer is the same as the total number of decimal places in the multiplicand *and* the multiplier. The "Concept" relates multiplication of decimals to multiplication of fractions. The Problem Solving section gives some everyday uses of multiplication of decimals.

1-27 (p. 162) Two main cases are considered here: 1) when the divisor is a whole number and 2) when the divisor is a decimal. Students will use the caret ($\wedge$) to indicate the new position (if necessary) of the decimal point in both the divisor and the dividend. Two procedures are given for finding a quotient correct to the nearest decimal place. The "Concept" relates division of decimals to division of fractions.

1-28 (p. 173) The skills developed in this exercise are used in converting units of U.S. money and units of the metric system. Annexing zeros (whole numbers) and moving the decimal point (decimals) are the two procedures presented.

1-29 (p. 175) The skills developed in this exercise are used in converting units of U.S. money and units of the metric system. Moving the decimal point is the main procedure.

1-30 (p. 178) Three methods are presented: 1) division; 2) recognizing the decimal equivalent ($\frac{1}{4} = .25$); 3) fractions with denominators of 10, 100, etc.

1-31 (p. 181) Two methods are presented: 1) using a power of ten and 2) recognizing a common fraction equivalent. A table of equivalents is given.

1-32 (p. 184) First students must find the common fraction name. (Review Exercises 1–18, "Finding What Fractional Part One Number Is of Another.") Then the fraction is changed to a decimal. (See Exercise 1-30.)

1-33 (p. 187) Skills developed in Exercise 1-27, "Division of Decimals," may be needed here. Two methods of solution are presented. (Note that in addition to the usual provision for review and testing, the optional section, "Estimating Answers," provides practice in estimating when working with decimals. "Check by Calculator" (optional) provides practice working with whole numbers and decimals on the calculators.)

UNIT FOUR SHORTENED NAMES FOR LARGE WHOLE NUMBERS

The main objective of the exercises in Unit Four is to help students become familiar with shortened names for large whole numbers, i.e. to write the shortened forms and to recognize the shortened forms and what they mean. The shortened form is very commonly used in newspapers, annual reports, and periodicals. (Shortened names for numbers have been introduced in Exercises 1-1, 1-2, 1-4 to 1-7, 1-20, 1-21, 1-23, 1-24, 1-26, and 1-27.)

Teaching Suggestions

1-34 (p. 197) Students will need to be familiar with the powers of ten for this exercise. Practical examples can be found in news magazines and local newspapers.

1-35 (p. 200) Students will need to be familiar with the names of the powers of ten for this exercise. Operations with shortened names are similar to operations with whole numbers and decimals.

UNIT FIVE   PERCENT

The main objective of the exercises in Unit Five is to help students master computations involving percentages. Students will learn how to change percents to decimals and vice versa, how to change percents to fractions and vice versa, and how to do computations with percents.

Teaching Suggestions

1-36 (p. 207) Percent means hundredths; it is a third way for representing hundredths. (Fractions and decimals are the other two.) Point out to students that 100% is all of an amount; 200% is twice the amount; and so on.

1-37 (p. 209) To change a percent to a decimal, the decimal point is moved two places to the left and the percent sign is eliminated. Practical examples of percent can be found in news magazines and newspapers.

1-38 (p. 211) To change a decimal to a percent, move the decimal point two places to the right and write the percent sign, %, after the numeral.

1-39 (p. 213) Two methods are presented: 1) writing the fraction in hundredths and reducing, and 2) recognizing the equivalent fraction. Point out to students that when a percent is greater than 100%, the answer is a mixed number.

1-40 (p. 215) Three methods are presented: 1) division, 2) recognizing the equivalent fraction, and 3) writing a fraction so that its denominator is 100. Mixed numbers are first changed to mixed decimals.

1-41 (p. 218) It is usually best to proceed by changing the percent to a decimal, although the student could change the percent to a common fraction. Then multiply the given number by the decimal (or fraction). Multiplication of decimals was presented in Exercise 1–26. The Problem Solving section shows some everyday uses of percent and provide more practice in finding percents.

1-42 (p. 223) Making a fraction and division are the two skills needed for this exercise. The Problem Solving section shows some everyday uses of finding what percent one number is of another.

1-43 (p. 228) The procedure presented here is to change the percent to a decimal or fraction and then divide. Solving percent problems by using ratio and proportion is also introduced (optional). (Note that in addition to the usual provision for review and testing, students are provided with practice in "Estimating Answers" (optional) and "Using the Calculator" (optional) at the end of this unit.)

UNIT SIX   SQUARES AND SQUARE ROOTS

The main objective of the exercises in Unit Six is to help students master computations involving squaring a number or finding the square root of a number. Irrational numbers are also introduced.

Teaching Suggestions

1-44 (p. 240) Squaring whole numbers, fractions, and decimals is covered in this exercise. Point out to students that a table may also be used to find the square of a number.

1-45 (p. 242) The standard method, using the largest square for each group of two digits, is presented first. Alternate methods of solution (optional) are also given; these include the estimation-division-average method and using a square root table. The notation for units squared is shown in a footnote to the Practical Applications.

UNIT SEVEN SUPPLEMENTARY TOPICS

The main objective of the exercises in Unit Seven is to introduce the student to various topics that are usually studied in a contemporary arithmetic program. Various concepts about numbers and number systems and several arithmetic computational skills are included in these exercises.

Teaching Suggestions

1-46 (p. 250) Have students write Roman numerals for the numbers representing years. The Roman numerals are read by reading the letter for each numeral, "eye" for I, "vee" for V, "ex" for X, "el" for L, "see" for C, "dee" for D, "em" for M, and so on. MDCCLXVI would read, "em, dee, see, see, el, ex, vee, eye."

1-47 (p. 252) The use of three dots is introduced in this exercise to indicate a continuing pattern, such as 1, 2, 3, 4, . . .

1-48 (p. 253) Students have seen the x^2 form in Exercise 1-44. Powers of 10 may have been used in Unit Three, "Decimal Fractions." Point out to students that the exponent represents the number of repeated factors.

1-49 (p. 254) Practical applications for this exercise can be found in science texts and periodicals. Point out to students that the exponent 10 represents the number of zeros when the complete numeral is written out. You may wish to refer to Unit Four, "Shortened Names for Large Whole Numbers," when presenting this exercise.

1-50 (p. 255) Students may need to review the ideas of place value (see the introductions to Units One and Three) for this exercise.

1-51 (p. 257) Students have already worked with the relations equal, less than, and greater than. The symbols for not equal to, not less than, and not greater than are introduced here. These relation signs are usually used as verbs in number sentences.

1-52 (p. 259) Generally only whole numbers are classified as even or odd.

1-53 (p. 261) Students used the ideas of this exercise in Unit Two when working with fractions.

1-54 (p. 263) Students will use the ideas developed in the preceding two exercises for this exercise.

1-55 (p. 265) The two methods presented here are 1) a factor tree and 2) repeated division until only the numeral 1 remains.

1-56 (p. 267) Factoring will be used in this exercise when determining the least common multiple. Point out to students the distinctions between the least common denominator and the least common multiple.

1-57 (p. 269) Students used these ideas in Exercises 1-10 and 1-11 when working with fractions. Equivalent fractions are useful in all four operations with fractions and in comparing fractions.

1-58 (p. 270) This topic was also presented in Exercise 1-14. Here the method of cross products is introduced to compare fractions. Symbols to represent numbers (a, b, etc.) are also introduced in this exercise.

1-59 (p. 272) Students used reciprocals in the division of fractions, Exercise 1-17. The reciprocal and the multiplicative inverse for any number are the same.

1-60 (p. 273) Students will use the skills developed in Exercises 1-48 and 1-49 in this exercise. Significant digits are introduced here; they are discussed more fully in Exercise 2-20. To change a numeral to scientific notation results when the decimal place is moved to the left; a negative exponent results when the decimal place is moved to the right. When changing from scientific notation to completely written out form, a positive exponent indicates that the decimal point should be moved to the right; a negative exponent indicates that the point should be moved to the left.

1-61 (p. 275) Students worked with repeating decimals in Exercise 1-30 when they changed common fractions to decimals. A repeating sequence is indicated by a horizontal bar over the sequence.

1-62 (p. 276) The method presented is an algebraic solution using number sentences and equations.

PART 2 BASIC UNITS OF MEASURE

The primary objective of Part 2 is to help students master the use of the basic units of measure. Coverage in this part includes conversion within a system to smaller or larger units; the metric system; the customary system; measures of time, angles, arcs, speed, and temperature; precision and accuracy in measurement; significant digits; and approximate numbers. The introduction to Part 2 discusses changing to smaller or larger units; multiplication and division skills are needed here. The Inventory Test in Part 2 can be used to diagnose general difficulties before beginning the section.

UNITS OF MEASURE

This short section discusses and gives examples of "Changing to a Smaller Unit of Measure" and "Changing to a Larger Unit of Measure."

METRIC SYSTEM OF MEASURE

The main objective of these exercises is to help students master the use of the five basic metric system measures: length, mass, capacity, area, and volume. The chart on page 289 shows the place values for numerals and the corresponding unit of measure.

Teaching Suggestions

2-1 (p. 290) A metric ruler is used in this exercise. Millimeters, centimeters, decimeters, and meters are discussed. To change to smaller or larger units, see the introduction. For short methods of computation, see Exercises 1-28 and 1-29.

2-2 (p. 294) Although the kilogram is used to measure mass, students will use it to measure weight, since the word "weight" in everyday use almost always means mass. The kilogram is considered the basic unit of weight. To change to smaller or larger units, see the introduction to Part 2. For short methods of computation, see Exercises 1-28 and 1-29.

2-3 (p. 296) The liter is the basic unit used here. The units for capacity are the same as the units for volume (Exercise 2-5). To change to smaller or larger units, see the introduction to Part 2. For short methods of computation, see Exercises 1-28 and 1-29.

2-4 (p. 298) Measures of area use units squared; the exponent raised, small 2 used as an exponent represents the word "square." Centare and hectare are defined. To change to smaller or larger units, see the introduction to Part 2. For short methods of computation, see Exercises 1-28 and 1-29.

2-5 (p. 300) Units of capacity (Exercise 2-3) are also used to measure volume. Volume is the number of units of cubic measure; the exponent small 3 used as an exponent is used to represent the word "cubic." The chart shows some equivalents for volume and capacity. To change to smaller or larger units, see the introduction to Part 2. For short methods of computation, see Exercises 1-28 and 1-29.

CUSTOMARY SYSTEM OF MEASURE

The main objective of these exercises is to help students master the use of six basic measures in the customary system: length, weight, liquid measure, dry measure, area, and volume. Volume, capacity, and weight relationships are also covered. "Fundamental Operations with Denominate Numbers" on page 304 gives examples and procedures for each operation (addition, subtraction, multiplication, and division) with denominate numbers.

Teaching Suggestions

2-6 (p. 305) The basic units used are inches, feet, yards, rods, and miles. Students should simplify their answers as shown in the models on page 304.

2-7 (p. 307) The basic units presented are ounces, pounds, short tons, and long tons. Students should simplify their answers as shown in the models on page 304.

2-8 (p. 308) The basic units presented are ounces, pints, quarts, and gallons. Students should simplify their answers as shown in the models on page 304.

2-9 (p. 310) The basic units presented are pints, quarts, pecks, and bushels. Students should simplify their answers as shown in the models on page 304.

2-10 (p. 311) The basic units presented are the squared units of Exercise 2-6.

2-11 (p. 312) The basic units presented are cubic inches, cubic feet, and cubic yards.

2-12 (p. 313) These relationships are considered: gallons to cubic inches and cubic feet; pounds to cubic feet; bushels to cubic feet and cubic yards.

OTHER BASIC MEASURES

The main objective of these exercises is to help students master the use of several other basic measures: time, angles and arcs, rates of speed, and temperature. Coverage also includes the twenty-four hour clock and time zones.

Teaching Suggestions

2-13 (p. 314) The basic units presented are seconds, minutes, hours, days, weeks, months, and years. Students should simplify their answers as shown in the models on page 304.

2-14 (p. 317) The units presented are minutes, seconds, and degrees. Students should simplify their answers as shown in the models on page 304.

2-15 (p. 318) The knot and the nautical mile are presented here. The symbols for rates of speed in the metric system are also presented. In the customary system use the word "per"; in the metric system use the slash (/) for the word "per."

2-16 (p. 319) Students will have to know how to use formulas to convert units.

2-17 (p. 320) A 4-digit numeral is used to tell time on a 24-hour clock; 1200 is noon, 0100 is 1 A.M., and 2400 is midnight.

2-18 (p. 321) The four basic time zones in the United States are Eastern (EST), Central (CST), Mountain (MST), and Pacific (PST). You might mention daylight-saving time for each zone. Also, encourage a discussion on why the time zones exist.

MEASUREMENT

The main objective of these exercises is to help students use precision and accuracy in measurement, determine significant digits, and compute with approximate numbers.

Teaching Suggestions

2-19 (p. 323) Point out the distinction between precision and accuracy. Accuracy is determined by relative error. Students could perform some length measurements to develop their skills in determining precision and accuracy of measurements.

2-20 (p. 326) Significant digits usually need to be used with scientific notation (Exercise 1-60) and in determining precision in measurements.

2-21 (p. 327) Counting numbers are exact numbers. Numbers determined by measuring are approximate numbers. The result of computations with approximate numbers cannot be more accurate than the least accurate approximate number involved in the computation.

PART 3 FACTS AND SKILLS IN GEOMETRY

The primary objective of Part 3 is to help students master facts and skills of geometry. Techniques for Problem Solving (p. 356) will assist students in their solving of Geometric problems. Coverage in this part includes applications with units of length, applications with angles and arcs, constructions of various geometric figures, indirect measurement using trigonometry, and applications involving area and volume. The introduction to Part 3 points out that geometry means "earth measure." Thus geometry is the study of points, lines, planes, and space. The Inventory Test in Part 3 can be used to diagnose general difficulties before beginning the section.

APPLICATIONS USING UNITS OF LENGTH

The main objective of these exercises is to help students master the work with applications that use units of length. The topics covered include points and lines, drawing and measuring line segments, drawing to scale, geometric figures, perimeter, and the circumference of a circle.

Teaching Suggestions

3-1 (p. 336) Definitions of a point, a line, and a half-line are given. To name lines we use capital letters or a single small letter. Points are named by capital letters. Lines may be vertical, horizontal, or slanting. The identifying characteristics of intersecting, parallel, and perpendicular lines are given.

3-2 (p. 343) Three instruments can be used to draw and/or measure line segments: the straightedge, the ruler, and the compass. Instructions are given for using all three tools.

3-3 (p. 345) Scale definitions may be written as equality statements or as customary or metric ratios. Multiplication and division are used to determine actual distances, scale distances, or the scale. No actual drawing is required for this exercise. A road map would be a useful teaching tool for this exercise.

3-4 (p. 349) Coverage includes planes, space, angles, closed figures, triangles, quadrilaterals, circles, solid figures, and the Euler formula. It would be helpful to students to point out physical objects having the shapes covered in this exercise.

3-5 (p. 357) Three types of figures are covered: rectangles, squares, and triangles. Formulas are given for each type. Students may find it helpful to draw and label appropriate figures.

3-6 (p. 360) Various parts of a circle are defined and relationships given. The Greek lowercase letter "pi" is introduced.

APPLICATIONS USING ANGLES AND ARCS

The main objective of these exercises is to help students master the work with applications using angles and arcs. Coverage includes naming, reading, measuring, and drawing angles; angles in navigation; sums of angles of polygons; pairs of angles; parallel lines and angle relationships; congruent triangles; similar triangles; longitude, latitude, and nautical miles; longitude and time; and the magnetic compass.

Teaching Suggestions

3-7 (p. 363) Angles are formed by rays. Angles can be used to show direction. Three ways of naming angles are presented. Angle measure is discussed and then five kinds of angles are defined.

3-8 (p. 367)   The protractor can be used to measure angles. It is also used in drawing angles. Degree measure is used in this exercise (see Exercise 3-7).

3-9 (p. 370)   Direction of motion and wind direction are covered in this exercise. Vectors are introduced to show direction.

3-10 (p. 372)   Coverage includes angles of triangles, quadrilaterals, and other polygons. Students should recall the material presented in Exercise 3-4, "Geometric Figures," as well as that in Exercise 3-8. The rules for the sum of the measures of the angles in the three types of figures named above are given.

3-11 (p. 374)   Several different kinds of pairs of angles are discussed: complementary angles, supplementary angles, opposite or vertical angles, and the exterior angle of a triangle.

3-12 (p. 377)   This exercise hinges on the work of the previous exercise. The various pairs of angles formed when a transversal cuts two parallel lines are discussed.

3-13 (p. 379)   The symbol for "is congruent to" is introduced in this exercise. Three methods for identifying congruent triangles are presented.

Two of the practice problems involve geometric proofs.

3-14 (p. 381)   Three methods for identifying similar triangles are presented. In the problems students are asked to find the missing parts of two similar triangles when other parts of those triangles are given.

3-15 (p. 383)   This exercise provides practical applications of the ideas and skills taught in the previous exercises. Latitude, longitude, and related ideas are defined here. The nautical mile which was first introduced in Exercise 2-15, "Rates of Speed," is taken up again.

3-16 (p. 386)   This exercise again relates to the field of navigation. Time zones are reviewed; they were first discussed in Exercise 2-18. Then the conversion factors for changing longitude to time and time to longitude are presented. Solar or sun time is introduced here.

3-17 (p. 389)   This exercise again relates to navigation; this time the field is air navigation. The course of an airplane is the direction it flies over the earth and is given as an angle. Three different types of courses are presented: the true course, the magnetic course, and the compass course. Students learn how to find each type.

CONSTRUCTIONS

The main objective of these exercises is to help students master constructions of various geometric figures. Coverage includes constructions of line segments, triangles, polygons, angles, perpendicular lines, bisectors of line segments and angles, and parallel lines. Students will need a ruler and a compass for the exercises here.

Teaching Suggestions

3-18 (p. 393)   Students may use either a ruler or a compass for this exercise. The copies made with the compass may be more accurate. For additional practice, have students copy line segments from figures in the previous exercises of this text.

3-19 (p. 394)   Three parts, including at least one side, (out of six parts) of a triangle must be known in order to construct it. There are three ways that three such parts may be known; each of the combinations is presented. Students will need compass, ruler, and protractor for this exercise.

3-20 (p. 397)   Regular polygons can most easily be constructed by inscribing them in a circle, using the fact that equal central angles of a circle intercept equal arcs and equal chords. Review the

special names of polygons with the students (see Exercise 3-4). Students will need compass and protractor for this exercise.

3-21 (p. 399)   The definition of perpendicular lines, first seen in Exercise 3-1, is given again. The symbol for "is perpendicular to" is introduced. Students will need compass and protractor for this exercise.

3-22 (p. 400)   Students will need a compass for this exercise. You may wish to combine Exercises 3-21 and 3-22.

3-23 (p. 401)   Students may use either a ruler or a compass to bisect a line. For additional practice students may be referred to some of the figures in the previous exercises of Part 3.

3-24 (p. 402) Point out that the sides of the angle do not have to be equal in length to those of the given angle when students copy a given angle. Students may use a protractor or a compass for this exercise. For additional practice refer students to figures in the previous exercises of Part 3.

3-25 (p. 403) Students may use compass or protractor for this exercise. For additional practice refer students to figures in the previous exercises of Part 3 or to the angles they copied in Exercise 3-24.

3-26 (p. 405) The definition of parallel lines is given again here (see Exercise 3-1). The symbol for parallel lines is introduced. Students may use protractor or compass for this exercise. For additional practice have students draw some polygons with parallel sides.

INDIRECT MEASUREMENT

The main objective of these exercises is to help students master indirect measurements. Coverage includes the rule of Pythagoras, similar triangles, and numerical trigonometry (for right angles).

Teaching Suggestions

3-27 (p. 407) Have students measure the sides of triangles to verify the rule of Pythagoras. Students will use concepts they have learned about triangles in Part 3, squaring a number (Exercise 1-44), and finding square roots (Exercise 1-45).

3-28 (p. 410) Similar triangles were presented in Exercise 3-14; students may need to review that exercise before beginning this exercise. Ratios are used in the problems in this exercise; ratio was first introduced at the end of Exercise 1-18.

The three conditions for similar triangles are presented here.

3-29 (p. 412) The sine, cosine, and tangent are introduced in this exercise. Three steps are shown for solving problems using trigonometric ratios. Review the concept of complement of an angle with students since this idea is often used in numerical trigonometry. The boxes in the text give detailed solutions for different types of problems. A table of trigonometric values for the sine, cosine, and tangent (from 0° through 90°) is given on page 417.

APPLICATIONS MEASURING AREA AND VOLUME

The main objective of these exercises is to help students master work in the applications measuring area and volume. Coverage includes area of two-dimensional figures, surface area of solids, and volume of solids.

Teaching Suggestions

3-30 (p. 419) This exercise is a lengthy one and includes area measure of rectangles, squares, parallelograms, triangles, trapezoids, circles, rectangular solids, cubes, right circular cylinders, and spheres. Formulas for each type of figure are given. Students should be reminded to draw figures when figures are not given in the problems to help them find the solutions.

3-31 (p. 429) Physical representations of the solid geometric figures considered in this exercise could be used to help students find volumes. Coverage in this exercise includes volume measure of rectangular solids, cubes, right circular cylinders, spheres, right circular cones, and pyramids. You may wish to have students draw pictures of the solids to help them in their work.

PART 4 ELEMENTS OF ALGEBRA

The primary objective of Part 4 is to help students master the basic elements of algebra. Coverage in this part includes positive and negative numbers, computations with positive and negative numbers, the language of algebra, and open sentences in one variable including solutions of sentences and graphing. The Inventory Test in Part 4 can be used to diagnose general difficulties before beginning the section.

POSITIVE AND NEGATIVE NUMBERS; EXTENSION OF THE NUMBER SYSTEM

The main objective of these exercises is to help students master the concepts of positive and negative numbers, rational numbers, absolute value, the real number line, and additive inverse. Also, the student will learn about vectors, opposite meanings, and comparing integers.

Teaching Suggestions

4-1 (p. 446) The number line (introduced in Unit One of Part 1) is extended in this exercise to locate negative numbers, which are also introduced in this exercise. The idea of opposites is discussed and the integers are defined. Have students read the " − " sign as "negative" so that they do not confuse it with the operation sign designating minus. Also, have students read the " + " sign as "positive."

4-2 (p. 449) Rational numbers, irrational numbers, and real numbers are covered in this exercise. Fractional numbers are rational numbers; non-terminating and non-repeating decimals (see Exercise 1-61) are irrational numbers, as are square roots of any positive number other than a perfect square. All rational and irrational numbers together make up the real numbers.

4-3 (p. 452) Vertical bars are used to designate absolute value. The magnitude of a distance or a vector represented by a signed number are examples of the use of absolute value.

4-4 (p. 453) This exercise introduces students to the idea of graphing. Remind students that points are named by capital letters (see Exercise 3-1).

4-5 (p. 456) Have students make up their own examples that show opposite meanings. Also, students can represent opposite meanings on a number line.

4-6 (p. 457) Students will use skills developed in Exercises 1-14 and 1-53 for comparisons in this exercise. If students have trouble with comparisons involving negative numbers, have them graph the numbers on the real number line and point out that numbers to the left are always less than numbers found to the right.

4-7 (p. 459) Vectors have been used in previous exercises to show positive direction. Practical examples for this exercise include velocity of vehicles and wind and concepts such as force. A physics textbook would be a source of some practical applications.

4-8 (p. 462) The opposite of a signed number is sometimes called the additive inverse of the number. The sum of a number and its additive inverse is zero. Students will get practice in reading and writing expressions involving opposites, such as − (−4), in this exercise.

COMPUTATION WITH POSITIVE AND NEGATIVE NUMBERS

The main objective of these exercises is to help students master computations with positive and negative numbers. Students will learn how to do the four basic operations with integers and rational numbers.

Teaching Suggestions

4-9 (p. 464) Students may find it helpful to use vectors in finding sums of integers and rational numbers. Point out that the sum of two positive numbers is positive; the sum of two negative numbers is negative. The sum of a positive number and a negative number will have the sign of the number with the greater absolute value.

4-10 (p. 468) Opposites (see Exercises 4-5, 4-7, and 4-8) can be used to an advantage in subtracting integers and rational numbers. If students have trouble with subtraction here, review those earlier exercises. Again, the use of vectors may be helpful to students in subtracting integers and rational numbers.

4-11 (p. 470)   Point out that the product of two positive numbers is positive; the product of two negative numbers is positive. The product of a positive number and a negative number is negative. Addition facts can be used to develop multiplication. Students may have trouble multiplying two negative numbers. The steps on page 471 of the text show a logical development of this operation. For rational numbers, the idea of multiplicative inverse should be reviewed.

4-12 (p. 472)   The sign of a quotient in a division problem is determined in the same way as the sign of a product. Positive divided by positive is positive; negative divided by negative is positive. The quotient of a positive and a negative number is negative. The text suggests using absolute value for the division of integers and rational numbers and then affixing the sign to the quotient per the rules above.

LANGUAGE OF ALGEBRA

The main objective of these exercises is to help students master the language of algebra. Coverage includes symbols; mathematical expressions, phrases, and sentences; formulas; and evaluating algebraic expressions and formulas. The introduction reviews the signs of operation.

Teaching Suggestions

4-13 (p. 475)   This exercise reviews several common symbols used in algebra. The relation signs such as $=$ and $\neq$, exponents, square root signs, and parentheses are discussed. The terms "constant" and "variable" are defined and examples of variables are given. For additional practice in identifying variables, have students name the variables in the problems of Exercise 4-19.

4-14 (p. 476)   Numerical expressions, algebraic expressions, and mathematical expressions are explained. For additional practice, have students make up their own examples.

4-15 (p. 478)   Coverage in this exercise includes open sentences, equations and inequalities, and compound sentences. All open sentences contain one or more variables. (That is, $5 + 3 = 8$ is not an open sentence.) For additional practice refer students to the formulas for area and volume in Exercises 3-30 and 3-31.

4-16 (p. 480)   Formulas usually relate two or more quantities. Have students make up some of their own formulas. For additional practice refer students to the formulas for area and volume given in Exercises 3-30 and 3-31.

4-17 (p. 482)   Be sure students write the algebraic expression and the numerical values that substitute for each variable before they evaluate the expression. This will help them to avoid mistakes caused by carelessness. For additional practice refer students to the problems in Exercise 4-14 and let them choose their own values for substitution.

4-18 (p. 485)   Be sure students copy the formula, then substitute the given values in the formula, and then perform the operations. For extra practice refer students to the problems of Exercise 4-16 and have them use their own values for substitution.

OPEN SENTENCES IN ONE VARIABLE

The main objective of these exercises is to help students master work with open sentences in one variable. Coverage includes inverse operations; properties of equality; solving equations in one variable; solving equations when the replacements for the variable are restricted; solving equations with absolute values; using formulas; graphing an equation in one variable on the number line; solving equations involving fraction, decimal, and percent number relationships; ratio and proportion; solving inequalities in one variable; and graphing an inequality in one variable on the number line.

Teaching Suggestions

4-19 (p. 486)   Students will substitute numbers into a given equation to see if the resulting numerical sentence is true. When it is, that number is the solution to the equation. This exercise depends on skills developed in Exercises 4-17 and 4-18. For extra practice refer back to the problems in those exercises.

4-20 (p. 489) Addition and subtraction are associated inverse operations; multiplication and division are associated inverse operations. Inverse operations will be used in later exercises to solve equations. The properties of equality state that adding, subtracting, multiplying, or dividing (except by zero) both sides of an equation by equals gives equal results. These properties are used in later exercises to solve equations. For extra practice refer students to the problems in Exercise 4-18.

4-21 (p. 492) This exercise is a lengthy one; you may wish to devote two or more class periods to its discussion. Four types of equations are presented: equations involving (1) addition, (2) subtraction, (3) multiplication, and (4) division. Each type can be solved by inverse operations and by the additive or multiplicative inverse. Point out that equations may be given in forms slightly different from the models. Have students write out each individual step in the solution process to help them avoid mistakes due to carelessness. The last topic in the exercise deals with equations that must be solved by using more than one axiom (such as $5x + 14 = 25$).

4-22 (p. 499) The replacements for the variable could be restricted to a certain group of numbers, such as whole numbers, integers, real numbers, prime numbers, even numbers, odd numbers, positive integers, and so on. In these cases the solution to the equation may not exist.

4-23 (p. 501) Absolute value has been used repeatedly in this part. However, students may need to review Exercise 4-3, "Absolute Value," before proceeding with this exercise. The solutions of the equation $|x| = a$ (where a is any constant) and $+a$ and $-a$. For extra practice, refer students to the problems in Exercise 4-21, replacing the variables with the absolute values of the variable.

4-24 (p. 501) Students will use skills developed in Exercises 4-18 and 4-21 in this exercise. For extra practice refer students to the problems in Exercise 4-18.

4-25 (p. 503) The graph of most equations in one variable will be a single point on the number line; the graph of an equation with absolute value will be two points. Students may find it helpful to review Exercise 4-4 before proceeding with this exercise. For extra practice refer students to the problems in Exercise 4-21.

4-26 (p. 505) Corresponding previous exercises that dealt with these kinds of number relationships are Exercises 1-18, 1-19, 1-32, 1-33, 1-42, and 1-43. Refer students to these exercises and their problems for additional practice and review. Setting up the equation may be difficult for some students; the models in the boxes in the text give detailed solutions.

4-27 (p. 507) Ratios were first presented in Exercise 1-18; students may need to review the material there before proceeding with this exercise. A ratio is a comparison of two quantities by division. The colon may be used to show a ratio. 3 to 8 is 3:8. Rate of speed and the number of items selling at a certain price are two common applications of ratio.

4-28 (p. 510) There are four terms in a proportion. If any three of the terms are known, the fourth term may be found by solving the equation. Multiplicative inverse will be used in the solution. To determine if the proportion is true, find the cross products.

4-29 (p. 513) The solution of an equation in one variable is one number, but the solution of an inequality in one variable is more than one number. Inequalities are solved by using transformations in much the same way that equations are solved. Point out that multiplying or dividing both sides of an inequality by the same *negative* number *reverses* the order of the inequality. The axioms used in solving inequalities are classified in three parts, and the first group of practice problems are then classified into corresponding parts. Three dots are used in writing the solutions for some inequalities; see Exercise 1-47, "Reading and Writing Shortened Lists of Numerals."

4-30 (p. 518) The graph of an inequality in one variable may be a half-line, a ray, a line segment, or an interval (see Exercise 3-1). Open dots are used to show the exclusion of endpoints on the graph. For extra practice refer students to the problems in Exercise 4-29.

PART 5 GRAPHS, STATISTICS, AND PROBABILITY

The primary objective of Part 5 is to help students master the basic ideas of graphs, statistics, and probability. Coverage in this part includes bar, line, and circle graphs; frequency distributions, histograms, and frequency polygons; averages and measures of central tendency; percentiles and quartiles; and probability and odds. The Inventory Test in Part 5 can be used to diagnose general difficulties before beginning the section.

GRAPHS

The main objective of these exercises is to help students master work with various types of statistical graphs, namely, bar graphs, line graphs, and circle graphs.

Teaching Suggestions

5-1 (p. 527) Bar graphs are often used to compare population statistics and changes in prices, cost of living, wages, and so on. Practical examples of bar graphs can be found in news magazines, especially those devoted to the business interests, and local newspapers. Additional problems could be made up from these sources or from statistics given in the various world almanacs.

5-2 (p. 528) Line graphs are often used to show relationships between quantities, such as between income and amount of education, temperature and month, stock prices and day or month, and so on. Practical examples of line graphs can be found in magazines, especially those devoted to business and education interests, and local newspapers. Additional problems could be made up from these sources or from statistics given in the various world almanacs.

5-3 (p. 530) Circle graphs are often used to compare percentages, such as percentages of oil reserves in the world, percentages of money spent for various items, and percentages of the federal budget devoted to various expenditures. Practical examples of circle graphs can be found in news magazines, especially those devoted to business and economic interests, and local newspapers. Additional problems could be made up from these sources or from statistics given in the various world almanacs.

STATISTICS

The main objective of these exercises is to help students master the ideas of statistics, such as frequency distributions and polygons, histograms, averages, percentiles, and quartiles.

Teaching Suggestions

5-4 (p. 531) A table of data showing frequency is a frequency distribution. The graphs of the tabular data are histograms (a bar graph) and frequency polygons (a line graph). Practical examples of these kinds of data can be found in news magazines and local newspapers. Additional problems could be made up from these sources or from statistics given in the various world almanacs.

5-5 (p. 532) Three common types of measures of central tendency are presented: mean, median, and mode. The mean is the measure that is commonly associated with the word "average." The mean can be used to find the average number of students in a class; the median can be used to find the average income of a group of people; the mode can be used to find the average shoe size purchased in a store. The range is a measure of data variation; it is the difference between the highest and lowest numbers in the data. Practical examples might be found in educational journals; other examples to use for additional practice can be taken from statistics textbooks.

5-6 (p. 535) A percentile is a percentage rank of a score. Quartiles divide a group of scores into four parts. For additional practice refer students to the problems in Exercise 5-5.

5-7 (p. 536) Probability is used in statistics to determine the accuracy of a prediction or an estimation. It is also used to determine the chance of error for a statistical method. Probability is the ratio of favorable outcomes to the total number of possible outcomes. Odds are ratios. Numerous practical examples are shown in the practice problems. Other examples can be found in a probability textbook.

PART 6 EVERYDAY APPLICATIONS—PROBLEM SOLVING

The primary objective of Part 6 is to help students master everyday applications of mathematics, especially problems they may encounter in the future as consumers. Coverage includes income, various types of taxes, applications involving spending money and managing money, and applications in business and travel.

INCOME, TAKE-HOME PAY, INCOME TAXES

The main objective of these exercises is to help students master applications involving income and taxes, including state and local income taxes.

Teaching Suggestions

6-1 (p. 542) Various types of incomes are presented in this exercise, including hourly wages; weekly, monthly, and annual salaries; piece work; commission and bonuses; profit; fees, tips, and pensions; and interest and dividends. You may wish to discuss the various occupations or ways that each of these types might be earned. News magazines and business magazines would be sources for published income figures.

6-2 (p. 548) Two methods are shown here: the percentage method and the table method. You might wish to obtain information and tables from the local internal revenue office for single persons' taxes since those rates are different from the married persons' rates shown in the text.

6-3 (p. 551) Additional information about social security taxes can be obtained from the local social security office. Also news magazines frequently publish articles about social security taxes that might be of interest to your students.

6-4 (p. 551) Students may be interested in their own local tax rates; local and state tax rates in your area can be obtained from local government offices.

6-5 (p. 552) Published figures for some average take-home pays can be found in business magazines and news magazines. These figures might be of interest to your students, especially for those careers which the students might be considering.

6-6 (p. 554) The method shown here is the use of the tax tables. You may wish to obtain the tax tables for single persons since those figures are different from the figures for married persons given in the text. The local internal revenue office can supply you with more information and additional tables.

SPENDING MONEY

The main objective of these exercises is to help students master applications involving spending money, including buying food, unit pricing, making change, sales tax, utilities, owning an automobile, insurance, property tax, and buying and renting a home.

Teaching Suggestions

6-7 (p. 557) Students might visit a local grocery or supermarket to find current prices on items of food in which they are interested. Or grocery or supermarket ads in the local newspaper can be consulted for additional practice problems.

6-8 (p. 560)   Consult supermarket ads in the local newspapers for additional practice. Or have some students compare prices at a local grocery.

6-9 (p. 564)   For additional practice problems consult a local newspaper for sales in local stores. Also, for similar items on sale at several stores, you might have students determine the better buy.

6-10 (p. 566)   Use the prices given in some of the previous exercises for additional practice problems; have students specify the cash tendered for the purchase. Also, it may be helpful to have students state the bills and coins they expect to receive in change as well as the amount in dollars and cents figures.

6-11 (p. 567)   Use the tax rate for your area and some ads in a local newspaper to determine total selling price of items in the ads. Have students find out if there is also a local meal tax at restaurants, a tax for staying in a motel or hotel, or any other specific taxes in your area.

6-12 (p. 568)   Have students read the meters in their own homes. Then use the local rates (obtain them from your local utility companies) to create a fictitious bill for a month of service. Local utility companies may also have typical usages for various sizes of apartments and homes which are of interest to students.

6-13 (p. 571)   Current prices of new and used cars can be found in local newspapers. Other expenses, such as local taxes, gasoline costs, insurance premiums, and so on, can be obtained from appropriate local business and government offices. Average repair costs can also be obtained, along with an estimate of the average number and kinds of repairs typical car owners experience.

6-14 (p. 572)   Three types of loans are discussed here: installment buying, annual percentage rate loans, and deferred payment purchase plan. Local stores and banks might be contacted to find out current credit arrangements. Additional information could also be obtained from nationwide credit card companies.

6-15 (p. 577)   Local insurance agents can be contacted to obtain more up-to-date information or information specific to your area. A typical fire insurance policy can be obtained and some of the details of the policy can be explored by the class.

6-16 (p. 578)   Local insurance agents may be contacted to obtain current information about policies. Information may also be obtained from the home offices of nationwide life insurance companies. Copies of typical policies may be obtained and their terms discussed by the class.

6-17 (p. 580)   Contact the local government office to find out local property tax rates on different types of buildings and real estate. Typical assessed values on houses in the area can also be obtained. You might wish to discuss assessed value versus true market value of homes if these figures differ in your locality.

6-18 (p. 582)   Consult local newspapers or contact local real estate agents to find out prices of typical homes in your area and costs of renting houses. Banks can be contacted to find out about the terms of mortgages in your areas. Real estate offices or local government and federal government offices might have information on the typical expenses that are encountered in owning a home, such as insurance, repairs, depreciation, improvements, and so on.

MANAGING MONEY

The main objective of these exercises is to help students master applications involving managing money, including banking, simple interest, borrowing money, compound interest, and investments such as stocks and bonds.

Teaching Suggestions

6-19 (p. 584)   Deposit slips and blank check forms can be obtained from a local bank. Various forms can be shown to the class and discussed with them.

6-20 (p. 586)   The formula $i = prt$ is used in this exercise. Encourage students to write the formula each time they solve a problem and to write an equation (such as $r = 4\%$) for each item of information they are given.

6-21 (p. 589)   Three topics are discussed in this exercise: discount and add-on interest loans, financing a car, and amortizing a loan. Local banks may be contacted to find out current terms and rates of interest in your area which can be discussed with the class. You might also wish to find out what other types of general loans are available (home improvement loans, collateral loans, etc.) and discuss these with the class.

6-22 (p. 594) Most savings accounts offer compound interest. Contact a local bank to find out the different types of savings plans available and discuss these with your class. Notice from the table that a person's money is doubled in 12 years at 6% interest, in 15 years at 5% interest, and in a little over 17 years at 4% interest.

6-23 (p. 598) Banks and stock brokerages can be contacted for more information on stocks, bonds, mutual funds, and money rates. Several other types of investments are also available (such as money markets); you might wish to discuss these briefly with your class. Local or major city newspapers can be consulted for current prices on the stock exchange and for additional items of interest about investments. Financial magazines are other sources of additional information.

APPLICATIONS IN BUSINESS AND TRAVEL

The main objective of these exercises is to help students master applications in business and travel, including discount, commission, profit and loss, reading a road map, and reading a timetable.

Teaching Suggestions

6-24 (p. 604) The problems in this exercise are similar to those in Exercise 6-9. As in that exercise, you might wish to consult a local newspaper for sales in local stores for additional practice problems. Discount stores and catalog stores are other sources of applications of the ideas presented in this exercise.

6-25 (p. 607) Commissions are usually found in the field of sales; most salespersons work on a commission basis. You might wish to discuss the many types of items that may be sold on a commission and contrast the commission rates (selling a house versus selling a car, for example).

6-26 (p. 609) You might wish to discuss with your students the types of stores that would have relatively high markups and those that would have relatively low ones. (For example, small specialty stores might have high markups but most large grocery chains have low markups on most goods.)

6-27 (p. 614) For additional practice problems, bring a local road map to class and have students read it. Refer also to Exercises 3-3, "Scale."

6-28 (p. 615) Local train and bus terminals may be contacted to obtain various types of timetables. These can then be discussed with the class. Airlines may also be contacted for their timetables.

ACTIVITIES

(1-7, 5-5) Calorie Counting

Page References: pp. 68–70, 532–534
Time Required: one class period, one week of homework
Materials Needed: calorie-counting booklets

◆

Invite the school nurse, a dietician, or a physician to address the class on nutrition.

Ask the students (or volunteers) to keep records of everything they eat each day for a week.

Make available to the students copies of a calorie-counting booklet. Have each assess his/her daily, weekly, and average daily calorie consumption. The class can then determine the total consumption and compute an average.

Using the calorie booklets, students should try to determine individual calorie needs to maintain, increase, or decrease weight, and thus establish suitable diets.

(1-13, 5-2, 6-23) Investing in the Stock Market

Page References: pp. 86–97, 528–529, 598–603
Time Required: one class period, one month's homework
Materials Needed: stock market page from newspaper

◆

Allocate to each student $1,000 to invest in shares in the stock market.

Have each student choose a stock from the New York Stock Exchange or the American Stock Exchange. Require each student to plot on a line graph the stock movement for one month.

At the beginning of each class the students will add to their line graphs, using newspapers brought into class by students and/or the teacher.

At the end of the month, each student should discuss with the rest of the class his or her graph, indicating the profit or loss made on the investment if the stock were "sold" at this time. Make sure deductions are made for brokers' commissions.

Invite a businessperson to address the class, allowing for follow-up questions by the class.

(1-24, 6-5) Adding and Subtracting Decimals

Page References: pp. 133–152, 542–556
Time Required: one class period
Materials Needed: one 1040A tax form for each member of the class, available at your local post office

◆

Divide the class into groups (perhaps by rows). Allocate different amounts by group for wages, exemptions, and taxes withheld. Ask the class to fill out the form. Stress the need for accuracy and allow sufficient time for completion.

If there is enough time, groups can exchange forms and check for accuracy.

(1-36) Percent Designs

Page References: pp. 206–208
Time Required: one–two class periods
Materials Needed: ditto master of a grid measuring 10 cm × 10 cm, divided into 100 congruent squares; in each square both digonals are drawn so that the grid contains 400 congruent triangles

◆

Provide a copy of the grid to each student.

Then have each student shade triangles on his/her grid to form any creative design of his/her choosing. (You may want to require a symmetric design.) No partial shading of any triangle is allowed.

Have each student record the percentage of his/her grid that is shaded and the percentage that is not. Then circulate the grids and have each student record the percentages of every other students' grids beside their names. Collect the designs and the student sheets of percentages for correction.

(2-3, 2-5) Metric Volume (Capacity)

Page References: pp. 296–298, 300–302
Time Required: one class period
Materials Needed: several 25 mL or 100 mL graduated cylinders, common household containers such as teaspoons, tablespoons, cups, tumblers, 1 qt. or 2 qt. milk or beverage bottles or containers, tap water (or large containers filled with water) and provision for disposing of waste water

◆

Instruct students to:

Partially fill a 25 mL or 100 mL graduated cylinder with water. Set the cylinder down on a level surface where you can read the lowest level of the curved liquid surface with your eyes at the same level as the liquid. Read the volume of water in the cylinder to the nearest milliliter.

Carefully pour just enough water into the teaspoon to fill it level with the top. Be careful not to spill any water. Read the volume of water left in the graduated cylinder. By subtraction determine the volume in milliliters of water poured into the teaspoon. Repeat the measurements for the table-spoon or any other small-volume container you may have available.

Refill the graduated cylinder almost up to its maximum capacity and record the volume of water in it. Pour all of the water into a cup or tumbler which holds more than 200 mL. Refill the cylinder as many times as necessary and empty into the larger container. Each time record the volume of water poured out to the nearest mL. The last liquid added to the container will be less than the volume of the graduated cylinder. Fill the cup or tumbler level full, but not overflowing with water. By addition, find the volume of the larger container to the nearest milliliter.

If you are using a 100 mL graduated cylinder and have a 1 qt. or 2 qt. bottle or container, use the same method to determine the volume of the large container.

Have the students make a chart showing their findings.

(2-5) Measuring Your Textbook

Page References: pp. 300–302
Time Required: one class period
Materials Needed: *Refresher Mathematics*
  textbook, metric ruler

◆

Have students use a metric ruler to determine the length, width, and thickness of the pages of their textbook, excluding the covers. Measurements should be made to the nearest millimeter.

From the pages numbered with both Roman and Arabic numerals, plus unnumbered pages, students should find the total number of pages, excluding the covers. From this, find the total number of sheets in the book. (How many pages to a sheet?) Continuing their calculations in measurement, have students:

  (a) calculate the area of each page,
  (b) find the volume of the book, excluding the covers,
  (c) calculate the thickness of each sheet to the nearest hundredth of a millimeter,
  (d) express this last figure in meters; in micrometers.

(2-15, 6-13, 6-27) Planning a Motor Trip

Page References: pp. 288–318, 571, 614
Time Required: one class period
Materials Needed: road map of home state

◆

Supply, or have students obtain road maps of their home state or region. (An atlas that shows major highways and distances between major cit-ies will suffice.) Have students plan a motor trip from their home town to the state capital, from there to another major city 200 or 300 miles away, and then back to their home town. (If students live in the state capital, choose another city.) Have them:

—Lay out the journey from city to city, estimating distance and time required for each leg
—Estimate the driving time for each leg by assuming that they average 50 miles per hour
—Find the total distance and total driving time for the three legs of the journey
—Estimate the amount of gasoline required for each leg, assuming that the car averages 20 miles per gallon (Or have students use the actual rate of fuel consumption of their family's car, if known)
—Use the prevailing cost locally of the kind of gasoline used in the family car to estimate the total cost of fuel for the journey

Have the students convert their estimates of total distances, driving times, gasoline consumption, and fuel costs into kilometers, hours, liters, and dollars, assuming there are 1.6 kilometers per mile, and 3.8 liters per gallon. Explain why two of the totals change while two of the totals remain the same.

(3-3, 3-14) Scale Drawing

Page References: pp. 345–348, 379–382
Time Required: three class periods
Materials Needed: graph paper, poster board,
  colored pencils or magic markers

◆

Instruct groups of students to prepare wall posters advertising a product or event of the group's choosing. The poster should first be designed in a scale drawing on graph paper. The scales should be given.

Encourage students to plan the layout carefully, advising them to determine the central message and how to best effect it by clever use of large and small printing and open space.

After the scale drawing is complete, the full-sized poster should be made in accordance with the scale established in the scale drawing.

(3-6) Finding the Value of π (pi)

Page References: pp. 360–361
Time Required: one class period
Materials Needed: metric tape measure, meter
  stick or metric ruler, several cylindrical objects
  inside and outside the classroom

◆

Have available in the classroom (or have students bring in) several cylindrical containers of different diameters, such as tin cans, tumblers, pots, etc. Send some students outside the classroom to measure a flagpole, telephone pole, cylindrical trash container, etc. For each object have students measure to the nearest millimeter the circumference (with metric tape measure) and diameter (with meter stick or ruler). Each student should be asked to measure at least five objects of as widely different diameters as possible. For each object, divide the circumference by the diameter, carrying it out to the nearest two decimal places. The ratio of the circumference to the diameter of a circle is called π (pi). Find the average value of π for all the measurements made. Have students look up the accepted value of π in a reference book. If their measured values differ from the accepted value (to two decimal places), have them determine the percent difference.

(5-4) Probability and Statistics—Hat Sizes

Page References: pp. 531–540
Time Required: two class periods and optional out-of-class activity
Materials Needed: string, centimeter ruler

◆

Have students determine their hat sizes to the nearest centimeter using a piece of string and a centimeter ruler, and write the numbers on the chalkboard. From this data have students construct a frequency table and calculate the percentage of each size. (See textbook p. 531)

Ask the students to design a simple hat which could actually or hypothetically be made and sold to the other students in the school as promotion for some school function. They would need to decide the following:
(a) the number of hats needed (an estimated percentage of the student body),
(b) the percentage of each size as indicated by the class sample,
(c) the number of hats in each size,
(d) the amount of material needed for the hats,
(e) the price of the material,
(f) the price charged for each hat for the project to break even.

Conduct a class discussion on the usefulness and many applications of sampling and polls in society, e.g.: manufacturing decisions which must relate to disparate market needs; political predictions; quality control, i.e., expected defective parts.

Allow interested students to research one or more of these areas and report to the class.

(5-5, 6-8) Planning a Luncheon

Page References: pp. 532–534, 560–563
Time Required: three class periods, one homework assignment
Materials Needed: none

◆

Plan a lunch for senior citizens. Set up groups to:
—Arrange for school accommodations (establishing the number to be invited), invitations, and transportation
—Decide on the menu and the necessary quantity of each item
—(Two groups) comparatively price commodities in competing stores
—Keep careful records of hours spent planning and actual work time spent on the project
Following the planning, the class should develop a chart showing:
(a) the number of pupil hours spent,
(b) the individual and total projected costs of the project,
(c) the average cost per guest.

(5-7) A Demonstration of Probability Theory

Page References: pp. 536–539
Time Required: one class period, one homework assignment
Materials Needed: two pennies (or other coins of the same kind) for each pair of students

◆

Have each pair of students flip two similar coins simultaneously at least 50 times and record the results under the following categories: 2 heads; 1 head and 1 tail; 2 tails. Add up the results from each pair of students to get a total for each category for the entire class. Ask the students to analyze these results to discover the approximate ratio expressed in small whole numbers. (Rounding will be necessary to arrive at the 1:2:1 ratio.)

As a follow-up application ask students to make an outside investigation into genetics, the study of how characteristics are passed on from parents to offspring. Refer students to encyclopedias or biology textbooks for information on Gregor Mendel's law of genetics. Have them look for specific applications of the 1:2:1 ratio in the inheritance from individual parents of contrasting pure traits.

As an alternative, invite a biology teacher or eager biology student to explain to the class how the 1:2:1 ratio applies to Mendel's laws of genetics, as found in blending inheritance.

(6-1, 6-13) Spending Money and Earning Money

Page References: pp. 542–547, 557–571
Time Required: one–two class periods
Materials Needed: *Dictionary of Occupational Titles, Occupational Outlook Handbook*

◆

Have each student:
—Select a hobby or activity of his/her choice
—Determine the initial involvement costs and the continuing costs for a year
—Determine the amount of hours per day, per week, per month, per year the hobby/activity has taken or is likely to require
—Make a list of jobs associated with the hobby/activity (have students consult the *Dictionary of Occupational Titles* and/or the *Occupational Outlook Handbook*), giving an estimate of what gach job pays
—Choose one job, ascertain the approximate salary per year, and compute the hourly, weekly, monthly wage
—Determine accordingly how much his/her time already invested in the hobby or activity would have been worth monetarily
—State all of the ways in which math is used in the hobby or activity, demonstrating at least three specifically

(6-1) Producing a Play

Page References: pp. 542–547
Time Required: two class periods, one homework assignment
Materials Needed: the production costs for a recent school theatrical production

◆

Invite a member of the theatre arts department to speak to the class about the costs of producing a play. List these on the board, being sure to include the following:
  (a) royalty fees, if any, for all performances,
  (b) cost of scripts,
  (c) cost of make-up and costumes,
  (d) costs for scenery, including paint, props, boards for backdrops, and any labor charges involved in constructing the staging,
  (e) cost of sheet music, if used,
  (f) cost of advertising, including raw materials for display posters and tickets.
When the costs have been estimated, have the class determine how many tickets at what price must be sold in order to make $300 profit. If $300 is an unrealistic figure, have students select a more reasonable amount. In computing the number of tickets that must be sold, students should determine the number of seats in the auditorium, the number of performances, and the average

number of spectators they can expect for each performance.
    As an added activity have students determine the number, by category, of students who will contribute to the production in any capacity. Lastly, have them estimate the hours spent by the cast and crew for the entire production.

(6-7, 6-13) Spending Money

Page References: pp. 542–547, 557–571
Time Required: one class period
Materials Needed: none

◆

As an exercise in budgeting, instruct students to:
—List several items which they would realistically expect or desire to purchase over the next year
—List the separate costs for each item
—Establish their average weekly income or use $10 as a hypothetical weekly allowance if their actual allowance or earnings is less than $10
—List weekly amounts used for fixed and estimated expenditures
—Determine how much of their weekly income could be saved toward the purchase of these items
—Determine if any income can be set aside for savings in general and not toward the purchase of the desired items
At the conclusion of this exercise, students should be able to show whether or not any or all of their desired items could be purchased at some specific time during the year.

(6-8) Comparative Shopping

Page References: pp. 560–563
Time Required: one class period, one homework assignment
Materials Needed: none

◆

Have students choose hobbies requiring several pieces of equipment (e.g., skiing, photography, oil painting, skin diving) and make a list of necessary items. Then, using catalogs or stores, comparatively price each item to establish at least four prices for each.
    Have students compute and decide upon three packages of needed equipment: the least expensive, the most expensive, and the package of personal choice.

(6-8, 6-18) Repairing a Home

Page References: pp. 560–563, 582–583
Time Required: two class periods, one home-work assignment
Materials Needed: none

◆

Have a student with parental permission volunteer the family's home to establish basic dimensions necessary to make the following repairs:
—All new gutters
—New wallpaper in the living room
—New paint on the exterior of the house
—New linoleum on kitchen floor

Divide the class into four groups, assigning one repair to each group. Each group should research the following:
(a) the quantity of material needed for the repair,
(b) the unit cost of 3 competing materials,
(c) the total cost of repairs using each competing material.

Each group should report to the class on its findings, including the relative merits of competing material. The class as a whole then votes on which material to use.

(6-9) Comparing Food Prices

Page References: pp. 557–565
Time Required: one class period, one home-work assignment
Materials Needed: none

◆

As an exercise in price comparison shopping, divide the class into groups of two. Instruct each pair of students to visit one supermarket with the following list of food:

    canned tomatoes
    hamburger
    salad dressing
    margarine
    tuna
    5 lbs. potatoes
    vegetable soup
    canned green beans
    coffee

One student will select and record the most expensive product on the shelf for each item. The second student will select and record the least expensive item. Each student should find the same weight and amount of each product. Have students total the results of their selections. Compare the difference in cost between the expensive and inexpensive brands.

Compute the savings on the same items if they were bought each week for one year.

As a follow-up activity conduct a class discussion on the real value of coupons and sale prices. Have groups of students validate conclusions at local markets.

(6-12) Family Energy Costs

Page References: pp. 568–570
Time Required: one class period, one home-work assignment
Materials Needed: graph paper (optional)

◆

Have volunteers make a line graph, comparing their family costs over one year for electricity, gas, and heat for a one-family home. Place the months of the year on the horizontal line and the cost in dollars on the vertical line. Then have students determine the average cost per year for each utility.

As an additional exercise, have students list at least ten specific ways they can save on energy costs.

(6-13, 6-22) Buying a Car

Page References: pp. 571–576, 589–597
Time Required: two class periods, one home-work assignment
Materials Needed: several copies of the automotive section of a Sunday paper, bank information on car loans

◆

Divide the class into groups and ask each group to choose from the newspaper a used car of their choice. Stipulate that they have $500 to use as a down payment.

Each group will have to ascertain from a local bank how much money the bank will loan based on the make, year, and model of their choice and what the interest rate will be. Each group should list the cash price, down payment, interest rate, and the number of monthly payments they decide upon. Based on the amount and number of monthly payments, each group will then determine the annual percentage rate, using the chart on page 575. Finally, each group should compute the difference between the cash price and the actual cost including the finance charge.

Have each group present their results to the rest of the class for discussion.

As a bonus activity, the group may:
(a) calculate the depreciation of the car by the time the loan has been repaid,
(b) find the cost of insurance for one year,
(c) make a chart showing the fixed and variable costs of owning a car. Include under fixed costs insurance, license fees, and depreciation. List gasoline and oil, maintenance, and tires under variable costs.

(6-18) Housing—Buying vs. Renting

Page References: pp. 568–570, 572–577, 580–583

Time Required: four class periods, three homework assignments

Materials Needed: real estate section of a local newspaper

◆

Have students:

—Look through the real estate ads in their local newspapers for selling prices of homes

—Ascertain a current interest rate on a home loan from a local bank, and determine how much a 20-year mortgage would cost for a house of their selection assuming the minimum down payment.

—Break down the mortgage loan into monthly payments

—Determine the cost of property taxes for their locale

—Find out the cost of insurance needed for the value of their house

—Calculate the approximate costs per year for utilities

—Adding all costs, determine the expected yearly costs of home ownership

When students have completed their analysis of home ownership costs, have them calculate and compare the costs per year to rent a house of comparable size and value. Be sure they include utilities if not included in the rent. Discuss with the class the advantages and disadvantages of home ownership.

As an additional activity in the study of housing costs, have the students compare the costs of renting comparable furnished and unfurnished apartments.

Have them determine the costs of furnishing the unfurnished apartment including finance charges and add the results to the rent. Compare the total yearly cost to that of the furnished apartment.

(6-19, 6-26) Investing Money

Page References: pp. 584–613

Time Required: two class periods, one homework assignment

Materials Needed: bank information on various types of accounts

◆

Allocate $10,000 to each student to be "invested" in any of the local banks. Have them investigate the various savings plans, e.g., regular accounts, special notice accounts, and term deposit certificates.

Have each student "invest" his/her money in a combination of plans that will account for:

(a) an amount needed for emergency,

(b) an amount probably needed in a few years,

(c) an amount for investment only.

Then have students compute the results of their plans in a ten-year period with no prior withdrawals.

(6-26) Recycling Project

Page References: pp. 557–583

Time Required: two–three class periods, three homework assignments

Materials Needed: none

◆

Ask volunteers to find out how much scrap dealers in aluminum, glass, or paper will pay for scrap. Students may use the Yellow Pages to locate the dealers.

Based on availability of waste product and price they will receive per ton, students should choose one of the scrap products and plan a collection drive. Students should plan:

(a) the gasoline costs used for vehicles in the collection process,

(b) the rental costs for trucks if needed,

(c) the cost of advertising,

(d) the number of hours required for collection,

(e) the number of workers needed.

By checking with parents and neighbors, students should be able to determine a realistic figure for the amount they can expect to collect. (The entire class can pool their findings and come up with an average.) Based on the expected size of collection, the initial investment costs, and the price offered by the scrap dealer, students should then determine their expected profit.

(6-28) Planning a Trip

Page References: pp. 557–572, 614–617

Time Required: one class period, one homework assignment

Materials Needed: travel page from newspaper, travel agencies

◆

Have each student plan a trip for one week to a city at least three hundred miles away from home. Have them:

—Obtain information conerning traveling costs from newspaper ads or travel agencies

—Compare cost for airplane, bus, train, and car

—Plan the cost of accommodations at a reasonably priced hotel

—Estimate transportation cost within the city

—Include new travel clothing if needed for the trip

Then have each student account for individual and total costs.

COMPUTER ACTIVITIES

The activities below exemplify the use of the computer with specific Exercises in RE-FRESHER MATHEMATICS. The type of computer, student ability, and time will dictate modifications and expansions. Familiarity with your specific computer and its BASIC language is necessary. Individual *computer syntax* must be verified in order to use the examples expressed below.

Activity Aim: To use the computer as a calculator.

SAMPLES

Addition:

PRINT 357 + 563
The output is 920.

Subtraction:

PRINT 803 − 259
The output is 544.

Multiplication:

PRINT 1728 * 563
The output is 972,864.

Division:

PRINT 4563/9
The output is 507.

Use the computer as a calculator for any calculation involving whole numbers and decimals. See especially "Check By Calculator" on pp. 71 and 190.

Activity Aim: To use the computer to program basic arithmetic operations.

Addition:

```
10   CLS
20   FOR X = 1 TO 5
30   INPUT "ENTER FIRST ADDEND"; A
40   INPUT "ENTER SECOND ADDEND"; B
50   C = A + B
60   PRINT; A; "+"; B; "="; C: PRINT
70   NEXT X
80   END
```

See pp. 30–40 for addition of whole number problems.

Subtraction:

```
10   CLS
20   FOR X = 1 TO 5
30   INPUT "ENTER MINUEND"; A
40   INPUT "ENTER SUBTRAHEND"; B
50   IF B > A GOTO 90
60   C = A − B
70   PRINT A; "−"; B; "="; C: PRINT
80   NEXT X
85   GOTO 110
90   PRINT "SUBTRAHEND TOO LARGE!
     REENTER"
100  GOTO 30
110  END
```

See pp. 41–48 for subtraction of whole number problems.

Note the program changes that are underlined in red. Have the class now program multiplication (pp. 49–57) and division (pp. 58–70) problems.

Activity Aim: To use the computer to solve problems using random selection of addends.

```
10   CLS
20   PRINT "ADDITION PROBLEM":
     PRINT
30   W = W + 1
40   A = RND (99)
50   B = RND (99)
60   C = A + B
70   PRINT A; "+"; B; "="
80   INPUT "ANSWER"; X
90   IF C = X THEN 110
100  GOTO 150
110  Q = 0
120  CLS
130  IF W = 5 THEN 250
140  GOTO 20
150  Q = Q + 1
160  IF Q = 2 THEN 190
170  PRINT "TRY IT ONE MORE TIME"
180  GOTO 70
190  PRINT "THE CORRECT ANSWER IS"
200  PRINT A; "+"; B; "="; C: PRINT
210  Q = 0
220  C = X
230  If W = 5 THEN 250
240  GOTO 20
250  END
```

You can expand this program to all operations with whole numbers. See pp. 30–70.

Activity Aim: To use the computer to solve rectangular area and perimeter problems using formulas.

Area:

```
10   CLS
20   PRINT "FINDING THE AREA OF A
     RECTANGLE": PRINT
30   PRINT "AREA = LENGTH × WIDTH"
40   FOR X = 1 TO 5
50   INPUT "ENTER LENGTH"; L
60   INPUT "ENTER WIDTH"; W
70   A = L * W
80   PRINT "AREA ="; L; "X"; W; "="; A;
     "SQUARE MEASURE": PRINT
90   NEXT X
100  END
```

See p. 419 for rectangular area problems.

Perimeter:

```
10   CLS
20   PRINT "FINDING THE PERIMETER
     OF A RECTANGLE": PRINT
30   PRINT "P = 2L + 2W"
40   FOR X = 1 TO 5
50   INPUT "ENTER LENGTH"; L
60   INPUT "ENTER WIDTH"; W
70   P = 2 * (L + W)
80   PRINT "PERIMETER ="; P; "LINEAR
     MEASURE": PRINT
90   NEXT X
100  END
```

See pp. 357 and 358 for rectangular perimeter problems.

Activity Aim: To use the computer to convert from Celsius to Fahrenheit using the formula.

```
10   CLS
20   INPUT "ENTER CELSIUS"; C
30   F = (9/5) * C + 32
40   PRINT C; "DEGREES CELSIUS ="; F;
     "DEGREES FAHRENHEIT"
```

Rewrite the formula and program to convert Fahrenheit to Celsius.

See pp. 319–320 for Celsius and Fahrenheit conversion problems.

Activity Aim: To use the computer to express exponents.

SAMPLE

$4 \times 4 \times 4$ may be written 4^3.
The computer receives the information in this form:
PRINT 4 ↑ 3
The output is 64.

See p. 253 for exponential expression problems.

Activity Aim: To use the computer to solve basic formulas.

SAMPLE

$$d = 2r \qquad r = \frac{d}{2}$$

Have the students write separate programs to find the diameter and radius using the formulas above.

See p. 361 for diameter and radius conversion problems.

Activity Aim: To have the students write a program to solve a percentage problem.

SAMPLE

A store is going to have a year-end sale. All prices are to be marked down 25%. Write a program to help the sales manager. As a second part of your program, include a 5% sales tax to the new price.

See pp. 221, 222, 564, 565, 604, 605, and 606 for percentage problems.

Similarly, programs can be written for other word problems in the text.

NAME _____

Basic Skills—Inventory Test for Part 1

1. Read the numeral 7,621,034 or write it in words.
Seven million, six hundred twenty-one thousand, thirty-four

2. Write the numeral naming: Fifteen million, six hundred thousand, four hundred fifty-one
15,600,451

3. Round 12,209,125 to the nearest thousand. 12,209,000

4. Add: 322
419
876   1617

5. Subtract: 10,109
7,926   2,183

6. Multiply: 64 × 5,107
326,848

7. Divide: 423)56,682   134

8. Express $\frac{105}{...}$ in lowest terms. $\frac{7}{7}$

9. Change $\frac{12}{5}$ to a mixed number. $2\frac{2}{5}$

10. Raise to higher terms: $\frac{5}{12} = \frac{20}{48}$ 48

11. Find the lowest common denominator of the fractions: $\frac{3}{4}, \frac{1}{8}, \frac{7}{24}$ 48

12. Add: $3\frac{1}{4}$
$+5\frac{2}{3}$ $8\frac{11}{12}$

13. Subtract: $7\frac{3}{10}$
$-4\frac{4}{5}$ $2\frac{1}{2}$

14. Which is greater: $\frac{7}{11}$ or $\frac{7}{12}$? $\frac{7}{11}$

15. Change $12\frac{1}{3}$ to an improper fraction. $\frac{37}{3}$

16. Multiply: $\frac{7}{8} \times \frac{4}{5}$

17. Divide: $5\frac{1}{3} \div 1\frac{1}{3}$ 4

18. 22 is what part of 55? $\frac{2}{5}$

19. $\frac{2}{3}$ of what number is 30?   45

20. Read the numeral 29.085 or write it in words.
Twenty-nine and eighty-five thousandths

21. Write as a decimal numeral: Four hundred sixteen thousandths   .416

22. Find correct to the nearest hundredth: 17.859   17.86

23. Add: 1.06 + .73   1.79

24. Subtract: 12 − 8.2   3.8

25. Which is smaller: .9 or .105?   .105

26. Multiply: 7.5 × .32   2.4

27. Divide: .178)2.43148   13.66

28. Multiply by short method: 1,000 × 5.2   5,200

29. Divide by short method: .05 ÷ 100   .0005

30. Change $\frac{3}{5}$ to a decimal.   .6

31. Change $.66\frac{2}{3}$ to a common fraction. $\frac{2}{3}$

32. What decimal part of 20 is 7?   .35

33. .9 of what number is 27?   30

34. Write 302.16 *billion* completely as a numeral.   302,160,000,000

35. Write the shortened name for 52,900,000,000 in billions.   52.9 billion

36. Write *eighty-nine hundredths* as a percent, as a decimal, and as a common fraction.
89%, .89, $\frac{89}{100}$

37. Express 27% as a decimal.   .27

38. Express 1.32 as a percent.   132%

39. Express 15% as a common fraction. $\frac{3}{20}$

40. Express $\frac{...}{20}$ as a percent.   85%

41. Find 65% of 325.   211.25

42. What percent of 36 is 27?   75%

43. 25% of what number is 114?   456

44. Square 5.3   28.09

45. Find the square root of 5,041.   71

1

NAME _____

Problem Solving—Inventory Test

Solve each of the following problems:

1. If a car averages 82 kilometers per hour, how long will it take to drive a distance of 574 kilometers? 7 hours

2. The attendance at the Funland Amusement Park on Friday was 17,842, Saturday, 19,791, and Sunday, 16,444. Find the total attendance for the 3-day period. 54,077

3. The scouts of Troop 62 hiked an average of 13 miles a day on a 17-day camping trip. How many miles did the scouts hike in all? 221 miles

4. The number of shares of the Killington Corporation increased from 1,716,842 to 2,319,729. Find the amount of increase. 602,887 shares

5. Snowfall in December was $14\frac{1}{4}$ inches, in January $19\frac{3}{4}$ inches, and in February $22\frac{1}{4}$ inches. What was the total snowfall for the three months? $56\frac{1}{4}$ inches

6. The width of a rectangle is one fourth as long as the length. How long is the width if the length is $9\frac{1}{8}$ centimeters? $2\frac{9}{32}$ centimeters

7. When Meg started her diet she measured $32\frac{3}{4}$ inches around her waist. At the end of her diet Meg measured $26\frac{7}{8}$ inches. How many inches did Meg lose? $5\frac{7}{8}$ inches

8. Nancy bought $3\frac{1}{4}$ yards of quilted material to make 8 placemats for the dining room table. How much material will she use for each? $\frac{13}{32}$ yards

9. If 36 out of 50 students passed the Algebra test, what part of the students passed? $\frac{36}{50}$ or $\frac{18}{25}$

10. On a fishing trip, Louise caught 6 fish or $\frac{2}{9}$ of the total catch. How many fish were caught altogether? 27

11. Richard is 187.4 centimeters tall. Sara is 26.8 centimeters shorter than Richard. How tall is Sara? 160.6 centimeters

12. If Bill bought a fly fishing rod for $161.98 and reel for $42.29, how much did he spend for his rod and reel? $204.27

13. Sascha earns $5.95 an hour as a night supervisor at the Key Market. How much money does she earn for a 32-hour work week? $190.40

14. Donna bought some pizza dough for $4.95. If the dough costs $.55 per pound, how many pounds did she buy? 9

15. Aileen made 7 and missed 12 baskets in the basketball game Friday night. What was her percentage of baskets made? 36.8%

16. A soccer ball which regularly sells for $29.95 is reduced 25% on sale. What is the sales price? $22.46

17. How much must Earl invest at 14% annual interest to earn $5,000 per year? $35,714.29

2

T41

Test for Unit Two

1. Express each of the following fractions in lowest terms:
   a. $\frac{20}{25}$ $\frac{4}{5}$
   b. $\frac{18}{36}$ $\frac{1}{2}$
   c. $\frac{55}{88}$ $\frac{5}{8}$
   d. $\frac{96}{144}$ $\frac{2}{3}$
   e. $\frac{108}{180}$ $\frac{3}{5}$

2. Change each of the following improper fractions to a whole number or a mixed number:
   a. $\frac{7}{2}$ $3\frac{1}{2}$
   b. $\frac{40}{3}$ $13\frac{1}{3}$
   c. $\frac{24}{7}$ 3
   d. $\frac{58}{5}$ $11\frac{3}{5}$
   e. $\frac{60}{12}$ $5\frac{5}{6}$

3. Change each of the following mixed numbers to simplest form:
   a. $5\frac{1}{4}$ 6
   b. $4\frac{8}{9}$ 9
   c. $8\frac{5}{6}$ $8\frac{3}{5}$
   d. $6\frac{1}{4}$ $10\frac{2}{3}$
   e. $11\frac{19}{18}$ $13\frac{1}{4}$

4. Change each of the following to equivalent fractions having denominators as specified:
   a. $\frac{1}{2} = \frac{5}{10}$
   b. $\frac{5}{6} = \frac{20}{24}$
   c. $\frac{13}{16} = \frac{35}{80}$
   d. Change $\frac{3}{16}$ to 32nds. $\frac{6}{32}$

5. Find the lowest common denominator of each of the following groups of fractions:
   a. $\frac{1}{4}$ and $\frac{11}{12}$ 12
   b. $\frac{2}{3}$ and $\frac{3}{8}$ 24
   c. $\frac{3}{10}$ and $\frac{1}{6}$ 30
   d. $\frac{1}{2}$, $\frac{5}{15}$, and $\frac{2}{9}$ 18

6. Add:
   a. $\frac{1}{2}$ $\frac{2}{3}$ $\frac{1}{6}$
   b. $2\frac{3}{4}$ $\frac{3}{4}$ $3\frac{11}{20}$
   c. $4\frac{3}{5}$ $6\frac{3}{8}$ $10\frac{17}{20}$
   d. $5\frac{1}{2}$ $\frac{7}{10}$ $3\frac{3}{5}$ $9\frac{4}{5}$

7. Find the missing numbers:
   a. $4 = 3\frac{2}{3}$ 5
   b. $1\frac{7}{10} = \frac{7}{10}$ 17

8. Subtract:
   a. $\frac{13}{18}$ $\frac{5}{9}$ $\frac{7}{12}$ $3\frac{11}{20}$
   b. $7\frac{5}{8}$ $4\frac{3}{16}$ $3\frac{7}{16}$
   c. 10 $3\frac{3}{5}$ $6\frac{2}{5}$
   d. $6\frac{3}{8}$ $2\frac{5}{6}$ $3\frac{13}{24}$

9. a. Which is smaller: $\frac{2}{3}$ or $\frac{3}{4}$? $\frac{1}{4}$ or $\frac{1}{5}$? $3\frac{1}{4}$, $\frac{1}{7}$
   b. Which is greater: $\frac{1}{6}$ or $\frac{1}{11}$? $\frac{3}{5}$ or $\frac{5}{9}$? $\frac{1}{6}$, $\frac{3}{5}$

10. Change the following mixed numbers to improper fractions:
    a. $2\frac{3}{4}$ $\frac{11}{4}$
    b. $9\frac{1}{5}$ $\frac{46}{5}$
    c. $1\frac{3}{5}$ $\frac{13}{5}$

11. Multiply:
    a. $\frac{1}{2} \times \frac{5}{8}$ $\frac{5}{16}$
    b. $15 \times 2\frac{4}{5}$ 42
    c. $4\frac{1}{8} \times \frac{7}{16}$ $2\frac{11}{16}$
    d. $1\frac{1}{2} \times 2\frac{2}{3}$ $2\frac{9}{10}$
    e. $5\frac{1}{4}$ 112

12. Divide:
    a. $\frac{3}{4} \div \frac{5}{16}$ $2\frac{2}{5}$
    b. $4\frac{1}{2} \div 10\frac{5}{12}$
    c. $15 \div 3\frac{1}{3}$ $4\frac{1}{2}$
    d. $10\frac{1}{4} \div 2\frac{1}{4}$ $3\frac{9}{11}$

13. a. What part of 8 is 5? $\frac{5}{8}$
    b. 34 is what part of 85? $\frac{2}{5}$

14. a. 15 is $\frac{1}{2}$ of what number? 90
    b. $\frac{2}{3}$ of what number is 24? 36

Test for Unit One

1. Write each of the following numerals in words:
   a. 1701 One thousand seven hundred one (or seventeen hundred one)
   b. 7,907,561 Seven million, nine hundred seven thousand, five hundred sixty-one
   c. 152,695 One hundred fifty-two thousand, six hundred ninety-five
   d. 21,691,000 Twenty-one million, six hundred ninety-one thousand

2. Write the numeral naming each of the following:
   a. Fifteen thousand, twenty-nine 15,029
   b. Six million, five hundred five thousand, two hundred seventy-two 6,505,272
   c. Two hundred twenty-two thousand, two hundred two 222,202
   d. Eighty-four billion, six hundred fifty million 84,650,000,000

3. Round these:
   a. 4,975 to the nearest hundred 5,000
   b. 20,731 to the nearest thousand 21,000
   c. 138,862,497 to the nearest ten thousand 138,860,000
   d. 784 to the nearest ten 780

4. Add:
   a. 910 296 624 105 1935
   b. 7320 8090 1251 4679 21,340
   c. 21,370 72,234 54,630 18,698 166,932
   d. 476,463 150,000 851,237 341,985 1,819,685

5. Subtract:
   a. 2416 1952 464
   b. 91,778 46,119 45,659
   c. 396,539 119,268 277,271
   d. 2,022 697 1,325

6. Multiply:
   a. 49 91 4,459
   b. 639 387 247,293
   c. 4,731 574 2,715,594
   d. 56,745 1,071 60,773,895

7. Divide:
   a. 45)4410 98
   b. 13)52988 4,076
   c. 463)75,006 162
   d. 2,984)3,357,193 1,125 R193

8. The Mississippi River is 2,348 miles long and the Missouri River is 3,710 miles long. How many miles longer is the Missouri River? 1,362 miles

9. An airplane flies nonstop from Chicago to Hong Kong, an air distance of about 13,080 kilometers, in 12 hours. What speed did the airplane average for the trip? 1090 kilometers per hour

10. Henrietta bought 15 radios at $32 each, 12 irons at $16 each, 24 toasters at $8 each and 10 blenders at $20 each, for her hardware store. What was the total cost? $1,064

Test for Unit Three

1. Write each of the following as a decimal numeral:
a. Twenty-two hundredths **.22**
b. Eight hundred eighty-seven thousandths **0.887**
c. Fifty and nine tenths **50.9**
d. Four thousand two hundred fifteen millionths **.004215**
e. Six ten thousandths **.0006**

2. Write each of the following numerals as a word statement:
a. .2 **Two tenths**
b. 0.793 **Seven hundred ninety-three thousandths**
c. 11.01 **Eleven and one hundredth**
d. .00623 **Six hundred twenty-three hundred thousandths**
e. 91.1991 **Ninety-one and one thousand nine hundred ninety-one ten thousandths**

3. Round:
a. .85 correct to nearest tenth **.9**
b. 2.9225 correct to nearest hundredth **2.92**
c. 17.2749 correct to nearest thousandth **17.275**
d. $21.553 correct to nearest cent **$21.55**

4. Add:
a. 9.1
4.3
2.7 **16.1**

b. 2.03
0.05
6.01
9.09 **17.18**

c. 3.4 + .28 + 13 **16.68**

d. Find the sum of: 56.1, .253, and 1.96 **58.313**
e. $5.99 + $19.27 + $.82 + $3.14 + $150.05 **$179.27**

5. Subtract:
a. 96.431
75.912 **20.519**

b. .711 − .3 **.411**

c. .8 − .425 **.375**

d. 10 − .003 **9.997**
e. $5.75 from $27 **$21.25**

6. Which is greater:
a. .9 or .75? **.9**
b. 2.15 or .512? **2.15**

Which is smaller:
c. .115 or .21? **.115**
d. 7.6 or 67? **.67**

7. Multiply:
a. .05
.1 **.005**

b. 80.15
.191 **15.30865**

c. .022 × .09 **.00198**

Find correct to nearest cent:
d. $\frac{1}{3}$ of $6.13 **$2.04**
e. $25.12 × .75$\frac{1}{4}$ **$18.90**

Test for Unit Three (cont.)

8. Divide:
a. 52)9152 **.0176**
b. 2.1)24.78 **11.8**
c. 43 ÷ .215 **200**
d. $.05)$3 **$60**
e. Find correct to nearest tenth: 51.92)695 **13.4**

9. Multiply:
a. 100 × 60 **6,000**
b. 10 × 8.73 **87.3**
c. 1,000 × .5 **500**
d. 100 × .02 **2**
e. 1,000 × 67.34 **67,340**

10. Divide:
a. 800 ÷ 10 **80**
b. 560 ÷ 1,000 **.56**
c. .2 ÷ 100 **.002**
d. 49.4 ÷ 1,000 **.0494**
e. 392 ÷ 100 **3.92**

11. Change each of the following fractions to decimals:
a. $\frac{3}{8}$ **.375**
b. $\frac{17}{100}$ **.17**
c. $\frac{21}{25}$ **.84**
d. $\frac{43}{65}$ (to 3 decimal places) **.661**

12. Change each of the following decimals to fractions or mixed numbers:
a. .4 **$\frac{2}{5}$**
b. 0.75 **$\frac{3}{4}$**
c. .625 **$\frac{5}{8}$**
d. 2.33$\frac{1}{3}$ **2$\frac{1}{3}$**
e. .875 **$\frac{7}{8}$**

13. a. What decimal part of 65 is 52? **.8**
b. 7 is what decimal part of 16? **.4375**

14. a. .07 of what amount is $420? **6000**
b. 18 is .4 of what number? **45**

Test for Unit Four

1. Read, or write in words, each of the following number names:
   a. 523.5 billion __Five hundred twenty-three point five billion__
   b. 21.963 million __Twenty-one point nine six three million__
   c. $9.47 trillion __Nine point four seven trillion dollars__
   d. 739.1 thousand __Seven hundred thirty-nine point one thousand__

2. Read, or write in words, each shortened name in the following:
   a. The United States produces around 7.8 billion bushels of corn a year. __Seven point eight billion__
   b. The announcer said that volume on the New York Stock Exchange reached 34.29 million shares today, down from 34.32 million the day before. __Thirty-four point two nine million; thirty-four point three two million__
   c. The city's budget for this year is $72.5 million; its budget was $48.6 million only two years ago. __Seventy-two point five million dollars; forty-eight point six million dollars__
   d. Youth employment programs this year will cost $592 million for 119.5 thousand public jobs. __Five hundred ninety-two million dollars; one hundred nineteen point five thousand__

3. Write the complete numeral for each of the following:
   a. $29.64 million __$29,640,000__
   b. 617.9 thousand __617,900__
   c. 935.12 billion __935,120,000,000__
   d. $1.936 trillion __$1,936,000,000,000__

4. Write the shortened number name for each of the following:
   a. In millions: 38,600,000 __38.6 million__
   b. In billions: 902,560,000,000 __902.56 billion__
   c. In thousands: 2,958,000 __2.958 thousand__
   d. In trillions: 7,983,000,000,000 __7.983 trillion__

5.
   a. Add: 321.4 million / 17.6 million / 922.8 million __1,261.8 million__
   b. Subtract: $162.5 billion / 145.3 billion __$17.2 billion__
   c. Multiply: $7 \times \$3.81$ trillion __$26.67 trillion__
   d. Divide: 40.32 million ÷ 7.2 __5.6 million__

6.
   a. Add as indicated: $7.24 billion + $356 million __$7.596 billion (or $7,596 million)__
   b. Subtract as indicated: 2.5 trillion − 618 billion __1.882 trillion (or 1,882 billion)__

7. This year American taxpayers will spend $2.2 billion for youth work programs. Last year they spent $1.56 billion. What is the increase in the amount that taxpayers will spend? __$640 million (or $.64 billion)__

8. Retail food sales this quarter are expected to increase to $14.2 million from $13.7 million last quarter. What is the amount of expected increase? __$500 thousand (or $.5 million)__

9. The number of Americans receiving food stamps this year is 19.2 million, which is three times the number who received food stamps ten years ago. How many Americans were receiving food stamps ten years ago? __6.4 million__

10. Revenues last year for a company were $6.3 billion. Revenues this year should be four times that amount. What will the revenues be this year? __$25.2 billion__

Test for Unit Five

1. Write each of the following as a percent, decimal, and fraction:
   a. Two hundredths __2%; .02; $\frac{2}{100}$, $\frac{1}{50}$__
   b. Seventy-nine hundredths __79%; .79; $\frac{79}{100}$__
   c. 27 out of 100 __27%; .27; $\frac{27}{100}$__
   d. 84 out of 100 __84%; .84; $\frac{84}{100}$, $\frac{21}{25}$__
   e. Sixty-seven and two-thirds hundredths __$67\frac{2}{3}\%$; $.67\frac{2}{3}$; $\frac{67\frac{2}{3}}{100}$__

2. Express each of the following percents as a decimal:
   a. 7% __.07__
   b. 61% __.61__
   c. 180% __1.8__
   d. $100\frac{1}{2}\%$ __1.05__
   e. 9.18% __.0918__

3. Express each of the following decimals as a percent:
   a. .09 __9%__
   b. .9 __90%__
   c. 1.36 __136%__
   d. .829 __82.9%__
   e. $.33\frac{1}{3}$ __$33\frac{1}{3}\%$__

4. Express each of the following percents as a common fraction or a mixed number:
   a. 17% __$\frac{17}{100}$__
   b. 75% __$\frac{75}{100} = \frac{3}{4}$__
   c. 8% __$\frac{2}{25}$__
   d. $2\frac{1}{4}\%$ __$\frac{9}{40}$__
   e. 350% __$3\frac{1}{2}$__

5. Express each of the following fractions or mixed numbers as a percent:
   a. $\frac{1}{5}$ __20%__
   b. $\frac{18}{25}$ __72%__
   c. $\frac{9}{4}$ __225%__
   d. $\frac{23}{23}$ __100%__
   e. $\frac{1}{3}$ __$33\frac{1}{3}\%$__

6. Find the following:
   a. 91% of 260 __236.6__
   b. 8% of 50 __4__
   c. 175% of 320 __560__
   d. 6% of $56.25 __$3.66__
   e. 12.47% of 5000 __623.5__

7. Find the following:
   a. 35 is what percent of 42? __$83\frac{1}{3}\%$__
   b. What percent of 75 is 30? __40%__
   c. $.98 is what percent of $4.90? __20%__
   d. 9 is what percent of 4? __225%__
   e. What percent of 11 is 7? (correct to nearest tenth) __63.6%__

8. Find the missing numbers:
   a. 36% of what number is 9? __25__
   b. 84 is 12% of what number? __700__
   c. 72 is 2.4% of what number? __3000__
   d. $62\frac{1}{2}\%$ of what number is 35? __56__
   e. $5\frac{3}{8}\%$ of what number is 69? __8__

9. A raincoat is marked down to $68 from a regular price of $85. What percent reduction is this markdown? __20%__

10. A used motorcycle is advertised at $1,020 and the owner claims that this is a 40% reduction from what it originally cost. What was the original cost of the motorcycle? __$2550__

Test for Unit Six

1. Square each of the following numbers:
   a. 17   289 ____   b. 150   22,500 ____
   c. .009   .000081 ____   d. 6.3   39.69 ____
   e. $\frac{4}{5}$ $\frac{16}{25}$ ____   f. $2\frac{1}{2}$ $\frac{25}{4} = 6\frac{1}{4}$ ____

2. Find the value of each of the following:
   a. 7^2 49 ____   b. $(4.1)^2$ 16.81 ____
   c. $(\frac{2}{5})^2$ $\frac{4}{25}$ ____   d. 20^2 400 ____
   e. $(.03)^2$ .0009 ____   f. $(1\frac{1}{2})^2$ $5\frac{1}{4}$ ____

3. Find the square root of each of the following numbers. If there is a remainder, find the answer correct to the nearest hundredth:
   a. 7,396   86 ____   b. 69,169   263 ____
   c. 46,308,025   6,805 ____   d. 15,848,361   3,981 ____
   e. 30   5.48 ____   f. 720.6   26.84 ____

4. Find the square root of each of the following numbers by estimation, division, and average (3 decimal places):
   a. 15   3.873 ____   b. 6   2.449 ____
   c. 83   9.110 ____   d. 42   6.481 ____
   e. 12   3.464 ____   f. 30   5.477 ____

5. Find the square root of each of the following numbers by using the table of squares and square roots on page 621:
   a. 48   6.928 ____   b. 32   5.657 ____
   c. 91   9.539 ____   d. 9,216   96 ____
   e. 289   17 ____   f. 2,916   54 ____

6. Find the area of a square that measures 21 cm on each side.   441 cm² ____

7. Find the length of each side of a square whose area is 225 square feet.   15 feet ____

8. A number times itself is 576. What is the number?   24 ____

9. A rug measures 79 inches on each side. How much floor area would that rug cover?   6241 square inches ____

10. Jennifer is making a new tabletop for her card table. She would like the new top to be square, with an area of 729 square inches. What should each side of her tabletop measure?   27 inches ____

Test for Unit Seven

1. What year does MCMLXXXIV represent?   1984 ____

2. Read, or write in words, each of the following:
   a. 0, 3, 6, 9, . . .   Zero, three, six, nine, and so on without end ____
   b. 10, 20, 30, . . . , 1,000   Ten, twenty, thirty, and so on up to one thousand ____

3. a. Find the value of 5^4.   625 ____   b. Express 32 as a power of 2. 2^5 ____

4. Express 500,000,000,000 as a product of a digit and a power of ten. 5×10^{11} ____

5. Write 3,578.06 in expanded form as a polynomial.
   $(3 \times 10^3) + (5 \times 10^2) + (7 \times 10^1) + (8 \times 10^0) + (0 \times 10^{-1}) + (6 \times 10^{-2})$ ____

6. Read, or write in words:
   a. $21 + 4 = 4 + 21$ Twenty-one plus four equals four plus twenty-one ____
   b. $15 - 3 < 3 \times 15$ Fifteen minus three is less than three times fifteen ____
   c. $6 > \frac{1}{6} \times 10$ Six is greater than one-sixth times ten ____
   d. $14 \div 2 \not< 28 \div 7$ Fourteen divided by two is not less than twenty-eight divided by seven ____

7. Determine which of the following sentences are true and which are false:
   a. $62 - 12 < 18 + 23$ False ____
   b. $56 - 13 = 5^2$ False ____
   c. $200 > 144 \div 12$ True ____
   d. $6 + 6 \not> 6 \times 6$ True ____

8. a. Is 243 an odd number?   Yes ____   b. Is 421 an even number?   No ____

9. What is the greatest common factor of 64, 192, and 240?   16 ____

10. a. Is 97 a prime number?   Yes ____   b. Is 81 a composite number?   Yes ____

11. Factor 306 as a product of prime numbers. $2 \times 3 \times 51$ ____

12. Find the least common multiple of 8, 12, and 18.   72 ____

13. Are $\frac{7}{12}$ and $\frac{35}{60}$ equivalent fractions?   Yes ____

14. Write the simplest form and three other fractions equivalent to $\frac{16}{24}$. $\frac{2}{3}, \frac{4}{6}, \frac{6}{9}, \frac{8}{12}$ ____

15. Determine by the cross-product test which of the following statements are true:
    a. $\frac{3}{4} > \frac{2}{3}$ True ____   b. $\frac{5}{21} > \frac{6}{19}$ False ____

16. Write the multiplicative inverse or reciprocal of each of the following:
    a. 9 $\frac{1}{9}$ ____   b. $\frac{1}{11}$ 11 ____   c. $\frac{3}{8}$ $\frac{8}{3}$ ____   d. $\frac{22}{3}$ $\frac{3}{22}$ ____   e. $4\frac{5}{6}$ $\frac{6}{29}$ ____

17. Express 3,750,000,000 in scientific notation. 3.75×10^9 ____

18. Express .00000000106 in scientific notation. 1.06×10^{-9} ____

19. Express $\frac{4}{11}$ as a repeating decimal. $.\overline{36}$ ____

20. Express .55 as a numeral naming a common fraction. $\frac{11}{20}$ ____

Basic Skills—Achievement Test for Part 1

1. Read the numeral 198,376,521 or write it in words.
One hundred ninety-eight million, three hundred seventy-six thousand, five hundred twenty-one

2. Write the numeral naming: Six hundred eight million, nine hundred sixty-two thousand, one hundred seventy-three. 608,962,173

3. Round 9,216,053 to the nearest ten thousand. 9,220,000

4. Add: 219 + 3,025 + 29,428 + 17 + 390 33,079

5. Take 5,296 from 13,904. 8,608

6. Multiply: 7 × 26 × 3,914 712,348

7. Divide: 119)30,702 258

8. Express $\frac{36}{48}$ in lowest terms. $\frac{3}{4}$

9. Change $\frac{65}{14}$ to a mixed number. $4\frac{9}{14}$

10. Raise to higher terms: $\frac{5}{8} = \frac{?}{48}$ $\frac{40}{48}$

11. Find the lowest common denominator of the fractions: $\frac{3}{5}, \frac{7}{18}, \frac{11}{24}$ 72

12. Add: $5\frac{7}{8} + 9\frac{5}{6}$ $15\frac{17}{24}$

13. Subtract: $9\frac{5}{6} - 5\frac{3}{4}$ $4\frac{1}{12}$

14. Which is smaller: $\frac{1}{3}$ or $\frac{3}{7}$? $\frac{1}{3}$

15. Change $3\frac{5}{6}$ to an improper fraction. $\frac{23}{6}$

16. Multiply: $6\frac{2}{3} \times 4\frac{4}{5}$ 32

17. Divide: $2\frac{2}{3} \div 4\frac{1}{2}$ $\frac{16}{27}$

18. What fractional part of 70 is 42? $\frac{3}{5}$

19. $\frac{3}{4}$ of what number is 48? 64

20. Read the numeral 15.261 or write it in words. Fifteen and two hundred sixty-one thousandths

21. Write as a decimal numeral: One and seventy-two hundredths 1.72

22. Find correct to the nearest cent: $31.953 $31.95

23. Add: 13.021
4.176
8.922
.759 26.878

24. Subtract: 113.16 − 98.89 14.27

25. Which is greater: 1.05 or 1.105? 1.105

26. Multiply: 4.07 × 16.93 68.9051

27. Divide: 1.9)39.14 20.6

28. Use short method: 100 × .002 .2

29. Use short method: 93.7 ÷ 1,000 .0937

30. Change $\frac{5}{24}$ to a decimal (2 places). .21

31. Change .072 to a common fraction. $\frac{9}{125}$

32. What decimal part of 16 is 5? .3125

33. .12 of what number is 84? 700

34. Write 923.5 million completely as a numeral. 923,500,000

35. Write the shortened name for 6,700,000,000,000 in trillions. 6.7 trillion

36. Write twenty-three hundredths as a percent, as a decimal, and as a common fraction. 23%, .23, $\frac{23}{100}$

37. Express 92% as a decimal. .92

38. Express .03 as a percent. 3%

39. Express $37\frac{1}{2}$% as a common fraction. $\frac{3}{8}$

40. Express $\frac{3}{4}$ as a percent. 75%

41. Find 42% of 350. 147

42. 73 is what percent of 50? 146%

43. $33\frac{1}{3}$% of what number is 70? 210

44. Square 5.12. 26.21

45. Find the square root of 10 to the nearest hundredth. 3.16

Test for Metric System of Measure

1. Change: a. 42.7 cm to cm 4,270 cm
b. 923 mm to cm 92.3 cm
c. 19 km to m 19,000 m
d. 4,176 m to m 41.76 m

2. What metric unit of length does each of these symbols represent?
a. m meter
b. cm centimeter
c. dam dekameter
d. km kilometer

3. Add: 3 cm + 5 m + 22 mm
5052 mm = 505.2 cm = 5.052 m

4. Subtract: 5 m − 34 cm
466 cm = 4.66 m

5. Change:
a. 6.3 cg to mg 63 mg
b. 79 g to cg 7,900 cg
c. 6,210 mg to dg 62.1 dg
d. 540 g to kg 0.54 kg

6. What metric unit of mass (weight) does each of these symbols represent?
a. kg kilogram
b. g gram
c. dg decigram
d. hg hectogram

7. Add: 15 mg + 15 g + 15 kg 15,015,015 mg = 15,015.015 g = 15.015015 kg

8. Subtract: 895 g − 418 mg 894,582 mg = 894.582 g

9. Change: a. .563 L to mL 563 mL
b. 2,356 mL to L 2.356 L
c. 29.05 kL to hL 290.5 hL
d. 594 L to kL 0.594 kL

10. What metric unit of capacity does each of these symbols represent?
a. mL milliliter
b. dL deciliter
c. L liter
d. cL centiliter

11. Add: 5 kL + 600 L + 4,991 mL
5,604,991 mL = 5,604.991 L = 5.604991 kL

12. Subtract: 792 L − 416 mL
791,584 mL = 791.584 L

13. Change: a. 2.4 dm² to mm² 24,000 mm²
b. 17.3 cm² to m² .00173 m²
c. 34 km² to m² 34,000,000 m²
d. 1,491 mm² to cm² 14.91 cm²

14. What metric unit of area does each of these symbols represent?
a. m² square meters
b. cm² square centimeters
c. hm² square hectometers
d. km² square kilometers

15. Multiply: 30 m × 17 m 510 m²

16. Multiply: 142 km × 15 km 2,130 km²

17. Change:
a. 15 cm³ to mm³ 15,000 mm³
b. 115 m³ to cm³ 115,000,000 cm³
c. 1,783 cm³ to mL 1,783 mL
d. 45,990 cm³ to m³ .04599 m³

18. What metric unit of volume does each of these symbols represent?
a. cm³ cubic centimeter
b. km³ cubic kilometer
c. dam³ cubic dekameter
d. mm³ cubic millimeter

19. Multiply: 14 mm × 16 mm × 8 mm 1,792 mm³

20. Multiply: 121 cm × 9 cm × 176 cm 191,664 cm³

Test for Customary Units of Measure

1. Change:
   a. 5¼ ft. to in. __63 in.__
   b. 2.5 mi to ft. __13,200 ft.__
   c. 5 mi. to rd. __1,600 rd.__
   d. 44,880 ft. to mi. __8½ mi.__
   e. 81 in. to ft. __6¾ ft.__

2. Add and simplify: 4 yd. 5 ft.
   6 yd. 2 ft.
   __12 yd. 1 ft.__

3. Subtract: 3 mi. 1,500 ft.
   1 mi. 4,500 ft.
   __1 mi. 2,280 ft.__

4. Change:
   a. 4.4 lb. to oz. __17.4 oz.__
   b. 3.5 s.t. to lb. __7,000 lb.__
   c. 10 oz. to lb. __.625 lb.__
   d. 6.1 l.t. to lb. __13,664 lb.__
   e. 19,000 lb. to s.t. __9.5 s.t.__

5. Add and simplify: 6 lb. 5 oz.
   8 lb. 3 oz.
   7 lb. 14 oz. __22 lb. 6 oz.__

6. Subtract: 10 lb. 3 oz.
   7 lb. 12 oz.
   __2 lb. 7 oz.__

7. Change: a. 2.2 qt. to oz. __70.4 oz.__
   b. 20 qt. to pt. __40 pt.__
   c. 150 gal. to qt. __600 qt.__
   d. 640 oz. to gal. __5 gal.__
   e. 240 oz. to qt. __7½ qt.__

8. Add and simplify: 5 qt. 1 pt. 12 oz.
   3 qt. 1 pt. 8 oz.
   __9 qt. 1 pt. 4 oz.__

9. Subtract: 17 gal. 8 oz.
   11 gal. 13 oz.
   __5 gal. 123 oz.__

10. Change:
    a. 15 qt. to pt. __30 pt.__
    b. 13 pk. to qt. __104 qt.__
    c. 1¼ bu. to pk. __6 pk.__
    d. 144 qt. to bu. __4½ bu.__
    e. 32 pt. to qt. __16 qt.__

11. Multiply and simplify: 12 bu. 3 pk. × 3   __38 bu. 1 pk.__

12. Divide: 5 pk. 5 qt. ÷ 5   __1 pk. 1 qt.__

13. Change:
    a. 6 sq. yd. to sq. in. __7,776 sq. in.__
    b. 4 acres to sq. ft. __174,240 sq. ft.__
    c. 1.25 sq. mi. to sq. yd. __3,872,000 sq. yd.__
    d. 130,680 sq. ft. to acres   __3 acres__
    e. 489.6 sq. in. to sq. ft. __3.4 sq. ft.__

14. Change:
    a. 40 cu. ft. to cu. in. __69,120 cu. in.__
    b. 6,048 cu. in. to cu. ft. __3.5 cu. ft.__
    c. 150 cu. ft. to cu. yd. __5⅝ cu. yd.__
    d. 576 cu. in. to cu. ft. __⅓ cu. ft.__
    e. 2.5 cu. yd. to cu. ft. __67.5 cu. ft.__

15. Find the equivalent: 112.5 gal.
    a. 15 cu. ft. in gal. __112.5 gal.__
    b. 6 gal. in cu. in. __1,386 cu. in.__
    c. 3 bu. in cu. ft. __3¾ cu. ft.__
    d. 120 gal. in cu. ft. __16 cu. ft.__
    e. 31.25 cu. ft. in bu. __25 bu.__

13

Test for Other Basic Measures

1. Change:
   a. 10.3 min. to sec. __618 sec.__
   b. 16 min. 30 sec. to sec. __990 sec.__
   c. 210 sec. to min. __3½ min.__
   d. 4½ hr. to min. __270 min.__
   e. 720 min. to hr. __12 hr.__

2. Change:
   a. ½ yr. to days __182½ days__
   b. 8⅓ wk. to days __58 days__
   c. 294 days to wk. __42 wk.__
   d. 78 mo. to yr. __6.5 yr.__
   e. 1,825 days to yr. __5 yr.__

3. Add and simplify: 16 hr. 20 min. 19 sec.
   7 hr. 39 min. 12 sec.
   4 hr. 25 min. 48 sec.
   __28 hr. 25 min. 19 sec.__

4. Subtract: 17 days 16 hr.
   12 days 21 hr.
   __4 days 19 hr.__

5. Write as A.M. or P.M. time:
   a. 7 o'clock at night. __7 P.M.__
   b. 8 o'clock in the morning. __8 A.M.__
   c. Half past ten in the morning. __10:30 A.M.__
   d. Quarter to two in the afternoon. __1:45 P.M.__
   e. One minute past midnight. __12:01 A.M.__

6. Change:
   a. 15° to minutes __900'__
   b. 48' to seconds __2,880"__
   c. 240" to minutes __4'__
   d. 360' to degrees __6°__
   e. 14° 5' to seconds __50,700"__

7. Multiply and simplify: 52° 25' 55" × 4
   __109° 43' 40"__

8. Divide: 180° 45' ÷ 3
   __60° 15'__

9. Change:
   a. 300 km/h to m/h __300,000 m/h__
   b. 180 cm/min to m/min __1.8 m/min__
   c. 1,080 mm/s to cm/s __108 cm/s__
   d. 100 m/h to km/h __0.1 km/h__
   e. 5,160 m/h to km/h __5.16 km/h__

10. Change:
    a. 200 stat. m.p.h. to ft. per sec. __293⅓ ft. per sec.__
    b. 3,600 ft. per min. to stat. mi. per min. __12/17 stat. mi. per min.__
    c. 180 knots to stat. m.p.h. __207 stat. m.p.h.__
    d. 690 stat. m.p.h. to knots __600.3 knots__
    e. 1,408 ft. per sec. to stat. m.p.h. __960 stat. m.p.h.__

11. Change:
    a. 17°C to K __290 K__
    b. 112°C to K __385 K__
    c. 412 K to °C __139°C__
    d. 292 K to °C __19°C__

14

T47

NAME

Test for Other Basic Measures (cont.)

12. Change:
   a. 30°C to °F   86°F
   b. 5°C to °F   41°F
   c. 64°F to °C (to nearest hundredth)   17.78°C
   d. 96°F to °C (to nearest tenth)   35.6°C

13. Using the 24-hour system, express each of the following:
   a. 8:00 A.M.   0800
   b. 12:15 P.M.   1215
   c. 6:30 P.M.   1830
   d. 10:08 A.M.   1008

14. Express each of the following in A.M. or P.M. time:
   a. 0206   2:06 A.M.
   b. 1230   12:30 P.M.
   c. 1716   5:16 P.M.
   d. 2357   11:57 P.M.

15. If a nationwide program is telecast at 1:00 P.M. from New York City, what time is it seen in:
   a. Los Angeles?   10:00 A.M.
   b. Boston?   1:00 P.M.
   c. Chicago?   12:00 noon
   d. Denver?   11:00 A.M.
   e. New Orleans?   12:00 P.M.
   f. Miami?   1:00 P.M.

NAME

Test for Measurement

1. To what unit of measure are these measurements precise?
   a. .06 cm   0.01 cm
   b. 152 km   1 km
   c. 116.1 g   0.1 g
   d. 16.5 ft.   .1 ft.
   e. $1\frac{1}{4}$ hr. $\frac{1}{5}$ hr.
   f. $8\frac{1}{4}$ lb. $\frac{1}{16}$ lb.

2. Find the greatest possible error in each of these measurements:
   a. 32 kg   0.5 kg
   b. 5.2 L   0.05 L
   c. 0.14 mm   0.005 mm
   d. $8\frac{1}{4}$ hr. $\frac{1}{4}$ hr.
   e. $3\frac{1}{4}$ lb. $\frac{1}{16}$ lb.
   f. $4\frac{7}{16}$ in. $\frac{1}{32}$ in.

3. Find the relative error in each of these measurements:
   a. 5.073 cm   .0000986
   b. 932 gal. $\frac{1}{1864}$
   c. 5,280 ft. $\frac{1}{5280}$
   d. 149.6 g   .0003342
   e. .043 mL   .0116
   f. 25,500 tons $\frac{1}{510}$

4. Which measurement in each of the following is more accurate:
   a. 32.2 cm or 7.21 cm?   7.21 cm
   b. 40 mm or .004 mm?   same
   c. .09 m or 340 km?   340 km
   d. 55,000 km or .6 mm?   55,000 km
   e. 72 ft. or 7.2 ft.?   same
   f. $5\frac{1}{4}$ in. or $12\frac{5}{8}$ mi.? $12\frac{5}{8}$ mi.

5. Determine the number of significant digits in each of the following:
   a. 43   2
   b. 900   1
   c. 5,685   4
   d. 14,000   2
   e. 2   1
   f. .06   2

6. Name the significant digits in each of the following:
   a. 56   5, 6
   b. 900   9
   c. 17,281   1, 7, 2, 8, 1
   d. .073   0, 7, 3
   e. 5,070,000   5, 0, 7
   f. 89.04   8, 9, 0, 4

7. Add these approximate numbers:
   a. 8.4 + 4.66   13.0
   b. 22 + .16 + 1.7   24

8. Subtract these approximate numbers:
   a. 42.117 − 14.16   27.96
   b. 34.1 − .052   34.0

9. Multiply these approximate numbers:
   a. 7.1 × 4.3   31
   b. 29 × .038   1.1

10. Divide these approximate numbers:
   a. 14 ÷ 12.61   1.1
   b. 4 ÷ .043   90

Test for Applications Using Units of Length

1. Read, or write in words, each of the following: $\overleftrightarrow{AB}$; $\overline{DF}$; $\overrightarrow{OP}$
   line AB; segment DF; ray OP

2. Name each of the following:

   line XY; ray RQ

3. Name the points in the above drawings. **X, Y, R, Q**

4. Name the line segments in the following triangle: **JH, JM, HM**

5. At what point do $\overleftrightarrow{HJ}$ and $\overleftrightarrow{JM}$ meet? **point J**

6. Are points H, M, and J collinear? **No**

7. Are $\overleftrightarrow{HM}$ and $\overleftrightarrow{HJ}$ concurrent? **Yes**

8. Identify which of these figures shows parallel lines. Perpendicular lines. Intersecting lines.

   a. b. c.

   parallel   intersecting   perpendicular

9. Draw a representation of a line in a:
   a. vertical position   b. slanting position   c. horizontal position

10. If a scale is 1 in. = 50 mi, what actual distance is represented by the scale distance of $4\frac{1}{4}$ in.? **418 cm** $212\frac{1}{2}$ mi.

11. What is the perimeter of a rectangle 122 cm long and 87 cm wide? **418 cm**

12. Find the circumference of a circle whose diameter is 36 meters. **113.04 m**

13. What is the perimeter of a square whose side measures $5\frac{1}{2}$ feet? **22 ft.**

14. What is the circumference of a circle whose radius is 14 inches, using $\pi = \frac{22}{7}$? **88 in.**

15. Find the perimeter of a triangle whose sides measure 2.9 centimeters, 1.7 centimeters, and 2.4 centimeters. **7 cm**

Test for Applications Using Units of Length (cont.)

16. How many yards of wooden molding are needed for a bedroom floor 11 feet 6 inches by 12 feet 8 inches? **48 ft. 4 in.**

17. Which is longer, the perimeter of a card table measuring 27 inches on a side or the circumference of a circular table measuring 33 inches in diameter? **Perimeter of card table 27 in. on a side**

18. Allowing 1 centimeter for the overlap, how long should a paper label be in order to go around a metal container having a diameter of 15 centimeters? **48.1 cm**

19. Name each of the following figures:

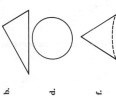

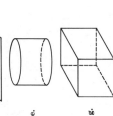

   a. **rectangle**   b. **triangle**
   c. **parallelogram**   d. **circles**
   e. **cylinder**   f. **cone**
   g. **cube**   h. **pyramid**

20. a. Find the perimeter of a rectangle whose sides measure 11 cm and 8 cm. **38 cm**

    b. Find the circumference of a circle whose diameter is 21 m, using $\pi = \frac{22}{7}$. **66 m**

T50

Test for Applications Using Angles and Arcs

1. Write the symbol used to designate the word "angle." ∠

2. Name the sides and vertex of this angle: Sides *JQ* and *JT*, vertex *J*

3. Which of the following are measures of an obtuse angle? 108°
   a. 108°   b. 15°   c. 198°

4. Draw an angle measuring 26°.

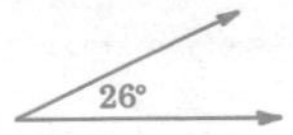

5. Draw:
   a. a straight angle   b. a reflex angle   c. a right angle

6. Find the measure of the third angle of a triangle when the other two angles measure 34° and 95°. 51°

7. If you face west, then turn clockwise to face northeast, how many degrees do you turn? 135°

8. What is the complement of an angle measuring 13°? 77°

9. Find the supplement of an angle measuring 126° 30'? 53°30'

10. In the following figure find the measures of ∠2, ∠1 and ∠3 when m∠4 = 112°.

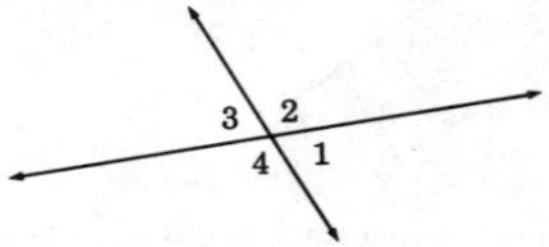

m∠2 = 112°
m∠1 = 68°
m∠3 = 68°

11. In the following figure find the measures of ∠s 2 and 4 when m∠1 = 55° and m∠3 = 61°.

m∠2 = 64°
m∠4 = 119°

Test for Applications Using Angles and Arcs (cont.)

12. Parallel lines *AB* and *EF* are cut by transversal *ST*. If m∠3 = 119°, what is the measure of ∠1? ∠2? ∠4? ∠6? ∠5? ∠8? ∠7?

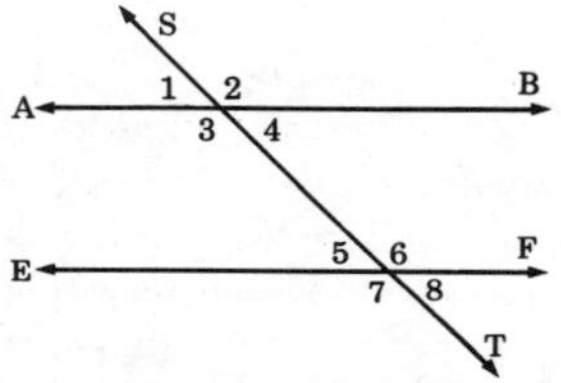

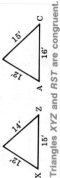

m∠1 = 61°
m∠2 = 119°
m∠4 = 61°
m∠6 = 119°
m∠5 = 61°
m∠8 = 61°
m∠7 = 119°

13. Select from the following group two triangles that are congruent.

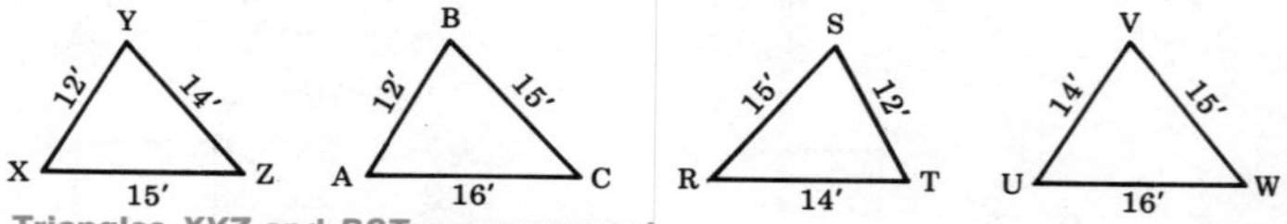

Triangles *XYZ* and *RST* are congruent.

14. Select from the following group two triangles that are similar.

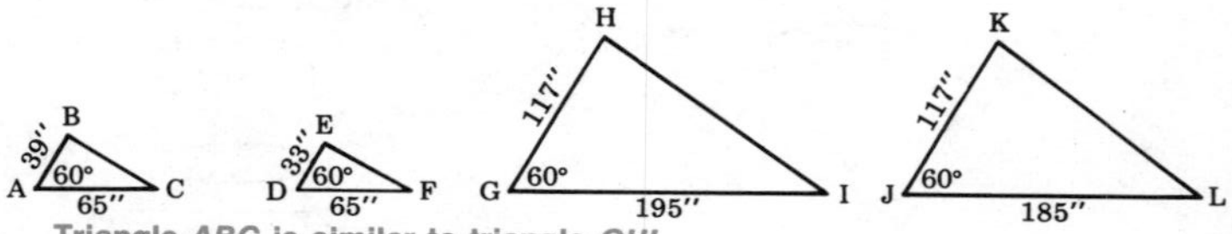

Triangle *ABC* is similar to triangle *GHI*.

15. How many nautical miles apart are two ships if one ship is located at 21°N. 72°E. and the other ship is at 21°N. 57°E.?
900 nautical miles

16. a. Change 56° 30' arc of longitude to units of time. 3 hr. 46 min.
    b. Change 4 hr. 32 min. to units measuring longitude. 68°

17. What compass course should a pilot steer if the true course is 92°, variation is 12°E., and deviation is 2°E.? 78°

NAME

Test for Constructions (cont.)

8. Draw a line segment that measures 86 millimeters long and draw the perpendicular bisector of the line segment.

43mm   43mm

9. Draw with a protractor an angle of 67°. Then construct with a compass an angle equal to it. Check your copy of the angle with a protractor.

67°

10. Draw an angle measuring 130°. Bisect it using a compass. Check with a protractor.

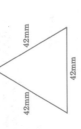

65°   65°

11. Draw a line 62 millimeters long. Select a point A outside this line. Through this point construct a line parallel to the line you have drawn. Make the parallel line 62 millimeters long.

62mm
A
62mm

12. Inscribe a regular pentagon inside a circle that has a diameter of 44 millimeters. Measure the sides of the pentagon and write this measure.

22mm   27mm

13. Construct a rectangle 2¼ inches long and 1¾ inches wide.

$1\frac{3}{4}$ in. $2\frac{1}{4}$ in.

22

NAME

Test for Constructions

1. Make a copy of the following line segment, using a compass and a straightedge.

2. Construct a triangle with sides measuring 3 centimeters, 5.5 centimeters, and 4 centimeters.

4cm   3cm   5.5cm

3. Construct a triangle with two angles measuring 80° and 60° and an included side 25 millimeters long.

80°   60°   25mm

4. Construct a triangle with two sides measuring 1¼ inches and 1½ inches and an included angle of 50°.

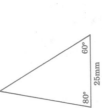

50° $1\frac{1}{2}$ in. $1\frac{7}{8}$ in.

5. Draw an equilateral triangle, each side measuring 42 mm.

42mm   42mm   42mm

6. Draw a square, each side measuring 3 cm.

3cm   3cm   3cm   3cm

7. Draw a line 4 inches long. Using a compass, construct a perpendicular to it:
   a. At a point on the line.
   b. From a point not on the line.

a.   4″   b.

21

T51

Test for Indirect Measurement   (Use table on page 417 in text when appropriate.)

1. In the figure shown here, name the hypotenuse, the altitude, and the base.
   Hypotenuse is f, altitude is e, base is g.

2. Find the hypotenuse of a right triangle if the altitude is 24 cm and the base is 18 cm. **30 cm**

3. What is the base of a right triangle if the hypotenuse is 135 mm and the altitude is 81 mm?
   108 mm

4. Find the altitude of a right triangle if the base is 96 yards and the hypotenuse is 204 yards.
   180 yd.

5. Find the height of a tree that casts a shadow of 45 feet at the time that a 12-foot telephone pole casts a shadow of 9 feet. **60 ft.**

6. Show by the rule of Pythagoras that a 39-inch ski pole can be fitted into a rectangular packing box having inside dimensions of 16 inches by 36 inches.
   $16^2 + 36^2 = 1552$; $39^2 = 1521$; $1552 > 1521$

7. In right triangle ABC shown here, find the indicated parts:

   a. Find side a (to nearest hundredth) when angle $A = 24°$ and side $b = 25$.   **11.13**

   b. Find side c (to nearest hundredth) when angle $B = 80°$ and side $a = 12$.   **69.12**

   c. Find angle B when side $b = 4$ and side $c = 5$.   **53°**

   d. Find side b (to nearest tenth) when side $a = 76$ and angle $A = 35°$.
   108.5

   e. Find angle A when side $b = 12$ and side $a = 9$.   **37°**

8. A sail for a sailboat has the shape and dimensions shown in the following figure:
   Find the altitude x (to nearest hundredth) of the sail.
   10.25 ft.

9. A ramp has the shape and dimensions shown in the following figure:
   Find the length x (to the nearest tenth) of the ramp.
   12.4 m

Test for Applications Measuring Area and Volume

Find the area of:

1. A rectangle 23 feet long and 85 feet wide.   **1955 sq. ft.**

2. A circle whose radius is 91 meters.   **26,002.34 m²**

3. A square whose side measures 102 centimeters.   **10,404 cm²**

4. A parallelogram with an altitude of 13 inches and a base of 27 inches.
   351 sq. in.

5. A circle whose diameter is 5.3 millimeters.   **22.05065 mm²**

6. A triangle with a base of 42 yards and a height of 75 yards.   **1575 sq. yd.**

7. A trapezoid with bases of 25 kilometers and 15 kilometers and a height of 20 kilometers.
   400 km²

Find the total area of:

8. A cube whose edge measures 12 meters.   **864 m²**

9. A rectangular solid 52 feet long, 22 feet wide, and 13 feet high.
   4,212 sq. ft.

10. A right circular cylinder 77 centimeters in diameter and 65 centimeters high.
    25,024.23 cm²

11. A sphere whose radius is 15 millimeters.
    2,826 mm²

12. How many 12 inch by 12 inch square carpet tiles are needed for a floor that measures 12 feet long and 15 feet wide?
    180 carpet tiles

13. The surface area of a circular skating rink having a diameter of 20 meters is how many times as large as the area of a rink having a diameter of 15 meters?
    $\frac{16}{9} = 1.7$ times as large

Find the volume of:

14. A cube whose edge measures 54 inches.   **157,464 cu. in.**

15. A cylinder 88 millimeters high with the radius of its base 22 millimeters.
    133,738.88 mm³

16. A rectangular solid 45 feet long, 32 feet wide, and 28 feet high.   **40,320 cu. ft.**

17. A sphere whose diameter is 8 yards.   **267.94667 cu. yd.**

18. A right circular cone 42 meters high with the diameter of its base 30 meters.
    9,891 m³

19. A square pyramid 22.8 centimeters on each side and 25 centimeters high.   **4,332 cm³**

20. A circular dish is 8¼ inches in diameter and 2 inches deep. How many cubic inches of food will it hold?
    106.858125 cu. in.

Test for Positive and Negative Numbers

1. Read, or write in words, each of the following:

a. -5 negative five
b. +12 positive twelve
c. -2 negative two
d. +17 positive seventeen
e. -46 negative forty-six
f. +72 positive seventy-two

2. What number is the opposite of the following?

a. +8 -8
b. -4 +4
c. -23 +23
d. +65 -65
e. -29 +29

3. Read, or write in words, each of the following:

a. $+\frac{7}{8}$ positive seven-eighths
b. $-\frac{1}{12}$ negative one-twelfth
c. $+5\frac{1}{4}$ positive five and one-fourth
d. -.75 negative point seven five
e. +12.6 positive twelve point six
f. $-3\frac{5}{6}$ negative 3 and five-sixths

4. a. Is $-\frac{4}{5}$ an integer? A rational number? A real number? no; yes; yes
b. Is $+5\frac{3}{11}$ an integer? A rational number? A real number? no; yes; yes
c. Is .63 an integer? A rational number? A real number? no; yes; yes
d. Is -6.7 an integer? A rational number? A real number? no; yes; yes
e. Is -7 an integer? A rational number? A real number? yes; yes; yes
f. Is $+\sqrt{43}$ an integer? A rational number? A real number? no; no; yes

5. Read, or write in words, each of the following:

a. |+13| absolute value of positive thirteen
b. |-5| absolute value of negative five
c. |+67| absolute value of positive sixty-seven
d. $|-\frac{4}{5}|$ absolute value of negative four-fifths
e. |-.87| absolute value of negative point eight seven
f. $|+4\frac{5}{8}|$ absolute value of positive four and five-eights

6. Find the absolute value of each of the following:

a. |-9| 9
b. |+1| 1
c. |0| 0
d. |-14| 14
e. $|+\frac{2}{3}|$ $\frac{2}{3}$

7. Use the number line shown here to name the number that corresponds to each of the following points.

```
   C       A  D E                            B
-10 -9 -8 -7 -6 -5 -4 -3 -2 -1 0 1 2 3 4 5 6 7 8 9 10
```

a. A -4
b. B +10
c. C -7
d. D -2
e. E +1

8. If -16° means 16° below zero, how can 32° above zero be represented? +32°

9. If an increase of $17 in the cost of living index can be represented by +$17, how can a decrease of $3 in the cost of living be represented? -$3

10. If -12 km/h indicates a speed of 12 km/h in the easterly direction, +24 km/h represents: 24 km/h in the westerly direction

---

Test for Positive and Negative Numbers (cont.)

11. Which number is greater?

a. +8 or +3 +8
b. -8 or -3 -3
c. 0 or +5 +5
d. -10 or 0 0
e. +5 or -5 +5
f. -7 or +3 +3

12. Which of the following sentences are true?

a. +2 > -6 True
b. 0 < -7 False
c. +4 > +6 False
d. -16 > +11 False
e. -7 < -2 True
f. +9 < +4 True

13. A movement of how many units of distance and in what direction (right or left) is represented by each of the following numbers?

a. -6 six units to the left
b. +8 eight units to the right
c. -1 one unit to the left
d. +14 fourteen units to the right
e. +6.7 six point seven units to the right
f. $-3\frac{1}{2}$ three and one-half units to the left

14. Using the number line as a scale, draw a vector representing:

a. -9
b. $+4\frac{1}{4}$
c. +10
d. +6.75
e. -7.5
f. $-2\frac{1}{2}$

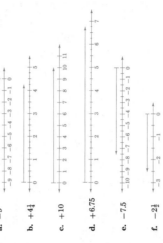

15. What number is the additive inverse of each of the following:

a. -3 +3
b. +28 -28
c. 0 0
d. $-\frac{3}{16}$ $+\frac{3}{16}$
e. +5.17 -5.17
f. $-\frac{13}{4}$ $+\frac{13}{4}$

16. Read, or write in words, each of the following:

a. -4 negative four
b. -(+12) the opposite of positive twelve
c. -(-1.11) the opposite of negative one point eleven
d. -(-37) the opposite of negative thirty-seven
e. $-(-\frac{7}{8})$ the opposite of negative seven-eighths
f. -(+1) the opposite of positive one

T53

Test for Computation with Positive and Negative Numbers

1. Add on the number line, using vectors:

a.
$$\begin{array}{r} +7 \\ +4 \\ \hline \end{array}$$

b.
$$\begin{array}{r} -6 \\ -3 \\ \hline \end{array}$$

c.
$$\begin{array}{r} +8 \\ -6 \\ \hline \end{array}$$

d.
$$\begin{array}{r} -5 \\ +1 \\ \hline \end{array}$$

2. Select the word that correctly completes the sentence:

a. The sum of a positive number and a positive number is (*positive, negative*). **positive**

b. The sum of a positive number and a negative number is determined by the sign of the number that has the (*greater, smaller*) absolute value. **greater**

c. The sum of a negative number and a negative number is (*positive, negative*). **negative**

3. Add:

a.
$$\begin{array}{r} +7 \\ +9 \\ \hline +16 \end{array}$$
b.
$$\begin{array}{r} -8 \\ -15 \\ \hline -23 \end{array}$$
c.
$$\begin{array}{r} -25 \\ +14 \\ \hline -11 \end{array}$$
d.
$$\begin{array}{r} +67 \\ -29 \\ \hline +38 \end{array}$$

4. Add:

a.
$$\begin{array}{r} -\tfrac{5}{8} \\ -\tfrac{23}{24} \\ \hline \end{array}$$
b.
$$\begin{array}{r} +\tfrac{4}{5} \\ +\tfrac{7}{10} \\ +\tfrac{1}{2} \\ \hline \end{array}$$
c.
$$\begin{array}{r} -7.6 \\ +2.1 \\ +5.5 \\ \hline \end{array}$$
d.
$$\begin{array}{r} +7.86 \\ -3.74 \\ +4.12 \\ \hline \end{array}$$

5. Subtract on the number line, using vectors:

a.
$$\begin{array}{r} +3 \\ +8 \\ \hline \end{array}$$

Test for Computation with Positive and Negative Numbers (cont.)

b.
$$\begin{array}{r} -4 \\ -5 \\ \hline \end{array}$$

c.
$$\begin{array}{r} +2 \\ -7 \\ \hline \end{array}$$

d.
$$\begin{array}{r} -1 \\ +9 \\ \hline \end{array}$$

6. Subtract:

a.
$$\begin{array}{r} +16 \\ +10 \\ \hline +6 \end{array}$$
b.
$$\begin{array}{r} +3 \\ +17 \\ \hline -14 \end{array}$$
c.
$$\begin{array}{r} -53 \\ +25 \\ \hline -78 \end{array}$$
d.
$$\begin{array}{r} +76 \\ -92 \\ \hline +168 \end{array}$$

7. Subtract:

a.
$$\begin{array}{r} -4\tfrac{3}{4} \\ -6\tfrac{1}{2} \\ \hline +15 \end{array}$$
b.
$$\begin{array}{r} -\tfrac{3}{4} \\ +1\tfrac{3}{8} \\ \hline \tfrac{5}{8} \end{array}$$
c.
$$\begin{array}{r} +7.9 \\ +4.6 \\ \hline +3.3 \end{array}$$
d.
$$\begin{array}{r} +.81 \\ -.62 \\ \hline +.19 \end{array}$$

8. Multiply:

a.
$$\begin{array}{r} +4 \\ +8 \\ \hline +32 \end{array}$$
b.
$$\begin{array}{r} -11 \\ -7 \\ \hline +77 \end{array}$$
c.
$$\begin{array}{r} +13 \\ -6 \\ \hline -78 \end{array}$$
d.
$$\begin{array}{r} -15 \\ +10 \\ \hline -150 \end{array}$$

9. Select the word that correctly completes the sentence:

a. The product of a negative number and a negative number is (*positive, negative*). **positive**

b. The product of a positive number and a positive number is (*positive, negative*). **positive**

c. The product of a positive number and a negative number is (*positive, negative*). **negative**

10. Multiply:

a.
$$\begin{array}{r} +3\tfrac{1}{2} \\ -4 \\ \hline +12\tfrac{4}{8} \end{array}$$
b.
$$\begin{array}{r} -3\tfrac{1}{4} \\ -2\tfrac{1}{4} \\ \hline +7\tfrac{7}{16} \end{array}$$
c.
$$\begin{array}{r} -.3 \\ +.9 \\ \hline -0.27 \end{array}$$
d.
$$\begin{array}{r} +7.6 \\ +2.8 \\ \hline +21.28 \end{array}$$

11. Name the multiplicative inverse of each of the following:

a. -9 $-\tfrac{1}{9}$
b. $+22$ $+\tfrac{1}{22}$
c. $-\tfrac{5}{4}$ $-\tfrac{4}{5}$
d. $+6\tfrac{1}{3}$ $+\tfrac{3}{19}$

12. Divide:

a.
$$\tfrac{+18}{+6} \quad +3$$
b.
$$\tfrac{-56}{-8} \quad +7$$
c.
$$\tfrac{-60}{+15} \quad -4$$
d.
$$\tfrac{+42}{-7} \quad -6$$

13. Divide:

a. $-5)\overline{+10.5}$ -7
b. $+11)\overline{-1.21}$ $-.11$
c. $-1.2)\overline{-15.6}$ $+13$
d. $+.9)\overline{+1.17}$ $+1.3$

Test for Language of Algebra

1. Read, or write in words, each of the following:
   a. $(+6) - (-17)$ positive six minus negative seventeen
   b. $(23)(12)$ Twenty-three times twelve.
   c. $\sqrt{9}$ The square root of nine.
   d. $\frac{56}{7}$ Fifty-six divided by seven.
   e. 8^3 Eight cubed.
   f. $5(10 + 34)$ Five times the sum of ten and thirty-four.

2. What symbol represents the variable in each of the following:
   a. $8 + x = 15$ x
   b. $52 = m + 6$ m
   c. $6y = 42$ y
   d. $12 = 2d$ d
   e. $\frac{s}{5} = 15$ c
   f. $33 - A = 11$ A

3. Write each of the following as an algebraic expression:
   a. The difference between twenty-nine and seventeen. $29 - 17$
   b. The product of three and six. 3×6
   c. The sum of A and sixteen. $A + 16$
   d. Fifteen divided by five. $\frac{15}{5}$
   e. Seven times the sum of six and two. $7(6 + 2)$
   f. The square root of the diameter d. $\sqrt{d}$

4. Read, or write in words, each of the following:
   a. $9x + 4y$ The sum of nine times x and four times y.
   b. $\frac{2\pi}{3}$ Two times pi divided by three.
   c. $12 - 4$ The difference between twelve and four.
   d. $\frac{1}{4}\pi d^2 h$ One-fourth times pi times d squared times h.
   e. $x - 15$ The difference between x and fifteen.
   f. $2lw$ Two times l times w.

5. Read, or write in words, each of the following:
   a. $g \leq 4$ Some number g is less than or equal to four.
   b. $4x + 3 > 3x - 7$ The sum of four times some number x and three is greater than the difference between three times some number x and seven.
   c. $4d \not> 18$ Four times some number d is not greater than eighteen.
   d. $15 \leq 2p \leq 0$ Fifteen is less than or equal to two times some number p which is less than or equal to zero.
   e. $2x + 6 \neq 21$ The sum of two times some number x and six is not equal to twenty-one.
   f. $3m = 18$ Three times some number m is equal to eighteen.

6. Write each of the following as an open sentence symbolically:
   a. Nine times some number x is greater than or equal to seventy-two. $9x \geq 72$
   b. Zero is less than some number x which is less than seven. $0 < x < 7$
   c. The sum of some number y and four is greater than twelve. $y + 4 > 12$

29

Test for Language of Algebra (cont.)

   d. The sum of five times some number a and two is not less than the difference between two times some number a and three. $5a + 2 \not< 2a - 3$
   e. The sum of two times some number p plus four is not equal to nine. $2p + 4 \neq 9$
   f. Each number A is equal to each number s squared. $A = s^2$

7. Express each of the following as a formula:
   a. The circumference of a circle (c) is equal to two times pi (π) times the diameter (d). $C = 2\pi d$
   b. The volume (V) of a rectangular solid is equal to the length (l) times the width (w) times the height (h). $V = lwh$
   c. The perimeter (p) of an equilateral triangle is equal to three times the length of the side (s). $p = 3s$
   d. The temperature (K) in kelvins is equal to the sum of the temperature (C) in degrees Celsius and two hundred seventy-three. $K = C + 273$
   e. The distance (d) a moving object travels is equal to its average rate of speed (r) times the time of travel (t). $d = rt$
   f. Work done (W) in moving an object is equal to the force (F) exerted times the distance (d) the object traveled. $W = Fd$

8. Find the value of each of the following:
   a. $A + 16$ when $A = 3$ 19
   b. $9x + 4y$ when $x = 2$ and $y = 3$ 30
   c. $\frac{1}{4}\pi d^2 h$ when $d = 4$, $h = 2$, and $\pi = 3.14$ 25.12
   d. $x - 15$ when $x = 45$ 30
   e. $2lw$ when $l = 6$ and $w = 9$ 4
   f. $\frac{m}{n}$ when $m = 12$, and $n = 3$ 27

9. Find the value of each of the following:
   a. B when $A = 47$. Formula: $B = 90 - A$ 43
   b. F when $W = 56$ and $d = 3$. Formula: $W = Fd$ $18\frac{2}{3}$
   c. C when $K = 292$. Formula: $K = C + 273$ 19
   d. V when $l = 5$, $w = 2\frac{1}{2}$, and $h = 7$. Formula: $V = lwh$ $87\frac{1}{2}$
   e. r when $d = 100$ and $t = 4$. Formula: $r = \frac{d}{t}$ 25
   f. p when $s = 16$. Formula: $p = 3s$ 48

30

NAME _____

Test for Open Sentences In One Variable

Underline which of the numbers when substituted for x will make the sentence true?

1. $x + 6 = 13$
   a. 6 b. <u>7</u> c. 19 d. $\frac{1}{3}$

2. $x - 9 = 2$
   a. 1 b. 7 c. $\frac{2}{9}$ d. <u>11</u>

3. $7x = 56$
   a. <u>8</u> b. $\frac{1}{56}$ c. 9 d. 49

4. $\frac{4}{4} = 24$
   a. 6 b. <u>96</u> c. 48 d. 8

Find the missing number.

5. $(38 - 25) + ? = 38$ **25**

6. $(x + 6) - ? = x$ **6**

7. $(4 \times y) \div ? = y$ **4**

8. $\frac{9}{10} \times ? = 9$ **10**

Solve and check.

9. $5 + a = 26$ **a = 21**
10. $17 = b + 4$ **b = 13**
11. $c - 28 = 9$ **c = 37**
12. $0 = d - 51$ **d = 51**
13. $8f = 72$ **f = 9**
14. $132 = 11g$ **g = 12**
15. $\frac{h}{3} = 27$ **h = 81**
16. $\frac{4}{5}j = 1.6$ **j = 2**
17. $3k - 9 = 30$ **k = 13**
18. $m + \frac{1}{2}m = 60$ **m = 40**

19. Find the solution of the equation $2n - 3 = 19$ when the replacements for the variable are:
   a. The whole numbers. **n = 11**
   b. The prime numbers. **n = 11**
   c. The negative integers. **no solutions**
   d. The even integers. **no solutions**

20. Find the solutions of each of the following equations:
   a. $|p| = 3$ **+3, -3**
   b. $|r| + 6 = 9$ **+3, -3**
   c. $7 \times |s| = 105$ **+9, -9**
   d. $\frac{|t|}{4} = 16$ **+8, -8**

21. Find the value of h when $A = 40$ and $b = 5$. Formula: $A = bh$ **h = 8**
22. Find the value of r when $A = 880$, $\pi = \frac{22}{7}$, and $h = 14$. Formula: $A = 2\pi rh$ **r = 10**
23. Find the value of t when $v = 224$, $V = 48$, and $g = 32$. Formula: $v = V + gt$ **t = 5½**

24. For each of the following equations draw an appropriate number line, then graph its solution. The replacements for the variable are all the real numbers.
   a. $x + 3 = 8$ [number line 0 ... 5]
   b. $9n - n = 48$ [number line 0 ... 6]

NAME _____

Test for Open Sentences In One Variable (cont.)

c. $|y| = 1$ [number line -1 ... 1] d. $\frac{1}{2}a + 10 = 9$ [number line -2 ... -1]

25. By the equation method, find the required number in each of these:
   a. $\frac{7}{12}$ of 36 **21**
   b. 19 is what part of 57? **1/3**
   c. .82 of 109 **89.38**
   d. What percent of 40 is 8? **20%**
   e. $12\frac{1}{2}\%$ of 140 **17.5**
   f. 12% of what number is 6? **50**

26. Find the ratio of:
   a. 6 to 11 $\frac{6}{11}$
   b. 63 to 56 $\frac{63}{56} = \frac{9}{8}$
   c. x to 17 $\frac{x}{17}$
   d. 27 to 84 $\frac{27}{84} = \frac{9}{28}$
   e. 15 to 90 $\frac{15}{90} = \frac{1}{6}$
   f. 52 to 78 $\frac{52}{78} = \frac{2}{3}$

Solve and check:

27. $\frac{x}{4} = \frac{48}{?}$ **x = 3**
28. $\frac{9}{20} = \frac{y}{80}$ **y = 36**
29. $\frac{5}{9} = \frac{5}{c}$ **c = 6**
30. $\frac{1.3}{?} = \frac{.7}{n}$ **n = .5**

Find the solutions of each of the following inequalities when the replacements for the variable are all the real numbers.

31. $y + 6 < 16$ **All numbers less than 10.**
32. $x - 2 > 10$ **All numbers greater than 12.**
33. $\frac{n}{3} \geq 14$ **42 and all numbers greater than 42.**
34. $12b \neq 252$ **All numbers except 21.**
35. $-9t \leq 45$ **-5 and all numbers greater than -5.**
36. $-9r \not< -36$ **4 and all numbers less than 4.**

For each of the following inequalities draw an appropriate number line, then graph its solutions. The replacements for the variable are all the real numbers.

37. $a \geq 5$ [number line 0 ... 5]
38. $-2 \leq b \leq 0$ [number line -2 ... 2]
39. $c < -3$ [number line -3 ... 0]
40. $-9d \not> -18$ [number line 0 ... 2]

Test for Graphs

Results of Test A

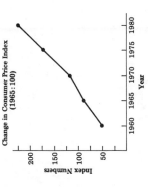

Use the graph above to answer the following questions regarding the exam grades of a class.

1. What quantity does the vertical scale represent? The horizontal scale? **Number of students. Exam grades.**
2. How many students are represented by four scale marks on the vertical axis? **20**
3. Which grade was received by the most number of students? **B**
4. Which grade was received by the least number of students? **D**
5. How many students received a grade of D? a grade of C? a grade of B? a grade of A? **D: 10; C: 15; B: 30; A: 20**
6. Construct a bar graph for the following data which gives the average weekly salaries for the employees of a firm.

Receptionists $147.00 Secretaries $168.00
Filing Clerks $135.00 Assistant Managers $250.00
Typists $152.00 Managers $350.00

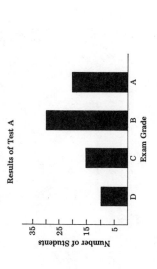

33

Test for Graphs (cont.)

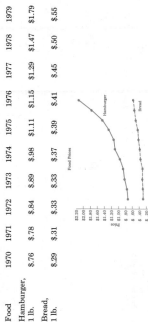

Use the line graph above to answer the following questions about the consumer price index.

7. What quantity does the vertical scale represent? The horizontal scale? **Consumer price index. Year.**
8. What value of the index is represented by five scale marks on the vertical axis? What year is represented by the third scale mark on the horizontal axis? **125; 1970**
9. In what year was the consumer price index the lowest? In what year was it the highest? **1960; 1980**
10. What was the (approximate) value of the index in 1960? In 1970? In 1980? **50; 120; 225**
11. Construct line graphs for the following data which give the increases in the cost of bread and hamburger from the years 1970 through 1980. Use a solid line for the price of hamburger and a dotted line for the price of bread.

Food	1970	1971	1972	1973	1974	1975	1976	1977	1978	1979	1980
Hamburger, 1 lb.	$.76	$.78	$.84	$.89	$.98	$1.11	$1.15	$1.29	$1.47	$1.79	$2.15
Bread, 1 lb.	$.29	$.31	$.33	$.37	$.39	$.41	$.45	$.50	$.55	$.59	

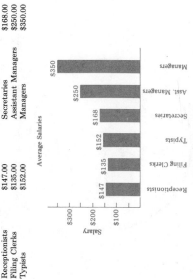

34

Test for Graphs (cont.)

U.S. Building Fire Losses by Causes

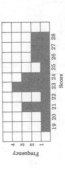

Use the circle graph above to answer the following questions about U.S. building fire losses and their causes.

12. What fractional part of fire losses are caused by heating and cooking equipment? By suspected arson? By electrical equipment? $\frac{1}{5}$; $\frac{7}{100}$; $\frac{17}{100}$

13. Compare the percentage of fire losses due to children and matches to the percentage due to smoking and matches. $\frac{5}{12}$

14. Which three categories cause the largest percentages of losses? **Heating and cooking equipment, other, and electrical.**

15. If the total number of losses by fire is 996,000, what are the number of building losses for each cause?
Heating and cooking equipment 199,200; **Electrical** 169,320;
Smoking and Matches 119,520; **Unknown** 109,560; **Suspected arson** 69,720;
Open flames and sparks 99,600; **Children and matches** 49,800; **Other** 179,280; **Total 996,000**

16. Construct a circle graph of the following data which show the percentages of employed persons by major occupational groups.

Percent of Employed by Occupational Group

Occupational Group	Percent of Total U.S. Workers
Clerical Workers	20%
Other White-collar Workers	30%
Blue-collar Workers	35%
Service Workers	10%
Farm Workers	5%

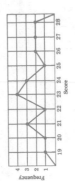

Percent of Total U.S. Workers

35

Test for Statistics and Probability

1. Make:
   a. A frequency distribution table   **b.** A histogram
   c. A frequency polygon
   for the following list of scores:
   26, 23, 21, 24, 26, 28, 22, 21, 24, 23, 27, 20, 19, 23, 27, 25, 21, 24, 28, 23

Frequency Distribution Table

Score	Tally	Frequency
19	1	1
20	1	1
21	111	3
22	1	1
23	1111	4
24	111	3
25	1	1
26	11	2
27	11	2
28	11	2
		20

2. Find the mean for these scores:
   74, 80, 71, 72, 75, 77, 76 **75**

3. Find the median for these scores:
   57, 55, 56, 54, 58, 59, 52, 49, 54 **55**

4. Find the mode for these scores:
   81, 85, 80, 89, 85, 82, 81, 85, 84, 85, 83, 81 **85**

5. Find the range for these scores:
   63, 81, 77, 56, 66, 71, 48, 98, 75, 91, 67, 77, 88, 83, 65, 70 **50**

36

Test for Statistics and Probability (cont.)

6. Tally and arrange the following list of scores in a frequency distribution table; then find the mean, median, mode, and range:
39, 36, 33, 40, 35, 37, 35, 39, 34, 38, 37, 40, 37, 33, 35, 38, 36, 37, 41, 39, 32, 35, 40, 36, 38, 41, 40, 38, 39, 34, 35

Frequency Distribution Table

Score	Tally	Frequency
32	1	1
33	11	2
34	11	2
35	4HT	5
36	111	3
37	1111	4
38	1111	4
39	1111	4
40	1111	4
41	11	2
		31

mean = 37; median = 37; mode = 35; range = 9

7. Find the percentile rank of each of the following:
   a. Tom ranked eighth in a class of 40 students. 80th
   b. Sarah ranked seventieth in a class of 175 students. 60th
   c. Anthony ranked ninth in a class of 45 students. 80th
   d. Bonny ranked fifteenth in a class of 50 students. 70th

8. Find the lower quartile, upper quartile, and 50th percentile for each of the following lists of scores:
   a. 23, 24, 26, 30, 29, 27, 28, 25
   upper quartile = 29; lower quartile = 25; 50th percentile = 27
   b. 77, 77, 78, 78, 80, 81, 81, 84, 84, 86
   upper quartile = 84; lower quartile = 78; 50th percentile = 81

9. What is the probability of drawing at random on the first draw:
   a. A blue bead from a box containing 12 beads of which 5 are blue? $\frac{5}{12}$
   b. A white block from a box containing 18 blocks of which 7 are white? $\frac{7}{18}$
   c. A vanilla caramel from a bag of 50 caramels of which 24 are vanilla? $\frac{12}{25}$
   d. A wooden chip from a bag containing 30 chips of which 18 are wooden? $\frac{3}{5}$
   e. A lime gum drop from a jar containing 64 gum drops of which 12 are lime? $\frac{3}{16}$

10. A standard deck of 52 playing cards consists of an ace, king, queen, jack, 10, 9, 8, 7, 6, 5, 4, 3, and 2 in four different suits: clubs (black), diamonds (red), hearts (red), and spades (black). The king, queen, and jack cards are called picture cards.
    From a well-shuffled deck and drawn at random, what is the probability that the first draw will be:
    a. a black card? $\frac{1}{2}$
    b. a 9? $\frac{1}{13}$
    c. the king of hearts? $\frac{1}{52}$
    d. a picture card? $\frac{3}{13}$
    What are the odds in favor of the first draw being:
    e. a 3? 1 to 13
    f. a red card? 1 to 2
    g. a picture card? 3 to 13
    h. the ace of spades? 1 to 52

Test for Income, Take-Home Pay, Income Tax

1. What is the weekly wage of a person who works 20 hours per week at each of the following rates?
   a. $6.85 $137
   b. $9.60 $192
   c. $4.50 $90
   d. $5.95 $119
   e. $10.25 $205

2. Find the weekly earnings of a person who works:
   a. 25 hours at $8.00 per hour $200
   b. 22 hours at $6.10 per hour $134.20
   c. 30 hours at $4.25 per hour $127.50
   d. 40 hours at $7.50 per hour $300
   e. 25½ hours at $9.40 per hour $239.70
   f. 18½ hours at $12.50 per hour $231.25

3. Find the hourly rate of pay when a person's weekly earnings are:
   a. $226 for 40 hours of work $5.65
   b. $98 for 20 hours of work $4.90
   c. $145 for 29 hours of work $5.00
   d. $297.50 for 35 hours of work $8.50

4. Find the annual salary of a person earning each week:
   a. $125 $6,500
   b. $275 $14,300
   c. $57.75 $3,003
   d. $325.25 $16,913
   e. $452.75 $23,530

5. A farm worker picked 76 boxes of lettuce per day at 43¢ a box. How much did she earn for the day? $32.68

6. A machinist makes 1,342 bolts a day at 6¢ a bolt. What are his weekly earnings (5 days)? $402.60

7. At the rate of 6% commission on sales, how much will a real estate salesperson receive for selling a house at $64,900? $3,894.

8. A craft store, charging 11% commission, sold goods amounting to $2,540. How much is the commission? $279.40

9. A car mechanic charges $45.50 for a regular tune-up and $35 for tire alignment. What were her earnings for the day when she did 5 tune-ups and 7 tire alignments? $357.50

10. How much interest should a person receive for the year if his savings account in the bank averages over the year:
    a. $4,250 and the bank pays 5½% annual interest? $233.75
    b. $1,200 and the bank pays 5% annual interest? $60
    c. $15,000 and the bank pays 5¾% annual interest? $862.50

NAME _____

Test for Income, Take-Home Pay, Income Tax (cont.)

d. $10,000 and the bank pays $6\frac{1}{2}\%$ annual interest? **$650**

e. $5,000 and the bank pays $7\frac{1}{2}\%$ annual interest? **$375**

11. Use the percentage method (text page 548) to compute the amount of income tax to be withheld each week if a married person's weekly wages are:

a. $122 and 1 withholding allowance is claimed __**$7.07**__

b. $158 and 2 withholding allowances is claimed __**$10.96**__

c. $280 and 3 withholding allowances are claimed __**$30.88**__

d. $390 and 2 withholding allowances are claimed __**$60.93**__

12. Find the amount deducted from the employee's wages for social security tax if an employee earns each week:

a. $80 __**$5.36**__   b. $232 __**$15.54**__   c. $96 __**$6.43**__

d. $416 __**$27.87**__   e. $573 __**$38.39**__

13. If the wage tax rate is $2\frac{3}{4}\%$ of the wages, find the tax a person owes for the week if the person earns weekly:

a. $92 __**$2.53**__   b. $129 __**$3.55**__   c. $246 __**$6.77**__

d. $183.50 __**$5.05**__   e. $387.25 __**$10.65**__

14. Use the table of withholding tax (text page 549) for all wages between $88 and $330; for all other wages use the rates given on text page 548. The social security tax rate is 6.7%. Find the take-home pay for each of the following weekly wages:

a. $206 with 2 withholding allowances claimed __**$172.70**__

b. $145 with 1 withholding allowance claimed __**$123.58**__

c. $365 with 2 withholding allowances claimed __**$286.05**__

d. $412 with 3 withholding allowances claimed __**$321.35**__

15. Use the earnings given as the taxable income. Use the tax table on text page 556. Find the federal income tax owed by a married person filing a joint return who earns annually:

a. $9,720 and claims 3 exemptions __**$567**__

b. $10,575 and claims 2 exemptions __**$859**__

c. $11,120 and claims 3 exemptions __**$825**__

d. $9,975 and claims 4 exemptions __**$442**__

NAME _____

Test for Spending Money

1. Find the cost of the following, using these grocery prices:

Orange juice, 1 qt. for 77¢   Beans, 5 cans for $1.69
Coffee, $2.60 for 10-oz. jar   Soup, 6 bars for $1.08
Beets, 4 cans for 95¢   Soup, 4 cans for 95¢
Flour, 5 lb. for 99¢   Sugar, 10 lb. for $3.35
Paper towels, 2 pkgs. for $1.89   Bread, 2 loaves for $1.49
Pears, 2 cans for $1.59   Peanut butter, 3-lb. jar for $3.88

a. 1 can of beets __**$.25**__   b. 1 can of pears __**$.80**__

c. 1 can of soup __**$.24**__   d. 1 pkg. of paper towels __**$.95**__

e. 12 bars of soap __**$2.16**__   f. 8 cans of beets __**$1.98**__

g. 3 cans of beans __**$1.02**__   h. 2 bars of soap __**$.36**__

i. 2 qts. of orange juice __**$1.54**__   j. 20 lb. of flour __**$3.96**__

k. 4 loaves of bread __**$2.98**__   l. 2 jars of peanut butter __**$7.76**__

2. Find the unit price (cost per pound) to the nearest tenth of a cent for each of the following:

a. 5-lb. pkg. hamburger at $6.95 __**$1.39**__   b. 12-oz. pkg. cheese at $1.29 __**$1.72**__

c. 10-oz. jar peanut butter at 99¢ __**$1.584**__   d. 1-lb. 8-oz. pie at $2.29 __**$1.53**__

3. In each of the following the same item is priced at two rates. Which is the better buy?

a. Oranges: 6 for 75¢ or 12 for $1.39 __**12 for $1.39**__

b. Cookies: 10 oz. for 89¢ or 16 oz. for $1.35 __**16 oz. for $1.35**__

c. Bread: 5 loaves for $2.59 or 2 loaves for $1.05 __**5 loaves for $2.59**__

d. Jelly: 1 for 59¢ or 3 for $1.75 __**3 lb. for $1.75**__

4. Find the sale price of each of the following articles:

a. Bed sheet; regular price, $4.59; reduced 15%. __**$3.90**__

b. Radio; regular price, $69.95; reduced 25%. __**$52.47**__

c. Tape recorder; regular price, $199; $\frac{1}{3}$ off. __**$132.67**__

d. Desk; regular price, $259; reduced 35%. __**$168.35**__

5. Find how much change you should receive from $1 if your purchases cost:

a. $.89 __**11¢**__   b. $.52 __**48¢**__   c. $.18 __**82¢**__

d. $.27 __**73¢**__   e. $.74 __**26¢**__

6. Find how much change you should receive from $5 if your purchases cost:

a. $1.57 __**$3.43**__   b. $2.36 __**$2.64**__   c. $4.71 __**29¢**__

d. $.69 __**$4.31**__   e. $3.85 __**$1.15**__

7. Find the tax on each of the following sales when the tax rate is:

a. 3%: (1) $4.00 __**12¢**__   (2) $12.50 __**38¢**__   (3) $89.75 __**$2.69**__

b. 6%: (1) $22.00 __**$1.32**__   (2) $.59 __**4¢**__   (3) $9.87 __**59¢**__

c. 5%: (1) $2.00 __**10¢**__   (2) $10.25 __**51¢**__   (3) $75.90 __**$3.80**__

d. 4%: (1) $889.00 __**$3.56**__   (2) $7.98 __**32¢**__   (3) $572.50 __**$22.90**__

Test for Spending Money (cont.)

8. Given the reading at the beginning and at the end of the period, use the rates shown below the readings in each of the following to find the cost of the electricity used during the period.

Reading		Rates
At Beginning	At End	First 10 kWh or less $5.50
5837	5884	Next 40 kWh @ $.08
5064	5155	Next 50 kWh @ $.07
1292	1508	Over 100 kWh @ $.05
3296	3462	

a. __$8.46__
b. __$11.57__
c. __$18.46__
d. __$15.76__

9. a. Mrs. Tomaso bought a new car for $7,575. After using it for 3 years, she sold it for $3,050. What was the average yearly depreciation of the car? __$1508.33⅓__

b. At the end of the first year the odometer read 11,416 miles; at the end of the second year 25,065 miles; and at the end of the third year 37,847 miles. How many miles did Mrs. Tomaso drive during the first year? During the second year? During the third year? __11,416; 13,649; 12,782__

c. During the first year her expenses, excluding the average yearly depreciation, were: gasoline, 400 gallons at $1.38\frac{3}{10}$ per gal.; oil, 25 quarts at $1.60 per qt.; insurance, $400 per yr.; license fees, $24 per yr.; miscellaneous expenses, $416. (1) How much did it cost her to run her car for the first year, excluding depreciation? (2) What was the average cost per month? Per week? Per mile?
1) __$1190.60__; 2) __$9.22; $24.80; $.10__

10. Find the carrying charge when each article is purchased on the installment plan:

	Cash Price	Down Payment	Monthly Payment	Number of Monthly Payments	
a. Stereo system	$775	$100	$39.00	20	a. $105
b. Television	$495	$50	$43.75	12	b. $80
c. Winter coat	$150	$20	$16.25	9	c. $16.25
d. Skis	$259	$75	$43.50	5	d. $33.50
e. Bicycle	$229	$25	$15.64	15	e. $30.60

11. What is the cost of $45,000 fire insurance for 3 years at the rate of $.22 per $100? __$267.30__

12. How much must you pay for taxes on a house assessed for $32,500 if the tax rate is $55.40 per $1,000? __$1800.50__

13. If no more than 25% of a person's income should be spent for the rent of a house or apartment, find the highest monthly rent someone earning each of the following incomes can afford to pay:
a. $21,000 per year __$437.50__
b. $920 per month __$230__
c. $280 per week __$303.34__

14. If a family should not buy a house costing more than 2½ times its annual income, what is the highest price a family can afford to pay if its income is:
a. $32,750 per year __$81,875__
b. $1,875 per week __$45,500__

Test for Managing Money

1. Find the interest on:
a. $400 for 1 yr. at 6% __$24__
b. $1,000 for 1 yr. at 8% __$80__
c. $5,000 for 2 yr. at 5½% __$550__
d. $3,000 for 1 yr. at 10¾% __$322.50__

2. Find the interest on:
a. $500 for 3 yr. at 7% __$105__
b. $2,000 for 5 yr. at 11¼% __$1125__
c. $10,000 for 10 yr. at 5¾% __$5750__
d. $1,000 for 10 yr. at 6.5% __$650__

3. What is the annual rate of interest you are paying when you:
a. Borrow $108 and pay back $126 at the end of a month? __200%__
b. Borrow $50 and pay back $53.75 at the end of a week? __390%__
c. Borrow $500 and pay back $503 the next day? __219%__

4. What is the actual amount of money (net proceeds) you receive and how much is the monthly payment for a discount loan when the amount of the note is $5,000, the interest rate is 6%, and the period of the loan is 12 months? __$4700; $416.67__

5. What is the actual amount of money you receive and how much is the monthly payment for an add-on-interest loan if the principal is $5,000, the interest rate is 6%, and the period of the loan is 12 months? __$5000; $441.66__

6. Use the table in the text on page 592 to find the monthly payment to finance a car if you borrow:
a. $3,500 for 36 months __$111.80__
b. $2,500 for 36 months __$79.86__
c. $1,800 for 30 months __$67.50__
d. $2,000 for 24 months __$91.66__

7. Use the table in the text on page 593 to find the monthly payment on a loan of $50,000 at 8½% annual interest when the loan is to be amortized in:
a. 20 years __$433.92__
b. 35 years __$373.44__
c. 10 years __$619.93__
d. 25 years __$402.62__
e. 30 years __$384.46__
f. 40 years __$366.55__

8. For these problems use the table in the text on page 595. Find the amount and the interest earned when the interest is compounded annually on:
a. $1,000 for 8 yr. at 5% __$1477.46; $477.46__
b. $10,000 for 12 yr. at 6% __$20,121.96; $10,121.96__
c. $2,000 for 20 yr? at 16% __$38,921.20; $36,921.20__
d. $5,000 for 15 yr. at 6% __$11,982.80; $6982.80__

9. For these problems use the table in the text on page 595. Find the amount and the interest earned when the interest is compounded quarterly on:
a. $500 for 2 yr. at 6% __$796.92; $296.92__
b. $100 for 3 yr. at 10% __$313.84; $213.84__
c. $3,000 for 5 yr. at 5% __$7959.89; $4959.89__
d. $20,000 for 1 yr. at 20% __$41,472.00; $21,472.00__

NAME

Test for Managing Money (cont.)

10. Using the closing prices listed in the stock quotations in the text on page 600, find the total cost of each of the following purchases, including the commission (use the fee schedule on text page 598).

a. 100 shares of IBM $5,587.00

b. 100 shares of IdahoP $2550.50

c. 100 shares of IntHarv $3900.04

d. 25 shares of InIndStl $902.38

e. 50 shares of IntTT $1427.87

f. 15 shares of INACp $627.60

g. 500 shares of IllPowr $11,253.13

h. 1000 shares of IowaPS $24,737

i. 50 shares of ItekCp $1047.99

j. 50 shares of IngerR $2535.84

11. Find the date of maturity and the interest rate for each of the following bonds:

a. Alcoa 5¼s 91 1991, 5¼%

b. St O Ind 9.2s 04 2004, 9.2%

c. Exxon 6¼s 98 1998, 6¼%

d. GMA 8s 93 1993, 8%

e. Digit 4¼s 02 2002, 4¼%

f. ATT 8⅝s 07 2007, 8⅝%

12. Find the amount earned the first year by each of the following:

a. A 1-yr. certificate of deposit for $10,000 at 11.72% interest $1172

b. A 8-yr. certificate of deposit for $100,000 at 8.83% interest $8830

c. A 6-yr. certificate of deposit for $20,000 at 9.06% interest $1812

d. A 4-yr. certificate of deposit for $5,000 at 10.79% interest $539.50

NAME

Test for Applications in Business and Travel

1. Find the discount and the net price:

List price	$72	$159.95	75¢	$55.50	$225
Rate of discount	9%	15%	10%	33%	45%

$6.48, $65.52; $23.99, $135.96; 8¢, 67¢; $18.32, $37.18; $101.25; $123.75

2. Find the rate of discount when a stereo, listed for $259, is sold for $189.07. 27%

3. The net price of a refrigerator is $455 when a 30% discount is allowed. What is its list price? $650

4. A car salesperson sold a new car for $8,390 and was paid a commission of 11%. What was the amount of the salesperson's commission? What were the net proceeds the owner of the dealership received? $922.90, $7467.10

5. Find the commission and net proceeds:

Sales	$75,500	$79	$12,500	$185.95	$775
Rate of commission	8%	4%	9%	12%	15%

$6040, $69,460; $3.16, $75.84; $1125, $11,375; $22.31, $163.64; $116.25, $658.75

6. Find the rate of commission if the commission is $7,962.50 on sales amounting to $122,500. 6.5%

7. What must the sales be for a 9% rate of commission to bring a commission of $315? $3500

8. Find the selling price:

Cost	$122	$72	$1,345	$687.25	$22.90
Markup on cost	50%	125%	75%	35%	42%

$183; $2353.75; $162; $927.79; $32.52

9. What is the selling price of a used car if it costs the dealer $2,400 and the rate of markup on the selling price is 48%? $4615.38

10. Find the rate of markup on cost:

Cost	$220	$18.80	$1,650	$575	$1.62
Selling price	$253	$21.15	$1,980	$776.25	$2.51

15%; 20%; 12½%; 35%; 55%

11. What does the route symbol ⑨ mean on a road map? A federal highway

12. How would you interpret the markings found on a road map? Mileage between the two points is 2 miles.

Test for Applications in Business and Travel (cont.)

13. Use the train timetable in the text on page 615 to answer these questions:
   a. When does train 177 leave New York? When does it arrive in Baltimore?
   7:15 P.M.; 10:17 P.M.
   b. How many miles is it from New York to Baltimore? What is train 177's rate of speed for the trip from New York to Baltimore?
   184 miles, 60.66 m.p.h.

14. Use the bus timetable in the text on page 616 to answer these questions:
   a. When does bus 824 leave Memphis? When does it arrive in Oklahoma City? In Fresno?
   1:10 A.M.; 3:30 P.M.; 3:30 A.M. on the next day
   b. How long is the trip from Flagstaff to Little Rock on bus 823? On bus 827?
   33 hr. 15 min.; 30 hr. 55 min.

15. Use the airline schedule in the text on page 617 to answer these questions:
   a. What is the flight number of the airplane leaving Dallas-Fort Worth for Chicago at 7:15 P.M.?
   Are there any stops on this flight? How long does the flight take?
   120; yes; 3½ hr.
   b. When does flight 182 leave Dallas-Fort Worth for Tacoma? When does it arrive in Tacoma?
   How long is the flight?
   6:40 P.M.; 8:20 P.M.; 1 hr. 40 min.

Problem Solving Competency Check Test

Solve each problem and select the letter corresponding to your answer.

1. The monthly earnings of a person who works 35 hours at $7.95 per hour are:
   (a.) $1,113.00   b. $278.25   c. $1,213.00   d. $14,469.00

2. A realty receives 6% commission on the sale of a house. If the realty sold a house for $85,000, what would be the commission?
   a. $51,000   b. $8,500   **(c.)** $5,100   d. $6,000

3. If the withholding tax is $27.50 and social security tax is $15.75 on your weekly salary of $235, your take-home pay is:
   a. $207.50   b. $219.25   c. $191.25   **(d.)** $191.75

4. Which is the best buy of the same kind of apples?
   a. 4 for $1   **(b.)** 2 for 59¢   c. 8 for $1.75   d. 12 for $2.79

5. What is the unit price (cost per liter) to the nearest tenth of a cent of a 2-L bottle of beverage costing $1.09?
   (a.) 54.5¢   b. 55.4¢   c. $2.18   d. 27.3¢

6. A football which regularly sells for $28.50 is reduced 16%. Its sale price is:
   a. $4.56   **(b.)** $23.94   c. $24.94   d. $24.23

7. What is the regular price of a freezer that sold for $189.85 at a 20% reduction sale?
   (a.) $236.88   b. $949.25   c. $37.97   d. $169.85

8. If a purchase is $11.46, the change from a $20 bill should be:
   a. $8.64   **(b.)** $8.54   c. $9.54   d. $9.64

9. What is the selling price, including sales tax of 7%, of a television which sells for $498.95?
   a. $464.02   b. $533.77   c. $598.95   **(d.)** $533.88

10. Jason can purchase a bicycle for the cash price of $189.98 or $20 down and 6 equal monthly payments of $33.27. By paying cash he can save:
   a. $9.67   b. $20.00   c. $26.97   **(d.)** $29.67

11. What is the cost of $80,000 fire insurance for 1 year at $.32 per $100?
   a. $2,560   b. $25.60   **(c.)** $256   d. $320

12. How much must you pay for taxes on a house assessed for $42,800 if the tax rate is $5.80 per $100?
   a. $248.24   b. $580   c. $2,482.40   **(d.)** $2,482.40

13. Mr. Freedman bought a house for $92,400. He paid $23,100 down and gave a mortgage for the balance. What is the amount of his mortgage?
   (a.) $69,300   b. $79,300   c. $69,100   d. $79,100

Problem Solving Competency Check Test (cont.)

14. Edward Swider paid the Collegiate Racketball Club membership by check in the amount of $80 on April 3, 1982.

```
FIRST STATE BANK                    650
                              19 _____
PAY TO THE
ORDER OF ① _____   $ _____
_____ DOLLARS
MEMO _____
95 646 718
```

On line ① should be written:
a. Edward Swider   b. Collegiate Racketball Club
c. April 3, 1982   (d.) Eighty and 00/100

15. If you have $1,219 in your checking account and make a withdrawal of $33.86, your new balance is:
a. $1,252.86   (b.) $1,185.14   c. $1,085.14   d. $1,152.86

16. The interest on $4,400 for 5 years at 18% annual interest is:
a. $396   (b.) $792   c. $3,960   d. $2,200

17. Ernie borrowed $7,000 for 48 months to buy a car and paid $250.83 per month. What was his finance charge?
a. $250.83   (b.) $5,039.84   c. $4,039.84   d. $700

18. What is the monthly payment on an add-on-interest loan, if $5,000 is borrowed at 14% for 36 months?
(a.) $197.22   b. $158.33   c. $700   d. $138.88

19. If you purchase 150 shares of XYZ at 16 and sell at 22½, your profit, excluding commissions, is:
a. $875   b. $150   c. $2,400   (d.) $975

20. Mrs. Moulton owns a $10,000 Treasury Note that pays an annual rate of 8¾% interest. She receives semiannually:
(a.) $437.50   b. $875   c. $800   d. $400

21. Mr. Sawyer bought a picture at a 20% discount. If the regular price was $120, what price did he pay?
a. $24   b. $100   (c.) $96   d. $106

22. What is the selling price of a calculator if it costs $58 and the rate of markup on the selling price is 40%?
a. $23.20   (b.) $81.20   c. $34.80   d. $92.80

23. A bus leaves Boston at 2:30 P.M. and arrives in New York, 339 km away, at 6:15 P.M. The average speed of the bus is:
(a.) 90.4 km/h   b. 84.75 km/h   c. 123.27 km/h   d. 79.76 km/h

24. When the monthly finance charge rate is 1.6%, the annual rate is:
a. 17.2%   b. 18.4%   c. 16%   (d.) 19.2%

25. At the average cost of 5.8¢ per kilowatt hour, the weekly cost of operating 3 hours each day a stereo that uses 350 watts of power is:
(a.) 42.6¢   b. 4.2¢   c. $4.26   d. 52.6¢

Competency Check Test I

Multiple Choice—Choose the letter of the correct answer.

1. Add: 42,561 / 97,183 / 65,919 / 76,240
a. 281,902   b. 279,903   (c.) 281,903   d. 280,803

2. Subtract: 427,632 − 268,549
(a.) 159,083   b. 159,081   c. 159,183   d. 259,082

3. Multiply: 422 × 968
a. 408,490   (b.) 408,496   c. 418,492   d. 418,496

4. Divide: 28)10,276
a. 376   b. 267   (c.) 367   d. 366

5. A floor is constructed of 17 boards, each 48 inches wide. About how wide is the floor?
a. 10,000 inches   b. 100 inches   c. 10 feet   (d.) 1,000 inches

6. The principal of Center School paid $368 for 12 chairs. Approximately how much did each chair cost?
a. $100   b. $5   (c.) $30   d. $20

7. The Anderson family drives from their home in Cincinnati to visit relatives in Los Angeles. The trip log shows that they drive 2,295 miles in a total driving time of 50 hours. What speed, in miles per hour, did they average for their trip?
a. 48 m.p.h.   b. 40 m.p.h.   (c.) 46 m.p.h.   d. 42 m.p.h.

8. When listed in order from greatest to smallest, which number would be first?
a. 36,845   b. 36,485   (c.) 36,854   d. 36,584

9. Change the improper fraction $\frac{29}{3}$ to a mixed number.
a. $4\frac{2}{3}$ b. $4\frac{1}{3}$ (c.) $9\frac{2}{3}$ d. $4\frac{2}{3}$

10. Add: $5\frac{9}{10}$ / $3\frac{3}{5}$
a. $8\frac{11}{15}$ b. $9\frac{3}{4}$ (c.) $9\frac{1}{2}$ d. $4\frac{3}{4}$

11. Subtract: 12 − $9\frac{3}{4}$
a. $13\frac{1}{4}$ b. $2\frac{3}{4}$ c. $3\frac{1}{4}$ (d.) $2\frac{1}{4}$

12. Multiply: $\frac{2}{5} \times \frac{9}{10}$
a. $\frac{2}{3}$ (b.) $\frac{3}{4}$ c. $\frac{1}{4}$ d. $\frac{18}{40}$

13. Divide: $36 \div 2\frac{1}{4}$
(a.) 16   b. $18\frac{1}{4}$ c. 81   d. $17\frac{3}{4}$

14. There are three stacks of dimes on the table. One stack is $\frac{3}{4}$ inch high; the second is $1\frac{3}{8}$ inch high, and the third is $\frac{7}{8}$ inch high. If all three stacks were placed one on top of the other, how high would that one stack be?
a. $2\frac{9}{8}$ (b.) $2\frac{7}{16}$ c. $2\frac{1}{4}$ d. $1\frac{3}{16}$

15. Round $72.568 to the nearest cent.
a. $72.56   (b.) $72.57   c. $72.50   d. $72.58

16. Add: 87.2, 461, and 3.89
a. 91.441   (b.) 91.371   c. 91.551   d. 95.7

17. Subtract: 6.87 − .6
a. 6.81   b. 0.87   (c.) 6.27   d. .087

18. Multiply: .216 × .23
a. .446   (b.) .05   c. 49.68   d. .5

19. Divide: .04)‾.002
a. 5   b. 50   (c.) .05   d. .5

Competency Check Test I (cont.)

35. How many students received a grade of B?
a. 10
b. 15
c. 7
d. 5

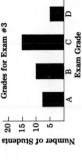

Grades for Exam #3
Number of Students — 20, 15, 10, 5
Exam Grade — A B C D

36. What is the perimeter of a square lamp table whose sides measure 30 inches?
a. 90 inches
b. 60 inches
c. 120 inches
d. 900 inches

37. In five different tests Carlo received grades of 80, 86, 75, 90 and 89. What is Carlo's average grade?
a. 85
b. 84
c. 88
d. 83

38. Find the product of −5 and −3.
a. −15
b. 8
c. 15
d. −8

39. Which of these equations has −4 as its solution?
a. $7x = -14$
b. $x - 1 = 3$
c. $x + 9 = 5$
d. $\frac{8}{6} = 2$

40. How much change should you receive from $10 when your purchases cost $8.37?
a. $2.63
b. $1.63
c. $1.73
d. $2.37

41. Tom worked 25 hours at $4.50 per hour. How much did he earn?
a. $11.25
b. $112.50
c. $1,125
d. $25

42. Find the value of A using the formula $A = \pi r^2$ when $\pi = 3.14$ and $r = 10$ cm.
a. 31.4 cm²
b. 314 cm²
c. 314 cm
d. 3,140 cm²

43. Jim earns $278 a week. If the Federal Withholding Tax is $39.94 and the Social Security Tax is $18.63, what is his take-home pay?
a. $260.96
b. $238.06
c. $219.70
d. $255.10

44. Joan has a bank balance of $396.59. If she deposits a check in the amount of $25.86 what is her new balance?
a. $370.73
b. $422.45
c. $655.19
d. $399.17

45. Find the volume of a rectangular box that is 15 cm long, 8 cm wide and 6 cm high.
a. 7.20 m
b. 720 cm³
c. 29 cm³
d. 7,200 cm

46. Which is the best buy: 10 pears for $.89, 12 pears for $1.00, or 8 pears for $.69?
a. 8 pears for $.69
b. 12 pears for $1.00
c. 10 pears for $.89
d. all the same price

47. Find the interest on $1,500 invested at 5¼% for 2 years.
a. $1.58
b. $157.50
c. $1,575
d. $15.75

48. Clara bought a skirt for $23.97. If the sales tax rate is 4%, how much altogether will Clara pay for the skirt?
a. $.96
b. $.95
c. $24.93
d. $24.92

49. Lisa bought a stereo system paying $50 down and 12 monthly payments of $35. What was the total cost?
a. $475
b. $420
c. $400
d. $470

50. Todd's father bought a house at $58,000. He paid 25% down and gave a mortgage for the remainder. What is the amount of the mortgage?
a. $14,000
b. $14,500
c. $14,500
d. $43,500

Competency Check Test I (cont.)

20. What percent is equivalent to .3?
a. .3%
b. 3%
c. 30%
d. 300%

21. Find 20% of $72.25
a. $1,445
b. $14.45
c. $144.50
d. $1,445

22. What percent of $56 is $35?
a. 19½%
b. 62½%
c. 160%
d. 21%

23. 6% of what amount is $23.40?
a. $140.40
b. $256
c. $390
d. $39

24. When a tennis pro shop offers a 35% reduction on all tennis rackets, how much would you pay for a racket which regularly sells for $32?
a. $112
b. $2.80
c. $11.20
d. $20.80

25. Find the rate of reduction allowed when a camera that regularly sells for $75 can be purchased for $60.
a. 15%
b. 20%
c. 12½%
d. 2%

26. The school baseball team won 75% of the games it played. If it won 18 games, how many games did it lose?
a. 4
b. 24
c. 6
d. 7

27. Jason wants to put a row of old bricks along the edge of the garden. Each brick is 20 cm long and the edge of the garden measures 10 m. How many bricks should Jason buy?
a. 55
b. 5
c. 50
d. 500

28. Lillian mixes 500 mL of frozen orange juice concentrate with 1,500 mL of water. How many liters of orange juice mixture results?
a. 0.2 L
b. 20 L
c. 1 L
d. 2 L

29. Clingmans Dome, a mountain in Tennessee, has an elevation of about 6,600 feet. How many miles high is Clingmans Dome?
a. 1¼ miles
b. 1¼ miles
c. 1¼ miles
d. 1¾ miles

30. Kathy left for a trip on Saturday at 9:45 A.M. and returned home the next day at 1:20 P.M. How long was she away?
a. 3 hours 35 minutes
b. 15 hours 35 minutes
c. 8 hours 25 minutes
d. 27 hours 35 minutes

31. What is the freezing point of water on the Celsius scale?
a. −10°C
b. 273°C
c. 32°C
d. 0°C

32. How many millimeters long is the line?
a. .025 mm
b. 25 mm
c. 2.5 mm
d. .25 mm

33. The distance on a map between Chicago and Cleveland is 9 inches. If the scale of the map is 1 inch = 10 miles, what is the actual distance between Chicago and Cleveland?
a. 36 miles
b. 900 miles
c. 90 miles
d. 360 miles

34. What is the best estimate of the width of this page?
a. 20 mm
b. 22 cm
c. 25 dm
d. 1 m

Competency Check Test II

Multiple Choice—Choose the letter of the correct answer.

1. Add: 7,126
   1,493
   2,057
   5,739
   9,814

   a. 26,238   b. 26,228   c. 25,229   **d. 26,229**

2. Subtract: 15,360 − 9,279

   a. 5,081   **b. 6,081**   c. 6,091   d. 6,181

3. Multiply: 783 × 506

   a. 396,198   b. 395,198   c. 395,198   d. 43,848

4. Divide: 22,848 ÷ 56

   a. 428   b. 448   **c. 408**   d. 418

5. The Brown family, on their vacation trip, drove 322 miles the first day, 462 miles the second day, 298 miles the third day, and 357 miles the fourth day. How many miles did they drive altogether?

   a. 1,439 miles   b. 1,438 miles   c. 1,429 miles   d. 1,339 miles

6. Add: $3\frac{1}{4}$
   $2\frac{5}{6}$

   a. $1\frac{5}{6}$ b. 6   c. $5\frac{7}{6}$ **d. $5\frac{13}{36}$**

7. Subtract: $1\frac{3}{4} - \frac{3}{4}$

   a. $2\frac{3}{4}$ b. $\frac{3}{4}$ **b. $1\frac{1}{4}$**   c. $\frac{1}{2}$ d. $2\frac{1}{12}$

8. Which of the following improper fractions are equivalent to $2\frac{4}{9}$?

   a. $2\frac{4}{9}$ b. $\frac{9}{8}$ **c. $3\frac{4}{9}$**   d. $\frac{10}{4}$

9. Multiply: $\frac{5}{12} \times \frac{3}{8}$

   a. $\frac{45}{96}$ b. $\frac{15}{12}$ c. $\frac{8}{96}$ d. $\frac{8}{20}$

10. Divide: $3\frac{1}{2} \div 4$

    a. 7   **b. $\frac{7}{8}$**   c. 14   d. $\frac{7}{4}$

11. If a piece $2\frac{1}{4}$ inches long is cut off a board measuring $15\frac{5}{8}$ inches, the remaining length of the board is:

    a. 13 inches   b. $13\frac{3}{8}$ inches   **c. $12\frac{7}{8}$ inches**   d. $13\frac{3}{8}$ inches

12. The number named twenty and seven tenths is represented by:

    a. .27   **b. 20.7**   c. 2.07   d. .207

13. Add: 5.92 + .318 + 74.3

    a. 83.4   b. 56.4   c. 1,653   **d. 80.538**

14. Subtract: 8.5 − 7.296

    a. 1.204   b. 12.04   c. 77.704   d. 7.7704

15. Multiply: .03 × .002

    a. .032   b. 6   c. .0006   **d. .00006**

16. Divide: $.08 \overline{)\$72}$

    a. $.09   b. $.90   **c. $900**   d. $9

17. If a taxi cab charges $1.19 for the first $\frac{1}{4}$ mile and $.35 for each additional $\frac{1}{4}$ mile, the charge for a ride of 2 miles is:

    a. $4.64   b. $2.75   **c. $3.64**   d. $2.45

18. The decimal equivalent of 45% is:

    a. 45   b. 4.5   c. .045   **d. .45**

Competency Check Test II (cont.)

19. Find 87% of 560.

    a. 4,872   **b. 487.2**   c. 48.72   d. 4.872

20. What percent of $75 is $27?

    a. 36%   b. 48%   c. 54%   d. 75.27%

21. 4% of what number is 60?

    a. 240   **b. 1,500**   c. 24   d. 150

22. A ski jacket that regularly sells for $79 is reduced 15%. Its sale price is:

    a. $59.25   **b. $67.15**   c. $77.82   d. $11.85

23. The number of grams of hamburger in 1.5 kilograms is:

    a. 1,500 g   b. 150 g   c. 15 g   d. 15,000 g

24. Kevin fell asleep at 10:40 P.M. and woke up the next morning at 7:10 A.M. How long did Kevin sleep?

    a. 7 hr. 40 min.   **b. $8\frac{1}{2}$ hr.**   c. 8 hr. 10 min.   d. 8 hr. 20 min.

25. How long is the line?

    inches
    1

    a. $1\frac{1}{2}$ inches   **b. $1\frac{1}{4}$ inches**   c. $1\frac{1}{4}$ inches   d. $1\frac{1}{16}$ inches

26. If the scale is 1 cm = 50 km, what actual distance is represented by 4.2 cm?

    a. 2,100 km   **b. 210 km**   c. 21 km   d. .21 km

27. What is the surface area of a rectangular coffee table whose sides are 12 inches and 27 inches?

    a. 324 square inches   b. 78 square inches
    c. 39 square inches   d. 1,296 square inches

28. Find the circumference of a circle whose radius is 70 cm when $\pi = \frac{22}{7}$.

    a. 22 cm   b. 44 cm   c. 220 cm   **d. 440 cm**

29. The area of a triangle whose altitude is 18 mm and base is 21 mm is:

    a. 37.8 mm²   **b. 189 mm²**   c. 39 mm²   d. 378 mm²

30. The volume of a cube whose side measures 4 m is:

    a. 16 m³   b. 128 m³   **c. 64 m³**   d. 256 m³

31. The measure of the third angle of a triangle, when the other two angles measure 68° and 53°, is:

    a. 239°   b. 121°   **c. 59°**   d. 15°

32. Which number is smallest?

    a. −9   b. +3   c. 0   d. −7

Competency Check Test II (cont.)

33. When -28 is divided by $+7$, the quotient is:

   a. -21 b. $+21$ c. -4 d. $+4$

34. The value of $\frac{m^2 - n^2}{m + n}$ when $m = 6$ and $n = 4$ is:

   a. 10   b. 2   c. 1   d. 5.2

35. Which of the following is the solution of the equation $N - 8 = 8$?

   a. -16 b. 1   c. 0   d. 16

36. Find the value of V when $v = 143$, $g = 32$ and $t = 4$ using the formula $v = V + gt$:

   a. 107   b. 271   c. 15   d. 179

37. How much of the budget is allowed for food and rent?

[pie chart: Clothing 8%, Food 37%, Rent 25%, Operating expenses 13%, Miscellaneous 17%]

   a. 38%
   b. 37%
   c. 62%
   d. 57%

38. Find the average (arithmetic mean) of the following scores: 9, 10, 5, 8, 7, 9

   a. 10   b. 8   c. 5   d. 9

39. There are 144 girls and 126 boys enrolled in a school. The ratio of the number of girls to the total enrollment is:

   a. 8 to 7   b. 15 to 8   c. 7 to 8   d. 8 to 15

40. The probability of drawing at random on the first draw a white tennis ball from a box containing 10 white balls and 15 yellow balls is:

   a. $\frac{2}{5}$ b. $\frac{1}{10}$ c. $\frac{1}{25}$ d. $\frac{3}{10}$

41. Which is the better buy: a dozen cookies for \$1.29, a half-dozen cookies for \$.69, or 8 cookies for \$.88?

   a. a half-dozen for \$.69
   b. all the same price
   c. 8 for \$.88
   d. a dozen for \$1.29

42. The Golden family had dinner at the local restaurant and the bill came to \$22.50. If the meal tax rate is 5%, what was the total bill?

   a. \$23.63   b. \$23.60   c. \$33.75   d. \$22.61

Competency Check Test II (cont.)

43. If the Federal Withholding Tax is \$44.60, Social Security Tax is \$21.78 and City Wage Tax is \$7.15, when you earn \$325 weekly, your take-home pay is:

   a. \$241.47   b. \$251.47   c. \$271.25   d. \$280.40

44. If the tax rate is \$7.20 per \$100, the property tax on a house costing \$63,000 and assessed for \$48,000 is:

   a. \$4,536   b. \$453.60   c. \$3,456   d. \$345.60

45. Mrs. Carroll owns a \$10,000 bond that pays an annual rate of $11\frac{1}{2}\%$ interest. She receives semi-annually:

   a. \$1,150   b. \$57.50   c. \$2,300   d. \$575

46. If you borrow \$3,500 for 30 months to buy a car and pay \$131.25 per month, the finance charge is:

   a. \$3,937.50   b. \$437.50   c. \$3,368.75   d. \$306.25

47. Frank bought a basketball at a 30% reduction paying \$22.40 for it. What was the regular price?

   a. \$15.68   b. \$52.40   c. \$32.00   d. \$22.70

48. Barbara's mother can purchase a refrigerator for the cash price of \$599 or for \$60 down and eight equal payments of \$76.40. By paying cash she can save:

   a. \$7.20   b. \$61.12   c. \$12.20   d. \$72.20

49. The amount of money that must be invested at the annual rate of 12% to earn \$18,000 per year is:

   a. \$1,500,000   b. \$150,000   c. \$15,000   d. \$1,500

50. Lisa Lieberman paid Jaffe Stores by check in the amount of \$86.39 on June 29, 1980.

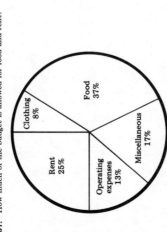

On line ① should be written:

   a. Jaffe Stores
   b. June 29, 1980
   c. Lisa Lieberman
   d. Eighty-Six and $\frac{39}{100}$

INDIVIDUALIZED PUPIL PERFORMANCE RECORD

For Allyn and Bacon's REFRESHER MATHEMATICS by Edwin I. Stein   Class

Name _____

When the book is used as a diagnostic and remedial textbook, each pupil should:

1. Work each *Inventory Test*; **cross** the numerals of all incorrect problems in the appropriate column at the left of the sheet.
2. Turn to the *Exercises* which correspond to the crossed numerals. Study the instructions and sample solutions.
3. Work the *Diagnostic Test in the Exercise*. **Cross** the numerals of all incorrect problems in the appropriate row under the heading *Diagnostic Tests and Related Set Numbers*.
4. Work Sets that correspond to the numerals crossed in step 3. **Circle** these numerals.
5. Work each *Achievement Test*; **cross** the numerals of any incorrect problems in the appropriate column at the right. Repeat steps 2, 3, and 4.

When the book is used as a basic textbook and the scoresheet is used as a record of the completed work, each pupil should:

1. **Circle** the numeral in the appropriate column at left when problem is correct on *Inventory Test*.
2. When problem is correct on *Diagnostic Test*, **circle** numeral in appropriate row under the heading *Diagnostic Tests and Related Set Numbers*.
3. When all problems in Set are correct, **check** the corresponding numeral under the heading *Diagnostic Test and Related Set Numbers*.
4. **Circle** numeral in appropriate column at right when *Achievement Test* problem is correct.

Exercise	Inventory Tests I	II	III	IV	Diagnostic Test and Related Set Numbers	Achievement Tests I	II	III	IV
1-1	1	1	1	1	1 2 3 4 5 6 7 8 9 10 11 12 13 14 15 16 17 18 19 20 21 22 23 24 25	1	1	1	1
1-2	2	2	2	2	1 2 3 4 5 6 7 8 9 10 11	2	2	2	2
1-3	3	3	3	3	1a b 2a b 3a b 4a b 5a b 6a b 7a b 8a b	3	3	3	3
1-4	4	4	4	4	1 2 3 4 5 6 7 8 9 10 11 12 13 14 15 16 17 18 19 20 21 22a b 23a b 24 25a b c 26	4	4	4	4
1-5	5	5	5	5	1 2 3 4 5 6 7 8 9 10 11 12 13 14 15 16 17 18 19 20 21 22 23 24 25 26 27 28 29 30 31a b 32 33 34	5	5	5	5
1-6A					1 2 3 4 5 6 7 8 9 10 11 12 13 14 15 16 17 18 19 20 21				
1-6B	6	6	6	6	1 2 3 4 5 6 7 8 9 10 11 12 13 14 15 16 17 18 19 20 21 22 23 24 25	6	6	6	6
1-7A					1 2 3 4 5 6 7 8 9 10 11 12 13 14 15 16 17 18 19 20 21 22 23 24 25 26				
1-7B	7	7	7	7	1 2 3 4 5 6 7 8 9 10 11 12 13 14 15 16 17 18 19 20 21 22 23 24 25 26 27 28 29 30 31 32 33	7	7	7	7
1-8	8	8	8	8	1 2 3 4 5 6 7		8	8	8	8
1-9	9	9	9	9	1 2 3 4 5 6 7 8 9 10	9	9	9	9
1-10	10	10	10	10	1 2a b 3 4a b 5	10	10	10	10
1-11	11	11	11	11	1 2 3 4	11	11	11	11
1-12	12	12	12	12	1 2 3 4 5 6 7 8 9 10 11 12 13 14 15 16 17 18 19 20 21 22 23 24 25 26 27	12	12	12	12
1-13	13	13	13	13	1 2 3 4 5 6 7 8 9 10 11 12 13 14 15 16 17 18 19 20 21 22 23 24	13	13	13	13
1-14	14	14	14	14	1 2 3 4 5 6 7 8	14	14	14	14
1-15	15	15	15	15	1 2 3 4a b c	15	15	15	15
1-16	16	16	16	16	1 2 3 4 5 6 7 8 9 10 11 12 13 14 15 16 17 18 19 20 21 22 23 24 25 26 27 28 29 30 31 32 33 34 35 36 37 38a b c d e f g h i j	16	16	16	16
1-17	17	17	17	17	1 2 3 4 5 6 7 8 9 10 11 12 13 14 15 16 17 18 19 20 21 22 23 24 25 26 27 28 29a b 30a b c 31a b c d e f	17	17	17	17
1-18	18	18	18	18	1 2 3a b c 4 5 6 7 8 9 10	18	18	18	18
1-19	19	19	19	19	1a b c d e 2a b c d e	19	19	19	19

Exercise	Inventory Tests I	II	III	IV	Diagnostic Test and Related Set Numbers	Achievement Tests I	II	III	IV
1-20	20	20	20	20	1 2 3 4 5 6 7 8 9 10 11 12 13 14 15 16 17	20	20	20	20
1-21	21	21	21	21	1 2 3 4 5 6 7 8 9 10 11 12 13 14 15 16 17	21	21	21	21
1-22	22	22	22	22	1a b 2a b 3a b 4a 5a b 6a b 7a b c d 8a b	22	22	22	22
1-23	23	23	23	23	1 2 3 4 5 6 7 8 9 10 11 12 13 14 15 16 17 18 19 20 21 22a b 23 24 25	23	23	23	23
1-24	24	24	24	24	1 2 3 4 5 6 7 8 9 10 11 12 13 14 15 16 17 18 19 20 21 22 23 24a b 25 26 27 28 29 30 31 32	24	24	24	24
1-25	25	25	25	25	1 2 3 4 5 6 7 8	25	25	25	25
1-26	26	26	26	26	1 2 3 4 5 6 7 8 9 10 11 12 13 14 15 16 17 18 19 20 21 22a b 23 24 25 26 27 28 29 30 31 32	26	26	26	26
1-27A					1 2 3 4 5 6 7 8 9 10 11 12a b c 13a b 14				
1-27B					1 2 3 4 5 6 7 8 9 10 11a b c				
1-27C					1 2 3 4 5 6 7 8 9 10a b c 11a b				
1-27D	27	27	27	27	1 2 3 4 5 6 7 8 9 10a b c	27	27	27	27
1-28	28	28	28	28	1 2 3 4 5 6 7 8 9 10 11 12 13 14 15 16 17 18	28	28	28	28
1-29	29	29	29	29	1 2 3 4 5 6 7 8 9 10 11 12 13 14 15 16 17	29	29	29	29
1-30	30	30	30	30	1 2 3 4 5 6 7 8 9 10 11 12 13 14 15 16 17 18	30	30	30	30
1-31	31	31	31	31	1 2 3 4 5 6 7 8 9 10 11 12 13 14	31	31	31	31
1-32	32	32	32	32	1 2 3 4 5 6 7 8 9 10 11 12 13 14	32	32	32	32
1-33	33	33	33	33	1 2 3 4 5 6 7 8 9 10 11 12 13 14 15 16	33	33	33	33
1-34	34	34	34	34	1 2 3 4 5 6 7 8 9 10 11 12 13 14 15 16 17 18 19 20	34	34	34	34
1-35	35	35	35	35	1 2 3 4 5 6 7 8 9 10 11 12 13 14 15 16 17 18 19 20 21 22 23	35	35	35	35
1-36	36	36	36	36	1 2 3 4	36	36	36	36
1-37	37	37	37	37	1 2a b 3 4 5 6 7 8 9 10 11 12 13 14 15 16 17	37	37	37	37
1-38	38	38	38	38	1 2 3 4 5 6 7 8 9 10 11 12 13 14 15 16	38	38	38	38
1-39	39	39	39	39	1 2 3 4	39	39	39	39
1-40	40	40	40	40	1 2 3 4 5 6 7 8 9 10 11 12 13 14	40	40	40	40
1-41	41	41	41	41	1 2 3 4 5 6 7 8 9 10 11 12 13 14 15a b 16a b	41	41	41	41
1-42	42	42	42	42	1a b 2a b 3 4a b 5 6a b 7 8 9 10 11 12a b 13 14 15 16	42	42	42	42
1-43	43	43	43	43	1 2 3 4 5 6 7 8 9 10 11 12 13 14 15	43	43	43	43
1-44	44	44	44	44	1 2 3 4 5 6	44	44	44	44
1-45	45	45	45	45	1 2 3 4 5 6 7 8 9 10 11 12 13 14 15	45	45	45	45

EXTENDED ANSWERS

Diagnostic Test, p. 20

3. Sixty-eight   4. Five hundred eighty-two   5. Four thousand, nine hundred seventy-five
6. Five thousand, three   7. Ten thousand, five hundred
8. Eight hundred twenty-three thousand, six hundred fifty-nine
9. Four million, nine hundred thousand
10. Seven million, eight hundred forty-seven thousand
11. Three million, five hundred eighty-two thousand, nine hundred forty-two
12. Eighteen million, four hundred sixty-seven thousand, one hundred twenty-five
13. Nine hundred ninety-five million, twenty-eight thousand
14. Sixty-two billion, six hundred sixty-nine million, four thousand, seven hundred sixteen
15. Nine hundred four trillion, sixty billion, three hundred eighty-one million, two hundred sixty-nine thousand, eight
16. Seventeen trillion, five hundred billion
17. Seven hundred forty thousand, five hundred nine
18. Sixty-three million, eighty-six thousand, four hundred ninety-three
19. Two hundred eighteen billion, nine hundred four million, two hundred twenty-six thousand, seven
20. Five trillion, six hundred ninety-three billion, one hundred sixty-nine million, forty thousand
21. Ninety-four million   22. Five hundred seventy-eight billion   23. Sixteen trillion
24. One hundred seven million dollars 25. a. Five thousand, three hundred b. Fifty-three hundred

Related Practice Examples, pp. 21–22

Set 3: 1. Ninety-six   2. Forty-seven   3. Fifty-three   4. Eighty-four   5. Seventy-nine
Set 4: 1. Four hundred twenty-six   2. Seven hundred eighty-four 3. Two hundred sixty-nine
        4. Nine hundred five   5. Three hundred eighty
Set 5: 1. Eight thousand, two hundred seventy-eight
        2. Three thousand, nine hundred twenty-six
        3. Two thousand, four hundred seventy-four 4. Four thousand, two hundred six
        5. Seven thousand, nine hundred fifty-nine
Set 6: 1. Six thousand, four   2. Four thousand, forty-five   3. Nine thousand, eighty
        4. Seven thousand, thirty-two 5. Three thousand, seven
Set 7: 1. Twenty-seven thousand, four hundred thirty-two
        2. Fifty-nine thousand, one hundred seventeen
        3. Twenty-thousand, seven hundred thirty
        4. Forty-five thousand sixty-three
        5. Eighty-one thousand, eight hundred ninety-six
Set 8: 1. One hundred fifty thousand
        2. Two hundred ninety-one thousand, four hundred twenty-nine
        3. Seven hundred twenty-three thousand, eight hundred five
        4. Nine hundred thousand, two hundred eighty-one
        5. Six hundred seventy-four thousand, three hundred sixty-four
Set 9: 1. Three million   2. Eight million   3. Two million, five hundred thousand
        4. Seven million, six hundred thousand   5. Four million, two hundred thousand
Set 10: 1. Nine million, two hundred fifty thousand
        2. Three million, seven hundred forty-six thousand
        3. Four million, eight hundred ninety-one thousand
        4. Two million, six thousand   5. Five million, six hundred seventy thousand
Set 11: 1. One million, five hundred ninety-six thousand, two hundred seventy-one
        2. Seven million, one hundred twenty-two thousand, eight hundred forty-three
        3. Eight million, nine hundred thousand, five hundred twenty-seven
        4. Five million, four hundred sixteen thousand, eighty-four
        5. Six million, four thousand, nine
Set 12: 1. Thirty-two million, four hundred twenty-nine thousand seven hundred eighty-four
        2. Fifty-seven million, one hundred five thousand, nine hundred thirty-three

3. Twenty-eight million, two hundred forty-six thousand, two hundred nineteen
4. Ninety-six million, eight hundred sixty-three thousand, eight hundred forty-eight
5. Twenty-five million, nine hundred seventy-five

Set 13:
1. Seven hundred million
2. Two hundred forty-one million, eight hundred forty-nine thousand
3. Nine hundred twenty-eight million, three hundred seventy-six thousand, five hundred
4. Four hundred six million, nine hundred twenty-three thousand, six hundred eighty-five
5. Eight hundred forty-three million, two hundred ninety-seven thousand, eight hundred sixty-one

Set 14:
1. Five billion 2. Six billion, four hundred fifty million
3. Ninety-three billion, eight hundred eighty-four million, five hundred twenty-six thousand, seven hundred sixty-seven
4. Two hundred eighty-six billion, eight million, four hundred thirty-one thousand
5. Eight hundred forty-two billion, six hundred ninety-nine million, nineteen thousand, five hundred eight

Set 15:
1. Nine trillion
2. Twenty-three trillion, five hundred eighty-seven billion, nine hundred fourteen million, sixty thousand
3. Sixty-seven trillion, three hundred forty-six billion, five hundred twenty-eight million, nine hundred fifteen thousand, eight hundred ninety-eight
4. One hundred thirty-nine trillion, five billion, four hundred seven million, eight hundred sixty-two thousand, one hundred
5. Seven hundred sixty-eight trillion, eight hundred forty-one billion, two hundred fifty-three million, one hundred ninety-six thousand, three hundred sixty-nine

Set 16:
1. Seven million, five hundred ninety-three thousand
2. Eighty-two million, one hundred thousand, four hundred seventy-six
3. Four billion, seven hundred sixty-three million, two hundred eighty-five thousand, one hundred
4. Three hundred sixty billion, five million, four hundred thirty thousand
5. Sixty-seven trillion, eight hundred billion

Set 17:
1. Twenty-eight thousand, nine hundred
2. Nine hundred eighty-three thousand, two hundred twenty-five
3. Seven thousand, three
4. Four hundred fifty-five thousand, six hundred seven
5. Sixty-one thousand, four hundred fifty-nine

Set 18:
1. Six million, five hundred eighty-two thousand
2. Thirty-four million, eight hundred nine thousand, five hundred
3. Eight hundred three million, five hundred seventy-eight thousand, six hundred thirty-two
4. Fifty-nine million, two hundred thousand, forty-eight
5. Seven hundred eighty-five million, four hundred six thousand, nine hundred seventeen

Set 19:
1. Seventy-five billion
2. Eight billion, four hundred ninety-seven million, five hundred thousand
3. Four hundred sixty-billion, one hundred twenty-two million, eight hundred forty-nine thousand
4. Ninety-three billion, six hundred seventy-four million, three hundred seventeen thousand, five hundred five
5. Two hundred forty-five billion, three hundred ninety million, six thousand, seven hundred

Set 20:
1. Nine trillion 2. Twenty-seven trillion, five hundred forty-billion
3. One hundred twenty-one trillion, three hundred twenty-five billion, nine hundred thirteen million, six thousand, two hundred
4. Two trillion, six hundred seven billion, five hundred fifty-nine million, forty thousand, one hundred seventy-two
5. Fifty-three trillion, four hundred eighty-two billion, eighty million

Set 21:
1. Forty-seven million 2. Nine million 3. Five hundred sixty million
4. Eighty-eight million 5. One billion, seven hundred twenty-five million

Set 22:
1. Four billion 2. Six hundred three billion 3. Fifty-one billion
4. Two trillion, eight hundred fifty billion 5. Ninety-nine billion

Set 23:
1. Ten trillion 2. Eight trillion 3. Two hundred trillion
4. Eighty-seven trillion 5. Three hundred seventy-five trillion

Set 24: 1. Seventy-three billion dollars   2. Four hundred five million dollars
        3. Fourteen trillion dollars   4. Fifty-nine million dollars
        5. Six hundred billion dollars
Set 25: 1. a. Six thousand, nine hundred   b. Sixty-nine hundred
        2. a. Four thousand, one hundred   b. Forty-one hundred
        3. a. Two thousand, eight hundred   b. Twenty-eight hundred
        4. a. Eight thousand, seven hundred b. Eighty-seven hundred
        5. a. Seven thousand, four hundred b. Seventy-four hundred

Practical Applications, p. 22

1. a. One hundred sixty-five million, two hundred forty-six thousand, three hundred twenty square kilometers
   b. Eighty-two million, four hundred forty-one thousand, five hundred sixty square kilometers
   c. Seventy-three million, four hundred forty-two thousand, seven hundred fifty-five square kilometers
   d. Fourteen million, ninety thousand, one hundred ten square kilometers
2. Forty-nine million, eight hundred two thousand, one hundred thirty shares; Two billion, nine hundred twenty-one million, six hundred eighty-six thousand, six hundred twenty-six shares
3. a. Ten million, eight hundred twenty thousand   b. Nine million, five thousand
   c. Seven million, eight hundred ninety-five thousand, five hundred forty-three
   d. Seven million, fifty thousand
   e. Five million, nine hundred forty-eight thousand, five hundred forty-six
4. Thirteen billion, three hundred seventy-seven million, six hundred fifty-nine thousand, seven hundred forty-seven coins; Seven hundred sixty-nine million, four hundred sixty-six thousand, two hundred nine dollars; Eighty-one billion, one hundred ninety-six million dollars
5. a. Four hundred ninety-nine billion dollars   b. Nineteen billion dollars

Diagnostic Test, p. 128

1. Two tenths   2. Six hundredths   3. Fifty-eight hundredths   4. One and five tenths
5. Three and seventy-three hundredths   6. Four thousandths
7. Seventy-six thousandths   8. Two hundred eighty-nine thousandths
9. Fourteen and seven hundred eight thousandths 10. Thirty-seven ten-thousandths
11. Seventeen thousand nine hundred twenty-five hundred-thousandths
12. Four hundred fifty-six millionths
13. One hundred twenty-nine and four thousand two hundred sixty-one ten-thousandths
14. Eight hundred thirty-six thousandths; Point eight three six
15. Twenty-four and seventy-eight hundredths; Twenty-four point seventy-eight
16. Three hundred eighty-four and five tenths million
17. Sixteen trillion, nine hundred twenty-seven billion dollars

Related Practice Examples, p. 129

Set 1: 1. Eight tenths   2. One tenth   3. Five tenths   4. Four tenths   5. Nine tenths
Set 2: 1. Three hundredths   2. Seven hundredths   3. Two hundredths
        4. Eight hundredths   5. Five hundredths
Set 3: 1. Twenty-four hundredths   2. Eighty-five hundredths   3. Ninety-one hundredths
        4. Sixty hundredths   5. Thirty-seven hundredths
Set 4: 1. One and six tenths   2. Two and nine tenths   3. Fourteen and three tenths
        4. Thirty-eight and five tenths   5. One hundred twenty-six and four tenths
Set 5: 1. Two and fifty-one hundredths   2. Seven and thirty-seven hundredths
        3. Thirty and forty-six hundredths   4. Eighty-nine and three hundredths
        5. Two hundred forty-eight and nineteen hundredths
Set 6: 1. Five thousandths   2. Eight thousandths
        3. One thousandth   4. Seven thousandths   5. Three thousandths
Set 7: 1. Twenty-four thousandths   2. Sixty-three thousandths
        3. Eighty thousandths   4. Ninety-two thousandths   5. Thirty-five thousandths

Set 8: 1. Eight hundred thirty-two thousandths   2. Nine hundred forty-six thousandths
3. Two hundred fifty-three thousandths   4. Seven hundred ninety-eight thousandths
5. Four hundred sixty-five thousandths

Set 9: 1. Six and five thousandths   2. Twenty-one and seven hundred sixty-nine thousandths
3. Thirty-four and ninety-four thousandths
4. One hundred eighty-six and five hundred twenty-eight thousandths
5. Two hundred and forty-two thousandths

Set 10: 1. Seven ten-thousandths   2. Eighty-nine ten-thousandths
3. Five hundred seventy-four ten-thousandths
4. Three thousand nine hundred sixty-two ten-thousandths
5. Nine thousand three hundred fifty ten-thousandths

Set 11: 1. Six hundred-thousandths   2. Three hundred ninety-two hundred thousandths
3. Nine thousand four hundred thirteen hundred-thousandths
4. Twenty-five thousand four hundred-thousandths
5. Seventy-two thousand eight hundred fifteen hundred-thousandths

Set 12: 1. One millionth   2. Five hundred thirty-four millionths
3. Eighty thousand seventy-six millionths
4. One hundred seventy-five thousand two hundred eighty-three millionths
5. Eight hundred thousand five hundred sixty millionths

Set 13: 1. Eight and twenty-five ten-thousandths
2. Twenty-three and nine thousand three hundred seventeen ten-thousandths
3. Forty-nine and eight thousand three hundred twenty-nine hundred-thousandths
4. Three hundred fifty-four and three thousand six hundred twenty-five ten-thousandths
5. Five hundred seventy-one and five thousand eight hundred seventy-five hundred thousandths

Set 14: 1. Point nine three; Ninety-three hundredths
2. Point six one nine; Six hundred nineteen thousandths   3. Point two; Two tenths
4. Point three four two eight; Three thousand four hundred twenty-eight ten-thousandths
5. Point four seven six five one; Forty-seven thousand six hundred fifty-one hundred-thousandths

Set 15: 1. Thirty-six point eight nine; Thirty-six and eighty-nine hundredths
2. Seventeen point six; Seventeen and six tenths
3. Five point nine two five; Five and nine hundred twenty-five thousandths
4. Two hundred ninety-four point zero one three; Two hundred ninety-four and thirteen thousandths
5. Eighty-seven point four six six two; Eighty-seven and four thousand six hundred sixty-two ten-thousandths

Set 16: 1. Six trillion, eight hundred billion
2. Ninety-seven million, nine hundred twenty thousand
3. Eighty-four billion, four hundred thirty-six million
4. Four hundred six million, seven hundred fifty-three thousand
5. Two trillion, five hundred sixty-nine billion, forty million

Set 17: 1. Thirty-three point four billion dollars   2. Sixty-one point five six million dollars
3. Two point eight nine eight trillion dollars
4. One thousand, seven hundred fifty point zero eight billion dollars
5. Nine hundred twenty-six point eight seven two million dollars

Review Exercises, p. 195

1. **a.** Eighty-five million   **b.** Six hundred ninety-one billion dollars
   c. Three million four hundred eighty-seven thousand   **d.** Fourteen trillion
2. **a.** One point three trillion   **b.** Sixteen point seven thousand
   c. Seventy-nine point six million
   d. Twenty-five point eight billion
3. **a.** Four point ninety-five billion.   **b.** Twenty-three point zero nine million dollars
   c. One point forty-two trillion dollars   **d.** Seventy-four point thirty-eight thousand
4. **a.** Five hundred eighty-seven point four million
   b. Nine hundred fourteen point two two trillion

c. Six hundred nine point one thousand

d. Four hundred forty-five point six nine billion dollars

5. a. Eight point six eight six billion   **b.** Three hundred eighty-one point two nine nine million

c. Eight point zero five six trillion dollars

d. Seven hundred thirty-six point eight four seven billion

Diagnostic Test, p. 197

1. 725,000,000   **2.** 51,900,000   **3.** 483,280,000

4. 6,071,000   **5.** $34,560,000   **6.** 49,000,000,000

7. 98,400,000,000   **8.** 2,060,000,000   **9.** 350,742,000,000

10. $691,180,000,000   **11.** 31,000,000,000,000   **12.** 85,300,000,000,000

13. 408,570,000,000,000   **14.** 9,529,000,000,000   **15.** $70,040,000,000,000

16. 616,000   **17.** 18,947   **18.** $5,090   **19.** 182,600   **20.** 6,230

Related Practice Examples, p. 198–199

Set 1:   **1.** 83,000,000   **2.** 9,000,000   **3.** 460,000,000   **4.** 2,805,000,000   **5.** 589,000,000

Set 2:   **1.** 6,800,000   **2.** 84,700,000   **3.** 358,300,000   **4.** 1,098,500,000   **5.** 49,400,000

Set 3:   **1.** 17,450,000   **2.** 5,070,000   **3.** 227,930,000   **4.** 1,540,160,000   **5.** 24,820,000

Set 4:   **1.** 8,562,000   **2.** 13,468,000   **3.** 479,704,000   **4.** 3,872,835,000   **5.** 93,696,000

Set 5:   **1.** $4,500,000   **2.** $27,400,000   **3.** 75,090,000   **4.** 408,263,000   **5.** 667,850,000

Set 6:   **1.** 78,000,000,000   **2.** 5,000,000,000   **3.** 609,000,000,000
       4. 3,570,000,000,000   **5.** 91,000,000,000

Set 7:   **1.** 9,400,000,000   **2.** 72,800,000,000   **3.** 403,700,000,000
       4. 2,576,300,000,000   **5.** 80,200,000,000

Set 8:   **1.** 12,060,000,000   **2.** 6,950,000,000   **3.** 820,730,000,000
       4. 3,505,810,000,000   **5.** 49,240,000,000

Set 9:   **1.** 7,568,000,000   **2.** 31,047,000,000   **3.** 564,382,000,000
       4. 1,067,509,000,000   **5.** 81,635,000,000

Set 10:   **1.** $8,300,000,000   **2.** $30,700,000,000   **3.** $54,610,000,000
       4. $935,054,000,000   **5.** $819,370,000,000

Set 11:   **1.** 4,000,000,000,000   **2.** 17,000,000,000,000   **3.** 106,000,000,000,000
       4. 58,000,000,000,000   **5.** 292,000,000,000,000

Set 12:   **1.** 3,900,000,000,000   **2.** 18,500,000,000,000   **3.** 207,800,000,000,000
       4. 83,100,000,000,000   **5.** 6,700,000,000,000

Set 13:   **1.** 5,920,000,000,000   **2.** 60,070,000,000,000   **3.** 139,780,000,000,000
       4. 17,050,000,000,000   **5.** 308,590,000,000,000

Set 14:   **1.** 8,227,000,000,000   **2.** 41,058,000,000,000   **3.** 9,464,000,000,000
       4. 250,829,000,000,000   **5.** 11,256,000,000,000

Set 15:   **1.** $9,600,000,000,000   **2.** $3,280,000,000,000   **3.** $87,920,000,000,000
       4. $40,071,000,000,000   **5.** $156,435,000,000,000

Practice Problems, p. 252

1. a. zero, one, two, three and so on without end   **b.** zero, two, four, six and so on without end

c. one, three, five, seven and so on without end   **d.** two, three, five, seven and so on without end

e. zero, five, ten, fifteen and so on without end   **f.** one, four, nine, sixteen and so on without end

2. a. zero, one, two, three and up to and including 25

b. zero, four, eight, twelve and up to and including sixty-four

c. one, three, five, seven and up to and including sixty-one

d. zero, ten, twenty, thirty and up to and including two hundred

e. zero, eight, sixteen, twenty-four and up to and including ninety-six

f. two, four, six, eight and up to and including fifty

3. a. 0, 3, 6, 9, . . .   **b.** 0, 6, 12, 18, . . . , 72   **c.** 2, 3, 5, 7, . . . , 89

d. 0, 20, 40, 60, . . .   **e.** 1, 2, 3, 4, . . . , 100   **f.** 3, 9, 27, 81, . . .

1. a. Three to the fourth power **b.** Seven to the eighth power **c.** Two to the eleventh power
d. Ten to the fifth power **e.** Six to the first power **f.** Eleven to the ninth power
g. Five to the third power or five cubed **h.** Twenty to the second power or twenty squared
2. a. 5^7 **b.** 9^3 **c.** 50^2
3. a. 7 **b.** 9 **c.** 1 **d.** 12 **e.** 4
4. a. 5 **b.** 6 **c.** 2 **d.** 12 **e.** 21
5. a. 8 **b.** 12 **c.** 7 **d.** 15 **e.** 30
6. a. 2^4 **b.** 6^{10} **7. a.** $7^3 = 7 \times 7 \times 7$ **b.** $6^4 = 6 \times 6 \times 6 \times 6$
c. $3^{10} = 3 \times 3 \times 3 \times 3 \times 3 \times 3 \times 3 \times 3 \times 3 \times 3$ **d.** $8^7 = 8 \times 8 \times 8 \times 8 \times 8 \times 8 \times 8$
e. $12^9 = 12 \times 12 \times 12 \times 12 \times 12 \times 12 \times 12 \times 12 \times 12$
f. $9^{12} = 9 \times 9 \times 9 \times 9 \times 9 \times 9 \times 9 \times 9 \times 9 \times 9 \times 9 \times 9$
8. a. 8^2 **b.** 3^4
9. a. $36 = 6 \times 6 = 6^2$ **b.** $25 = 5 \times 5 = 5^2$ **c.** $8 = 2 \times 2 \times 2 = 2^3$
d. $81 = 9 \times 9 = 9^2$ **e.** $121 = 11 \times 11 = 11^2$ **f.** $625 = 25 \times 25 = 25^2$
10. a. 81 **b.** 256 **c.** 256 **d.** 2,187 **e.** 512 **f.** 15,625

1. a. 67,294 **b.** 356,827
2. a. $(5 \times 10^1) + (8 \times 1)$ **b.** $(4 \times 10^2) + (9 \times 10^1) + (2 \times 1)$
c. $(3 \times 10^3) + (5 \times 10^2) + (6 \times 10^1) + (4 \times 1)$
d. $(8 \times 10^5) + (9 \times 10^4) + (2 \times 10^3) + (1 \times 10^2) + (8 \times 10^1) + (7 \times 1)$
e. $(3 \times 10^4) + (7 \times 10^3) + (0 \times 10^2) + (2 \times 10^1) + (6 \times 1)$
f. $(2 \times 10^7) + (9 \times 10^6) + (1 \times 10^5) + (4 \times 10^4) + (2 \times 10^3) + (3 \times 10^2) + (5 \times 10^1) + (8 \times 1)$
g. $(5 \times 10^6) + (6 \times 10^5) + (8 \times 10^4) + (8 \times 10^3) + (2 \times 10^2) + (4 \times 10^1) + (1 \times 1)$
h. $(8 \times 10^8) + (2 \times 10^7) + (3 \times 10^6) + (9 \times 10^5) + (1 \times 10^4) + (7 \times 10^3) + (0 \times 10^2) +$
$(8 \times 10^1) + (4 \times 1)$
i. $(4 \times 10^7) + (6 \times 10^6) + (2 \times 10^5) + (1 \times 10^4) + (2 \times 10^3) + (5 \times 10^2) + (9 \times 10^1) + (7 \times 1)$
j. $(2 \times 10^{10}) + (5 \times 10^9) + (3 \times 10^8) + (8 \times 10^7) + (4 \times 10^6) + (5 \times 10^5) + (6 \times 10^4) +$
$(9 \times 10^3) + (7 \times 10^2) + (0 \times 10^1) + (3 \times 1)$
k. $(6 \times 10^9) + (7 \times 10^8) + (3 \times 10^7) + (9 \times 10^6) + (1 \times 10^5) + (5 \times 10^4) + (8 \times 10^3) +$
$(2 \times 10^2) + (4 \times 10^1) + (5 \times 1)$
l. $(9 \times 10^{11}) + (3 \times 10^{10}) + (7 \times 10^9) + (1 \times 10^8) + (7 \times 10^7) + (5 \times 10^6) + (4 \times 10^5) +$
$(8 \times 10^4) + (6 \times 10^3) + (2 \times 10^2) + (9 \times 10^1) + (8 \times 1)$
3. a. .734 **b.** .6175 **c.** 93.68 **d.** 207.49
e. 48,368.027 **f.** 6,492.50381
4. a. $(2 \times 10^{-1}) + (9 \times 10^{-2})$ **b.** $(4 \times 10^{-1}) + (7 \times 10^{-2}) + (6 \times 10^{-3})$
c. $(3 \times 10^{-1}) + (5 \times 10^{-2}) + (0 \times 10^{-3}) + (9 \times 10^{-4})$
d. $(6 \times 10^{-1}) + (8 \times 10^{-2}) + (1 \times 10^{-3}) + (4 \times 10^{-4}) + (5 \times 10^{-5})$
e. $(9 \times 10^{-1}) + (7 \times 10^{-2}) + (5 \times 10^{-3}) + (2 \times 10^{-4}) + (6 \times 10^{-5}) + (3 \times 10^{-6})$
f. $(8 \times 10^0) + (2 \times 10^{-1})$ **g.** $(8 \times 10^1) + (9 \times 10^0) + (4 \times 10^{-1}) + (6 \times 10^{-2})$
h. $(2 \times 10^2) + (1 \times 10^1) + (5 \times 10^0) + (8 \times 10^{-1}) + (3 \times 10^{-2})$
i. $(5 \times 10^1) + (3 \times 10^0) + (9 \times 10^{-1}) + (2 \times 10^{-2}) + (7 \times 10^{-3})$
j. $(7 \times 10^2) + (5 \times 10^1) + (9 \times 10^0) + (5 \times 10^{-1}) + (0 \times 10^{-2}) + (4 \times 10^{-3})$
k. $(3 \times 10^3) + (1 \times 10^2) + (4 \times 10^1) + (6 \times 10^0) + (2 \times 10^{-1}) + (5 \times 10^{-2})$
l. $(8 \times 10^3) + (3 \times 10^2) + (7 \times 10^1) + (2 \times 10^0) + (4 \times 10^{-1}) + (9 \times 10^{-2}) + (1 \times 10^{-3}) +$
(7×10^{-4})

1. a. Thirty-four equals thirty-four **b.** Eight times nine is equal to nine times eight
c. Fifty-three is greater than forty-nine **d.** Forty-eight minus nine is greater than thirty-seven
e. Sixteen is less than twenty **f.** Four is less than twenty-one minus sixteen
g. Seventeen is not equal to ten plus eight
h. Six times seven is not equal to ninety-eight divided by two
i. Fifteen is not greater than twenty-three

j. Forty-five times four is not greater than two hundred

k. Two plus three is not less than three plus two

l. Nine squared is not less than four plus seventy-seven

m. Fifteen times ten is equal to ten times fifteen

n. Six times nine is less than forty-nine plus seven

o. Eleven minus four is greater than two times three

p. Twenty-seven divided by three is not equal to three divided by twenty-seven

q. Thirty-seven minus eighteen is not less than fifteen minus fifteen

r. Sixteen times two equals sixty-four divided by two

s. Sixty-three divided by seven is less than sixty-three minus seven

t. Four plus zero is greater than seven times zero

u. One times one is equal to one divided by one

v. Seventy-two divided by nine is not greater than two times four

w. Point six minus point six is less than point six divided by point six

x. One half plus one half is greater than one half times one half

Practice Problems, pp. 261–262

3. a. 1, 2, 13, 26 **b.** 1, 2, 3, 6, 9, 18 **c.** 1, 17
  d. 1, 3, 5, 9, 15, 45 **e.** 1, 2, 3, 6, 9, 18, 27, 54 **f.** 1, 2, 5, 7, 10, 14, 35, 70
  g. 1, 2, 3, 4, 6, 8, 12, 16, 24, 48 **h.** 1, 2, 3, 4, 6, 7, 12, 14, 21, 28, 42, 84
  i. 1, 2, 5, 10, 25, 50, 125, 250 **j.** 1, 2, 3, 4, 6, 8, 9, 12, 18, 24, 36, 72
  k. 1, 2, 4, 5, 8, 10, 16, 20, 25, 40, 50, 80, 100, 200, 400
  l. 1, 2, 3, 4, 6, 11, 12, 22, 33, 44, 66, 132
4. a. 1, 2, 3, 4, 6, 12 **b.** 1, 2, 4, 8, 16 **c.** 1, 2, 4
5. a. 1, 2, 3, 4, 6, 9, 12, 18, 36 **b.** 1, 2, 3, 4, 6, 9, 12, 18, 27, 36, 54, 108
  c. 1, 3, 9, 27, 81 **d.** 1, 3, 9
6. a. 1, 2, 4, 8; 1, 2, 3, 4, 6, 12; 1, 2, 4 **b.** 1, 2, 4; 1, 5; 1
  c. 1, 2, 3, 4, 6, 8, 12, 16, 24, 48; 1, 2, 4, 7, 8, 14, 28, 56; 1, 2, 4, 8
  d. 1, 2, 3, 6, 9, 18, 27, 54; 1, 2, 3, 4, 6, 8, 9, 12, 18, 24, 36, 72; 1, 2, 3, 6, 9, 18
  e. 1, 3, 7, 9, 21, 63; 1, 2, 3, 4, 6, 7, 12, 14, 21, 28, 42, 84; 1, 3, 7, 21
  f. 1, 5, 9, 25, 45, 225; 1, 2, 3, 5, 6, 10, 15, 25, 30, 50, 75, 150; 1, 5, 25
  g. 1, 2, 3, 4, 6, 8, 12, 24; 1, 2, 3, 4, 5, 6, 10, 12, 15, 20, 30, 60; 1, 2, 3, 4, 6, 8, 12, 16, 24, 32, 48, 96;
    1, 2, 3, 4, 6, 12
  h. 1, 2, 4, 7, 8, 14, 28, 56; 1, 2, 4, 5, 7, 10, 14, 20, 28, 35, 70, 140;
    1, 2, 3, 4, 6, 7, 8, 12, 14, 21, 24, 28, 42, 56, 84, 168; 1, 2, 4, 7, 14, 28
  i. 1, 2, 3, 4, 6, 7, 12, 14, 21, 28, 42, 84; 1, 2, 3, 4, 6, 8, 9, 12, 16, 18, 24, 36, 48, 72, 144;
    1, 2, 3, 4, 6, 8, 9, 10, 12, 15, 18, 20, 24, 30, 36, 40, 45, 60, 90, 120, 180, 360; 1, 2, 3, 4, 6, 12
7. a. 1, 2 **b.** 1, 3 **c.** 1, 2, 4, 8
  d. 1, 3, 5, 9, 15, 45 **e.** 1, 2 **f.** 1, 2, 3, 4, 6, 12
  g. 1, 13 **h.** 1, 2, 3, 4, 6, 7, 12, 14, 21, 28, 42, 84 **i.** 1, 3, 5, 15, 25, 75
8. a. 1, 2, 3, 6, 7, 9, 14, 18, 21, 42, 63, 126 **b.** 1, 2, 3, 6, 9, 18, 27, 54, 81, 162
  c. 1, 2, 3, 6, 9, 18 **d.** 18
9. a. 1, 2, 3, 4, 6, 8, 12, 16, 24, 48 **b.** 1, 2, 4, 5, 8, 10, 16, 20, 40, 80
  c. 1, 2, 4, 7, 8, 14, 16, 28, 56, 112 **d.** 1, 2, 4, 8, 16
  e. 16
10. a. 4 **b.** 1 **c.** 18 **d.** 7 **e.** 50
  f. 24 **g.** 18 **h.** 17 **i.** 125

Practice Problems, p. 266

1. a. 1 × 24, 24 × 1, 2 × 12, 12 × 2, 3 × 8, 8 × 3, 4 × 6, 6 × 4 **b.** 1 × 19, 19 × 1
  c. 1 × 42, 42 × 1, 2 × 21, 21 × 2, 3 × 14, 14 × 3, 6 × 7, 7 × 6
  d. 1 × 144, 144 × 1, 2 × 72, 72 × 2, 3 × 48, 48 × 3, 4 × 36, 36 × 4, 6 × 24, 24 × 6, 8 × 18, 18 × 8,
    9 × 16, 16 × 9, 12 × 12

e. 1×108, 108×1, 2×54, 54×2, 3×36, 36×3, 4×27, 27×4, 6×18, 18×6, 9×12, 12×9

f. 1×400, 400×1, 2×200, 200×2, 4×100, 100×4, 5×80, 80×5, 8×50, 50×8, 10×40, 40×10, 16×25, 25×16, 20×20

g. 1×360, 360×1, 2×180, 180×2, 3×120, 120×3, 4×90, 90×4, 5×72, 72×5, 6×60, 60×6, 9×40, 40×9, 10×36, 36×10, 12×30, 30×12, 15×24, 24×15, 18×20, 20×18

2. a. $2 \times 2 \times 7$ arranged in different orders   **b.** $2 \times 3 \times 5$ arranged in different orders

c. $3 \times 5 \times 5$ arranged in different orders

d. $2 \times 2 \times 16$, $2 \times 4 \times 8$ and $4 \times 4 \times 4$ arranged in different orders

e. $2 \times 2 \times 30$, $2 \times 3 \times 20$, $2 \times 4 \times 15$, $2 \times 5 \times 12$, $2 \times 6 \times 10$, $3 \times 4 \times 10$, $3 \times 5 \times 8$, and $4 \times 5 \times 6$ arranged in different orders

f. $2 \times 2 \times 125$, $2 \times 5 \times 50$, $2 \times 10 \times 25$, $4 \times 5 \times 25$, $5 \times 5 \times 20$ and $5 \times 10 \times 10$ arranged in different orders

g. $2 \times 7 \times 7$ arranged in different orders

3. a. $2 \times 2 \times 2 \times 5$ arranged in different orders

b. $2 \times 2 \times 2 \times 9$ or $2 \times 2 \times 3 \times 6$ or $2 \times 3 \times 3 \times 4$ arranged in different orders

c. $2 \times 3 \times 3 \times 3$ arranged in different orders   **d.** $3 \times 3 \times 5 \times 5$ arranged in different orders

e. $2 \times 2 \times 2 \times 20$ or $2 \times 2 \times 4 \times 10$ or $2 \times 2 \times 5 \times 8$ or $2 \times 4 \times 4 \times 5$ arranged in different orders

f. $2 \times 2 \times 3 \times 35$ or $2 \times 2 \times 5 \times 21$ or $2 \times 2 \times 7 \times 15$ or $2 \times 3 \times 5 \times 14$ or $2 \times 3 \times 7 \times 10$ or $2 \times 5 \times 6 \times 7$ or $3 \times 4 \times 5 \times 7$ arranged in different orders

g. Any of the following arranged in different orders:
$2 \times 2 \times 2 \times 150$, $2 \times 2 \times 3 \times 100$, $2 \times 2 \times 4 \times 75$, $2 \times 2 \times 5 \times 60$, $2 \times 2 \times 6 \times 50$, $2 \times 2 \times 10 \times 30$, $2 \times 2 \times 12 \times 25$, $2 \times 2 \times 15 \times 20$, $2 \times 3 \times 4 \times 50$, $2 \times 3 \times 5 \times 40$, $2 \times 3 \times 8 \times 25$, $2 \times 3 \times 10 \times 20$, $2 \times 4 \times 5 \times 30$, $2 \times 4 \times 6 \times 25$, $2 \times 4 \times 10 \times 15$, $2 \times 5 \times 5 \times 24$, $2 \times 5 \times 6 \times 20$, $2 \times 5 \times 8 \times 15$, $2 \times 5 \times 10 \times 12$, $2 \times 6 \times 10 \times 10$, $3 \times 4 \times 4 \times 25$, $3 \times 4 \times 5 \times 20$, $3 \times 4 \times 10 \times 10$, $3 \times 5 \times 5 \times 16$, $3 \times 5 \times 8 \times 10$, $4 \times 4 \times 5 \times 15$, $4 \times 5 \times 5 \times 12$, $4 \times 5 \times 6 \times 10$, $5 \times 5 \times 6 \times 8$

4. a. 2×7 **b.** $2 \times 3 \times 3 \times 3$ **c.** $2 \times 2 \times 2 \times 3 \times 3$

d. $3 \times 3 \times 3 \times 5$ **e.** $2 \times 2 \times 2 \times 3 \times 7$ **f.** $2 \times 2 \times 2 \times 3 \times 5 \times 5$

g. $2 \times 2 \times 2 \times 2 \times 2 \times 2 \times 3 \times 5$

5. a. $2 \times 2 \times 5$ **b.** $2 \times 3 \times 3 \times 3$ **c.** 3×7 **d.** $2 \times 2 \times 2 \times 2 \times 3$

e. $2 \times 2 \times 5 \times 5$ **f.** $2 \times 2 \times 2 \times 3 \times 3$ **g.** $2 \times 3 \times 5 \times 5$

Practice Problems, p. 268

1. a. 28; 52; 148   **b.** 84; 91; 119   **c.** 42; 51; 102

2. a. 0, 5, 10, 15, 20   **b.** 0, 2, 4, 6, 8   **c.** 0, 6, 12, 18, 24

    d. 0, 9, 18, 27, 36   **e.** 0, 11, 22, 33, 44   **f.** 0, 14, 28, 42, 56

    g. 0, 27, 54, 81, 108   **h.** 0, 59, 118, 177, 236   **i.** 0, 100, 200, 300, 400

    j. 0, 225, 450, 675, 900   **3.** Yes   **4.** Yes

5. a. 0, 8, 16, 24, 32, . . .   **b.** 0, 10, 20, 30, 40, . . .   **c.** 0, 11, 22, 33, 44, . . .

    d. 0, 32, 64, 96, 128, . . .   **e.** 0, 40, 80, 120, 160, . . .   **f.** 0, 17, 34, 51, 68, . . .

    g. 0, 75, 150, 225, 300, . . .   **h.** 0, 96, 192, 288, 384, . . .   **i.** 0, 120, 240, 360, 480, . . .

    j. 0, 275, 550, 825, 1,100, . . .

6. 1, 2, 3, 4, 6, 8, 12, 24; Yes

7. Yes, Yes, Yes

8. 0, 8, 16, 24, 32, 40, 48, 56, 64, 72, 80, 88, 96, 104, 112, 120, 128, 136, 144, 152;
0, 12, 24, 36, 48, 60, 72, 84, 96, 108, 120, 132, 144, 156, 168, 180, 192, 204, 216, 228;
0, 24, 48, 72, 96, 120

9. a. 0, 15, 30, 45, . . .   **b.** 0, 8, 16, 24, . . .   **c.** 0, 60, 120, 180, . . .

    d. 0, 144, 288, 432, . . .   **e.** 0, 100, 200, 300, . . .   **f.** 0, 40, 80, 120, . . .

    g. 0, 16, 32, 48, . . .   **h.** 0, 180, 360, 540, . . .   **i.** 0, 210, 420, 630, . . .

10. Yes; No; 20

11. a. 24   **b.** 12   **c.** 60   **d.** 48   **e.** 100   **f.** 168

    g. 672   **h.** 300   **i.** 1,050   **j.** 12   **k.** 120   **l.** 600

12. a. 2 and 24   **b.** 5 and 50   **c.** 4 and 48   **d.** 6 and 72

The product of the G.C.F. and the L.C.M. is equal to the product of the two given numbers.

Practice Problems, p. 270

1. a. $\frac{1}{2}, \frac{2}{4}, \frac{3}{6}, \frac{4}{8}, \ldots$ **b.** $\frac{1}{9}, \frac{2}{18}, \frac{3}{27}, \frac{4}{36}, \ldots$ **c.** $\frac{3}{5}, \frac{6}{10}, \frac{9}{15}, \frac{12}{20}, \ldots$ **d.** $\frac{2}{3}, \frac{4}{6}, \frac{6}{9}, \frac{8}{12}, \ldots$

e. $\frac{5}{8}, \frac{10}{16}, \frac{15}{24}, \frac{20}{32}, \ldots$ **f.** $\frac{13}{16}, \frac{26}{32}, \frac{39}{48}, \frac{52}{54}, \ldots$ **g.** $\frac{11}{12}, \frac{22}{24}, \frac{33}{36}, \frac{44}{48}, \ldots$ **h.** $\frac{5}{6}, \frac{10}{12}, \frac{15}{18}, \frac{20}{24}, \ldots$

i. $\frac{3}{4}, \frac{6}{8}, \frac{9}{12}, \frac{12}{16}, \ldots$ **j.** $\frac{7}{10}, \frac{14}{20}, \frac{21}{30}, \frac{28}{40}, \ldots$ **k.** $\frac{9}{13}, \frac{18}{26}, \frac{27}{39}, \frac{36}{52}, \ldots$ **l.** $\frac{8}{15}, \frac{16}{30}, \frac{24}{45}, \frac{32}{60}, \ldots$

m. $\frac{4}{11}, \frac{8}{22}, \frac{12}{33}, \frac{16}{44}, \ldots$ **n.** $\frac{19}{25}, \frac{38}{50}, \frac{57}{75}, \frac{76}{100}, \ldots$ **o.** $\frac{17}{20}, \frac{34}{40}, \frac{51}{60}, \frac{68}{80}, \ldots$ **p.** $\frac{31}{50}, \frac{62}{100}, \frac{93}{150}, \frac{124}{200}, \ldots$

2. a. $\frac{1}{6}, \frac{2}{12}, \frac{3}{18}, \frac{4}{24}$ **b.** $\frac{4}{5}, \frac{8}{10}, \frac{12}{15}, \frac{16}{20}$ **c.** $\frac{1}{4}, \frac{2}{8}, \frac{3}{12}, \frac{4}{16}$ **d.** $\frac{7}{8}, \frac{14}{16}, \frac{21}{24}, \frac{28}{32}$

e. $\frac{5}{12}, \frac{10}{24}, \frac{15}{36}, \frac{20}{48}$ **f.** $\frac{6}{7}, \frac{12}{14}, \frac{18}{21}, \frac{24}{28}$ **g.** $\frac{8}{9}, \frac{16}{18}, \frac{24}{27}, \frac{32}{36}$ **h.** $\frac{7}{12}, \frac{14}{24}, \frac{21}{36}, \frac{28}{48}$

i. $\frac{11}{16}, \frac{22}{32}, \frac{33}{48}, \frac{44}{64}$ **j.** $\frac{19}{24}, \frac{38}{48}, \frac{57}{72}, \frac{76}{96}$ **k.** $\frac{3}{10}, \frac{6}{20}, \frac{9}{30}, \frac{12}{40}$ **l.** $\frac{13}{18}, \frac{26}{36}, \frac{39}{54}, \frac{52}{72}$

m. $\frac{17}{20}, \frac{34}{40}, \frac{51}{60}, \frac{68}{80}$ **n.** $\frac{29}{50}, \frac{58}{100}, \frac{87}{150}, \frac{116}{200}$ **o.** $\frac{8}{75}, \frac{16}{150}, \frac{24}{225}, \frac{32}{300}$ **p.** $\frac{10}{11}, \frac{20}{22}, \frac{30}{33}, \frac{40}{44}$

Practice Problems, p. 274

1. a. 5×10; 8.7×10; 4.00×10^2; 9.14×10^2; 8.0×10^3; 3.06×10^3
b. 6×10^4; 7.9×10^4; 5.2×10^5; 3.6×10^7; 4.875×10^6
c. 2×10^9; 3.08×10^{11}; 8.45×10^{13} **d.** 9.6×10^8; 5.1×10^{11}; 4.23×10^{17}
e. 7×10^{18}; 6.7×10^{21}

2. a. 4×10^{-1}; 9×10^{-1}; 3×10^{-1}; 3.4×10^{-1}; 2×10^{-2}; 8.8×10^{-1}; 6.9×10^{-1}
b. 5.7×10^{-2}; 7×10^{-3}; 9.83×10^{-1}; 5.42×10^{-2}; 9.6×10^{-3}; 8.167×10^{-1}; 2.005×10^{-1}
c. 1.8×10^{-3}; 2.49×10^{-3}; 6.356×10^{-2}; 7.84×10^{-4}; 9.1×10^{-6}
d. 3.65×10^{-5}; 7.9×10^{-7}; 8.01×10^{-7}; 3.6×10^{-9}
e. 4.3×10^{-11}; 5×10^{-14}; 6.82×10^{-16}

3. a. 5×10^{23} horsepower   **b.** 3.86×10^9 kilometers   **c.** 6.6×10^{21} tons
d. 7.79×10^8 kilometers   **e.** (1) 4.9×10^{16} (2) 5.28×10^{20}

Inventory Test, p. 284

1. a. 420 mm   **b.** .690 km   **2. a.** 1,800 g   **b.** 5.5 g
3. a. 700 cL   **b.** .665 L   **4. a.** 4,900 ha   **b.** 8,000,000 cm^2
5. a. 8,300,000 cm^3 **b.** 9.250 dm^3 **c.** 6,750 cm^3; 6.75 kg
6. 22 ft.   **7.** $\frac{3}{4}$ **8.** 72 pt.   **9.** 24   **10.** $\frac{3}{4}$ sq. ft.
11. 189 cu. ft.   **12.** $\frac{5}{6}$ **13.** 76° 18′   **14.** 1,200 m/min.   **15.** No
16. a. 9:05 P.M.   **b.** 12:30 A.M.   **17.** 8 A.M.
18. a. (1) 5 cm; $6\frac{1}{4}$%   (2) 500 kg; 2%   (3) .05 L; .676%   (4) $\frac{1}{6}$ hr.; 3.57%
b. 0.060 mm; 0.060 mm
c. Precision: 8.2 mm, 69 cm, 40 m, 750,000 km   Accuracy: 8.2 mm, 750,000 km, 69 cm, 40 m
19. a. 4 significant digits; 9, 0, 0, 8   **b.** 2 significant digits; 4, 0
c. 8 significant digits; 8, 9, 0, 2, 6, 5, 7, 1   **d.** 5 significant digits; 6, 7, 1, 3, 9
e. 2 significant digits; 3, 8   **f.** 3 significant digits; 5, 6, 3
20. a. 10.11   **b.** 37.932   **c.** 1.2567   **d.** .2408333

Practice Problems, pp. 301–302

3. 8,000 mm^3; 15,000,000,000 mm^3; 6,100,000 mm^3; 25,800 mm^3; 53,600,000,000 mm^3
4. 27,000,000 cm^3; 4,950 cm^3; 6.226 cm^3; 8,500,000 cm^3; .74 cm^3
5. 63,000 dm^3; 2.930 dm^3; 17,900 dm^3; .073250 dm^3; .0864 dm^3
6. 17 m^3; .0368 m^3; .0094 m^3; .575 m^3; .6 m^3 **a.** 21 cm^3
b. 9,000 cm^3 **c.** 430 mL   **d.** 18.6 L   **e.** 7.9 L   **f.** 5 kg
g. 67 g   **h.** 820 g   **i.** 9.1 kg   **j.** 615 g   **k.** 4 L
l. 75 mL   **m.** .545 L   **n.** 26 dm^3 **o.** 19 cm^3 **p.** 8,800 cm^3

Practice Problems, p. 325

4. a. $\frac{1}{4}$ kg   **b.** .05 L   **c.** .0005 cm   **d.** .005 mm   **e.** $\frac{1}{2}$ km
f. $\frac{1}{8}$ hr.   **g.** $\frac{1}{16}$ lb.   **h.** $\frac{1}{2}$ gal.   **i.** .05 ft.   **j.** $\frac{1}{64}$ in.

k. 5 m   **l.** 50 tons   **m.** 5 km   **n.** 5,000 mi.   **o.** .0005 mg

p. $\frac{1}{2}$ cm   **q.** $\frac{1}{2}$ in.   **r.** $\frac{1}{2}$ oz.   **s.** $\frac{1}{2}$ min.   **t.** 500 mm

5. a. $\frac{1}{130}$ **b.** $\frac{1}{92}$ **c.** $\frac{1}{4,114}$ **d.** $\frac{1}{16}$ **e.** $\frac{1}{946}$

    f. $\frac{1}{74}$ **g.** $\frac{1}{94}$ **h.** $\frac{1}{1,734}$ **i.** $\frac{1}{190,536}$ **j.** $\frac{1}{114}$

    k. $\frac{1}{1,520}$ **l.** $\frac{1}{800}$ **m.** $\frac{1}{54,000}$ **n.** $\frac{1}{1,160}$ **o.** $\frac{1}{78}$

    p. $\frac{1}{1,218}$ **q.** $\frac{1}{70}$ **r.** $\frac{1}{40}$ **s.** $\frac{1}{278}$ **t.** $\frac{1}{4,120}$

6. a. $5\frac{5}{8}\%$ **b.** 10%   **c.** $3\frac{4}{7}\%$ **d.** $\frac{2}{3}\%$ **e.** $4\frac{1}{6}\%$

7. a. 6.52 cm   **b.** same   **c.** 250 km   **d.** 75,000 km

    e. same   **f.** $10\frac{7}{8}$ mi.

8. .004 mm is more precise; 65,000 m is more accurate

9. a. Precision: .009 mm, .09 mm, .9 mm, 9 mm; 90 mm; accuracy: all the same

    b. Precision: 7.26 in., 9.375 yd., 63.4 ft., 83.0 mi., 500 mi.;

       accuracy: 9.375 yd., 83.0 mi., 7.26 in., 63.4 ft., 500 mi.

10. a. 8.11 mm to 8.21 mm inclusive   **b.** $2\frac{5}{8}''$ to $2\frac{7}{8}''$ inclusive

    c. 4.775 cm to 4.825 cm inclusive   **d.** 2.995'' to 3.145'' inclusive

    e. 1.999 mm to 2.001 mm inclusive   **f.** $9\frac{13}{32}''$ to $9\frac{19}{32}''$ inclusive

Practice Problems, p. 376

6. Yes. The measure of an exterior angle, when it is a right angle, will equal the measure of the adjacent interior angle.

No. Since the measure of an exterior angle of a triangle is equal to the sum of the measures of the opposite two interior angles and each interior angle must have a measure greater than zero, the measures of an exterior angle and one opposite interior angle will not be the same.

Practice Problems, pp. 377-378

7. a. $\angle 6$ and $\angle 7$, and $\angle 3$ and $\angle 2$ are opposite angles and hence equal. $\angle 3$ and $\angle 6$ are alternate-interior angles and hence are equal. Therefore, if $\angle 3$ and $\angle 6$ are equal, their opposite angles, $\angle 2$ and $\angle 7$ are equal.

    b. $\angle 2$ and $\angle 6$ are corresponding angles and hence are equal. $\angle 1$ and $\angle 2$ are supplementary angles. Therefore, since $\angle 2 = \angle 6$ and $\angle 1$ and $\angle 2$ are supplementary, $\angle 1$ and $\angle 6$ are supplementary.

    c. $\angle 5$ and $\angle 4$ are alternate-interior angles and hence are equal. $\angle 1$ and $\angle 4$, and $\angle 5$ and $\angle 8$ are opposite angles and hence are equal. If $\angle 5 = \angle 4$, $\angle 1 = \angle 4$ and $\angle 5 = \angle 8$, then $\angle 1 = \angle 8$.

    d. $\angle 3$ and $\angle 6$ are opposite-interior angles and therefore are equal. Since $\angle 3 = \angle 6$ and $\angle 5$ and $\angle 6$ are supplementary, $\angle 3$ and $\angle 5$ must be supplementary.

    e. $\angle 6$ and $\angle 8$ are supplementary angles. $\angle 2$ and $\angle 6$ are equal because they are corresponding angles. Since $\angle 2 = \angle 6$ and $\angle 6$ and $\angle 8$ are supplementary, $\angle 2$ and $\angle 8$ are supplementary.

8. a. $m\angle 6 = 130°$ **b.** $m\angle 7 = 106°$ **c.** $m\angle 5 = 75°$ **d.** $m\angle 2 = 93°$

    e. $m\angle 3 = 133°$ **f.** $m\angle 8 = 64°$ **g.** $m\angle 2 = 97°$ **h.** $m\angle 4 = 65°$

    i. $m\angle 1 = 48°$; $m\angle 2 = 132°$; $m\angle 3 = 132°$; $m\angle 4 = 48°$; $m\angle 5 = 48°$; $m\angle 6 = 132°$; $m\angle 7 = 132°$

    j. $m\angle 1 = 55°$; $m\angle 2 = 125°$; $m\angle 4 = 55°$; $m\angle 5 = 55°$; $m\angle 6 = 125°$; $m\angle 7 = 125°$; $m\angle 8 = 55°$

Practice Problems, pp. 379-380

1. Triangles ABC and JKL are congruent because three sides of triangle ABC are equal to three sides of triangle JKL.

2. Triangles GHI and JKL are congruent because two angles and an included side of triangle GHI are equal respectively to two angles and an included side of triangle JKL.

3. Triangles DEF and JKL are congruent because two sides and an included angle of triangle DEF are equal respectively to two sides and an included angle of triangle JKL.

6. Given $AB = BC$, $\angle ABD = \angle DBC$ and $BD = BD$, triangles ABD and DBC are congruent because two sides and an included angle of triangle ABD are equal respectively to two sides and an included angle of DBC.

7. It is given that side EF = side HG and side EH = side FG. Since both triangles share a common side EG, the two triangles EGH and EFG have three equal sides and are therefore congruent.

T79

Set 9:   1. 60°   2. 105°   3. 30°   4. 345°   5. 150°
Set 10:   1. 7° 30′   2. 11° 15′   3. 5°   4. 9°   5. 3° 45′
Set 11:   1. 3′ 45″   2. 10′   3. 3′   4. 7′ 30″   5. 12′ 15″
Set 12:   1. 33° 45′   2. 21°   3. 72° 30′   4. 232° 45′   5. 136° 30′
Set 13:   1. 56° 22′ 30″ 2. 33° 12″   3. 22° 17′ 30″ 4. 120° 58′ 30″   5. 84° 21′ 15″
Set 14:   1. (a) 1000   (b) 1324   (c) 2100   (d) 0657   (e) 1617
           2. (a) 1755   (b) 0655   (c) 1053   (d) 20 sec. past 1812   (e) 32 sec past 0644
           3. 0604   4. 1917   5. 12 sec. past 1023
Set 15:   1. 15° W.; 45° E.; 90° W.; 80° E.; 123° W.   2. 10° E.; 26° W.; 130° W.; 113° E.; 46° W.
           3. 18° 45′ E.   4. 175° 52′ W.   5. 90° 31′ W.
Set 16:   1. 26 min. 16 sec.; 15 min. 44 sec.; 9 min. 32 sec.; 7 min. 4 sec.; 27 min. 32 sec.;
              14 min. 44 sec.; 9 min. 44 sec.
           2. 36 sec. past 1617; 24 sec. past 1652; 24 sec. past 1351; 52 sec. past 1500;
              44 sec. past 1454; 48 sec. past 1704; 20 sec. past 1542

Inventory Test, pp. 442-445

8. a. 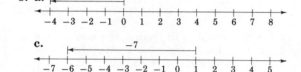   b.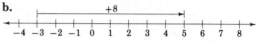

   c.   d.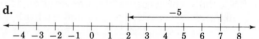

28. a. all real numbers greater than 17   b. all real numbers less than 8
    c. −2 and all real numbers less than −2   d. −3 and all real numbers greater than −3
    e. all real numbers greater than 0   f. all real numbers except −3
    g. 3 and all real numbers less than 3   h. all real numbers less than 5
    i. 70 and all real numbers less than 70

29. a. all integers greater than 5   b. all multiples of 3 that are 12 or greater
    c. no solution

30. a. 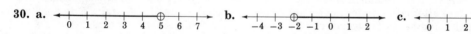   b. 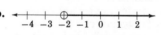   c.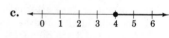

    d. 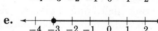   e. 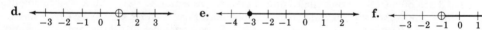   f.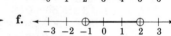

Practice Problems, p. 451

1. a. positive five sixths   b. negative thirteen fourths
   c. negative six and one third   d. positive eighty-nine hundredths
   e. negative seventeen point three   f. positive two and two fifths
   g. negative eleven sixteenths   h. negative four thousandths

2. a. $-\frac{1}{5}$ b. $+\frac{7}{12}$ c. $+\frac{9}{100}$ d. $-4\frac{3}{10}$ e. $-9\frac{48}{100}$ f. $+\frac{161}{1,000}$

3. a. $-\frac{5}{8}$ b. $+\frac{9}{4}$ c. $-.07$ d. $+2.75$
   e. $+1\frac{3}{5}$ f. $-.102$ g. $-6\frac{4}{7}$ h. $+58.3$

4. a. No; yes; yes   b. No; yes; yes   c. No; yes; yes
   d. No; yes; yes   e. Yes; yes; yes   f. No; no; yes

5. a. Integers: -17, $\frac{16}{2}$ Rational: -17, $+\frac{3}{4}$, $-.05$, $\frac{16}{2}$, $-7\frac{1}{3}$
      Irrational: $-\sqrt{29}$ Real: -17, $+\frac{3}{4}$, $-.05$, $-\sqrt{29}$, $\frac{16}{2}$, $-7\frac{1}{3}$
   b. Integers: $+100$, $-\frac{18}{6}$ Rational: $-\frac{4}{7}$, $+100$, -5.08, $-2\frac{3}{4}$, $-\frac{18}{6}$
      Irrational: $+\sqrt{13}$ Real: $-\frac{4}{7}$, $+100$, -5.08, $-2\frac{3}{4}$, $+\sqrt{13}$, $-\frac{18}{6}$
   c. Integers: $-\sqrt{16}$, -69, $\frac{28}{4}$ Rational: $-.375$, $-\frac{5}{8}$, $-\sqrt{16}$, -69, $+\frac{28}{4}$, $+3\frac{5}{6}$
      Irrational: None   Real: $-.375$, $-\frac{5}{8}$, $-\sqrt{16}$, -69, $+\frac{28}{4}$, $+3\frac{5}{6}$

d. Integers: $+97$ Rational: $-\frac{15}{16}, +97, +.18, -4\frac{7}{12}, +\frac{21}{4}$

Irrational: $-\sqrt{5}$ Real: $-\sqrt{5}, -\frac{15}{16}, +97, +.18, -4\frac{7}{12}, +\frac{21}{4}$

Practice Problems, p. 455

1. a. $-2; +2; -7; +5; 0; -9; +8; +3; -6; +10$ **b.** $O; U; B; D; F; I; N; Q; A; T$

2. a. **b.**

c. **d.**

e.

f.

g.

h.

i.

j.

3. a. **b.**

c.

Practice Problems, pp. 460–461

6. a. **b.**

c.

d.

e.

f.

g.

h.

i. -8

j. $+19$

k. -18

l. $+9$

m. -8

n. $+5$

o. -6

p. $+22$

q. -11

r. -6

Practice Problems, p. 462

3. a. The opposite of nine or negative nine   **b.** The opposite of positive two
  c. The opposite of negative five sixths   **d.** The opposite of negative eleven
  e. The opposite of negative eight is equal to positive eight
  f. The opposite of positive ten is equal to negative ten
  g. The opposite of fifteen is equal to negative fifteen
  h. The opposite of negative six point four is equal to positive six point four

Practice Problems, p. 475

1. a. Negative five plus positive seven   **b.** Thirty-six times twenty-eight
  c. Fifty-three squared   **d.** Forty-two divided by six
  e. The square root of sixty-five   **f.** Positive twelve minus positive thirty
  g. Twenty-four times nineteen
  h. Seventeen cubed or seventeen to the third power
  i. Negative fifteen times negative four   **j.** Six to the power of eight
  k. The absolute value of negative thirteen   **l.** Seven times the sum of twenty-one and nine
  m. Eighteen minus the difference of positive twenty minus positive seven
  n. Nine times the product of twelve times five   **o.** The negative of negative twenty
  p. The absolute value of negative six plus negative three

Practice Problems, p. 479

1.
  a. Each number n is less than sixteen
  b. Each number x is greater than twenty-nine
  c. Three times each number y is not equal to seventeen
  d. Each number b plus six is greater than twenty-five
  e. Each number a minus five is less than fifty-three
  f. Seven times each number t is not less than negative fourteen
  g. Nine times some number m plus four is equal to fifteen
  h. Eight times each number c minus nine is not greater than the number c plus seven
  i. Two times each number n plus six is less than three times the number n minus eight
  j. Each number b divided by five is greater than twenty
  k. Fifteen times each number x minus four is not greater than eighteen minus seven times the number x
  l. Each number h is greater than or equal to negative six
  m. Nineteen times each number w is less than or equal to thirty-eight
  n. Five times each number d plus seven is less than or equal to twelve
  o. Each number n minus sixteen is greater than or equal to four times the number n plus nine
  p. Each number n is greater than negative three and less than eight
  q. Each number x is less than eighteen and greater than zero
  r. Each number b is greater than or equal to four and less than or equal to twenty-seven
  s. Seven times each number r is less than or equal to twelve and greater than or equal to negative six
  t. Each number a is greater than two and less than or equal to forty-five

2.   **a.** $x + 10 = 41$ **b.** $n - 4 < 18$ **c.** $4t > 12$ **d.** $9y + 2 \neq 50$ **e.** $a + 5 \nless 9$
  f. $\frac{m}{7} \ngtr 14$ **g.** $7x \geq 30$ **h.** $c + 9 \leq 10$ **i.** $-5 < t < -1$ **j.** $4 \leq 12y \leq 15$

Practice Problems, pp. 480–481

2.
  a. The perimeter of an equilateral triangle equals three times the length of a side
  b. Electromotive force in volts equals the current in amperes times the resistance in ohms
  c. Time of travel equals distance traveled divided by the average rate of speed
  d. Fahrenheit temperature reading equals one and eight tenths times the Celsius temperature reading plus 32 degrees
  e. Distance a freely falling body drops equals one-half the product of the acceleration due to gravity and the square of the time of falling

Practice Problems, p. 504

1.   **a.**    **b.** 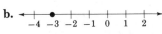   **c.**

  d. 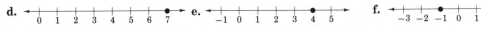   **e.** 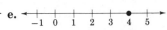   **f.**

  g. 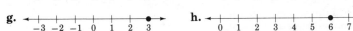   **h.**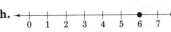

  i.    **j.**

  k. no solution   **l.**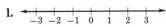

Practice Problems—Type I, p. 515

1.   **a.** $n > 3$, or all real numbers greater than 3   **b.** $a > 0$, or all real numbers greater than 0
  c. $d > -11$, or all real numbers greater than -11
2.   **a.** $c > 23$, or all real numbers greater than 23   **b.** $x > -1$, or all real numbers greater than -1
  c. $r > 4$, or all real numbers greater than 4

3. a. $b < 6$, or all real numbers less than 6   **b.** $g < -7$, or all real numbers less than -7
  c. $m < 1.9$, or all real numbers less than 1.9
4. a. $d < 15$, or all real numbers less than 15   **b.** $t < 28$, or all real numbers less than 28
  c. $h < -2$, or all real numbers less than -2
5. a. $m \geq 4$, or 4 and all real numbers greater than 4
  b. $f \geq -17$, or -17 and all real numbers greater than -17
  c. $w \geq -16$, or -16 and all real numbers greater than -16
6. a. $x \geq 15$, or 15 and all real numbers greater than 15
  b. $k \geq 12.2$, or 12.2 and all real numbers greater than 12.2
  c. $a \geq 3$, or 3 and all real numbers greater than 3
7. a. $y \leq 7$, or 7 and all real numbers less than 7
  b. $e \leq \frac{1}{2}$, or $\frac{1}{2}$ and all real numbers less than $\frac{1}{2}$
  c. $v \leq -3$, or -3 and all real numbers less than -3
8. a. $r \leq 14$, or 14 and all real numbers less than 14
  b. $c \leq -2$, or -2 and all real numbers less than -2
  c. $x \leq 18$, or 18 and all real numbers less than 18
9. a. $s \neq 8$, or all real numbers except 8   **b.** $n \neq 1.3$, or all real numbers except 1.3
  c. $b \neq -11$, or all real numbers except -11
10. a. $w \neq 17$, or all real numbers except 17   **b.** $y \neq 2\frac{3}{8}$, or all real numbers except $2\frac{3}{8}$
  c. $r \neq 0$, or all real numbers except 0
11. a. $x > 9$, or all real numbers greater than 9   **b.** $y < -5$, or all real numbers less than -5
  c. $n \geq -6$, or -6 and all real numbers greater than -6
  d. $a \leq 20$, or 20 and all real numbers less than 20
  e. $t \neq 0$, or all real numbers except 0
  f. $b < -4$, or all real numbers less than -4
  g. $r > -4$, or all real numbers greater than -4
  h. $z \geq 19$, or 19 and all real numbers greater than 19
  i. $w \leq -10$, or -10 and all real numbers less than -10

Type II, pp. 515–516

1. a. $x > 2$, or all real numbers greater than 2   **b.** $y > -8$, or all real numbers greater than -8
  c. $a > \frac{1}{4}$, or all real numbers greater than $\frac{1}{4}$
2. a. $m > 8$, or all real numbers greater than 8   **b.** $c > 70$, or all real numbers greater than 70
  c. $r > -54$, or all real numbers greater than -54
3. a. $z < 5$, or all real numbers less than 5   **b.** $s < -8$, or all real numbers less than -8
  c. $b < 1\frac{1}{2}$, or all real numbers less than $1\frac{1}{2}$
4. a. $h < 24$, or all real numbers less than 24   **b.** $p < -15$, or all real numbers less than -15
  c. $w < 35$, or all real numbers less than 35
5. a. $n \geq 5$, or 5 and all real numbers greater than 5
  b. $a \geq \frac{2}{3}$, or $\frac{2}{3}$ and all real numbers greater than $\frac{2}{3}$
  c. $x \geq 32$, or 32 and all real numbers greater than 32
6. a. $b \geq 0$, or 0 and all real numbers greater than 0
  b. $r \geq -24$, or -24 and all real numbers greater than -24
  c. $z \geq 15$, or 15 and all real numbers greater than 15
7. a. $s \leq 9$, or 9 and all real numbers less than 9
  b. $h \leq 2$, or 2 and all real numbers less than 2
  c. $n \leq 400$, or 400 and all real numbers less than 400
8. a. $y \leq 42$, or 42 and all real numbers less than 42
  b. $t \leq 28$, or 28 and all real numbers less than 28
  c. $b \leq -30$, or -30 and all real numbers less than -30
9. a. $c \neq 16$, or all real numbers except 16   **b.** $x \neq 128$, or all real numbers except 128
  c. $f \neq 0$, or all real numbers except 0
10. a. $n \neq 60$, or all real numbers except 60   **b.** $d \neq -12$, or all real numbers except -12
  c. $y \neq 90$, or all real numbers except 90
11. a. $b < 4$, or all real numbers less than 4
  b. $a > -16$, or all real numbers greater than -16
  c. $x \geq -\frac{3}{4}$, or $-\frac{3}{4}$ and all real numbers greater than $-\frac{3}{4}$
  d. $y \leq 4\frac{1}{3}$, or $4\frac{1}{3}$ and all real numbers less than $4\frac{1}{3}$

e. $m \neq -8$, or all real numbers except -8 **f.** $t < 49$, or all real numbers less than 49

g. $n > 0$, or all real numbers greater than 0

h. $z > 75$, or 75 and all real numbers greater than 75

i. $x \leq 128$, or 128 and all real numbers less than 128

Type III, p. 516

1. a. $d < -6$, or all real numbers less than -6 **b.** $c < -14$, or all real numbers less than -14

  c. $h < -3$, or all real numbers less than -3

2. a. $b < 7$, or all real numbers less than 7   **b.** $m < 13\frac{1}{3}$, or all real numbers less than $13\frac{1}{3}$

  c. $z < 40$, or all real numbers less than 40

3. a. $x > -9$, or all real numbers greater than -9 **b.** $r > -6$, or all real numbers greater than -6

  c. $a > -4\frac{1}{2}$, or all real numbers greater than $-4\frac{1}{2}$

4. a. $t > 8$, or all real numbers greater than 8   **b.** $y > 3$, or all real numbers greater than 3

  c. $m > 6$, or all real numbers greater than 6

5. a. $y \leq -7$, or -7 and all real numbers less than -7

  b. $r \leq 0$, or 0 and all real numbers less than 0

  c. $x \leq -2$, or -2 and all real numbers less than -2

6. a. $n \leq 9$, or 9 and all real numbers less than 9

  b. $g \leq 162$, or 162 and all real numbers less than 162

  c. $t \leq 4\frac{1}{4}$, or $4\frac{1}{4}$ and all real numbers less than $4\frac{1}{4}$

7. a. $s \geq -20$, or -20 and all real numbers greater than -20

  b. $y \geq 0$, or 0 and all real numbers greater than 0

  c. $x \geq -5$, or -5 and all real numbers greater than -5

8. a. $z \geq 36$, or 36 and all real numbers greater than 36

  b. $n \geq -9$, or -9 and all real numbers greater than -9

  c. $y \geq 90$, or 90 and all real numbers greater than 90

9. a. $n > -36$, or all real numbers greater than -36

  b. $b < 12$, or all real numbers less than 12

  c. $y \geq -48$, or -48 and all real numbers greater than -48

  d. $c \leq -90$, or -90 and all real numbers less than -90

  e. $a > 21$, or all real numbers greater than 21

  f. $x < -40$, or all real numbers less than -40

  g. $s \leq 60$, or 60 and all real numbers less than 60

  h. $m < -36$, or all real numbers less than -36

  i. $d \geq 4$, or 4 and all real numbers greater than 4

10. a. $m > -8$, or all real numbers greater than -8

  b. $z < 11$, or all real numbers less than 11

  c. $x \geq -23$, or -23 and all real numbers greater than -23

  d. $a \leq -4\frac{1}{2}$, or $-4\frac{1}{2}$ and all real numbers less than $-4\frac{1}{2}$

  e. $r > 0$, or all real numbers greater than 0   **f.** $b < 56$, or all real numbers less than 56

  g. $y < -7$, or all real numbers less than -7 **h.** $n \leq \frac{1}{2}$, or $\frac{1}{2}$ and all real numbers less than $\frac{1}{2}$

  i. $g \geq 7$, or 7 and all real numbers greater than 7

Miscellaneous, p. 517

1. a. $x < 4$, or all real numbers less than 4   **b.** $a > -5$, or all real numbers greater than -5

  c. $m \leq 2$, or 2 and all real numbers less than 2   **d.** $s \geq 2$, or 2 and all real numbers greater than 2

  e. $n \neq -8$, or all real numbers except -8 **f.** $y < 9$, or all real numbers less than 9

  g. $b > -\frac{1}{2}$, or all numbers greater than $-\frac{1}{2}$

  h. $r \geq 3$, or 3 and all real numbers greater than 3

  i. $t < -7$, or all real numbers less than -7

  j. $z \leq 5\frac{1}{2}$, or $5\frac{1}{2}$ and all real numbers less than $5\frac{1}{2}$

  k. $w > 14$, or all real numbers greater than 14   **l.** $x \neq \frac{1}{2}$, or all real numbers except $\frac{1}{2}$

2. a. $x > 3$, or all real numbers greater than 3   **b.** $c < -12$, or all real numbers less than -12

  c. $d \leq -7$, or -7 and all real numbers less than -7

  d. $y \neq -15$, or all real numbers except -15

  e. $m \leq 0$, or 0 and all real numbers less than 0

f. $t < 17$, or all real numbers less than 17   **g.** $b < -4$, or all real numbers less than -4
h. $x \leq -9$, or -9 and all real numbers less than -9
i. $n \geq 5$, or 5 and all real numbers greater than 5
j. $r \neq 5$, or all real numbers except 5   **k.** $w > -12$, or all real numbers greater than -12
l. $a > -18$, or all real numbers greater than -18

Restricted Replacement, p. 517

1. 4, 5, 6, 7, 8, 9; 0, 1, 2, 3, 4, 5, 6; 0, 1, 2, 3; 0, 1, 2, 3, 4
2. $-3, -2, -1, 0, 1, 3$; $-3, -2, -1, 0, 1$; $-3, -2, -1, 0, 1, 2, 3$; $-3, -2, -1, 0, 1, 2$
3. 2, 3, 5; 2, 3, 5, 7, 11, 13, 17, 19, 23; all prime numbers except 11; 2, 3
4. $\ldots, -10, -9, -8; \ldots, -7, -6, -5$; 20, 21, 22, $\ldots$; all integers except -2
5. 0, 2, 4, 6, 8, 10, 12, 14, $\ldots$; 0, 2; 36, 38, 40, $\ldots$
6. 0, $-1, -2, -3, -4$; no replacements; all negative integers; $-1, 0$
7. no replacements; no replacements; $-2, -1, 0, 1, 2$; $-2, -1, 0, 1, 3$
8. 5, 7; no replacements; 2, 3, 7; 7
9. 16, 20, 24, 28; 16, 20, 24, 28; no replacements; 12, 16, 20, 24

Practice Problems, p. 520

1. a. **b.** **c.**

d. **e.** **f.**

g. **h.** **i.**

j. **k.** **l.**

m. **n.** **o.**

2. a. **b.**

c. **d.**

e. **f.**

3. a. $a < 1$ or any equivalent inequality   **b.** $b > -4$ or any equivalent inequality
c. $-1 \leq c \leq 2$ or any equivalent inequality   **d.** $d \geq 5$ or any equivalent inequality
e. $-4 < e < 0$ or any equivalent inequality   **f.** $f \leq -1$ or any equivalent inequality

Review of Part 4, pp. 521–522

15. a. **b.**

c.

16. 9 games   **17. a.** $\frac{3}{4}$ **b.** $x = 24$
18. a. $n < 12$, or all real numbers less than 12   **b.** $b > 13$, or all real numbers greater than 13
 c. $b \geq -5$, or -5 and all real numbers greater than -5
 d. $x \neq -2$, or all real numbers except -2
 e. $y \leq -2$, or -2 and all real numbers less than -2
 f. $n \geq -30$, or -30 and all real numbers greater than -30

19. 2, 3, 5, 7

20. a.

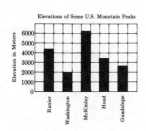

Number line from −3 to 9, open circle at 4, bold ray extending left.

b.

Number line from −9 to 3, open circle at −5, bold ray extending left.

c.

Number line from −7 to 5, closed dot at −1, bold ray extending left.

d.

Number line from −6 to 7, open circle at 3.

e.

Number line from −8 to 5, closed dot at −3.

f.

Number line from −6 to 7, open circles at −1 and 4.

Inventory Test, p. 526

1.

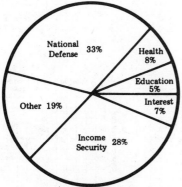

Elevations of Some U.S. Mountain Peaks

2.

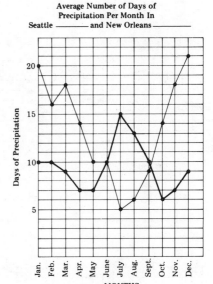

Average Number of Days of Precipitation Per Month In
Seattle ———— and New Orleans ————

3.

Average Annual Outlays of the Federal Government

National Defense 33%
Health 8%
Education 5%
Interest 7%
Income Security 28%
Other 19%

4.

Frequency Distribution Table

Score	Tally	Frequency
75	11	2
74	1111	5
73	1	1
72	111	3
71	111	3
70	111	3
69	111	3
68	1111	4
67	11	2
66	111	3
65	11	2

Practice Problems, p. 529

2.

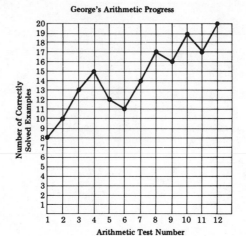

George's Arithmetic Progress

3.

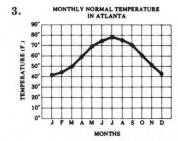

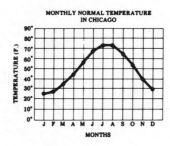

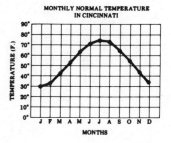

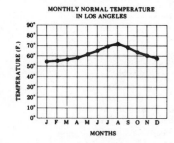

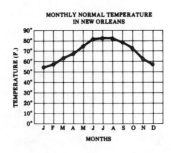

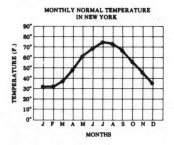

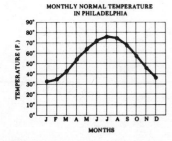

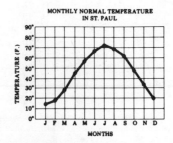

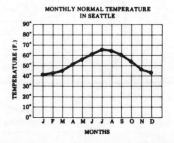

Practice Problems, p. 530

1. c. Food: $4,650; $387.50; $89.42   Savings: $1,860; $155; $35.77   Clothing: $2,232; $186; $4̶
Family Car: $1,860; $155; $35.77   Shelter: $3,720; $310; $71.54
Other Expenses: $2,790; $232.50; $53.65   Operating Expenses: $1,488; $124; $28.62

2.

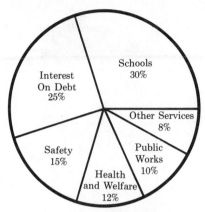

Planned City Budget

3.

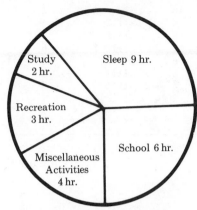

How Charles Spends His Day

4.

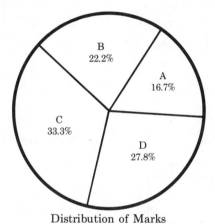

Distribution of Marks

Practice Problems, p. 532

1. a.

Frequency Distribution Table

Score	Tally	Frequency
100	1	1
95	11	2
90	1111	4
85	1111	4
80	111	3
75	1111	4
70	111	3
65	1	1
60	11	2
40	1	1

b. Histogram

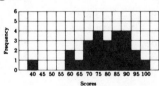

c. Frequency Polygon

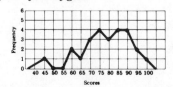

2. a.

Frequency Distribution Table

Score	Tally	Frequency
10	1	1
9	1111	4
8	1111	5
7	1111 1	6
6	1111	4
5	1111	4
4	11	2
3	1	1
2	11	2
1	1	1

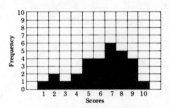

c. Frequency Polygon

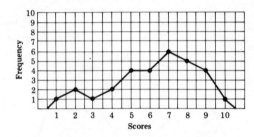

3. a.

b. Histogram

Frequency Distribution Table

Score	Tally	Frequency
52	11	2
51	111	3
50	1111	4
49	1111	5
48	111	3
47	111	3
46	11	2
45	1111 11	7
44	111	3
43	111	3

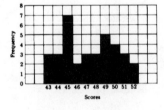

c. Frequency Polygon

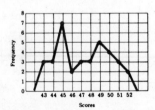

4. a.

Frequency Distribution Table

Score	Tally	Frequency
19	1111	4
18	1111	4
17	1111	5
16	111	3
15	1111	5
14	1111 111	8
13	1111	4
12	1111	5
11	11	2
10	1	1

b. Histogram

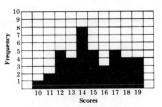

c. Frequency Polygon

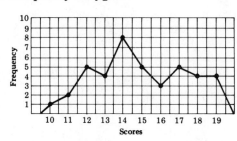

5. a.

Frequency Distribution Table

Score	Tally	Frequency
10	111	3
9	1111 1	6
8	1111 1111	9
7	1111 1	6
6	1111 11	7
5	1111 1	6
4	1111	5
3	111	3
2	1111	4
1	1	1

b. Histogram

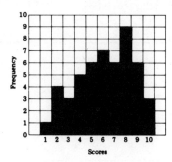

c. Frequency Polygon

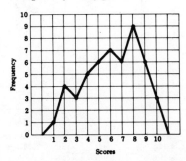

5. a.

Score	Tally	Frequency
9	11111 1	6
8	11111 1	6
7	1111	4
6	1111	4
5	11	2
4	11	2
3	1	1

Mean   7
Median 7
Mode   8 and 9

b.

Score	Tally	Frequency
50	1	1
49	11	2
48	111	3
47	11111	5
46	1	1
45	1	1
44	11	2
43	11	2
42	1	1
41		0
40	1	1

Mean   46
Median 47
Mode   47

c.

Score	Tally	Frequency
20	11	2
19	1111	4
18	1111	4
17	1111	4
16	1111	4
15	111	3
14	11111	5
13	11	2
12	11	2
11	1	1

Mean   16
Median 16
Mode   14

d.

Score	Tally	Frequency
64	1111	4
63	111	3
62	1111	4
61	1111	4
60	1111	4
59	11111 11	7
58	1111	4
57	111	3
56	11	2
55	1	1

Mean   60
Median 60
Mode   59

e.

Score	Tally	Frequency
10	11	2
9	11111 111	8
8	11111 1	6
7	11111	5
6	11111	5
5	11111 11	7
4	11111	5
3	11111	5
2	111	3
1	1	1

Mean   6
Median 6
Mode   9

1.

Family Income Spending

2. a.

Frequency Distribution Table

Score	Tally	Frequency
9	1	1
8	1111	4
7	111	3
6	1111	5
5	1111 1	6
4	1111	4
3	111	3
2	1111	4
1	11	2

b. Histogram

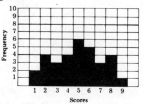

c. Frequency Polygon

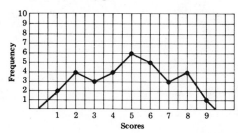

3.

Score	Tally	Frequency
20	1	1
19	1111	5
18	111	3
17	1111	5
16	1111 111	8
15	1	1
14	11	2

Mean 17   Mode 16
Median 17   Range 6

T93

Practice Problems, p. 560–563

3. **a. (1)** Thirty-four and six tenths cents   **(2)** Five and four tenths cents
(3) Seventy-one and three tenths cents   **(4)** Eight and two tenths cents
(5) Sixty-nine and twenty-five hundredths cents
  b. (1) Two dollars and thirty-seven and four tenths cents   **(2)** One dollar and thirty-two and nine tenths cents   **(3)** Five dollars and six and seven tenths cents   **(4)** Three dollars and ten and eight tenths cents   **(5)** One dollar and forty-three and nine tenths cents

Practice Problems, pp. 581–582

1. $.007; $.019; $.050; $.132; $.084; $.175; $.060; $.092
2. 2¢; .9¢; 3.5¢; 1.4¢; 14.6¢; 7¢; 9.5¢; 18¢; 5.8¢; 25¢
3. 40 mills; 80 mills; 5 mills; 26 mills; 61 mills; 105 mills
4. 4¢ per $1.00; 2.8¢ per $1.00; 4.3¢ per $1.00; 3.4¢ per $1.00
5. 50 mills per $1; 33 mills per $1; 45 mills per $1; 190 mills per $1

Practice Problems, pp. 600–603

4. **a.** 1 point above low and $2\frac{1}{2}$ points below high
  b. $1\frac{1}{2}$ points above low and $6\frac{1}{2}$ points below high
  c. $12\frac{1}{8}$ points above low and $6\frac{1}{2}$ points below high
  d. $7\frac{5}{8}$ points above low and $1\frac{7}{8}$ points below high
  e. $73\frac{1}{2}$ points above low and 8 points below high
  f. $\frac{3}{8}$ point above low and 21.75 points below high
5. INA Cp, $.65; Idaho P, $.57; Ill Powr, $.57; Inger R, $.79; Inlnd Stl, $.70; IBM, $.86;
Int Harv, $.575; Int Min, $.65; Int Paper, $.55; Int T&T, $.55; Iowa PS, $.60; Itek Cp, $---
6. INA Cp, $8.05; Idaho P, $3.13; Ill Powr, $2.78; Inger R, $7.11; Inlnd Stl, $7.00; IBM, $5.67;
Int Harv, $6.40; Int Min, $5.95; Int Paper, $4.69; Int T&T, $4.65; Iowa PS, $3.50; Itek Cp, $1.57

Stein's
REFRESHER MATHEMATICS

Eighth Edition

Edwin I. Stein

Allyn and Bacon, Inc.

Boston   Rockleigh, N.J.   Atlanta   Dallas   San Jose
London   Sydney   Toronto

Editor: Andrew P. Mastronardi
Senior Designer: L. Christopher Valente
Designers: Janet Hobbs, Dorothy Spence
Buyers: Martha Ballentine, Roger Powers

Technical art: Anco/Boston; Parts 1, 2, 3

Graphics Etcetera; Parts 4, 5, 6, Answer Section

Illustrations: John Ham; 9, 14, 18, 127, 206, 250, 278, 288, 314, 336, 394, 420, 446, 474, 527, 557, 604

Charles Shaw; 2, 74, 195, 239, 285, 303, 323, 363, 408, 464, 486, 525, 542, 584

Photo credits: Allyn & Bacon Library; 1, 541
Taurus Photo/Mike Wannemacher; 283
Picture Cube/Margaret Thompson; 331
Andrew Brilliant; 332
Picture Cube/Frank Siteman; 441
Nasa; 442
United States Government; 525

Cover design: John Martucci

ISBN   0-205-07771-4

Library of Congress Catalog Card Number   81–71790

Printed in the United States of America

3   4   5   6   7   8   9   90   89   88   87   86   85   84

Preface

REFRESHER MATHEMATICS is the ideal textbook for any class where initial or remedial instruction in the basic skills of mathematics is required.

This edition of REFRESHER MATHEMATICS retains all the features of the preceding edition including its specially designed individualized program in the basic skills of arithmetic and its comprehensive treatment of the metric system, informal geometry, introductory algebra, statistics, probability, and basic consumer applications of earning, spending, borrowing, and investing money.

Several new features have been added. *Competency Check Tests* are found in each section of the book, giving students the experience of taking multiple choice tests. Most importantly, in view of the developing everyday use of shortened names to represent large whole numbers, an extensive treatment of this subject is included. *Check by Calculator* reviews, exercises in reading a road map and railroad, bus, and airline timetables, and illustrations relating careers and applications to mathematics are all new. Problem-solving strategies are provided to help students develop their problem solving techniques.

The arithmetic program in Part 1 is so organized that each pupil finds his or her deficiencies in arithmetic by a simple testing procedure and is directed quickly to specific assignments based on individual needs. Since pupils do not waste time drilling on examples that they can solve, this system of individualized assignments accelerates individual achievement.

The *Basic Skills—Inventory Tests* in arithmetic that open the text are used to diagnose general difficulties and are keyed to the *Exercises*. A keyed *Problem Solving—Inventory Test* is also included to diagnose each student's problem solving weaknesses. For each *Exercise* the text gives (I) the aim, (II) the procedure to be followed, and (III) completely worked-out sample solutions. Following this developmental material come the diagnostic tests, keyed to the immediately succeeding practice examples.

The examples in the diagnostic tests and the related practice examples are closely calibrated in difficulty. Step by step, because one example is not perceptibly more difficult than the preceding one, the pupil is led to think clearly, to develop a full understanding of mathematical concepts, and to experience the pride of success. The final examples in each exercise provide an automatic end test. The final keyed *Achievement Tests* serve as mastery tests to close this section.

The inventory and achievement tests now contain more questions on the reading and writing of decimal numerals and include the new shortened name concept for the development of greater mathematical literacy. A new unit on *Shortened Names for Large Whole Numbers* has been added. Operations with shortened names are also included in the text.

Both metric and customary units of measure are studied in Part 2.

The section on informal geometry contains geometric measurement, constructions, relationships, and facts using both metric and customary units.

The introductory algebra section includes positive and negative numbers, language of algebra, solutions of equations and inequalities, and graphing on the number line.

The section dealing with statistics includes bar, line, and circle graphs, concepts of frequency distribution, measures of central tendency, percentiles and quartiles. The study of probability is also included.

Each of the above areas has a keyed *Inventory Test* so that a student may locate specific deficiencies and may be directed to individualized help. This makes the book adaptable to the specific needs of each individual pupil. The practice material is varied and usually graded by difficulty.

While most exercises contain practical applications, the last section of the text is a complete unit on everyday consumer problems. Here the techniques of problem solving may be used. The *Competency Check Test* at the end of this section is an excellent review of the basic essentials of the mathematics required in everyday living.

Maintenance tests, called *Refresh Your Skills,* and *Reviews* are provided throughout the book.

The carefully planned, flexible organization of REFRESHER MATHEMATICS allows for individual differences and at the same time provides maximum and minimum materials for modern group methods of instruction. Practice materials for class, home, and optional assignments are included. REFRESHER MATHEMATICS is a complete program in basal and remedial mathematics, all in a single text.

The author acknowledges the assistance of his wife, Elaine, and of Charlotte Jaffe and Marilyn Lieberman.

Edwin I. Stein

Contents

PART 1 BASIC SKILLS IN ARITHMETIC AND PROBLEM SOLVING 1

UNIT One WHOLE NUMBERS   18

UNIT Two COMMON FRACTIONS   74

PART 2 BASIC UNITS OF MEASURE 283

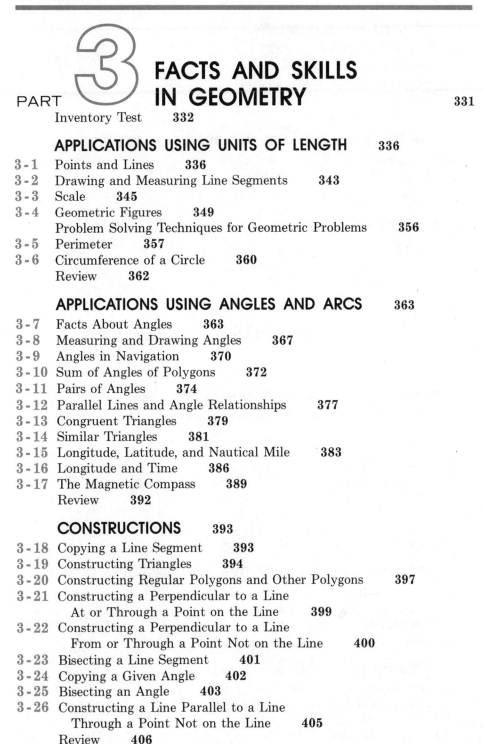

3 FACTS AND SKILLS
IN GEOMETRY

PART   **331**

PART 5 GRAPHS, STATISTICS, AND PROBABILITY   525

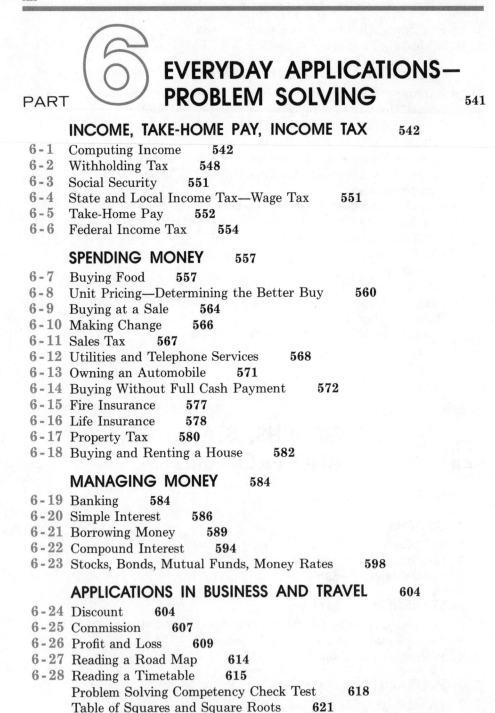

PART **6** **EVERYDAY APPLICATIONS— PROBLEM SOLVING** 541

Basic Skills in Arithmetic and Problem Solving

INTRODUCTION

COUNTING

The primitive people who lived many, many years ago did not know how to count. They did not write number symbols or speak number names.

These primitive people used to match animals and objects one-by-one with pebbles or with their fingers. Although primitive shepherds could not name or indicate by symbols how many sheep they had, they could tell by the corresponding tally of pebbles or fingers whether any sheep were missing upon their return from pasture.

It is interesting to note that the word *calculate* is derived from the Latin word *calculus,* meaning pebble, and the word *digit* from *digitus,* meaning finger.

Later number names were developed to represent numbers. Sometimes names of things containing the same quantity were used. The word for two in Tibet is the same as the word meaning wings and in China it is the same as the word meaning ears.

When people began to trade, a system of counting was developed. Counting devices were used. One was a wax- or sand-covered board on which symbols were marked with a stylus. The Romans used a table marked with lines. To count they placed counters between these lines. The Chinese and Japanese still use a counting device, called an abacus, consisting of a frame with movable beads on wires.

In counting there is exact pairing or matching or one-to-one correspondence between objects that are being counted and the counting numbers. See page 18.

NUMBER SYMBOLS

After people learned to count, they developed number symbols so that they could make permanent records. Clay tablets have been found containing "cuneiform" number symbols which were written more than 5,000 years ago by the Babylonians. The British Museum has in its possession an arithmetic manuscript on a roll of papyrus, said to have been written about 4,000 years ago by Ahmes, an Egyptian.

Number symbols were expressed in many ways.

The Babylonians used repeated wedge-shaped symbols (▼◀).

The Egyptians used different grouping symbols: | (1), ∩ (10),

⊃ (100), ⚷ (1000), ((10,000), ⌇ (100,000), and ⚱ (1,000,000).

The Mayans used combinations of dots and horizontal bars (⠇≡).

The Romans used the symbols I, V, X, L, C, D, and M.

The Greeks and Hebrews used the letters of their respective alphabets.

The Chinese used their traditional number symbols: – (1), = (2),

三 (3), 四 (4), 五 (5), 六 (6), 七 (7), 八 (8), 九 (9), 十 (10),

百 (100), and 千 (1000).

We use the number symbols of the Hindu-Arabic System. These symbols were introduced in Europe during the 12th century and are now used generally throughout the world. They were brought to the United States by the colonists. In this system there are ten symbols: 0, 1, 2, 3, 4, 5, 6, 7, 8, and 9. Each of these number symbols is called a digit or figure.

The number symbols we write to represent numbers are not numbers but names for numbers which we call numerals. Number is an abstract idea. We cannot see or write a number. Although it is usually called a number, technically 25 is a numeral or group of number symbols which represent the number named *twenty-five*. A number may be represented by symbols in many ways; it has many names. The number named *twenty-five* may be represented by 25 or XXV or 13 + 12 or 5^2, etc. It should be noted however, that number symbols are generally used to denote both numbers and numerals, as "Add the numbers 35 and 54" and "Write the numeral 796."

WORD NAMES FOR NUMBERS

Most civilized people throughout the world use the same number symbols 0, 1, 2, 3, 4, 5, 6, 7, 8, and 9, but they differ in writing and speaking word names for numbers because of their respective languages.

Here are the word names for whole numbers from one to ten inclusive in several different languages:

	English	French	Spanish	Italian	German
1	one	un	uno	uno	eins
2	two	deux	dos	due	zwei
3	three	trois	tres	tre	drei
4	four	quatre	cuatro	quattro	vier
5	five	cinq	cinco	cinque	fünf
6	six	six	seis	sei	sechs
7	seven	sept	siete	sette	sieben
8	eight	huit	ocho	otto	acht
9	nine	neuf	nueve	nove	neun
10	ten	dix	diez	dieci	zehn

DECIMAL SYSTEM OF NUMERATION

A system of numeration is a method of naming numbers by writing numerals.

In our system of numeration, called the decimal system, ten number symbols, 0, 1, 2, 3, 4, 5, 6, 7, 8, and 9, are used to represent all numbers. There is no single number symbol in our notation system for the number ten or for numbers greater than ten. The numerals representing numbers greater than nine are formed by writing two or more number symbols next to each other in different positions or places.

Our system of numeration is built on the base ten. The base of a system of numeration is the number it takes in any one place to make 1 in the next higher place. In the decimal system it takes ten in any one place to make 1 in the next higher place. It takes 10 ones to make 1 ten, 10 tens to make 1 hundred, 10 hundreds to make 1 thousand, and so forth.

The decimal system is a positional system. Instead of a special symbol to represent each power of ten, it uses place value. The value of each place in the decimal number scale is ten times the value of the next place to the right. (See page 18.) Thus in a numeral the value of each symbol depends not only on what the symbol is but also on its position in the numeral. The symbol 4 in the decimal numeral 47 means 4 tens but in a decimal numeral 427 it means 4 hundreds.

A numeral may be written as a sum of terms called a polynomial. Each digit is expressed as a product of the digit and a power of ten. (See Exercise 1-50.)

OPERATIONS

In arithmetic we use the operations of addition, subtraction, multiplication, and division. In each case we are operating with two numbers to get a third number. This is called a binary operation.

Operations that undo each other are called inverse operations. Addition and subtraction are inverse operations. Also multiplication and division are inverse operations because they undo each other. Addition and multiplication are considered to be principal operations. Subtraction is the inverse operation of addition and division is the inverse operation of multiplication.

PROPERTIES

Operations have certain characteristics or properties.

1. Commutative Property of Addition

Adding 5 and 6 gives the same sum as adding 6 and 5.

$$\text{That is, } 5 + 6 = 6 + 5$$

The order we use in adding two numbers does not affect the sum. The commutative property of addition states that when we add one number to a second number we get the same sum as when we add the second number to the first.

2. Associative Property of Addition

Only two addends may be added at any one time. When there are three addends, we must first select two addends, find their sum, and add the third addend to this sum.

$$5 + 6 + 4 \text{ may be thought of as either:}$$

$$(5 + 6) + 4 \quad \text{or} \quad 5 + (6 + 4)$$

The associative property of addition permits us to group or associate the first and second numbers and add their sum to the third number or to group or associate the second and third numbers and add their sum to the first number. Either way we get the same final sum.

3. Commutative Property of Multiplication

Multiplying 5 by 6 is the same as multiplying 6 by 5.

$$\text{That is, } 5 \times 6 = 6 \times 5$$

The order we use in multiplying two numbers does not affect the product. The commutative property of multiplication states that when we multiply one number by a second number we get the same product as when we multiply the second number by the first number.

4. Associative Property of Multiplication

Only two factors may be multiplied at any one time. When there are three factors, we must first select two factors, find their product, and then multiply this product by the third factor.

$$5 \times 6 \times 4 \text{ may be thought of as either:}$$

$$(5 \times 6) \times 4 \quad \text{or} \quad 5 \times (6 \times 4)$$

The associative property of multiplication permits us to group or associate the first and second numbers and multiply their product by the third number or to group or associate the second and third numbers and multiply their product by the first number. We get the same final product in each case.

5. Distributive Property of Multiplication over Addition

To find the product of $5 \times (3 + 7)$ we may find the sum of 3 and 7 and then multiply this sum (10) by 5.

$$5 \times (3 + 7) = 5 \times 10 = 50$$

Or we may multiply the 3 by 5 and the 7 by 5 and add the products.

$$5 \times (3 + 7) = (5 \times 3) + (5 \times 7) = 15 + 35 = 50$$

Either way we get the same final result.

The distributive property of multiplication over addition states that when we multiply one number by the sum of a second and a third number we get the same result as when we add the product of the first and second numbers to the product of the first and third numbers. Multiplication is being distributed over addition.

We use the distributive property in computation such as 2×23.

$$2 \times 23 = 2 \times (20 + 3) = (2 \times 20) + (2 \times 3) = 40 + 6 = 46$$

or
$$\frac{23}{\times 2} = \frac{20 + 3}{\times 2} = \frac{20}{\times 2} + \frac{3}{\times 2}$$

We also use the distributive property in the following:

$$(9 \times 6) + (9 \times 4) = 9 \times (6 + 4)$$

or
$$(6 \times 9) + (4 \times 9) = (6 + 4) \times 9$$

6. Closure

When we add the whole numbers 5 and 4, the sum is the whole number 9. When we multiply the whole numbers 6 and 7, the product is the whole number 42. However, when we subtract the whole number 8 from the whole number 3, the answer is the integer -5. See Exercise 4-10. When we divide the whole number 6 by the whole number 4, the answer is the mixed number $1\frac{1}{2}$.

If we, using all numbers in a given group, add any two numbers (or subtract or multiply or divide) and get as our answer in every case one of the numbers described in the given group, we say the group is *closed* under that operation. This property is called closure.

Thus we see above that the whole numbers are closed under addition and multiplication but not under subtraction and division.

7. Properties of Zero

a. When we add zero (0) to any number, the number remains unchanged. This addition property of zero is illustrated by the following:

The sum of 5 and 0 is 5. $5 + 0 = 5$

The sum of 0 and 8 is 8. $0 + 8 = 8$

A number which, when added to a given number, does not change the given number is called the additive identity (or identity element for addition). Therefore zero (0) is the additive identity.

b. Zero subtracted from any number is the number. $6 - 0 = 6$

c. The difference between any number and itself is zero. $9 - 9 = 0$

d. The product of any number and zero is zero. $3 \times 0 = 0$; $0 \times 8 = 0$; $0 \times 0 = 0$

e. If the product of two numbers is zero, then one of the factors is zero or both factors are zero.

f. If zero is divided by any number other than zero, the quotient is zero. $0 \div 2 = 0$

g. In arithmetic division by zero is excluded.

8. Properties of One

a. When we multiply any number by one (1), the number remains unchanged. This multiplicative property of one is illustrated by the following:

$$1 \text{ times } 7 \text{ is } 7. \qquad 1 \times 7 = 7$$

$$5 \text{ times } 1 \text{ is } 5. \qquad 5 \times 1 = 5$$

A number which, when multiplied by a given number, does not change the given number is called the multiplicative identity (or the identity element for multiplication). Therefore one (1) is the multiplicative identity.

b. When any number, except zero, is divided by itself, the quotient is one. Thus one (1) may be expressed in symbols as any of the following:

$$\tfrac{1}{1}, \tfrac{2}{2}, \tfrac{3}{3}, \tfrac{4}{4}, \tfrac{5}{5}, \tfrac{6}{6}, \text{ etc.}$$

c. One (1) raised to any power is one. For example:

$$1^5 = 1 \times 1 \times 1 \times 1 \times 1 = 1$$

d. The multiplicative identity one (1) is used to change the form of fractions to lower terms (Exercise 1-8), and to higher terms (Exercise 1-10); to divide fractions and mixed numbers (Exercise 1-17), and to divide by a decimal (Exercise 1-27).

e. When one (1) is added to any whole number, the sum is the next higher whole number.

f. When one (1) is added to any even number, the sum is an odd number. When one (1) is added to any odd number, the sum is an even number.

How well do you know your **arithmetic?**

Try the examples in each of the *Basic Skills—Inventory Tests* on the following pages.

If your difficulty is with the same kind of example in all tests, you have located a **weakness.**

If you have difficulty with any examples in these tests, you need help.

To get this help, see page 14.

1-1. Read the numeral 95,237 or write it in words.

1-2. Write the numeral naming: Eight hundred seven thousand, forty-eight 807,048

1-3. Round 6,849 to the nearest hundred. 6,800

1-4. Add: 578
     946
     825
   2,349

1-5. Subtract: 9,856
       3,682
      6,174

1-6. Multiply: 59×83 4,897   1-7. Divide: $34\overline{)918}$ 27

1-8. Express $\frac{10}{25}$ in lowest terms. $\frac{2}{5}$

1-9. Change $\frac{11}{8}$ to a mixed number. $1\frac{3}{8}$

1-10. Raise to higher terms: $\frac{3}{5} = \frac{?}{20}$ $\frac{12}{20}$

1-11. Find the lowest common denominator of the fractions: $\frac{1}{2}$ and $\frac{2}{3}$ 6

1-12. Add: $4\frac{2}{5}$
    $+ 7\frac{3}{10}$ $11\frac{7}{10}$

1-13. Subtract: $5\frac{3}{4}$
    $- 2\frac{1}{3}$ $3\frac{5}{12}$

1-14. Which is greater: $\frac{5}{12}$ or $\frac{5}{8}$? $\frac{5}{8}$

1-15. Change $4\frac{1}{2}$ to an improper fraction. $\frac{9}{2}$

1-16. Multiply: $\frac{2}{3} \times \frac{4}{5}$ $\frac{8}{15}$ 1-17. Divide: $2\frac{1}{4} \div \frac{1}{2}$ $4\frac{1}{2}$

1-18. 8 is what part of 20? $\frac{2}{5}$ 1-19. $\frac{1}{4}$ of what number is 36? 144

1-20. Read the numeral .375 or write it in words. three eighths

1-21. Write as a decimal numeral: Ninety-six hundredths .96

1-22. Find correct to the nearest tenth: 6.79 6.8

1-23. Add: $.39 + 2.5$ 2.89   1-24. Subtract: $9 - 4.6$ 4.4

1-25. Which is smaller: .7 or .65? .65

1-26. Multiply: $5.4 \times .13$ .702   1-27. Divide: $.2\overline{)6.8}$ 34

1-28. Multiply by short method: 100×38 3,800

1-29. Divide by short method: $4.9 \div 10$ .49

1-30. Change $\frac{3}{4}$ to a decimal. .75   1-31. Change .8 to a common fraction. $\frac{4}{5}$

1-32. What decimal part of 25 is 16? .64   1-33. .3 of what number is 12? 40

1-34. Write 2.7 *million* completely as a numeral. 2,700,000

1-35. Write the shortened name for 16,300,000 in millions. 16.3 million

1-36. Write *seventeen-hundredths* as a percent, as a decimal, and as a common fraction. 17%, .17, $\frac{17}{100}$

1-37. Express 9% as a decimal. .09   1-38. Express .12 as a percent. 12%

1-39. Express 40% as a common fraction. $\frac{4}{10}$

1-40. Express $\frac{9}{10}$ as a percent. 90%   1-41. Find 25% of 93. 23.25

1-42. What percent of 60 is 45? 75%   1-43. 80% of what number is 72? 90

1-44. Square 16. 256   1-45. Find the square root of 3,844. 62

1-1. Ninety-five thousand, two hundred thirty-seven

1-1. Read the numeral 5,374,089 or write it in words.

1-2. Write the complete numeral naming: 638 *million* **638,000,000**

1-3. Round 4,826,678 to the nearest thousand. **4,827,000**

1-4. Add: 279   1-5. Subtract: 1-6. Multiply: 5,280 × 97 **512,160**
  3,482
  68,926   427,534   1-7. Divide: 13,968 ÷ 144 **97**
  8,597   372,895
  19,308   ‾‾‾‾‾‾‾   1-8. Express $\frac{60}{96}$ in lowest terms. $\frac{5}{8}$
  ‾‾‾‾‾‾   54,639
  100,592   1-9. Change $\frac{25}{12}$ to a mixed number. $2\frac{1}{12}$

                         1-10. Change $\frac{3}{4}$ to 24ths. $\frac{18}{24}$

1-11. Find the lowest common denominator of the fractions: $\frac{1}{4}, \frac{2}{3}, \frac{5}{6}$ **12**

1-12. Add: $6\frac{1}{2}$ 1-13. Subtract: 1-14. Which is greater: $\frac{3}{5}$ or $\frac{5}{8}$? $\frac{5}{8}$
  $5\frac{3}{8}$ $7\frac{1}{4}$
  $4\frac{7}{16}$ $16\frac{5}{16}$ $6\frac{7}{8}$ $\frac{3}{8}$

1-15. Change $3\frac{1}{3}$ to an improper fraction. $\frac{10}{3}$

1-16. Multiply: $1\frac{2}{3} \times 2\frac{1}{4}$ $3\frac{3}{4}$ 1-17. Divide: $6\frac{3}{4} \div 1\frac{5}{16}$ $5\frac{1}{7}$

1-18. 60 is what part of 72? $\frac{5}{6}$ 1-19. $\frac{2}{3}$ of what number is 18? **27**

1-20. Read the numeral 9.46 *billion* or write it in words.

1-21. Write as a decimal numeral: Five hundred four thousandths **.504**

1-22. Find correct to the nearest hundredth: 6.8345 **6.83**

1-23. Add: .36   1-24. Subtract: 26 − .14 **25.86**
  5.2
  4.   1-25. Which is greater: .03 or .029? **.03**
  1.02
  ‾‾‾‾   1-26. Multiply: .06 × 1.35 **.081**
  10.58
                         1-27. Divide: .36)‾18 **50**

1-28. Multiply by short method: 1,000 × 34.27 **34,270**

1-29. Divide by short method: 6.89 ÷ 100 **.0689**

1-30. Change $\frac{3}{20}$ to a decimal. **.15**

1-31. Change 0.25 to a common fraction. $\frac{1}{4}$

.30 1-32. What decimal part of 60 is 18?   1-33. 0.7 of what number is 21? **30**

1-34. Write 24.9 *billion* completely as a numeral. **24,900,000,000**

1-35. Write the shortened name for 6,400,000,000,000 in trillions. **6.4 trillion**

1-36. Write *forty-one hundredths* as a percent, as a decimal, and as a common fraction. **41%, .41,** $\frac{41}{100}$

.64 1-37. Express 64% as a decimal.   1-38. Express 0.05 as a percent. **5%**

1-39. Express $83\frac{1}{3}$% as a common fraction. $\frac{5}{6}$

1-40. Express $\frac{17}{20}$ as a percent. **85%** 1-41. Find 29% of 67. **19.43**

45% 1-42. What percent of 80 is 36?   1-43. 15% of what number is 3? **20**

1-44. Square 62. **3,844**   1-45. Find the square root of 24,336. **156**

1-1. Five million, three hundred seventy-four thousand, eighty-nine
1-20. Nine billion, four hundred sixty million

1-1. Read the numeral 163,000,987 or write it in words.

1-2. Write the numeral naming: Seventeen billion, four hundred seven million 17,407,000,000

1-3. Round 18,268,395 to the nearest hundred thousand. 18,300,000

1-4. Find the sum of 4,132; 375; 81,580; 69; and 5,364 91,520

1-5. Subtract 259 from 3,120. 2,861

4,474,880 **1-6.** Multiply: 736 × 6,080   **1-7.** Divide: $693\overline{)30,185}$ 43.56

1-8. Express $\frac{52}{64}$ in lowest terms. $\frac{13}{16}$ **1-9.** Change $\frac{8}{3}$ to a mixed number. $2\frac{2}{3}$

1-10. Raise to higher terms: $\frac{7}{8} = \frac{?}{64}$ 56

1-11. Find the lowest common denominator of the fractions: $\frac{1}{6}$ and $\frac{3}{8}$ 24

1-12. Add: $2\frac{3}{4} + 6\frac{5}{8}$ $9\frac{3}{8}$ **1-13.** Subtract: 8

1-14. Which is smaller: $\frac{3}{4}$ or $\frac{7}{10}$? $\frac{7}{10}$ $3\frac{2}{5}$

1-15. Change $1\frac{5}{6}$ to an improper fraction. $\frac{11}{6}$ $4\frac{3}{5}$

1-16. Find $\frac{7}{12}$ of 84. 49   **1-17.** Divide: $2\frac{9}{16} \div 2$ $1\frac{9}{32}$

1-18. What part of 50 is 35? $\frac{7}{10}$ **1-19.** $\frac{3}{8}$ of what number is 27? 72

1-20. Read the numeral .0839 or write it in words. **Point zero eight three nine**

1-21. Write as a decimal numeral: Two hundred and seventy-two thousandths 200.072

1-22. Find correct to the nearest cent: $8.3674 $8.37

1-23. Add: 6.8 + .24 + 16 23.04   **1-24.** Subtract: 18 − 1.5 16.5

1-25. Which is smaller: 1.05 or .247? .247 **1-26.** Find .75 of 2.60 1.95

1-27. Divide: $3.6\overline{)51.12}$ 14.2

1-28. Multiply by short method: 100 × .09 9

1-29. Divide by short method: 576.5 ÷ 1,000 .5765

.875 **1-30.** Change $\frac{7}{8}$ to a decimal.   **1-31.** Change 4.75 to a mixed number. $4\frac{3}{4}$

1-32. 54 is what decimal part of 72? .75

1-33. 9 is 0.5 of what number? 18

1-34. Write 9.14 *million* completely as a numeral. 9,140,000

1-35. Write the shortened name for 75,280,000,000 in billions. 75.28 billion

1-36. Write *seven hundredths* as a percent, as a decimal, and as a common fraction. 7%, .07, $\frac{7}{100}$

1-37. Express 30% as a decimal. .3 or .30 **1-38.** Express $.08\frac{1}{2}$ as a percent. $8\frac{1}{2}$%

1-39. Express $12\frac{1}{2}$% as a common fraction. $\frac{1}{8}$

1-40. Express $\frac{27}{36}$ as a percent. 75% **1-41.** Find 8% of $8.27 (nearest cent). $.66

133% **1-42.** 16 is what percent of 12? **1-43.** 42 is 7% of what number? 600

1-44. Square .08 .0064   **1-45.** Find the square root of .7396 .86

1-1. One hundred sixty-three million, nine hundred eighty-seven

1-1. Read the numeral 5,493,268,003,751 or write it in words.

1-2. Write the numeral naming: Forty-three million, nine hundred five thousand, twenty-six **43,905,026**

1-3. Round 839,599,696 to the nearest million. **840,000,000**

1-4. Add: 583,289
698,785
835,869
549,397
897,298 **3,564,638**

1-5. Subtract:
3,210,050
1,839,096
——————
1,370,954

1-6. Multiply: 6,905
9,087
62,745,735

1-7. Divide:
5,022,702 ÷ 5,937 **846**

1-8. Express $\frac{48}{108}$ in lowest terms. $\frac{4}{9}$ **1-9.** Change $\frac{29}{9}$ to a mixed number. $3\frac{2}{9}$

1-10. Raise to higher terms: $\frac{5}{6} = \frac{?}{72}$ **$\frac{60}{72}$**

1-11. Find the lowest common denominator of the fractions: $\frac{1}{10}, \frac{1}{4},$ and $\frac{1}{6}$ **60**

1-12. Add: $1\frac{2}{3} + 5\frac{7}{12} + 3\frac{1}{6}$ **$10\frac{5}{12}$** **1-13.** Subtract: $14\frac{1}{3} - 9\frac{11}{16}$ **$4\frac{31}{48}$**

1-14. Is the statement $\frac{7}{9} < \frac{10}{13}$ true? **No**

1-15. Change $6\frac{5}{7}$ to an improper fraction. **$\frac{47}{7}$**

1-16. Multiply: $2\frac{3}{4} \times 8\frac{4}{5}$ **$24\frac{1}{5}$** **1-17.** Divide: $(6 \div 8\frac{1}{4}) \div 1\frac{3}{5}$ **$\frac{5}{11}$**

1-18. What part of 192 is 54? **$\frac{9}{32}$** **1-19.** 220 is $\frac{11}{16}$ of what number? **320**

1-20. Read the numeral 42.6 *million* or write it in words. **(see above)**

1-21. Write as a decimal numeral: Sixty and nine ten-thousandths **60.0009**

1-22. Find correct to nearest cent: $68.5349 **$68.53**

1-23. Add: 3.24 + 53.7 + .938 **57.878** **1-24.** Subtract: 4.2 − .351 **3.849**

1-25. Is the statement 7.5 > .750 true? **Yes**

1-26. Multiply: .013 × .07 **.00091** **1-27.** Divide: .05)‾6‾ **120**

1-28. Multiply by short method: 100 × 90.235 **9,023.5**

1-29. Divide by short method: 6,340.7 ÷ 10,000 **.63407**

1-30. Change $\frac{21}{32}$ to a decimal (2 places). **.66**

1-31. Change $.16\frac{2}{3}$ to a common fraction. **$\frac{1}{6}$**

.875 1-32. 91 is what decimal part of 104? **1-33.** 120 is .625 of what number? **192**

1-34. Write 10.25 trillion completely as a numeral. **10,250,000,000,000**

1-35. Write the shortened name for 1,839,000,000,000 in trillions. **1.839 trillion**

1-36. Write *one-fourth hundredth* as a percent, as a decimal, and as a common fraction. **$\frac{1}{4}$%, .0025, $\frac{1}{100}$ or $\frac{1}{400}$**

1-37. Express 103% as a decimal. **1.03** **1-38.** Express .0875 as a percent. **8.75%**

1-39. Express $233\frac{1}{3}$% as a mixed number. **$2\frac{1}{3}$**

1-40. Express $\frac{27}{75}$ as a percent. **36%** **1-41.** Find $100\frac{1}{2}$% of $9,000. **$9,045**

$116\frac{2}{3}$% 1-42. $.56 is what percent of $.48? **1-43.** $83\frac{1}{3}$% of what number is 30? **36**

1-44. Square .09 **.0081** **1-45.** Find the square root of .000225 **.015**

Do you need help in arithmetic?

Have **Inventory Tests** I, II, III and IV shown you the kind of examples on which you need help?

For help turn to the EXERCISES that have the same numerals as the examples you could not do.

For example:

Did you have trouble with example 1-7 (long division) in every test?

If you did, turn to EXERCISE 1-7, page 58.

Maybe you did example 1-16 incorrectly.

Then turn to EXERCISE 1-16, page 102.

In each EXERCISE, study the instructions and the sample solution. Do the examples in the **Diagnostic Test.**

Concentrate on the practice work you need most. The **Practice Examples** in each EXERCISE are keyed to those in the **Diagnostic Test** so that you can quickly find the ones you need.

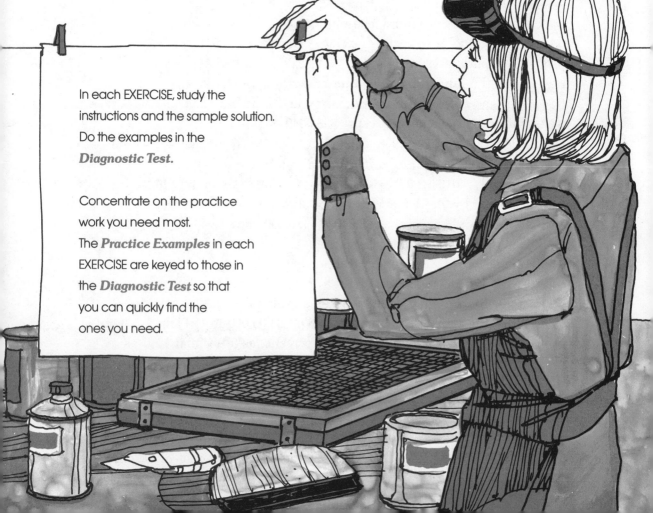

Solve each of the following problems:

1. If an airplane averages 785 kilometers per hour, how long will it take to fly a distance of 2,355 kilometers? (68) 3 hours

2. Find the total enrollment when the township school system has 5,827 high school students, 6,988 middle school students, and 9,379 elementary school students. (38) 22,194

3. With the EPG rating of 34 miles per gallon (mpg), how far can you expect to drive the car if its tank holds 19 gallons of gasoline? (56) 646 miles

4. The population of Houston increased in 10 years from 1,233,535 to 1,594,086. Find the amount of increase. (46) 360,551

5. Rainfall in April was $1\frac{3}{4}$ inches, in May $2\frac{1}{2}$ inches, and in June $\frac{3}{4}$ inch. What was the total rainfall for the three months? (90) 5 inches

6. The radius of a circle is one half the diameter. How long is the radius of a circle whose diameter is $5\frac{3}{8}$ inches? (107) $2\frac{11}{16}$ inches

7. Jerry's mother bought *XYZ* stock at the price of $30\frac{5}{8}$ per share. How much did the price per share fall when it sold for $27\frac{3}{4}$? (96) $2\frac{7}{8}$

8. Find the length of each piece (disregarding waste) when a board, $15\frac{3}{4}$ feet long, is cut into 9 pieces of equal length. (114) $1\frac{3}{4}$ feet

9. The new penny has about 2 parts of copper and about 98 parts zinc. What part of the penny is copper? (119) $\frac{2}{100}$ or $\frac{1}{50}$

10. Rita failed to do correctly 8 test problems or $\frac{2}{5}$ of the test. How many problems were on the test? (123) 20

11. The Dow Jones stock averages opened at 905.06 and during the day lost 7.14 points. What was the closing stock average? (151) 897.93

12. If your bank balance is $984.09 and you deposit $25.95, what is your new bank balance? (144) $1,010.04

13. If the rate is $.53 for the first minute and $.36 for each additional minute, find the charges for a telephone call 9 minutes long? (160) $3.41

14. Tony bought a small computer for $989.99, paying $119.99 down and $72.50 per month. How many monthly payments must he make? (170) 12

15. Susan spelled 39 words correctly and 11 words incorrectly. What percent of the words did she spell incorrectly? (226) 22%

16. A tennis racket which regularly sells for $49 is reduced 25%. What is its sales price? (221) $36.75

17. How much must you invest at 12% annual interest to earn $6,000 per year? (231) $50,000

First study pages 16 and 17. Then turn to the pages indicated by the red numerals for more practice. If you are having difficulty in performing the operations to solve problems, go to the development section of the respective Exercises for assistance.

For an alternate Problem Solving-Inventory Test see p. T41 or *Testing Program*, p. 2.

Problem Solving Strategy

Use the following procedure in solving problems:

1. *Read* the problem carefully to find:
  a. The facts that are given.
  b. What is being asked.

2. *Decide* what to do to solve the problem.
  a. Which given facts should be used with which operations?
  b. If more than one step is needed, in what order should the operations be used?
  c. Select only the facts you need. Sometimes more facts than are needed are given. Other times not enough facts are given.
  d. Write or think of a mathematical sentence (or equation) that relates the given facts to what is asked.

3. *Solve* the problem according to your plan. Perform the operation or operations accurately:
  a. First with rounded numbers to estimate the answer.
  b. Then with the actual numbers to find your answer.
  c. Use a diagram or drawing when necessary.

4. *Answer*
  a. Write your answer.
  b. Check your answer.
     Compare it with your estimated answer for reasonableness.

Problem: A merchant purchased 48 radios at $79 each. What was the total cost?

1. *Read and find:*
Given facts   48 radios were purchased; each radio costs $79.
Question asked   What was the total cost?

2. *Decide on plan:*   Multiply the cost of a radio ($79) by the number of radios (48) to find total cost.
                     Think: $48 \times \$79 = $ total cost

3. *Solve:*

Estimate: 50 × $80 = $4,000

Actual: $79 cost per radio

  48 radios
  ___

  632

  316
  ___

  $3,792 total cost

4. *Answer:*

Write answer: Answer, $3,792 total cost

Check answer: The answer $3,792 compares reasonably with the $4,000 estimate.

Problem: Mrs. Gomez can buy an electric sweeper for the cash price of $149.95 or pay $15 down and $16.09 a month for 10 months. How much can she save by paying cash?

1. *Read and find:*

Given facts Cash price is $149.95.

Installment price is $15 down payment and $16.09 per month for 10 months.

Question asked How much can be saved by paying cash?

2. *Decide on plan:*

a. Multiply monthly payment ($16.09) by number of payments (10) and add $15 to the product to find the installment price.

b. Then subtract the cash price ($149.95) from the installment price to find the amount that could be saved.

Think: (10 × $16.09 + $15) − ($149.95) = savings

3. *Solve:*

Estimate: (10 × $16 + $15) − ($150)

  $160 + $15 − $150 = $25

Actual: 10 × $16.09 = $160.90

  $160.90 + $15 = $175.90 installment price

  $175.90 installment price

  − $149.95 cash price
  _____

  $25.95 savings

4. *Answer:*

Write answer: Answer, $25.95 can be saved.

Check answer: The answer $25.95 compares reasonably with the $25 estimate.

UNIT **One**

WHOLE NUMBERS

The numbers 0, 1, 2, 3, 4, 5, 6, 7, 8, 9, 10, 11, 12, 13, 14, 15, etc. are called whole numbers. The whole numbers beginning with 1 that are used in counting are called counting or natural numbers. Whole numbers have order. Every whole number is followed by another whole number. Thus there is no last whole number and no greatest whole number.

A number that tells how many things are in a group or the size of a group is called a cardinal number. A number that tells the order as first, second, third, fourth, and so forth is called an ordinal number. The same number symbols are used for both the cardinal and ordinal numbers.

A line may be thought of as an endless set of points. A straight line that has its points labeled with numerals so that each point is associated with a number is called a number line. The numerals are arranged in a definite order on the number line so that they correspond one-to-one with the points on the line.

Each point, equally-spaced, is given a corresponding whole number (*coordinate*) in consecutive order. Observe that the number corresponding to the point on the line farther to the right is the greater number. Sometimes a point is indicated by a capital letter. On the number line above, the coordinate of point A is 3.

18

Reading Numerals
Naming Whole Numbers

VOCABULARY: whole numbers
natural numbers
ordinal number
counting numbers
cardinal number
number line

OBJECTIVE:

I. Aim   To read numerals naming whole numbers.

II. Procedure

1. Starting from the ones place and going to the left, separate by commas the given numeral; when it contains four or more digits, separate into as many groups (or periods) of three digits as possible. (See sample solution 1.)

Sometimes spaces are used instead of commas to separate these groups. (See sample solution 2.)

2. Starting from the left and going to the right, read each period of digits separately, applying the name of the period indicated by each comma or space as it is reached.

The word "and" is not used between the names of the periods.

3. A 4-digit numeral may appear without a comma or a space. (To read it, see sample solution 7.)

4. Shortened names consisting of a numeral and a period name are sometimes used to indicate large whole numbers whose numerals end in a series of zeros. (See sample solution 8.) The $ symbol indicates *dollars*.

From the right,

the first comma or space is read *"thousand;"*
the second comma or space is read *"million;"*
the third comma or space is read *"billion;"*
the fourth comma or space is read *"trillion;"*
the fifth comma or space is read *"quadrillion."*

	Trillions			Billions			Millions			Thousands			Ones		
quadrillions	hundred trillions	ten trillions	trillions	hundred billions	ten billions	billions	hundred millions	ten millions	millions	hundred thousands	ten thousands	thousands	hundreds	tens	ones
6,	6	6	6,	6	6	6,	6	6	6,	6	6	6,	6	6	6

Related Resources: *Practical Applications in Mathematics:* p. 3, Example 1; Diagnostic Test, p. 3, items 1–15

In the number scale the *ones* (sometimes called *units*) are located on the right. One place to the left of the ones position is the *tens* place; one place to the left of the tens position is the *hundreds* place. These three places form a *group* or *period*.

The scale illustrates that the periods in increasing order are: *ones, thousands, millions, billions, trillions,* and *quadrillions.* The names of other periods in increasing order are *quintillion, sextillion, septillion, octillion, nonillion,* and *decillion.* One decillion is written: 1,000,000,000,000,000,000,000,000,000,000,000

III. Sample Solutions

1. Separate 9258076 into periods, using commas.   *Answer:* 9,258,076
2. Separate 60759241378 into periods, using spaces.

   Answer: 60 759 241 378
3. Read 476,923.   *Answer:* Four hundred seventy-six thousand, nine hundred twenty-three
4. Read 34,008,290.   *Answer:* Thirty-four million, eight thousand, two hundred ninety
5. Read 2,540,250,000.   *Answer:* Two billion, five hundred forty million, two hundred fifty thousand
6. Read 83,700,000,000,000.   *Answer:* Eighty-three trillion, seven hundred billion
7. Read 9100 in two ways.   *Answer:* (a) Ninety-one hundred
   (b) Nine thousand one hundred
8. Read $56 *billion.*   *Answer:* Fifty-six billion dollars

DIAGNOSTIC TEST

Separate the following numerals into proper periods:

1. By using commas: 29740059265040   29,740,059,265,040
2. By using spaces: 817604009000   817 604 009 000

For ex. 3–25
see Extended
Answers, p. T70.
Read, or write in words, each of the following:

3. 68	9. 4,900,000	15. 904,060,381,269,008
4. 582	10. 7,847,000	16. 17500000000000
5. 4,975	11. 3,582,942	17. 740 509
6. 5,003	12. 18,467,125	18. 63 086 493
7. 10,500	13. 995,028,000	19. 218 904 226 007
8. 823,659	14. 62,669,004,716	20. 5 693 169 040 000

Read, or write in words, each of the following:

21. 94 million 22. 578 billion 23. 16 trillion 24. $107 million
25. Read in two ways: 5300

RELATED PRACTICE EXAMPLES

Separate the following numerals into proper periods:

SET 1	SET 2
By using commas:	By using spaces:
1. 793058 793,058	**1.** 61437 61 437
2. 96400127 96,400,127	**2.** 483050095 483 050 095
3. 592473085604 592,473,085,604	**3.** 90225006873 90 225 006 873
4. 23100000000000 23,100,000,000,000	**4.** 216038823700000
5. 8335000000000000	**5.** 8903250000000000000

8,335,000,000,000,000   **4.** 216 038 823 700 000   **5.** 890 325 000 000 000 000

Read, or write in words, each of the following:
For sets 3–16 see Extended Answers, pp. T70–T71.

SET 3	SET 4	SET 5	SET 6
1. 96	**1.** 426	**1.** 8,278	**1.** 6,004
2. 47	**2.** 784	**2.** 3,926	**2.** 4,045
3. 53	**3.** 269	**3.** 2,474	**3.** 9,080
4. 84	**4.** 905	**4.** 4,206	**4.** 7,032
5. 79	**5.** 380	**5.** 7,959	**5.** 3,007

SET 7	SET 8	SET 9	SET 10
1. 27,432	**1.** 150,000	**1.** 3,000,000	**1.** 9,250,000
2. 59,117	**2.** 291,429	**2.** 8,000,000	**2.** 3,746,000
3. 20,730	**3.** 723,805	**3.** 2,500,000	**3.** 4,891,000
4. 45,063	**4.** 900,281	**4.** 7,600,000	**4.** 2,006,000
5. 81,896	**5.** 674,364	**5.** 4,200,000	**5.** 5,670,000

SET 11	SET 12	SET 13
1. 1,596,271	**1.** 32,429,784	**1.** 700,000,000
2. 7,122,843	**2.** 57,105,933	**2.** 241,849,000
3. 8,900,527	**3.** 28,246,219	**3.** 928,376,500
4. 5,416,084	**4.** 96,863,848	**4.** 406,923,685
5. 6,004,009	**5.** 25,000,975	**5.** 843,297,861

SET 14	SET 15	SET 16
1. 5,000,000,000	**1.** 9,000,000,000,000	**1.** 7593000
2. 6,450,000,000	**2.** 23,587,914,060,000	**2.** 82100476
3. 93,884,526,767	**3.** 67,346,528,915,898	**3.** 4763285100
4. 286,008,431,000	**4.** 139,005,407,862,100	**4.** 360005430000
5. 842,699,019,508	**5.** 768,841,253,196,369	**5.** 67800000000000

For sets 17–25 see Extended Answers, pp. T71–T72.

SET 17	SET 18	SET 19
1. 28 900	1. 6 582 000	1. 75 000 000 000
2. 983 225	2. 34 809 500	2. 8 497 500 000
3. 7 003	3. 803 578 632	3. 460 122 849 000
4. 455 607	4. 59 200 048	4. 93 674 317 505
5. 61 459	5. 785 406 917	5. 245 390 006 700

SET 20	SET 21	SET 22
1. 9 000 000 000 000	1. 47 million	1. 4 billion
2. 27 540 000 000 000	2. 9 million	2. 603 billion
3. 141 325 913 006 200	3. 560 million	3. 51 billion
4. 2 607 559 040 172	4. 88 million	4. 2,850 billion
5. 53 482 080 000 000	5. 1,725 million	5. 99 billion

SET 23	SET 24	SET 25
1. 10 trillion	1. $73 billion	Read in two ways:
2. 8 trillion	2. $405 million	1. 6900
3. 200 trillion	3. $14 trillion	2. 4100
4. 87 trillion	4. $59 million	3. 2800
5. 375 trillion	5. $600 billion	4. 8700
		5. 7400

PRACTICAL APPLICATIONS

For ex. 1–5 see Extended Answers, p. T72.

Read, or write in words, each of the numerals that appear in the following:

1. The Pacific Ocean has an area of 165,246,320 square kilometers; Atlantic Ocean, 82,441,560 square kilometers; Indian Ocean, 73,442,755 square kilometers; and Arctic Ocean, 14,090,110 square kilometers.

2. The trading volume on the New York Stock Exchange one day was 49,802,130 shares making the total volume so far for the year 2,921,686,626 shares.

3. Recently Shanghai had a population of 10,820,000; Tokyo, 9,005,000; New York, 7,895,543; Moscow, 7,050,000; and Bombay, 5,948,546.

4. The total number of United States coins minted during the year was 13,377,659,747 valued at $769,466,209 while the total money in circulation was valued at $81,196,000,000.

5. Congress approved a $499 billion budget allowing a $19 billion tax cut.

Writing Numerals
Naming Whole Numbers

OBJECTIVE:

I. Aim To write numerals naming whole numbers.

II. Procedure

1. Write each period of digits and use commas (or spaces when required) to represent the names of the periods. (See sample solutions 1 and 2.)

2. Where the number is named by a numeral prefix and a period name, replace the period name by the appropriate number of zeros:

thousand — 3 zeros (,000)   billion — 9 zeros (,000,000,000)
million — 6 zeros (,000,000) trillion — 12 zeros (,000,000,000,000)
(See sample solutions 3 and 4.)

III. Sample Solutions

1. Write the numeral naming: Seven million, five hundred four thousand, twenty-nine   *Answer: 7,504,029*
2. Write the numeral naming: Twelve billion, nine hundred twenty-six million, eight hundred thousand   *Answer: 12 926 800 000*
3. Write the complete numeral for 125 million. *Answer: 125,000,000*
4. Write the complete numeral for $37 trillion.
          Answer: $37,000,000,000,000

DIAGNOSTIC TEST

Write, using commas, the numeral naming each of the following:

1. Six hundred fifty-two 652
2. Five hundred nineteen thousand, eight hundred forty-seven 519,847
3. Thirty million, four hundred five thousand, eighteen 30,405,018
4. Seven billion, nine hundred thirty million, two hundred thousand 7,930,200,000
5. Ten trillion, twenty-one billion, three hundred million, three hundred fifty 10,021,300,000,350

Write the complete numeral that names each of the following:

6. 38 million   8. 5 trillion   10. 67 hundred 6700
7. 503 billion   9. 608 thousand 608,000
11. Write, using spaces, the numeral: Four hundred sixty million

---

6. 38,000,000 8. 5,000,000,000,000
7. 503,000,000,000 11. 460 000 000

Basic: Sets 1–8, Misc. Ex. 1–2, Appl. 1–2; *Average:* Sets 1–10, Misc. Ex. 1–4, Appl. 1–3; *Enriched:*
Sets 1–11, Misc. Ex. 1–5, Appl. 1–4

24   **WHOLE NUMBERS**

RELATED PRACTICE EXAMPLES

Write, using commas, the numeral that names each of the following:

SET 1

1. Four hundred 400
2. Eight hundred thirty-three 833
3. Seven hundred nine 709
4. Nine hundred seventeen 917
5. Two hundred forty-eight 248

SET 2

1. Four thousand, six hundred ten 4,610
2. Ninety thousand, fifty 90,050
3. Seven hundred eighty-one thousand 781,000
4. Twenty-two thousand, nine 22,009
5. Five hundred sixteen thousand, one hundred seventy-three 516,173

SET 3

1. 3,900,000 2. 85,006,240
3. 608,150,092

1. Three million, nine hundred thousand
2. Eighty-five million, six thousand, two hundred forty
3. Six hundred eight million, one hundred fifty thousand, ninety-two
4. Thirty-five million, seventy-seven thousand, nine hundred sixty-eight
5. Nine hundred fourteen million, seven hundred thousand, fifty 914,700,050

4. 35,077,968

SET 4

1. Forty-eight billion, eighteen million, seven thousand 48,018,007,000
2. Two hundred billion, sixty-two thousand, nine hundred 200,000,062,900
3. Eight hundred seventeen billion, four hundred million, seventy-six thousand 817,400,076,000
4. Seven billion, five hundred four million, seven hundred thousand, fifty-nine 7,504,700,059
5. Nineteen billion, forty million, three hundred thousand, six hundred seventy-five 19,040,300,675

SET 5

1. 4,950,000,000,000 2. 11,800,236,000,000

1. Four trillion, nine hundred fifty billion
2. Eleven trillion, eight hundred billion, two hundred thirty-six million
3. Three hundred trillion, forty billion, five million, eighty-nine thousand, two hundred one 300,040,005,089,201
4. One trillion, six hundred eighty-two billion, four hundred seven million, sixteen thousand 1,682,407,016,000
5. Five hundred trillion, ninety-three billion, four million, nine hundred eighty-eight 500,093,004,000,988

Write the complete numeral for each of the following:

SET 6	SET 7	SET 8
1. 75 million 75,000,000	**1.** 3 billion 3,000,000,000	**1.** 11 trillion
2. 430 million 430,000,000	**2.** $500 billion $500,000,000,000	**2.** 60 trillion
3. $6 million $6,000,000	**3.** 18 billion 18,000,000,000	**3.** 100 trillion
4. $89 million $89,000,000	**4.** $1,490 billion	**4.** $8 trillion
5. 3,908 million	**5.** $67 billion	**5.** $326 trillion

3,908,000,000   4. $1,490,000,000,000   5. $67,000,000,000

1. 11,000,000,000,000
2. 60,000,000,000,000
3. 100,000,000,000,000
4. $8,000,000,000,000
5. $326,000,000,000,000

SET 9	SET 10
675,000 **1.** 675 thousand	500 **1.** 5 hundred
54,000 **2.** 54 thousand	$3600 **2.** $36 hundred
$5,000 **3.** $5 thousand	9,000 **3.** 90 hundred
70,000 **4.** 70 thousand	$12,800 **4.** $128 hundred
$830,000 **5.** $830 thousand	76,900 **5.** 769 hundred

SET 11

Write, using spaces, the numeral that names each of the following:

1. Fourteen thousand, six hundred 14 600
2. Three million, eighty thousand, nine hundred twenty-three 3 080 923
3. Nineteen billion, seventy-five million, six hundred forty thousand
4. Thirteen trillion 13 000 000 000 000　　　　　　　　　19 075 640 000
5. Eight trillion, one hundred billion, fifty million 8 100 050 000 000

MISCELLANEOUS EXAMPLES

1. What new group is formed by 1 more than:
a. 9,999?　　**b.** 999,999?　　**c.** 99,999,999?　　**d.** 9,999,999,999?
ten thousands　　millions　　hundred millions　　ten billions
2. a. Name the greatest possible 12-place whole number. 999,999,999,999
b. Name the smallest possible 7-place whole number. 1,000,000

3. Write the numeral that names (1) the greatest possible number, and (2) the smallest possible number, using the digits:
　　a. 5, 2, 9　　**b.** 6, 3, 0, 7　　**c.** 4, 0, 5, 0, 8, 2　　**d.** 3, 1, 9, 0, 1, 6, 8

4. Arrange the following numbers according to size, writing the numeral for the greatest number first: 759,863 759,836 759,638 759,368
　　　759,368　　　759,638　　　759,863　　　759,836

5. Arrange the following numbers according to size, writing the numeral for the smallest number first:
　　　3,582,745　　　3,578,542　　　3,587,425　　　3,578,524

3.a. 952; 259   b. 7,630; 367   c. 854,200; 2,458   d. 9,863,110; 113,689
5. 3,578,524   3,578,542   3,582,745   3,587,425

PRACTICAL APPLICATIONS

1. From the newspaper—

Write the complete numeral naming each number in the following news items: $2,000,000,000

a. The Treasury stated that it will sell $2 billion in two-year notes.

b. The Congressional committee rejected a rollback in Social Security tax for 106 million workers and their employees. 106,000,000

c. The city banks agreed to invest $500 million in local bonds. $500,000,000

d. The federal deficit for the new fiscal year was projected to be about $53 billion. $53,000,000,000

e. The local utility approved an $85 million increase in its rates.

2. From the stock page— $85,000,000

Observe in the sales column of the stock report that the numerals represent hundreds.

Write the complete numeral naming the number of shares sold of each of the following stocks:

a. Ala Gas 600   **d.** Aetna Lf 109,000   **g.** ACF 8,900   **j.** Airb Frt 2,000

b. APL 3,900   **e.** ASA 43,900   **h.** Abbt Lb   **k.** AMF 15,400

c. Air Prd 23,200 **f.** Acme C 6,200   **i.** Addrsg 76,900 **l.** Ada Ex 6,800

                                       h. 155,000

52 Weeks High	Low	Stock	Div.	Yld %	P-E Ratio	Sales 100s	High	Low	Close	Net Chg.
36⅞	28⅞	ACF	2.10	5.8	9	89	36⅜	36¼	36⅜	+ ⅛
21	15⅝	AMF	1.24	6.4	9	154	19½	19⅛	19½	+ ⅛
15¾	9⅜	APL	1	7.5	9	39	13½	13¼	13¼	− ⅜
45	32⅜	ARA	1.45	3.2	10	173	u45¾	43⅞	45	+1
27½	17⅞	ASA	1	3.8	..	439	26⅝	25½	26¼	+ ½
13¾	7¾	ATO	.48	3.5	8	116	13¾	13⅜	13⅝	+ ⅛
40	29	AbbtLb	.72	1.8	17	1550	39⅜	38½	39	+ ½
21	11	AcmeC	.80	3.9	8	62	20⅝	20¼	20⅝	.....
5½	2¾	AdmDg	.04	.8	7	55	4⅞	4¾	4¾	.....
13	11¼	AdaEx	1.11	9.1	..	68	12¼	12⅛	12¼	.....
7½	3⅞	AdmMl	.20e	3.1	8	27	6¾	6½	6½	− ¼
27¼	12¼	Addrsg	.20	.7	..	769	u27⅞	26⅞	27½	+ ½
44⅛	31	AetnaLf	2.20	5.2	5	1090	42⅛	41	42⅛	+ ⅝
24¾	15⅞	Ahmans	1	4.0	5	131	u24⅞	24½	24⅞	+ ⅛
3⅞	2¼	Aileen	..	..	58	30	3	2⅞	2⅞	− ⅛
31⅞	21⅜	AirPrd	.60	1.9	12	232	31⅝	31⅛	31⅜	.....
25¼	12⅜	AirbFrt	1	4.0	14	20	25⅛	24⅞	25⅛	+ ⅛
16¾	11¼	Akzona	.80	5.9	16	17	13½	13½	13½	.....
18⅜	15¾	AlaGas	1.40	8.4	5	6	17	16¾	16¾	+ ⅛
9¾	8½	AlaP	dpf.87	9.7	..	37	9	9	9	.....

3. From statistical reports—

Write the complete numeral naming each of the numbers in the following reports:

a.

World Production of Crude Petroleum (In thousands of barrels)			Add three zeros to each numeral.
North America	3,788,426	Africa	2,131,314
South America	1,301,421	Asian Area	921,068
Western Europe	311,464	Eastern Europe and China	4,615,621
Middle East	8,049,566		

b.

United States Foreign Trade (In millions of dollars)	Add six zeros to each numeral.	
	Exports	Imports
Western Hemisphere	$41,082	$43,356
Europe	35,903	23,604
Asia	30,541	39,459
Oceania	2,690	1,671
Africa	4,396	12,547

4. From financial reports—

Write the complete numeral naming each of the numbers in the following reports: **Add three zeros to each numeral.**

a.   **Balance Sheet**
(Dollars in thousands)

Assets

Cash	$ 25,360
Accounts Receivable	194,525
Inventories	321,837
Property and Equipment	273,548
Other assets	69,856
Total	$885,126

Liabilities and Stockholders' Equity

Notes Payable	$ 73,418
Accounts Payable	138,734
Other Liabilities	41,583
Long-term Debt	210,895
Stockholders' Equity	420,496
Total	$885,126

b.   **Statement of Income**
(Dollars in thousands)

Sales and Other Income

Sales	$1,239,476
Other Income	56,805
Total	$1,296,281

Costs and Expenses

Operating Expenses	$826,993
Selling Expenses	49,238
Depreciation	65,784
Interest Expense	5,477
Total	947,492
Income before taxes	348,789
Taxes	149,979
Net Income	$198,810

1-3 Rounding Whole Numbers

OBJECTIVE:
I. Aim   To round whole numbers.

II. Procedure

1. Locate the place in the numeral to which the number is to be rounded. Rewrite the given figures to the left of the required place and write a zero in place of each figure to the right of the required place.

2. If the first figure dropped is 5 or more, increase the figure in the required place by 1, otherwise write the same figure as given.

III. Sample Solutions

1. 287 to nearest ten.   *Answer: 290*
2. 5,249 to nearest hundred.   *Answer: 5,200*
3. 48,628 to nearest thousand.   *Answer: 49,000*
4. 283,175 to nearest ten thousand.   *Answer: 280,000*
5. 654,927 to nearest hundred thousand.   *Answer: 700,000*
6. 26,385,000 to nearest million.   *Answer: 26,000,000*
7. 381,874,059,900 to nearest billion.   *Answer: 382,000,000,000*
8. 9,298,845,763,000 to nearest trillion.   *Answer: 9,000,000,000,000*

DIAGNOSTIC TEST

Round each of the following numbers to the nearest:

1. Ten: a. 57 60   b. 394 390

2. Hundred: a. 5,626 5,600   b. 890 900

3. Thousand: a. 4,501 5,000   b. 63,289 63,000

4. Ten thousand: a. 71,989 70,000   b. 149,000 150,000

5. Hundred thousand: a. 243,762 200,000   b. 1,650,059 1,700,000

6. Million: a. 3,728,409 4,000,000 b. 41,371,000 41,000,000

7. Billion: a. 24,492,568,200   b. 170,964,135,000 171,000,000,000
                      24,000,000,000

8. Trillion: a. 1,602,373,068,040   b. 43,089,967,000,000
                      2,000,000,000,000   43,000,000,000,000

RELATED PRACTICE EXAMPLES

Round each of the following numbers to the

Nearest ten:

SET 1(a)
1. 28 30
2. 45 50
3. 97 100
4. 416 420
5. 2,539 2,540

SET 1(b)
1. 31 30
2. 84 80
3. 152 150
4. 683 680
5. 3,844 3,840

Nearest hundred:

SET 2(a)
1. 230 200
2. 947 900
3. 2,529 2,500
4. 3,814 3,800
5. 5,408 5,400

SET 2(b)
1. 389 400
2. 551 600
3. 7,863 7,900
4. 9,092 9,100
5. 15,654 15,700

Nearest thousand:

SET 3(a)
1. 8,726 9,000
2. 3,867 4,000
3. 24,539 25,000
4. 57,603 58,000
5. 10,954 11,000

SET 3(b)
1. 9,485 9,000
2. 5,094 5,000
3. 36,320 36,000
4. 40,159 40,000
5. 878,208 878,000

Nearest ten thousand:

SET 4(a)
1. 23,400 20,000
2. 94,872 90,000
3. 80,056 80,000
4. 142,625 140,000
5. 6,531,009 6,530,000

SET 4(b)
1. 56,000 60,000
2. 37,160 40,000
3. 295,928 300,000
4. 528,050 530,000
5. 7,849,323 7,850,000

Nearest hundred thousand:

SET 5(a)
1. 120,000 100
2. 345,946 300
3. 1,510,214 1,500
4. 3,835,837 3,800
5. 22,403,541 22,400

SET 5(b)
1. 560,500 600
2. 283,475 300
3. 2,375,908 2,400
4. 5,481,481 5,500
5. 19,792,624 19,800

Nearest million:

SET 6(a)
1. 4,500,000 5
2. 9,742,820 10
3. 8,906,437 9
4. 32,671,352 33
5. 84,827,466 85

SET 6(b)
1. 2,450,000 2
2. 7,398,250 7
3. 14,487,095 14
4. 462,264,128 462
5. 8,250,175,000 8,250

Nearest billion:

SET 7(a)
1. 6,150,000,000 6
2. 17,428,819,000 17
3. 39,076,524,692 39
4. 253,264,800,000 253
5. 598,337,060,489 598

SET 7(b)
1. 8,832,000,000 9
2. 21,584,427,000 22
3. 67,940,839,148 68
4. 820,756,922,000 821
5. 908,671,595,286 909

Nearest trillion:

SET 8(a)
1. 5,592,000,000,000 6
2. 8,709,362,570,000 9
3. 46,860,813,289,400 47
4. 130,914,325,674,149 131
5. 815,637,453,116,588 816

SET 8(b)
1. 7,459,540,000,000 7
2. 6,187,969,486,725 6
3. 83,356,708,232,947 83
4. 95,019,831,645,103 95
5. 604,231,993,584,270 604

VOCABULARY: sum
addends
plus sign

OBJECTIVE:

I. Aim To add whole numbers.

II. Procedure

1. If it is necessary to write the example, place units under units, tens under tens, etc.

2. When there are two or more columns of numerals, add each column, starting from the units column. If the sum of any column is ten or more, write the last figure of the sum in the answer and carry the other figures to the next column.

3. Check by adding the columns in the opposite direction or by adding the sum of each column as shown in sample solution 1.

   Note: See page 5 for the commutative and associative properties.

4. Numbers named by shortened numerals may be added. (See sample solution 2.)

III. Sample Solutions

1. Add as indicated: $984 + 1,292 + 2,484 + 337 + 4,069$

   Solution: *Check:*

984	
1,292	
2,484 ←addends	
337	
4,069	
9,166 ←sum	

26	sum of units column
34	sum of tens column
18	sum of hundreds column
7	sum of thousands column
9,166	

   Answer: 9,166

2. Add: 65 billion
           7 billion
        286 billion
        358 billion   *Answer:* 358 billion

IV. Definitions

1. The sum is the answer in addition.
2. The addends are the numbers that are added.
3. A plus (+) sign indicates addition.

Related Resources: *Practical Applications in Mathematics:* Diag. Test and Prac. Ex., pp. 5–6

30

PRELIMINARY EXAMPLES (all students)

Basic Addition Combinations

Add:

1. 2	3	7	1	0	5	4	6	8	9
1	5	6	8	3	2	4	9	0	7
3	8	13	9	3	7	8	15	8	16
2. 4	7	5	6	9	3	0	8	1	2
6	1	0	7	9	4	2	5	3	8
10	8	5	13	18	7	2	13	4	10
3. 5	2	9	0	6	7	3	1	4	8
3	0	2	4	5	7	9	1	8	6
8	2	11	4	11	14	12	2	12	14
4. 0	6	3	9	1	2	8	7	4	5
1	8	0	3	4	6	7	2	9	5
1	14	3	12	5	8	15	9	13	10
5. 3	4	8	2	0	7	5	9	6	1
8	5	2	7	9	3	6	1	4	0
11	9	10	9	9	10	11	10	10	1
6. 6	0	2	5	4	8	3	1	9	7
6	0	9	8	3	1	2	7	5	4
12	0	11	13	7	9	5	8	14	11
7. 1	8	4	7	2	9	6	0	3	5
5	4	0	8	2	6	3	7	1	9
6	12	4	15	4	15	9	7	4	14
8. 9	5	1	8	3	4	2	7	0	6
0	7	9	3	6	1	4	5	8	2
9	12	10	11	9	5	6	12	8	8
9. 4	1	0	7	8	6	9	2	5	3
2	6	5	9	8	0	4	3	1	7
6	7	5	16	16	6	13	5	6	10
10. 8	4	6	3	5	1	7	0	2	9
9	7	1	3	4	2	0	6	5	8
17	11	7	6	9	3	7	6	7	17

DIAGNOSTIC TEST

Add and check:

1. 32
45 77

2. 56
17 73

3. 67
59 126

4. 9
5
8 22

5. 15
26
87 128

6. 6
2
4
9 21

7. 25
94
84
39 242

8. 5
7
6
8
9 35

9. 34
90
23
65
57 269

10. 389
459
 848

11. 3,592
2,738 6,330

12. 86,056
44,598 130,654

13. 435
599
796 1,830

14. 3,598
6,487
5,739 15,824

15. 24,673
12,762
37,857 75,292

16. 556
479
628
493 2,156

17. 3,962
6,109
2,854
6,875 19,800

18. 79,459
68,417
75,388
91,754 315,018

19. 258
584
845
207
396 2,290

20. 4,973
9,282
3,970
2,639
9,789 30,653

21. 83,625
94,774
87,146
32,753
64,838
 363,136

22. 9,651
78
83,795
206
5,184
16,745
 115,659

23. 1,483,297
3,915,485
20,500
372,173
5,283,094
7,848
608,556 11,690,953

24. 276 million
35 million
659 million
 970 million

25. 3,845 + 928 + 63,847 + 795 + 1,356 70,771

26. Find the sum of: 2,381; 967; 29; 8,406; 750 12,533

Basic: Sets 1–19,
Misc. Ex. 1–7,
Prob. 1–10

Average: Sets 1–23,
Misc. Ex. 1–9, Prob. 1–15

Enriched: Sets 1–26,
Misc. Ex. 1–10, Prob. 1–20

RELATED PRACTICE EXAMPLES

Add and check:

SET 1

34	67	20	42	63
25	32	58	44	26
59	99	78	86	89

SET 2

59	36	37	59	28
8	45	57	24	62
67	81	94	83	90

SET 3

48	24	83	56	95
94	76	60	58	99
142	100	143	114	194

SET 4

7	8	9	4	9
5	6	7	6	8
8	8	7	9	9
20	22	23	19	26

SET 5

42	25	86	65	34
27	50	92	7	79
19	83	58	94	88
88	158	236	166	201

SET 6

3	8	9	4	7
5	4	7	0	6
7	9	7	9	2
4	6	8	8	9
19	27	31	21	24

SET 7

82	45	94	28	86
43	79	36	8	59
92	63	49	7	97
30	42	78	14	95
247	229	257	57	337

SET 8

5	8	2	8	9
1	6	5	9	9
3	9	9	4	6
9	4	7	7	8
2	7	4	8	9
20	34	27	36	41

SET 9

12	82	73	91	76
26	64	9	30	89
21	37	24	54	97
13	56	8	79	88
22	28	87	68	53
94	267	201	322	403

SET 10

231	573	267	974	608
457	119	385	566	96
688	692	652	1,540	704

SET 11

6142	2,358	6,834	5,705	7,478
3756	4,135	8,879	986	9,757
9898	6,493	15,713	6,691	17,235

SET 12

37454	75,633	946	35,878	21,973
41345	83,698	30,306	89,646	92,027
78799	159,331	31,252	125,524	114,000

SET 13

582	861	779	926	639
143	495	698	15	852
214	827	579	658	543
939	2,183	2,056	1,599	2,034

SET 14

4214	2,616	4,043	8,588	6,957
1152	3,942	3,285	765	9,799
3403	4,585	9,772	89	7,186
8769	11,143	17,100	9,442	23,942

SET 15

13581	62,583	45,496	68,432	73,858
61132	17,914	27,383	257	59,576
15273	28,619	31,576	9,746	38,643
89986	109,116	104,455	78,435	172,077

SET 16

637	458	682	39	787
131	173	745	896	957
528	649	384	9	236
300	326	973	372	598
1,596	1,606	2,784	1,316	2,578

SET 17

2105	2,374	4,966	7,581	6,479
3216	4,834	3,859	470	8,642
1143	1,496	5,573	28	9,736
2319	3,587	8,426	6	6,857
8783	12,291	22,824	8,085	31,714

SET 18

34832	56,834	42,763	6,445	76,542
11913	67,509	96,833	95,214	81,276
22741	31,915	47,485	87	998
10420	47,628	92,739	1,756	5,823
79906	203,886	279,820	103,502	164,639

SET 19

129	289	646	7	688
325	769	53	226	734
356	427	8	589	597
218	892	24	36	905
497	976	593	328	799
1,525	3,353	1,324	1,186	3,723

SET 20

1673	2,765	46	7,328	6,494
2191	4,497	315	478	3,849
1814	2,086	2	9,663	8,589
1271	2,199	1,597	98	6,355
2353	7,385	221	8,976	2,914
9302	18,932	2,181	26,543	28,201

SET 21

15924	54,683	45,618	75	96,545
24872	82,759	4,973	5,491	32,678
29644	38,742	87	67,987	49,924
18873	81,599	564	446	85,397
15692	76,847	9	51,743	26,785
105005	334,630	51,251	125,742	291,329

SET 22(a)

6	16	387	529	6,583
5	8	596	63	3,476
2	29	347	784	4,285
8	9	482	8	3,842
4	65	574	29	5,273
2	22	108	685	2,646
27	149	2,494	2,098	26,105

SET 22(b)

95329	82,541	4,953	362,474	642,566
458	31,796	58,275	67,580	255,887
7509	28,865	87	85,030	974,464
26	49,496	893,574	250,545	850,952
84531	22,738	64,705	414,650	527,327
727	57,633	9,268	2,647	741,969
188580	273,069	1,030,862	1,182,926	3,993,165

SET 23(a)

8	48	653	4,591	46,398
3	36	28	8,216	127
9	53	3	9,137	3,874
7	29	792	1,615	58,462
4	74	89	2,293	95
2	83	246	3,748	4,267
6	98	36	5,157	67,839
39	421	1,847	34,757	181,062

SET 23(b)

167450	698,523	487,944	2,459,630	4,369,565
32184	34,769	921,593	25,400	7,436,859
500863	178	678,105	580,575	3,050,762
200	9,874	637,050	243,125	4,598,083
89157	968,452	986,967	8,549	6,452,136
8	789,347	369,725	9,783,484	7,587,327
38524	59,791	917,862	15,338	3,967,009
828386	2,560,934	4,999,246	13,116,101	37,461,741

SET 24

26 trillion	327 billion	$273 billion	438 million	$ 407 million
9 trillion	95 billion	486 billion	161 million	519 million
42 trillion	158 billion	958 billion	85 million	6,248 million
77 trillion	63 billion	567 billion	79 million	3,639 million
	643 billion	$2,284 billion	229 million	794 million
			992 million	$11,607 million

Add as indicated: **SET 25(a)**

1. 6 + 2 + 7 15
2. 8 + 5 + 6 + 4 + 9 32
3. 72 + 37 + 84 193
4. 69 + 75 + 29 + 82 + 51 306
5. 24 + 33 + 49 + 53 + 38 + 84 + 70 351
6. 18 + 3 + 5 + 26 52
7. 89 + 7 + 16 + 8 + 19 139
8. 47 + 4 + 19 + 72 + 5 + 27 174
9. 6 + 21 + 78 + 7 + 36 + 81 229
10. 85 + 15 + 3 + 30 + 9 + 28 + 6 176

SET 25(b)

1. 903 + 412 1,315
2. 328 + 139 + 426 893
3. 475 + 189 + 88 + 423 1,175
4. 264 + 593 + 34 + 766 + 872 2,529
5. 502 + 74 + 365 + 82 + 171 1,194
6. 97 + 227 + 56 + 5 + 764 1,149
7. 639 + 422 + 91 + 587 + 36 + 58 1,833
8. 597 + 329 + 649 + 864 + 975 + 800 4,214
9. 278 + 42 + 764 + 95 + 556 + 476 2,211
10. 886 + 529 + 793 + 438 + 927 + 572 + 106 4,251

SET 25(c)

1. 6,545 + 7,182 13,727
2. 92,114 + 345 + 21,533 113,992
3. 69,483 + 24,752 + 80,599 + 2,843 177,677
4. 9,787 + 1,578 + 463 + 9,242 + 89 21,159
5. 50,422 + 1,525 + 396 + 85 + 4,152 56,580
6. 253 + 9,052 + 59,628 + 4 + 786 69,723
7. 20,500 + 2,483 + 15,814 + 350 + 9,453 + 87 48,687
8. 14,208 + 932 + 5,919 + 28 + 36,545 + 15,970 73,602
9. 29,751 + 34,462 + 45,432 + 382 + 2,456 + 3,084 115,567
10. 37,442 + 42,596 + 807 + 6,638 + 52,800 + 51 + 9,046 149,380

SET 26

Find the sum of:

 1. 86; 153; 128; 249 616
 2. 425; 836; 595; 26; 84 1,966
 3. 762; 158; 32; 17; 4; 365 1,338
 4. 200; 175; 335; 95; 480; 45 1,330
 5. 731; 453; 846; 924; 877; 382; 502 4,715
 6. 1,469; 8,355; 2,076; 3,416; 123; 4,125 19,564
 7. 8,500; 250; 960; 63,075; 407; 2,155 75,347
 8. 24,057; 6,799; 4,276; 9,341; 8,223; 37,165 89,861
 9. 83; 9; 472; 1,857; 343; 8,158; 48,494 59,416
10. 53,196; 9,324; 4,197; 3,283; 44,915; 6,258; 1,970 123,143

MISCELLANEOUS EXAMPLES

Find the missing numbers, adding horizontally and vertically:

1. $8 + 7 + 4 =$ 19
 $3 + 5 + 9 =$ 17
 $6 + 3 + 7 =$ 16
 $9 + 8 + 6 =$ 23
 ————————
 $26 + 23 + 26 =$ 75

2. $62 + 25 + 32 =$ 119
 $13 + 7 + 15 =$ 35
 $6 + 18 + 48 =$ 72
 $35 + 9 + 4 =$ 48
 ————————
 $116 + 59 + 99 =$ 274

3. $153 + 42 + 29 =$ 224
 $35 + 51 + 7 =$ 93
 $24 + 143 + 62 =$ 229
 $109 + 84 + 46 =$ 239
 ————————
 $321 + 320 + 144 =$ 785

4. $16 + 95 + 84 + 7 =$ 202
 $8 + 23 + 2 + 91 =$ 124
 $27 + 8 + 30 + 25 =$ 90
 $70 + 16 + 53 + 6 =$ 145
 ————————
 $121 + 142 + 169 + 129 =$ 561

5. $5 + 8 + 6 =$ 19
 $9 + 7 + 8 =$ 24
 $3 + 9 + 9 =$ 21
 $4 + 5 + 9 =$ 18
 $7 + 8 + 4 =$ 19
 ————————
 $28 + 37 + 36 =$ 101

6. $9 + 5 + 7 + 8 =$ 29
 $4 + 9 + 8 + 4 =$ 25
 $7 + 9 + 4 + 9 =$ 29
 $2 + 8 + 2 + 5 =$ 17
 $8 + 9 + 5 + 8 =$ 30
 ————————
 $30 + 40 + 26 + 34 =$ 130

7. $15 + 58 + 97 =$ 170
 $28 + 9 + 34 =$ 71
 $14 + 73 + 81 =$ 168
 $55 + 61 + 48 =$ 164
 $74 + 26 + 64 =$ 164
 ————————
 $186 + 227 + 324 =$ 737

8. $38 + 49 + 2 =$ 89
 $57 + 27 + 95 =$ 179
 $21 + 54 + 7 =$ 82
 $68 + 15 + 38 =$ 121
 $95 + 6 + 14 =$ 115
 ————————
 $279 + 151 + 156 =$ 586

9. 152 + 69 + 200 =   421
   827 + 143 + 92 = 1,062
   507 + 258 + 181 =   946
    26 + 400 +   6 =   432
   242 + 75 + 47 =   364
  1,754 + 945 + 526 = 3,225

10. 1,500 +   938 + 4,926 = 7,364
    284 +   29 + 1,385 = 1,698
     65 +   300 +   50 =   415
   146 +   8 +   267 =   421
 3,650 + 1,825 +   928 = 6,403
 5,645 + 3,100 + 7,556 = 16,301

PROBLEM SOLVING

See page 16, *Problem Solving Strategy.*
For problems 1–5, select the letter corresponding to your answer.

1. The Washington High School presented this year three performances of their dramatic show. 1,058 persons saw the opening performance. On the succeeding two nights 993 and 1,196 persons respectively attended. What was the total attendance at the school show?

  a. 3,147 persons   **c.** 3,437 persons
  b. 3,247 persons   **d.** Answer not given

2. How many calories are contained in the following meal: vegetable soup, 86; roast lamb, 175; fresh peas, 66; boiled potato, 117; 2 slices of white bread, 134; apple, 81; glass of milk, 170?

  a. 819 calories   **c.** 929 calories
  b. 739 calories   **d.** 829 calories

3. During the year Mr. Santos bought the following amounts of fuel oil to heat his house: 665 liters, 787 liters, 936 liters, 905 liters, 738 liters, and 946 liters. How many liters of fuel oil did he buy in all?

  a. 4,877 liters   **c.** 4,977 liters
  b. 4,887 liters   **d.** 5,067 liters

4. What is the total area of the Great Lakes if the area of each lake is as follows: Lake Superior, 82,414 square kilometers; Lake Erie, 25,745 square kilometers; Lake Michigan, 58,016 square kilometers; Lake Ontario, 19,529 square kilometers; and Lake Huron, 59,596 square kilometers?

  a. 246,300 square kilometers   **c.** 245,300 square kilometers
  b. 245,290 square kilometers   **d.** 246,290 square kilometers

5. What is the population of New England if Maine has a population of 993,663; New Hampshire, 737,681; Vermont, 444,732; Massachusetts, 5,689,170; Rhode Island, 949,723; and Connecticut, 3,032,217?

  a. 11,848,186 people   **c.** 10,848,186 people
  b. 11,847,186 people   **d.** Answer not given

6. A 2-story factory building with office and showroom is available for rent. The office has 87 square meters of floor space; showroom, 224 square meters; first floor, 1,986 square meters; and second floor, 1,468 square meters. What is the total floor space of the building? 3,765 square meters

7. In the election for president of the student body at the Fernwood High School, 146 freshmen, 198 sophomores, 127 juniors, and 208 seniors voted for Jack and 179, 124, 218, and 134 respectively voted for Maria. Who received the greater number of votes? Jack

8. Find the total enrollment at the Central Junior High School if in grade 7A there are 225 pupils; 7B, 193 pupils; 8A, 249 pupils; 8B, 216 pupils; 9A, 368 pupils; and 9B, 257 pupils. 1,508 pupils

9. What was the total attendance at the 4 school league football games played by Crispus Attucks High School if the first game was witnessed by 2,973 spectators, the second by 2,684, the third by 3,189, and the last game by 2,708 spectators? 11,554 spectators

10. There are 135 boys and 146 girls in the freshman class at the Carson High School, 118 boys and 139 girls in the sophomore class, 169 boys and 129 girls in the junior class, and 107 boys and 158 girls in the senior class. How many boys are enrolled at the school? How many girls? What is the total enrollment? 529 boys; 572 girls; 1,101 pupils

11. Find the final scores of the following high school games:

Football

| Washington | 7 | 14 | 0 | 13 = 34 | Carson | 6 | 19 | 7 | 14 = 46 |
| Fernwood | 13 | 9 | 16 | 6 = 44 | Greenville | 0 | 13 | 12 | 6 = 31 |

Baseball

| Carson | 0 | 3 | 0 | 2 | 4 | 0 | 2 | 1 | 0 = 12 |
| Springdale | 1 | 2 | 1 | 0 | 2 | 1 | 0 | 3 | 0 = 10 |

Basketball

| Greenville | 12 | 21 | 14 | 19 = 66 | Washington | 15 | 8 | 24 | 16 = 63 |
| Jefferson | 17 | 15 | 9 | 26 = 67 | Hilton | 13 | 18 | 17 | 10 = 58 |

12. Bowling—3-game match—Find (*a*) which player scored the most points, (*b*) which team won the most games, and (*c*) which team had the greater total match score.

Washington High				Kingston High			
Benson	152	217	169	Richards	154	131	176
Allen	163	150	178	Carter	159	182	198
Young	148	176	214	Turner	165	157	188

12. a. Carter, 539 points   b. Kingston High, 2 games   c. Washington High, 1,567 points

13. Find the winner of a 9-hole golf match (the smaller number of strokes wins):

Jones wins

| Jones, Carson H. S. | 4 | 4 | 5 | 6 | 3 | 4 | 5 | 4 | 5 = | 40 |
| Williams, Dalton H. S. | 3 | 5 | 4 | 5 | 4 | 5 | 6 | 5 | 4 = | 41 |

14. A streamlined passenger train goes from Miami to Jacksonville, a distance of 652 kilometers; from Jacksonville to Savannah, 225 kilometers; from Savannah to Raleigh, 552 kilometers; from Raleigh to Richmond, 253 kilometers; from Richmond to Washington, 187 kilometers; and from Washington to New York City, 365 kilometers. What is the total distance by train from Miami to New York City? 2,234 kilometers

15. What is the total seating capacity at the county baseball stadium if there are 978 box seats, 8,496 reserved seats, 19,564 general admission seats, and 3,825 bleacher seats? 32,863 seats

16. The township public library has 2,309 fiction books, 1,894 nonfiction books, 195 reference books, and 275 magazines. What is the total number of books and magazines? 4,673 books and magazines

17. Find the total area of the Pacific States if the area of California is 404,975 square kilometers; Oregon, 249,117 square kilometers; and Washington, 172,416 square kilometers. 826,508 square kilometers

60,040,000 **18.** In the United States recently there were 3,263,000 pupils enrolled in kindergarten, 33,507,000 pupils in the elementary schools (grades 1 through 8), 15,183,000 pupils in the senior high schools (grades 9 through 12), and 8,087,000 students in colleges. Find the total school enrollment.

19. How many farms are in the East North Central States (Ohio, Indiana, Illinois, Michigan, Wisconsin) if there are 111,332 farms in Ohio, 101,479 farms in Indiana, 123,565 in Illinois, 77,946 in Michigan, and 98,973 in Wisconsin? 513,295 farms

20. What was the total attendance at the New York–Los Angeles 6-game World Series if 56,668 persons saw the first game, 56,691 saw the second, 55,992 the third, 55,995 the fourth, 55,955 the fifth, and 56,407 the sixth game? 337,708 people

---

REFRESH YOUR SKILLS

1. Write 6,862,925 as a word statement.
Six million, eight hundred sixty-two thousand, nine hundred twenty-five

2. Round 129,484,056 to the nearest million.
129,000,000

3. Add:
16834
32605
57428
45887
84065
236819

4. Add:
92,153
8,449
376
9,849
75,698
186,525

5. Add:
38,669
81,767
63,898
89,948
56,877
331,159

For a Computer Activity see p. T39.

Subtraction of Whole Numbers

VOCABULARY: minuend
subtrahend
remainder
difference
minus sign

I. Aim
OBJECTIVE:
To subtract whole numbers.

II. Procedure

1. If it is necessary to write the example, place units under units, tens under tens, etc.

2. Subtract the figures in the subtrahend from the corresponding figures in the minuend. Start from the units place and work to the left. If any figure in the subtrahend is greater than the corresponding figure in the minuend, increase the figure in the minuend by 10 by taking 1 from the preceding figure in the next higher place.

3. Check by adding the remainder to the subtrahend. Their sum should equal the minuend.

4. Numbers named by shortened numerals may be subtracted. (See sample solution 5.)

III. Sample Solutions

Subtract in the following examples:

1. 7,485 − 2,648

Solution:		*Check:*
7,485	← minuend	2,648
2,648	← subtrahend	4,837
4,837	← remainder	7,485
	or difference	

Answer: 4,837

2. 94,500 − 93,264

Solution:	*Check:*
94,500	93,264
93,264	1,236
1,236	94,500

Answer: 1,236

3. From 8,337 subtract 6,828.

Solution:	*Check:*
8,337	6,828
6,828	1,509
1,509	8,337

Answer: 1,509

4. Take 279 from 3,456.

Solution:	*Check:*
3,456	279
279	3,177
3,177	3,456

Answer: 3,177

5. 950 million
187 million
763 million

Answer: 763 million

Related Resources: *Practical Applications in Mathematics:* Diag. Test and Prac. Ex., pp. 7–8

IV. Definitions

1. The minuend is the number from which you subtract.
2. The subtrahend is the number you subtract.
3. The remainder or difference is the answer in subtraction.
4. The minus (−) sign indicates subtraction.

PRELIMINARY EXAMPLES (all students)

Basic Subtraction Combinations

Subtract:

1.	5	7	2	6	17	16	11	9	8	7
	2	7	0	4	8	9	6	5	1	3
	3	0	2	2	9	7	5	4	7	4
2.	7	10	13	5	9	3	8	6	10	11
	4	5	9	0	1	2	8	3	6	7
	3	5	4	5	8	1	0	3	4	4
3.	6	11	9	10	5	15	3	12	8	6
	5	9	6	1	4	8	3	7	0	2
	1	2	3	9	1	7	0	5	8	4
4.	1	5	6	7	16	9	11	7	9	14
	0	3	1	6	7	4	5	2	9	8
	1	2	5	1	9	5	6	5	0	6
5.	10	11	15	12	8	4	17	3	8	10
	3	8	7	5	6	1	9	0	2	4
	7	3	8	7	2	3	8	3	6	6
6.	6	4	13	8	10	14	7	16	5	4
	6	2	4	3	9	7	1	8	5	0
	0	2	9	5	1	7	6	8	0	4
7.	5	12	2	15	11	8	0	12	10	9
	1	4	2	6	3	5	0	9	7	8
	4	8	0	9	8	3	0	3	3	1
8.	15	3	4	10	8	14	7	4	13	7
	9	1	3	2	7	6	0	4	8	5
	6	2	1	8	1	8	7	0	5	2
9.	13	9	12	18	14	12	9	13	11	1
	7	0	8	9	5	3	2	6	4	1
	6	9	4	9	9	9	7	7	7	0
10.	10	12	13	9	11	6	14	2	9	8
	8	6	5	7	2	0	9	1	3	4
	2	6	8	2	9	6	5	1	6	4

DIAGNOSTIC TEST

Subtract and check:

1. 28
  5 *23*
	6. 59
	  51 *8*
		11. 465
		  359 *106*
			16. 6,841
			  2,327 *4,514*
				21. 85,924
				  23,713 *62,211*

2. 31
  6 *25*
	7. 50
	  29 *21*
		12. 800
		  698 *102*
			17. 7,653
			  4,725 *2,928*
				22. 92,045
				  21,924 *70,121*

3. 76
  14 *62*
	8. 496
	  263 *233*
		13. 604
		  589 *15*
			18. 3,962
			  1,487 *2,475*
				23. 68,247
				  44,395 *23,852*

4. 87
  68 *19*
	9. 562
	  228 *334*
		14. 255
		  37 *218*
			19. 5,000
			  3,792 *1,208*
				24. 89,463
				  15,684 *73,779*

5. 32
  12 *20*
	10. 643
	  367 *276*
		15. 4,965
		  1,842 *3,123*
			20. 8,429
			  64 *8,365*
				25. 70,571
				  39,782 *30,789*

26. 35,823
    19,341 *16,482*

27. 45,936
    798 *45,138*

28. 845,094
    384,276 *460,818*

29. 4,575,000
    1,395,463 *3,179,537*

438 billion **30.** 527 billion − 89 billion   **31.** 98,370 − 84,697 *13,673*

7,884 **32.** From 8,463 subtract 579.   **33.** Take 3,582 from 9,348. *5,766*

34. Find the difference between 17,947 and 13,799. *4,148*

Basic: Sets 1–25, Average: Sets 1–30, Prob. 1–15   Enriched: Sets 1–34,
Prob. 1–10   Prob. 1–19

RELATED PRACTICE EXAMPLES

Subtract and check:

SET 1

59	84	96	77	69
4	2	3	6	5
55	82	93	71	64

SET 2

23	86	94	47	52
8	7	6	9	3
15	79	88	38	49

SET 3

85	66	98	49	57
61	35	26	15	23
24	31	72	34	34

SET 4

36	71	64	93	85
19	23	25	58	37
17	48	39	35	48

SET 5

78	53	95	29	87
48	23	35	9	17
30	30	60	20	70

SET 6

26	37	86	98	79
23	34	82	90	72
3	3	4	8	7

SET 7

40	60	30	40	90
17	36	23	8	65
23	24	7	32	25

SET 8

475	978	362	549	853
132	642	150	325	721
343	336	212	224	132

	SET 9					SET 10			
891	582	439	918	647	842	516	975	482	923
379	166	284	627	493	358	179	587	295	476
512	416	155	291	154	484	337	388	187	447

	SET 11					SET 12			
428	785	972	964	545	609	750	500	300	600
218	380	672	357	138	392	273	307	160	485
210	405	300	607	407	217	477	193	140	115

	SET 13					SET 14			
538	972	784	901	735	253	467	526	884	573
453	965	697	838	686	26	58	72	9	83
85	7	87	63	49	227	409	454	875	490

SET 15

8654	4,296	6,582	9,728	7,849
2431	3,003	4,172	6,215	3,812
6223	1,293	2,410	3,513	4,037

SET 16

9642	8,657	2,938	3,463	9,185
6425	5,846	1,682	2,702	6,345
3217	2,811	1,256	761	2,840

SET 17

7194	5,371	8,362	5,774	3,680
3457	2,925	4,554	2,968	2,962
3737	2,446	3,808	2,806	718

SET 18

9426	8,335	4,900	5,371	8,796
3258	6,452	3,825	4,691	6,099
6168	1,883	1,075	680	2,697

SET 19

8324	6,153	8,050	6,000	9,000
5975	2,896	4,584	2,505	8,427
2349	3,257	3,466	3,495	573

SET 20

4823	6,752	3,058	1,971	3,000
8	127	65	986	479
4815	6,625	2,993	985	2,521

SET 21

38762	93,596	55,706	45,383	79,648
26521	61,593	52,304	32,162	36,136
12241	32,003	3,402	13,221	43,512

SET 22

45962	79,614	83,253	90,349	52,767
12535	48,532	52,028	70,281	28,343
33427	31,082	31,225	20,068	24,424

SET 23

37424	56,382	80,545	76,200	38,509
15258	42,491	39,644	41,350	32,754
22166	13,891	40,901	34,850	5,755

SET 24

61847	97,325	54,000	18,508	79,406
25952	83,459	23,475	10,809	25,837
35895	13,866	30,525	7,699	53,569

SET 25

97437	42,662	50,583	62,471	90,000
28769	25,896	48,794	42,587	82,575
68668	16,766	1,789	19,884	7,425

SET 26

24593	57,924	67,500	29,070	82,957
10638	39,819	59,045	14,924	53,498
13955	18,105	8,455	14,146	29,459

SET 27

65925	46,000	21,507	86,500	95,374
386	897	4,742	95	8,298
65539	45,103	16,765	86,405	87,076

SET 28

645987	705,961	658,235	450,000	532,654
314265	536,809	621,791	328,542	2,867
331722	169,152	36,444	121,458	529,787

SET 29

8935299	2,963,475	7,694,251	2,500,000	4,000,000
3411274	1,724,133	86,206	1,758,625	2,950,966
5524025	1,239,342	7,608,045	741,375	1,049,034

SET 30

15 trillion	533 million	$287 billion
9 trillion	475 million	68 billion
6 trillion	58 million	$219 billion

$900 million	1,056 billion	$2,304 billion
409 million	897 billion	1,086 billion
$491 million	159 billion	$1,218 billion

Subtract as indicated:

SET 31(a)	**SET 31(b)**

SET 31(a)

1. 25 − 18 7
2. 645 − 382 263
3. 476 − 387 89
4. 500 − 298 202
5. 4,563 − 3,275 1,288
6. 5,694 − 4,987 707
7. 78,000 − 54,092 23,908
8. 95,375 − 92,186 3,189
9. 636,059 − 278,770 357,289
10. 3,540,000 − 3,269,145 270,855

SET 31(b)

1. 63 − 8 55
2. 427 − 76 351
3. 300 − 84 216
4. 4,517 − 28 4,489
5. 3,788 − 909 2,879
6. 1,500 − 6 1,494
7. 58,625 − 9,775 48,850
8. 32,743 − 814 31,929
9. 545,611 − 39,768 505,843
10. 750,000 − 7,050 742,950

SET 32

1. From 300 subtract 152. 148
2. From 798 subtract 476. 322
3. From 237 subtract 95. 142
4. From 4,583 subtract 3,178. 1,405
5. From 3,489 subtract 2,183. 1,306
6. From 2,000 take 504. 1,496
7. From 9,153 take 2,846. 6,307
8. From 15,000 take 3,825. 11,175
9. From 37,450 take 15,968. 21,482
10. From 84,073 take 28,536. 55,537

SET 33

1. Take 63 from 324. 261
2. Take 574 from 1,582. 1,008
3. Take 497 from 520. 23
4. Take 38 from 1,000. 962
5. Take 7,352 from 9,514. 2,162
6. Subtract 633 from 1,981. 1,348
7. Subtract 200 from 1,582. 1,382
8. Subtract 4,156 from 8,915. 4,759
9. Subtract 347 from 59,784. 59,437
10. Subtract 2,059 from 37,685. 35,626

SET 34

Find the difference between:

1. 942 and 368 574
2. 450 and 199 251
3. 8,256 and 8,147 109
4. 3,794 and 900 2,894
5. 45,836 and 21,862 23,974
6. 68,593 and 40,500 28,093
7. 300,000 and 171,938 128,062
8. 821,575 and 390,695 430,880
9. 761,350 and 385,025 376,325
10. 5,000,000 and 2,750,000 2,250,000

PROBLEM SOLVING

See page 16, *Problem Solving Strategy.*

For problems 1–5, select the letter corresponding to your answer.

1. The present enrollment at the Township Junior High School is 849 pupils. There are 675 pupils registered at the senior high school. What is the difference in their enrollments?

a. 185 pupils   b. 164 pupils   c. 174 pupils   d. Answer not given

2. The seating capacity at the new Jackson High School athletic field is 4,120. The stands at the old field held only 1,584 persons. How many more spectators can now be seated?

    a. 2,436 spectators   **c.** 2,446 spectators

    <u>**b.**</u> 2,536 spectators   **d.** 2,546 spectators

3. A salesperson's automobile mileage for the past year was 41,266 kilometers. If 11,291 kilometers represents pleasure driving, how many miles was the car driven for business purposes?

    a. 30,875 kilometers   **c.** 29,985 kilometers

    b. 29,875 kilometers   <u>**d.**</u> 29,975 kilometers

4. How much larger is the state of Texas with an area of 688,617 square kilometers than the state of Rhode Island with an area of 3,232 square kilometers?

    a. 684,439 square kilometers   **c.** 685,439 square kilometers

    b. 695,439 square kilometers   <u>**d.**</u> Answer not given

5. How many fewer farms are there in the United States if the number of farms decreased from 5,859,169 to 2,730,242?

    a. 3,028,927 farms   **c.** 3,027,927 farms

    b. 3,138,927 farms   <u>**d.**</u> 3,128,927 farms

6. During the football season Southwest High School scored 219 points against their opponents' 136 points and gained 1,246 yards against 891 yards in running plays, and 932 yards against 798 yards in passing. How many more points did Southwest score? How many more yards did they gain running? Passing? 83 points; 355 yards; 134 yards

7. In the election for secretary of the student association at the Rowell Junior High School, Jeff received 1,206 votes and Mark, 978 votes. How many more votes did Jeff receive? 228 votes

8. The total attendance at all athletic events at the Amerigo Vespucci High School for the school year was 48,547. The attendance for the previous year was 39,498. Find the amount of increase in attendance. 9,049 persons

9. Tom's father is interested in buying a new house. He likes two houses, one priced at $49,600 and the other at $50,575. How much more does the second house cost? If he expects to make a down payment of $12,750, what would be the amount of the mortgage (the balance) in each case? $975; $36,850; $37,825

10. Mother asked John to check the electric and gas bills. If the previous month's reading of the electric meter was 7,095 kilowatt-hours and the present reading is 7,264 kilowatt-hours, how many kilowatt-hours of electricity were used during the month? He found the present reading of the gas meter to be 5,105 cubic meters and the previous reading 4,937 cubic meters. How many cubic meters of gas were used?

169 kilowatt-hours; 168 cubic meters

$27,254 **11.** Ms. Schmidt paid $29,450 for her house 4 years ago. If the house depreciated in the amount of $2,196, what would its present value be?

12. A poultry farmer handled 129,344 chickens during the year. His sales totaled 97,872 chickens. Find the number of chickens that remained unsold. 31,472 chickens

13. The deepest place thus far discovered in the world is 10,863 meters in the Marianas Trench near the island of Guam. The greatest depth in the Atlantic Ocean is 9,219 meters, located near Puerto Rico. Find the difference in these depths. 1,644 meters

14. The highest point in Asia is Mount Everest with an elevation of 8,848 meters. Mount McKinley in Alaska with an elevation of 6,186 meters is the highest point in North America. What is the difference in their altitudes? 2,662 meters

15. Just off the coast of California is a submarine mountain called San Juan Seamount. How far below the surface of the water is the top of the mountain if the ocean floor depth is 3,658 meters and the mountain rises 3,105 meters? 553 meters

16. The area of the earth is 510,100,500 square kilometers. If there are 361,149,600 square kilometers of water, how many square kilometers of land are on the earth? 148,950,900 square kilometers

17. The seating capacity of the Rose Bowl in Pasadena, California, is 100,188. How many more people can be seated in the Philadelphia municipal stadium if its seating capacity is 105,000? 4,812

18. The distance from the moon to the earth is 384,393 kilometers and from the sun to the earth is 149,499,812 kilometers. How much farther away is the sun from the earth than the moon? 149,115,419 kilometers

19. How many more square kilometers of territory did the United States acquire by the Louisiana Purchase of 2,142,427 square kilometers than by the purchase of Alaska with 1,518,776 square kilometers?
623,651 square kilometers

REFRESH YOUR SKILLS

1. Write 208,070,009 as a word statement.

Two hundred eight million, seventy thousand, nine

2. Round 7,583,284 to the nearest thousand.

7,583,000

3. Add:
45,836
29,797
75,086
18,925
56,498
―――――
226,142

4. Subtract:
21673
4689
――― 16,984

5. Subtract:
965,425
829,517
―――――
135,908

For a Computer Activity see p. T39.

Multiplication of Whole Numbers

I. Aim OBJECTIVE: To multiply whole numbers.

VOCABULARY: multiplicand   partial product
multiplier   times sign
product   factor

II. Procedure

1. To write the example, place units under units, tens under tens, etc.

2. If the multiplier has only one figure, multiply each and every figure in the multiplicand by this figure. Start from the right and work in order to the left. (See sample solution 1.)

3. If the multiplier has more than one figure, find the partial products by multiplying all the figures in the multiplicand by each figure in the multiplier, starting from the right. Write the partial products under each other, placing the numbers so that the right-hand figure of each partial product is directly under its corresponding figure in the multiplier. Then add the partial products. (See sample solutions 2 and 3.)

4. Check by interchanging the multiplier and multiplicand and multiplying again (see sample solution 2), or by dividing the product by either the multiplier or the multiplicand. (See sample solutions 1 and 3. See pages 5 and 6 for commutative, associative, and distributive properties).

5. Numbers named by shortened numerals may be multiplied. (See sample solution 4.)

III. Sample Solutions

1. Multiply 647 by 8.

Solution: *Check:*

$$
\begin{array}{r}
647 \\
8 \\
\hline
5{,}176
\end{array}
\qquad
8\overline{)5{,}176}^{\,647}
$$

Answer: 5,176

2. Multiply 36 by 45.

Solution: *Check:*

$$
\begin{array}{r}
36 \leftarrow \text{multiplicand} \\
45 \leftarrow \text{multiplier} \\
\hline
180 \quad\text{partial} \\
144 \quad\text{products} \\
\hline
1{,}620 \leftarrow \text{product}
\end{array}
\qquad
\begin{array}{r}
45 \\
36 \\
\hline
270 \\
135 \\
\hline
1{,}620
\end{array}
$$

Answer: 1,620

3. Multiply 54 by 29.

Solution: *Check:*

$$
\begin{array}{r}
54 \\
29 \\
\hline
486 \\
108 \\
\hline
1{,}566
\end{array}
\qquad
54\overline{)1{,}566}
\begin{array}{r}
29 \\
108 \\
486 \\
486
\end{array}
\qquad
29\overline{)1{,}566}
\begin{array}{r}
54 \\
145 \\
116 \\
116
\end{array}
$$

Answer: 1,566 $\cdots$ $\cdots$

4. Multiply: 5×18 billion

$$
\begin{array}{r}
18 \text{ billion} \\
\times \; 5 \\
\hline
90 \text{ billion}
\end{array}
$$

Answer: 90 billion

IV. Definitions

1. The multiplicand is the number you multiply.
2. The multiplier is the number by which you multiply.
3. The product is the answer in multiplication.
4. A partial product is the product obtained by multiplying the multiplicand by any figure in the multiplier.
5. The symbol $\times$ (times sign) indicates multiplication.
6. A factor is any one of the numbers used in multiplication to form a product.

PRELIMINARY EXAMPLES (all students)

Basic Multiplication Combinations

Multiply:

1.	5	8	2	9	7	1	6	3	4	0
	3	1	5	6	9	2	7	0	4	8
	15	8	10	54	63	2	42	0	16	0
2.	1	7	4	5	8	6	0	9	2	3
	7	6	3	8	2	4	5	9	0	1
	7	42	12	40	16	24	0	81	0	3
3.	0	2	7	9	4	3	6	5	1	8
	2	9	7	3	0	5	8	6	1	4
	0	18	49	27	0	15	48	30	1	32
4.	4	8	6	7	9	5	0	1	3	2
	8	7	9	0	5	1	4	3	6	2
	32	56	54	0	45	5	0	3	18	4
5.	7	3	2	8	6	9	1	4	0	5
	4	3	1	9	0	8	6	5	7	2
	28	9	2	72	0	72	6	20	0	10
6.	6	2	0	3	7	9	5	8	4	1
	1	4	6	2	3	7	0	8	9	5
	6	8	0	6	21	63	0	64	36	5
7.	5	7	6	1	0	8	2	3	9	4
	4	2	5	9	1	3	8	7	0	6
	20	14	30	9	0	24	16	21	0	24
8.	8	3	9	6	5	1	0	4	7	2
	6	9	4	2	5	0	3	1	8	7
	48	27	36	12	25	0	0	4	56	14
9.	0	8	1	6	2	4	7	9	3	5
	0	5	8	6	3	7	1	2	4	9
	0	40	8	36	6	28	7	18	12	45
10.	7	3	4	8	2	0	9	6	5	1
	5	8	2	0	6	9	1	3	7	4
	35	24	8	0	12	0	9	18	35	4

DIAGNOSTIC TEST A

Multiply and check:

1. 23
 3 _69_

2. 72
 4 _288_

3. 24
 3 _72_

4. 39
 7 _273_

5. 231
 2 _462_

6. 319
 3 _957_

7. 874
 6 _5,244_

8. 1,728
 9 _15,552_

9. 34,267
 6 _205,602_

10. 8
 2,532 _20,256_

11. 60
 3 _180_

12. 400
 5 _2,000_

13. 302
 2 _604_

14. 5,208
 8 _41,664_

15. 3,000
 5 _15,000_

16. 4,006
 7 _28,042_

17. 50,800
 4 _203,200_

18. 6 × 985   _5,910_

19. 197 × 65 billion   _12,805 billion_

20. Multiply 48 by 9.   _432_

21. Find the product of 314 and 8.   _2,512_

Basic: Sets 1–15(A), 1–18(B), Prob. 1–10   Average: Sets 1–18(A), 1–22(B), Prob. 1–15   Enriched: Sets 1–21(A), 1–25(B), Prob. 1–20

RELATED PRACTICE EXAMPLES

Multiply and check:

SET 1

32	11	43	22	21
2	8	2	4	3
64	88	86	88	63

SET 2

63	94	52	82	71
3	2	3	2	6
189	188	156	164	426

SET 3

23	28	19	15	46
4	3	5	6	2
92	84	95	90	92

SET 4

48	65	36	64	59
6	8	7	4	9
288	520	252	256	531

SET 5

321	144	542	823	912
3	2	2	3	4
963	288	1,084	2,469	3,648

SET 6

217	928	816	191	649
4	3	5	6	2
868	2,784	4,080	1,146	1,298

SET 7

376	593	842	938	697
2	7	5	9	8
752	4,151	4,210	8,442	5,576

SET 8

2143	1,728	5,914	7,896	4,789
2	4	5	7	9
4286	6,912	29,570	55,272	43,101

SET 9

24239	69,382	49,663	39,145	84,574
4	9	7	6	8
96956	624,438	347,641	234,870	676,592

SET 10

7	8	9	3	6
13	36	537	2182	72,589
91	288	4,833	6,546	435,534

SET 11   ## SET 12

20	60	4	5	50	700	9	200	100	6
8	5	70	80	9	3	900	5	8	800
160	300	280	400	450	2,100	8,100	1,000	800	4,800

SET 13   ## SET 14

104	203	2	3,012	34,032	208	405	1,806	8	46,084
2	3	304	3	2	3	4	7	9,072	5
208	609	608	9,036	68,064	624	1,620	12,642	72,576	230,420

SET 15

5280	1,760	9,500	82,000	6
8	4	7	5	30,000
42240	7,040	66,500	410,000	180,000

SET 16

2004	4,008	30,049	80,005	90,026
2	7	3	6	9
4008	28,056	90,147	480,030	810,234

SET 17

3040	4,070	6,080	50,700	40,030
7	6	5	9	8
21280	24,420	30,400	456,300	320,240

SET 18

1. 8×21 168   **5.** $7 \times 13,758$ 96,306   **9.** 605×4 2,420
2. 4×923 3,692   **6.** 5×700 3,500   **10.** $5 \times 2,006$ 10,030
3. 6×709 4,254   **7.** $8 \times 5,030$ 40,240
4. $3 \times 5,762$ 17,286   **8.** 351×9 3,159

SET 19

1. 4×7 trillion   28 trillion   **3.** 6×152 million   912 million   **5.** $3 \times \$238$ billion   \$714 billion
2. 8×29 billion   232 billion   **4.** $9 \times \$84$ billion   \$756 billion   **6.** $5 \times 4,596$ million   22,980 million

SET 20

Multiply:

1. 26 by 4 104   **5.** 5,975 by 8 47,800   **9.** 5,280 by 7 36,960
2. 752 by 7 5,264   **6.** 5,000 by 3 15,000   **10.** 16,000 by 5 80,000
3. 6 by 401 2,406   **7.** 3,002 by 9 27,018
4. 3,600 by 5 18,000   **8.** 1,728 by 4 6,912

SET 21

Find the product of:

1. 144 and 9 1,296
2. 6 and 231 1,386
3. 640 and 8 5,120
4. 5 and 320 1,600

5. 4 and 1,000 4,000
6. 5,175 and 3 15,525
7. 9 and 2,446 22,014
8. 6,080 and 7 42,560

150,300
9. 25,050 and 6
10. 81,041 and 2
162,082

DIAGNOSTIC TEST B

11. 727,665,698
12. 2,921,270,634

Multiply and check:

1. 37 24 888	**6.** 36 9,967 358,812	**11.** 74,686 9,743	**16.** 693 907 628,551
2. 78 56 4,368	**7.** 592 231 136,752	**12.** 38,457 75,962	**17.** 5,009 69 345,621
3. 485 92 44,620	**8.** 6,342 358 2,270,436	**13.** 8,500 54 459,000	**18.** 40,603 28 1,136,884
4. 6,948 89 618,372	**9.** 16,959 786 13,329,774	**14.** 700 500 350,000	**19.** 8,001 306 2,448,306
5. 45,847 65 2,980,055	**10.** 4,574 1,728 7,903,872	**15.** 208 144 29,952	**20.** 6,080 705 4,286,400

21. 384 × 597 229,248
22. 36 × 407 × 743 10,886,436
23. Multiply 75 by 49. 3,675

24. Find the product of 144 and 24. 3,456
25. 36 × 50 million 1,800 million

RELATED PRACTICE EXAMPLES

Multiply and check:

SET 1

23	28	42	19	53
12	25	21	37	14
276	700	882	703	742

SET 2

72	93	35	57	74
18	27	49	68	96
1,296	2,511	1,715	3,876	7,104

SET 3

144	526	967	231	897
23	42	36	47	88
3,312	22,092	34,812	10,857	78,936

SET 4

4113	6,374	8,439	1,728	6,582
21	35	78	93	46
86,373	223,090	658,242	160,704	302,772

SET 5

93153	38,642	68,459	84,696	54,587
24	85	47	63	72
2235672	3,284,570	3,217,573	5,335,848	3,930,264

SET 6

24	53	32	25	48
312	659	2,978	68,426	76,383
7,488	34,927	95,296	1,710,650	3,666,384

SET 7   SET 8

144	975	347	583	886	9645	2,786	149	8,327	4,784
324	638	231	963	697	529	231	9,687	524	379
46,656	622,050	80,157	561,429	617,542	5,102,205	643,566	1,443,363	4,363,348	1,813,136

SET 9

21462	78,356	65,815	764	15,647
344	492	873	34,687	989
7382928	38,551,152	57,456,495	26,500,868	15,474,883

SET 10

2467	6,374	8,622	9,675	4,569
1236	2,462	7,393	8,326	6,271
3049212	15,692,788	63,742,446	80,554,050	28,652,199

SET 11

12345	64,632	56,397	3,846	96,749
1728	4,436	8,457	86,798	8,795
21332160	286,707,552	476,949,429	333,825,108	850,907,455

SET 12

23814	49,279	57,625	91,352	42,387
16523	82,536	34,719	73,858	29,998
393478722	4,067,291,544	2,000,682,375	6,747,076,016	1,271,525,226

SET 13   SET 14

50	300	6,000	2,240	3,600	59	8,963	365	640	5,280
62	27	24	38	375	30	40	200	3,000	15,000
3,100	8,100	144,000	85,120	1,350,000	1,770	358,520	73,000	1,920,000	79,200,000

SET 15   SET 16

501	409	603	803	3,205	529	69	2,534	59,735	3,842
26	32	59	144	475	706	608	109	203	8,067
13,026	13,088	35,577	115,632	1,522,375	373,474	41,952	276,206	12,126,205	30,993,414

SET 17

3006	4,009	6,008	9,005	80,006
34	87	365	2,183	176
102204	348,783	2,192,920	19,657,915	14,081,056

SET 18

2050	6,080	60,506	40,702	329
16	53	125	6,713	7,070
32800	322,240	7,563,250	273,232,526	2,326,030

SET 19

4003	3,009	7,006	20,058	80,002
205	6,082	7,006	1,009	74,004
820615	18,300,738	49,084,036	20,238,522	5,920,468,008

SET 20

6080	50,306	73,050	46,050	50,903
203	504	8,007	2,500	60,704
1234240	25,354,224	584,911,350	115,125,000	3,090,015,712

SET 21

1. 24 × 35 840
2. 87 × 832 72,384
3. 56 × 4,973 278,488
4. 68 × 13,764 935,952
5. 156 × 849 132,444

6. 523 × 6,942 3,630,666
7. 709 × 52,723 37,380,607
8. 408 × 2,050 836,400
9. 200 × 503 100,600
10. 4,520 × 3,006 13,587,120

SET 22

1. 42 × 37 × 23 35,742
2. 63 × 524 × 201 6,635,412
3. 30 × 605 × 178 3,230,700
4. 264 × 671 × 9 1,594,296
5. 829 × 72 × 485 28,948,680

6. 273 × 100 × 126 3,439,800
7. 3,000 × 20 × 50 3,000,000
8. 10 × 6,020 × 200 12,040,000
9. 845 × 172 × 674 97,959,160
10. 3,568 × 4,123 × 9,035
132,912,656,240

SET 23

Multiply:
1. 32 by 16 512
2. 50 by 29 1,450
3. 18 by 45 810
4. 296 by 83 24,568
5. 144 by 97 13,968

6. 314 by 52 16,328
7. 400 by 125 50,000
8. 871 by 408 355,368
9. 693 by 834 577,962
10. 1,265 by 200 253,000

SET 24

Find the product of:
1. 56 and 17 952
2. 48 and 70 3,360
3. 253 and 85 21,505
4. 72 and 481 34,632
5. 953 and 249 237,297

6. 405 and 840 340,200
7. 9,700 and 82 795,400
8. 762 and 300 228,600
9. 649 and 206 133,694
10. 5,280 and 487 2,571,360

SET 25 $28,288 million

980 trillion
1. 70 × 14 trillion
2. 25 × 98 billion
2,450 billion

3. 68 × $416 million
4. 100 × $39 billion
$3,900 billion

45,167 million
5. 31 × 1,457 million
6. 475 × 620 billion
294,500 billion

PROBLEM SOLVING

See page 16, *Problem Solving Strategy.*
For problems 1–5, select the letter corresponding to your answer.

1. How far can a car go on a tankful of gasoline if it averages 6 kilometers on a liter and the tank holds 80 liters?

a. 470 kilometers   c. 480 kilometers

b. 380 kilometers   d. Answer not given

2. Each of the 26 major baseball teams has 25 players on its team roster. How many major league players are there in all?

a. 640 players   b. 650 players   c. 550 players   d. 560 players

3. David's brother wants to buy an automobile costing $4,000. If he pays $500 in cash, he could pay off the balance in 36 monthly installments of $113 each. How much could he save by paying the entire sum at the time of purchase?

a. $468   b. $486   c. $648   d. $568

4. At a speed of 299,851 kilometers per second, how far does light travel in 1 hour?

a. 18,001,960 kilometers   c. 17,901,960 kilometers

b. 17,991,060 kilometers   d. Answer not given

5. In the Martin Luther King High School there are 8 classes with 40 pupils on roll, 15 classes with 41 pupils, and 13 classes with 39 pupils. What is the total enrollment of the school?

a. 1,542 pupils   b. 1,432 pupils   c. 1,442 pupils   d. 1,532 pupils

6. Find the total number of books needed to make 29 class sets of 36 books each? 1,044 books

7. At an average ground speed of 725 kilometers per hour, how far can an airplane fly in 14 hours? 10,150 kilometers

8. If a box contains 144 envelopes, how many envelopes will there be in 26 boxes? 3,744 envelopes

9. How many sheets of writing paper are in 19 packages, each containing 500 sheets? 9,500 sheets

10. The auditorium of Franklin High School has 26 rows of 24 seats each. In the lunch room there are 32 tables each accommodating 17 pupils. What is the seating capacity of the auditorium? Of the lunch room? Which is greater? How much greater? 624; 544; auditorium by 80 seats

11. Find the leading scorer in the school football league. Thomas made 13 touchdowns, 2 field goals, and 5 points after touchdown. Wilson made 14 touchdowns and 4 points after touchdown. Harrison made 12 touchdowns, 3 field goals, and 9 points after touchdown. Harrison with 90 points

12. Lisa's mother saves $250 every month out of her monthly salary of $1,325. What is her annual salary? How much does she save in a year? $15,900; $3,000

13. A merchant bought 48 radios at $59 each. He sold them at $76 each. How much profit did he make altogether? $816

14. If 1 centimeter represents 125 kilometers, how many kilometers apart are two cities 12 centimeters apart on the chart? 1,500 kilometers

15. A farmer has 39 rows of pear trees each containing 26 trees. He expects each tree to produce on an average 29 bushels of pears. How many bushels of pears does the farmer expect from his trees? 29,406 bushels

16. A merchant purchased 18 dozen shirts at $59 per dozen, 16 dozen ties at $35 per dozen, 30 dozen pairs of socks at $11 per dozen, and 9 dozen belts at $27 per dozen. Find the total cost of the merchandise. $2,195

17. If sound travels at a speed of 332 meters per second, how far does it travel in 1 hour? 1,195,200 meters per hour

18. Marie's average reading rate is 205 words per minute. How many words can she read in half an hour? 6,150 words

19. Andrew can type an average of 48 words per minute. His term report contains 2,136 words. Will he be able to type it in 45 minutes? yes

20. If there are 179 days in the average school year, what is the total number of days one attends the elementary and secondary schools (12 years) when one's attendance is perfect? 2,148 days

For a Computer Activity see p. T39.

REFRESH YOUR SKILLS

1. Write 19,207,485,000,000 in words. Nineteen trillion, two hundred seven billion, four hundred eighty-five million

2. Write the numeral naming:

Four million, twenty-six thousand, seven hundred ninety-six 4,026,796

3. Write the complete numeral naming 75 *billion*. 75,000,000,000

4. Round 5,963,499,716 to the nearest million. 5,963,000,000

5. Add:	**6.** Subtract:	**7.** Multiply:	**8.** Multiply:
9,832	105,000	807	1,728
17,496	98,634	609	853
50,957	6,366	491,463	1,473,984
8,268			
594			
87,147			

EXERCISE

1-7 Division of Whole Numbers*

VOCABULARY: dividend
divisor
quotient
remainder
partial dividend
division signs

OBJECTIVE:

I. Aim To divide whole numbers.

II. Procedure

1. Find the quotient figure by dividing the one-figure divisor or the trial divisor, when it contains more than one figure, into the first figure or, if necessary, into the first two figures of the dividend. When the divisor contains two or more figures, use as the trial divisor the first figure of the divisor if the next figure on the right is 0, 1, 2, 3, 4, or 5, and increase the first figure of the divisor by one (1) if the next figure on the right is 6, 7, 8, or 9. The largest trial quotient figure that can be used at any one time is 9. (See step 1 of sample solution 2.)
2. Multiply the whole divisor by this quotient figure. Write this product under the corresponding figures in the dividend. If this product is greater than the partial dividend, then use as the trial quotient a figure one less than the figure first tried. (See step 2 of sample solution 2.)
3. Subtract this product from the corresponding numbers in the dividend. If the remainder is greater than the whole divisor, then use as the trial quotient figure one more than the figure tried. (See step 3 of sample solution 2.)
4. Bring down the next figure of the dividend and annex it to the remainder, if any. (See step 4 of sample solution 2.)
5. Using the remainder and the annexed numbers as partial dividends, repeat steps 1 to 4 until the remainder is no longer divisible.
6. Check by multiplying the quotient by the divisor and adding the remainder, if any, to the product. The result should equal the dividend. (See sample solutions 3 and 4.)
7. Numbers named by shortened numerals may be divided. (See sample solution 5.)

III. Sample Solutions

1. Divide 42,574 by 7.

Solution:

$$\begin{array}{r} 6{,}082 \leftarrow \text{quotient} \\ \text{divisor} \rightarrow 7\overline{)42{,}574} \leftarrow \text{dividend} \end{array}$$

Check:

$$\begin{array}{r} 6{,}082 \\ \times 7 \\ \hline 42{,}574 \end{array}$$

Answer: 6,082

*See "Tests for Divisibility" page 66.

Related Resources: *Practical Applications in Mathematics:* Diag. Tests A and B, and Prac. Ex., pp. 11–13

2. Divide 986 by 29.

Step 1	*Step 2*	*Step 3*	*Step 4*
$\dfrac{3}{29)\overline{986}}$	$\dfrac{3}{29)\overline{986}}$ 87	$\dfrac{3}{29)\overline{986}}$ $\dfrac{87}{11}$	$\dfrac{3}{29)\overline{986}}$ $\dfrac{87}{116}$ ← partial dividend

Repeat Step 1.	*Repeat Step 2.*	*Repeat Step 3.*	*Check:*
34 29)986 87 — 116	34 29)986 87 — 116 116	34 29)986 87 — 116 116 . . .	34 × 29 — 306 68 — 986

Answer: 34

3. Divide 45,034 by 89.

Solution:	*Check:*
506 89)45,034 44 5 — 534 534 . . .	506 89 — 4 554 40 48 — 45,034

Answer: 506

4. Divide 70,861 by 165.

Solution:	*Check:*
429 165)70,861 66 0 — 4 86 3 30 — 1 561 1 485 — Remainder → 76	429 165 — 2 145 25 74 42 9 — 70,785 76 — 70,861

Answer: 429 R76 or $429\frac{76}{165}$

5. Divide 72 million by 24.

$$\frac{3 \text{ million}}{24)\overline{72 \text{ million}}}$$

Answer: 3 million

IV. Definitions

1. The dividend is the number you divide.
2. The divisor is the number by which you divide.
3. The quotient is the answer in division.
4. When the division is not exact, the number left is the remainder
5. The partial dividend is the part of the dividend annexed to the remainder.
6. The symbols ÷ and $\overline{\ )}$ indicate division

PRELIMINARY EXAMPLES (all students)

Basic Division Facts

Divide:

1. $3\overline{)6}$ 2 $4\overline{)28}$ 7 $8\overline{)8}$ 1 $2\overline{)18}$ 9 $9\overline{)0}$ 0 $5\overline{)20}$ 4 $1\overline{)1}$ 1 $6\overline{)24}$ 4 $7\overline{)42}$ 6

2. $2\overline{)14}$ 7 $3\overline{)15}$ 5 $7\overline{)35}$ 5 $1\overline{)9}$ 9 $5\overline{)10}$ 2 $4\overline{)32}$ 8 $6\overline{)0}$ 0 $8\overline{)32}$ 4 $9\overline{)63}$ 7

3. $5\overline{)35}$ 7 $2\overline{)2}$ 1 $9\overline{)45}$ 5 $6\overline{)36}$ 6 $7\overline{)21}$ 3 $1\overline{)0}$ 0 $3\overline{)21}$ 7 $4\overline{)12}$ 3 $8\overline{)24}$ 3

4. $6\overline{)48}$ 8 $9\overline{)18}$ 2 $3\overline{)0}$ 0 $5\overline{)45}$ 9 $1\overline{)7}$ 7 $2\overline{)4}$ 2 $8\overline{)16}$ 2 $7\overline{)63}$ 9 $4\overline{)4}$ 1

5. $4\overline{)0}$ 0 $6\overline{)18}$ 3 $5\overline{)5}$ 1 $8\overline{)72}$ 9 $2\overline{)6}$ 3 $3\overline{)27}$ 9 $9\overline{)81}$ 9 $1\overline{)5}$ 5 $7\overline{)14}$ 2

6. $1\overline{)3}$ 3 $8\overline{)48}$ 6 $4\overline{)24}$ 6 $7\overline{)7}$ 1 $6\overline{)42}$ 7 $9\overline{)54}$ 6 $3\overline{)9}$ 3 $2\overline{)16}$ 8 $5\overline{)0}$ 0

7. $9\overline{)36}$ 4 $5\overline{)40}$ 8 $1\overline{)2}$ 2 $8\overline{)0}$ 0 $3\overline{)3}$ 1 $4\overline{)20}$ 5 $2\overline{)10}$ 5 $7\overline{)28}$ 4 $6\overline{)54}$ 9

8. $7\overline{)56}$ 8 $1\overline{)4}$ 4 $4\overline{)16}$ 4 $9\overline{)72}$ 8 $8\overline{)64}$ 8 $6\overline{)6}$ 1 $5\overline{)15}$ 3 $2\overline{)12}$ 6 $3\overline{)18}$ 6

9. $8\overline{)40}$ 5 $4\overline{)36}$ 9 $6\overline{)30}$ 5 $3\overline{)12}$ 4 $5\overline{)25}$ 5 $1\overline{)8}$ 8 $7\overline{)0}$ 0 $9\overline{)27}$ 3 $2\overline{)8}$ 4

10. $4\overline{)8}$ 2 $2\overline{)0}$ 0 $7\overline{)49}$ 7 $5\overline{)30}$ 6 $6\overline{)12}$ 2 $9\overline{)9}$ 1 $3\overline{)24}$ 8 $8\overline{)56}$ 7 $1\overline{)6}$ 6

DIAGNOSTIC TEST A

Divide and check:

1. $2\overline{)68}$ 34 6. $7\overline{)427}$ 61 11. $8\overline{)3,928}$ 491 15. $4\overline{)804}$ 201

2. $3\overline{)78}$ 26 7. $6\overline{)552}$ 92 12. $4\overline{)96,852}$ 24,213 16. $6\overline{)642}$ 107

3. $2\overline{)846}$ 423 8. $4\overline{)8,448}$ 2,112 13. $7\overline{)36,533}$ 5,219 17. $2\overline{)6,008}$ 3,004

4. $4\overline{)928}$ 232 9. $3\overline{)8,526}$ 2,842 14. $3\overline{)6,900}$ 2,300 18. $5\overline{)5,030}$ 1,006

5. $6\overline{)834}$ 139 10. $6\overline{)9,852}$ 1,642

Write quotient and remainder:

19. $7\overline{)156}$ 22 R2   20. $4\overline{)6,923}$ 1,730 R3

Write remainder as fraction in lowest terms:

21. $6\overline{)935}$ $155\frac{5}{6}$ 22. $8\overline{)4,668}$ $583\frac{1}{2}$ 23. $7\overline{)843}$ $120\frac{3}{7}$

Divide as indicated:

24. $8,795 \div 5$ 25. Divide 3,870 by 9.   26. 54 billion $\div$ 6
     1,759   430   9 billion

RELATED PRACTICE EXAMPLES

Divide and check:
Basic: Sets 1–19(A), 1–24(B), Practice 1, Prob. 1–2
Average: Sets 1–23(A), 1–30(B), Practice 1–2, Prob. 1–4
Enriched: Sets 1–26(A), 1–33(B), Practice 1–2, Prob. 1–5

SET 1

$\overset{31}{3\overline{)93}}$ $\overset{23}{2\overline{)46}}$ $\overset{}{3\overline{)66}}$ 22 $\overset{21}{4\overline{)84}}$ $\overset{14}{2\overline{)28}}$

SET 2

$\overset{17}{2\overline{)34}}$ $\overset{29}{3\overline{)87}}$ $\overset{}{5\overline{)85}}$ 17 $\overset{16}{6\overline{)96}}$ $\overset{16}{4\overline{)64}}$

SET 3

$\overset{231}{3\overline{)693}}$ $\overset{341}{2\overline{)682}}$ $\overset{}{4\overline{)488}}$ 122 $\overset{323}{2\overline{)646}}$ $\overset{122}{3\overline{)366}}$

SET 4

$\overset{141}{6\overline{)846}}$ $\overset{124}{4\overline{)496}}$ $\overset{}{3\overline{)819}}$ 273 $\overset{114}{7\overline{)798}}$ $\overset{121}{8\overline{)968}}$

SET 5

$\overset{236}{4\overline{)944}}$ $\overset{149}{5\overline{)745}}$ $\overset{}{6\overline{)918}}$ 153 $\overset{287}{3\overline{)861}}$ $\overset{126}{7\overline{)882}}$

SET 6

$\overset{41}{6\overline{)246}}$ $\overset{81}{9\overline{)729}}$ $\overset{}{2\overline{)128}}$ 64 $\overset{72}{3\overline{)216}}$ $\overset{92}{4\overline{)368}}$

SET 7

$\overset{72}{7\overline{)504}}$ $\overset{87}{4\overline{)348}}$ $\overset{}{6\overline{)468}}$ 78 $\overset{63}{5\overline{)315}}$ $\overset{76}{8\overline{)608}}$

SET 8

$\overset{3224}{2\overline{)6448}}$ $\overset{2,121}{4\overline{)8,484}}$ $\overset{}{3\overline{)9,636}}$ 3,212 $\overset{1,243}{2\overline{)2,486}}$ $\overset{2,331}{3\overline{)6,993}}$

SET 9

$\overset{1651}{5\overline{)8255}}$ $\overset{1,241}{8\overline{)9,928}}$ $\overset{}{6\overline{)6,894}}$ 1,149 $\overset{3,246}{3\overline{)9,738}}$ $\overset{2,392}{4\overline{)9,568}}$

SET 10

$\overset{2497}{3\overline{)7491}}$ $\overset{2,289}{4\overline{)9,156}}$ $\overset{}{6\overline{)8,748}}$ 1,458 $\overset{4,787}{2\overline{)9,574}}$ $\overset{1,498}{5\overline{)7,490}}$

SET 11

$\overset{749}{7\overline{)5243}}$ $\overset{359}{6\overline{)2,154}}$ $\overset{}{8\overline{)6,792}}$ 849 $\overset{848}{4\overline{)3,392}}$ $\overset{764}{6\overline{)4,584}}$

SET 12

$\overset{23292}{3\overline{)69876}}$ $\overset{13,792}{7\overline{)96,544}}$ $\overset{}{4\overline{)95,860}}$ 23,965 $\overset{38,476}{2\overline{)76,952}}$ $\overset{14,299}{6\overline{)85,794}}$

$$\overset{9122}{8\overline{)72976}} \qquad \overset{7,492}{5\overline{)37,460}} \qquad \textbf{SET 13}\ \overset{5,856}{7\overline{)40,992}} \qquad \overset{4,865}{3\overline{)14,595}} \qquad \overset{7,496}{4\overline{)29,984}}$$

$$\overset{210}{4\overline{)840}} \qquad \overset{4,840}{2\overline{)9,680}} \qquad \textbf{SET 14}\ \overset{1,200}{3\overline{)3,600}} \qquad \overset{400}{5\overline{)2,000}} \qquad \overset{15,000}{6\overline{)90,000}}$$

$$\overset{304}{2\overline{)608}} \qquad \overset{301}{3\overline{)903}} \qquad \textbf{SET 15}\ \overset{2,041}{2\overline{)4,082}} \qquad \overset{2,102}{4\overline{)8,408}} \qquad \overset{3,020}{3\overline{)9,060}}$$

$$\overset{105}{5\overline{)525}} \qquad \overset{209}{4\overline{)836}} \qquad \textbf{SET 16}\ \overset{1,052}{8\overline{)8,416}} \qquad \overset{11,203}{6\overline{)67,218}} \qquad \overset{40,553}{2\overline{)81,106}}$$

$$\overset{3001}{3\overline{)9003}} \qquad \overset{4,002}{2\overline{)8,004}} \qquad \textbf{SET 17}\ \overset{30,012}{2\overline{)60,024}} \qquad \overset{10,002}{4\overline{)40,008}} \qquad \overset{10,001}{5\overline{)50,005}}$$

$$\overset{2004}{3\overline{)6012}} \qquad \overset{4,006}{6\overline{)24,036}} \qquad \textbf{SET 18}\ \overset{20,008}{4\overline{)80,032}} \qquad \overset{5,005}{2\overline{)10,010}} \qquad \overset{30,090}{2\overline{)60,180}}$$

SET 19

Write quotient and remainder: $\overset{763\ R5}{6\overline{)4,583}} \qquad \overset{3,949\ R2}{7\overline{)27,645}} \qquad \overset{24,655\ R2}{4\overline{)98,622}}$

$5\overline{)14}$ 2 R4 $3\overline{)553}$ 184 R1

SET 20

Write quotient and remainder: $\overset{2,860\ R2}{3\overline{)8,582}} \qquad \overset{7,680\ R3}{5\overline{)38,403}} \qquad \overset{16,400\ R5}{6\overline{)98,405}}$

$6\overline{)65}$ 10 R5 $8\overline{)243}$ 30 R3

SET 21

Write remainder as a fraction in lowest terms: $\overset{776\frac{1}{6}}{6\overline{)4,657}} \qquad \overset{11,218\frac{5}{8}}{8\overline{)89,749}}$

$9\overline{)35}$ $3\frac{8}{9}$ $5\overline{)74}$ $14\frac{4}{5}$ $4\overline{)685}$ $171\frac{1}{4}$

SET 22

Write remainder as a fraction in lowest terms: $\overset{4,697\frac{1}{2}}{8\overline{)37,580}} \qquad \overset{9,895\frac{2}{3}}{6\overline{)59,374}}$

$4\overline{)78}$ $19\frac{1}{2}$ $6\overline{)940}$ $156\frac{2}{3}$ $8\overline{)7,534}$ $941\frac{3}{4}$

SET 23

Write remainder as a fraction in lowest terms: $\overset{1,090\frac{5}{9}}{9\overline{)9,815}} \qquad \overset{12,180\frac{4}{7}}{7\overline{)85,264}}$

$6\overline{)365}$ $60\frac{5}{6}$ $8\overline{)644}$ $80\frac{1}{2}$ $3\overline{)812}$ $270\frac{2}{3}$

SET 24

1. 438 ÷ 3 146
2. 8,694 ÷ 2 4,347
3. 69,376 ÷ 9 7,708$\frac{4}{9}$
4. 42,965 ÷ 6 7,160$\frac{5}{6}$
5. 94,500 ÷ 4 23,625

6. 37,624 ÷ 8 4,703
7. 23,583 ÷ 7 3,369
8. 84,215 ÷ 5 16,843
9. 72,708 ÷ 9 8,078$\frac{2}{3}$
10. 91,862 ÷ 6 15,310$\frac{1}{3}$

SET 25

1. Divide 932 by 4. 233
2. Divide 5,795 by 8. 724$\frac{3}{8}$
3. Divide 3,482 by 9. 386$\frac{8}{9}$
4. Divide 83,007 by 3. 27,669
5. Divide 94,075 by 2. 47,037$\frac{1}{2}$

6. Divide 42,840 by 5. 8,568
7. Divide 51,382 by 6. 8,563$\frac{2}{3}$
8. Divide 65,424 by 8. 8,178
9. Divide 38,971 by 9. 4,330$\frac{1}{9}$
10. Divide 70,056 by 7. 10,008

SET 26

1. 45 trillion ÷ 5 9 trillion
2. 675 million ÷ 9 75 million
3. $208 billion ÷ 8 $26 billion

4. $772 million ÷ 4 $193 million
5. 2,814 million ÷ 7 million 402
6. 9,174 billion ÷ 6 billion 1,529

DIAGNOSTIC TEST B

Divide and check:

1. 24)96 4
2. 38)228 6
3. 27)567 21
4. 14)1,022 73
5. 26)8,164 314
6. 85)71,995 847
7. 48)61,584 1,283
8. 54)349,272 6,468
9. 60)28,980 483
10. 57)34,656 608
11. 36)230,400 6,400
12. 42)126,252 3,006
13. 63)27,044 429 R17

Write remainder as fraction in Ex. 14:

14. 72)45,018 625$\frac{1}{4}$
15. 144)864 6
16. 174)7,482 43
17. 298)11,026 37
18. 357)44,982 126
19. 946)810,722 857
20. 907)61,676 68
21. 506)405,306 801
22. 843)552,660 655 R495
23. 400)338,000 845
24. 391)185,805 475 R80

Write remainder as fraction in Ex. 25:

25. 144)75,984 527$\frac{2}{3}$
26. 1,760)14,080 8
27. 1,728)101,952 59
28. 8,526)3,939,012 462
29. 5,280)23,876,160 4,522
30. 8,005)4,306,690 538
31. 21,714 ÷ 231 94
32. Divide 300,960 by 5,280. 57
33. 945 million ÷ 45 21 million

RELATED PRACTICE EXAMPLES

Divide and check:

SET 1

$\overset{5}{16)\overline{80}}$
$\qquad$
$\overset{4}{23)\overline{92}}$
$\qquad$
$\overset{3}{27)\overline{81}}$ 3
$\qquad$
$\overset{3}{24)\overline{72}}$
$\qquad$
$\overset{4}{19)\overline{76}}$

SET 2

$\overset{7}{54)\overline{378}}$
$\qquad$
$\overset{7}{43)\overline{301}}$
$\qquad$
$\overset{}{48)\overline{432}}$ 9
$\qquad$
$\overset{4}{83)\overline{332}}$
$\qquad$
$\overset{8}{67)\overline{536}}$

SET 3

$\overset{28}{32)\overline{896}}$
$\qquad$
$\overset{14}{48)\overline{672}}$
$\qquad$
$\overset{}{29)\overline{667}}$ 23
$\qquad$
$\overset{27}{12)\overline{324}}$
$\qquad$
$\overset{41}{24)\overline{984}}$

SET 4

$\overset{65}{48)\overline{3120}}$
$\qquad$
$\overset{76}{24)\overline{1,824}}$
$\qquad$
$\overset{}{72)\overline{2,664}}$ 37
$\qquad$
$\overset{73}{95)\overline{6,935}}$
$\qquad$
$\overset{45}{41)\overline{1,845}}$

SET 5

$\overset{123}{37)\overline{4551}}$
$\qquad$
$\overset{343}{26)\overline{8,918}}$
$\qquad$
$\overset{}{34)\overline{7,344}}$ 216
$\qquad$
$\overset{198}{17)\overline{3,366}}$
$\qquad$
$\overset{186}{52)\overline{9,672}}$

SET 6

$\overset{548}{72)\overline{39456}}$
$\qquad$
$\overset{492}{85)\overline{41,820}}$
$\qquad$
$\overset{}{94)\overline{73,508}}$ 782
$\qquad$
$\overset{653}{46)\overline{30,038}}$
$\qquad$
$\overset{839}{77)\overline{64,603}}$

SET 7 2,876

$\overset{2415}{29)\overline{70035}}$
$\qquad$
$\overset{1,562}{32)\overline{49,984}}$
$\qquad$
$\overset{}{17)\overline{48,892}}$
$\qquad$
$\overset{3,253}{25)\overline{81,325}}$
$\qquad$
$\overset{1,562}{48)\overline{74,976}}$

SET 8 3,319

$\overset{5219}{43)\overline{224417}}$
$\qquad$
$\overset{9,758}{54)\overline{526,932}}$
$\qquad$
$\overset{}{64)\overline{212,416}}$
$\qquad$
$\overset{1,957}{71)\overline{138,947}}$
$\qquad$
$\overset{689\ R44}{89)\overline{61,365}}$

SET 9

$\overset{189}{30)\overline{5670}}$
$\qquad$
$\overset{749}{50)\overline{37,450}}$
$\qquad$
$\overset{}{70)\overline{61,040}}$ 872
$\qquad$
$\overset{523}{80)\overline{41,840}}$
$\qquad$
$\overset{926}{40)\overline{37,040}}$

SET 10

$\overset{705}{36)\overline{25380}}$
$\qquad$
$\overset{603}{69)\overline{41,607}}$
$\qquad$
$\overset{}{37)\overline{29,933}}$ 809
$\qquad$
$\overset{406}{55)\overline{22,330}}$
$\qquad$
$\overset{304}{83)\overline{25,232}}$

SET 11 4,800

$\overset{220}{15)\overline{3300}}$
$\qquad$
$\overset{900}{57)\overline{51,300}}$
$\qquad$
$\overset{}{30)\overline{144,000}}$
$\qquad$
$\overset{7,800}{46)\overline{358,800}}$
$\qquad$
$\overset{8,700}{66)\overline{574,200}}$

SET 12 4,003

$\overset{2004}{12)\overline{24048}}$
$\qquad$
$\overset{5,007}{24)\overline{120,168}}$
$\qquad$
$\overset{}{69)\overline{276,207}}$
$\qquad$
$\overset{7,008}{56)\overline{392,448}}$
$\qquad$
$\overset{3,004}{35)\overline{105,140}}$

SET 13 463 R15

$\overset{862\ R25}{56)\overline{48297}}$
$\qquad$
$\overset{258\ R16}{75)\overline{19,366}}$
$\qquad$
$\overset{}{82)\overline{37,981}}$
$\qquad$
$\overset{769\ R11}{48)\overline{36,923}}$
$\qquad$
$\overset{388\ R37}{64)\overline{24,869}}$

Write remainders as fractions:

$172\frac{2}{7}$ 14)2412 $473\frac{1}{2}$ 38)17,993   **SET 14** $916\frac{7}{52}$ 52)47,639 $2,450\frac{19}{24}$ 24)58,819 $8,347\frac{1}{3}$ 78)651,092

$3\frac{18}{71}$ 213)693   7   128)896   **SET 15**   427)3,416 8   3   640)1,920   6   569)3,414

52   173)8996   47   144)6,768   **SET 16**   321)8,025 25   23   234)5,382   34   285)9,690

48   229)10992   58   231)13,398   **SET 17**   929)52,024 56   69   745)51,405   75   144)10,800

629   152)95608   146   289)42,194   **SET 18**   223)66,008 296   159   475)75,525   284   337)95,708

754   863)650702   584   629)367,336   **SET 19** 399   847)337,953   565   768)433,920   746   487)363,302

37   208)7696   78   706)55,068   **SET 20**   302)28,086 93   62   405)25,110   98   504)49,392

301   106)31906   608   204)124,032   **SET 21** 407   903)367,521   708   502)355,416   409   805)329,245

320   478)152960   820   640)524,800   **SET 22** 590   372)219,480   810   679)549,990   420   725)304,500

426   200)85200   967   300)290,100   **SET 23** 584   700)408,800   609   500)304,500   820   600)492,000

$528\frac{113}{692}$ 692)365489 $862\frac{217}{375}$ 375)323,467   **SET 24** $463\frac{26}{927}$ 927)429,227 $736\frac{200}{589}$ 589)433,704 $674\frac{75}{866}$ 866)583,759

Write remainders as fractions:

$467\frac{1}{4}$ 320)149520 $816\frac{2}{21}$ 231)188,518   **SET 25** $285\frac{1}{3}$ 144)41,088 $471\frac{2}{5}$ 625)294,625 $740\frac{1}{2}$ 258)191,049

8   5280)42240   8   2,774)22,192   **SET 26** 9   2,240)20,160   9   8,304)74,736   5   4,568)22,840

SET 27

$$\overset{42}{6080\overline{)255360}} \qquad \overset{67}{1,728\overline{)115,776}} \qquad \overset{57}{9,286\overline{)529,302}} \qquad \overset{85}{3,974\overline{)337,790}}$$

$$\overset{}{5,280\overline{)153,120}} \; 29$$

SET 28

$$\overset{467}{8366\overline{)3906922}} \qquad \overset{470}{5,280\overline{)2,481,600}} \qquad \overset{490}{5,863\overline{)2,872,870}} \qquad \overset{942}{1,728\overline{)1,627,776}}$$

$$\overset{}{6,080\overline{)1,824,000}} \; 300$$

SET 29

$$\overset{5485}{2240\overline{)12286400}} \qquad \overset{3,756}{8,267\overline{)31,050,852}} \qquad \overset{9,842}{1,760\overline{)17,321,920}} \qquad \overset{4,582}{2,794\overline{)12,802,108}}$$

$$\overset{}{3,246\overline{)24,377,460}} \; 7,510$$

SET 30

$$\overset{654}{7006\overline{)4581924}} \qquad \overset{752}{6,080\overline{)4,572,160}} \qquad \overset{586}{3,600\overline{)2,109,600}} \qquad \overset{206}{9,004\overline{)1,854,824}}$$

$$\overset{}{3,007\overline{)1,202,800}} \; 400$$

SET 31	SET 32
1. $408 \div 24$ 17	**1.** Divide 5,616 by 144. 39
2. $93,940 \div 35$ 2,684	**2.** Divide 13,398 by 231. 58
3. $287,246 \div 41$ 7,006	**3.** Divide 153,600 by 320. 480
4. $173,712 \div 231$ 752	**4.** Divide 342,720 by 2,240. 153
5. $401,280 \div 5,280$ 76	**5.** Divide 1,487,200 by 1,760. 845

SET 33

1. 216 billion $\div$ 18 12 billion

2. \$713 million $\div$ 31 23 million

3. 4,002 million $\div$ 87 46 million

4. \$8,295 billion $\div$ 105 79 billion

5. 2,074 billion $\div$ 34 billion 61

6. 13,520 million $\div$ 26 million 520

TESTS FOR DIVISIBILITY

The following tests may be used to determine whether a given number is divisible (that is, can be divided exactly so that there is no remainder) by 2, 3, 4, 5, 6, 8, 9, or 10.

a. A number is divisible by 2 only if it ends in 0, 2, 4, 6, or 8. All even numbers are divisible by 2.

b. A number is divisible by 3 only if the sum of its digits is divisible by 3.

To test whether 4,971 is divisible by 3, first find the sum of the digits of 4,971 (4 + 9 + 7 + 1). This sum is 21, which is divisible by 3. Therefore, 4,971 is divisible by 3.

c. A number is divisible by 4 only if it is an even number* and the number represented by the last two digits (tens and units digits) is divisible by 4. Numbers ending in two zeros are divisible by 4.

To test whether the even number 92,136 is divisible by 4, check whether the number formed by the last two digits (36) is divisible by 4. Since 36 is divisible by 4, therefore 92,136 is divisible by 4.

d. A number is divisible by 5 only if it ends in 5 or 0.
The number 3,645 ends in a 5; therefore it is divisible by 5. The number 430 ends in a 0; therefore it is divisible by 5.

e. A number is divisible by 6 only if it is an even number and the sum of its digits is divisible by 3.

To test whether 9,558 is divisible by 6, first check whether 9,558 is an even number. Then check whether the sum of the digits of 9,558 (9 + 5 + 5 + 8) is divisible by 3. The sum of the digits is 27 which is divisible by 3. Therefore 9,558 is divisible by 6.

f. A number is divisible by 8 only if it is an even number and the number represented by the last three digits (hundreds, tens, and units digits) is divisible by 8. Numbers ending in three zeros are divisible by 8. Use this test only when the number is 1,000 or larger.

The even number 39,712 is divisible by 8. Observe that the number formed by the last three digits (712) is divisible by 8.

g. A number is divisible by 9 only if the sum of its digits is divisible by 9.

50,382 is divisible by 9. Observe that the sum of the digits of 50,382 (5 + 0 + 3 + 8 + 2) is 18 and 18 is divisible by 9.

h. A number is divisible by 10 only if it ends in 0.
9,670 is divisible by 10 because it ends in 0.

PRACTICE PROBLEMS

1. a. Is 68,532 divisible by 2? Yes   by 3? Yes   by 9? No   by 6? Yes
   b. Is 384,670 divisible by 4? No   by 2? Yes   by 5? Yes   by 10? Yes
   c. Is 165,501 divisible by 6? No   by 9? Yes   by 3? Yes   by 2? No
   d. Is 483,000 divisible by 10? Yes   by 5? Yes   by 8? Yes   by 3? Yes
   e. Is 759,258 divisible by 3? Yes   by 8? No   by 4? No   by 6? Yes

---
* See Ex. 1-52.

2. Determine whether the following numbers are divisible:

a. By 3:	417 Yes	2,853 Yes	7,415 No	29,538 Yes	593,618 No
b. By 5:	251 No	8,970 Yes	1,565 Yes	38,006 No	686,400 Yes
c. By 8:	4,794 No	1,352 Yes	86,727 No	75,000 Yes	827,223 No
d. By 2:	518 Yes	6,359 No	4,506 Yes	83,000 Yes	179,244 Yes
e. By 9:	837 Yes	5,706 Yes	89,784 Yes	40,389 No	779,895 Yes
f. By 4:	536 Yes	9,252 Yes	1,874 No	92,425 No	954,768 Yes
g. By 6:	914 No	8,349 No	6,858 Yes	27,291 No	906,702 Yes
h. By 10:	470 Yes	3,000 Yes	20,105 No	90,790 Yes	813,060 Yes

PROBLEM SOLVING

See page 16, *Problem Solving Strategy.*
For problems 1–3, select the letter corresponding to your answer.

1. Ms. Watson earns an annual salary of $18,468. What is her monthly salary?

    a. $1,549   **b.** $1,449   <u>**c.**</u> $1,539   **d.** Answer not given

2. Joan's mother bought a washing machine costing $395. If she paid $35 in cash and arranged to pay off the balance in 8 equal monthly payments, what is the amount of each payment?

    a. $48   **b.** $52   **c.** $55   <u>**d.**</u> $45

3. Two cities, 350 kilometers apart, are to be plotted on a chart with the scale 1 centimeter = 25 kilometers. How many centimeters apart should they be plotted?

    a. 12 centimeters   **c.** 16 centimeters
    <u>**b.**</u> 14 centimeters   **d.** 18 centimeters

4. How long will it take a person, driving at a speed of 80 kilometers per hour, to travel from Los Angeles to San Francisco, a distance of 680 kilometers? If the car averages 5 kilometers on a liter of gasoline, how many liters are required to make the trip? $8\frac{1}{2}$ hours; 136 liters

5. A ship reached port after steaming for a distance of 1,121 nautical miles at a speed of 19 knots (1 knot = 1 nautical mile per hour). How many hours did it take? 59 hours

Averages

1. In the last football game of the season the Hartville High School team gained 168 yards from scrimmage in 28 tries. How many yards gained did they average per try? 6 yards

2. The total attendance at the Bartram Junior High School for the 21 school days in March was 19,467. Find the average daily attendance. _{927 pupils}

3. An airplane left airport A at 1 P.M. and reached airport B, 956 kilometers away, at 4 P.M. Find the average speed of the airplane. $318\frac{2}{3}$ km/hr

4. In 16 basketball games, Ted Harris scored 83 field goals (2 points each) and 26 foul goals (1 point each). What is his average in points per game? 12 points

5. When Mr. Conroy had the tank of his car filled with gasoline, he noticed that the speedometer read 29,762. The next time he stopped for gasoline, it required 52 liters to fill the tank. If the speedometer then read 30,074, how many kilometers was the car averaging on a liter of gasoline?

6 km/L

To find the average of 2 or more numbers, add the given numbers, then divide the sum by the number of given numbers. If there are 2 numbers given, divide their sum by 2; if there are 3 numbers given, divide their sum by 3, etc.

6. Find the average of the following numbers:

a. 1,389 women and 2,563 women. 1,976 women

b. 46 pupils, 35 pupils, 38 pupils, 41 pupils, and 45 pupils. 41 pupils

c. 225 kilometers, 199 kilometers, and 206 kilometers. 210 km

d. 8 centimeters, 15 centimeters, 9 centimeters, and 17 centimeters.

e. $28, $19, $23, $34, $20, and $44. $28 $12\frac{1}{4}$ cm

7. What is Lucia's average in arithmetic if she receives the following marks: 87, 94, 72, 65, 81, 100, 75, and 90? 83

8. A class with an enrollment of 43 pupils had 1 absentee on Monday, 2 on Tuesday, 3 on Wednesday, 1 on Thursday, and 3 on Friday. Find the average daily attendance for the week. 41 pupils

9. What was the average daily temperature for a week in a western city if the temperatures were as follows: Monday, 15°; Tuesday, 16°; Wednesday, 19°; Thursday, 14°; Friday, 18°; Saturday, 14°; Sunday, 16°? 16°

10. The seven linemen on the State College football team starting with the left end weigh 77, 89, 100, 98, 106, 102, and 79 kilograms respectively. The backfield men weigh 76, 80, 82, and 90 kilograms. What is the average weight per man in the line? In the backfield? Of the eleven starting players? 93 kg; 82 kg; 89 kg

11. In the championship football game the four backs of the Jane Addams High School team played as follows: Johnson in 9 tries gained 108 yards, Bell in 13 tries gained 104 yards, Leonard in 4 tries gained 20 yards, and Tompkins in 3 tries gained 36 yards. Find each player's average gain per try. Find the team's average gain per try.

Johnson, 12 yd.; Bell, 8 yd.; Leonard, 5 yd.; Tompkins, 12 yd.; team's average, $9\frac{7}{29}$ yd.

12. The total enrollment in the 5 freshman sections at the Fairmount High School is 185 pupils; in 4 sophomore sections, 152 pupils; in 4 junior sections, 164 pupils; and in 3 senior sections, 123 pupils. What is the average pupil enrollment per section for each grade? For the school?
Fresh., 37; soph., 38; junior, 41; senior, 41; entire school, 39 pupils

---

REFRESH YOUR SKILLS

1. Round 409,261 to the nearest hundred. **409,300**

2. Add:	**3.** Subtract:	**4.** Multiply:	**5.** Divide:
916	483,251	6,080	708
3,827	479,628	905	875⟌619,500
54,296	——	——	
8,689	3,623	5,502,400	
26,778			
——			
94,506			

For a Class Activity see p. T33.   For a Computer Activity see p. T39.

ESTIMATING ANSWERS

To estimate a sum, a difference, a product or a quotient, first round the given numbers to convenient place values (see page 28) and then perform the operation or operations.

For each of the following select your nearest estimate:

1. 639 + 795 + 389 is approximately: 1,600   1,800   2,000

2. 8,960 − 3,012 is approximately: 5,000   4,000   6,000

3. 73 × 48 is approximately: 3,500   4,500   2,500

4. 810 ÷ 39 is approximately: 40   30   20

5. 605 × 98 is approximately: 60,000   600,000   6,000,000

6. 4,392 − 2,509 is approximately: 2,500   2,100   1,900

7. 345 + 927 + 775 is approximately: 1,800   1,900   2,000

8. 4,131 ÷ 51 is approximately: 85   80   90

9. 20,000 − 12,063 is approximately: 7,000   7,500   8,000

10. 406 × 9,117 is approximately: 37,000,000   3,700,000   370,000

11. 64,898 ÷ 103 is approximately: 63,000   6,300   630

12. 945 + 793 + 666 is approximately: 2,400   2,200   2,000

13. 1,493 × 704 is approximately: 100,000   10,000   1,000,000

14. 8,217 + 5,399 + 6,016 is approximately: 15,000   20,000   25,000

15. 39,758 ÷ 198 is approximately: 200   250   300

16. 56,302 − 9,294 is approximately: 50,000   47,000   53,000

17. 2,941 + 6,128 + 4,290 is approximately: 10,000   12,000   13,000

18. 594,849 ÷ 851 is approximately: 800   600   700

19. 8,523 × 6,973 is approximately: 60,000,000   50,000,000   48,000,000

20. 79,698 − 47,715 is approximately: 30,000   35,000   40,000

CHECK BY CALCULATOR

Use a calculator to check the following answers. Indicate the wrong answers. If a calculator is not available, check by computation.

1. Add: b and c are incorrect

	a.	b.	c.	d.	e.
	846	2,984	83,455	91,548	397,426
	5,863	5,979	6,908	28,307	586,979
	9,596	4,898	56,568	45,279	8,395
	688	3,419	9,675	74,486	65,074
	2,599	9,382	63,398	29,158	858,987
	19,592	26,762	219,004	268,778	1,916,861

2. Subtract: c and d are incorrect

a.	b.	c.	d.	e.
5,439	81,506	68,042	791,000	926,183
2,148	9,408	38,569	295,044	799,287
3,291	72,098	29,573	505,956	126,896

3. Multiply: a and d are incorrect

a. $326 \times 284 = 91,584$
b. $657 \times 469 = 308,133$
c. $5,309 \times 6,058 = 32,161,922$
d. $9,845 \times 7,699 = 75,696,655$

4. Divide: a and c are incorrect

a. $29,743 \div 49 = 67$
b. $45,105 \div 485 = 93$
c. $410,892 \div 706 = 592$
d. $785,519 \div 943 = 833$

For a Computer Activity see p. T39.

COMPETENCY CHECK TEST

Solve each problem and select the letter corresponding to your answer. The numerals in red boxes indicate Exercises where help may be found.

1. Write the numeral naming: Twelve thousand, six hundred forty ⬛1-2
   a. 12,640   **b.** 1264   **c.** 12,600,40   **d.** Answer not given

2. Write the complete numeral naming 49 *billion*. ⬛1-2
   a. 49,000   **b.** 49,000,000   **c.** 49,000,000,000,000   d. Answer not given

3. Round 87,493,056 to the nearest hundred thousand. ⬛1-3
   a. 87,000,000   **b.** 87,490,000   c. 87,500,000   **d.** 87,000,000,000

4. Add: 5,967   **a.** 33,465
           8,458   b. 33,565
           9,389   **c.** 32,456
           6,554   **d.** Answer
   ⬛1-4 3,197   not given

5. Subtract: 53,046   **a.** 21,057
               32,989   **b.** 20,157
                           c. 20,057
                           d. Answer
        ⬛1-5   not given

6. Multiply: 798×907 **a.** 723,876   b. 723,786   **c.** 732,786   **d.** 732,876   ⬛1-6

7. Divide: 72$\overline{)43,848}$ **a.** 69   **b.** 604   c. 609   **d.** 64   ⬛1-7

1. a. Four million, nine hundred ninety-eight thousand, two hundred eighty-one
b. Sixty-seven million, fifty thousand, three

72   **WHOLE NUMBERS**

REVIEW OF UNIT ONE

c. Seven hundred three million, four hundred twenty-eight thousand, nine hundred eighty-six

1. Write each of the following numerals in words:

a. 4,998,281   **b.** 67,050,003   **c.** 703,428,986

d. 2,000,083,000   **e.** 401,575,700,000   d. Two billion, eighty-three thousand

e. Four hundred one billion, five hundred seventy-five million, seven hundred thousand

2. Write the numeral naming each of the following:

a. Nine hundred six thousand, twenty-five 906,025

b. Seven million, eleven thousand, four hundred seventy-three 7,011,473

c. Sixteen billion, three hundred fifteen million, two hundred fifty thousand 16,315,250,000

3. Write the complete numeral naming:

a. 928 billion   **b.** 17 trillion   **c.** 2,509 million
928,000,000,000   17,000,000,000,000   2,509,000,000

4. Round: **a.** 275,648 to the nearest hundred; 275,600

           b. 19,841,007 to the nearest million; 20,000,000

           c. 386,248,500 to the nearest hundred thousand; 386,200,000

           d. 7,009,989 to the nearest thousand; 7,010,000

           e. 4,952,426 to the nearest ten thousand. 4,950,000

5. Add:

a.	**b.**	**c.**	**d.**
5894	68,299	82,625	659,693
7579	85,067	49,836	429,774
6658	3,188	76,257	787,309
8784	59,083	34,967	169,497
9679	4,348	25,814	858,935
38594	219,985	269,499	589,788
			3,494,996

e. Find the sum of: 462; 8,395; 1,984; 26; 3,572 14,439

6. Subtract:

a.	**b.**	**c.**	**d.**
8392	25,703	610,324	7,800,000
5487	9,698	280,719	5,975,008
2905	16,005	329,605	1,824,992

e. Subtract 438,794 from 647,600. 208,806

7. Multiply:

a.	**b.**	**c.**	**d.**
48	207	896	1,509
93	900	759	3,084
4,464	186,300	680,064	4,653,756

e. Multiply 685 by 45. 30,825

8. Divide:

a. $67\overline{)3953}$ 59   **b.** $75\overline{)29{,}850}$ 398   **c.** $905\overline{)8{,}913{,}345}$ 9,849

d. $5{,}280\overline{)4{,}609{,}440}$ 873   **e.** Divide 561,660 by 138. 4,070

For problems 9 and 10, solve each problem and select the letter in red corresponding to your answer.

9. The Amazon River is 6,276 kilometers long and the St. Lawrence is 3,130 kilometers long. How many kilometers longer is the Amazon River?

    a. 3,246 km   b. 3,156 km   c. 3,256 km   d. 3,146 km

10. A boat sailed from Panama to Jacksonville, a distance of 2,550 kilometers in 75 hours. What speed did it average for the trip?

    a. 36 km/h   b. 26 km/h   c. 34 km/h   d. Answer not given

11. Last year Joan's mother received $12,350 salary, $4,785 commission, and $495 bonus. Find the total amount of her earnings for the year. $17,630

12. An electric appliance dealer during the last month purchased 15 refrigerators at $398 each, 36 electric irons at $15 each, 20 washing machines at $265 each, 12 clothes dryers at $179 each, 24 television sets at $307 each, and 48 digital clock radios at $34 each. What was the total cost of the merchandise? $22,958

13. Find the increase in population in the Northeast region of the United States if recently the population was 48,999,999 and ten years earlier it was 44,667,819. 4,332,180

14. Alaska has a land area of 1,530,858 square kilometers and an inland water area of 39,718 square kilometers while Hawaii has a land area of 16,615 square kilometers and an inland water area of 23 square kilometers. Find the total area of each state and the difference in their total areas. Alaska 1,570,576 km²; Hawaii 16,638 km²; 1,553,938 km²

KEYED ACHIEVEMENT TEST

For additional practice turn to the Exercises indicated by the numerals in the red boxes.

1. Write 4,010,001 in words. 1-1 Four million, ten thousand, one

2. Write the numeral naming: Three million, seven hundred thousand 1-2 3,700,000

3. Write the complete numeral naming 535 *million*. 1-2 535,000,000

4. Round 1,290,075,986 to the nearest million. 1-3 1,290,000,000

5. Add:

68,426
99,587
42,686
87,967
95,989 1-4
394,655

6. Subtract:

724,005
296,486 1-5
427,519

7. Multiply:

8,004
609 1-6
4,874,436

8. Divide: 9,060
869)7,873,140 1-7

For a Unit Test see p. T42 or *Testing Program,* p. 3.

UNIT **Two**

COMMON FRACTIONS

When a thing or unit is divided into equal parts, the number expressing the relation of one or more of the equal parts to the total number of equal parts is called a fraction

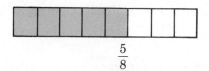

$$\frac{5}{8}$$

When a group of things is divided into equal parts, the number expressing the relation of one or more of the equal parts of the group to the total number of equal parts in the group is also considered to be a fraction.

Fractions that are expressed by a symbol consisting of a pair of numerals, one written above the other, with a horizontal bar between them, are called common fractions. In the fraction $\frac{5}{8}$, the numbers 5 and 8 are the terms of the fraction. The number above the fraction bar is called the numerator. The number below the fraction bar is called the denominator. In the fraction $\frac{5}{8}$, 5 is the numerator and 8 is the denominator. The denominator tells us the number of equal parts into which an object is divided. The numerator tells us how many equal parts are being used. The fraction $\frac{5}{8}$ means 5 parts of 8 equal parts. The denominator cannot be zero.

74

There is a point on the number line corresponding to each fraction.

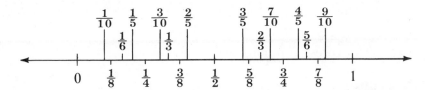

Although "fraction" is generally used to mean both the fractional number and the numeral written for the number, some mathematicians use the term "fraction" to mean the symbol and "rational number" to mean the fractional number. We shall use the word "fraction" to mean both the number and the numeral except in situations where it may be misunderstood.

A fraction may be used to indicate that one number is divided by another number. The fraction $\frac{5}{8}$ means 5 divided by 8 representing $5 \div 8$ or $8\overline{)5}$. The numerator is the number that is divided and the denominator is the number by which you divide. The whole number 2 may be considered as $\frac{2}{1}$, 3 as $\frac{3}{1}$, 4 as $\frac{4}{1}$, etc.

A fraction may be used to compare two things or two groups of things. For example: a pint is $\frac{1}{2}$ of a quart.

Fractions that name the same number are called equivalent fractions The fractions $\frac{3}{4}$, $\frac{6}{8}$, and $\frac{12}{16}$ are equivalent fractions. They all name the same number which in simplest form is $\frac{3}{4}$.

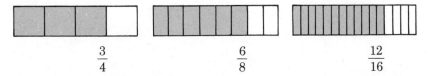

$$\frac{3}{4} \qquad \frac{6}{8} \qquad \frac{12}{16}$$

When we change a given fraction to lower terms or to higher terms or change an improper fraction to a whole number or mixed number or change a whole number to a fraction with a specified denominator or change a mixed number to an improper fraction, we use equivalent fractions. We shall see that when the numerator and denominator of any fraction are each divided by the same number (see page 76) or multiplied by the same number (see page 81), except by zero, the result is an equivalent fraction.

Observe that the larger the denominator, the smaller the size of the part—that is, $\frac{1}{4}$ is smaller than $\frac{1}{3}$, $\frac{1}{5}$ is smaller than $\frac{1}{4}$, etc.

Also we shall see in Exercise 2-19 that the smaller the size of the part, the more precise is the measurement.

Expressing Fractions in Lowest Terms

VOCABULARY: fraction
lowest terms
simplest form

OBJECTIVE:

I. Aim   To express fractions in lowest terms.

II. Procedure

1. Divide the numerator and denominator of the fraction by the largest number that can be divided exactly into both.

2. Or, factor (see Exercise 1-53) and use the multiplicative identity (see page 8).

III. Sample Solution

Express $\frac{63}{72}$ in lowest terms.

$$\frac{63}{72} = \frac{63 \div 9}{72 \div 9} = \frac{7}{8} \quad \text{or} \quad \frac{63}{72} = \frac{7 \cdot 9}{8 \cdot 9} = \frac{7}{8} \times 1 = \frac{7}{8}$$

Answer: $\frac{7}{8}$

IV. Definition

A fraction is in its lowest terms or in simplest form when its numerator and denominator cannot be divided exactly by the same number, except by 1.

DIAGNOSTIC TEST

Express the following fractions in lowest terms:

1. $\frac{3}{27}$ $\frac{1}{9}$ **3.** $\frac{6}{9}$ $\frac{2}{3}$ **5.** $\frac{26}{39}$ $\frac{2}{3}$ **7.** $\frac{250}{1000}$ $\frac{1}{4}$

2. $\frac{18}{36}$ $\frac{1}{2}$ **4.** $\frac{48}{64}$ $\frac{3}{4}$ **6.** $\frac{648}{852}$ $\frac{54}{71}$

Basic: Sets 1–5;   *Average:* Sets 1–6; Misc. Ex. 1–3   *Enriched:* Sets 1–7;
Misc. Ex. 1–2   Misc. Ex. 1–5

RELATED PRACTICE EXAMPLES

Express the following fractions in lowest terms:

	SET 1				SET 2	
1. $\frac{2}{4}$ $\frac{1}{2}$		**6.** $\frac{3}{18}$ $\frac{1}{6}$		**1.** $\frac{8}{16}$ $\frac{1}{2}$		**6.** $\frac{12}{36}$ $\frac{1}{3}$
2. $\frac{5}{15}$ $\frac{1}{3}$		**7.** $\frac{2}{12}$ $\frac{1}{6}$		**2.** $\frac{4}{40}$ $\frac{1}{10}$		**7.** $\frac{15}{45}$ $\frac{1}{3}$
3. $\frac{7}{21}$ $\frac{1}{3}$		**8.** $\frac{11}{88}$ $\frac{1}{8}$		**3.** $\frac{8}{48}$ $\frac{1}{6}$		**8.** $\frac{32}{96}$ $\frac{1}{3}$
4. $\frac{3}{12}$ $\frac{1}{4}$		**9.** $\frac{23}{46}$ $\frac{1}{2}$		**4.** $\frac{6}{42}$ $\frac{1}{7}$		**9.** $\frac{21}{84}$ $\frac{1}{4}$
5. $\frac{7}{35}$ $\frac{1}{5}$		**10.** $\frac{31}{93}$ $\frac{1}{3}$		**5.** $\frac{9}{54}$ $\frac{1}{6}$		**10.** $\frac{27}{81}$ $\frac{1}{3}$

SET 3

1. $\frac{8}{10}$ $\frac{4}{5}$
2. $\frac{4}{6}$ $\frac{2}{3}$
3. $\frac{6}{20}$ $\frac{3}{10}$
4. $\frac{10}{25}$ $\frac{2}{5}$
5. $\frac{9}{12}$ $\frac{3}{4}$
6. $\frac{6}{15}$ $\frac{2}{5}$
7. $\frac{10}{16}$ $\frac{5}{8}$
8. $\frac{14}{18}$ $\frac{7}{9}$
9. $\frac{21}{28}$ $\frac{3}{4}$
10. $\frac{18}{32}$ $\frac{9}{16}$

SET 4

1. $\frac{36}{60}$ $\frac{3}{5}$
2. $\frac{12}{16}$ $\frac{3}{4}$
3. $\frac{8}{20}$ $\frac{2}{5}$
4. $\frac{18}{24}$ $\frac{3}{4}$
5. $\frac{24}{32}$ $\frac{3}{4}$
6. $\frac{20}{36}$ $\frac{5}{9}$
7. $\frac{32}{48}$ $\frac{2}{3}$
8. $\frac{50}{75}$ $\frac{2}{3}$
9. $\frac{42}{60}$ $\frac{7}{10}$
10. $\frac{56}{64}$ $\frac{7}{8}$

SET 5

1. $\frac{38}{57}$ $\frac{2}{3}$
2. $\frac{68}{85}$ $\frac{4}{5}$
3. $\frac{58}{87}$ $\frac{2}{3}$
4. $\frac{57}{76}$ $\frac{3}{4}$
5. $\frac{69}{92}$ $\frac{3}{4}$
6. $\frac{52}{91}$ $\frac{4}{7}$
7. $\frac{62}{93}$ $\frac{2}{3}$
8. $\frac{34}{119}$ $\frac{2}{7}$
9. $\frac{91}{104}$ $\frac{7}{8}$
10. $\frac{85}{102}$ $\frac{5}{6}$

SET 6

1. $\frac{70}{112}$ $\frac{5}{8}$
2. $\frac{28}{400}$ $\frac{7}{100}$
3. $\frac{84}{192}$ $\frac{7}{16}$
4. $\frac{135}{144}$ $\frac{15}{16}$
5. $\frac{180}{216}$ $\frac{5}{6}$
6. $\frac{256}{352}$ $\frac{8}{11}$
7. $\frac{420}{756}$ $\frac{5}{9}$
8. $\frac{405}{567}$ $\frac{5}{7}$
9. $\frac{680}{765}$ $\frac{8}{9}$
10. $\frac{360}{456}$ $\frac{15}{19}$

SET 7

1. $\frac{10}{40}$ $\frac{1}{4}$
2. $\frac{20}{60}$ $\frac{1}{3}$
3. $\frac{80}{100}$ $\frac{4}{5}$
4. $\frac{40}{200}$ $\frac{1}{5}$
5. $\frac{150}{600}$ $\frac{1}{4}$
6. $\frac{300}{1000}$ $\frac{3}{10}$
7. $\frac{1200}{2000}$ $\frac{3}{5}$
8. $\frac{750}{2500}$ $\frac{3}{10}$
9. $\frac{3500}{5000}$ $\frac{7}{10}$
10. $\frac{2000}{3600}$ $\frac{5}{9}$

MISCELLANEOUS EXAMPLES

1. Express the following measurements in lowest terms:

SET 1

a. $\frac{2}{4}$ inch $\frac{1}{2}$ in.
b. $\frac{14}{16}$ inch $\frac{7}{8}$ in.
c. $\frac{6}{8}$ inch $\frac{3}{4}$ in.
d. $\frac{20}{32}$ inch $\frac{5}{8}$ in.
e. $\frac{36}{64}$ inch $\frac{9}{16}$ in.

SET 2

a. $\frac{8}{12}$ foot $\frac{2}{3}$ ft.
b. $\frac{3}{12}$ foot $\frac{1}{4}$ ft.
c. $\frac{30}{36}$ yard $\frac{5}{6}$ yd.
d. $\frac{220}{1760}$ mile $\frac{1}{8}$ mi.
e. $\frac{3960}{5280}$ mile $\frac{3}{4}$ mi.

SET 3

a. $\frac{10}{16}$ pound $\frac{5}{8}$ lb.
b. $\frac{600}{2000}$ ton $\frac{3}{10}$ ton
c. $\frac{840}{2240}$ ton $\frac{3}{8}$ ton
d. $\frac{24}{32}$ quart $\frac{3}{4}$ qt.
e. $\frac{4}{8}$ peck $\frac{1}{2}$ pk.

SET 4

a. $\frac{24}{60}$ hour $\frac{2}{5}$ hr.
b. $\frac{45}{60}$ minute $\frac{3}{4}$ min.
c. $\frac{9}{24}$ day $\frac{3}{8}$ day
d. $\frac{10}{12}$ year $\frac{5}{6}$ yr.
e. $\frac{73}{365}$ year $\frac{1}{5}$ yr.

a. $\frac{9}{16}$ in. **2.** Which is greater: **a.** $\frac{9}{16}$ inch or $\frac{28}{64}$ inch?   **b.** $\frac{12}{16}$ inch or $\frac{7}{8}$ inch? b. $\frac{7}{8}$ in.

a. $\frac{26}{32}$ in. **3.** Which is smaller: **a.** $\frac{26}{32}$ inch or $\frac{15}{16}$ inch?   **b.** $\frac{3}{8}$ inch or $\frac{40}{64}$ inch? b. $\frac{3}{8}$ in.

4. A student read the length of a metal block as $\frac{56}{64}$ of an inch. In what other way could this measurement have been expressed? $\frac{7}{8}$ in.

5. Select the fractions which are not in lowest terms:

a. $\frac{7}{10}$ $\frac{16}{21}$ $\frac{18}{81}$ **b.** $\frac{21}{25}$ $\frac{63}{72}$ $\frac{27}{58}$

Changing Improper Fractions and Simplifying Mixed Numbers

OBJECTIVE:

I. Aim

To change (1) an improper fraction to a whole number or a mixed number;

(2) a mixed number to simplest form.

II. Procedure

1. To change an improper fraction to a whole number or a mixed number:

a. Divide the numerator by the denominator.

b. Write the remainder, if any, as a fraction expressed in lowest terms.

2. To change a mixed number to simplest form:

a. Simplify the fraction of the given mixed number.

b. Add the result to the given whole number wherever possible.

III. Sample Solutions

1. Change $\frac{6}{3}$ to a whole number.

$$\frac{6}{3} = 2$$

Answer: 2

2. Change $\frac{8}{5}$ to a mixed number.

$$\frac{8}{5} = 1\frac{3}{5}$$

Answer: $1\frac{3}{5}$

3. Change $\frac{14}{6}$ to a mixed number.

$$\frac{14}{6} = 2\frac{2}{6} = 2\frac{1}{3}$$

Answer: $2\frac{1}{3}$

4. Simplify $4\frac{8}{8}$.

$$4\frac{8}{8} = 5$$

since $\frac{8}{8} = 1$ and $4 + 1 = 5$

Answer: 5

5. Simplify $7\frac{4}{6}$.

$$7\frac{4}{6} = 7\frac{2}{3}$$

since $\frac{4}{6} = \frac{2}{3}$

Answer: $7\frac{2}{3}$

6. Simplify $6\frac{14}{9}$.

$$6\frac{14}{9} = 7\frac{5}{9}$$

since $\frac{14}{9} = 1\frac{5}{9}$ and

$6 + 1\frac{5}{9} = 7\frac{5}{9}$

Answer: $7\frac{5}{9}$

7. Simplify $3\frac{6}{4}$.

$$3\frac{6}{4} = 4\frac{1}{2}$$

since $\frac{6}{4} = 1\frac{2}{4} = 1\frac{1}{2}$

and $3 + 1\frac{1}{2} = 4\frac{1}{2}$

Answer: $4\frac{1}{2}$

IV. Definitions

1. A proper fraction is a fraction whose numerator is smaller than its denominator, as the fraction $\frac{3}{8}$. The value of a proper fraction is always less than one (1).

2. An improper fraction is a fraction whose numerator is equal to or larger than its denominator, as $\frac{3}{3}$ or $\frac{8}{3}$. The value of an improper fraction is one (1) or more than one (1).

3. A mixed number is a number containing a whole number and a fraction, as $2\frac{3}{8}$.

PRELIMINARY EXAMPLES (all students)

1. Select the mixed numbers in the following:

$\frac{13}{16}$ 16 $8\frac{1}{4}$ $\frac{5}{6}$ $3\frac{11}{12}$ 25 $1\frac{4}{5}$ $\frac{7}{8}$

2. Select the proper fractions in the following:

$\frac{3}{4}$ $\frac{9}{8}$ $\frac{7}{16}$ $\frac{5}{5}$ $\frac{19}{20}$ $\frac{11}{6}$ $\frac{8}{12}$ $\frac{21}{32}$

3. Select the improper fractions in the following:

$\frac{1}{8}$ $\frac{7}{6}$ $\frac{4}{4}$ $\frac{3}{16}$ $\frac{5}{2}$ $\frac{18}{20}$ $\frac{8}{8}$ $\frac{45}{24}$

DIAGNOSTIC TEST

Change each of the following improper fractions to a whole number or a mixed number:

1. $\frac{18}{6}$ 3   **2.** $\frac{5}{3}$ $1\frac{2}{3}$ **3.** $\frac{12}{8}$ $1\frac{1}{2}$ **4.** $\frac{19}{4}$ $4\frac{3}{4}$ **5.** $\frac{35}{10}$ $3\frac{1}{2}$

Change each of the following mixed numbers to simplest form:

6. $5\frac{6}{6}$ 6   **7.** $6\frac{4}{2}$ 8   **8.** $3\frac{12}{16}$ $3\frac{3}{4}$ **9.** $9\frac{13}{10}$ $10\frac{3}{10}$ **10.** $7\frac{14}{8}$ $8\frac{3}{4}$

Basic: Sets 1–7,   *Average:* Sets 1–9, Misc. Ex. 1–4   *Enriched:* Sets 1–10,
Misc. Ex. 1–2   Misc. Ex. 1–5

RELATED PRACTICE EXAMPLES

Change each of the following improper fractions to a whole number or a mixed number:

SET 1	**SET 2**	**SET 3**	**SET 4**	**SET 5**
1. $\frac{8}{8}$ 1	**1.** $\frac{3}{2}$ $1\frac{1}{2}$	**1.** $\frac{6}{4}$ $1\frac{1}{2}$	**1.** $\frac{9}{2}$ $4\frac{1}{2}$	**1.** $\frac{10}{4}$ $2\frac{1}{2}$
2. $\frac{5}{5}$ 1	**2.** $\frac{7}{5}$ $1\frac{2}{5}$	**2.** $\frac{10}{6}$ $1\frac{2}{3}$	**2.** $\frac{7}{3}$ $2\frac{1}{3}$	**2.** $\frac{28}{8}$ $3\frac{1}{2}$
3. $\frac{12}{4}$ 3	**3.** $\frac{4}{3}$ $1\frac{1}{3}$	**3.** $\frac{12}{9}$ $1\frac{1}{3}$	**3.** $\frac{12}{5}$ $2\frac{2}{5}$	**3.** $\frac{34}{6}$ $5\frac{2}{3}$
4. $\frac{4}{2}$ 2	**4.** $\frac{13}{8}$ $1\frac{5}{8}$	**4.** $\frac{14}{8}$ $1\frac{3}{4}$	**4.** $\frac{13}{4}$ $3\frac{1}{4}$	**4.** $\frac{60}{16}$ $3\frac{3}{4}$
5. $\frac{15}{5}$ 3	**5.** $\frac{11}{6}$ $1\frac{5}{6}$	**5.** $\frac{28}{16}$ $1\frac{3}{4}$	**5.** $\frac{25}{8}$ $3\frac{1}{8}$	**5.** $\frac{45}{20}$ $2\frac{1}{4}$
6. $\frac{12}{3}$ 4	**6.** $\frac{25}{16}$ $1\frac{9}{16}$	**6.** $\frac{18}{10}$ $1\frac{4}{5}$	**6.** $\frac{22}{7}$ $3\frac{1}{7}$	**6.** $\frac{46}{16}$ $2\frac{7}{8}$
7. $\frac{6}{2}$ 3	**7.** $\frac{14}{9}$ $1\frac{5}{9}$	**7.** $\frac{18}{12}$ $1\frac{1}{2}$	**7.** $\frac{41}{6}$ $6\frac{5}{6}$	**7.** $\frac{33}{9}$ $3\frac{2}{3}$
8. $\frac{18}{3}$ 6	**8.** $\frac{19}{12}$ $1\frac{7}{12}$	**8.** $\frac{38}{24}$ $1\frac{7}{12}$	**8.** $\frac{32}{9}$ $3\frac{5}{9}$	**8.** $\frac{100}{32}$ $3\frac{1}{8}$
9. $\frac{16}{2}$ 8	**9.** $\frac{7}{4}$ $1\frac{3}{4}$	**9.** $\frac{52}{32}$ $1\frac{5}{8}$	**9.** $\frac{55}{16}$ $3\frac{7}{16}$	**9.** $\frac{50}{12}$ $4\frac{1}{6}$
10. $\frac{36}{9}$ 4	**10.** $\frac{13}{10}$ $1\frac{3}{10}$	**10.** $\frac{120}{64}$ $1\frac{7}{8}$	**10.** $\frac{67}{12}$ $5\frac{7}{12}$	**10.** $\frac{80}{24}$ $3\frac{1}{3}$

Change each of the following mixed numbers to simplest form:

SET 6

1. $2\frac{3}{3}$ 3
2. $7\frac{2}{2}$ 8
3. $5\frac{7}{7}$ 6
4. $9\frac{12}{12}$ 10
5. $6\frac{16}{16}$ 7

SET 7

1. $3\frac{12}{3}$ 7
2. $9\frac{6}{2}$ 12
3. $8\frac{18}{9}$ 10
4. $12\frac{32}{4}$ 20
5. $15\frac{30}{5}$ 21

SET 8

1. $4\frac{6}{9}$ $4\frac{2}{3}$
2. $9\frac{10}{16}$ $9\frac{5}{8}$
3. $3\frac{5}{20}$ $3\frac{1}{4}$
4. $6\frac{21}{24}$ $6\frac{7}{8}$
5. $7\frac{35}{42}$ $7\frac{5}{6}$
6. $28\frac{8}{12}$ $28\frac{2}{3}$
7. $11\frac{16}{30}$ $11\frac{8}{15}$
8. $25\frac{24}{32}$ $25\frac{3}{4}$
9. $14\frac{25}{40}$ $14\frac{5}{8}$
10. $18\frac{28}{49}$ $18\frac{4}{7}$

SET 9

1. $6\frac{8}{5}$ $7\frac{3}{5}$
2. $5\frac{7}{4}$ $6\frac{3}{4}$
3. $9\frac{5}{2}$ $11\frac{1}{2}$
4. $8\frac{8}{3}$ $10\frac{2}{3}$
5. $2\frac{13}{8}$ $3\frac{5}{8}$
6. $11\frac{7}{6}$ $12\frac{1}{6}$
7. $23\frac{19}{10}$ $24\frac{9}{10}$
8. $40\frac{16}{9}$ $41\frac{7}{9}$
9. $17\frac{35}{16}$ $19\frac{3}{16}$
10. $35\frac{29}{12}$ $37\frac{5}{12}$

SET 10

1. $4\frac{10}{4}$ $6\frac{1}{2}$
2. $3\frac{12}{8}$ $4\frac{1}{2}$
3. $7\frac{8}{8}$ $8\frac{1}{3}$
4. $9\frac{32}{12}$ $11\frac{2}{3}$
5. $6\frac{15}{10}$ $7\frac{1}{2}$
6. $12\frac{10}{4}$ $13\frac{1}{4}$
7. $16\frac{15}{6}$ $18\frac{1}{2}$
8. $13\frac{40}{12}$ $16\frac{1}{3}$
9. $21\frac{26}{16}$ $22\frac{5}{8}$
10. $45\frac{52}{24}$ $47\frac{1}{6}$

MISCELLANEOUS EXAMPLES

1. Express each of the following measurements as a mixed number:

SET 1

a. $\frac{10}{8}$ inches $1\frac{1}{4}$ in.
b. $\frac{9}{4}$ inches $2\frac{1}{4}$ in.
c. $\frac{20}{16}$ inches $1\frac{1}{4}$ in.
d. $\frac{7}{2}$ inches $3\frac{1}{2}$ in.
e. $\frac{41}{32}$ inches $1\frac{9}{32}$ in.

SET 2

a. $\frac{27}{12}$ feet $2\frac{1}{4}$ ft.
b. $\frac{35}{12}$ feet $2\frac{11}{12}$ ft.
c. $\frac{43}{36}$ yards $1\frac{7}{36}$ yd.
d. $\frac{3080}{1760}$ miles $1\frac{3}{4}$ mi.
e. $\frac{7920}{5280}$ miles $1\frac{1}{2}$ mi.

SET 3

a. $\frac{23}{16}$ pounds $1\frac{7}{16}$ lb.
b. $\frac{3500}{2000}$ tons $1\frac{3}{4}$ t.
c. $\frac{5040}{2240}$ tons $2\frac{1}{4}$ t.
d. $\frac{17}{8}$ gallons $2\frac{1}{8}$ gal.
e. $\frac{14}{4}$ bushels $3\frac{1}{2}$ bu.

SET 4

a. $\frac{75}{60}$ hours $1\frac{1}{4}$ hr.
b. $\frac{108}{60}$ minutes $1\frac{4}{5}$ min.
c. $\frac{42}{24}$ days $1\frac{3}{4}$ days
d. $\frac{15}{12}$ years $1\frac{1}{4}$ yr.
e. $\frac{91}{52}$ years $1\frac{3}{4}$ yr.

2. Which is larger: $6\frac{3}{8}$ inches or $\frac{100}{16}$ inches? $6\frac{3}{8}$ in.

3. Which is smaller: $\frac{72}{32}$ inches or $2\frac{3}{4}$ inches? $\frac{72}{32}$ in.

4. Leslie wrote $\frac{28}{15}$ as her answer to a multiplication example. How should it have been expressed? $1\frac{13}{15}$

5. Express each of the following measurements in simplest form:

SET 1

a. $5\frac{2}{8}$ inches $5\frac{1}{4}$ in.
b. $3\frac{6}{16}$ inches $3\frac{3}{8}$ in.
c. $1\frac{14}{32}$ inches $1\frac{7}{16}$ in.
d. $6\frac{5}{4}$ inches $7\frac{1}{4}$ in.
e. $9\frac{2}{2}$ inches 10 in.

SET 2

a. $6\frac{4}{12}$ feet $6\frac{1}{3}$ ft.
b. $1\frac{17}{12}$ feet $2\frac{5}{12}$ ft.
c. $4\frac{30}{36}$ yards $4\frac{5}{6}$ yd.
d. $2\frac{440}{1760}$ miles $2\frac{1}{4}$ mi.
e. $3\frac{1980}{5280}$ miles $3\frac{3}{8}$ mi.

SET 3

a. $4\frac{8}{16}$ pounds $4\frac{1}{2}$ lb.
b. $5\frac{1400}{2000}$ tons $5\frac{7}{10}$ t.
c. $2\frac{1960}{2240}$ tons $2\frac{7}{8}$ t.
d. $8\frac{11}{4}$ gallons $10\frac{3}{4}$ gal.
e. $7\frac{6}{32}$ bushels $7\frac{3}{16}$ bu.

SET 4

a. $2\frac{18}{60}$ hours $2\frac{3}{10}$ hr.
b. $9\frac{72}{60}$ minutes $10\frac{1}{5}$ min.
c. $5\frac{16}{24}$ days $5\frac{2}{3}$ days
d. $3\frac{4}{12}$ years $3\frac{1}{3}$ yr.
e. $1\frac{13}{52}$ years $1\frac{1}{4}$ yr.

OBJECTIVE:

I. Aim   To change a fraction to higher terms.

II. Procedure

1. Divide the new denominator by the denominator of the given fraction.

2. Then multiply both the numerator and denominator of the given fraction by the quotient. This is the same as multiplying the fraction by the multiplicative identity one (1). See page 8.

III. Sample Solutions

Change the following fractions to equivalent fractions having denominators as specified:

1. $\frac{9}{25} = \frac{?}{100}$

   100 divided by 25 equals 4

   $$\frac{9}{25} = \frac{9 \times 4}{25 \times 4} = \frac{36}{100} \quad \text{or} \quad \frac{9}{25} = \frac{9}{25} \times 1 = \frac{9 \times 4}{25 \times 4} = \frac{36}{100}$$

   Answer: $\frac{36}{100}$

2. Change $\frac{1}{6}$ to 24ths.

   24 divided by 6 equals 4

   $$\frac{1}{6} = \frac{1 \times 4}{6 \times 4} = \frac{4}{24}$$

   Answer: $\frac{4}{24}$

3. Change $\frac{7}{8}$ to 16ths.

   16 divided by 8 equals 2

   $$\frac{7}{8} = \frac{7 \times 2}{8 \times 2} = \frac{14}{16}$$

   Answer: $\frac{14}{16}$

DIAGNOSTIC TEST

Change the following fractions to equivalent fractions having denominators as specified:

1. $\frac{1}{4} = \frac{?}{12}$ $\frac{3}{12}$

2. $\frac{5}{8} = \frac{?}{32}$ $\frac{20}{32}$

3. $\frac{9}{10} = \frac{?}{100}$ $\frac{90}{100}$

4. Change $\frac{3}{5}$ to 100ths. $\frac{60}{100}$

5. Change 6 to 10ths. $\frac{60}{10}$

Basic: Sets 1–3, Misc. Ex. 1–3; *Average:* Sets 1–4(a), Misc. Ex. 1–4; *Enriched:* Sets 1–5, Misc. Ex. 1–6

82

COMMON FRACTIONS

RELATED PRACTICE EXAMPLES

Change the following fractions to equivalent fractions having denominators as specified:

SET 1	SET 2(a)	SET 2(b)	SET 3
1. $\frac{1}{2} = \frac{?}{8}$ 4	1. $\frac{2}{3} = \frac{?}{12}$ 8	1. $\frac{5}{8} = \frac{?}{24}$ 15	1. $\frac{7}{10} = \frac{?}{100}$ 70
2. $\frac{1}{3} = \frac{?}{15}$ 5	2. $\frac{3}{4} = \frac{?}{24}$ 18	2. $\frac{7}{9} = \frac{?}{45}$ 35	2. $\frac{3}{4} = \frac{?}{100}$ 75
3. $\frac{1}{5} = \frac{?}{20}$ 4	3. $\frac{7}{8} = \frac{?}{32}$ 28	3. $\frac{13}{16} = \frac{?}{32}$ 26	3. $\frac{2}{5} = \frac{?}{100}$ 40
4. $\frac{1}{4} = \frac{?}{28}$ 7	4. $\frac{5}{6} = \frac{?}{12}$ 10	4. $\frac{7}{12} = \frac{?}{48}$ 28	4. $\frac{9}{20} = \frac{?}{100}$ 45
5. $\frac{1}{2} = \frac{?}{18}$ 9	5. $\frac{3}{16} = \frac{?}{48}$ 9	5. $\frac{3}{4} = \frac{?}{16}$ 12	5. $\frac{2}{25} = \frac{?}{100}$ 8
6. $\frac{1}{8} = \frac{?}{48}$ 6	6. $\frac{5}{12} = \frac{?}{60}$ 25	6. $\frac{9}{32} = \frac{?}{96}$ 27	6. $\frac{43}{50} = \frac{?}{100}$ 86
7. $\frac{1}{12} = \frac{?}{72}$ 6	7. $\frac{17}{32} = \frac{?}{64}$ 34	7. $\frac{17}{24} = \frac{?}{48}$ 34	7. $\frac{21}{25} = \frac{?}{100}$ 84
8. $\frac{1}{6} = \frac{?}{42}$ 7	8. $\frac{3}{8} = \frac{?}{40}$ 15	8. $\frac{11}{16} = \frac{?}{80}$ 55	8. $\frac{3}{8} = \frac{?}{100}$ $37\frac{1}{2}$
9. $\frac{1}{16} = \frac{?}{64}$ 4	9. $\frac{5}{9} = \frac{?}{72}$ 40	9. $\frac{7}{12} = \frac{?}{36}$ 21	9. $\frac{2}{3} = \frac{?}{100}$ $66\frac{2}{3}$
10. $\frac{1}{10} = \frac{?}{70}$ 7	10. $\frac{5}{7} = \frac{?}{21}$ 15	10. $\frac{15}{16} = \frac{?}{64}$ 60	10. $\frac{17}{50} = \frac{?}{100}$ 34

SET 4(a)

1. $\frac{1}{8}$ to 64ths $\frac{8}{64}$
2. $\frac{1}{4}$ to 32nds $\frac{8}{32}$
3. $\frac{1}{2}$ to 12ths $\frac{6}{12}$
4. $\frac{11}{16}$ to 64ths $\frac{44}{64}$

5. $\frac{2}{3}$ to 24ths $\frac{16}{24}$
6. $\frac{4}{5}$ to 100ths $\frac{80}{100}$
7. $\frac{5}{8}$ to 16ths $\frac{10}{16}$
8. $\frac{19}{32}$ to 64ths $\frac{38}{64}$

9. $\frac{5}{6}$ to 30ths $\frac{25}{30}$
10. $\frac{7}{9}$ to 36ths $\frac{28}{36}$
11. $\frac{2}{1}$ to 10ths $\frac{20}{10}$
12. 5 to 8ths $\frac{40}{8}$

SET 4(b)

1. $\frac{3}{8}$ to 32nds $\frac{12}{32}$
2. $\frac{9}{10}$ to 40ths $\frac{36}{40}$
3. $\frac{2}{3}$ to 18ths $\frac{12}{18}$
4. $\frac{3}{5}$ to 20ths $\frac{12}{20}$

5. $\frac{1}{6}$ to 48ths $\frac{8}{48}$
6. $\frac{11}{12}$ to 24ths $\frac{22}{24}$
7. $\frac{3}{4}$ to 64ths $\frac{48}{64}$
8. $\frac{15}{16}$ to 32nds $\frac{30}{32}$

9. $\frac{19}{24}$ to 96ths $\frac{76}{96}$
10. $\frac{23}{25}$ to 100ths $\frac{92}{100}$
11. $\frac{4}{1}$ to 12ths $\frac{48}{12}$
12. 9 to 90ths $\frac{810}{90}$

SET 5

1. $\frac{2}{1}$ to 10ths $\frac{20}{10}$
2. $\frac{4}{1}$ to 12ths $\frac{48}{12}$
3. $\frac{5}{1}$ to 24ths $\frac{120}{24}$
4. $\frac{7}{1}$ to 6ths $\frac{42}{6}$
5. $\frac{3}{1}$ to 60ths $\frac{180}{60}$
6. 5 to 8ths $\frac{40}{8}$
7. 3 to 5ths $\frac{15}{5}$
8. 7 to 100ths $\frac{700}{100}$
9. 12 to 10ths $\frac{120}{10}$
10. 9 to 72nds $\frac{648}{72}$

MISCELLANEOUS EXAMPLES

Use a ruler to find the answer to each of the following:

1. $\frac{1}{8}'' = \frac{?}{16}''$ 2

2. $\frac{5}{8}'' = \frac{?}{16}''$ 10

3. $\frac{1}{4}'' = \frac{?}{8}'' = \frac{?}{16}''$ 2; 4

4. $\frac{3}{4}'' = \frac{?}{8}'' = \frac{?}{16}''$ 6; 12

5. $\frac{1}{2}'' = \frac{?}{4}'' = \frac{?}{8}'' = \frac{?}{16}''$ 2; 4; 8

6. $3'' = \frac{?}{2}'' = \frac{?}{4}'' = \frac{?}{8}''$ 6; 12; 24

Finding the Lowest Common Denominator and Changing Fractions to Equivalent Fractions

VOCABULARY: Lowest common denominator (LCD)

OBJECTIVE:

I. Aim Given fractions with unlike denominators,
(1) to find their lowest common denominator;
(2) to change the given fractions to equivalent fractions with the lowest common denominator as their denominators.

II. Procedure

1. Find the smallest possible natural number that can be divided exactly by the denominators of all the given fractions. This number is the lowest common denominator (LCD).

Usually the LCD can be found by inspection when the LCD is the denominator of one of the given fractions (see sample solution 1) or when it is the product of the denominators of the given fractions. (See sample solution 2.)

When the LCD is greater than any given denominator but smaller than the product of all the denominators (see sample solutions 3 and 4), use either the following method when inspection fails, or you can use the method which is discussed in Exercise 1-56 to find the least common multiple.

a. Arrange the denominators in a row. Divide by the smallest number that will divide exactly into two or more of the denominators. Bring down to the next row the quotients and any denominator which is not exactly divisible.

b. Continue this process until no two of the remaining numbers can be divided exactly by any number other than by one (1).

c. Multiply together all the divisors and numbers remaining in the last row. The product found in this way is called the lowest common denominator.

2. To change each given fraction to an equivalent fraction having the lowest common denominator as its denominator, use the same procedure as that used to change fractions to higher terms.

III. Sample Solutions

Find the LCD of the given fractions, then change the given fractions to equivalent fractions having the LCD as their denominators.

1. $\frac{1}{3}$ and $\frac{5}{6}$

LCD = 6

(6 is the smallest number that can be divided exactly by 3 and 6.)

$$\frac{1}{3} = \frac{1 \times 2}{3 \times 2} = \frac{2}{6}$$

$$\frac{5}{6} = \frac{5}{6}$$

Answer: 6;
$\frac{2}{6}$ and $\frac{5}{6}$

2. $\frac{1}{2}$ and $\frac{3}{5}$

LCD = 10

(10 is the smallest number that can be divided exactly by 2 and 5.)

$$\frac{1}{2} = \frac{1 \times 5}{2 \times 5} = \frac{5}{10}$$

$$\frac{3}{5} = \frac{3 \times 2}{5 \times 2} = \frac{6}{10}$$

Answer: 10;
$\frac{5}{10}$ and $\frac{6}{10}$

3. $\frac{5}{12}$ and $\frac{7}{16}$

LCD = 48

(48 is the smallest number that can be divided exactly by 12 and 16.)

$$\frac{5}{12} = \frac{5 \times 4}{12 \times 4} = \frac{20}{48}$$

$$\frac{7}{16} = \frac{7 \times 3}{16 \times 3} = \frac{21}{48}$$

Answer: 48;
$\frac{20}{48}$ and $\frac{21}{48}$

4. $\frac{2}{3}$, $\frac{7}{8}$, and $\frac{1}{6}$

$$3\overline{)3\ \ 8\ \ 6}$$
$$2\overline{)1\ \ 8\ \ 2}$$
$$\ \ \ \ 1\ \ 4\ \ 1$$

LCD = $3 \times 2 \times 1 \times 4 \times 1 = 24$

Answer: 24; $\frac{16}{24}$, $\frac{21}{24}$, and $\frac{4}{24}$

$$\frac{2}{3} = \frac{2 \times 8}{3 \times 8} = \frac{16}{24}$$

$$\frac{7}{8} = \frac{7 \times 3}{8 \times 3} = \frac{21}{24}$$

$$\frac{1}{6} = \frac{1 \times 4}{6 \times 4} = \frac{4}{24}$$

IV. Definition

The lowest common denominator is the smallest possible natural number that can be divided exactly by the denominators of all the given fractions.

DIAGNOSTIC TEST

Find the lowest common denominator (LCD) of the given fractions, then change the given fractions to equivalent fractions having the lowest common denominator (LCD) as their denominators:

1. $\frac{1}{2}$ and $\frac{3}{8}$
8, $\frac{4}{8}$, $\frac{3}{8}$

2. $\frac{3}{4}$ and $\frac{2}{3}$
12, $\frac{9}{12}$, $\frac{8}{12}$

3. $\frac{7}{10}$ and $\frac{9}{16}$
80, $\frac{56}{80}$, $\frac{45}{80}$

4. $\frac{5}{6}$, $\frac{4}{5}$, and $\frac{1}{3}$
30, $\frac{25}{30}$, $\frac{24}{30}$, $\frac{10}{30}$

Basic: Sets 1–2, Misc. Ex. 1–2; *Average:* Sets 1–3, Misc. Ex. 1–4; *Enriched:* Sets 1–4, Misc. Ex. 1–5

1-11 FINDING THE LOWEST COMMON DENOMINATOR   **85**

RELATED PRACTICE EXAMPLES

Find the lowest common denominator (LCD) of the given fractions, then change the given fractions to equivalent fractions having the lowest common denominator (LCD) as their denominators:

SET 1

1. $\frac{1}{2}$ and $\frac{1}{4}$ $4, \frac{2}{4}, \frac{1}{4}$

2. $\frac{3}{5}$ and $\frac{9}{10}$ $10, \frac{6}{10}, \frac{9}{10}$

3. $\frac{1}{4}$ and $\frac{9}{16}$ $16, \frac{4}{16}, \frac{9}{16}$

4. $\frac{17}{20}$ and $\frac{4}{5}$ $20, \frac{17}{20}, \frac{16}{20}$

5. $\frac{2}{3}$ and $\frac{1}{6}$ $6, \frac{4}{6}, \frac{1}{6}$

6. $\frac{1}{2}$ and $\frac{13}{16}$ $16, \frac{8}{16}, \frac{13}{16}$

7. $\frac{7}{8}$ and $\frac{3}{4}$ $8, \frac{7}{8}, \frac{6}{8}$

8. $\frac{11}{24}$ and $\frac{5}{8}$ $24, \frac{11}{24}, \frac{15}{24}$

9. $\frac{1}{4}$ and $\frac{21}{32}$ $32, \frac{8}{32}, \frac{21}{32}$

10. $\frac{5}{6}$ and $\frac{7}{18}$ $18, \frac{15}{18}, \frac{7}{18}$

SET 2

1. $\frac{1}{3}$ and $\frac{1}{2}$ $6, \frac{2}{6}, \frac{3}{6}$

2. $\frac{2}{5}$ and $\frac{5}{6}$ $30, \frac{12}{30}, \frac{25}{30}$

3. $\frac{2}{3}$ and $\frac{1}{4}$ $12, \frac{8}{12}, \frac{3}{12}$

4. $\frac{7}{8}$ and $\frac{3}{5}$ $40, \frac{35}{40}, \frac{24}{40}$

5. $\frac{1}{3}$ and $\frac{5}{8}$ $24, \frac{8}{24}, \frac{15}{24}$

6. $\frac{4}{5}$ and $\frac{3}{4}$ $20, \frac{16}{20}, \frac{15}{20}$

7. $\frac{13}{15}$ and $\frac{1}{4}$ $60, \frac{52}{60}, \frac{15}{60}$

8. $\frac{3}{5}$ and $\frac{5}{12}$ $60, \frac{36}{60}, \frac{25}{60}$

9. $\frac{8}{9}$ and $\frac{3}{4}$ $36, \frac{32}{36}, \frac{27}{36}$

10. $\frac{1}{6}$ and $\frac{4}{7}$ $42, \frac{7}{42}, \frac{24}{42}$

SET 3

1. $\frac{1}{4}$ and $\frac{1}{6}$ $12, \frac{3}{12}, \frac{2}{12}$

2. $\frac{7}{8}$ and $\frac{9}{10}$ $40, \frac{35}{40}, \frac{36}{40}$

3. $\frac{5}{6}$ and $\frac{3}{8}$ $24, \frac{20}{24}, \frac{9}{24}$

4. $\frac{3}{4}$ and $\frac{7}{10}$ $20, \frac{15}{20}, \frac{14}{20}$

5. $\frac{1}{6}$ and $\frac{15}{16}$ $48, \frac{8}{48}, \frac{45}{48}$

6. $\frac{1}{10}$ and $\frac{9}{32}$ $160, \frac{16}{160}, \frac{45}{160}$

7. $\frac{5}{14}$ and $\frac{1}{4}$ $28, \frac{10}{28}, \frac{7}{28}$

8. $\frac{13}{20}$ and $\frac{7}{12}$ $60, \frac{39}{60}, \frac{35}{60}$

9. $\frac{19}{24}$ and $\frac{13}{18}$ $72, \frac{57}{72}, \frac{52}{72}$

10. $\frac{11}{12}$ and $\frac{27}{32}$ $96, \frac{88}{96}, \frac{81}{96}$

SET 4

1. $\frac{1}{2}, \frac{1}{3}$, and $\frac{1}{4}$ $12, \frac{6}{12}, \frac{4}{12}, \frac{3}{12}$

2. $\frac{3}{4}, \frac{1}{2}$, and $\frac{5}{8}$ $8, \frac{6}{8}, \frac{4}{8}, \frac{5}{8}$

3. $\frac{2}{3}, \frac{7}{12}$, and $\frac{5}{6}$ $12, \frac{8}{12}, \frac{7}{12}, \frac{10}{12}$

4. $\frac{4}{5}, \frac{1}{4}$, and $\frac{2}{3}$ $60, \frac{48}{60}, \frac{15}{60}, \frac{40}{60}$

5. $\frac{5}{6}, \frac{3}{4}$, and $\frac{11}{12}$ $12, \frac{10}{12}, \frac{9}{12}, \frac{11}{12}$

6. $\frac{3}{10}, \frac{5}{8}$, and $\frac{7}{16}$ $80, \frac{24}{80}, \frac{50}{80}, \frac{35}{80}$

7. $\frac{7}{8}, \frac{5}{6}$, and $\frac{3}{4}$ $24, \frac{21}{24}, \frac{20}{24}, \frac{18}{24}$

8. $\frac{9}{16}, \frac{3}{8}$, and $\frac{13}{24}$ $48, \frac{27}{48}, \frac{18}{48}, \frac{26}{48}$

9. $\frac{7}{8}, \frac{9}{10}$, and $\frac{1}{12}$ $120, \frac{105}{120}, \frac{108}{120}, \frac{10}{120}$

10. $\frac{3}{5}, \frac{5}{16}$, and $\frac{9}{20}$ $80, \frac{48}{80}, \frac{25}{80}, \frac{36}{80}$

MISCELLANEOUS EXAMPLES

1. Of which group of fractions is 24 the LCD?

   a. $\frac{3}{4}$ and $\frac{5}{6}$ **b.** $\frac{5}{8}$ and $\frac{2}{3}$ **c.** $\frac{1}{2}, \frac{11}{12}$, and $\frac{1}{3}$

2. Of which group of fractions is 18 the LCD?

   a. $\frac{1}{6}$ and $\frac{2}{3}$ **b.** $\frac{7}{9}$ and $\frac{1}{2}$ **c.** $\frac{7}{12}, \frac{5}{6}$, and $\frac{2}{9}$

3. Of which group of fractions is 12 the LCD?

   a. $\frac{5}{9}$ and $\frac{3}{4}$ **b.** $\frac{1}{2}$ and $\frac{5}{6}$ **c.** $\frac{2}{3}, \frac{3}{4}$, and $\frac{1}{6}$

4. Of which group of fractions is 32 the LCD?

   a. $\frac{11}{16}$ and $\frac{7}{8}$ **b.** $\frac{3}{4}$ and $\frac{15}{32}$ **c.** $\frac{3}{8}, \frac{1}{2}$, and $\frac{3}{4}$

5. Of which group of fractions is 48 the LCD?

   a. $\frac{5}{6}$ and $\frac{3}{8}$ **b.** $\frac{7}{12}$ and $\frac{3}{4}$ **c.** $\frac{11}{16}, \frac{1}{6}$, and $\frac{5}{8}$

Addition of Fractions and Mixed Numbers

OBJECTIVE:

I. Aim — To add fractions and mixed numbers.

II. Procedure

1. If the fractions have like denominators, add the numerators and write the sum over the common denominator. Change the answer to simplest form. (See sample solutions 1, 2, and 3.)

2. If the fractions have unlike denominators, find their lowest common denominator and change the fractions to equivalent fractions having a common denominator. Then add as explained in step 1. (See sample solution 4 and Exercise 1-11.)

3. To add mixed numbers, first add the fractions, then add this sum to the sum of the whole numbers. (See sample solutions 5, 7, and 8.)

4. Check by going over the work again.

III. Sample Solutions

Add in the following examples:

1. $\dfrac{1}{5}$
$\dfrac{1}{5}$
$\dfrac{2}{5}$
Answer: $\dfrac{2}{5}$

2. $\dfrac{5}{8}$
$\dfrac{1}{8}$
$\dfrac{6}{8} = \dfrac{3}{4}$
Answer: $\dfrac{3}{4}$

3. $\dfrac{11}{16}$
$\dfrac{8}{16}$
$\dfrac{19}{16} = 1\dfrac{3}{16}$
Answer: $1\dfrac{3}{16}$

4. $\dfrac{4}{5} = \dfrac{8}{10}$
$\dfrac{7}{10} = \dfrac{7}{10}$
$\dfrac{15}{10} = 1\dfrac{5}{10} = 1\dfrac{1}{2}$
Answer: $1\dfrac{1}{2}$

5. $2\dfrac{1}{2} = 2\dfrac{3}{6}$
$5\dfrac{2}{3} = 5\dfrac{4}{6}$
$7\dfrac{7}{6} = 8\dfrac{1}{6}$
Answer: $8\dfrac{1}{6}$

6. 4
$6\dfrac{5}{8}$
$10\dfrac{5}{8}$
Answer: $10\dfrac{5}{8}$

7. $9\dfrac{7}{8} + \dfrac{13}{16}$
Solution:
$9\dfrac{7}{8} = 9\dfrac{14}{16}$
$\dfrac{13}{16} = \dfrac{13}{16}$
$9\dfrac{27}{16} = 10\dfrac{11}{16}$
Answer: $10\dfrac{11}{16}$

8. $13\dfrac{1}{6} = 13\dfrac{1}{6}$
$5\dfrac{1}{3} = 5\dfrac{2}{6}$
$9\dfrac{1}{2} = 9\dfrac{3}{6}$
$27\dfrac{6}{6} = 28$
Answer: 28

Related Resources: *Practical Applications in Mathematics:* Diag. Test and Prac. Ex., pp
21–22

86

DIAGNOSTIC TEST

Add:

1. $\frac{1}{3}$
$\frac{1}{3}$ $\frac{2}{3}$

2. $\frac{5}{16}$
$\frac{7}{16}$ $\frac{3}{4}$

3. $\frac{5}{6}$
$\frac{1}{6}$ 1

4. $\frac{3}{5}$
$\frac{4}{5}$ $1\frac{2}{5}$

5. $\frac{3}{4}$
$\frac{3}{4}$ $1\frac{1}{2}$

6. $\frac{1}{2}$
$\frac{3}{8}$ $\frac{7}{8}$

7. $\frac{2}{5}$
$\frac{3}{4}$ $1\frac{3}{20}$

8. $\frac{7}{8}$
$\frac{5}{6}$ $1\frac{17}{24}$

9. $\frac{3}{10}$
$\frac{1}{2}$
$\frac{4}{5}$ $1\frac{3}{5}$

10. 3
$\frac{1}{8}$ $3\frac{1}{8}$

11. $2\frac{3}{4}$
5 $7\frac{3}{4}$

12. $3\frac{1}{5}$
$4\frac{3}{5}$ $7\frac{4}{5}$

13. $2\frac{3}{10}$
$3\frac{1}{10}$ $5\frac{2}{5}$

14. $16\frac{5}{8}$
$23\frac{3}{8}$ 40

15. $7\frac{2}{5}$
$9\frac{4}{5}$ $17\frac{1}{5}$

16. $5\frac{7}{12}$
$6\frac{11}{12}$ $12\frac{1}{2}$

17. $8\frac{1}{6}$
$\frac{5}{6}$ 9

18. $12\frac{2}{3}$
$5\frac{1}{4}$ $17\frac{11}{12}$

19. $6\frac{1}{12}$
$8\frac{1}{6}$ $14\frac{1}{4}$

20. $24\frac{9}{10}$
$17\frac{5}{8}$ $42\frac{21}{40}$

21. $15\frac{2}{3}$
$9\frac{5}{6}$ $25\frac{1}{2}$

22. $6\frac{7}{8}$
$\frac{11}{12}$ $7\frac{19}{24}$

23. $\frac{1}{2}$
$3\frac{3}{4}$
$\frac{7}{8}$ $5\frac{1}{8}$

24. $4\frac{2}{3}$
$6\frac{1}{2}$
$5\frac{5}{8}$ $16\frac{19}{24}$

25. $\frac{7}{8} + \frac{7}{12} + \frac{1}{6}$ $1\frac{5}{8}$

26. $2\frac{5}{6} + 3\frac{1}{10} + 4\frac{1}{2}$ $10\frac{13}{30}$

27. Find the sum of $6\frac{1}{4}$, $3\frac{13}{16}$, and $7\frac{3}{8}$. $17\frac{7}{16}$

Basic: Sets 1–20, Prob. 1–9
Average: Sets 1–24, Prob. 1–12
Enriched: Sets 1–27, Prob. 1–15

RELATED PRACTICE EXAMPLES

Add:

SET 1

$\frac{2}{5}$	$\frac{3}{7}$	$\frac{6}{25}$	$\frac{3}{6}$	$\frac{5}{16}$
$\frac{2}{5}$	$\frac{2}{7}$	$\frac{8}{25}$ $\frac{14}{25}$	$\frac{2}{6}$	$\frac{4}{16}$
$\frac{4}{5}$	$\frac{5}{7}$		$\frac{5}{6}$	$\frac{9}{16}$

SET 2

$\frac{1}{4}$	$\frac{3}{8}$	$\frac{7}{12}$	$\frac{13}{32}$	$\frac{27}{64}$
$\frac{1}{4}$	$\frac{1}{8}$	$\frac{1}{12}$ $\frac{2}{3}$	$\frac{15}{32}$	$\frac{21}{64}$
$\frac{1}{2}$	$\frac{1}{2}$		$\frac{7}{8}$	$\frac{3}{4}$

SET 3

$\frac{1}{2}$	$\frac{1}{3}$	$\frac{5}{8}$	$\frac{27}{32}$	$\frac{11}{20}$
$\frac{1}{2}$	$\frac{2}{3}$	$\frac{3}{8}$	$\frac{5}{32}$	$\frac{9}{20}$
1	1	1	1	1

SET 4

$\frac{2}{3}$	$\frac{4}{9}$	$\frac{3}{4}$	$\frac{9}{16}$	$\frac{4}{10}$
$\frac{2}{3}$	$\frac{7}{9}$	$\frac{2}{4}$ $1\frac{1}{4}$	$\frac{10}{16}$	$\frac{9}{10}$
$1\frac{1}{3}$	$1\frac{2}{9}$		$1\frac{3}{16}$	$1\frac{3}{10}$

SET 5

$\frac{5}{6}$	$\frac{11}{16}$	$\frac{3}{8}$	$\frac{17}{24}$	$\frac{15}{32}$
$\frac{5}{6}$	$\frac{9}{16}$	$\frac{7}{8}$ $1\frac{1}{4}$	$\frac{19}{24}$	$\frac{29}{32}$
$1\frac{2}{3}$	$1\frac{1}{4}$		$1\frac{1}{2}$	$1\frac{3}{8}$

SET 6

$\frac{1}{2}$	$\frac{5}{8}$	$\frac{5}{6}$	$\frac{1}{10}$	$\frac{13}{32}$
$\frac{1}{4}$	$\frac{3}{64}$	$\frac{5}{12}$ $1\frac{1}{4}$	$\frac{2}{5}$	$\frac{3}{4}$
$\frac{3}{4}$	$\frac{43}{64}$		$\frac{1}{2}$	$1\frac{5}{32}$

SET 7

$\frac{2}{3}$	$\frac{1}{2}$	$\frac{2}{5}$	$\frac{11}{16}$	$\frac{3}{5}$
$\frac{3}{4}$	$\frac{1}{3}$	$\frac{1}{6}$	$\frac{4}{5}$	$\frac{7}{8}$
$1\frac{5}{12}$	$\frac{5}{6}$	$\frac{17}{30}$	$1\frac{39}{80}$	$1\frac{19}{40}$

SET 8

$\frac{7}{10}$	$\frac{5}{12}$	$\frac{3}{4}$	$\frac{1}{8}$	$\frac{13}{24}$
$\frac{11}{16}$	$\frac{3}{10}$	$\frac{5}{6}$	$\frac{7}{12}$	$\frac{9}{16}$
$1\frac{31}{80}$	$\frac{43}{60}$	$1\frac{7}{12}$	$\frac{17}{24}$	$1\frac{5}{48}$

SET 9

$\frac{3}{8}$	$\frac{2}{3}$	$\frac{3}{16}$	$\frac{7}{24}$	$\frac{5}{6}$
$\frac{4}{8}$	$\frac{3}{4}$	$\frac{1}{8}$	$\frac{3}{16}$	$\frac{3}{10}$
$\frac{1}{8}$	$\frac{5}{6}$	$\frac{1}{4}$	$\frac{5}{12}$	$\frac{7}{16}$
1	$2\frac{1}{4}$	$\frac{9}{16}$	$\frac{43}{48}$	$1\frac{137}{240}$

SET 10

4	8	12	$\frac{3}{16}$	$\frac{4}{7}$
$\frac{3}{4}$	$\frac{7}{8}$	$\frac{1}{6}$	7	18
$4\frac{3}{4}$	$8\frac{7}{8}$	$12\frac{1}{6}$	$7\frac{3}{16}$	$18\frac{4}{7}$

Col 1	Col 2	SET (middle)	Col 4	Col 5
$6\frac{3}{5}$	$9\frac{5}{8}$	**SET 11**	10	23
7	4	8	$15\frac{2}{3}$	$8\frac{9}{32}$
$13\frac{3}{5}$	$13\frac{5}{8}$	$7\frac{1}{2}\quad 15\frac{1}{2}$	$25\frac{2}{3}$	$31\frac{9}{32}$
		SET 12		
$4\frac{1}{4}$	$7\frac{2}{9}$	$10\frac{4}{8}$	$15\frac{5}{16}$	$32\frac{3}{10}$
$3\frac{2}{4}$	$2\frac{5}{9}$	$9\frac{3}{8}$	$12\frac{8}{16}$	$14\frac{6}{10}$
$7\frac{3}{4}$	$9\frac{7}{9}$	$19\frac{7}{8}$	$27\frac{13}{16}$	$46\frac{9}{10}$
		SET 13		
$5\frac{1}{8}$	$6\frac{7}{32}$	$10\frac{7}{16}$	$17\frac{1}{12}$	$43\frac{9}{64}$
$8\frac{5}{8}$	$3\frac{9}{32}$	$16\frac{7}{16}$	$11\frac{7}{12}$	$18\frac{23}{64}$
$13\frac{3}{4}$	$9\frac{1}{2}$	$26\frac{7}{8}$	$28\frac{2}{3}$	$61\frac{1}{2}$
		SET 14		
$6\frac{3}{4}$	$15\frac{1}{2}$	$7\frac{5}{8}$	$13\frac{1}{6}$	$17\frac{3}{16}$
$5\frac{1}{4}$	$9\frac{1}{2}$	$7\frac{3}{8}$	$12\frac{5}{6}$	$22\frac{13}{16}$
12	25	15	26	40
		SET 15		
$1\frac{2}{3}$	$4\frac{3}{5}$	$32\frac{5}{9}$	$16\frac{4}{8}$	$4\frac{7}{16}$
$2\frac{2}{3}$	$7\frac{4}{5}$	$13\frac{8}{9}$	$19\frac{5}{8}$	$5\frac{12}{16}$
$4\frac{1}{3}$	$12\frac{2}{5}$	$46\frac{4}{9}$	$36\frac{1}{8}$	$10\frac{3}{16}$
		SET 16		
$8\frac{3}{4}$	$9\frac{8}{10}$	$5\frac{5}{6}$	$28\frac{23}{32}$	$6\frac{3}{8}$
$4\frac{3}{4}$	$8\frac{7}{10}$	$24\frac{5}{6}$	$12\frac{19}{32}$	$16\frac{7}{8}$
$13\frac{1}{2}$	$18\frac{1}{2}$	$30\frac{2}{3}$	$41\frac{5}{16}$	$23\frac{1}{4}$
		SET 17		
$6\frac{1}{5}$	$8\frac{5}{8}$	$\frac{1}{4}$	$\frac{7}{12}$	$\frac{9}{16}$
$\frac{1}{5}$	$\frac{7}{8}$	$7\frac{3}{4}$	$1\frac{11}{12}$	$18\frac{5}{16}$
$6\frac{2}{5}$	$9\frac{1}{2}$	8	$2\frac{1}{2}$	$18\frac{7}{8}$
		SET 18		
$8\frac{1}{8}$	$7\frac{1}{4}$	$6\frac{1}{6}$	$4\frac{3}{10}$	$15\frac{3}{4}$
$2\frac{1}{3}$	$6\frac{2}{5}$	$9\frac{5}{8}$	$8\frac{5}{12}$	$29\frac{3}{16}$
$10\frac{11}{24}$	$13\frac{13}{20}$	$15\frac{19}{24}$	$12\frac{43}{60}$	$44\frac{15}{16}$
		SET 19		
$3\frac{1}{3}$	$4\frac{1}{4}$	$5\frac{1}{6}$	$8\frac{1}{5}$	$22\frac{7}{12}$
$3\frac{1}{6}$	$8\frac{5}{12}$	$6\frac{1}{2}$	$6\frac{3}{10}$	$17\frac{1}{6}$
$6\frac{1}{2}$	$12\frac{2}{3}$	$11\frac{2}{3}$	$14\frac{1}{2}$	$39\frac{3}{4}$

SET 20

$9\frac{1}{2}$	$2\frac{2}{3}$	$8\frac{3}{4}$	$16\frac{11}{12}$	$32\frac{9}{16}$
$5\frac{3}{4}$	$6\frac{4}{5}$	$3\frac{7}{8}$ $\quad 12\frac{5}{8}$	$14\frac{5}{8}$	$15\frac{7}{12}$
$15\frac{1}{4}$	$9\frac{7}{15}$		$31\frac{13}{24}$	$48\frac{7}{48}$

SET 21

$4\frac{4}{5}$	$3\frac{3}{4}$	$4\frac{5}{6}$	$11\frac{17}{24}$	$21\frac{17}{20}$
$6\frac{7}{10}$	$7\frac{5}{12}$	$3\frac{1}{2}$ $\quad 8\frac{1}{3}$	$4\frac{5}{12}$	$37\frac{2}{5}$
$11\frac{1}{2}$	$11\frac{1}{6}$		$16\frac{1}{8}$	$59\frac{1}{4}$

SET 22

$7\frac{1}{4}$	$9\frac{2}{3}$	$5\frac{5}{8}$	$\frac{3}{8}$	$\frac{5}{6}$
$\frac{7}{16}$	$\frac{1}{2}$	$\frac{3}{4}$ $\quad 6\frac{3}{8}$	$14\frac{9}{10}$	$19\frac{2}{3}$
$7\frac{11}{16}$	$10\frac{1}{6}$		$15\frac{11}{40}$	$20\frac{1}{2}$

SET 23

$\frac{5}{6}$	$4\frac{3}{8}$	$7\frac{3}{4}$	$\frac{1}{3}$	$6\frac{7}{12}$
$\frac{2}{3}$	$9\frac{7}{16}$	$\frac{5}{8}$	$8\frac{2}{5}$	$\frac{3}{8}$
$2\frac{1}{2}$	$\frac{5}{8}$	$3\frac{13}{16}$ $\quad 12\frac{3}{16}$	$\frac{1}{4}$	$5\frac{1}{6}$
4	$14\frac{7}{16}$		$8\frac{59}{60}$	$12\frac{1}{8}$

SET 24

$5\frac{1}{8}$	$3\frac{2}{3}$	$4\frac{1}{3}$	$15\frac{7}{10}$	$18\frac{5}{8}$
$1\frac{1}{4}$	$2\frac{1}{6}$	$8\frac{3}{8}$	$2\frac{5}{6}$	$13\frac{11}{12}$
$4\frac{1}{2}$	$2\frac{7}{12}$	$2\frac{3}{4}$	$9\frac{1}{4}$	$42\frac{3}{16}$
$10\frac{7}{8}$	$8\frac{5}{12}$	$15\frac{11}{24}$	$27\frac{47}{60}$	$74\frac{35}{48}$

SET 25

Add as indicated:

1. $\frac{1}{4} + \frac{1}{8}$ $\quad \frac{3}{8}$
2. $\frac{2}{3} + \frac{1}{6}$ $\quad \frac{5}{6}$
3. $\frac{5}{16} + \frac{7}{8}$ $\quad 1\frac{3}{16}$
4. $\frac{2}{5} + \frac{3}{5}$ $\quad 1$
5. $\frac{1}{3} + \frac{1}{12} + \frac{1}{6}$ $\quad \frac{7}{12}$

6. $\frac{3}{4} + \frac{7}{8} + \frac{1}{2}$ $\quad 2\frac{1}{8}$
7. $\frac{9}{16} + \frac{3}{8} + \frac{3}{4}$ $\quad 1\frac{11}{16}$
8. $\frac{1}{2} + \frac{9}{10} + \frac{3}{5}$ $\quad 2$
9. $\frac{2}{3} + \frac{5}{6} + \frac{5}{8}$ $\quad 2\frac{1}{8}$
10. $\frac{3}{4} + \frac{1}{6} + \frac{7}{12}$ $\quad 1\frac{1}{2}$

SET 26

Add as indicated:

1. $2\frac{3}{8} + 1\frac{1}{4}$ $\quad 3\frac{5}{8}$
2. $7 + \frac{5}{16}$ $\quad 7\frac{5}{16}$
3. $1\frac{2}{5} + \frac{7}{10} + 9$ $\quad 11\frac{1}{10}$
4. $6\frac{1}{4} + 3\frac{3}{8} + 1\frac{7}{16}$ $\quad 11\frac{1}{16}$
5. $4\frac{3}{16} + 2\frac{7}{8} + 5\frac{1}{2}$ $\quad 12\frac{9}{16}$

6. $2\frac{1}{2} + 4\frac{9}{16} + 8\frac{1}{2}$ $\quad 15\frac{9}{16}$
7. $7\frac{2}{3} + 5\frac{1}{6} + 3\frac{1}{12}$ $\quad 15\frac{11}{12}$
8. $8\frac{5}{8} + 6\frac{3}{8} + \frac{5}{6}$ $\quad 15\frac{5}{6}$
9. $5\frac{1}{10} + 8\frac{2}{5} + 7\frac{1}{2}$ $\quad 21$
10. $23\frac{9}{10} + 12\frac{3}{8} + 9\frac{3}{4}$ $\quad 46\frac{1}{40}$

SET 27

Find the sum of:

1. $\frac{2}{3}$ and $\frac{7}{12}$ $1\frac{1}{4}$

2. $3\frac{1}{4}$ and $9\frac{3}{4}$ 13

3. 9, $\frac{4}{5}$, and $\frac{1}{2}$ $10\frac{3}{10}$

4. $4\frac{1}{3}$, $\frac{2}{3}$, and $6\frac{5}{8}$ $11\frac{5}{8}$

5. $9\frac{1}{5}$, $6\frac{3}{10}$, and $4\frac{1}{2}$ 20

6. $5\frac{3}{8}$, $8\frac{5}{16}$, and $10\frac{1}{4}$ $23\frac{15}{16}$

7. $3\frac{7}{8}$, $4\frac{3}{4}$, and $2\frac{9}{16}$ $11\frac{3}{16}$

8. $12\frac{1}{4}$, $5\frac{13}{16}$, and $4\frac{5}{8}$ $22\frac{11}{16}$

9. $8\frac{2}{5}$, $14\frac{7}{10}$, and $9\frac{9}{10}$ 33

10. $23\frac{5}{8}$, $18\frac{11}{16}$, and $32\frac{3}{4}$ $75\frac{1}{16}$

PROBLEM SOLVING

See page 16, *Problem Solving Strategy*.

For problems 1–5, select the letter corresponding to your answer.

1. A tailor made a 2-piece dress, requiring $2\frac{7}{8}$ yards for one part and $1\frac{3}{4}$ yards for the other. How much material did she use?

　　a. $5\frac{1}{8}$ yards　　b. $4\frac{5}{8}$ yards　　c. $4\frac{7}{8}$ yards　　d. Answer not given

2. Find the total thickness of two pieces of wood that Donald glued together if one is $\frac{5}{16}$ inch thick and the other $\frac{7}{8}$ inch thick.

　　a. $1\frac{5}{16}$ inches　　b. $1\frac{1}{2}$ inches　　c. $1\frac{3}{16}$ inches　　d. $1\frac{1}{4}$ inches

3. Carmen works after school. During a certain week she worked $3\frac{3}{4}$ hours on Monday, $2\frac{1}{2}$ hours on Wednesday, and 4 hours on Friday. How many hours did she work altogether?

　　a. $9\frac{3}{4}$ hours　　b. $10\frac{1}{4}$ hours　　c. $9\frac{1}{2}$ hours　　d. $10\frac{1}{2}$ hours

4. The running time of a train from Chicago to San Francisco was changed to $49\frac{1}{3}$ hours. If this schedule saves $13\frac{3}{4}$ hours, how long did the trip take before the change was made?

　　a. $64\frac{1}{4}$ hours　　b. $63\frac{2}{3}$ hours　　c. $63\frac{1}{12}$ hours　　d. $63\frac{5}{6}$ hours

5. The XYZ stock opened Monday morning at the price of $27\frac{5}{8}$. During the 5 business days of the week, the stock gained $\frac{7}{8}$ point, $1\frac{1}{4}$ points, $\frac{5}{8}$ point, $1\frac{1}{2}$ points, and $\frac{3}{4}$ point respectively. What was its closing price on Friday evening?

　　a. $32\frac{7}{8}$　　b. $32\frac{3}{4}$　　c. $32\frac{5}{8}$　　d. Answer not given

6. Mr. Kahn needed $2\frac{1}{4}$ cups of sifted cake flour to make a plain cake and $1\frac{1}{2}$ cups for a pineapple sponge cake. Find the total amount of flour he required. $3\frac{3}{4}$ cups

7. With the last event of the track meet yet to be run, the juniors had 27 points; seniors, $26\frac{1}{2}$ points; sophomores, 26 points; and freshmen, $22\frac{1}{2}$ points. In the last event, the seniors scored $1\frac{1}{2}$ points; juniors, $1\frac{1}{2}$ points; sophomores, 3 points; and freshmen, 5 points. Which class won the meet? sophomores with 29 points

8. What is the overall length of a certain machine part consisting of 3 joined pieces measuring $2\frac{9}{16}$ inches, $1\frac{27}{32}$ inches, and $\frac{7}{8}$ inch respectively. $5\frac{9}{32}$ in.

9. Team B is $4\frac{1}{2}$ games behind team A in the standings. Team C is $2\frac{1}{2}$ games behind team B. How many games behind team A is team C? 7 games

10. A heating engineer, in installing an oil burner, finds it necessary to use pieces of pipe measuring $4\frac{11}{16}$ inches, $7\frac{5}{8}$ inches, $3\frac{1}{2}$ inches, and 9 inches. What length of pipe does he need to be able to cut the 4 pieces, disregarding waste? $24\frac{13}{16}$ in.

11. A blueprint calls for 3 separate pieces of wood measuring $8\frac{3}{8}$ inches, $3\frac{5}{16}$ inches, and $5\frac{1}{4}$ inches in length. How long should the piece of wood be in order that the 3 pieces may be cut from it? Allow $\frac{1}{2}$ inch for waste. $17\frac{7}{16}$ in.

12. What is the outside diameter of tubing when the inside diameter is $2\frac{5}{8}$ inches and the wall thickness is $\frac{3}{16}$ inch? 3 in.

13. What are the outside dimensions of a closed box whose inside dimensions are $4\frac{15}{16}$ inches by $5\frac{5}{8}$ inches by 7 inches if the thickness of material is $\frac{1}{8}$ inch? $5\frac{3}{16}$ in. by $5\frac{7}{8}$ in. by $7\frac{1}{4}$ in.

14. What is the perimeter (distance around) of a triangle if its three sides measure $6\frac{3}{8}$ inches, $4\frac{11}{16}$ inches, and $5\frac{3}{4}$ inches respectively? $16\frac{13}{16}$ in.

15. Find the length of the bolt that should be used to go through a piece of tubing $\frac{1}{2}$ inch in diameter, a second piece of tubing $\frac{3}{4}$ inch in diameter, a washer $\frac{1}{16}$ inch thick, and a nut $\frac{1}{8}$ inch thick. $1\frac{7}{16}$ in.

REFRESH YOUR SKILLS

1. Round 6,280,501 to the nearest thousand. 6,281,000

2. Add:	**3.** Subtract:	**4.** Multiply:	**5.** Divide: 996
483	61,074	8,567	$144\overline{)143,424}$
98,276	38,294	473	
1,584	─────	─────	
67,791	22,780	4,052,191	
4,928			
─────			
173,062			

6. Express $\frac{72}{84}$ in lowest terms. $\frac{6}{7}$ **7.** Change $\frac{5}{6}$ to 24ths. $\frac{20}{24}$

8. Add:	**9.** Add:	**10.** Add: $7\frac{1}{2}$
$\frac{7}{8}$	$6\frac{3}{5}$	$2\frac{11}{16}$
$\frac{5}{6}$ $1\frac{17}{24}$	$4\frac{2}{3}$ $11\frac{4}{15}$	$3\frac{5}{8}$ $13\frac{13}{16}$

Subtraction of Fractions and Mixed Numbers

OBJECTIVE:

I. Aim   To subtract fractions and mixed numbers.

II. Procedure

1. If the fractions have like denominators, subtract the numerators and write the difference over the common denominator. Reduce the answer to lowest terms. (See sample solutions 1 and 2.)

2. If the fractions have unlike denominators, find their lowest common denominator and change the fractions to equivalent fractions having a common denominator. Then subtract as explained in step 1. (See sample solution 3 and Exercise 1-11.)

3. To subtract a fraction or mixed number from a whole number, first regroup by taking one (1) from the whole number and changing it into a fraction making the numerator and common denominator the same as the denominator of the given fraction. Then subtract fractions and subtract whole numbers. (See sample solution 6.)

4. To subtract mixed numbers or a fraction from a mixed number, first subtract the fractions, then subtract the whole numbers. (See sample solution 4.) If the fraction in the subtrahend is larger than the fraction in the minuend, regroup by taking one (1) from the whole number in the minuend and increasing the fraction. (See sample solutions 7 and 8.)

5. Check by going over the work again.

III. Sample Solutions

Subtract in the following examples:

1. $\frac{4}{5}$
$\frac{1}{5}$
$\overline{\frac{3}{5}}$
Answer: $\frac{3}{5}$

2. $\frac{11}{16}$
$\frac{5}{16}$
$\overline{\frac{6}{16}} = \frac{3}{8}$
Answer: $\frac{3}{8}$

3. $\frac{7}{8} = \frac{7}{8}$
$\frac{1}{4} = \frac{2}{8}$
$\overline{\frac{5}{8}}$
Answer: $\frac{5}{8}$

4. $6\frac{4}{5} = 6\frac{8}{10}$
$3\frac{1}{2} = 3\frac{5}{10}$
$\overline{3\frac{3}{10}}$
Answer: $3\frac{3}{10}$

5. $7\frac{2}{3}$
$\underline{4}$
$3\frac{2}{3}$
Answer: $3\frac{2}{3}$

6. $8 = 7\frac{16}{16}$
$\underline{1\frac{7}{16} = 1\frac{7}{16}}$
$6\frac{9}{16}$
Answer: $6\frac{9}{16}$

7. $6\frac{1}{8} - \frac{7}{8}$
Solution:
$6\frac{1}{8} = 5\frac{9}{8}$
$\underline{\frac{7}{8} = \frac{7}{8}}$
$5\frac{2}{8} = 5\frac{1}{4}$

$6\frac{1}{8} = 5\frac{9}{8}$ since the 1 taken from the 6 equals $\frac{8}{8}$ and $\frac{1}{8} + \frac{8}{8} = \frac{9}{8}$

Answer: $5\frac{1}{4}$

8. $9\frac{1}{3} - 4\frac{5}{6}$
Solution:
$9\frac{1}{3} = 9\frac{2}{6} = 8\frac{8}{6}$
$\underline{4\frac{5}{6} = 4\frac{5}{6} = 4\frac{5}{6}}$
$4\frac{3}{6} = 4\frac{1}{2}$
Answer: $4\frac{1}{2}$

PRELIMINARY EXAMPLES—REGROUPING (All students)

Find the missing numbers:

SET 1	SET 2	SET 3
1. $4 = 3\frac{?}{2}$ 2	**1.** $1\frac{1}{2} = \frac{?}{2}$ 3	**1.** $3\frac{2}{3} = 2\frac{?}{3}$ 5
2. $3 = 2\frac{?}{6}$ 6	**2.** $1\frac{1}{4} = \frac{?}{4}$ 5	**2.** $8\frac{7}{8} = 7\frac{?}{8}$ 15
3. $7 = 6\frac{?}{4}$ 4	**3.** $1\frac{1}{3} = \frac{?}{3}$ 4	**3.** $2\frac{1}{2} = 1\frac{?}{2}$ 3
4. $9 = 8\frac{?}{3}$ 3	**4.** $1\frac{5}{8} = \frac{?}{8}$ 13	**4.** $4\frac{4}{5} = 3\frac{?}{5}$ 9
5. $12 = 11\frac{?}{8}$ 8	**5.** $1\frac{2}{5} = \frac{?}{5}$ 7	**5.** $9\frac{3}{4} = 8\frac{?}{4}$ 7
6. $8 = 7\frac{?}{10}$ 10	**6.** $1\frac{9}{10} = \frac{?}{10}$ 19	**6.** $15\frac{5}{6} = 14\frac{?}{6}$ 11
7. $6 = 5\frac{?}{24}$ 24	**7.** $1\frac{11}{16} = \frac{?}{16}$ 27	**7.** $12\frac{8}{9} = 11\frac{?}{9}$ 17
8. $5 = 4\frac{?}{16}$ 16	**8.** $1\frac{5}{6} = \frac{?}{6}$ 11	**8.** $6\frac{3}{10} = 5\frac{?}{10}$ 13
9. $14 = 13\frac{?}{32}$ 32	**9.** $1\frac{4}{7} = \frac{?}{7}$ 11	**9.** $7\frac{9}{16} = 6\frac{?}{16}$ 25
10. $2 = 1\frac{?}{20}$ 20	**10.** $1\frac{7}{12} = \frac{?}{12}$ 19	**10.** $10\frac{11}{12} = 9\frac{?}{12}$ 23

SET 4

1. $4\frac{1}{3} = 4\frac{?}{9} = 3\frac{?}{9}$ 3; 12

2. $7\frac{2}{5} = 7\frac{?}{10} = 6\frac{?}{10}$ 4; 14

3. $9\frac{3}{4} = 9\frac{?}{16} = 8\frac{?}{16}$ 12; 28

4. $5\frac{1}{2} = 5\frac{?}{16} = 4\frac{?}{16}$ 8; 24

5. $6\frac{3}{8} = 6\frac{?}{24} = 5\frac{?}{24}$ 9; 33

6. $3\frac{1}{7} = 3\frac{?}{35} = 2\frac{?}{35}$ 5; 40

7. $8\frac{3}{10} = 8\frac{?}{100} = 7\frac{?}{100}$ 30; 130

8. $2\frac{5}{6} = 2\frac{?}{30} = 1\frac{?}{30}$ 25; 55

9. $10\frac{7}{16} = 10\frac{?}{64} = 9\frac{?}{64}$ 28; 92

10. $15\frac{1}{12} = 15\frac{?}{36} = 14\frac{?}{36}$ 3; 39

DIAGNOSTIC TEST

Subtract:

1. $\frac{2}{3}$ − $\frac{1}{3}$ $\frac{1}{3}$

2. $\frac{5}{8}$ − $\frac{1}{8}$ $\frac{1}{2}$

3. $\frac{5}{6}$ − $\frac{1}{2}$ $\frac{1}{3}$

4. $\frac{7}{8}$ − $\frac{1}{5}$ $\frac{27}{40}$

5. $\frac{3}{4}$ − $\frac{7}{10}$ $\frac{1}{20}$

6. $4\frac{4}{5}$ − $3\frac{2}{5}$ $1\frac{2}{5}$

7. $8\frac{13}{16}$ − $5\frac{3}{16}$ $3\frac{5}{8}$

8. $45\frac{1}{3}$ − $32\frac{2}{3}$ $12\frac{2}{3}$

9. $12\frac{3}{8}$ − $7\frac{7}{8}$ $4\frac{1}{2}$

10. $38\frac{9}{16}$ − $36\frac{9}{16}$ 2

11. $5\frac{15}{32}$ − 3 $2\frac{15}{32}$

12. 6 − $2\frac{3}{4}$ $3\frac{1}{4}$

13. 9 − $\frac{4}{5}$ $8\frac{1}{5}$

14. $8\frac{11}{32}$ − $5\frac{1}{6}$ $3\frac{17}{96}$

15. $9\frac{3}{4}$ − $3\frac{1}{3}$ $6\frac{5}{12}$

16. $13\frac{7}{16}$ − $7\frac{5}{12}$ $6\frac{1}{48}$

17. $14\frac{5}{8}$ − $5\frac{3}{4}$ $8\frac{7}{8}$

18. $8\frac{1}{5}$ − $2\frac{1}{3}$ $5\frac{13}{15}$

19. $6\frac{3}{10}$ − $3\frac{9}{16}$ $2\frac{59}{80}$

20. $14\frac{1}{8}$ − $13\frac{1}{2}$ $\frac{5}{8}$

21. $1\frac{1}{4}$ − $\frac{5}{6}$ $\frac{5}{12}$

22. $\frac{15}{16}$ − $\frac{3}{4}$ $\frac{3}{16}$

23. $14\frac{1}{4}$ − $5\frac{2}{3}$ $8\frac{7}{12}$

24. Subtract $1\frac{5}{8}$ from 6. $4\frac{3}{8}$

RELATED PRACTICE EXAMPLES

Subtract:

SET 1

$\frac{4}{5}$ $\frac{5}{7}$ $\frac{5}{8}$ $\frac{3}{4}$ $\frac{27}{32}$

$\frac{3}{5}$ $\frac{3}{7}$ $\frac{2}{8}$ $\frac{2}{4}$ $\frac{18}{32}$

$\frac{1}{5}$ $\frac{2}{7}$ $\frac{3}{8}$ $\frac{1}{4}$ $\frac{9}{32}$

SET 2

$\frac{3}{4}$ $\frac{9}{16}$ $\frac{5}{6}$ $\frac{7}{10}$ $\frac{53}{64}$

$\frac{1}{4}$ $\frac{3}{16}$ $\frac{1}{6}$ $\frac{3}{10}$ $\frac{29}{64}$

$\frac{1}{2}$ $\frac{3}{8}$ $\frac{2}{3}$ $\frac{2}{5}$ $\frac{3}{8}$

SET 3

$\frac{7}{8}$ $\frac{2}{3}$ $\frac{3}{4}$ $\frac{17}{20}$ $\frac{15}{16}$

$\frac{3}{16}$ $\frac{1}{6}$ $\frac{5}{12}$ $\frac{3}{5}$ $\frac{21}{32}$

$\frac{11}{16}$ $\frac{1}{2}$ $\frac{1}{3}$ $\frac{1}{4}$ $\frac{9}{32}$

SET 4

$\frac{1}{2}$ $\frac{3}{4}$ $\frac{7}{8}$ $\frac{2}{3}$ $\frac{3}{5}$

$\frac{1}{3}$ $\frac{2}{5}$ $\frac{2}{3}$ $\frac{7}{16}$ $\frac{7}{12}$

$\frac{1}{6}$ $\frac{7}{20}$ $\frac{5}{24}$ $\frac{11}{48}$ $\frac{1}{60}$

SET 5

$\frac{1}{4}$ $\frac{5}{6}$ $\frac{7}{12}$ $\frac{5}{8}$ $\frac{11}{16}$

$\frac{1}{6}$ $\frac{3}{8}$ $\frac{3}{16}$ $\frac{1}{12}$ $\frac{9}{20}$

$\frac{1}{12}$ $\frac{11}{24}$ $\frac{19}{48}$ $\frac{13}{24}$ $\frac{19}{80}$

SET 6

$6\frac{3}{5}$ $8\frac{8}{9}$ $15\frac{2}{4}$ $34\frac{13}{16}$ $53\frac{7}{8}$

$2\frac{1}{5}$ $7\frac{3}{9}$ $4\frac{1}{4}$ $21\frac{8}{16}$ $25\frac{2}{8}$

$4\frac{2}{5}$ $1\frac{5}{9}$ $11\frac{1}{4}$ $13\frac{5}{16}$ $28\frac{5}{8}$

SET 7

$3\frac{5}{8}$ $6\frac{11}{16}$ $9\frac{7}{12}$ $11\frac{9}{10}$ $31\frac{17}{20}$

$1\frac{3}{8}$ $3\frac{5}{16}$ $7\frac{5}{12}$ $6\frac{3}{10}$ $19\frac{13}{20}$

$2\frac{1}{4}$ $3\frac{3}{8}$ $2\frac{1}{6}$ $5\frac{3}{5}$ $12\frac{1}{5}$

SET 8

$9\frac{2}{5}$ $5\frac{3}{7}$ $16\frac{9}{16}$ $18\frac{1}{4}$ $36\frac{1}{5}$

$4\frac{4}{5}$ $1\frac{6}{7}$ $7\frac{12}{16}$ $14\frac{2}{4}$ $18\frac{4}{5}$

$4\frac{3}{5}$ $3\frac{4}{7}$ $8\frac{13}{16}$ $3\frac{3}{4}$ $17\frac{2}{5}$

SET 9

$7\frac{1}{6}$ $15\frac{5}{8}$ $13\frac{3}{10}$ $8\frac{13}{16}$ $17\frac{5}{32}$

$1\frac{5}{6}$ $6\frac{7}{8}$ $10\frac{9}{10}$ $6\frac{15}{16}$ $13\frac{29}{32}$

$5\frac{1}{3}$ $8\frac{3}{4}$ $2\frac{2}{5}$ $1\frac{7}{8}$ $3\frac{1}{4}$

SET 10

$6\frac{1}{2}$ $11\frac{3}{4}$ $29\frac{2}{5}$ $43\frac{3}{10}$ $32\frac{11}{12}$

$5\frac{1}{2}$ $4\frac{3}{4}$ $16\frac{2}{5}$ $28\frac{3}{10}$ $23\frac{11}{12}$

1 7 13 15 9

SET 11

$4\frac{1}{2}$ $9\frac{3}{4}$ $15\frac{3}{8}$ $43\frac{9}{10}$ $39\frac{25}{32}$

2 6 9 27 14

$2\frac{1}{2}$ $3\frac{3}{4}$ $6\frac{3}{8}$ $16\frac{9}{10}$ $25\frac{25}{32}$

SET 12

9 3 13 42 27

$2\frac{1}{4}$ $1\frac{3}{5}$ $6\frac{5}{8}$ $17\frac{7}{20}$ $20\frac{13}{16}$

$6\frac{3}{4}$ $1\frac{2}{5}$ $6\frac{3}{8}$ $24\frac{13}{20}$ $6\frac{3}{16}$

SET 13

5	9	4	12	10
$\frac{3}{8}$	$\frac{7}{12}$	$\frac{1}{2}$ $3\frac{1}{2}$	$\frac{9}{10}$	$\frac{2}{3}$
$4\frac{5}{8}$	$8\frac{5}{12}$		$11\frac{1}{10}$	$9\frac{1}{3}$

SET 14

$8\frac{5}{6}$	$11\frac{11}{12}$	$14\frac{3}{4}$	$18\frac{9}{10}$	$42\frac{7}{8}$
$6\frac{1}{3}$	$8\frac{1}{4}$	$3\frac{1}{8}$ $11\frac{5}{8}$	$15\frac{2}{5}$	$9\frac{3}{16}$
$2\frac{1}{2}$	$3\frac{2}{3}$		$3\frac{1}{2}$	$33\frac{11}{16}$

SET 15

$7\frac{1}{3}$	$20\frac{3}{5}$	$11\frac{2}{3}$	$23\frac{13}{16}$	$40\frac{4}{5}$
$4\frac{1}{5}$	$12\frac{7}{16}$	$5\frac{1}{2}$ $6\frac{1}{6}$	$17\frac{2}{3}$	$8\frac{3}{4}$
$3\frac{2}{15}$	$8\frac{13}{80}$		$6\frac{7}{48}$	$32\frac{1}{20}$

SET 16

$8\frac{3}{4}$	$3\frac{7}{12}$	$7\frac{15}{16}$	$13\frac{9}{10}$	$32\frac{13}{20}$
$5\frac{1}{6}$	$1\frac{3}{8}$	$2\frac{5}{6}$ $5\frac{5}{48}$	$9\frac{3}{4}$	$27\frac{5}{8}$
$3\frac{7}{12}$	$2\frac{5}{24}$		$4\frac{3}{20}$	$5\frac{1}{40}$

SET 17

$6\frac{1}{2}$	$9\frac{3}{8}$	$10\frac{5}{16}$	$16\frac{3}{10}$	$17\frac{5}{32}$
$3\frac{3}{4}$	$5\frac{9}{16}$	$7\frac{3}{4}$ $2\frac{9}{16}$	$8\frac{1}{2}$	$12\frac{9}{16}$
$2\frac{3}{4}$	$3\frac{13}{16}$		$7\frac{4}{5}$	$4\frac{19}{32}$

SET 18

$10\frac{1}{3}$	$13\frac{1}{4}$	$9\frac{1}{6}$	$36\frac{2}{3}$	$25\frac{3}{4}$
$4\frac{1}{2}$	$8\frac{2}{3}$	$2\frac{3}{5}$ $6\frac{17}{30}$	$17\frac{7}{8}$	$21\frac{4}{5}$
$5\frac{5}{6}$	$4\frac{7}{12}$		$18\frac{19}{24}$	$3\frac{19}{20}$

SET 19

$9\frac{3}{4}$	$7\frac{1}{6}$	$18\frac{5}{6}$	$23\frac{3}{8}$	$24\frac{7}{12}$
$1\frac{9}{10}$	$5\frac{3}{8}$	$12\frac{15}{16}$ $5\frac{43}{48}$	$11\frac{5}{12}$	$14\frac{7}{10}$
$7\frac{17}{20}$	$1\frac{19}{24}$		$11\frac{23}{24}$	$9\frac{53}{60}$

SET 20

$8\frac{11}{16}$	$4\frac{7}{8}$	$7\frac{1}{3}$	$13\frac{3}{4}$	$19\frac{2}{5}$
$8\frac{7}{16}$	$4\frac{1}{2}$	$6\frac{2}{3}$ $\frac{2}{3}$	$12\frac{5}{6}$	$18\frac{9}{10}$
$\frac{1}{4}$	$\frac{3}{8}$		$\frac{11}{12}$	$\frac{1}{2}$

SET 21

$5\frac{1}{2}$	$3\frac{9}{16}$	$1\frac{7}{10}$	$1\frac{1}{3}$	$1\frac{3}{4}$
$\frac{3}{4}$	$\frac{7}{8}$	$\frac{5}{6}$ $\frac{13}{15}$	$\frac{1}{2}$	$\frac{15}{16}$
$4\frac{3}{4}$	$2\frac{11}{16}$		$\frac{5}{6}$	$\frac{13}{16}$

SET 22

Subtract as indicated:

1. $\frac{3}{5} - \frac{1}{5}$ $\frac{2}{5}$
2. $\frac{5}{8} - \frac{1}{4}$ $\frac{3}{8}$
3. $\frac{1}{2} - \frac{1}{6}$ $\frac{1}{3}$
4. $\frac{4}{5} - \frac{3}{4}$ $\frac{1}{20}$
5. $\frac{13}{16} - \frac{7}{10}$ $\frac{9}{80}$
6. $2\frac{3}{8} - 1$ $1\frac{3}{8}$
7. $2 - 1\frac{3}{8}$ $\frac{5}{8}$
8. $4 - \frac{9}{16}$ $3\frac{7}{16}$
9. $6\frac{1}{2} - \frac{1}{2}$ 6
10. $3\frac{5}{6} - \frac{1}{3}$ $3\frac{1}{2}$

SET 23

Subtract as indicated:

1. $6\frac{7}{8} - 2\frac{3}{8}$ $4\frac{1}{2}$ 6. $24\frac{9}{16} - 8\frac{3}{4}$ $15\frac{13}{16}$

2. $2\frac{1}{2} - 1\frac{3}{4}$ $\frac{3}{4}$ 7. $21\frac{3}{10} - 13\frac{1}{4}$ $8\frac{1}{20}$

3. $8\frac{1}{4} - 4\frac{2}{5}$ $3\frac{17}{20}$ 8. $10\frac{5}{6} - 10\frac{7}{12}$ $\frac{1}{4}$

4. $11\frac{1}{2} - 2\frac{1}{3}$ $9\frac{1}{6}$ 9. $8\frac{7}{8} - 7\frac{15}{16}$ $\frac{15}{16}$

5. $14\frac{3}{8} - 9\frac{1}{10}$ $5\frac{11}{40}$ 10. $4\frac{1}{6} - 3\frac{2}{3}$ $\frac{1}{2}$

SET 24

1. From $\frac{7}{8}$ subtract $\frac{5}{16}$. $\frac{9}{16}$ 6. Take $5\frac{3}{4}$ from $7\frac{1}{8}$. $1\frac{3}{8}$

2. From $1\frac{4}{5}$ take $\frac{9}{10}$. $\frac{9}{10}$ 7. Find the difference between $6\frac{1}{3}$ and $5\frac{5}{6}$. $\frac{1}{2}$

3. From 7 subtract $2\frac{11}{16}$. $4\frac{5}{16}$ 8. Find the difference between $4\frac{5}{8}$ and $2\frac{9}{10}$. $1\frac{29}{40}$

4. Take 3 from $4\frac{5}{8}$. $1\frac{5}{8}$ 9. Find the difference between $15\frac{1}{4}$ and $8\frac{1}{2}$. $6\frac{3}{4}$

5. Subtract $2\frac{1}{2}$ from $9\frac{1}{2}$. 7   10. Find the difference between 9 and $3\frac{3}{4}$. $5\frac{1}{4}$

PROBLEM SOLVING

See page 16, *Problem Solving Strategy.*
For problems 1–5, select the letter corresponding to your answer.

1. Two months ago Marian weighed $123\frac{1}{4}$ pounds. Now she weighs $116\frac{3}{4}$ pounds. How many pounds did she lose?

   a. $7\frac{1}{2}$ pounds   b. $6\frac{3}{4}$ pounds   c. $6\frac{1}{2}$ pounds   d. Answer not given

2. If normal body temperature is $98\frac{3}{5}$ degrees Fahrenheit, how many degrees above normal is a temperature of 101 degrees?

   a. $3\frac{2}{5}$ degrees   b. $2\frac{2}{5}$ degrees   c. $2\frac{4}{5}$ degrees   d. $3\frac{1}{5}$ degrees

3. Find the net change in a stock if it opened at $63\frac{7}{8}$ and closed at $65\frac{1}{4}$.

   a. $2\frac{3}{8}$ points gain   c. $1\frac{3}{8}$ points gain
   b. $1\frac{7}{8}$ points gain   d. Answer not given

4. A merchant sold $7\frac{5}{8}$ yards of cloth to a customer. If it was cut from a bolt that contained $18\frac{2}{3}$ yards, what length remained on the bolt?

   a. $10\frac{11}{24}$ yards   b. $11\frac{1}{24}$ yards   c. $11\frac{1}{6}$ yards   d. $10\frac{5}{8}$ yards

5. In arranging an $8\frac{1}{2}$ by 11 inch piece of paper in the drawing class, pupils were directed to draw a line $\frac{1}{4}$ inch from each edge. What are the inside dimensions between the lines?

   a. $8\frac{1}{4} \times 10\frac{3}{4}$ inches   c. $8 \times 10\frac{1}{2}$ inches
   b. $8 \times 10\frac{3}{4}$ inches   d. Answer not given

6. Helen bought $4\frac{3}{8}$ yards of material for a dress. If her mother plans to use only $3\frac{3}{4}$ yards, how much extra material did Helen buy? $\frac{5}{8}$ yd.

7. A $4\frac{1}{4}$-pound chicken weighed $3\frac{1}{8}$ pounds when dressed. Find the loss in weight. **$1\frac{1}{8}$ lb.**

8. During the baseball season a newspaper reported that the 6th place team was 16 games behind the leader while the 5th place team was $13\frac{1}{2}$ games behind the leader. How many games behind the 5th place team was the 6th place team? **$2\frac{1}{2}$ games**

9. The running time of a crack streamliner from Philadelphia to Chicago is $14\frac{3}{4}$ hours. Another train takes $15\frac{2}{3}$ hours to make the same trip. How much faster is the first train? **55 min. or $\frac{11}{12}$ hr.**

10. A pilot finds that winds slow her progress in reaching her destination. Her trip out takes $2\frac{3}{4}$ hours. How long should her return trip take if she must return at the end of $4\frac{1}{2}$ hours flying time? **$1\frac{3}{4}$ hr.**

11. If one mechanic can assemble a motor in $6\frac{1}{2}$ hours while another mechanic can do the same job in $7\frac{5}{6}$ hours, how much more quickly can the first mechanic do the job? **$1\frac{1}{3}$ hr.**

12. A $2\frac{1}{2}$-inch nail is driven through a piece of wood, $1\frac{1}{8}$ inches thick, supporting a joist. How far into the joist did the nail extend? **$1\frac{3}{8}$ in.**

13. After flying for $3\frac{1}{4}$ hours, a pilot was forced to land at an alternate airport, 960 kilometers from his destination. Following a delay of $\frac{1}{2}$ hour, the airplane took off. If the pilot wishes to arrive on schedule, $5\frac{3}{4}$ hours after his original take-off, how fast must he fly? **480 km/hr.**

14. A plumber, in installing water pipes, used pieces measuring $5\frac{1}{2}$ feet, $3\frac{3}{4}$ feet, and $1\frac{2}{3}$ feet. If they were cut from a 15-foot length of pipe, how many feet of pipe remained? Disregard waste. **$4\frac{1}{12}$ ft.**

15. The wall thickness of certain tubing is $\frac{5}{16}$ inch and its outside diameter is $2\frac{7}{8}$ inches. What is the inside diameter? **$2\frac{1}{4}$ in.**

REFRESH YOUR SKILLS

1. Add:
68,322
87,947
48,279
70,976
91,385
366,909

2. Subtract:
832,506
95,708
736,798

3. Multiply:
869
798
693,462

4. Divide: **689**
$2{,}240\overline{)1{,}543{,}360}$

5. Round 269,826,140 to the nearest million. **270,000,000**

6. Express $\frac{54}{90}$ in lowest terms. **$\frac{3}{5}$**

7. Change $\frac{3}{4}$ to 64ths. **$\frac{48}{64}$**

8. Add: $4\frac{5}{6}$
$2\frac{3}{4}$
$5\frac{7}{12}$ **$13\frac{1}{6}$**

9. Subtract:
$10\frac{5}{6}$
$8\frac{1}{3}$ **$2\frac{1}{2}$**

10. Subtract:
9
$2\frac{13}{16}$ **$6\frac{3}{16}$**

For a Class Activity see p. T33.

OBJECTIVE:

I. Aim To compare fractions.

VOCABULARY: is less than
is greater than

II. Procedure

1. Change the given fractions to fractions having a common denominator.

2. Then take the given fraction that is equal to the fraction having the greater numerator and common denominator as the greater fraction.

3. The symbol $<$ means "is less than." (See sample solution 4.)

4. The symbol $>$ means "is greater than." (See sample solution 5.)

III. Sample Solutions

1. Which is greater: $\frac{2}{5}$ or $\frac{1}{2}$?

$\frac{2}{5} = \frac{4}{10}$

$\frac{1}{2} = \frac{5}{10}$

Answer: $\frac{1}{2}$ is greater

2. Which is smaller: $\frac{3}{4}$ or $\frac{5}{6}$?

$\frac{3}{4} = \frac{9}{12}$

$\frac{5}{6} = \frac{10}{12}$

Answer: $\frac{3}{4}$ is smaller

3. Arrange in order of size (greatest first): $\frac{3}{4}$, $\frac{7}{8}$, and $\frac{2}{3}$

$\frac{3}{4} = \frac{18}{24}$, $\frac{7}{8} = \frac{21}{24}$, and $\frac{2}{3} = \frac{16}{24}$

Answer: $\frac{7}{8}$, $\frac{3}{4}$, and $\frac{2}{3}$

4. Is the statement $\frac{2}{3} < \frac{3}{5}$ true?

$\frac{2}{3} = \frac{10}{15}$ $\frac{3}{5} = \frac{9}{15}$

$\frac{2}{3}$ is not less than $\frac{3}{5}$

Answer: No, false

5. Is the statement $\frac{3}{10} > \frac{1}{4}$ true?

$\frac{3}{10} = \frac{6}{20}$ $\frac{1}{4} = \frac{5}{20}$

$\frac{3}{10}$ is greater than $\frac{1}{4}$

Answer: Yes, true

DIAGNOSTIC TEST

Which is greater:

1. $\frac{1}{3}$ or $\frac{1}{2}$? $\frac{1}{2}$

2. $\frac{5}{6}$ or $\frac{7}{8}$? $\frac{7}{8}$

Which is smaller:

3. $\frac{1}{4}$ or $\frac{1}{10}$? $\frac{1}{10}$

4. $\frac{5}{8}$ or $\frac{11}{16}$? $\frac{5}{8}$

Arrange in order of size (greatest first):

5. $\frac{7}{32}$, $\frac{3}{8}$, and $\frac{13}{16}$. $\frac{13}{16}$, $\frac{3}{8}$, $\frac{7}{32}$

Arrange in order of size (smallest first):

6. $\frac{3}{5}$, $\frac{7}{12}$, and $\frac{1}{2}$. $\frac{1}{2}$, $\frac{7}{12}$, $\frac{3}{5}$

7. Is the statement $\frac{4}{5} < \frac{5}{6}$ true? Yes

8. Is the statement $\frac{1}{3} > \frac{3}{8}$ true? No

* Also see Exercise 1-58.

Related Resources: *Practical Applications in Mathematics:* Diag. Test and Prac. Ex., p. 26 **98**

Basic: Sets 1–5, Misc. Ex. 1; *Average:* Sets 1–7, Misc. Ex. 1–2; *Enriched:* Sets 1–8, Misc. Ex. 1–2

1-14 COMPARING FRACTIONS   **99**

RELATED PRACTICE EXAMPLES

SET 1	**SET 2**	**SET 3**	**SET 4**	**SET 5**

SET 1 / SET 2
Which is greater in each of the following?

SET 3 / SET 4
Which is smaller in each of the following?

SET 5
Arrange in order of size (greatest first):

$\frac{1}{3}$ **1.** $\frac{1}{4}$ or $\frac{1}{3}$?   **1.** $\frac{2}{3}$ or $\frac{5}{8}$? $\frac{2}{3}$ **1.** $\frac{1}{2}$ or $\frac{1}{6}$?   **1.** $\frac{5}{16}$ or $\frac{1}{4}$? $\frac{1}{6}$ **1.** $\frac{1}{2}$, $\frac{1}{5}$, and $\frac{1}{3}$ $\frac{1}{2}, \frac{1}{3}, \frac{1}{5}$

$\frac{1}{6}$ **2.** $\frac{1}{6}$ or $\frac{1}{10}$?   **2.** $\frac{4}{5}$ or $\frac{3}{4}$? $\frac{4}{5}$ **2.** $\frac{1}{8}$ or $\frac{1}{10}$?   **2.** $\frac{5}{8}$ or $\frac{7}{12}$? $\frac{7}{12}$ **2.** $\frac{5}{8}$, $\frac{2}{3}$, and $\frac{3}{5}$ $\frac{2}{3}, \frac{5}{8}, \frac{3}{5}$

$\frac{1}{2}$ **3.** $\frac{1}{2}$ or $\frac{1}{16}$?   **3.** $\frac{7}{16}$ or $\frac{2}{5}$? $\frac{7}{16}$ **3.** $\frac{1}{16}$ or $\frac{1}{12}$?   **3.** $\frac{2}{5}$ or $\frac{3}{8}$? $\frac{1}{16}$ **3.** $\frac{3}{4}$, $\frac{9}{16}$, and $\frac{7}{12}$ $\frac{3}{4}, \frac{7}{12}, \frac{9}{16}$

$\frac{1}{5}$ **4.** $\frac{1}{5}$ or $\frac{1}{8}$?   **4.** $\frac{3}{8}$ or $\frac{1}{4}$? $\frac{3}{8}$ **4.** $\frac{1}{3}$ or $\frac{1}{5}$?   **4.** $\frac{2}{3}$ or $\frac{11}{16}$? $\frac{1}{5}$ **4.** $\frac{7}{8}$, $\frac{5}{6}$, and $\frac{4}{5}$ $\frac{7}{8}, \frac{5}{6}, \frac{4}{5}$

$\frac{1}{10}$ **5.** $\frac{1}{12}$ or $\frac{1}{10}$?   **5.** $\frac{13}{16}$ or $\frac{5}{6}$? $\frac{5}{6}$ **5.** $\frac{1}{8}$ or $\frac{1}{4}$?   **5.** $\frac{7}{10}$ or $\frac{3}{4}$? $\frac{1}{8}$ **5.** $\frac{1}{4}$, $\frac{5}{16}$, and $\frac{3}{8}$ $\frac{3}{8}, \frac{5}{16}, \frac{1}{4}$

SET 6	**SET 7**	**SET 8**

SET 6
Arrange in order of size (smallest first):

SET 7
Which of the following statements are true?

SET 8
Which of the following statements are true?

1. $\frac{1}{4}$, $\frac{1}{2}$, and $\frac{1}{6}$ $\frac{1}{6}, \frac{1}{4}, \frac{1}{2}$

1. $\frac{1}{10} < \frac{1}{8}$ True

1. $\frac{3}{5} > \frac{1}{2}$ True

2. $\frac{1}{2}$, $\frac{2}{5}$, and $\frac{3}{10}$ $\frac{3}{10}, \frac{2}{5}, \frac{1}{2}$

2. $\frac{1}{2} < \frac{3}{5}$ True

2. $\frac{11}{16} > \frac{3}{4}$ False

3. $\frac{11}{16}$, $\frac{5}{8}$, and $\frac{3}{4}$ $\frac{5}{8}, \frac{11}{16}, \frac{3}{4}$

3. $\frac{2}{3} < \frac{7}{12}$ False

3. $\frac{3}{10} > \frac{2}{7}$ True

4. $\frac{2}{3}$, $\frac{11}{12}$, and $\frac{5}{6}$ $\frac{2}{3}, \frac{5}{6}, \frac{11}{12}$

4. $\frac{5}{6} < \frac{8}{9}$ True

4. $\frac{3}{8} > \frac{5}{12}$ False

5. $\frac{5}{8}$, $\frac{3}{5}$, and $\frac{3}{4}$ $\frac{3}{5}, \frac{5}{8}, \frac{3}{4}$

5. $\frac{13}{16} < \frac{19}{24}$ False

5. $\frac{21}{25} > \frac{5}{6}$ True

MISCELLANEOUS EXAMPLES

1. To round a mixed number to the nearest whole number, drop the fraction. However, add 1 to the whole number if the fraction is one half or more. If the fraction is less than one half, do not add anything.

Round the following mixed numbers to the nearest whole numbers:

SET 1	**SET 2**	**SET 3**
a. $4\frac{1}{4}$ 4	**a.** $3\frac{2}{3}$ 4	**a.** $1\frac{5}{6}$ 2
b. $9\frac{3}{8}$ 9	**b.** $5\frac{4}{5}$ 6	**b.** $6\frac{5}{12}$ 6
c. $7\frac{2}{5}$ 7	**c.** $8\frac{1}{2}$ 9	**c.** $2\frac{9}{16}$ 3
d. $12\frac{7}{16}$ 12	**d.** $16\frac{17}{32}$ 17	**d.** $27\frac{3}{5}$ 28
e. $28\frac{13}{32}$ 28	**e.** $34\frac{7}{8}$ 35	**e.** $53\frac{27}{64}$ 53

2. a. Which is larger: $\frac{1}{2}$ of a gallon or $\frac{1}{2}$ of a quart? $\frac{1}{2}$ of a gallon

   b. Which is shorter: $\frac{3}{4}$ of a foot or $\frac{3}{4}$ of a yard? $\frac{3}{4}$ of a foot

   c. Which weighs more: $\frac{2}{3}$ of a ton or $\frac{2}{3}$ of a pound? $\frac{2}{3}$ of a ton

   d. Which is longer: $\frac{1}{6}$ of an hour or $\frac{1}{6}$ of a minute? $\frac{1}{6}$ of an hour

Changing Mixed Numbers to Improper Fractions

I. Aim   OBJECTIVE:
To change a mixed number to an improper fraction.

II. Procedure

1. Multiply the whole number by the denominator of the fraction. Add the numerator of the fraction to this product.
2. Write this sum over the denominator of the fraction.

III. Sample Solutions

1. Change $1\frac{3}{8}$ to an improper fraction. $1\frac{3}{8} = \frac{11}{8}$

Multiply 1 by 8, then add 3 to the product. Write this sum (11) over the denominator (8) of the given fraction. This is a brief way of saying:

$$1\frac{3}{8} = 1 + \frac{3}{8} = \frac{8}{8} + \frac{3}{8} = \frac{11}{8} \qquad \textit{Answer: } \frac{11}{8}$$

2. Change $4\frac{2}{5}$ to an improper fraction. $4\frac{2}{5} = \frac{22}{5}$

Multiply 4 by 5, then add 2 to the product. Write this sum (22) over the denominator (5) of the given fraction.

$$4\frac{2}{5} = 4 + \frac{2}{5} = \frac{20}{5} + \frac{2}{5} = \frac{22}{5} \qquad \textit{Answer: } \frac{22}{5}$$

DIAGNOSTIC TEST

Change the following mixed numbers to improper fractions:

1. $1\frac{1}{3}$ $\frac{4}{3}$ **2.** $1\frac{3}{4}$ $\frac{7}{4}$ **3.** $7\frac{1}{2}$ $\frac{15}{2}$ **4.** $5\frac{7}{8}$ $\frac{47}{8}$

Basic: Sets 1–3,   Average: Sets 1–4(b), Misc. Ex. 1–4, Appl. 1–4   Enriched: Sets 1–4(c),
Misc. Ex. 1–2, Appl. 1–2 **RELATED PRACTICE EXAMPLES** Misc. Ex. 1–5, Appl. 1–5

Change the following mixed numbers to improper fractions:

SET 1	SET 2	SET 3	SET 4(a)	SET 4(b)	SET 4(c)
1. $1\frac{1}{5}$ $\frac{6}{5}$	**1.** $1\frac{5}{8}$ $\frac{13}{8}$	**1.** $8\frac{1}{5}$ $\frac{41}{5}$	**1.** $4\frac{2}{3}$ $\frac{14}{3}$	**1.** $7\frac{4}{5}$ $\frac{39}{5}$	**1.** $3\frac{5}{6}$ $\frac{23}{6}$
2. $1\frac{1}{4}$ $\frac{5}{4}$	**2.** $1\frac{2}{3}$ $\frac{5}{3}$	**2.** $3\frac{1}{7}$ $\frac{22}{7}$	**2.** $3\frac{3}{4}$ $\frac{15}{4}$	**2.** $6\frac{2}{9}$ $\frac{56}{9}$	**2.** $8\frac{7}{10}$ $\frac{87}{10}$
3. $1\frac{1}{2}$ $\frac{3}{2}$	**3.** $1\frac{7}{12}$ $\frac{19}{12}$	**3.** $9\frac{1}{6}$ $\frac{55}{6}$	**3.** $2\frac{5}{8}$ $\frac{21}{8}$	**3.** $5\frac{4}{7}$ $\frac{39}{7}$	**3.** $4\frac{3}{8}$ $\frac{35}{8}$
4. $1\frac{1}{8}$ $\frac{9}{8}$	**4.** $1\frac{13}{16}$ $\frac{29}{16}$	**4.** $16\frac{1}{3}$ $\frac{49}{3}$	**4.** $3\frac{7}{12}$ $\frac{43}{12}$	**4.** $4\frac{13}{32}$ $\frac{141}{32}$	**4.** $10\frac{9}{16}$ $\frac{169}{16}$
5. $1\frac{1}{16}$ $\frac{17}{16}$	**5.** $1\frac{9}{32}$ $\frac{41}{32}$	**5.** $14\frac{1}{4}$ $\frac{57}{4}$	**5.** $5\frac{3}{16}$ $\frac{83}{16}$	**5.** $2\frac{11}{12}$ $\frac{35}{12}$	**5.** $13\frac{3}{5}$ $\frac{68}{5}$

Related Resources: *Practical Applications in Mathematics:*
Diag. Test and Prac. Ex., p. 27

MISCELLANEOUS EXAMPLES

1. How many half-inches in $5\frac{1}{2}$ inches? in $9\frac{1}{2}$ inches? 11; 19
2. How many eighths of an inch in $4\frac{7}{8}$ inches? in $2\frac{3}{8}$ inches? 39; 19
3. How many sixteenths of an inch in $2\frac{13}{16}$ inches? in $1\frac{9}{16}$ inches? 45; 25
4. How many quarters of an inch in $6\frac{3}{4}$ inches? in $7\frac{1}{4}$ inches? 27; 29
5. How many thirty-seconds of an inch in $1\frac{17}{32}$ inches? in $3\frac{5}{32}$ inches? 49; 101

PRACTICAL APPLICATIONS

1. A bag contains $37\frac{1}{2}$ pounds of sugar. How many $\frac{1}{2}$-pound bags can be filled? 75 bags

2. How many pieces $\frac{1}{4}$ inch long can be cut from a strip of metal $14\frac{3}{4}$ inches long? 59 pieces

3. If it takes $\frac{1}{8}$ yard of material to make a certain necktie, how many similar neckties can be made from a bolt of goods containing $15\frac{3}{8}$ yards? 123 neckties

4. How many pieces of wood $\frac{1}{3}$ foot long can be cut from a board $6\frac{2}{3}$ feet long? 20 pieces

5. How many $\frac{1}{10}$-ounce tea bags can be made from $9\frac{3}{10}$ ounces of tea? 93 tea bags

---

REFRESH YOUR SKILLS

1. Add:
59,864
46,239
67,403
76,216
85,945   335,667

2. Subtract:
968,325
898,296
70,029

3. Multiply:
5,280
  905   4,778,400

4. Divide: 698
$3{,}600\overline{)2{,}512{,}800}$

5. Round 7,899,846 to the nearest thousand. 7,900,000

6. Express $\frac{54}{96}$ in lowest terms. $\frac{9}{16}$

7. Change $\frac{3}{5}$ to 60ths. $\frac{36}{60}$

8. Arrange in order of size (greatest first): $\frac{2}{3}, \frac{7}{8}, \frac{9}{10}$ $\frac{9}{10}, \frac{7}{8}, \frac{2}{3}$

9. Add:
$8\frac{1}{4}$
$5\frac{3}{16}$
$2\frac{5}{8}$ $16\frac{1}{16}$

10. Subtract:
$1\frac{3}{10}$
$\frac{4}{5}$
$\frac{1}{2}$

Multiplication of Fractions and Mixed Numbers

OBJECTIVE:

I. Aim   To multiply fractions and mixed numbers.

II. Procedure

1. Change each mixed number to an improper fraction. (See Exercise 1-15.) A whole number may be expressed in fractional form by using the whole number as the numerator and the figure 1 as the denominator. (See sample solution 4.)

2. Wherever it is possible, first divide any numerator and denominator by the greatest possible number that is exactly contained in both (greatest common factor) to simplify computation.

3. Multiply the resulting numerators to obtain the numerator of the answer.

4. Multiply the resulting denominators to obtain the denominator of the answer.

5. Where necessary, express the answer in simplest form.

6. When multiplying a mixed number and a whole number, the vertical form may also be used. (See sample solution 7.)

7. Check by going over the work again.

Note that when a number is multiplied by a proper fraction, the product is always smaller than the given number because a fractional part of the given number is being found.

III. Sample Solutions

Multiply in the following examples:

1. $\frac{7}{8} \times \frac{3}{5} = \frac{21}{40}$

Answer: $\frac{21}{40}$

2. $\frac{2}{3} \times 5 = \frac{10}{3} = 3\frac{1}{3}$

Answer: $3\frac{1}{3}$

3. $\frac{4}{5} \times \frac{15}{16} = \frac{\overset{1}{\cancel{4}}}{\cancel{5}} \times \frac{\overset{3}{\cancel{15}}}{\underset{4}{\cancel{16}}} = \frac{3}{4}$

Answer: $\frac{3}{4}$

4. $6 \times \frac{3}{8} = \frac{\cancel{6}}{1} \times \frac{3}{\underset{4}{\cancel{8}}} = \frac{9}{4} = 2\frac{1}{4}$

Answer: $2\frac{1}{4}$

5. $2\frac{1}{2} \times 1\frac{1}{5} = \frac{\overset{1}{\cancel{5}}}{\cancel{2}} \times \frac{\overset{3}{\cancel{6}}}{\cancel{5}} = \frac{3}{1}$ or 3

Answer: 3

6. $2\frac{5}{8} \times 1\frac{1}{4} = \frac{21}{8} \times \frac{5}{4} = \frac{105}{32} = 3\frac{9}{32}$

Answer: $3\frac{9}{32}$

7.   21
$\underline{\times\ 3\frac{1}{2}}$ $\frac{1}{2}\times 21 = \frac{21}{2} = 10\frac{1}{2}$

63
$\underline{+\ 10\frac{1}{2}}$

$73\frac{1}{2}$ *Answer:* $73\frac{1}{2}$

8. $1\frac{4}{5}\times\frac{2}{3}\times 3\frac{1}{8} = \dfrac{\overset{3}{\cancel{9}}}{\underset{1}{\cancel{5}}}\times\dfrac{\overset{1}{\cancel{2}}}{\underset{1}{\cancel{3}}}\times\dfrac{\overset{5}{\cancel{25}}}{\underset{4}{\cancel{8}}} = \dfrac{15}{4} = 3\frac{3}{4}$ *Answer:* $3\frac{3}{4}$

9. $\dfrac{6}{100}\times 800 = \dfrac{6}{\cancel{100}}\times\cancel{800} = 48$ *Answer:* 48

10. To find a fractional part of a number, multiply as in multiplication of fractions.

Find $\frac{5}{8}$ of 48.

$\dfrac{5}{\underset{1}{\cancel{8}}}\times\overset{6}{\cancel{48}} = 30$

Answer: 30

11. Reduction may be used in examples involving both multiplication and division as follows:

Simplify:

$\dfrac{5,280\times 60}{3,600} = \dfrac{\overset{88}{\cancel{5,280}}\times\overset{1}{\cancel{60}}}{\underset{\underset{1}{\cancel{6}}}{\cancel{3,600}}} = \dfrac{88}{1}$ or 88

Answer: 88

DIAGNOSTIC TEST

Multiply as indicated:

1. $\frac{1}{5}\times\frac{1}{3}$ $\frac{1}{15}$	**9.** $\frac{2}{3}\times 6$ 4	**17.** $5\times\frac{3}{16}$ $\frac{15}{16}$	**25.** $6\frac{1}{4}\times\frac{3}{8}$ $2\frac{11}{32}$
2. $\frac{3}{4}\times\frac{5}{8}$ $\frac{15}{32}$	**10.** $\frac{3}{4}\times 2$ $1\frac{1}{2}$	**18.** $4\frac{1}{2}\times 4$ 18	**26.** $\frac{5}{16}\times 9\frac{3}{5}$ 3
3. $\frac{1}{2}\times\frac{2}{3}$ $\frac{1}{3}$	**11.** $\frac{5}{6}\times 10$ $8\frac{1}{3}$	**19.** $1\frac{7}{12}\times 8$ $12\frac{2}{3}$	**27.** $\frac{5}{6}\times 1\frac{9}{16}$ $1\frac{29}{96}$
4. $\frac{3}{8}\times\frac{4}{5}$ $\frac{3}{10}$	**12.** $\frac{3}{5}\times 7$ $4\frac{1}{5}$	**20.** $3\frac{1}{3}\times 5$ $16\frac{2}{3}$	**28.** $5\frac{1}{3}\times 1\frac{1}{8}$ 6
5. $\frac{3}{4}\times\frac{8}{15}$ $\frac{2}{5}$	**13.** $10\times\frac{9}{10}$ 9	**21.** $12\times 1\frac{5}{6}$ 22	**29.** $2\frac{5}{8}\times 2\frac{2}{5}$ $6\frac{3}{10}$
6. $\frac{9}{16}\times\frac{5}{6}$ $\frac{15}{32}$	**14.** $48\times\frac{7}{12}$ 28	**22.** $10\times 2\frac{9}{16}$ $25\frac{5}{8}$	**30.** $4\frac{1}{2}\times 2\frac{1}{4}$ $10\frac{1}{8}$
7. $\frac{5}{2}\times\frac{10}{3}$ $8\frac{1}{3}$	**15.** $4\times\frac{7}{8}$ $3\frac{1}{2}$	**23.** $7\times 3\frac{1}{4}$ $22\frac{3}{4}$	**31.** $\frac{1}{2}\times\frac{8}{15}\times\frac{5}{6}$ $\frac{2}{9}$
8. $\frac{7}{8}\times 8$ 7	**16.** $12\times\frac{5}{8}$ $7\frac{1}{2}$	**24.** $2\frac{1}{2}\times\frac{4}{5}$ 2	**32.** $1\frac{3}{4}\times 3\frac{1}{7}\times 1\frac{3}{5}$ $8\frac{4}{5}$

33. 18
$\underline{\quad 7\frac{1}{3}}$ 132

34. $12\frac{5}{6}$
$\underline{\quad 8}$ $102\frac{2}{3}$

35. Find the product of $2\frac{7}{8}$ and $1\frac{3}{4}$. $5\frac{1}{32}$

36. $\frac{5}{100}\times 900$ 45 **37.** Simplify: $\dfrac{132\times 3,600}{5,280}$ 90 **38.** Find $\frac{3}{4}$ of 18. $13\frac{1}{2}$

Basic: Sets 1–27, 38(a)–38(d), Prob. 1–10; *Average:* Sets 1–33, 38(a)–38(h), Prob. 1–15; *Enriched:*
Sets: 1–38(j), Prob. 1–20

104   **COMMON FRACTIONS**

RELATED PRACTICE EXAMPLES

Multiply as indicated:

SET 1

$\frac{1}{4} \times \frac{1}{2}$ $\frac{1}{8}$ $\frac{1}{8} \times \frac{1}{3}$ $\frac{1}{24}$ $\frac{1}{5} \times \frac{1}{4}$ $\frac{1}{20}$ $\frac{1}{2} \times \frac{1}{10}$ $\frac{1}{20}$ $\frac{1}{16} \times \frac{1}{4}$ $\frac{1}{64}$

SET 2

$\frac{1}{2} \times \frac{3}{5}$ $\frac{3}{10}$ $\frac{5}{6} \times \frac{7}{8}$ $\frac{35}{48}$ $\frac{7}{16} \times \frac{3}{4}$ $\frac{21}{64}$ $\frac{9}{10} \times \frac{3}{8}$ $\frac{27}{80}$ $\frac{5}{12} \times \frac{1}{3}$ $\frac{5}{36}$

SET 3

$\frac{1}{3} \times \frac{3}{5}$ $\frac{1}{5}$ $\frac{4}{5} \times \frac{3}{4}$ $\frac{3}{5}$ $\frac{3}{10} \times \frac{1}{3}$ $\frac{1}{10}$ $\frac{2}{3} \times \frac{1}{2}$ $\frac{1}{3}$ $\frac{5}{6} \times \frac{6}{7}$ $\frac{5}{7}$

SET 4

$\frac{5}{6} \times \frac{3}{8}$ $\frac{5}{16}$ $\frac{6}{7} \times \frac{11}{12}$ $\frac{11}{14}$ $\frac{4}{5} \times \frac{7}{8}$ $\frac{7}{10}$ $\frac{3}{10} \times \frac{5}{14}$ $\frac{3}{28}$ $\frac{1}{3} \times \frac{9}{10}$ $\frac{3}{10}$

SET 5

$\frac{2}{5} \times \frac{5}{12}$ $\frac{1}{6}$ $\frac{9}{10} \times \frac{2}{3}$ $\frac{3}{5}$ $\frac{5}{8} \times \frac{16}{25}$ $\frac{2}{5}$ $\frac{8}{9} \times \frac{3}{4}$ $\frac{2}{3}$ $\frac{4}{21} \times \frac{7}{8}$ $\frac{1}{6}$

SET 6

$\frac{5}{6} \times \frac{4}{5}$ $\frac{2}{3}$ $\frac{3}{16} \times \frac{6}{7}$ $\frac{9}{56}$ $\frac{7}{8} \times \frac{12}{21}$ $\frac{1}{2}$ $\frac{10}{12} \times \frac{14}{15}$ $\frac{7}{9}$ $\frac{15}{16} \times \frac{9}{10}$ $\frac{27}{32}$

SET 7

$\frac{5}{4} \times \frac{2}{3}$ $\frac{5}{6}$ $\frac{7}{2} \times \frac{4}{5}$ $2\frac{4}{5}$ $\frac{4}{3} \times \frac{3}{4}$ 1 $\frac{21}{10} \times \frac{15}{14}$ $2\frac{1}{4}$ $\frac{10}{9} \times \frac{15}{8}$ $2\frac{1}{12}$

SET 8

$\frac{1}{2} \times 2$ 1 $\frac{5}{6} \times 6$ 5 $\frac{11}{8} \times 8$ 11 $\frac{7}{16} \times 16$ 7 $\frac{3}{5} \times 5$ 3

SET 9

$\frac{3}{4} \times 8$ 6 $\frac{5}{8} \times 16$ 10 $\frac{7}{6} \times 72$ 84 $\frac{5}{12} \times 36$ 15 $\frac{9}{16} \times 64$ 36

SET 10

$\frac{3}{8} \times 4$ $1\frac{1}{2}$ $\frac{7}{16} \times 2$ $\frac{7}{8}$ $\frac{5}{12} \times 3$ $1\frac{1}{4}$ $\frac{11}{24} \times 12$ $5\frac{1}{2}$ $\frac{9}{10} \times 5$ $4\frac{1}{2}$

SET 11

$\frac{3}{8} \times 6$ $2\frac{1}{4}$ $\frac{9}{16} \times 20$ $11\frac{1}{4}$ $\frac{19}{12} \times 8$ $12\frac{2}{3}$ $\frac{7}{10} \times 26$ $18\frac{1}{5}$ $\frac{5}{6} \times 4$ $3\frac{1}{3}$

SET 12

$\frac{1}{3} \times 7$ $2\frac{1}{3}$ $\frac{3}{5} \times 9$ $5\frac{2}{5}$ $\frac{9}{8} \times 5$ $5\frac{5}{8}$ $\frac{13}{16} \times 3$ $2\frac{7}{16}$ $\frac{3}{10} \times 21$ $6\frac{3}{10}$

SET 13

$4 \times \frac{3}{4}$ 3 $8 \times \frac{5}{8}$ 5 $12 \times \frac{13}{12}$ 13 $32 \times \frac{29}{32}$ 29 $16 \times \frac{15}{16}$ 15

SET 14

$12 \times \frac{5}{6}$ 10 $24 \times \frac{3}{8}$ 9 $32 \times \frac{17}{16}$ 34 $8 \times \frac{3}{4}$ 6 $30 \times \frac{2}{5}$ 12

SET 15

$2 \times \frac{3}{8}$ $\frac{3}{4}$ $4 \times \frac{5}{16}$ $1\frac{1}{4}$ $2 \times \frac{7}{4}$ $3\frac{1}{2}$ $5 \times \frac{3}{10}$ $1\frac{1}{2}$ $6 \times \frac{13}{24}$ $3\frac{1}{4}$

SET 16

$12 \times \frac{7}{16}$ $5\frac{1}{4}$ $18 \times \frac{15}{32}$ $8\frac{7}{16}$ $16 \times \frac{19}{12}$ $25\frac{1}{3}$ $14 \times \frac{5}{8}$ $8\frac{3}{4}$ $15 \times \frac{3}{10}$ $4\frac{1}{2}$

SET 17

$4 \times \frac{2}{3}$ $2\frac{2}{3}$ $9 \times \frac{4}{5}$ $7\frac{1}{5}$ $3 \times \frac{9}{8}$ $3\frac{3}{8}$ $7 \times \frac{5}{12}$ $2\frac{11}{12}$ $13 \times \frac{9}{10}$ $11\frac{7}{10}$

SET 18

$2\frac{1}{8} \times 16$ 34 $3\frac{1}{4} \times 12$ 39 $1\frac{2}{3} \times 9$ 15 $7\frac{1}{2} \times 6$ 45 $4\frac{3}{5} \times 10$ 46

SET 19

$2\frac{5}{6} \times 3$ $8\frac{1}{2}$ $5\frac{7}{8} \times 6$ $35\frac{1}{4}$ $3\frac{11}{12} \times 8$ $31\frac{1}{3}$ $1\frac{9}{10} \times 12$ $22\frac{4}{5}$ $6\frac{13}{16} \times 20$ $136\frac{1}{4}$

SET 20

$2\frac{1}{5} \times 4$ $8\frac{4}{5}$ $3\frac{1}{2} \times 7$ $24\frac{1}{2}$ $5\frac{2}{3} \times 2$ $11\frac{1}{3}$ $4\frac{5}{8} \times 3$ $13\frac{7}{8}$ $3\frac{9}{16} \times 5$ $17\frac{13}{16}$

SET 21

$8 \times 5\frac{1}{4}$ 42 $16 \times 2\frac{5}{8}$ 42 $12 \times 3\frac{5}{6}$ 46 $48 \times 1\frac{13}{16}$ 87 $24 \times 5\frac{11}{12}$ 142

SET 22

$15 \times 1\frac{7}{10}$ $25\frac{1}{2}$ $8 \times 2\frac{9}{16}$ $20\frac{1}{2}$ $18 \times 5\frac{5}{8}$ $101\frac{1}{4}$ $4 \times 3\frac{1}{6}$ $12\frac{2}{3}$ $16 \times 4\frac{7}{12}$ $73\frac{1}{3}$

SET 23

$2 \times 6\frac{1}{3}$ $12\frac{2}{3}$ $5 \times 4\frac{3}{8}$ $21\frac{7}{8}$ $3 \times 9\frac{1}{2}$ $28\frac{1}{2}$ $9 \times 5\frac{7}{16}$ $48\frac{15}{16}$ $7 \times 1\frac{3}{4}$ $12\frac{1}{4}$

SET 24

$3\frac{1}{3} \times \frac{3}{5}$ 2 $2\frac{1}{5} \times \frac{5}{6}$ $1\frac{5}{6}$ $6\frac{3}{4} \times \frac{2}{3}$ $4\frac{1}{2}$ $3\frac{1}{8} \times \frac{8}{15}$ $1\frac{2}{3}$ $4\frac{1}{2} \times \frac{2}{9}$ 1

SET 25

$2\frac{1}{2} \times \frac{3}{4}$ $1\frac{7}{8}$ $5\frac{2}{3} \times \frac{5}{8}$ $3\frac{13}{24}$ $1\frac{3}{8} \times \frac{1}{2}$ $\frac{11}{16}$ $3\frac{7}{16} \times \frac{3}{4}$ $2\frac{37}{64}$ $4\frac{5}{6} \times \frac{11}{16}$ $3\frac{31}{96}$

SET 26

$\frac{1}{2} \times 3\frac{1}{5}$ $1\frac{3}{5}$ $\frac{5}{6} \times 2\frac{3}{10}$ $1\frac{11}{12}$ $\frac{7}{8} \times 3\frac{1}{7}$ $2\frac{3}{4}$ $\frac{9}{16} \times 1\frac{1}{3}$ $\frac{3}{4}$ $\frac{5}{12} \times 2\frac{1}{4}$ $\frac{15}{16}$

SET 27

$\frac{7}{8} \times 1\frac{1}{4}$ $1\frac{3}{32}$ $\frac{1}{2} \times 3\frac{3}{8}$ $1\frac{11}{16}$ $\frac{1}{5} \times 4\frac{1}{3}$ $\frac{13}{15}$ $\frac{9}{10} \times 1\frac{1}{2}$ $1\frac{7}{20}$ $\frac{3}{4} \times 2\frac{3}{16}$ $1\frac{41}{64}$

SET 28

$1\frac{1}{4} \times 1\frac{3}{5}$ 2 $5\frac{1}{3} \times 4\frac{1}{2}$ 24 $2\frac{5}{8} \times 1\frac{5}{7}$ $4\frac{1}{2}$ $3\frac{3}{4} \times 1\frac{1}{5}$ $4\frac{1}{2}$ $2\frac{2}{3} \times 3\frac{3}{8}$ 9

SET 29

$2\frac{1}{3} \times 1\frac{1}{5}$ $2\frac{4}{5}$ $2\frac{1}{6} \times 2\frac{2}{3}$ $5\frac{7}{9}$ $2\frac{2}{5} \times 1\frac{3}{16}$ $2\frac{17}{20}$ $1\frac{7}{16} \times 1\frac{5}{9}$ $2\frac{17}{72}$ $1\frac{3}{4} \times 3\frac{1}{3}$ $5\frac{5}{6}$

SET 30

$2\frac{1}{8} \times 1\frac{1}{2}$ $3\frac{3}{16}$ $3\frac{3}{4} \times 2\frac{7}{8}$ $10\frac{25}{32}$ $1\frac{9}{16} \times 4\frac{1}{3}$ $6\frac{37}{48}$ $2\frac{5}{6} \times 1\frac{3}{8}$ $3\frac{43}{48}$ $4\frac{1}{8} \times 3\frac{13}{16}$ $15\frac{93}{128}$

SET 31

$\frac{1}{4} \times \frac{5}{6} \times \frac{2}{5}$ $\frac{1}{12}$ $\frac{2}{3} \times \frac{5}{8} \times \frac{3}{10}$ $\frac{1}{8}$ $\frac{5}{12} \times \frac{3}{16} \times \frac{4}{5}$ $\frac{1}{16}$ $\frac{2}{5} \times \frac{3}{4} \times \frac{15}{16}$ $\frac{9}{32}$ $\frac{3}{4} \times \frac{3}{5} \times \frac{1}{2}$ $\frac{9}{40}$

SET 32

$2\frac{3}{4} \times 1\frac{1}{8} \times 3\frac{5}{6}$ $11\frac{55}{64}$ $1\frac{1}{2} \times \frac{4}{5} \times 2\frac{1}{6}$ $2\frac{3}{5}$ $3\frac{1}{5} \times 1\frac{1}{4} \times 1\frac{1}{3}$ $5\frac{1}{3}$ $1\frac{5}{16} \times 2\frac{2}{3} \times 3\frac{1}{7}$ 11

$4\frac{1}{6} \times 3\frac{1}{5} \times 1\frac{3}{10}$ $17\frac{1}{3}$

	SET 33			
28	16	35	6	29
$4\frac{1}{2}$	$5\frac{3}{4}$	$7\frac{5}{8}$	$8\frac{4}{5}$	$2\frac{2}{3}$
126	92	$266\frac{7}{8}$	$52\frac{4}{5}$	$77\frac{1}{3}$

	SET 34			
$32\frac{1}{4}$	$24\frac{3}{8}$	$17\frac{3}{5}$	$5\frac{5}{6}$	$14\frac{3}{16}$
8	6	10	9	8
258	$146\frac{1}{4}$	176	$52\frac{1}{2}$	$113\frac{1}{2}$

SET 35

1. Multiply $4\frac{7}{8}$ by 6. $29\frac{1}{4}$
2. Multiply $2\frac{3}{4}$ by $4\frac{1}{2}$. $12\frac{3}{8}$
3. Find the product of $1\frac{3}{4}$ and $3\frac{1}{7}$. $5\frac{1}{2}$
4. Find the product of $1\frac{1}{8}$ and $5\frac{1}{3}$. 6
5. Find the product of $2\frac{3}{16}$ and $2\frac{2}{5}$. $5\frac{1}{4}$

SET 36

1. $\frac{3}{100} \times 400$ 12
2. $\frac{4}{100} \times 500$ 20
3. $\frac{2}{100} \times 1,000$ 20
4. $\frac{6}{100} \times 2,500$ 150
5. $\frac{8}{100} \times 4,800$ 384

SET 37

Simplify:

1. $\frac{1,760 \times 30}{60}$ 880

2. $\frac{75 \times 32}{25 \times 24}$ 4

3. $\frac{22 \times 21 \times 21}{7 \times 2 \times 2}$ $346\frac{1}{2}$

4. $\frac{33 \times 6 \times 14}{231}$ 12

5. $\frac{18 \times 12 \times 16}{144}$ 24

6. $\frac{5,280 \times 120}{3,600}$ 176

7. $\frac{3,600 \times 66}{5,280}$ 45

8. $\frac{5,280 \times 4}{6,080}$ $3\frac{9}{19}$

9. $\frac{5,280 \times 100}{3,600}$ $146\frac{2}{3}$

10. $\frac{6,080 \times 12}{5,280}$ $13\frac{9}{11}$

Finding Fractional Parts of a Number

Find:

SET 38(a)	SET 38(b)	SET 38(c)	SET 38(d)
1. $\frac{1}{2}$ of 48 24	1. $\frac{1}{3}$ of 96 32	1. $\frac{1}{8}$ of 104 13	1. $\frac{1}{6}$ of 78 13
2. $\frac{1}{2}$ of 32¢ 16¢	2. $\frac{1}{3}$ of 8 $2\frac{2}{3}$	2. $\frac{1}{8}$ of $136 $17	2. $\frac{1}{6}$ of $120 $20
3. $\frac{1}{2}$ of $60 $30	3. $\frac{1}{4}$ of 52¢ 13¢	3. $\frac{1}{8}$ of $\frac{2}{5}$ $\frac{1}{20}$	3. $\frac{1}{6}$ of 4 $\frac{2}{3}$
4. $\frac{1}{2}$ of 17 $8\frac{1}{2}$	4. $\frac{1}{4}$ of $1,000 $250	4. $\frac{1}{16}$ of 288 18	4. $\frac{1}{12}$ of 84¢ 7¢
5. $\frac{1}{2}$ of $2\frac{1}{2}$ $1\frac{1}{4}$	5. $\frac{1}{4}$ of $1\frac{3}{5}$ $\frac{2}{5}$	5. $\frac{1}{16}$ of 24 $1\frac{1}{2}$	5. $\frac{1}{12}$ of $2\frac{1}{4}$ $\frac{3}{16}$

SET 38(e)	SET 38(f)	SET 38(g)	SET 38(h)
1. $\frac{1}{5}$ of $285 $57	1. $\frac{3}{4}$ of 16 12	1. $\frac{2}{5}$ of 45¢ 18¢	1. $\frac{3}{8}$ of 24¢ 9¢
2. $\frac{1}{5}$ of 19 $3\frac{4}{5}$	2. $\frac{3}{4}$ of $272 $204	2. $\frac{2}{5}$ of 12 $4\frac{4}{5}$	2. $\frac{5}{8}$ of $168 $105
3. $\frac{1}{10}$ of 80¢ 8¢	3. $\frac{3}{4}$ of $\frac{3}{8}$ $\frac{9}{32}$	3. $\frac{3}{5}$ of $235 $141	3. $\frac{7}{8}$ of 6 $5\frac{1}{4}$
4. $\frac{1}{20}$ of 45 $2\frac{1}{4}$	4. $\frac{2}{3}$ of 87¢ 58¢	4. $\frac{3}{5}$ of $1\frac{2}{3}$ 1	4. $\frac{15}{16}$ of $1,120 $1,050
5. $\frac{1}{100}$ of 1,400 14	5. $\frac{2}{3}$ of $1\frac{11}{16}$ $1\frac{1}{8}$	5. $\frac{4}{5}$ of 585 468	5. $\frac{9}{16}$ of $\frac{2}{3}$ $\frac{3}{8}$

SET 38(i)		SET 38(j)	
1. $\frac{7}{10}$ of 180 126	4. $\frac{17}{20}$ of 12 $10\frac{1}{5}$	1. $\frac{5}{6}$ of 72¢ 60¢	4. $\frac{5}{12}$ of $2,400 $1,000
2. $\frac{9}{100}$ of 250 $22\frac{1}{2}$	5. $\frac{3}{10}$ of $1\frac{1}{2}$ $\frac{9}{20}$	2. $\frac{5}{6}$ of 16 $13\frac{1}{3}$	5. $\frac{7}{12}$ of $2\frac{2}{5}$ $1\frac{2}{5}$
3. $\frac{41}{100}$ of $5,000 $2,050		3. $\frac{11}{12}$ of 132 121	

PROBLEM SOLVING

See page 16, *Problem Solving Strategy.*

For problems 1–5, select the letter corresponding to your answer.

1. A sewing class is making costumes for the school play. If each costume requires $3\frac{5}{8}$ yards of goods, how many yards are needed to make 26 costumes?

   a. $85\frac{3}{4}$ yards   **b.** $91\frac{7}{8}$ yards   <u>**c.**</u> $94\frac{1}{4}$ yards   **d.** Answer not given

2. How much wood is needed to make 15 shelves each $6\frac{2}{3}$ feet long?

   a. $96\frac{1}{3}$ feet   <u>**b.**</u> 100 feet   **c.** $98\frac{2}{3}$ feet   **d.** $101\frac{1}{2}$ feet

3. Paul wishes to buy a cassette recorder priced at $60. He pays $\frac{1}{5}$ of the price in cash and the rest in 6 equal monthly installments. How much must he pay each month?

   <u>**a.**</u> $8   **b.** $9   **c.** $10   **d.** Answer not given

4. Mr. Rodgers earns $5 per hour. If he gets time and a half for overtime, what is his hourly rate for overtime work?

   a. $6\frac{1}{2}$ **b.** $8\frac{1}{2}$ **c.** $7   <u>**d.**</u> $7\frac{1}{2}$

5. A house worth \$52,800 is assessed at $\frac{2}{3}$ of its value. What is the assessed value of the house?

     a. \$33,600   **b.** \$37,500   **c.** \$35,200   **d.** \$30,500

6. Find the cost of: **a.** $\frac{1}{2}$ dozen eggs at 96¢ per dozen. \$.48

b. $\frac{3}{4}$ pound of cake at 80¢ per pound. \$.60

c. $2\frac{1}{4}$ pounds of bananas at 32¢ per pound. \$.72

d. $2\frac{3}{16}$ pounds of string beans at 48¢ per pound. \$1.05

e. $1\frac{1}{8}$ pounds of candy at 88¢ per pound. \$.99

7. The cooking class is divided into 6 teams. If each team uses a recipe that requires $2\frac{1}{2}$ cups cake flour, $2\frac{1}{4}$ teaspoons baking powder, $\frac{1}{2}$ cup shortening, and $1\frac{1}{4}$ cups sugar, how much of each ingredient does the entire class need? 15 cups flour, $13\frac{1}{2}$ tea. baking powder, 3 cups short., $7\frac{1}{2}$ cups sugar

8. If the premium rate for a 3-year fire insurance policy is $2\frac{1}{2}$ times the annual rate, find the rate for a 3-year policy if the annual rate is: \$.45/\$100 **a.** 18¢ per \$100; **b.** 32¢ per \$100; **c.** 25¢ per \$100; **d.** 19¢ per \$100; \$.92$\frac{1}{2}$/\$100 **e.** 37¢ per \$100.   \$.80/\$100   \$.62$\frac{1}{2}$/\$100   \$.47$\frac{1}{2}$/\$100

9. How much will a trip over a distance of 12 kilometers cost at $11\frac{3}{4}$ cents per kilometer? \$1.41

10. A family budgets $\frac{1}{4}$ of its annual income of \$18,600 for food, $\frac{3}{10}$ for shelter including operating expenses and furnishings, $\frac{1}{10}$ for transportation, $\frac{1}{20}$ for clothing, $\frac{1}{8}$ for savings, and the remainder for miscellaneous expenses. How much is allowed for each item annually? Food, \$4,650; Shelter, \$5,580; Trans., \$1,860; Clothing, \$930; Sav., \$2,325; misc., \$3,255

11. a. If the scale on a chart is 1 inch = 30 miles, how many miles do $6\frac{1}{2}$ inches represent? 195 mi.

b. Using the scale 1 inch = 150 miles, find the distance represented by $\frac{3}{4}$ inch. 112.50 mi.

12. a. In constructing a vector diagram the scale 1 inch = 48 m.p.h. is used. What speed in miles per hour will a line $2\frac{3}{16}$ inches in length represent? 105 m.p.h.

b. What is the wind velocity if it is represented by a line $\frac{5}{8}$ inch and the scale 1 inch = 64 m.p.h. is used? 40 m.p.h.

c. What is the air speed if it is represented by a line $3\frac{1}{2}$ inches long and the scale 1 inch = 50 m.p.h. is used? 175 m.p.h.

d. What is the ground speed if it is represented by a line $4\frac{3}{4}$ inches long and the scale 1 inch = 60 m.p.h. is used? 285 m.p.h.

13. To change Fahrenheit temperature readings to Celsius readings, first subtract 32 degrees from the Fahrenheit reading, then take $\frac{5}{9}$ of the answer.

Change the following Fahrenheit readings to Celsius:

     a. 68°F   **b.** 104°F   **c.** 41°F   **d.** 32°F   **e.** 212°F
        20°C   40°C   5°C   0°C   100°C

14.a. 1. $3\frac{1}{2}$ in., 2. 35 ft., 3. $1\frac{1}{4}$ in., 4. $\frac{3}{8}$ in., 5. $12\frac{1}{2}$ ft.

14.b. 1. 9 ft., 2. $16\frac{1}{2}$ ft., 3. $2\frac{7}{16}$ in., 4. $\frac{3}{8}$ in., 5. $\frac{27}{32}$ in.

1-16 MULTIPLICATION OF FRACTIONS AND MIXED NUMBERS **109**

14. a. The diameter of a circle is twice the radius. Find the length of the diameter if the radius is: (1) $1\frac{3}{4}$ in. (2) $17\frac{1}{2}$ ft. (3) $\frac{5}{8}$ in. (4) $\frac{3}{16}$ in. (5) $6\frac{1}{4}$ ft.

b. The radius of a circle is one half the diameter. Find the length of the radius if the diameter is: (1) 18 ft. (2) 33 ft. (3) $4\frac{7}{8}$ in. (4) $\frac{3}{4}$ in. (5) $1\frac{11}{16}$ in.

c. The circumference (distance around) of a circle is $3\frac{1}{7}$ times the diameter. Find the circumference when the diameter is: (1) 56 ft. (2) 15 ft. (3) $10\frac{1}{2}$ in. (4) $8\frac{3}{4}$ in. (5) $\frac{7}{8}$ in. 1. 176 ft., 2. $47\frac{1}{7}$ ft., 3. 33 in., 4. $27\frac{1}{2}$ in., 5. $2\frac{3}{4}$ in.

15. The temperature of a parcel of air decreases at the rate of $5\frac{1}{2}$°F for each 1,000 feet it rises following the dry adiabat.

a. If the temperature at the surface is 68°F, find the temperature at an altitude of 4,000 feet. 46°F

b. What is the temperature at the surface if the temperature at an altitude of 5,000 feet is 42°F? $69\frac{1}{2}$°F

16. A plane encounters a wind which cuts its ground speed to $\frac{2}{3}$ of its scheduled speed. If the pilot planned to fly at a ground speed of 360 miles per hour and reach a point 720 miles distant, how much longer will it take him to get to his destination, flying at the reduced speed? 1 hr. longer

17. If 1 cubic foot holds about $7\frac{1}{2}$ gallons, how many gallons of oil will a tank hold if its volume is $24\frac{1}{2}$ cubic feet? $183\frac{3}{4}$ gal.

18. If 1 cubic foot of water weighs $62\frac{1}{2}$ pounds, find the weight of a column of water containing 32 cubic feet. 2,000 lb.

19. A ship steams at $18\frac{1}{2}$ knots. At that rate how many nautical miles does it go in $6\frac{1}{4}$ hours? (1 knot = 1 nautical m.p.h.) $115\frac{5}{8}$ naut. mi.

20. The speed of a certain submarine when submerged is $\frac{3}{5}$ of its surface speed. If its maximum surface speed is $17\frac{1}{2}$ knots, what is its maximum speed when submerged? $10\frac{1}{2}$ knots

---

REFRESH YOUR SKILLS

1. Round 687,473 to the nearest hundred. 687,500

2. Add:	**3.** Subtract:	**4.** Multiply:	**5.** Divide: 409
627	800,000	5,280	365$\overline{)149,285}$
4,963	706,094	5,280	
81,896	93,906	27,878,400	
97,988			
8,569			
194,043			

6. Express $\frac{36}{54}$ in lowest terms. $\frac{2}{3}$

9. Multiply: **10.** Multiply:

$\frac{7}{8} \times \frac{3}{4}$ $\frac{21}{32}$ $4\frac{3}{8} \times 3\frac{1}{5}$ 14

7. Add:

$3\frac{7}{10}$

$6\frac{5}{12}$ $10\frac{7}{60}$

8. Subtract:

$4\frac{1}{3}$

$3\frac{5}{6}$ $\frac{1}{2}$

Division of Fractions and Mixed Numbers

VOCABULARY: Complex fraction

OBJECTIVE:

I. Aim   To divide fractions and mixed numbers.

II. Procedure

1. Change each mixed number to an improper fraction. (See Exercise 1-15.) Express each whole number in fraction form by using the whole number as the numerator and the figure 1 as the denominator.

2. Since division by a number gives the same result as multiplication by the reciprocal of this number (see Exercise 1-59), invert the divisor to find the reciprocal and then multiply as in the multiplication of fractions, using reduction where possible. (See Exercise 1-16.)

3. To simplify a complex fraction, divide the fraction in the numerator by the fraction in the denominator.

4. Check by going over the work again.

   Note that when a number is divided by a proper fraction (less than 1), the quotient is always greater than the given number. (See sample solution 3.) When a number is divided by a mixed number, the quotient is always smaller than the given number. (See sample solution 6.)

III. Sample Solutions

Divide in the following examples:

1. $\dfrac{2}{5} \div \dfrac{5}{8} = \dfrac{2}{5} \times \dfrac{8}{5} = \dfrac{16}{25}$

3. $15 \div \dfrac{3}{8} = \dfrac{\overset{5}{\cancel{15}}}{1} \times \dfrac{8}{\underset{1}{\cancel{3}}} = \dfrac{40}{1}$ or 40

Answer: $\frac{16}{25}$

Answer: 40

2. $\dfrac{2}{3} \div \dfrac{5}{6} = \dfrac{2}{\underset{1}{\cancel{3}}} \times \dfrac{\overset{2}{\cancel{6}}}{5} = \dfrac{4}{5}$

4. $8\frac{3}{4} \div 7 = \dfrac{35}{4} \div 7 = \dfrac{\overset{5}{\cancel{35}}}{4} \times \dfrac{1}{\underset{1}{\cancel{7}}} = \dfrac{5}{4} = 1\frac{1}{4}$

Answer: $\frac{4}{5}$

Answer: $1\frac{1}{4}$

5. $2\frac{1}{2} \div \dfrac{3}{4} = \dfrac{5}{2} \div \dfrac{3}{4} = \dfrac{5}{\underset{1}{\cancel{2}}} \times \dfrac{\overset{2}{\cancel{4}}}{3} = \dfrac{10}{3} = 3\frac{1}{3}$

Answer: $3\frac{1}{3}$

6. $2\frac{3}{16} \div 1\frac{1}{4} = \frac{35}{16} \div \frac{5}{4} = \frac{\overset{7}{\cancel{35}}}{\cancel{16}} \times \frac{\overset{1}{\cancel{4}}}{\cancel{5}} = \frac{7}{4} = 1\frac{3}{4}$

Answer: $1\frac{3}{4}$

7. a. $\left(2\frac{1}{2} \div \frac{5}{6}\right) \div \frac{2}{3} = \left(\frac{\cancel{5}}{\cancel{2}} \times \frac{\overset{3}{\cancel{6}}}{\cancel{5}}\right) \div \frac{2}{3} = \frac{3}{1} \div \frac{2}{3} = \frac{3}{1} \times \frac{3}{2} = \frac{9}{2} = 4\frac{1}{2}$

Answer: $4\frac{1}{2}$

b. $2\frac{1}{2} \div \left(\frac{5}{6} \div \frac{2}{3}\right) = 2\frac{1}{2} \div \left(\frac{5}{\cancel{6}} \times \frac{\overset{1}{\cancel{3}}}{2}\right) = 2\frac{1}{2} \div \frac{5}{4} = \frac{\cancel{5}}{\cancel{2}} \times \frac{\overset{2}{\cancel{4}}}{\cancel{5}} = \frac{2}{1} = 2$

Answer: 2

8. Simplify the following complex fraction:

$$\frac{\frac{9}{16}}{\frac{3}{8}} = \frac{9}{16} \div \frac{3}{8} = \frac{\overset{3}{\cancel{9}}}{\cancel{16}} \times \frac{\overset{1}{\cancel{8}}}{\cancel{3}} = \frac{3}{2} = 1\frac{1}{2}$$

Answer: $1\frac{1}{2}$

IV. Definition

A complex fraction is a fraction in which the numerator or denominator or both have a fraction as a term.

DIAGNOSTIC TEST

Divide as indicated:

1. $\frac{1}{3} \div \frac{3}{4}$ $\frac{4}{9}$	**7.** $\frac{7}{8} \div 2$ $\frac{7}{16}$	**13.** $1\frac{2}{3} \div 4$ $\frac{5}{12}$	**19.** $2\frac{5}{8} \div \frac{3}{5}$ $4\frac{3}{8}$
2. $\frac{2}{3} \div \frac{5}{16}$ $2\frac{2}{15}$	**8.** $8 \div \frac{1}{2}$ 16	**14.** $15 \div 1\frac{7}{8}$ 8	**20.** $1\frac{7}{16} \div \frac{2}{3}$ $2\frac{5}{32}$
3. $\frac{3}{5} \div \frac{9}{10}$ $\frac{2}{3}$	**9.** $5 \div \frac{15}{16}$ $5\frac{1}{3}$	**15.** $6 \div 4\frac{1}{2}$ $1\frac{1}{3}$	**21.** $\frac{7}{8} \div 1\frac{3}{4}$ $\frac{1}{2}$
4. $\frac{3}{4} \div \frac{3}{8}$ 2	**10.** $2 \div \frac{3}{5}$ $3\frac{1}{3}$	**16.** $4 \div 5\frac{1}{3}$ $\frac{3}{4}$	**22.** $\frac{3}{4} \div 1\frac{3}{5}$ $\frac{15}{32}$
5. $\frac{5}{6} \div \frac{7}{12}$ $1\frac{3}{7}$	**11.** $4\frac{1}{2} \div 18$ $\frac{1}{4}$	**17.** $7 \div 2\frac{3}{4}$ $2\frac{6}{11}$	**23.** $11\frac{1}{3} \div 2\frac{5}{6}$ 4
6. $\frac{3}{4} \div 6$ $\frac{1}{8}$	**12.** $4\frac{2}{3} \div 2$ $2\frac{1}{3}$	**18.** $2\frac{1}{2} \div \frac{5}{6}$ 3	**24.** $3\frac{3}{16} \div 2\frac{1}{8}$ $1\frac{1}{2}$

25. $1\frac{1}{6} \div 9\frac{1}{3}$ $\frac{1}{8}$	**29.** $\left(1\frac{1}{8} \div 2\frac{1}{4}\right) \div 4\frac{2}{3}$ $\frac{3}{28}$
26. $5\frac{3}{8} \div 1\frac{2}{3}$ $3\frac{47}{56}$	**30.** $7\frac{1}{3} \div \left(3\frac{1}{7} \times 2\frac{4}{5}\right)$ $\frac{5}{6}$
27. $6\frac{3}{5} \div 6\frac{3}{5}$ 1	
28. Divide $9\frac{1}{3}$ by $3\frac{1}{7}$. $2\frac{32}{33}$	**31.** Simplify: $\dfrac{\frac{2}{3}}{\frac{5}{8}}$ $1\frac{1}{15}$

RELATED PRACTICE EXAMPLES

Divide as indicated:

SET 1

$\frac{1}{4} \div \frac{1}{3}$ $\frac{3}{4}$ $\frac{3}{4} \div \frac{4}{5}$ $\frac{15}{16}$ $\frac{1}{2} \div \frac{2}{3}$ $\frac{3}{4}$ $\frac{2}{5} \div \frac{9}{16}$ $\frac{32}{45}$ $\frac{3}{5} \div \frac{11}{12}$ $\frac{36}{55}$

SET 2

$\frac{1}{2} \div \frac{1}{5}$ $2\frac{1}{2}$ $\frac{7}{8} \div \frac{2}{3}$ $1\frac{5}{16}$ $\frac{9}{10} \div \frac{1}{3}$ $2\frac{7}{10}$ $\frac{11}{12} \div \frac{4}{7}$ $1\frac{29}{48}$ $\frac{4}{9} \div \frac{7}{16}$ $1\frac{1}{63}$

SET 3

$\frac{1}{4} \div \frac{3}{4}$ $\frac{1}{3}$ $\frac{5}{8} \div \frac{5}{6}$ $\frac{3}{4}$ $\frac{3}{16} \div \frac{5}{12}$ $\frac{9}{20}$ $\frac{2}{3} \div \frac{4}{5}$ $\frac{5}{6}$ $\frac{5}{6} \div \frac{7}{8}$ $\frac{20}{21}$

SET 4

$\frac{5}{6} \div \frac{5}{12}$ 2 $\frac{7}{8} \div \frac{7}{16}$ 2 $\frac{2}{3} \div \frac{2}{9}$ 3 $\frac{5}{8} \div \frac{5}{8}$ 1 $\frac{4}{5} \div \frac{4}{15}$ 3

SET 5

$\frac{1}{2} \div \frac{7}{16}$ $1\frac{1}{7}$ $\frac{2}{5} \div \frac{3}{10}$ $1\frac{1}{3}$ $\frac{7}{8} \div \frac{5}{12}$ $2\frac{1}{10}$ $\frac{4}{5} \div \frac{7}{10}$ $1\frac{1}{7}$ $\frac{5}{12} \div \frac{3}{16}$ $2\frac{2}{9}$

SET 6

$\frac{2}{3} \div 4$ $\frac{1}{6}$ $\frac{4}{5} \div 2$ $\frac{2}{5}$ $\frac{7}{8} \div 7$ $\frac{1}{8}$ $\frac{9}{10} \div 6$ $\frac{3}{20}$ $\frac{3}{5} \div 9$ $\frac{1}{15}$

SET 7

$\frac{3}{5} \div 8$ $\frac{3}{40}$ $\frac{1}{2} \div 10$ $\frac{1}{20}$ $\frac{7}{16} \div 3$ $\frac{7}{48}$ $\frac{11}{12} \div 5$ $\frac{11}{60}$ $\frac{5}{8} \div 4$ $\frac{5}{32}$

SET 8

$6 \div \frac{1}{3}$ 18 $7 \div \frac{1}{8}$ 56 $8 \div \frac{4}{5}$ 10 $10 \div \frac{5}{16}$ 32 $4 \div \frac{2}{5}$ 10

SET 9

$18 \div \frac{9}{10}$ 20 $6 \div \frac{4}{5}$ $7\frac{1}{2}$ $10 \div \frac{2}{3}$ 15 $12 \div \frac{15}{16}$ $12\frac{4}{5}$ $15 \div \frac{21}{32}$ $22\frac{6}{7}$

SET 10

$5 \div \frac{3}{4}$ $6\frac{2}{3}$ $13 \div \frac{2}{3}$ $19\frac{1}{2}$ $9 \div \frac{7}{8}$ $10\frac{2}{7}$ $14 \div \frac{5}{6}$ $16\frac{4}{5}$ $7 \div \frac{9}{10}$ $7\frac{7}{9}$

SET 11

$1\frac{1}{2} \div 3$ $\frac{1}{2}$ $4\frac{2}{3} \div 14$ $\frac{1}{3}$ $5\frac{3}{5} \div 7$ $\frac{4}{5}$ $2\frac{7}{16} \div 13$ $\frac{3}{16}$ $3\frac{3}{8} \div 9$ $\frac{3}{8}$

SET 12

$3\frac{3}{4} \div 3$ $1\frac{1}{4}$ $7\frac{1}{2} \div 5$ $1\frac{1}{2}$ $8\frac{2}{5} \div 6$ $1\frac{2}{5}$ $18\frac{3}{4} \div 10$ $1\frac{7}{8}$ $11\frac{7}{8} \div 5$ $2\frac{3}{8}$

SET 13

$1\frac{5}{8} \div 2$ $\frac{13}{16}$ $2\frac{1}{16} \div 4$ $\frac{33}{64}$ $8\frac{1}{2} \div 3$ $2\frac{5}{6}$ $5\frac{3}{4} \div 5$ $1\frac{3}{20}$ $3\frac{4}{5} \div 8$ $\frac{19}{40}$

SET 14

$6 \div 1\frac{1}{2}$ 4 $8 \div 1\frac{1}{3}$ 6 $27 \div 2\frac{1}{4}$ 12 $68 \div 3\frac{2}{5}$ 20 $57 \div 2\frac{3}{8}$ 24

SET 15

$8 \div 2\frac{2}{5}$ $3\frac{1}{3}$ $10 \div 1\frac{7}{8}$ $5\frac{1}{3}$ $14 \div 1\frac{5}{16}$ $10\frac{2}{3}$ $40 \div 5\frac{1}{3}$ $7\frac{1}{2}$ $6 \div 2\frac{1}{4}$ $2\frac{2}{3}$

SET 16

$5 \div 6\frac{2}{3}$ $\frac{3}{4}$ $8 \div 9\frac{3}{5}$ $\frac{5}{6}$ $3 \div 6\frac{3}{4}$ $\frac{4}{9}$ $6 \div 7\frac{7}{8}$ $\frac{16}{21}$ $2 \div 4\frac{4}{5}$ $\frac{5}{12}$

SET 17

$6 \div 1\frac{2}{3}$ $3\frac{3}{5}$ $4 \div 6\frac{3}{5}$ $\frac{20}{33}$ $8 \div 1\frac{4}{5}$ $4\frac{4}{9}$ $1 \div 2\frac{1}{3}$ $\frac{3}{7}$ $3 \div 1\frac{5}{8}$ $1\frac{11}{13}$

SET 18

$1\frac{1}{2} \div \frac{9}{16}$ $2\frac{2}{3}$ $1\frac{1}{8} \div \frac{3}{32}$ 12 $8\frac{3}{4} \div \frac{7}{8}$ 10 $1\frac{7}{8} \div \frac{3}{16}$ 10 $9\frac{1}{3} \div \frac{7}{24}$ 32

SET 19

$1\frac{3}{5} \div \frac{2}{3}$ $2\frac{2}{5}$ $2\frac{1}{3} \div \frac{7}{8}$ $2\frac{2}{3}$ $1\frac{5}{6} \div \frac{5}{12}$ $4\frac{2}{5}$ $3\frac{1}{4} \div \frac{5}{6}$ $3\frac{9}{10}$ $2\frac{1}{8} \div \frac{9}{10}$ $2\frac{13}{36}$

SET 20

$4\frac{1}{4} \div \frac{3}{5}$ $7\frac{1}{12}$ $2\frac{3}{5} \div \frac{5}{8}$ $4\frac{4}{25}$ $3\frac{1}{7} \div \frac{3}{4}$ $4\frac{4}{21}$ $2\frac{2}{3} \div \frac{9}{16}$ $4\frac{20}{27}$ $2\frac{3}{8} \div \frac{4}{5}$ $2\frac{31}{32}$

SET 21

$\frac{5}{12} \div 8\frac{1}{3}$ $\frac{1}{20}$ $\frac{5}{6} \div 1\frac{1}{9}$ $\frac{3}{4}$ $\frac{3}{5} \div 2\frac{2}{5}$ $\frac{1}{4}$ $\frac{7}{8} \div 3\frac{3}{4}$ $\frac{7}{30}$ $\frac{13}{16} \div 1\frac{7}{32}$ $\frac{2}{3}$

SET 22

$\frac{2}{3} \div 1\frac{1}{4}$ $\frac{8}{15}$ $\frac{7}{8} \div 3\frac{1}{3}$ $\frac{21}{80}$ $\frac{3}{16} \div 1\frac{3}{5}$ $\frac{15}{128}$ $\frac{1}{3} \div 2\frac{9}{16}$ $\frac{16}{123}$ $\frac{4}{5} \div 4\frac{5}{8}$ $\frac{32}{185}$

SET 23

$14\frac{3}{8} \div 2\frac{7}{8}$ 5 $7\frac{1}{2} \div 1\frac{1}{4}$ 6 $14\frac{1}{2} \div 3\frac{5}{8}$ 4 $18\frac{1}{3} \div 1\frac{5}{6}$ 10 $50\frac{1}{4} \div 4\frac{3}{16}$ 12

SET 24

$3\frac{1}{5} \div 1\frac{1}{3}$ $2\frac{2}{5}$ $1\frac{5}{8} \div 1\frac{7}{32}$ $1\frac{1}{3}$ $4\frac{2}{3} \div 1\frac{3}{5}$ $2\frac{11}{12}$ $11\frac{1}{4} \div 2\frac{1}{2}$ $4\frac{1}{2}$ $2\frac{1}{16} \div 1\frac{3}{8}$ $1\frac{1}{2}$

SET 25

$2\frac{1}{4} \div 3\frac{3}{8}$ $\frac{2}{3}$ $1\frac{3}{5} \div 3\frac{1}{5}$ $\frac{1}{2}$ $1\frac{13}{16} \div 2\frac{1}{4}$ $\frac{29}{36}$ $2\frac{1}{12} \div 3\frac{3}{4}$ $\frac{5}{9}$ $1\frac{17}{32} \div 2\frac{5}{8}$ $\frac{7}{12}$

SET 26

$1\frac{2}{5} \div 2\frac{2}{3}$ $\frac{21}{40}$ $4\frac{1}{4} \div 2\frac{4}{5}$ $1\frac{29}{56}$ $3\frac{3}{8} \div 3\frac{1}{5}$ $1\frac{7}{128}$ $1\frac{1}{2} \div 1\frac{7}{9}$ $\frac{27}{32}$ $1\frac{2}{3} \div 1\frac{7}{16}$ $1\frac{11}{69}$

SET 27

$\frac{3}{8} \div \frac{3}{8}$ 1 $1\frac{5}{12} \div 1\frac{5}{12}$ 1 $2\frac{2}{3} \div 2\frac{2}{3}$ 1 $9\frac{1}{2} \div 9\frac{1}{2}$ 1 $6\frac{13}{16} \div 6\frac{13}{16}$ 1

SET 28

1. Divide $4\frac{1}{2}$ by $\frac{3}{5}$. $7\frac{1}{2}$
2. Divide $6\frac{1}{4}$ by 8. $\frac{25}{32}$
3. Divide $2\frac{2}{3}$ by $7\frac{1}{2}$. $\frac{16}{45}$
4. Divide $3\frac{3}{4}$ by $4\frac{2}{5}$. $\frac{75}{88}$
5. Divide $1\frac{7}{8}$ by $1\frac{1}{3}$. $1\frac{13}{32}$

SET 29(a)

1. $(\frac{7}{8} \div \frac{3}{16}) \div \frac{21}{32}$ $7\frac{1}{9}$
2. $(\frac{3}{4} \div \frac{4}{5}) \div \frac{3}{8}$ $2\frac{1}{2}$
3. $(1\frac{1}{2} \div \frac{5}{8}) \div 1\frac{1}{5}$ 2
4. $(4\frac{1}{5} \div 1\frac{3}{4}) \div 2\frac{7}{10}$ $\frac{8}{9}$
5. $(6\frac{2}{3} \div 3\frac{1}{5}) \div 3\frac{3}{4}$ $\frac{5}{9}$

SET 29(b)

1. $\frac{9}{10} \div (\frac{3}{5} \div \frac{3}{4})$ $1\frac{1}{8}$
2. $\frac{3}{8} \div (\frac{5}{6} \div \frac{7}{12})$ $\frac{21}{80}$
3. $5\frac{1}{2} \div (\frac{11}{12} \div 2\frac{2}{3})$ 16
4. $1\frac{7}{8} \div (3\frac{3}{4} \div 2\frac{2}{5})$ $1\frac{1}{5}$
5. $4\frac{2}{3} \div (3\frac{1}{2} \div 1\frac{3}{4})$ $2\frac{1}{3}$

SET 30(a)

1. $(\frac{5}{16} \div \frac{3}{8}) \times \frac{4}{5}$ $\frac{2}{3}$
2. $(1\frac{1}{4} \div 4\frac{1}{2}) \times \frac{9}{16}$ $\frac{5}{32}$
3. $(\frac{2}{3} \times \frac{3}{4}) \div \frac{3}{5}$ $\frac{5}{6}$
4. $(1\frac{7}{8} \times 1\frac{1}{2}) \div 2\frac{1}{4}$ $1\frac{1}{4}$
5. $(2\frac{1}{5} \times 3\frac{1}{2}) \times 4\frac{2}{3}$ $35\frac{14}{15}$

SET 30(b)

1. $\frac{7}{8} \times (\frac{3}{4} \div \frac{15}{16})$ $\frac{7}{10}$
2. $\frac{25}{32} \times (\frac{9}{10} \div \frac{5}{8})$ $1\frac{1}{8}$
3. $6\frac{2}{3} \times (\frac{13}{16} \div 5\frac{1}{5})$ $1\frac{1}{24}$
4. $1\frac{1}{2} \times (1\frac{3}{4} \div 1\frac{1}{5})$ $1\frac{11}{24}$
5. $2\frac{5}{6} \times (5\frac{1}{4} \div 1\frac{5}{16})$ $11\frac{1}{3}$

SET 30(c)

1. $\frac{5}{12} \div (\frac{1}{6} \times \frac{2}{3})$ $3\frac{3}{4}$
2. $\frac{1}{2} \div (\frac{3}{4} \times \frac{7}{8})$ $\frac{16}{21}$
3. $8\frac{3}{4} \div (\frac{15}{16} \times 2\frac{4}{5})$ $3\frac{1}{3}$
4. $4\frac{1}{2} \div (2\frac{1}{4} \times 1\frac{2}{3})$ $1\frac{1}{5}$
5. $7\frac{4}{5} \div (8\frac{2}{3} \times 3\frac{3}{8})$ $\frac{4}{15}$

Simplify:

SET 31(a)

$\frac{3}{5}$	$\frac{1}{2}$	$\frac{7}{16}$	$\frac{4}{5}$	$\frac{9}{16}$
$\frac{3}{8}$	$\frac{3}{4}$	$\frac{5}{8}$	$\frac{7}{8}$	$\frac{2}{5}$
$1\frac{3}{5}$	$\frac{2}{3}$	$\frac{21}{40}$	$\frac{32}{35}$	$1\frac{13}{32}$

SET 31(b)

$2\frac{2}{3}$	$3\frac{3}{8}$	$3\frac{15}{16}$	$10\frac{5}{8}$	$4\frac{2}{5}$
$5\frac{1}{3}$	$4\frac{1}{2}$	$2\frac{5}{8}$	$4\frac{1}{4}$	$2\frac{3}{4}$
$\frac{1}{2}$	$\frac{3}{4}$	$1\frac{1}{2}$	$2\frac{1}{2}$	$1\frac{3}{5}$

SET 31(c)

$\frac{5}{8}$	$2\frac{9}{16}$	$1\frac{5}{6}$	$9\frac{1}{2}$	$2\frac{1}{8}$
2	6	3	4	10
$\frac{5}{16}$	$\frac{41}{96}$	$\frac{11}{18}$	$2\frac{3}{8}$	$\frac{17}{80}$

SET 31(d)

$87\frac{1}{2}$	$33\frac{1}{3}$	$12\frac{1}{2}$	$62\frac{1}{2}$	$16\frac{2}{3}$
100	100	100	100	100
$\frac{7}{8}$	$\frac{1}{3}$	$\frac{1}{8}$	$\frac{5}{8}$	$\frac{1}{6}$

SET 31(e)

4	6	5	1	8
$\frac{2}{5}$	$1\frac{1}{2}$	$3\frac{3}{4}$	$2\frac{2}{3}$	$5\frac{3}{8}$
10	4	$1\frac{1}{3}$	$\frac{3}{8}$	$1\frac{21}{43}$

SET 31(f)

$\frac{\frac{1}{2} + \frac{1}{4}}{\frac{3}{8}}$ 2	$\frac{\frac{5}{6}}{3 - 1\frac{1}{3}}$ $\frac{1}{2}$	$\frac{\frac{3}{4} + \frac{7}{8}}{\frac{11}{16} - \frac{1}{2}}$ $8\frac{2}{3}$	$\frac{4\frac{1}{2} + 1\frac{1}{2}}{7\frac{1}{4} + 2\frac{5}{8}}$ $\frac{48}{79}$	$\frac{8\frac{1}{6} - 2\frac{2}{3}}{6 - 3\frac{3}{5}}$ $2\frac{7}{24}$

PROBLEM SOLVING

See page 16, *Problem Solving Strategy*.

For problems 1–5, select the letter corresponding to your answer.

1. If each costume for the school show requires $3\frac{1}{3}$ yards of material, how many costumes can be made from a 30-yard bolt of material?

    a. 10 costumes   b. 9 costumes   c. 8 costumes   d. Answer not given

2. If an airplane flies 1,120 kilometers in $1\frac{3}{4}$ hours, what is its average ground speed in kilometers per hour?

    a. 600 km/h   b. 620 km/h   c. 640 km/h   d. 600 km/h

3. If a board $10\frac{1}{2}$ feet long was cut into 6 pieces of equal length, what would the length of each piece be? Disregard waste.

    a. $1\frac{7}{8}$ ft.   **b.** $2\frac{1}{4}$ ft.   **c.** $1\frac{1}{2}$ ft.   **d.** $1\frac{3}{4}$ ft.

4. What are the actual dimensions of a porch which, drawn to the scale of $\frac{1}{4}$ inch = 1 foot, measures $1\frac{3}{4}$ inches by $2\frac{1}{2}$ inches?

    a. 8 ft. by 10 ft.   **c.** 7 ft. by 11 ft.

    b. 7 ft. by 10 ft.   **d.** Answer not given

5. A bus is scheduled to go a distance of $87\frac{1}{2}$ miles in $2\frac{1}{2}$ hours. What average speed must be maintained to arrive on schedule?

    a. 30 m.p.h.   **b.** 40 m.p.h.   **c.** 35 m.p.h.   **d.** 38 m.p.h.

6. How many athletic association membership cards $1\frac{3}{8}$ inches wide can be cut from stock 22 inches wide? 16 cards

7. Barry wishes to use a recipe but it makes twice the quantity he needs. If the recipe requires $\frac{1}{4}$ cup cake flour, $2\frac{1}{2}$ tablespoons lemon juice, 1 cup milk, and $\frac{3}{4}$ cup sugar, how much of each ingredient should he take to make the quantity he wants? $\frac{1}{8}$ cup flour, $1\frac{1}{4}$ tb. juice, $\frac{1}{2}$ cup milk, $\frac{3}{8}$ cup sugar

8. How much does a pound of each of the following items cost if:

    a. $2\frac{1}{2}$ pounds of peas cost 70¢? 28¢

    b. $1\frac{3}{4}$ pounds of asparagus cost 98¢? 56¢

    c. $2\frac{5}{8}$ pounds of apples cost 84¢? 32¢

9. A ship steamed a distance of 90 nautical miles in $8\frac{1}{3}$ hours. Find its average speed in knots. (1 knot = 1 nautical m.p.h.) $10\frac{4}{5}$ knots

10. The scale on a blueprint of a factory is $\frac{1}{8}$ inch = 1 foot. Find the actual dimensions of a room measuring $2\frac{7}{8}$ by $3\frac{5}{16}$ inches. 23 ft. $\times$ $26\frac{1}{2}$ ft.

11. a. Two cities, 252 miles apart, are plotted on a chart $5\frac{1}{4}$ inches apart. Find the scale of the chart.   **b.** Find the scale if:

Actual distance	217 mi.	350 mi.	100 mi.	8,125 yd.	$268\frac{3}{4}$ mi.
Distance on chart	$3\frac{7}{8}$ in.	$8\frac{3}{4}$ in.	$1\frac{9}{16}$ in.	$6\frac{1}{2}$ in.	$5\frac{3}{8}$ in.

a. 1 in. = 48 mi. b. 1 in. = 56 mi.; 1 in. = 40 mi.; 1 in. = 64 mi.; 1 in. = 1,250 yd.; 1 in. = 50 mi.

12. a. On a map a distance of 84 miles is represented by $2\frac{5}{8}$ inches. How long should a line be to represent a distance of 96 miles? 3 in.

    b. How long should a line be to represent a distance of 25 miles if a distance of 350 miles on the chart is represented by $1\frac{3}{4}$ inches? $\frac{1}{8}$ in.

    c. If the distance between Chicago and Cleveland, 360 miles, is represented on a road map by a length of $4\frac{1}{2}$ inches, what is the distance between Detroit and Memphis represented on the same map by a length of $9\frac{1}{8}$ inches? 730 mi.

13. The temperature of a parcel of air decreases at the rate of $5\frac{1}{2}°$F for each 1,000 feet it rises following the dry adiabat.

a. Find the altitude where the temperature is $53\frac{1}{2}°$F when the surface temperature is $70°$F. 3,000 ft.

b. At what altitude is the temperature $45°$F when the surface temperature is $78°$F? 6,000 ft.

c. Find the altitude where the temperature is $40\frac{1}{2}°$F when the surface temperature is $62\frac{1}{2}°$F. 4,000 ft.

14. a. If the weight of a column of water is 1,125 pounds, how many cubic feet does it occupy? (1 cu. ft. of water weighs about $62\frac{1}{2}$ lb.) 18 cu. ft.

b. A railroad tank car has a capacity of 6,000 gallons of oil. What is its capacity in cubic feet? (1 cu. ft. = $7\frac{1}{2}$ gal.) 800 cu. ft.

c. A storage bin has a capacity of 1,190 cubic feet. How many bushels of wheat can it hold? (1 bu. = $1\frac{1}{4}$ cu. ft.) 952 bu.

15. a. How many lengths of pipe $3\frac{1}{2}$ feet long can be cut from a pipe 28 feet long? Disregard waste in cutting. 8 lengths

b. A floor is 24 feet wide. How many floor boards $2\frac{1}{4}$ inches wide are needed to cover it? 128 boards

c. How many pieces $3\frac{3}{8}$ inches long can be cut from 50 metal rods each 54 inches long? Disregard waste. 800 pieces

16. The diameter of a circle is equal to the circumference (distance around) of the circle divided by $3\frac{1}{7}$. Find the diameter when the circumference is: **a.** 88 ft.   **b.** 50 ft.   **c.** $\frac{11}{16}$ in.   **d.** $9\frac{5}{8}$ in.   **e.** $16\frac{1}{2}$ in.

28 ft. $15\frac{10}{11}$ ft. $\frac{7}{32}$ in. $3\frac{1}{16}$ in. $5\frac{1}{4}$ in.

REFRESH YOUR SKILLS

1. Add: **2.** Subtract: **3.** Multiply: **4.** Divide:

92,838   156,072   3,600 $1,728\overline{)1,394,496}$ 807
79,665   75,483   905
83,756   80,589   3,258,000
44,973
58,786   360,018

5. Round 18,639,708 to the nearest thousand. 18,640,000

6. Add: **7.** Subtract: **8.** Multiply: **9.** Divide:

$2\frac{3}{8}$ $11\frac{1}{2}$ $2\frac{7}{10} \times 3\frac{3}{4}$ $10\frac{1}{8}$ $9\frac{1}{6} \div 3\frac{1}{7}$ $2\frac{11}{12}$
$\frac{11}{12}$ $5\frac{2}{3}$
$9\frac{1}{6}$ $12\frac{11}{24}$ $5\frac{5}{6}$

10. Express $\frac{96}{128}$ in lowest terms. $\frac{3}{4}$

Finding What Fractional Part One Number Is of Another

VOCABULARY: ratio

OBJECTIVE:

I. Aim   1. To find what fractional part one number is of another;
2. to compare numbers by using fractions.

II. Procedure

1. To find what fractional part one number is of another, make a fraction taking the number of parts used as the numerator and the number of parts in the whole unit or group as the denominator. Reduce the fraction to lowest terms.

2. To compare numbers by using fractions, make a fraction taking the number that is being compared as the numerator and the number with which it is being compared as the denominator. Change the fraction to simplest form.

III. Sample Solutions

1. What part of 5 is 4?

Answer: $\frac{4}{5}$

2. 9 is what part of 12?

$$\frac{9}{12} = \frac{3}{4}$$

Answer: $\frac{3}{4}$

3. Compare 10 with 16.
$\frac{10}{16} = \frac{5}{8}$ indicating that
10 is $\frac{5}{8}$ of 16

Answer: $\frac{5}{8}$

4. Compare 9 with 6.
$\frac{9}{6} = \frac{3}{2}$ or $1\frac{1}{2}$ indicating that
9 is $1\frac{1}{2}$ times 6

Answer: $\frac{3}{2}$ or $1\frac{1}{2}$

DIAGNOSTIC TEST

Find the following:

1. What part of 6 is 1? $\frac{1}{6}$
2. 2 is what part of 3? $\frac{2}{3}$
3. What part of 56 is 35? $\frac{5}{8}$
4. What part of 100 is 40? $\frac{2}{5}$
5. Compare 3 with 5. $\frac{3}{5}$

6. Compare 12 with 16. $\frac{3}{4}$
7. Compare 11 with 8. $\frac{11}{8}$ or $1\frac{3}{8}$
8. Compare 15 with 9. $\frac{15}{9}$ or $1\frac{2}{3}$
9. Compare 8 with 1. 8
10. Compare 24 with 24. 1

117

Basic: Sets 1–7, Prob. 1–7, Ratio 1–5; *Average:* Sets 1–9, Prob. 1–10, Ratio 1–7; *Enriched:* Sets 1–10, Prob. 1–14, Ratio 1–8

118   COMMON FRACTIONS

RELATED PRACTICE EXAMPLES

Find the following:

SET 1

1. What part of 4 is 1? $\frac{1}{4}$
2. What part of 5 is 1? $\frac{1}{5}$
3. What part of 8 is 7? $\frac{7}{8}$
4. What part of 24 is 13? $\frac{13}{24}$
5. What part of 9 is 5? $\frac{5}{9}$

SET 2

1. 2 is what part of 5? $\frac{2}{5}$
2. 3 is what part of 4? $\frac{3}{4}$
3. 7 is what part of 10? $\frac{7}{10}$
4. 8 is what part of 15? $\frac{8}{15}$
5. 29 is what part of 32? $\frac{29}{32}$

SET 3(a)

1. What part of 12 is 6? $\frac{1}{2}$
2. What part of 18 is 3? $\frac{1}{6}$
3. What part of 64 is 16? $\frac{1}{4}$
4. What part of 80 is 5? $\frac{1}{16}$
5. What part of 75 is 25? $\frac{1}{3}$

SET 3(b)

1. 12 is what part of 18? $\frac{2}{3}$
2. 16 is what part of 20? $\frac{4}{5}$
3. 49 is what part of 56? $\frac{7}{8}$
4. 27 is what part of 36? $\frac{3}{4}$
5. 45 is what part of 54? $\frac{5}{6}$

SET 3(c)

1. What part of 72 is 30? $\frac{5}{12}$
2. 78 is what part of 90? $\frac{13}{15}$
3. What part of 112 is 21? $\frac{3}{16}$
4. 66 is what part of 108? $\frac{11}{18}$
5. What part of 144 is 135? $\frac{15}{16}$

SET 4

1. What part of 100 is 28? $\frac{7}{25}$
2. 75 is what part of 100? $\frac{3}{4}$
3. What part of 100 is 64? $\frac{16}{25}$
4. $37\frac{1}{2}$ is what part of 100? $\frac{3}{8}$
5. What part of 100 is $66\frac{2}{3}$? $\frac{2}{3}$

SET 5

Compare:

1. 1 with 8 $\frac{1}{8}$
2. 3 with 7 $\frac{3}{7}$
3. 5 with 6 $\frac{5}{6}$
4. 1 with 2 $\frac{1}{2}$
5. 12 with 25 $\frac{12}{25}$

SET 6

Compare:

1. 6 with 10 $\frac{3}{5}$
2. 21 with 28 $\frac{3}{4}$
3. 84 with 96 $\frac{7}{8}$
4. 25 with 30 $\frac{5}{6}$
5. 36 with 54 $\frac{2}{3}$

SET 7

Compare:

1. 7 with 6 $1\frac{1}{6}$
2. 13 with 8 $1\frac{5}{8}$
3. 9 with 4 $2\frac{1}{4}$
4. 5 with 3 $1\frac{2}{3}$
5. 17 with 5 $3\frac{2}{5}$

SET 8

Compare:

1. 10 with 8 $1\frac{1}{4}$
2. 18 with 15 $1\frac{1}{5}$
3. 21 with 12 $1\frac{3}{4}$
4. 24 with 10 $2\frac{2}{5}$
5. 40 with 16 $2\frac{1}{2}$

SET 9

Compare:

1. 2 with 1 2
2. 5 with 1 5
3. 10 with 1 10
4. 50 with 1 50
5. 28 with 1 28

SET 10

Compare:

1. 8 with 8 1
2. 10 with 5 2
3. 24 with 6 4
4. 150 with 150 1
5. 96 with 4 24

PROBLEM SOLVING

See page 16, *Problem Solving Strategy*.

For problems 1–5, select the letter corresponding to your answer.

1. What part of a class attended school on a certain day if 39 pupils of the 42 enrolled were present?

    a. $\frac{6}{7}$ b. $\frac{11}{12}$ c. $\frac{13}{14}$ d. Answer not given

2. Jane missed 5 examples in an arithmetic test of 25 examples. What part of the test did she get right?

    a. $\frac{1}{5}$ b. $\frac{1}{2}$ c. $\frac{4}{5}$ d. $\frac{3}{4}$

3. In 24 times at bat, Joe hit safely 9 times. What part of the time did he hit safely?

    a. $\frac{3}{4}$ b. $\frac{3}{7}$ c. $\frac{3}{10}$ d. $\frac{3}{8}$

4. The Harriet Tubman School team won 9 games and lost 6. What part of the games played did it win?

    a. $\frac{2}{3}$ b. $\frac{3}{5}$ c. $\frac{1}{3}$ d. $\frac{2}{5}$

5. What part of her annual income of $15,000 does Elsa save if she deposits $125 each month in her savings account?

    a. $\frac{1}{9}$ b. $\frac{2}{15}$ c. $\frac{1}{10}$ d. Answer not given

6. There are 16 boys and 20 girls in a class. What part of the class is boys? What part is girls? $\frac{4}{9}$; $\frac{5}{9}$

7. If 175 out of 200 freshmen passed their physical examination, what part of the freshmen passed? $\frac{7}{8}$

8. On the final report 6 pupils in the mathematics class received A. The teacher also announced there were 12 B's, 10 C's, 8 D's, and 4 E's. What part of the class received A? B? C? D? E? A, $\frac{3}{20}$; B, $\frac{3}{10}$; C, $\frac{1}{4}$; D, $\frac{1}{5}$; E, $\frac{1}{10}$

9. Meyer made two errors in 25 fielding chances. What part of his chances did he field the ball cleanly? $\frac{23}{25}$

10. Janet Miller, a pitcher, won 8 games and lost 4. What part of the games did she win? $\frac{2}{3}$

11. At bat 36 times, Thompson made 7 singles, 2 doubles, 1 triple, and 2 home runs. What part of the time did he hit safely? $\frac{1}{3}$

12. If brass contains 3 parts copper and 2 parts zinc, what part of brass is copper? What part is zinc? $\frac{3}{5}$; $\frac{2}{5}$

13. A certain hydrochloric acid solution contains 4 parts acid and 8 parts water. What part of the solution is acid? How many quarts of acid are there in 18 quarts of the solution? $\frac{1}{3}$; 6 quarts

14. a. What part of a dollar is a dime? **b.** 50 minutes is what part of an hour? **c.** 3 is what part of a dozen? **d.** 12 ounces is what part of a pound? **e.** What part of a bushel is a peck? **f.** 6 inches is what part of a foot?

a. $\frac{1}{10}$ b. $\frac{5}{6}$ c. $\frac{1}{4}$ d. $\frac{3}{4}$ e. $\frac{1}{4}$ f. $\frac{1}{2}$

Ratio

Ratio is the answer obtained when two numbers are compared by division. (See Exercise 4-27.)

The ratio of 4 to 12 is $\frac{4}{12}$ or $\frac{1}{3}$, read "1 to 3" which indicates that 4 when compared to 12 is $\frac{1}{3}$ of 12.

The ratio of 12 to 4 is $\frac{12}{4}$ or $\frac{3}{1}$, read "3 to 1" which indicates that 12 when compared to 4 is 3 times 4.

Quantities being compared must be expressed in the same units.

1. Find the ratio of:

SET 1	SET 2	SET 3
a. 4 to 6 $\frac{2}{3}$	**a.** 6 to 4 $\frac{3}{2}$ or $1\frac{1}{2}$	**a.** 2 mm to 8 mm $\frac{1}{4}$
b. 9 to 12 $\frac{3}{4}$	**b.** 8 to 5 $\frac{8}{5}$ or $1\frac{3}{5}$	**b.** 10 in. to 1 ft. $\frac{5}{6}$
c. 5 to 7 $\frac{5}{7}$	**c.** 24 to 3 $\frac{8}{1}$ or 8	**c.** 16 in. to 1 yd. $\frac{4}{9}$
d. 8 to 32 $\frac{1}{4}$	**d.** 15 to 9 $\frac{5}{3}$ or $1\frac{2}{3}$	**d.** 20 kg to 12 kg $\frac{5}{3}$ or $1\frac{2}{3}$
e. 25 to 40 $\frac{5}{8}$	**e.** 18 to 16 $\frac{9}{8}$ or $1\frac{1}{8}$	**e.** 18 min. to 1 hr. $\frac{3}{10}$

a. $\frac{1}{5}$ **2. a.** What is the ratio of a nickel to a quarter? **b.** Compare 2 things   b. $\frac{1}{6}$ to a dozen. **c.** What is the ratio of 35 minutes to 15 minutes? **d.** Compare 1 dollar to 3 quarters. **e.** Compare $\frac{1}{4}$ hour to 2 hours.   c. $\frac{7}{3}$ or $2\frac{1}{3}$; d. $\frac{4}{3}$ or $1\frac{1}{3}$; e. $\frac{1}{8}$

3. Find the ratio of two meshed gears if one gear has 48 teeth and the other 36 teeth. $\frac{4}{3}$ or $1\frac{1}{3}$

4. The pitch of a roof is the ratio of the rise to the span. What is the pitch of a roof if the rise is 2 meters and the span is 8 meters? $\frac{1}{4}$

5. The following recipe will make 6 servings of custard: 3 eggs, 1 pint milk, $\frac{1}{4}$ teaspoon salt, $\frac{1}{2}$ teaspoon vanilla, $\frac{1}{4}$ cup sugar. How much of each ingredient should be used to make 12 servings? 3 servings? 2 servings? 9 servings? 4 servings?

6. A mixture used to make concrete contains 1 part cement, 2 parts sand, 4 parts gravel, and water. If 4 bags of sand are used, how many bags of cement and how many bags of gravel are required? **2 bags cement, 8 bags gravel**

7. A bag of mixed grass seed contains 4 pounds of rye-grass, 3 pounds of fescue, 1 pound of clover, and 2 pounds of bluegrass seed. What is the ratio of rye-grass seed to the total mixture? $\frac{2}{5}$

8. The Lift-Drag Ratio (L/D) is the ratio of the lift of an airplane to its drag. Find the Lift-Drag Ratio in each of the following cases:

	6	14	6	$13\frac{1}{2}$	$16\frac{2}{3}$
Lift (lb.)	900	1,400	1,050	2,700	3,000
Drag (lb.)	150	100	175	200	180

5. 6 eggs, 1 qt. milk, $\frac{1}{2}$ ts. salt, 1 ts. van., $\frac{1}{2}$ cup sugar; $1\frac{1}{2}$ eggs, $\frac{1}{2}$ pt. milk, $\frac{1}{8}$ ts. salt, $\frac{1}{4}$ ts. vanilla, $\frac{1}{8}$ cup sugar; 1 egg, $\frac{1}{3}$ pt. milk, $\frac{1}{12}$ ts. salt, $\frac{1}{6}$ ts. van., $\frac{1}{12}$ cup sugar; $4\frac{1}{2}$ eggs, $1\frac{1}{2}$ pt. milk, $\frac{3}{8}$ ts. salt, $\frac{3}{4}$ ts. van., $\frac{3}{8}$ cup sugar; 2 eggs, $\frac{2}{3}$ pt. milk, $\frac{1}{6}$ ts. salt, $\frac{1}{2}$ ts. van., $\frac{1}{6}$ cup sugar.

Finding a Number when a Fractional Part of it Is Known

OBJECTIVE:

I. Aim To find a number when a fractional part of it is known.

II. Procedure

1. Divide the given number representing the fractional part of the unknown number by the given fraction, or multiply the given number by the reciprocal of the given fraction. (See Exercise 1-59.)

2. Or follow method 2 shown in the sample solutions.

III. Sample Solutions

1. $\frac{1}{6}$ of what number is 18?

Method 1

$18 \div \frac{1}{6} = 18 \times 6 = 108$

Method 2

Since $\frac{1}{6}$ of the number = 18

$\frac{6}{6}$ of the number = $6 \times 18 = 108$

Therefore, the number = 108

Answer: 108

2. $\frac{3}{4}$ of what number is 21?

Method 1

$$21 \div \frac{3}{4} = \overset{7}{\cancel{21}} \times \frac{4}{\cancel{3}} = 28$$
$$1$$

Method 2

Since $\frac{3}{4}$ of the number = 21

$\frac{1}{4}$ of the number = $21 \div 3 = 7$

$\frac{4}{4}$ of the number = $4 \times 7 = 28$

Therefore, the number = 28

Answer: 28

DIAGNOSTIC TEST

Find the following:

1. $\frac{1}{4}$ of what number is 12? 48

2. $\frac{5}{8}$ of what number is 45? 72

Basic: Sets 1(a)–2(b), Misc. Ex. Set 1, Prob. 1–4; *Average:* Sets 1(a)–2(c), Misc. Ex. Sets 1–2, Prob. 1–6; *Enriched:* Sets 1(a)–2(e), Misc. Ex. Sets 1–2, Prob. 1–7

122   COMMON FRACTIONS

RELATED PRACTICE EXAMPLES

Find the following:

SET 1(a)

1. $\frac{1}{2}$ of what number is 5? 10
2. $\frac{1}{2}$ of what number is 12? 24
3. $\frac{1}{2}$ of what number is 38? 76
4. $\frac{1}{2}$ of what number is 275? 550
5. $\frac{1}{2}$ of what number is 500? 1,000

SET 1(b)

1. $\frac{1}{3}$ of what number is 4? 12
2. $\frac{1}{3}$ of what number is 18? 54
3. $\frac{1}{4}$ of what number is 8? 32
4. $\frac{1}{4}$ of what number is 27? 108
5. $\frac{1}{4}$ of what number is 250? 1,000

SET 1(c)

1. $\frac{1}{8}$ of what number is 2? 16
2. $\frac{1}{8}$ of what number is 8? 64
3. $\frac{1}{8}$ of what number is 19? 152
4. $\frac{1}{16}$ of what number is 10? 160
5. $\frac{1}{16}$ of what number is 21? 336

SET 1(d)

1. $\frac{1}{6}$ of what number is 7? 42
2. $\frac{1}{6}$ of what number is 24? 144
3. $\frac{1}{6}$ of what number is 52? 312
4. $\frac{1}{12}$ of what number is 36? 432
5. $\frac{1}{12}$ of what number is 200? 2,400

SET 1(e)

1. $\frac{1}{5}$ of what number is 6? 30
2. $\frac{1}{5}$ of what number is 125? 625
3. $\frac{1}{10}$ of what number is 40? 400
4. $\frac{1}{100}$ of what number is 9? 900
5. $\frac{1}{20}$ of what number is 150? 3,000

SET 2(a)

1. $\frac{3}{4}$ of what number is 9? 12
2. $\frac{3}{4}$ of what number is 15? 20
3. $\frac{3}{4}$ of what number is 26? $34\frac{2}{3}$
4. $\frac{2}{3}$ of what number is 14? 21
5. $\frac{2}{3}$ of what number is 110? 165

SET 2(b)

1. $\frac{2}{5}$ of what number is 42? 105
2. $\frac{3}{5}$ of what number is 3? 5
3. $\frac{3}{5}$ of what number is 96? 160
4. $\frac{4}{5}$ of what number is 8? 10
5. $\frac{4}{5}$ of what number is 55? $68\frac{3}{4}$

SET 2(c)

1. $\frac{5}{8}$ of what number is 20? 32
2. $\frac{3}{8}$ of what number is 12? 32
3. $\frac{7}{8}$ of what number is 35? 40
4. $\frac{11}{16}$ of what number is 33? 48
5. $\frac{3}{16}$ of what number is 9? 48

SET 2(d)

1. $\frac{9}{10}$ of what number is 18? 20
2. $\frac{7}{10}$ of what number is 105? 150
3. $\frac{11}{20}$ of what number is 55? 100
4. $\frac{3}{100}$ of what number is 60? 2,000
5. $\frac{47}{100}$ of what number is 94? 200

SET 2(e)

1. $\frac{5}{6}$ of what number is 10? 12
2. $\frac{5}{6}$ of what number is 28? $33\frac{3}{5}$
3. $\frac{7}{12}$ of what number is 21? 36
4. $\frac{11}{12}$ of what number is 66? 72
5. $\frac{5}{12}$ of what number is 45? 108

MISCELLANEOUS EXAMPLES

Find the following:

SET 1

1. 16 is $\frac{1}{2}$ of what number? 32
2. 35 is $\frac{1}{3}$ of what number? 105
3. 23 is $\frac{1}{4}$ of what number? 92
4. 9 is $\frac{1}{8}$ of what number? 72
5. 40 is $\frac{1}{12}$ of what number? 480

SET 2

1. 6 is $\frac{3}{4}$ of what number? 8
2. 30 is $\frac{5}{8}$ of what number? 48
3. 44 is $\frac{11}{12}$ of what number? 48
4. 27 is $\frac{2}{3}$ of what number? 40$\frac{1}{2}$
5. 56 is $\frac{7}{16}$ of what number? 128

PROBLEM SOLVING

See page 16, *Problem Solving Strategy*.
For problems 1–3, select the letter corresponding to your answer.

1. Charlotte received $\frac{5}{6}$ of all the votes cast in the election for school treasurer. If she received 885 votes, how many students voted?
    a. 1,054 students   c. 1,065 students
    b. 1,062 students   d. Answer not given
2. Marilyn bought a pair of ice skates at a sale for $18. What was the regular price of the skates if they were reduced one third?
    a. $6   b. $12   c. $18   d. $27
3. If the school baseball team won 16 games or $\frac{2}{3}$ of the games played, how many games were lost?
    a. 6 games   b. 7 games   c. 8 games   d. 9 games
4. If 138 students or $\frac{3}{8}$ of the graduating class selected the college preparatory course, how many pupils were in the graduating class? 368 pupils
5. The school athletic association sold 1,295 student membership 1,480 tickets. If $\frac{7}{8}$ of the school became members, what is the school enrollment? students
6. Herbert, being paid at the rate of $\frac{3}{20}$ of his sales, received $51 commission. What was the amount of his sales? $340
7. How much does a pound of each of the following cost if:
    a. $\frac{1}{2}$ lb. of pretzels costs 49¢? 98¢
    b. $\frac{3}{4}$ lb. of margarine costs 69¢? 92¢
    c. $\frac{7}{8}$ lb. of chicken costs 84¢? 96¢
    d. $\frac{15}{16}$ lb. of peppers costs 45¢? 48¢

REVIEW OF UNIT TWO

1. Express each of the following fractions in lowest terms:

a. $\frac{18}{24}$ $\frac{3}{4}$ **b.** $\frac{16}{36}$ $\frac{4}{9}$ **c.** $\frac{45}{75}$ $\frac{3}{5}$ **d.** $\frac{84}{108}$ $\frac{7}{9}$ **e.** $\frac{105}{140}$ $\frac{3}{4}$

2. Change each of the following improper fractions to a whole number or a mixed number:

a. $\frac{9}{5}$ $1\frac{4}{5}$ **b.** $\frac{15}{3}$ 5 **c.** $\frac{20}{6}$ $3\frac{1}{3}$ **d.** $\frac{35}{8}$ $4\frac{3}{8}$ **e.** $\frac{76}{24}$ $3\frac{1}{6}$

3. Change each of the following mixed numbers to simplest form:

a. $8\frac{3}{3}$ 9 **b.** $9\frac{24}{8}$ 12 **c.** $2\frac{36}{60}$ $2\frac{3}{5}$ **d.** $4\frac{11}{5}$ $6\frac{1}{5}$ **e.** $12\frac{28}{16}$ $13\frac{3}{4}$

4. Change each of the following to equivalent fractions having denominators as specified:

a. $\frac{1}{4} = \frac{?}{64}$ $\frac{16}{64}$ **b.** $\frac{2}{3} = \frac{?}{45}$ $\frac{30}{45}$ **c.** $\frac{17}{20} = \frac{?}{100}$ $\frac{85}{100}$

d. Change $\frac{5}{8}$ to 32nds. $\frac{20}{32}$ **e.** Change $\frac{11}{12}$ to 60ths. $\frac{55}{60}$

5. Find the lowest common denominator of each of the following groups of fractions:

a. $\frac{3}{5}$ and $\frac{7}{10}$ 10ths **b.** $\frac{1}{2}$ and $\frac{2}{3}$ 6ths **c.** $\frac{5}{6}$ and $\frac{3}{4}$ 12ths

d. $\frac{7}{8}$, $\frac{1}{4}$, and $\frac{5}{16}$ 16ths **e.** $\frac{3}{10}$, $\frac{7}{12}$, and $\frac{5}{8}$ 120ths

6. Add:

a. $\frac{5}{16}$ **b.** $1\frac{7}{8}$ **c.** $3\frac{9}{10}$ **d.** $6\frac{3}{4}$ **e.** $4\frac{11}{12} + 3\frac{3}{8} + 1\frac{9}{16}$ $9\frac{41}{48}$

$\frac{3}{4}$ $2\frac{2}{3}$ $10\frac{5}{8}$ $\frac{5}{12}$

$1\frac{1}{16}$ $4\frac{13}{24}$ $14\frac{21}{40}$ $2\frac{1}{6}$

                                                    $9\frac{1}{3}$

7. Find the missing numbers:

a. $6 = 5\frac{?}{8}$ 8 **b.** $1\frac{5}{12} = \frac{?}{12}$ 17 **c.** $10\frac{1}{6} = 9\frac{?}{6}$ 7

d. $7\frac{3}{4} = 7\frac{?}{16} = 6\frac{?}{16}$ 12; 28 **e.** $5\frac{2}{3} = 5\frac{?}{24} = 4\frac{?}{24}$ 16; 40

8. Subtract:

a. $\frac{7}{10}$ **b.** $6\frac{4}{5}$ **c.** 9 **d.** $4\frac{1}{8}$ **e.** Subtract

$\frac{1}{2}$ $4\frac{2}{3}$ $2\frac{17}{32}$ $2\frac{5}{6}$ $11\frac{3}{4}$ from $12\frac{9}{16}$. $\frac{13}{16}$

$\frac{1}{5}$ $2\frac{2}{15}$ $6\frac{15}{32}$ $1\frac{7}{24}$

9. a. Which is smaller: $\frac{1}{3}$ or $\frac{1}{5}$? $\frac{1}{5}$ $\frac{3}{4}$ or $\frac{2}{3}$? $\frac{2}{3}$

b. Which is greater: $\frac{1}{8}$ or $\frac{1}{12}$? $\frac{1}{8}$ $\frac{7}{16}$ or $\frac{1}{2}$? $\frac{1}{2}$

c. Arrange in order of size, smallest first: $\frac{11}{16}$, $\frac{3}{4}$, $\frac{7}{12}$ $\frac{7}{12}$, $\frac{11}{16}$, $\frac{3}{4}$

10. Change the following mixed numbers to improper fractions:

a. $1\frac{5}{6}$ $\frac{11}{6}$ **b.** $8\frac{1}{2}$ $\frac{17}{2}$ **c.** $5\frac{2}{3}$ $\frac{17}{3}$ **d.** $15\frac{7}{10}$ $\frac{157}{10}$ **e.** $2\frac{13}{16}$ $\frac{45}{16}$

11. Multiply:

a. $\frac{3}{4} \times \frac{13}{16}$ $\frac{39}{64}$ **b.** $18 \times 4\frac{5}{6}$ 87 **c.** $6\frac{3}{8} \times \frac{4}{5}$ $5\frac{1}{10}$ **d.** $5\frac{1}{4} \times 2\frac{2}{3}$ 14 **e.** 29

                                                           $6\frac{1}{2}$

                                                    $188\frac{1}{2}$

12. Divide:

a. $\frac{9}{10} \div \frac{3}{5}$ $1\frac{1}{2}$ **b.** $6\frac{2}{3} \div 10$ $\frac{2}{3}$ **c.** $14 \div 3\frac{1}{7}$ $4\frac{5}{11}$

d. $4\frac{11}{16} \div 3\frac{3}{4}$ $1\frac{1}{4}$ **e.** Divide $\frac{5}{8}$ by $2\frac{1}{3}$. $\frac{15}{56}$

13. a. What part of 16 is 9? $\frac{9}{16}$ **b.** 45 is what part of 72? $\frac{5}{8}$

14. a. $\frac{1}{3}$ of what number is 20? 60   **b.** $\frac{3}{5}$ of what number is 27? 45

For problems 15 through 17, solve each problem and select the letter in red that corresponds to your answer.

15. Find the total weight if a metal can weighs $2\frac{7}{8}$ ounces and its contents $12\frac{3}{4}$ ounces.

    a. $15\frac{1}{8}$ oz.   **b.** $15\frac{3}{8}$ oz.   **c.** $15\frac{5}{8}$ oz.   **d.** Answer not given

16. Antonio typed 225 words in $7\frac{1}{2}$ minutes. How many words did he average per minute?

    a. 30 words   **b.** 35 words   **c.** 37 words   **d.** 42 words

17. A certain stock sold at $39\frac{5}{8}$ when the stock market opened for the day and $37\frac{3}{4}$ when it closed. How many points did the stock lose?

    a. $1\frac{5}{8}$ points   **b.** $1\frac{7}{8}$ points   **c.** $2\frac{1}{8}$ points   **d.** $2\frac{3}{8}$ points

18. Tom's mother wishes to make 4 shelves each $2\frac{3}{4}$ feet long and 3 shelves each $3\frac{1}{4}$ feet long. Will a 20-foot board be long enough? How much more or less will she need? No, she needs $\frac{3}{4}$ ft. more.

19. The directions tell you to add 3 cans of water to 1 can of orange juice concentrate. What is the ratio of the water to the concentrate? Of the concentrate to the water? $\frac{3}{1}$ or 3; $\frac{1}{3}$

20. The radiator of a certain automobile holds 15 liters. If this radiator contains 6 liters of antifreeze and the rest water, what part of the solution is antifreeze? What part is water? $\frac{2}{5}$; $\frac{3}{5}$

COMPETENCY CHECK TEST

Solve each problem and select the letter corresponding to your answer. The numerals in red boxes indicate Exercises where help may be found.

1. Add: $8,621 + 19,470 + 267,922 + 84,534 + 741$ 1-4

    a. 380,288   **b.** 381,288   **c.** 381,388   **d.** 380,388

2. Subtract: $390,000 - 290,070$ 1-5

    a. 100,070   **b.** 100,030   **c.** 100,930   **d.** 99,930

3. Multiply: 830×508 1-6

    a. 42,164   **b.** 421,640   **c.** 4,216,400   **d.** Answer not given

4. Divide: $909\overline{)7,280,181}$ 1-7

    a. 809   **b.** 89   **c.** 8,009   **d.** Answer not given

(Continued on page 126.)

5. Add: $3\frac{5}{8} + 6\frac{13}{16}$ **a.** $10\frac{7}{16}$ **b.** $9\frac{18}{24}$ **c.** $9\frac{3}{4}$ **d.** $10\frac{1}{2}$ 1-12

6. Subtract: $10 - 5\frac{5}{6}$ **a.** $5\frac{1}{6}$ **b.** $5\frac{5}{6}$ **c.** $4\frac{1}{6}$ **d.** $15\frac{5}{6}$ 1-13

7. Multiply: $7\frac{1}{2} \times 4\frac{2}{5}$ **a.** $28\frac{1}{5}$ **b.** 33 **c.** $11\frac{3}{7}$ **d.** 31 1-16

8. Divide: $9 \div 6\frac{3}{4}$ **a.** $\frac{3}{4}$ **b.** $\frac{5}{8}$ **c.** $1\frac{1}{3}$ **d.** $1\frac{1}{4}$ 1-17

9. Express $\frac{56}{98}$ in lowest terms: **a.** $\frac{53}{95}$ **b.** $\frac{2}{3}$ **c.** $\frac{3}{5}$ **d.** $\frac{4}{7}$ 1-8

10. What part of 48 is 30?   **a.** $\frac{3}{4}$ **b.** $\frac{5}{8}$ **c.** $\frac{2}{3}$ **d.** $\frac{7}{10}$ 1-18

11. $\frac{9}{10}$ of what number is 72?   **a.** 80   **b.** 84   **c.** 90   **d.** 96   1-19

12. Arrange in order of size, greatest first: $\frac{3}{4}, \frac{2}{3}, \frac{5}{6}, \frac{7}{8}$ 1-14

a. $\frac{3}{4}, \frac{7}{8}, \frac{2}{3}, \frac{5}{6}$ **b.** $\frac{7}{8}, \frac{5}{6}, \frac{3}{4}, \frac{2}{3}$ **c.** $\frac{5}{6}, \frac{7}{8}, \frac{3}{4}, \frac{2}{3}$ **d.** Answer not given

KEYED ACHIEVEMENT TEST

For additional practice turn to the Exercises indicated by the numerals in the red boxes.

1. Write 1,978,246,523 in words.   1-1   1. One billion, nine hundred seventy-eight million, two hundred forty-six thousand, five hundred twenty-three

2. Round 3,602,754 to the nearest thousand.   1-3   3,603,000

3. Add: **4.** Subtract: **5.** Multiply: **6.** Divide: 967

82,579   9,850,601   8,349   984)951,528
21,429   9,788,963   1-5   768   1-6   1-7
51,678   ‾‾‾‾‾‾‾‾‾   ‾‾‾‾‾
30,199   61,638   6,412,032
92,389   1-4
‾‾‾‾‾‾
278,274

7. Express $\frac{54}{60}$ in lowest terms. $\frac{9}{10}$ **8.** Change $\frac{5}{6}$ to 48ths. $\frac{40}{48}$
1-8   1-10

9. Which is greater: $\frac{5}{8}$ or $\frac{2}{3}$? $\frac{2}{3}$
1-14

10. Add: **11.** Subtract: **12.** Multiply: **13.** Divide:

$7\frac{3}{4}$ $20\frac{1}{2}$ $3\frac{1}{7} \times 8\frac{1}{6}$ $25\frac{2}{3}$ $\frac{9}{16} \div 6\frac{3}{4}$ $\frac{1}{12}$

$8\frac{2}{3}$ $6\frac{7}{8}$ 1-13   1-16   1-17

$1\frac{1}{6}$ 1-12 $13\frac{5}{8}$
$17\frac{7}{12}$

14. What part of 80 is 64? $\frac{4}{5}$ **15.** $\frac{5}{8}$ of what number is 65?   104
1-18   1-19

For a Unit Test see p. T42 or *Testing Program*, p. 4.

DECIMAL FRACTIONS

A decimal fraction is a fractional number whose denominator is some power of ten (10; 100; 1,000; etc.) and is named by a numeral in which the denominator is not written as it is in a common fraction but is expressed by place value. Only the numerators appear in decimal notation.

To express parts of a unit, we extend the scale to the right of the units place. The first place to the right of the units place has the value of $\frac{1}{10}$ of a whole number unit; it expresses "tenths." A decimal point is used to separate the whole number from the fractional part. Thus, .6 or 0.6 and $\frac{6}{10}$ are numerals which name the same number.

The second place to the right of the units place expresses hundredths ($.01 = \frac{1}{100}$), the third place expresses thousandths ($.001 = \frac{1}{1,000}$), the fourth place expresses ten-thousandths ($.0001 = \frac{1}{10,000}$), the fifth place expresses hundred-thousandths ($.00001 = \frac{1}{100,000}$), and the sixth place expresses millionths ($.000001 = \frac{1}{1,000,000}$), etc.

hundred millions	ten millions	millions	hundred thousands	ten thousands	thousands	hundreds	tens	units or ones	and	tenths	hundredths	thousandths	ten-thousandths	hundred-thousandths	millionths
4	2	7	5	1	8	2	0	3	.	9	6	4	2	8	7

A mixed decimal is a number containing a whole number and a decimal fraction. The numeral in the box names a mixed decimal.

Since multiples of one are expressed in the places to the left of the ones place and parts of one are expressed in the places to the right of the ones place, the ones place is the center of the numeration system and the decimal point is not.

1-20 Reading Decimals

OBJECTIVE:

I. Aim   To read numerals naming decimals.

II. Procedure

1. In reading a numeral naming a decimal fraction:

a. Read the numeral to the right of the decimal point as you would a whole number and use the name that applies to the place value of the last figure. Sometimes the numeral may have a zero written in the ones place just preceding the decimal point. (See sample solution 1.)

b. Or read in order each digit separately after calling the decimal point "point." (See sample solution 2.)

2. In reading a numeral naming a mixed decimal:

a. First read the numeral for the whole number and then the numeral for the decimal fraction. Use the word "and" to indicate the position of the decimal point. (See sample solution 3.)

b. Or first read the numeral for the whole number and then read each digit of the decimal fraction separately. (See sample solution 4.)

3. In reading a shortened number name having a decimal prefix, read the decimal prefix and then the given period name. Usually the decimal prefix is read as in above steps 1b and 2b. (See sample solutions 5 and 6.)

III. Sample Solutions

1. Read .734 (sometimes written as 0.734)

Answer: Seven hundred thirty-four thousandths

2. Read .5291   *Answer:* Point five two nine one
3. Read 84.06   *Answer:* Eighty-four and six hundredths
4. Read 632.7   *Answer:* Six hundred thirty-two point seven
5. Read 29.843 *million.*

Answer: Twenty-nine point eight four three million

6. Read $45.69 *billion.*

Answer: Forty-five point six nine billion dollars

DIAGNOSTIC TEST   For ex. 1–17 see Extended Answers, p. T72.

Read the following decimal numerals (or write them in words):

1. .2	**5.** 3.73	**8.** .289	**11.** 0.17925
2. .06	**6.** .004	**9.** 14.708	**12.** .000456
3. 0.58	**7.** 0.076	**10.** .0037	**13.** 129.4261
4. 1.5			

Read (or write) in two ways: **14.** .836   **15.** 24.78

Read (or write in words): **16.** 384.5 million   **17.** $16.927 trillion

RELATED PRACTICE EXAMPLES For sets 1–17 see
Extended Answers, pp. T72–T73.

Read the following decimal numerals (or write them in words):

SET 1	**SET 2**	**SET 3**	**SET 4**	**SET 5**
1. .8	1. .03	1. .24	1. 1.6	1. 2.51
2. .1	2. .07	2. .85	2. 2.9	2. 7.37
3. 0.5	3. 0.02	3. 0.91	3. 14.3	3. 30.46
4. 0.4	4. 0.08	4. 0.60	4. 38.5	4. 89.03
5. .9	5. .05	5. .37	5. 126.4	5. 248.19

SET 6	**SET 7**	**SET 8**	**SET 9**	**SET 10**
1. .005	1. .024	1. .832	1. 6.005	1. .0007
2. .008	2. .063	2. .946	2. 21.769	2. .0089
3. 0.001	3. 0.080	3. 0.253	3. 34.094	3. 0.0574
4. 0.007	4. 0.092	4. 0.798	4. 186.528	4. 0.3962
5. .003	5. .035	5. .465	5. 200.042	5. .9350

SET 11	**SET 12**	**SET 13**
1. .00006	1. .000001	1. 8.0025
2. .00392	2. .000534	2. 23.9317
3. 0.09413	3. 0.080076	3. 49.08329
4. 0.25004	4. 0.175283	4. 354.3625
5. .72815	5. .800560	5. 571.05875

Read (or write in words) in two ways:

SET 14	**SET 15**
1. .93	1. 36.89
2. .619	2. 17.6
3. 0.2	3. 5.925
4. 0.3428	4. 294.013
5. .47651	5. 87.4662

Read (or write in words) the shortened number names:

SET 16	**SET 17**
1. 6.8 trillion	1. $33.4 billion
2. 97.92 million	2. $61.56 million
3. 84.436 billion	3. $2.898 trillion
4. 406.753 million	4. $1,750.08 billion
5. 2,569.04 billion	5. $926.872 million

OBJECTIVE:

I. Aim   To write numerals naming decimals.

II. Procedure

1. In writing a numeral naming a decimal fraction:

a. Write the figures as you do to name a whole number, but insert a decimal point so that the name of the part corresponds to the place value of the last figure. Prefix as many zeros as are required between the decimal point and the first figure when it is necessary to make the name of the part and the place value of the last figure correspond. (See sample solution 1.)

b. When each digit is named separately, write each corresponding digit in order immediately following the decimal point. (See sample solution 2.)

2. In writing a numeral naming a mixed decimal:

a. Write the numeral for the whole number, insert a decimal point for the word "and," then write the numeral for the decimal fraction. (See sample solution 3.)

b. Write the numeral for the whole number, insert a decimal point for the word "point," then write for each digit named separately the corresponding figure in order immediately following the decimal point. (See sample solution 4.)

3. In writing a shortened name requiring a decimal prefix, write the decimal prefix and then the given period name. (See sample solutions 5 and 6.)

4. To write numerals by expanded notation. (See Exercise 1-50.)

III. Sample Solutions

Write as a decimal numeral:

1. Five hundredths   *Answer:* .05 or 0.05
2. Point four nine two   *Answer:* .492
3. One hundred and fifty-nine thousandths   *Answer:* 100.059
4. Six hundred ninety point eighty-one   *Answer:* 690.81

Write as a shortened number name:

5. Twenty-four point three *million*   *Answer:* 24.3 million
6. Nine hundred seventy-seven point five six *billion dollars*

   Answer: $977.56 billion

DIAGNOSTIC TEST

Write each of the following as a decimal numeral:

1. Three tenths .3
2. Four and eight tenths 4.8
3. Seven hundredths .07
4. Twenty-five hundredths .25
5. One hundred nine and eighty-four hundredths 109.84
6. Nine thousandths .009
7. Ninety-four thousandths .094
8. Five hundred twenty-seven thousandths .527
9. Seven hundred and ninety-three thousandths 700.093
10. Four thousand six hundred thirty-six ten-thousandths .4636
11. Eight hundred forty-two hundred-thousandths .00842
12. Four millionths .000004
13. Two hundred sixty and three hundred forty-seven ten-thousandths 260.0347
14. Point eight three .83
15. Five hundred nineteen point nine 519.9

Write each of the following as a shortened number name:

16. Eighteen point seven four million 18.74 million
17. Ninety point three two nine billion dollars $90.329 billion

Basic: Sets 1–13; *Average:* Sets 1–15; *Enriched:* Sets 1–17

RELATED PRACTICE EXAMPLES

Write each of the following as a decimal numeral:

SET 1

1. Four tenths .4
2. Eight tenths .8
3. Two tenths .2
4. Nine tenths .9
5. One tenth .1

SET 2

1. Six and five tenths 6.5
2. Three and four tenths 3.4
3. Nine and seven tenths 9.7
4. Five and one tenth 5.1
5. Twenty and six tenths 20.6

SET 3

1. Eight hundredths .08
2. Two hundredths .02
3. Six hundredths .06
4. Four hundredths .04
5. Five hundredths .05

SET 4

1. Thirty-six hundredths .36
2. Fifty-seven hundredths .57
3. Eighty-four hundredths .84
4. Seventeen hundredths .17
5. Forty-nine hundredths .49

SET 5

1. Six and four hundredths   6.04
2. Five and sixty-two hundredths   5.62
3. Seventy-three and eighteen hundredths   73.18
4. Two hundred and five hundredths   200.05
5. Four hundred seven and twenty-five hundredths   407.25

SET 6

1. Three thousandths   .003
2. Seven thousandths   .007
3. Five thousandths   .005
4. Eight thousandths   .008
5. One thousandth   .001

SET 7

1. Sixty-nine thousandths   .069
2. Forty-seven thousandths   .047
3. Twenty-four thousandths   .024
4. Sixteen thousandths   .016
5. Eighty-three thousandths   .083

SET 8

1. Two hundred seventy-four thousandths   .274
2. Four hundred thirty-nine thousandths   .439
3. Seven hundred twenty-one thousandths   .721
4. Three hundred six thousandths   .306
5. One hundred eighty thousandths   .180

SET 9

1. Two and seventeen thousandths   2.017
2. Eight hundred and thirty-five thousandths   800.035
3. Thirty-six and two hundred fifty-three thousandths   36.253
4. Fifty-seven and seven hundred eighty-four thousandths   57.784
5. Nineteen and one hundred twenty-two thousandths   19.122

SET 10

1. Eight ten-thousandths   .0008
2. Thirty-six ten-thousandths   .0036
3. Four hundred ninety-four ten-thousandths   .0494
4. Three thousand five hundred sixty ten-thousandths   .3560
5. Eighteen ten-thousandths   .0018

SET 11

1. Three hundred-thousandths .00003
2. Forty-two hundred-thousandths .00042
3. Four hundred fifty-six hundred-thousandths .00456
4. Six thousand eight hundred twenty-two hundred-thousandths .06822
5. Fifteen thousand seven hundred eighty-six hundred-thousandths .15786

SET 12

1. Six millionths .000006
2. Ninety-five millionths .000095
3. Three hundred seven millionths .000307
4. Seventy-two thousand one hundred forty-nine millionths .072149
5. Five thousand thirteen millionths .005013

SET 13

1. Five hundred and fifty-eight ten-thousandths 500.0058
2. Two and three hundred twenty-five ten-thousandths 2.0325
3. Seventy and two thousand five hundred twelve ten-thousandths
4. Eighty-five and seven hundred thirty-two hundred-thousandths
5. Six hundred forty-three and sixty-seven millionths
   3. 70.2512   4. 85.00732   5. 643.000067

SET 14

1. Point seven .7
2. Point four nine .49
3. Point zero six eight .068
4. Point five one zero nine .5109
5. Point eight five seven .857

SET 15

1. Sixty-one point two 61.2
2. Zero point six four one 0.641
3. Two hundred fifty-eight point three five 258.35
4. Eight hundred eighty point one six one three 880.1613
5. Three thousand ten point seven nine 3,010.79

Write each of the following as a shortened number name:

SET 16

1. Eight point nine six million 8.96 million
2. Forty point five one one trillion 40.511 trillion
3. Twenty-five point two zero eight million 25.208 million
4. Five hundred eighteen point three billion 518.3 billion
5. One thousand sixty-nine point nine seven million 1,069.97 million

SET 17

1. Twelve point four trillion dollars $12.4 trillion
2. Five point eight six four million dollars $5.864 million
3. Seventy point three eight billion dollars $70.38 billion
4. Three hundred eighty-nine point zero seven million dollars $389.07 million
5. Four thousand nine hundred thirty point three nine four billion dollars $4,930.394 billion

---

REFRESH YOUR SKILLS

1. Add:

439,378
192,106
359,281
846,092
399,164
$\overline{2,236,021}$

2. Subtract:

8,200,061
  410,858
$\overline{7,789,203}$

3. Multiply:

6,029
8,709
$\overline{52,506,561}$

4. Divide: 9,809
$725\overline{)7,111,525}$

5. Add:

$2\frac{4}{5} + 1\frac{3}{4} + 6\frac{7}{10}$ $11\frac{1}{4}$

6. Subtract:

$5\frac{1}{2} - 1\frac{11}{12}$ $3\frac{7}{12}$

7. Multiply:

$11\frac{2}{3} \times 3\frac{9}{10}$ $45\frac{1}{2}$

8. Divide:

$14 \div 8\frac{1}{6}$ $1\frac{5}{7}$

9. Round 805,822,149 to the nearest million. 806,000,000

10. Write as a decimal numeral: Four hundred and twenty-seven thousandths 400.027

Rounding Decimals | EXERCISE 1-22

OBJECTIVE:

I. Aim   To round decimals.

II. Procedure

1. Rewrite as many figures as are needed and omit the rest.

2. If the first figure dropped is 5 or more, increase the preceding figure by 1.

3. In finding an answer correct to the nearest cent when the answer contains a fractional part of a cent, drop the fraction. However, if the fraction represents one-half cent or more, add one cent. (See sample solutions 10 and 11.)

4. Some calculators can (1) round *off,* (2) round *up,* and (3) round *down* decimals. (See sample solutions 14 and 15.)

III. Sample Solutions

Find:

1. .26	correct to nearest tenth.	*Answer:* .3
2. 4.542	correct to nearest tenth.	*Answer:* 4.5
3. .633	correct to nearest hundredth.	*Answer:* .63
4. 7.2051	correct to nearest hundredth.	*Answer:* 7.21
5. 8.1584	correct to nearest thousandth.	*Answer:* 8.158
6. .00257	correct to nearest ten-thousandth.	*Answer:* .0026
7. .001834	correct to nearest hundred-thousandth.	*Answer:* .00183
8. .0000286	correct to nearest millionth.	*Answer:* .000029
9. $3.658	correct to nearest cent.	*Answer:* $3.66
10. 5.24\frac{3}{4}$	correct to nearest cent.	*Answer:* $5.25
11. $.49$\frac{1}{3}$	correct to nearest cent.	*Answer:* $.49
12. 23.8	correct to nearest whole number.	*Answer:* 24
13. 5.499	correct to nearest whole number.	*Answer:* 5

14. 5.3682   rounded *off* to nearest hundredth is 5.37
rounded *up* to hundredths is 5.37
rounded *down* to hundredths is 5.36

(Here, nothing is added; all figures are just dropped.)

15. 34.2817   rounded *off* to nearest hundredth is 34.28
rounded *up* to hundredths is 34.29

(Here, 1 is always added when figures dropped are 1 or higher.)

rounded *down* to hundredths is 34.28

DIAGNOSTIC TEST

Find the following correct to—

1. Nearest tenth:
  a. .68 .7
  b. 5.425 5.4

4. Nearest ten-thousandth:
  a. .42769 .4277
  b. 26.58634 26.5863

2. Nearest hundredth:
  a. 8.349 8.35
  b. .5146 .51

5. Nearest hundred-thousandth:
  a. .005924 .00592
  b. 54.632857 54.63286

3. Nearest thousandth:
  a. 1.9685 1.969
  b. 3.24728 3.247

6. Nearest millionth:
  a. .0000985 .000099
  b. 2.0500302 2.050030

7. Nearest cent: **a.** $2.386   **b.** $4.843   **c.** 7.23\frac{2}{3}$ **d.** 1.68\frac{3}{8}$
                      $2.39 $4.84 $7.24 $1.68

8. Nearest whole number: **a.** 57.18   **b.** 89.562
                               57   90

Basic: Sets 1–7(a), Appl. 1; *Average:* Sets 1–8(a), Appl. 1–2; *Enriched:* Sets 1–8(b), Appl. 1–3

RELATED PRACTICE EXAMPLES

Find the following correct to—

Nearest tenth:

SET 1(a)	SET 1(b)
1. .25 .3	**1.** .14 .1
2. .87 .9	**2.** .32 .3
3. .984 1.0	**3.** 1.83 1.8
4. 4.16 4.2	**4.** 3.718 3.7
5. 2.39 2.4	**5.** 5.205 5.2

Nearest hundredth:

SET 2(a)	SET 2(b)
1. .517 .52	**1.** .323 .32
2. .308 .31	**2.** .934 .93
3. 5.845 5.85	**3.** 6.544 6.54
4. 9.3794 9.38	**4.** 8.3025 8.30
5. 15.2263 15.23	**5.** 14.0412 14.04

Nearest thousandth:

SET 3(a)	SET 3(b)
1. .2146 .215	**1.** .5452 .545
2. .3998 .400	**2.** .9293 .929
3. 3.0815 3.082	**3.** 5.0341 5.034
4. 1.5929 1.593	**4.** 2.0084 2.008
5. 16.76876 16.769	**5.** 24.82509 24.825

Nearest ten-thousandth:

SET 4(a)	SET 4(b)
1. .20585 .2059	**1.** .34143 .3414
2. .932473 .9325	**2.** .528129 .5281
3. 4.00619 4.0062	**3.** 2.30751 2.3075
4. 7.451852	**4.** 18.781304
5. 43.020471	**5.** 39.56372
4. 7.4519	4. 18.7813
5. 43.0205	5. 39.5637

Nearest hundred-thousandth: Nearest millionth:

SET 5(a)	SET 5(b)	SET 6(a)	SET 6(b)
1. .000083 .00008	**1.** .000079 .00008	**1.** .0000029	**1.** .0000052 .000005
2. .005142 .00514	**2.** .030585 .03059	**2.** .0001538	**2.** .0000194 .000019
3. 1.079324	**3.** 2.000467 2.00047	**3.** .0254985	**3.** .0039481 ,003948
4. 25.415291	**4.** 23.075306 23,07531	**4.** 1.0038296	**4.** 5.0006373
5. 83.591243	**5.** 71.114248 71.11425	**5.** 4.1830547	**5.** 8.2751254

3. 1.07932   6a. 1. .000003   4. 1.003830   5. 4.183055

4. 25.41529   2. .000154   Nearest cent: 4. 5.000637

5. 83.59124   3. .025499

                                                                        5. 8.275125

SET 7(a)	SET 7(b)	SET 7(c)	SET 7(d)
1. $.267 $.27	**1.** $.643 $.64	**1.** $.62$\frac{1}{2}$ $.63	**1.** $.46$\frac{1}{4}$ $.46
2. $0.59 $.59	**2.** $.834 $.83	**2.** $.28$\frac{3}{5}$ $.29	**2.** $.73$\frac{1}{3}$ $.73
3. $1.316 $1.32	**3.** $6.572 $6.57	**3.** 3.84\frac{7}{12}$ $3.85	**3.** 1.57\frac{2}{5}$ $1.57
4. $5.8682 $5.87	**4.** $14.9615 $14.96	**4.** 9.20\frac{7}{8}$ $9.21	**4.** 4.89\frac{5}{12}$ $4.89
5. $24.6754 $24.68	**5.** $30.2137 $30.21	**5.** 15.14\frac{5}{6}$ $15.15	**5.** 12.30\frac{11}{24}$ $12.30

Nearest whole number:

SET 8(a)		SET 8(b)	
1. 6.3 6	**4.** 34.28 34	**1.** 9.6 10	**4.** 25.989 26
2. 19.4 19	**5.** 858.126 858	**2.** 68.5 69	**5.** 399.704 400
3. 2.09 2		**3.** 100.81 101	

PRACTICAL APPLICATIONS

1. Round the following conversion factors correct to the nearest hundredth:

a. 1 nautical mile = 1.1515 statute miles 1.15

b. 1 meter = 1.0936 yards 1.09

c. 1 kilogram = 2.2046 pounds 2.20

2. Round correct to the nearest tenth:

a. 1 centimeter = 0.3937 inch 0.4

b. 1 bushel = 2,150.42 cubic inches 2,150.4

c. 1 cubic inch = 16.3872 cubic centimeters 16.4

3. Round correct to the nearest thousandth:

a. 1 statute mile = 0.8684 nautical mile 0.868

b. 1 square inch = 6.4516 square centimeters 6.452

c. 1 liquid quart = 0.9463 liter 0.946

OBJECTIVE:

I. Aim   To add decimals.

II. Procedure

1. If it is necessary to write the numerals in columns, write each addend so that the decimal points are directly under each other. Zeros may be annexed to the numerals naming decimal fractions so that the addends may have the same number of decimal places. (See sample solutions 4 and 5.)

2. Add as in the addition of whole numbers. (See Exercise 1-4.)

3. Place the decimal point in the sum directly under the decimal points in the addends.

4. When a decimal answer ends in one or more zeros to the right of the decimal point, the zeros may be dropped unless it is necessary to show the exact degree of measurement. (See sample solution 3.)

5. Check by adding the columns in the opposite direction or by adding the sum of each column as shown in Exercise 1-4.

6. Decimals named by shortened numerals may be added. (See sample solution 7.)

Concept:

(a) $.8 + .5$

Since $.8 = \frac{8}{10}$ and $.5 = \frac{5}{10}$,

then $\begin{array}{r} \frac{8}{10} \\ + \frac{5}{10} \\ \hline \frac{13}{10} = 1\frac{3}{10} \end{array}$ or using decimals $\begin{array}{r} .8 \\ + .5 \\ \hline 1.3 \end{array}$

$\qquad$ 1.3 is the same as $1\frac{3}{10}$

(b) $.46 + .7$

Since $.46 = \frac{46}{100}$ and $.7 = \frac{7}{10}$,

then $\begin{array}{r} \frac{46}{100} = \frac{46}{100} \\ + \frac{7}{10} = \frac{70}{100} \\ \hline \frac{116}{100} = 1\frac{16}{100} \end{array}$ or using decimals $\begin{array}{r} .46 \\ + .7 \\ \hline 1.16 \end{array}$

$\qquad$ 1.16 is the same as $1\frac{16}{100}$

III. Sample Solutions

Add in the following examples:

1. .5
  .8
  .9
  ‾‾‾
  2.2

Answer: 2.2

2. 2.08
  4.07
  1.03
  ‾‾‾‾
  7.18

Answer 7.18

3. .26
  .38
  .06
  ‾‾‾
  .70 = .7

Answer: .70 or .7

4. 2.9 + .12

  2.9   2.90
  .12 or   0.12
  ‾‾‾   ‾‾‾‾
  3.02   3.02

Answer: 3.02

5. Find the sum of
  .523, 4.16, 3, and 2.2

  .523   0.523
  4.16 or 4.160
  3.   3.000
  2.2   2.200
  ‾‾‾‾   ‾‾‾‾‾
  9.883   9.883

Answer: 9.883

6. $2.58 + $34.74 + $.69 + $5.87

  $2.58
  34.74
  .69
  5.87
  ‾‾‾‾‾
  $43.88

Answer: $43.88

7. Add: 82.6 million
  95.8 million
  4.3 million
  ‾‾‾‾‾‾‾‾‾‾‾
  182.7 million   *Answer: 182.7 million*

DIAGNOSTIC TEST

Add:

1.	**2.**	**3.**	**4.**
.4	.6	.02	.08
.3	.9	.01	.03
.2	.4	.03	.02
‾‾ .9	‾‾ 1.9	‾‾ .06	.06 .19

5.	**6.**	**7.**	**8.**
0.15	.52	0.85	1.5
0.03	.43	0.10	2.1
0.48	.69	0.05	4.3
0.17 .83	.74 2.38	‾‾‾‾ 1.00	‾‾‾ 7.9

9.	**10.**	**11.**	**12.**
6.4	3.06	4.57	7.49
8.9	4.09	2.93	3.5
4.4	2.08	4.87	‾‾‾‾ 10.99
3.5 23.2	‾‾‾‾ 9.23	‾‾‾‾ 12.37	

(Continued on next page.)

13. 2.86
.7
.12 3.68

14. 8
.05
8.05

15. 2.103
4.839
3.542 10.484

16. 50.48
37.59
23.84 111.91

17. 326.04
183.75
225.39
491.26
1,226.44

18. 19.47
8.46
592.75
74.81
126.78
91.33 913.6

19. $.50
.28
.79
.84
.67
$3.08

20. $42.85
9.74
223.30
54.67
7.98
$338.54

21. 18.59 billion
9.28 billion
53.06 billion
80.93 billion

22. .08 + 1.5 1.58

23. .752 + 4.53 + 6 11.282

24. $1.43 + $.89 + $5.07 + $.36 + $9.58 $17.33

25. Find the sum of: 6.4, .976, and 2.87 10.246

Basic: Sets 1–20, Prob. 1–5, Micrometer 1–2; *Average:* Sets 1–22(b), Prob. 1–8, Micrometer 1–4;
Enriched: Sets 1–25, Prob. 1–10, Micrometer 1–5

RELATED PRACTICE EXAMPLES

SET 1	SET 2	SET 3	SET 4	SET 5	SET 6
.2	.9	.03	.06	.68	.43
.6 .8	.7 1.6	.04 .07	.09 .15	.26 .94	.89 1.32
.3	.5	.03	.05	.57	.38
.2	.8	.02	.05	.04	.45
.2 .7	.4 1.7	.03 .08	.07 .17	.18 .79	.22 1.05
0.2	0.4	0.01	0.02	0.12	0.14
0.3	0.8	0.03	0.07	0.23	0.32
0.1	0.7	0.02	0.04	0.45	0.51
0.2 .8	0.4 2.3	0.03 .09	0.05 .18	0.16 .96	0.28 1.25
0.1	0.9	0.02	0.04	0.31	0.82
0.2	0.6	0.01	0.07	0.17	0.59
0.3	0.9	0.01	0.08	0.09	0.77
0.3 .9	0.2 2.6	0.05 .09	0.03 .22	0.24 .81	0.31 2.49
.1	.5	.03	.09	.23	.27
.3	.7	.01	.06	.18	.74
.2	.4	.02	.08	.29	.51
.1	.8	.01	.04	.03	.93
.2 .9	.3 2.7	.02 .09	.09 .36	.25 .98	.89 3.34

SET 7		SET 8		SET 9		SET 10		SET 11		SET 12	
.53		2.4		7.8		8.05		3.26		6.25	
.47	1.00 or 1	3.5	5.9	6.5	14.3	4.08	12.13	2.15	5.41	4.6	10.85
.48		6.2		4.1		9.08		6.27		6.74	
.26		9.3		1.5		8.03		2.83		9.3	
.16	.90 or .9	8.3	23.8	7.4	13.0 or 13	1.07	18.18	5.76	14.86	1.87	17.91
0.52		4.1		2.8		1.04		2.19		9.03	
0.81		2.2		1.4		3.08		3.82		8.2	
0.49		1.1		8.2		4.01		5.27		4.19	
0.68	2.50 or 2.5	9.2	16.6	7.5	19.9	2.05	10.18	1.38	12.66	2.3	23.72
0.27		2.2		1.6		6.01		3.41		5.7	
0.95		1.4		5.3		8.07		2.54		4.3	
0.32		5.1		4.4		4.05		4.68		9.25	
0.66	2.20 or 2.2	1.1	9.8	3.2	14.5	2.09	20.22	5.27	15.90 or 15.9	8.75	28.00 or 28
.17		9.2		8.5		6.08		3.86		4.92	
.08		3.2		4.9		2.03		2.29		8.1	
.49		4.3		6.3		9.07		4.57		9.6	
.27		2.1		7.4		4.09		5.18		2.43	
.99	2.00 or 2	7.1	25.9	5.6	32.7	7.03	28.30 or 28.3	1.33	17.23	1.8	26.85

SET 13*		SET 14*		SET 15*		SET 16*		SET 17*	
.9		.6		5.01		20.56		138.35	
5.28	6.18	4.	4.6	2.999	8.009	14.27	34.83	253.42	391.77
9.5		.94		3.728		36.87		639.32	
.33		7.		2.517		8.26		182.08	
.19	10.02	.8	8.74	9.282	15.527	15.84	60.97	24.19	845.59
1.4		8.		7.053		1.38		8.3407	
.7		.53		.96		16.48		.0038	
.29		2.1		8.524		9.27		1.153	
2.45	4.84	.68	11.31	1.8	18.337	.84	27.97	7.4619	16.9594
.81		3.9		1.516		28.45		.00532	
.9		.49		6.24		42.83		.13847	
3.9		6.		.006		56.19		.32296	
4.62	10.23	.57	10.96	4.518	12.280 or 12.28	18.28	145.75	.48325	.95000 or .95

*Continued on next page.

5.27	.36	4.783	32.28	335.48
7.4	4.68	1.829	14.85	618.37
.35	.2	5.318	9.74	407.54
3.82	9.	2.175	98.42	250.36
.06	1.7	9.384	.94	195.83
16.90 or 16.9	15.94	23.489	156.23	1,807.58

SET 18

.08	1.16	16.91	426.26	596.74
.57	2.65	28.38	9.84	289.57
.13	4.27	19.43	16.59	193.86
.42	1.04	37.21	153.60	319.29
.83	3.53	4.82	74.75	475.43
.66	5.21	.48 107.23	8.68	628.58
2.69	17.86		689.72	2,503.47

SET 19

$.24	$.83	$.36	$.22	$.84
.59	.25	.88	.93	.27
.37	.03	.41	.79	.49
.73	.62	.29	.68	.56
.85	.54	.37	.85	.99
$2.78	$2.27	$2.31	.42	.78
			$3.89	$3.93

SET 20

$2.33	$8.25	$45.63	$297.82	$355.95
.96	3.70	5.96	4.08	109.82
4.28	9.64	.83	16.29	481.56
.51	8.23	14.71	.62	247.49
9.16	4.72	9.43	148.37	575.37
$17.24	$34.54	18.65	974.46	281.25
		$95.21	$1,441.64	$2,051.44

SET 21

75.934 billion	6.4 trillion	$8.42 million
8.759 billion	1.7 trillion	15.95 million
84.693 billion	2.8 trillion	6.69 million
	10.9 trillion	28.94 million $60 million
$21.467 billion	8.26 million	$74.8 billion
106.594 billion	685.78 million	105.9 billion
8.299 billion	9.57 million	528.4 billion
57.038 billion	16.98 million	6.7 billion
193.398 billion	420.69 million	47.2 billion
	1,141.28 million	$763 billion

Add as indicated:

SET 22(a)

1. .3 + .6 .9
2. .05 + .12 .17
3. 3.6 + 5.1 8.7
4. .275 + .38 .655
5. .62 + .573 1.193
6. .47 + .1785 .6485
7. 18 + .32 18.32
8. .9 + 4 4.9
9. .017 + 15 15.017
10. .16 + .7 .86

SET 22(b)

1. .8 + .5 1.3
2. .46 + .04 .50 or .5
3. 15.8 + 6.3 22.1
4. .94 + .065 1.005
5. .824 + .19 1.014
6. .1692 + .23 .3992
7. 25 + .6 25.6
8. .08 + 9 9.08
9. 1.56 + 4 5.56
10. .036 + .1 .136

SET 23

1. .17 + .38 + .53 1.08
2. .52 + 1.6 + 8.26 10.38
3. .83 + 7 + 4.45 + .049 12.329
4. .06 + 1.2 + 46 + 3.825 + .075 51.16
5. 1.5 + 0.18 + 6.84 + 0.016 + 0.27 8.806
6. .83 + .37 + 4 + 3.9 + .051 9.151
7. 4.92 + 1.853 + 9.7 + 60 + 42.6 119.073
8. 1.4 + 26 + .39 + 5.98 + 9 42.77
9. 3.3 + .07 + 6 + 2.63 + .174 12.174
10. 15.6 + .19 + 4.75 + .836 + 200 221.376

SET 24

1. $.23 + $.75 $.98
2. $.17 + $.49 + $.83 $1.49
3. $1.80 + $2.60 + $4.25 $8.65
4. $2.75 + $3.35 + $.96 + $1.45 $8.51
5. $0.60 + $0.38 + $1.50 + $3.25 + $9.70 $15.43
6. $32.40 + $4.80 + $2.62 + $.61 + $.89 $41.32
7. $100 + $8.42 + $93.75 + $6.83 + $.14 $209.14
8. $5.57 + $2.83 + $1.69 + $10.50 + $7.25 $27.84
9. $.74 + $1.60 + $.99 + $4.88 + $.04 $8.25
10. $3.42 + $6.51 + $12.54 + $9.49 + $8.68 $40.64

SET 25

Find the sum of:

1. 4.23, 6.832, and 4.4 15.462
2. 8.01, .684, and 5.9 14.594
3. .07, 19.3, and 7.86 27.23
4. .6, 8, and .24 8.84
5. 1.4, 3.8, and .87 6.07
6. 2.05, .156, 4.69, and .08 6.976
7. $7.26, $.85, $.42, $.94, and $2.35 $11.82
8. $12.59, $9.47, $1.27, $.56, and $3.46 $27.35
9. $8.41, $2.25, $2.50, $7.64, and $.85 $21.65
10. $1.20, $4.81, $.39, $.74, and $3.08 $10.22

PROBLEM SOLVING

See page 16, *Problem Solving Strategy*.

For problems 1–5, select the letter corresponding to your answer.

1. The school orchestra bought a violin for $169.95, a saxophone for $249.75, and a trumpet for $124.50. What is the total cost of the new instruments?

    a. $534.20   **b.** $544.20   **c.** $554.20   **d.** Answer not given

2. Marilyn bought her graduation outfit. Her dress cost $49.98; shoes, $21.49; hat, $15.75; and bag, $16.95. How much did she spend?

    a. $105.17   **b.** $106.17   **c.** $104.17   **d.** $96.17

3. The Dow Jones stock averages opened at 885.57 and during the day gained 3.46 points. What was the closing stock average?

    a. 889.03   **b.** 888.93   **c.** 898.93   **d.** 988.93

4. Find the total expenses in producing the Jacques Cartier School show if renting and making costumes cost $362.55; royalty fee, $75.00; properties, $143.66; tickets, $25.50; lighting, $52.25; and miscellaneous items, $157.19.

    a. $806.15   **b.** $796.15   **c.** $826.15   **d.** $816.15

5. The freshmen contributed $139.75 to the Red Cross; the sophomores, $98.40; the juniors, $106.15; and the seniors, $134.90. What was the total amount of their contributions?

    a. $459.20   **b.** $469.20   **c.** $479.20   **d.** Answer not given

6. A mutual fund reported a dividend of 8.954 shares to a stockholder who already owns 490.573 shares. Find the new total of shares owned by the stockholder. 499.527 shares

7. Charlotte's parents took the family on a motor trip. The expenses included gasoline, $142.90; oil, $6.36; lodging, $175.00; meals, $259.25; amusements, $64.83; miscellaneous expenses, $39.54. How much did the trip cost? $687.88

8. The weather bureau reported that the total rainfall for the first 7 months in a city was 30.93 inches. During the 5 times it rained in August, 1.02 in., 2 in., 1.6 in., 0.58 in., and 0.4 in. of rain fell. What was the total rainfall at the end of the first 8 months? 36.53 in.

9. a. Robert's brother is a salesman. Last week he received $294.50 salary and $59.68 commission. How much did he earn in all? $354.18
b. What is his new bank balance if the previous balance was $297.06 and he deposited a check for $36.97? $334.03

10. a. For how much should a dealer sell a CB radio if it cost him $68.75 and he wishes to make a profit of $36.75? $105.50
b. Nick paid $77.88 for a bicycle. What was the original price if the storekeeper gave Nick a reduction of $11.10? $88.98

Micrometer Caliper Readings

Measurement by a micrometer caliper involves the addition of decimals. On the sleeve there are two graduated sets of lines. The set above the center line indicates the reading in millimeters (major divisions). The set below this line indicates the reading in half millimeters, 0.5 mm (subdivision). The thimble scale is marked in fifty (50) equal divisions, each representing 0.01 mm.

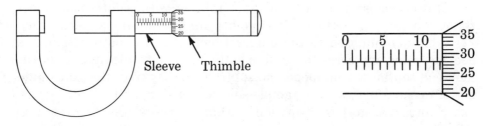

This setting reads 12.78 millimeters.

Find in each case the complete micrometer reading by adding the following given measurements:

	Major Divisions on Sleeve Indicating	Subdivision on Sleeve Indicating	Divisions on Thimble Indicating	
1. Piece A	5.00 mm	0.50 mm	0.14 mm	5.64 mm
2. Piece B	8.00 mm	0.50 mm	0.29 mm	8.79 mm
3. Piece C	14.00 mm	0.50 mm	0.48 mm	14.98 mm
4. Piece D	6.00 mm	0.00 mm	0.06 mm	6.06 mm
5. Piece E	11.00 mm	0.50 mm	0.37 mm	11.87 mm

REFRESH YOUR SKILLS

1. Add: **2.** Subtract: **3.** Multiply: **4.** Divide:

$4\frac{3}{5}$ 16 $5\frac{5}{8} \times 6\frac{2}{3}$ $37\frac{1}{2}$ $8 \div \frac{3}{4}$ $10\frac{2}{3}$

$\underline{5\frac{7}{10}}$ $10\frac{3}{10}$ $\underline{9\frac{11}{12}}$ $6\frac{1}{12}$

Forty-three point six; forty-three and

5. Read (or write in words) in two ways: 43.6   six-tenths

6. Write as a decimal numeral: Two hundred and forty-three thousandths   200.043

7. Round 8.276 to the nearest hundredth.   8.28

8. Add: 9.46 + .628 + 80.2   90.288

I. Aim   To subtract decimals.

II. Procedure

1. If it is necessary to write the numerals, write the subtrahend under the minuend so that the decimal points are directly under each other. Zeros may be annexed to the numeral naming a decimal fraction or a decimal point and zeros to the numeral naming a whole number in the minuend so that the minuend and subtrahend may have the same number of decimal places. (See sample solutions 4, 7, 8, and 9.)

2. Subtract as in the subtraction of whole numbers. (See Exercise 1-5.)

3. Place the decimal point in the remainder (or difference) directly under the decimal points of the subtrahend and minuend.

4. When a decimal answer ends in one or more zeros to the right of the decimal point, the zeros may be dropped unless it is necessary to show the exact degree of measurement. (See sample solution 5.)

5. Check by adding the remainder to the subtrahend. Their sum should equal the minuend. (See Exercise 1-5.)

6. Decimals named by shortened numerals may be subtracted. (See sample solution 11.)

Concept:

(*a*) $.9 - .2$

Since $.9 = \frac{9}{10}$ and $.2 = \frac{2}{10}$,

then $\begin{array}{r} \frac{9}{10} \\ - \frac{2}{10} \\ \hline \frac{7}{10} \end{array}$ or using decimals $\begin{array}{r} .9 \\ - .2 \\ \hline .7 \end{array}$

.7 is the same as $\frac{7}{10}$

(*b*) $4.5 - 1.6$

Since $4.5 = 4\frac{5}{10}$ and $1.6 = 1\frac{6}{10}$,

then $\begin{array}{r} 4\frac{5}{10} = 3\frac{15}{10} \\ - 1\frac{6}{10} = 1\frac{6}{10} \\ \hline 2\frac{9}{10} \end{array}$ or using decimals $\begin{array}{r} 4.5 \\ - 1.6 \\ \hline 2.9 \end{array}$

2.9 is the same as $2\frac{9}{10}$

III. Sample Solutions

Subtract in the following examples:

1. .59
.25
———
.34

Answer: .34

2. .47
.38
———
.09

Answer: .09

3. 16.2
4.6
———
11.6

Answer: 11.6

4. 5.61   5.61
.9   or   .90
————   ————
4.71   4.71

Answer: 4.71

5. 9.354
.854
—————
8.500 = 8.5

Answer: 8.500 or 8.5

6. $71.46
15.24
—————
$56.22

Answer: $56.22

7. .82 − .5
.82   .82
.5   or   .50
———   ———
.32   .32

Answer: .32

8. .6 − .524
.600
.524
————
.076

Answer: .076

9. $20 − $6.38
$20.00
6.38
—————
$13.62

Answer: $13.62

10. Subtract .27 from 2.7
2.70
.27
————
2.43

Answer: 2.43

11. Subtract:
21.87 trillion
11.78 trillion
———————————
10.09 trillion

Answer: 10.09 trillion

DIAGNOSTIC TEST

Subtract:

1. .8
.2
—— .6

7. 4.6
3.1
—— 1.5

13. .6593
.4978
————— .1615

19. 8.000
1.742
————— 6.258

25. 16.53 billion
9.67 billion
——————————— 6.86 billion

2. .38
.24
—— .14

8. 5.4
2.8
—— 2.6

14. 83.452
49.596
—————— 33.856

20. 3.56
.8
———— 2.76

26. .15 − .08 .07

27. .375 − .2 .175

3. .67
.48
—— .19

9. 6.5
2.5
—— 4.0

15. .34617
.14596
—————— .20021

21. .9
.735
———— .165

28. .7 − .625 .075

4. .05
.02
—— .03

10. .679
.398
———— .281

16. .4583
.2783
————— .1800 or .18

22. 5.
1.43
———— 3.57

29. 4.2 − .83 3.37

30. 9 − .05 8.95

5. .86
.79
—— .07

11. 3.72
1.95
———— 1.77

17. .85328
.84793
—————— .00535

23. 4.2
.372
———— 3.828

31. $16 − $1.50 $14.50

6. .57
.37
—— .20 or .2

12. 15.8
3.9
———— 11.9

18. 6.531
5.975
———— .556

24. $176.27
93.48
———— $82.79

32. Subtract
.08 from .3 .22

RELATED PRACTICE EXAMPLES

Subtract:

SET 1						SET 2			
.9	.4	.6	.7	.5	.29	.47	.73	.98	.64
.5	.3	.1	.2	.4	.03	.26	.32	.65	.21
.4	.1	.5	.5	.1	.26	.21	.41	.33	.43

SET 3						SET 4			
.36	.44	.25	.83	.58	.08	.06	.09	.07	.09
.18	.27	.09	.36	.29	.01	.04	.03	.05	.06
.18	.17	.16	.47	.29	.07	.02	.06	.02	.03

SET 5						SET 6			
.85	.92	.54	.25	.13	.58	.46	.25	.74	.92
.76	.84	.45	.17	.08	.28	.36	.05	.54	.42
.09	.08	.09	.08	.05	.30 or .3	.10 or .1	.20 or .2	.20 or .2	.50 or .5

SET 7						SET 8			
8.4	4.7	6.8	5.6	9.5	7.5	9.3	4.6	8.2	6.1
4.2	2.5	1.7	3.2	4.4	4.7	1.9	2.8	5.4	4.3
4.2	2.2	5.1	2.4	5.1	2.8	7.4	1.8	2.8	1.8

SET 9						SET 10			
5.3	7.6	9.2	2.8	8.4	.835	.594	.967	.076	.749
3.3	2.6	6.2	1.8	4.4	.214	.287	.378	.043	.729
2.0 or 2	5.0 or 5	3.0 or 3	1.0 or 1	4.0 or 4	.621	.307	.589	.033	.020 or .02

SET 11						SET 12			
7.64	6.51	9.68	4.29	9.01	18.5	17.3	24.5	38.4	42.7
3.53	4.32	5.95	1.76	7.84	6.2	5.8	3.5	7.1	5.9
4.11	2.19	3.73	2.53	1.17	12.3	11.5	21.0 or 21	31.3	36.8

SET 13				
.9355	.1327	60.07	375.3	5.724
.8492	.1219	42.38	190.4	3.856
.0863	.0108	17.69	184.9	1.868

SET 14				
.25683	34.645	9.4538	849.54	5,986.3
.14974	17.859	6.7425	258.46	2,894.7
.10709	16.786	2.7113	591.08	3,091.6

SET 15				
.5844	.4683	6.2937	.83572	43.596
.2837	.1678	4.2843	.43564	35.589
.3007	.3005	2.0094	.40008	8.007

SET 16

.7856	.3895	5.8362	.92041	82.475
.3256	.1995	3.7362	.52041	17.275
.4600 or .46	.1900 or .19	2.1000 or 2.1	.40000 or .4	65.200 or 65.2

SET 17

.6848	.04965	.00325	.97424	.03741
.6827	.04894	.00255	.96712	.03659
.0021	.00071	.00070 or .0007	.00712	.00082

SET 18

4.6	5.83	9.786	1.904	7.0352
3.8	4.96	8.895	1.737	6.9446
0.8	0.87	0.891	0.167	0.0906

SET 19

4.000	9.000	6.0000	10.000	24.000
1.753	4.068	5.9325	8.469	13.307
2.247	4.932	0.0675	1.531	10.693

SET 20

1.4	2.5	8.65	7.543	5.0046
.9	.7	.8	.28	4.307
.5	1.8	7.85	7.263	0.6976

SET 21

.36	.78	.07	.6	.4
.034	.1561	.0628	.49	.003
.326	.6239	.0072	.11	.397

SET 22

8	3	7	6	28
7.3	1.2	2.84	4.005	9.756
0.7	1.8	4.16	1.995	18.244

SET 23

5	9	3	1.6	4.9
.6	.09	.753	.54	.807
4.4	8.91	2.247	1.06	4.093

SET 24(a)

$4.85	$1.36	$36.80	$84.25	$25.00
2.60	.89	17.42	9.75	16.68
$2.25	$.47	$19.38	$74.50	$8.32

SET 24(b)

$193.45	$200.00	$476.13	$1,250.00	$5,000.00
108.91	127.18	85.75	975.80	3,752.69
$84.54	$72.82	$390.38	$274.20	$1,247.31

SET 25

42.8 million	$346.9 billion	$15.21 million
27.6 million	58.9 billion	13.95 million
15.2 million	**$288 billion**	**$1.26 million**
$529.03 billion	26.574 million	42.06 billion
488.95 billion	8.3 million	27.554 billion
$40.08 billion	**18.274 million**	**14.506 billion**

Subtract as indicated:

SET 26
1. .3 − .2 .1
2. .8 − .3 .5
3. .49 − .25 .24
4. .38 − .09 .29
5. .536 − .008 .528
6. .837 − .325 .512
7. .372 − .056 .316
8. .0045 − .0023 .0022
9. .1534 − .0976 .0558
10. .00316 − .00286 .00030 or .0003

SET 27
1. .45 − .4 .05
2. .84 − .3 .54
3. .97 − .6 .37
4. .384 − .2 .184
5. .518 − .46 .058
6. .039 − .01 .029
7. .6482 − .3 .3482
8. .8056 − .74 .0656
9. .5842 − .095 .4892
10. .0034 − .003 .0004

SET 28
1. .6 − .45 .15
2. .8 − .36 .44
3. .9 − .83 .07
4. .4 − .176 .224
5. .35 − .285 .065
6. .06 − .043 .017
7. .2 − .1356 .0644
8. .41 − .2538 .1562
9. .685 − .5903 .0947
10. .006 − .0005 .0055

SET 29
1. 3.6 − .24 3.36
2. 4.7 − .83 3.87
3. 23.4 − 1.75 21.65
4. 5.82 − .004 5.816
5. 6.03 − .307 5.723
6. 12.54 − 1.054 11.486
7. 81.2 − 4.18 77.02
8. 3.4 − .0056 3.3944
9. 2.875 − .375 2.5
10. 18.01 − .0002 18.0098

SET 30
1. 8 − .5 7.5
2. 2 − 1.7 0.3
3. 1 − .16 0.84
4. 3 − 1.38 1.62
5. 6 − .09 5.91
6. 4 − .875 3.125
7. 7 − 3.0625 3.9375
8. 10 − .375 9.625
9. 12 − 8.2003 3.7997
10. 27 − .0067 26.9933

SET 31
1. $.50 − $.25 $.25
2. $.78 − $.09 $.69
3. $2.74 − $.86 $1.88
4. $3.35 − $2.25 $1.10
5. $10 − $5.60 $4.40
6. $15.42 − $9 $6.42
7. $100 − $84.53 $15.47
8. $200 − $133.62 $66.38
9. $446.58 − $279.49 $167.09
10. $3,500 − $2,938.75 $561.25

SET 32
1. From .8 subtract .35 .45
2. From 1.4 subtract .07 1.33
3. From 45.9 take 6.38 39.52
4. From $2.84 take $1.75 $1.09
5. Subtract .004 from .08 .076
6. Subtract $.62 from $.79 $.17
7. Take .45 from 1.5 1.05
8. Take $8.40 from $12 $3.60
9. Find the difference between .03 and .003 .027
10. Find the difference between 4.81 and .481 4.329

PROBLEM SOLVING

See page 16, *Problem Solving Strategy.*
For problems 1–5, select the letter corresponding to your answer.

1. A girl ran the 100-meter dash in 15.4 seconds, then later ran the same distance in 13.7 seconds. How many seconds less did she take the second time?

    a. 2.7 seconds   **b.** 1.7 seconds   **c.** 2.3 seconds   **d.** 29.1 seconds

2. In addition to her weekly allowance of $8.50, Wanda earned $15.65 after school. How much money should she have left at the end of the week if her expenses were: bus fare, $3.50; school lunches and supplies, $6.97; movies, $1.85; church, $1.00; and savings, $3.75?

    a. $6.08   **b.** $5.08   **c.** $7.08   **d.** Answer not given

3. A stockholder owns 244.134 shares of stock in a mutual fund. If only 175 shares were issued to her, how many shares have not yet been issued?

    a. 89.034 shares   **c.** 89.134 shares
    b. 79.134 shares   **d.** 69.134 shares

4. The passenger ship, *United States,* on her maiden voyage established a record of 35.59 knots crossing the Atlantic Ocean. If the record speed of the *Queen Mary* is 30.99 knots, how much faster is the *United States?*

    a. 5.6 knots   **b.** 4.6 knots   **c.** 5.5 knots   **d.** Answer not given

5. During the month a merchant made deposits of $439.76, $180.53, $263.98, and $129.49. Checks and cash withdrawals were: $163.20, $248.00, $92.85, $310.94, and $8.52. If his previous monthly balance was $716.91, find his new bank balance.

    a. $897.16   **b.** $887.16   **c.** $907.16   **d.** Answer not given

6. How much change should you receive from a:
a. $1 bill if you owe $.59? $.83? $.46? $.11? $.78?
b. $5 bill if you owe $1.89? $2.98? $3.04? $.69? $4.44?
c. $10 bill if you owe $8.45? $5.09? $.97? $9.61? $2.28?

6.a. $.41; $.17; $.54; $.89; $.22 b. $3.11; $2.02; $1.96; $4.31; $.56 c. $1.55; $4.91; $9.03; $.39; $7.72

7. Harry bought a new suit costing $94.95; shoes, $21.75; hat, $12.50; shirt, $10.98; and tie, $3.50. If his mother gave him $150, how much money is left over after he pays for these articles? $6.32

8. John bought a baseball glove that regularly sold for $19.49 at a reduction of $2.98. How much did he pay? $16.51

9. He also left a deposit of $3.75 on a baseball uniform that cost $29.49. What balance does he owe? $25.74

10. In measuring a 1-inch block of metal by a precision instrument, a student found the average of all her readings to be 0.9996 inch. Find the amount of error. 0.0004 in.

11. The outside diameter of a piece of copper tubing is 2.375 millimeters and its wall thickness is .083 millimeters. What is the inside diameter? 2.209 mm

12. The net worth of a business is equal to its assets minus its liabilities. Find the net worth of a business whose assets are: cash, $982.75; merchandise, $12,369; accounts receivable, $536.89; other assets, $1,750; and liabilities are: notes payable, $400; accounts payable, $109.38; other liabilities, $327.67. $14,801.59

REFRESH YOUR SKILLS

1. Add: **2.** Subtract: **3.** Multiply: **4.** Divide: 8,695

149,526   821,427   1,760 $407)\overline{3,538,865}$

312,997   291,398   500

500,725   ———   ———

832,589   530,029   880,000

294,217

381,726

———

2,471,780

5. Round 30,246,701 to the nearest million. 30,000,000

6. Express $\frac{54}{72}$ in lowest terms. $\frac{3}{4}$

7. Change $\frac{3}{5}$ to 100ths. $\frac{60}{100}$

8. Which fraction is greater: $\frac{7}{8}$ or $\frac{4}{5}$? $\frac{7}{8}$

9. Add: **10.** Subtract: **11.** Multiply: **12.** Divide:

$6\frac{5}{12}$ $5\frac{1}{2}$ $3\frac{1}{7} \times 56$ 176 $5\frac{1}{4} \div \frac{7}{8}$ 6

$2\frac{1}{4}$ $3\frac{13}{16}$

$3\frac{5}{6}$ $1\frac{11}{16}$

———

$12\frac{1}{2}$

13. What part of 135 is 45? $\frac{1}{3}$ **14.** $\frac{3}{16}$ of what number is 27? 144

15. Write the numeral 800.08 in words. Eight hundred and eight hundredths

16. Write as a decimal numeral: Three and fifty-nine thousandths

17. Round $5.8273 to the nearest cent. $5.83   3.059

18. Add: **19.** Subtract: **20.** Subtract:

.693 + 4.82 + 37.6   4.37 $18 − $2.67

43.113   .6 $15.33

———

3.77

For a Class Activity, see p. T33.

Comparing Decimals

OBJECTIVE:

I. Aim   To compare decimals.

II. Procedure

1. If necessary, change the given numbers to decimals containing the same number of decimal places. Take the greater number as the greater decimal. (See sample solution 1.)

2. A mixed decimal or whole number is greater than a decimal fraction.

3. The symbol $<$ means "is less than." (See sample solution 4.)

4. The symbol $>$ means "is greater than." (See sample solution 5.)

III. Sample Solutions

1. Which is greater: .03 or .3?

.03 = .03
.3 = .30

Answer: .3 is greater

2. Which is smaller: 1.7 or .283?

1.7 = 1.700
.283 = .283

Answer: .283 is smaller

3. Arrange in order of size (smallest first):
.65, .635, 6.1, and .069

.65 = .650 and 6.1 = 6.100
.635 and .069 are already expressed in thousandths.

Answer: .069, .635, .65, and 6.1

4. Is the statement .16 $<$ .6 true?

.16 = .16   .6 = .60
.16 is less than .6

Answer: Yes, true

5. Is the statement .45 $>$ 4.3 true?

.45 = .45   4.3 = 4.30
.45 is not greater than 4.3

Answer: No, false

DIAGNOSTIC TEST

Which is greater:

1. .5 or .45? .5
2. .154 or 1.02? 1.02

Which is smaller:

3. .7 or .699? .699
4. 5.53 or .553? .553

5. Arrange in order of size (greatest first):
.06, 1.4, .19, and .388 1.4, .388, .19, .06

6. Arrange in order of size (smallest first):
4.72, .493, 4.8, and .465 .465, .493, 4.72, 4.8

7. Is the statement .725 $<$ .73 true? Yes

8. Is the statement .9 $>$ .19 true? Yes

RELATED PRACTICE EXAMPLES

Which is greater:

SET 1

1. .3 or .29? .3
2. .04 or .004? .04
3. .91 or .893? .91
4. .7 or .074? .7
5. .156 or .1561? .1561

SET 2

1. 1.47 or .278? 1.47
2. .63 or 4.5? 4.5
3. .28 or 2.8? 2.8
4. 2.9 or .899? 2.9
5. 4.504 or 4.5035? 4.504

Which is smaller:

SET 3

1. .89 or .9? .89
2. .2 or .21? .2
3. .50 or .05? .05
4. .36 or .4? .36
5. .0051 or .006? .0051

SET 4

1. 4.1 or 3.010? 3.010
2. 5.06 or 1.059? 1.059
3. 1.638 or .5376? .5376
4. 2.0439 or 2.04395? 2.0439
5. 9.3 or 8.754? 8.754

SET 5

Arrange in order of size
(greatest first):

1. .01, .001, .1, and .0001 .1; .01; .001; .0001
2. 2.25, .253, .2485, and 2.249 2.25; 2.249; .253;
3. .38, 1.5, .475, and .0506 1.5; 475; .38; .0506 .2485
4. .006, 5.02, .503, and .1483 5.02; .503; .1483;
5. .98, .89, .934, and .9 .98; .934; .9; .89 .006

SET 6

Arrange in order of size
(smallest first):
1. 19; .201; .21; 1.2   2. .4053; .465; .47; 4.5

1. .201, .19, 1.2, and .21
2. .465, .4053, .47, and 4.5
3. .51, .583, .60, and .5126
4. .04, 1.25, .156, and 2.3
5. .76, .7, .076, and .0710

Which of the following statements are true?

3. 51; .5126; .583; .60   4. .04; .156; 1.25;
2.3   5. .0710; .076; .7; .76

SET 7

1. .036 < 0.36 True
2. .3 < .27 False
3. .049 < .5 True
4. .9086 < .908 False
5. 6.7 < .675 False

SET 8

1. 4.8 > 5 False
2. 6.4 > .640 True
3. .087 > .1 False
4. 1.05 > .258 True
5. .0070 > .007 False

PRACTICAL APPLICATIONS

In the manufacture of a certain part 1 centimeter in length, a tolerance of ±.002 centimeter is permitted, that is, the part will be accepted if it measures anywhere from 0.998 centimeter to 1.002 centimeters.

Which of the following lengths fall within the acceptable range?

1. **a.** 1.015 cm No   **b.** 0.9991 cm Yes   **c.** 1.0031 cm No
2. **a.** 0.9989 cm Yes   **b.** 1.0012 cm Yes   **c.** 0.99 cm No
3. **a.** 1.02 cm No   **b.** 0.989 cm No   **c.** 1.0007 cm Yes

Multiplication of Decimals

I. Aim To multiply decimals.

II. Procedure

1. Write the given numerals and multiply as in the multiplication of whole numbers. The decimal point in the multiplier does not necessarily have to be under the decimal point in the multiplicand.

2. Find the total number of decimal places in the multiplicand and multiplier and point off in the product, counting from right to left, as many decimal places as there are in the multiplicand and multiplier together.

3. When the product contains fewer figures than the required number of decimal places, prefix as many zeros as are necessary. (See sample solutions 3, 4, and 6.)

4. When a decimal answer ends in one or more zeros to the right of the decimal point, the zeros may be dropped unless it is necessary to show the exact degree of measurement. (See sample solutions 4, 5, and 9.)

5. Check by interchanging the multiplier and multiplicand and multiplying again, or by dividing the product by either the multiplier or multiplicand. (See Exercise 1-6.)

6. Decimals named by shortened numerals may be multiplied. (See sample solution 12.)

Concept:

(*a*) $.2 \times .8$

Since $.2 = \frac{2}{10}$ and $.8 = \frac{8}{10}$,

then $\frac{2}{10} \times \frac{8}{10} = \frac{16}{100}$ or using decimals $.2 \times .8 = .16$

.16 is the same as $\frac{16}{100}$

(*b*) $.3 \times .02$

Since $.3 = \frac{3}{10}$ and $.02 = \frac{2}{100}$,

then $\frac{3}{10} \times \frac{2}{100} = \frac{6}{1000}$ or using decimals $.3 \times .02 = .006$

.006 is the same as $\frac{6}{1000}$

III. Sample Solutions

Multiply in the following examples:

1.   46
   .04
————
 1.84

Answer: 1.84

2. .2389
     2
————
 .4778

Answer: .4778

3.   .02
   .03
————
 .0006

Answer: .0006

4. 1.35
   .06
————
 .081Ø = .081

Answer: .081

---

5.   .625
     48
————
 5 000
25 00
————
30.000

Answer: 30

6. $2.6 \times .002 = .0052$

Answer: .0052

7. Find $\frac{3}{4}$ of \$1.68

  Solution:

        \$.42
$\frac{3}{4} \times$ \$1.68 = \$1.26

Answer: \$1.26

8. Find .56 of 2.43

  Solution:

    2.43
     .56
————
 1458
1 215
————
1.3608

Answer: 1.3608

9. Find $.37\frac{1}{2}$ of \$28.96

             1448
Solution: $\frac{1}{2} \times 2896 = 1448$

  \$28.96
   .37$\frac{1}{2}$
————
 20272
 8688
————
107152
 1448
————
\$10.86ØØ   *Answer:* \$10.86

10. Find $.16\frac{2}{3}$ of \$37.25 correct
    to nearest cent.

Solution: $\frac{2}{3} \times 3725 = \frac{7450}{3} = 2483\frac{1}{3}$

  \$37.25
   .16$\frac{2}{3}$
————
22350
 3725
————
59600
2483$\frac{1}{3}$
————
\$6.2083$\frac{1}{3}$ = \$6.21 *Answer:* \$6.21

11. Find $\frac{5}{6}$ of \$4.19 correct to nearest cent.

  Solution:

$$\frac{5}{6} \times \$4.19 = \frac{\$20.95}{6} = \$3.49\frac{1}{6} = \$3.49$$

Answer: \$3.49

12. 2.6×3.4 billion

Solution: 3.4 billion
          $\times$ 2.6
         ————
          204
           68
        ————
      8.84 billion

Answer: 8.84 billion

DIAGNOSTIC TEST

Multiply:

1. .3
  8 2.4

7. .03
  2 .06

13. .58
  .6 .348

19. .147
  .03 .00441

2. 43
  .24 10.32

8. 14
  .007 .098

14. .21
  .4 .084

20. 3.1416
  .75 2.356200 or 2.3562

3. .351
  86 30.186

9. .002
  4 .008

15. .56
  .37 .2072

21. .059
  .064 .003776

4. .6739
  7 4.7173

10. 3.14
  18 56.52

16. .05
  .01 .0005

22. .012 × .005
  .000060 or .00006

5. 75
  .48 36

11. .6
  .2 .12

17. 16.2
  .045 .729

23. 3.8 × 49.7 million
  188.86 million

6. 37
  .05 1.85

12. .3
  .3 .09

18. 34.89
  .875 30.52875

In examples involving money, find product correct to nearest cent:

24. $.25
  8 $2.00

25. $3.80
  24 $91.20

26. $3.62
  .06 $.22

27. $4.28
  .125 54¢

28. Find $\frac{1}{2}$ of $.84 $.42
29. Find $\frac{5}{8}$ of $3.45 $2.16
30. Find .46 of 150 69

31. $15.38
  .62$\frac{1}{2}$ $9.61

32. $23.89
  .33$\frac{1}{3}$ $7.96

RELATED PRACTICE EXAMPLES

Multiply:
Basic: Sets 1–25, Prob. 1–5
Average: Sets 1–29, Prob. 1–8
Enriched: Sets 1–35, Prob. 1–10

SET 1

4	6	12	.5	.9
.1	.7	.8	3	5
.4	4.2	9.6	1.5	4.5

SET 2

.23	7	84	.98	39
9	.56	.42	18	.61
2.07	3.92	35.28	17.64	23.79

SET 3

.247	.456	3	.572	28
5	34	.409	159	.717
1.235	15.504	1.227	90.948	20.076

SET 4

.9522	5	46	.1903	8725
8	.4673	.5034	156	.2839
7.6176	2.3365	23.1564	29.6868	2,477.0275

SET 5

25	.95	6	.125	246
.4	20	.35	8	.625
10.0 or 10	19.00 or 19	2.10 or 2.1	1.000 or 1	153.750 or 153.7

SET 6

.03	.05	10	200	.07
8	17	.04	.08	38
.24	.85	.4	16	2.66

SET 7

.04	.03	.02	3	5
2	3	4	.02	.01
.08	.09	.08	.06	.05

SET 8

.006	.012	.009	5	21
3	7	10	.013	.004
.018	.084	.09	.065	.084

SET 9

.001	.0015	.0007	13	4
9	2	5	.0006	.0014
.009	.003	.0035	.0078	.0056

SET 10

8.7	56.17	24	460	3.1416
6	75	3.6	4.8	32
52.2	4,212.75	86.4	2,208	100.5312

SET 11

.9	.5	3.6	3	5.7
.8	.6	.7	8.2	2.5
.72	.3	2.52	24.6	14.25

SET 12

.2	.3	.1	.4	.1
.2	.2	.1	.2	.8
.04	.06	.01	.08	.08

SET 13

.34	.95	.7	8.7	1.15
.3	.4	.66	.48	5.2
.102	.38	.462	4.176	5.98

SET 14

.24	.19	.3	.15	.07
.2	.5	.03	.6	1.3
.048	.095	.009	.09	.091

SET 15

.28	5.93	.45	12.52	256.79
.74	.87	4.91	.06	.03
.2072	5.1591	2.2095	.7512	7.7037

SET 16

.67	.05	.04	.02	2.14
.02	.76	.09	.02	.03
.0134	.038	.0036	.0004	.0642

SET 17

.003	.7	.375	28.2	.009
.4	.002	1.4	.064	36.6
.0012	.0014	.525	1.8048	.3294

SET 18

.368	32.49	70.84	95.26	453.40
.26	.625	.034	1.125	.375
.09568	20.30625	2.40856	107.1675	170.025

SET 19

.0002	.179	.057	3.14	.04
.2	.04	.38	.002	1.225
.00004	.00716	.02166	.00628	.049

SET 20

.268	3.1416	7.9582	8.504	2.423
.924	6.25	1.39	.015	9.146
.247632	19.635	11.061898	.12756	22.160758

SET 21

.00008	.003	.0052	.01	.02167
.6	.009	.05	.0007	1.8
.000048	.000027	.00026	.000007	.039006

SET 22(a)

1. $5 \times .7$ 3.5
2. $8 \times .04$ .32
3. $3 \times .28$ .84
4. $6 \times .005$ .03
5. $.06 \times 4$ .24
6. $9 \times .0001$ .0009
7. $7 \times .02$ .14
8. $10 \times .14$ 1.4
9. 1.2×12 14.4
10. $18 \times .05$ .9

SET 22(b)

1. $.2 \times .3$ .06
2. $.3 \times .04$ .012
3. $.4 \times .35$ .14
4. $.02 \times .09$ .0018
5. $.01 \times .005$ .00005
6. $.004 \times .02$ .00008
7. $.025 \times .008$ .0002
8. $.013 \times .01$ .00013
9. $.04 \times .029$ .00116
10. $.15 \times .060$ .009

SET 23

1. 8×3.1 trillion 24.8 trillion
2. 1.4×5.8 billion 8.12 billion
3. $2.9 \times \$4.53$ million $13.137 million
4. $1.75 \times \$16.5$ billion
5. 0.8×71.35 million
6. 4.25×50.84 billion

4. $28.875 billion 5. 57.08 million 6. 216.07 billion

In examples involving money, find products correct to nearest cent:

SET 24

$.42	$.69	$.80	$.75	$.36
4	18	7	48	15
$1.68	$12.42	$5.60	$36	$5.40

SET 25

$4.97	$5.76	$10.50	$16.31	$4.25
3	24	60	96	144
$14.91	$138.24	$630	$1,565.76	$612

SET 26

$89	$.75	$14.25	$152.80	$293.28
.04	.06	.19	.05	.63
$3.56	$.05	$2.71	$7.64	$184.77

SET 27

$840	$3.88	$15.61	$2500	$675.90
.625	.875	.045	.4375	.002
$525	$3.40	$.70	$1,093.75	$1.35

SET 28	**SET 29**	**SET 30**
Find:	Find:	Find:

SET 28

Find:

1. $\frac{1}{2}$ of $.98 $.49
2. $\frac{1}{2}$ of $4.63 $2.32
3. $\frac{1}{4}$ of .56 .14
4. $\frac{3}{4}$ of $5.20 $3.90
5. $\frac{1}{8}$ of $4.96 $.62
6. $\frac{7}{8}$ of 1.44 1.26
7. $7\frac{1}{3}$ of $.69 $5.06
8. $\frac{5}{6}$ of $12.48 $10.40
9. $2\frac{2}{3}$ of $38.16 $101.76
10. $4\frac{3}{5}$ of $51.40 $236.44

SET 29

Find:

1. $\frac{1}{2}$ of $.39 $.20
2. $\frac{3}{4}$ of .58 .435
3. $\frac{1}{6}$ of $1.22 $.20
4. $\frac{3}{8}$ of $2.58 $.97
5. $\frac{7}{12}$ of $10 $5.83
6. $1\frac{1}{2}$ of $.75 $1.13
7. $2\frac{5}{6}$ of $.83 $2.35
8. $4\frac{2}{5}$ of 3.64 16.016
9. $3\frac{1}{4}$ of $14.20 $46.15
10. $1\frac{2}{3}$ of $27.68 $46.13

SET 30

Find:

1. .25 of 60 15
2. .14 of .36 .0504
3. .04 of 9 .36
4. .08 of 1.5 .12
5. .75 of $4 $3
6. .06 of $93 $5.58
7. .39 of $3.40 $1.33
8. .86 of 9.57
9. .4 of 200
10. .13 of $15.64

8. 8.2302　9. 80　10. $2.03

SET 31

$80	$3,000	$.54	$2.80	$18.48
.12$\frac{1}{2}$	.04$\frac{1}{2}$	.37$\frac{1}{2}$	.03$\frac{1}{4}$	.05$\frac{3}{4}$
$10	$135	$.20	$.09	$1.06

SET 32

$246	$6,000	$.90	$1.15	$45.71
.33$\frac{1}{3}$	.08$\frac{1}{3}$	.04$\frac{2}{3}$	.66$\frac{2}{3}$	.83$\frac{1}{3}$
$82	$500	$.04	$.77	$38.09

PROBLEM SOLVING

See page 16, *Problem Solving Strategy.*

For problems 1–5, select the letter corresponding to your answer.

1. A refrigerator costs $625 cash or $80 down and 12 payments of $52.95 each. How much do you save by paying cash?

　　a. $80.00　　　**b.** $90.40　　　**c.** $52.95　　　**d.** Answer not given

2. A certain plane on a flight used 175.2 liters of gasoline per hour. If its flight lasted 4.5 hours, how many liters of gasoline were consumed?

　　a. 7,884 liters　　**b.** 78.84 liters　　**c.** 7.8840 liters　　**d.** 788.4 liters

3. A vessel heads N. 15° W. for 5 hours at 10.4 knots. Find the distance traveled in nautical miles. (1 knot = 1 nautical m.p.h.)

　　a. 52 nautical miles　　　　　　　**c.** 520 nautical miles
　　b. 5.20 nautical miles　　　　　　**d.** 50.2 nautical miles

4. Find the distance represented by 6.7 centimeters if the scale is 1 centimeter = 50 kilometers.

　　a. 325 kilometers　　　　　　　　**c.** 335 kilometers
　　b. 415 kilometers　　　　　　　　**d.** 300.7 kilometers

5. Two ports are 8.1 inches apart on a chart. If the scale is 1 inch = 40 nautical miles, how many hours will it take a ship going 18 knots to travel between ports?

    a. 16 hours   **b.** 18 hours   **c.** 17 hours   **d.** Answer not given

6. Find the cost of each of the following:

a. 5 kg of onions @ $.49 $2.45   **f.** 15 m² of linoleum @ $12.98 $194.70

b. 6 chairs @ $25.80 $154.80   **g.** 190 gallons of fuel oil @ $1.29 $245.10

c. 4 tires @ $59.90 $239.60   **h.** 3⅝ yd. of fabric @ $4.83 $17.51

d. 3 dozen eggs @ $.79 $2.37   **i.** 13 tons of coal @ $132.90 $1,727.70

e. 2¾ lb. of meat @ $2.29 $6.30   **j.** 24 cans of fruit @ $.69 $16.56

7. Find the cost per year of each of the following services:

a. Electricity, $29.60 per month, $355.20

b. Gas, $35.94 per month, $431.28

c. Telephone, $14.86 per month. $178.32

8. Find the earnings per week (time and a half rate over 40 hours):

Number of hours	32	46	29	43	51½
Rate per hour	$4.96	$8.20	$6.38	$5.15	$10.80
	$158.72	$401.80	$185.02	$229.18	$618.30

9. A finance company can be repaid on a loan of $100 in 6 monthly payments of $18.15, 12 monthly payments of $9.75, or 18 monthly payments of $6.97. Find the amounts paid back and the interest under each plan. $108.90, $8.90; $117, $17; $125.46, $25.46

10. An airplane has a ground speed of 325 knots. What is its ground speed in statute m.p.h. if 1 knot = 1.15 statute m.p.h.? 373.75 stat. m.p.h.

REFRESH YOUR SKILLS

1. Add: 28,104
625
9,302
29
75,412
828   114,300

2. Subtract:
1,500,020
690,175
809,845

3. Multiply:
876
938
821,688

4. Divide: 989
289)285,821

5. Add: $\frac{3}{8}$
$\frac{1}{4}$
$\frac{5}{6}$ $1\frac{11}{24}$

6. Subtract:
$8\frac{7}{12}$
$7\frac{2}{3}$ $\frac{11}{12}$

7. Multiply:
$12 \times 3\frac{7}{8}$
$46\frac{1}{2}$

8. Divide:
$1\frac{11}{16} \div 7\frac{1}{5}$
$\frac{15}{64}$

9. Add:
$.88 + .8 + .888$
2.568

10. Subtract:
$9.6 - .45$
9.15

11. Multiply:
3.1416
48
150.7968

12. Multiply:
$.004 \times 2.5$
.01

OBJECTIVE:
I. Aim   To divide decimals.

II. Procedure

1. If the divisor is a whole number: (See sample solutions 1 to 6.)

a. Divide as in division of whole numbers. (See Exercise 1-7.)

b. Place the decimal point in the quotient directly above the decimal point in the dividend.

2. If the divisor is a decimal: (See sample solutions 7 to 14.)

a. Make the divisor a whole number by moving its decimal point to the right of the last figure, indicating its new position by a caret (∧).

b. Move the decimal point in the dividend to the right as many places as you moved the decimal point in the divisor and indicate its new position by a caret (∧).

c. Divide as in the division of whole numbers and place the decimal point in the quotient directly above the caret (∧) in the dividend.

Observe in the following how the multiplicative identity one (1)* is used to make the divisor a whole number and to allow us to use steps a, b, and c. Both the divisor and dividend are multiplied by the power of ten which makes the divisor a whole number. The caret symbols are used in the sample solutions to indicate the new positions of the decimal points.

$$.12\overline{)0.492} = \frac{.492}{.12} = \frac{.492}{.12} \times 1 = \frac{.492 \times 100}{.12 \times 100} = \frac{49.2}{12} = 12\overline{)49.2}$$

Thus, $.12\overline{)0.492}$ becomes $12\overline{)49.2}$

or using carets, $.12_\wedge\overline{)0.49_\wedge 2}$ becomes $.12_\wedge\overline{)0.49_\wedge 2}$

d. When the dividend contains fewer decimal places than required, annex as many zeros as are necessary to a decimal dividend and a decimal point and the required zeros to a dividend containing a whole number. (See sample solutions 9, 11, 12, 13, and 14.)

3. Check by multiplying the quotient by the divisor and adding the remainder, if any, to the product. The result should equal the dividend. (See Exercise 1-7.)

4. To find the quotient correct to the nearest required decimal place, find the quotient to one more than the required number of decimal places, then round it off as outlined in Exercise 1-22. (See sample solution 13.) An alternate method is to find the quotient to the required number of decimal places, adding 1 to the last figure of the quotient if the remainder

---

* See page 8

162

is equal to one half, or more than one half, of the divisor. (See sample solution 14.)

5. Decimals named by shortened numerals may be divided. (See sample solution 15.)

Concept:

.468 ÷ .02

Since $.468 = \dfrac{468}{1000}$ and $.02 = \dfrac{2}{100}$,

then   or using decimals

$$\frac{468}{1000} \div \frac{2}{100} = \frac{\overset{234}{\cancel{468}}}{\cancel{1000}} \times \frac{\cancel{100}}{\cancel{2}} = \frac{234}{10} = 23\tfrac{4}{10}$$

$$.02_\wedge\overline{)\,.46_\wedge8}\quad\overset{23.4}{}$$

$$23.4 \text{ is the same as } 23\tfrac{4}{10}$$

III. Sample Solutions

1. 37.8 ÷ 6

$$6)\overline{37.8}\quad\overset{6.3}{}$$

Answer: 6.3

2. 9.12 ÷ 8

$$8)\overline{9.12}\quad\overset{1.14}{}$$

Answer: 1.14

3. 3.456 ÷ 4

$$4)\overline{3.456}\quad\overset{.864}{}$$

Answer: .864

4. Divide .015 by 5

$$5)\overline{.015}\quad\overset{.003}{}$$

Answer: .003

5. Divide 21 by 56. Find quotient to 3 decimal places.

$$56)\overline{21.000}\quad\overset{.375}{}$$
$$\underline{16\,8}$$
$$4\,20$$
$$\underline{3\,92}$$
$$280$$
$$\underline{280}$$

Answer: .375

6. Divide $8.50 by 6. Find answer correct to nearest cent.

$$6)\overline{\$8.50}\quad\overset{\$1.41\tfrac{2}{3} = \$1.42}{}$$

Answer: $1.42

7. Divide 35.6 by .4

$$.4_\wedge\overline{)\,35.6_\wedge}\quad\overset{8\,9.}{}$$

Answer: 89

8. .0119 ÷ .7

$$.7_\wedge\overline{)\,.0_\wedge119}\quad\overset{.017}{}$$

Answer: .017

9. $7 ÷ $1.75

$$\$1.75_\wedge\overline{)\,\$7.00_\wedge}\quad\overset{4.}{}$$
$$\underline{7.00}$$

Answer: 4

10. Divide .0006 by .012

$$\begin{array}{r} .05 \\ .012_\wedge\overline{).000_\wedge60} \end{array}$$

Answer: .05

11. 15 ÷ .625

$$\begin{array}{r} 24. \\ .625_\wedge\overline{)15.000_\wedge} \\ 12\ 50 \\ \hline 2\ 500 \\ 2\ 500 \\ \hline \end{array}$$

Answer: 24

12. Divide 122.8 by .004

$$\begin{array}{r} 30\ 700. \\ .004_\wedge\overline{)122.800_\wedge} \end{array}$$

Answer: 30,700

13. Divide 46 by 16.1 and find quotient correct to nearest thousandth.

$$\begin{array}{r} 2.8571 \\ 16.1_\wedge\overline{)46.0_\wedge0000} \\ 32\ 2 \\ \hline 13\ 80 \\ 12\ 88 \\ \hline 920 \\ 805 \\ \hline 1150 \\ 1127 \\ \hline 230 \\ 161 \\ \hline 69 \end{array}$$

Answer: 2.857

14. Divide 28.5 by .87 and find quotient correct to nearest tenth.

$$\begin{array}{r} 32.7 \\ .87\overline{)28.50_\wedge0} \\ 26\ 1 \\ \hline 2\ 40 \\ 1\ 74 \\ \hline 660 \\ 609 \\ \hline 51 \end{array}$$

Answer: 32.8

Since the remainder (51) is more than one half of the divisor (87), 1 is added to the last figure of the quotient.

15. 148.54 billion ÷ 2

$$\begin{array}{r} 74.27\ \text{billion} \\ 2\overline{)148.54\ \text{billion}} \end{array}$$

Answer: 74.27 billion

DIAGNOSTIC TEST A

Divide:

1. $4\overline{)9.2}$ → 2.3

2. $7\overline{)8.96}$ → 1.28

3. $2\overline{)5.328}$ → 2.664

4. $3\overline{).8226}$ → .2742

5. $6\overline{)5.124}$ → .854

6. $8\overline{).736}$ → .092

7. $36\overline{)91.44}$ → 2.54

8. $8\overline{)5.000}$ → .625

9. $200\overline{)4}$ → .02

10. Divide 3.6 by 6 .6

11. Divide 67.12 billion by 4 16.78 billion

12. Find quotient correct to nearest thousandth: $29\overline{)24}$ .828

13. Find answer correct to nearest cent: $12\overline{)\$2.57}$ \$.21

14. 2 ÷ 16 (Find quotient to 3 decimal places.) .125

1-27 DIVISION OF DECIMALS

RELATED PRACTICE EXAMPLES

Divide:

SET 1

$3\overline{)6.9}$ 2.3 $7\overline{)86.1}$ 12.3 $4\overline{)62.4}$ 15.6 $5\overline{)746.5}$ 149.3 $6\overline{)6765.6}$ 1127.6

SET 2

$2\overline{).86}$ .43 $4\overline{)5.88}$ 1.47 $7\overline{)92.96}$ 13.28 $6\overline{)88.02}$ 14.67 $8\overline{)971.44}$ 121.43

SET 3

$8\overline{).968}$ .121 $5\overline{).865}$ .173 $2\overline{)7.942}$ 3.971 $9\overline{)9.486}$ 1.054 $3\overline{)73.914}$ 24.638

SET 4

$4\overline{).8936}$ .2234 $6\overline{).6432}$ .1072 $3\overline{).5814}$ .1938 $5\overline{)6.9185}$ 1.3837 $7\overline{)8.9789}$ 1.2827

SET 5

$3\overline{)2.52}$ .84 $8\overline{)5.216}$ .652 $6\overline{)2.418}$ .403 $4\overline{)3.0612}$ .7653 $9\overline{)7.5519}$ .8391

SET 6

$6\overline{).228}$ .038 $2\overline{).1346}$ .0673 $8\overline{).024}$ .003 $7\overline{).0133}$ .0019 $5\overline{).00375}$ .00075

SET 7

$12\overline{)3.36}$ .28 $48\overline{)158.4}$ 3.3 $24\overline{)11.688}$ .487 $75\overline{)315.75}$ 4.21 $144\overline{)112.32}$ .78

SET 8

$5\overline{)2.0}$ .4 $4\overline{)5.00}$ 1.25 $8\overline{)7.000}$ .875 $16\overline{)9.0000}$ .5625 $32\overline{)13.00000}$ .40625

SET 9

$60\overline{)3}$ .05 $52\overline{)13}$ .25 $200\overline{)6}$ .03 $48\overline{)54}$ 1.125 $7,000\overline{)84}$ .012

SET 10

Divide:

4.2 by 7	.616 by 4	.0474 by 6	.56 by 8	$10.50 by 10
.6	.154	.0079	.07	$1.05

SET 11

1. 125.04 million ÷ 8   15.63 million

2. 16.5 trillion ÷ 3   5.5 trillion

3. $6.945 billion ÷ 5 $1.389 billion

4. $59.49 million ÷ 10 $5.949 million

5. $60.3 billion ÷ $9 billion   6.7

6. 125.8 million ÷ 100 million   1.258

SET 12(a)

Find quotient correct to nearest tenth:

$12\overline{)7}$ .6 $8\overline{)43}$ 5.4 $90\overline{)375}$ 4.2 $57\overline{)845.2}$ 14.8 $108\overline{)20,000}$ 185.2

SET 12(b)

Find quotient correct to nearest hundredth:

$9\overline{)5}$ .56 $6\overline{)38}$ 6.33 $46\overline{)2.8}$ .06 $156\overline{)108}$ .69 $270\overline{)60}$ .22

SET 12(c)

Find quotient correct to nearest thousandth:

$7\overline{)285}$ 40.714 $15\overline{)46}$ 3.067 $12\overline{)365}$ 30.417 $24\overline{)13.59}$ .566 $39\overline{)258.4}$ 6.626

SET 13(a)

Find answer correct to nearest cent:

$3\overline{)\$.72}$ \$.24 $6\overline{)\$5.40}$ \$.90 $12\overline{)\$9.60}$ \$.80 $24\overline{)\$8.16}$ \$.34 $144\overline{)\$11.52}$ \$.08

SET 13(b)

$4\overline{)\$.63}$ \$.16 $6\overline{)\$1.25}$ \$.21 $12\overline{)\$2.80}$ \$.23 $72\overline{)\$234}$ \$3.25 $144\overline{)\$1,056}$ \$7.33

SET 14

.87 ÷ 3 .29   1.072 ÷ 8 .134   9 ÷ 12 (2 decimal places) .75

8 ÷ 7 (nearest thousandth) 1.143   \$2.16 ÷ 5 (nearest cent) \$.43

DIAGNOSTIC TEST B

Divide:

1. $.3\overline{)247.8}$ 826   **4.** $.6\overline{)2.6898}$ 4.483   **7.** $.6\overline{)12.0}$ 20

2. $.5\overline{)9.25}$ 18.5   **5.** $.2\overline{).0034}$ .017   **8.** $.7\overline{)42}$ 60

3. $.8\overline{).896}$ 1.12   **6.** $1.2\overline{)108.72}$ 90.6   **9.** $.4\overline{)2}$ 5

10. 98.6 billion ÷ 2.9   34 billion

11. Find quotient correct to nearest tenth: $2.7\overline{)18}$ 6.7

RELATED PRACTICE EXAMPLES

Divide:

SET 1

$.4\overline{)7.6}$ 19 $.5\overline{)89.5}$ 179 $.3\overline{)176.7}$ 589 $.6\overline{)1804.2}$ 3007 $.2\overline{)6934.6}$ 34,673

$$\overset{2.8}{.3)\overline{.84}} \qquad \overset{23.8}{.2)\overline{4.76}} \qquad \overset{\textbf{SET 2}}{.5)\overline{32.15}}\ \text{64.3} \qquad \overset{125.6}{.7)\overline{87.92}} \qquad \overset{1,832.7}{.4)\overline{733.08}}$$

$$\overset{.42}{.7)\overline{.294}} \qquad \overset{7.56}{.4)\overline{3.024}} \qquad \overset{\textbf{SET 3}}{.8)\overline{9.792}}\ \text{12.24} \qquad \overset{209.76}{.3)\overline{62.928}} \qquad \overset{37.45}{.5)\overline{18.725}}$$

$$\overset{.841}{.5)\overline{.4205}} \qquad \overset{2.376}{.3)\overline{.7128}} \qquad \overset{\textbf{SET 4}}{.6)\overline{5.2524}}\ \text{8.754} \qquad \overset{10.427}{.9)\overline{9.3843}} \qquad \overset{9.246}{.7)\overline{6.4722}}$$

$$\overset{.002}{.8)\overline{.0016}} \qquad \overset{.032}{.7)\overline{.0224}} \qquad \overset{\textbf{SET 5}}{.4)\overline{.0028}}\ \text{.007} \qquad \overset{.003}{.3)\overline{.0009}} \qquad \overset{.092}{.6)\overline{.0552}}$$

$$\overset{24.3}{1.8)\overline{43.74}} \qquad \overset{5.76}{2.6)\overline{14.976}} \qquad \overset{\textbf{SET 6}}{3.5)\overline{3041.5}}\ \text{869} \qquad \overset{5.2}{19.6)\overline{101.92}} \qquad \overset{.364}{24.3)\overline{8.8452}}$$

$$\overset{35}{.4)\overline{14.0}} \qquad \overset{25}{.8)\overline{20.0}} \qquad \overset{\textbf{SET 7}}{2.8)\overline{49.00}}\ \text{17.5} \qquad \overset{16.25}{6.4)\overline{104.000}} \qquad \overset{9.375}{1.6)\overline{15.0000}}$$

$$\overset{30}{.5)\overline{15}} \qquad \overset{40}{.6)\overline{24}} \qquad \overset{\textbf{SET 8}}{.4)\overline{72}}\ \text{180} \qquad \overset{20}{1.8)\overline{36}} \qquad \overset{30}{4.2)\overline{126}}$$

$$\overset{5}{.6)\overline{3}} \qquad \overset{2.5}{.8)\overline{2}} \qquad \overset{\textbf{SET 9}}{5.6)\overline{14}}\ \text{2.5} \qquad \overset{1.25}{7.2)\overline{9}} \qquad \overset{3.125}{12.8)\overline{40}}$$

SET 10

1. 3.44 trillion ÷ .8 4.3 trillion
2. $48.5 billion ÷ .5 $97 billion
3. $7.75 trillion ÷ 3.1 $2.5 trillion

4. 9.259 million ÷ 4.7 1.97 million
5. 5.76 trillion ÷ 9.6 trillion .6
6. $73.6 billion ÷ $2.3 billion 32

SET 11(a)

Find quotient correct to nearest tenth:

$$\overset{}{8)\overline{9}}\ \text{1.1} \qquad \overset{}{6.3)\overline{15}}\ \text{2.4} \qquad \overset{}{7.5)\overline{456.2}}\ \text{60.8} \qquad \overset{3.2}{25.4)\overline{82.25}} \qquad \overset{2,272.7}{2.2)\overline{5,000}}$$

SET 11(b)

Find quotient correct to nearest hundredth:

$$\overset{}{.3)\overline{2}}\ \text{6.67} \qquad \overset{}{1.4)\overline{6}}\ \text{4.29} \qquad \overset{}{5.4)\overline{.48}}\ \text{.09} \qquad \overset{2.73}{13.2)\overline{36}} \qquad \overset{.58}{8.4)\overline{4.9}}$$

SET 11(c)

Find quotient correct to nearest thousandth:

$$\overset{}{.8)\overline{.45}}\ \text{.563} \qquad \overset{}{2.6)\overline{740}}\ \text{284.615} \qquad \overset{}{3.9)\overline{85.3}}\ \text{21.872} \qquad \overset{35.088}{5.7)\overline{200}} \qquad \overset{435.733}{7.5)\overline{3,268}}$$

DIAGNOSTIC TEST C

Divide:

1. $.02\overline{)521.56}$ 26,078

4. $.07\overline{).89789}$ 12.827

7. $.16\overline{)48.00}$ 300

2. $.79\overline{)4.661}$ 5.9

5. $.03\overline{).0009}$ .03

8. $.39\overline{)265.2}$ 680

3. $.56\overline{)2.4472}$ 4.37

6. $1.44\overline{)135.072}$ 93.8

9. $.25\overline{)50}$ 200

10. Find quotient correct to nearest thousandth: $.96\overline{)8.8}$ 9.167

11. $\$.04\overline{)\$1.84}$ 46

RELATED PRACTICE EXAMPLES

Divide:

		SET 1		
17	251		53	648
$.04\overline{).68}$	$.06\overline{)15.06}$	$.65\overline{)17.55}$ 27	$.32\overline{)16.96}$	$.57\overline{)369.36}$

		SET 2		
11.6	258.3		96.7	54.2
$.07\overline{).812}$	$.03\overline{)7.749}$	$.96\overline{)6.912}$ 7.2	$.25\overline{)24.175}$	$.81\overline{)43.902}$

		SET 3		
234.77	9.02		4.58	.63
$.02\overline{)4.6954}$	$.08\overline{).7216}$	$.14\overline{).4732}$ 3.38	$.43\overline{)1.9694}$	$.75\overline{).4725}$

		SET 4		
.992	3.085		.949	3.671
$.09\overline{).08928}$	$.05\overline{).15425}$	$.22\overline{).21692}$ .986	$.87\overline{).82563}$	$.36\overline{)1.32156}$

		SET 5		
.03	.05		.062	.009
$.06\overline{).0018}$	$.01\overline{).0005}$	$.03\overline{).00012}$ .004	$.68\overline{).04216}$	$.42\overline{).00378}$

		SET 6		
57	38.9		48.6	3.48
$3.65\overline{)208.05}$	$2.27\overline{)88.303}$	$4.24\overline{)7.0808}$ 1.67	$8.32\overline{)404.352}$	$5.39\overline{)18.7572}$

		SET 7		
1,900	50		87.5	62.5
$.04\overline{)76.00}$	$.36\overline{)18.00}$	$.52\overline{)39.00}$ 75	$.64\overline{)56.000}$	$1.92\overline{)120.000}$

		SET 8		
720	70		2,530	1,750
$.08\overline{)57.6}$	$.65\overline{)45.5}$	$.78\overline{)670.8}$ 860	$1.47\overline{)3,719.1}$	$2.83\overline{)4,952.5}$

		SET 9		
300	500		37.5	87.5
$.09\overline{)27}$	$.18\overline{)90}$	$.64\overline{)16}$ 25	$.32\overline{)12}$	$1.36\overline{)119}$

SET 10(a)

Find correct to nearest tenth:

$1.15\overline{)825}$ 717.4　$.87\overline{)79.4}$ 91.3　$3.14\overline{)853.76}$ 271.9　$.09\overline{)56.2}$ 624.4　$39.37\overline{)432}$ 11.0

SET 10(b)

Find correct to nearest hundredth:

$.06\overline{)4}$ 66.67　$.84\overline{)70}$ 83.33　$.37\overline{)2.9}$ 7.84　$1.53\overline{)34}$ 22.22　$2.21\overline{)13.67}$ 6.19

SET 10(c)

Find correct to nearest thousandth:

$.54\overline{)98}$ 181.481　$.69\overline{)8.45}$ 12.246　$.26\overline{)42.7}$ 164.231　$5.03\overline{)61}$ 12.127　$2.43\overline{).162}$ .067

SET 11(a)

Find correct to nearest cent:

$\$.02\overline{)\$3.46}$ 173　$\$.05\overline{)\$4}$ 80　$\$.38\overline{)\$27.36}$ 72　$\$.40\overline{)\$57.60}$ 144　$\$1.25\overline{)\$30}$ 24

(To nearest hundredth)　　　**SET 11(b)**

$\$.08\overline{)\$1.24}$ 15.5　$\$.25\overline{)\$3.90}$ 15.6　$\$.49\overline{)\$10.43}$ 21.29　$\$1.80\overline{)\$44.20}$ 24.56　$\$2.67\overline{)\$97.90}$ 36.67

DIAGNOSTIC TEST D

Divide:

1. $.006\overline{)74.898}$ 12,483　　4. $.231\overline{).00924}$ .04　　7. $.125\overline{)53.75}$ 430

2. $.018\overline{).4554}$ 25.3　　5. $4.375\overline{)11.8125}$ 2.7　　8. $.052\overline{)452.4}$ 8,700

3. $.007\overline{)6.53912}$ 934.16　　6. $.048\overline{)60.000}$ 1,250　　9. $.014\overline{)112}$ 8,000

10. Find quotient correct to nearest hundredth: $.333\overline{)249}$ 747.75

RELATED PRACTICE EXAMPLES

Divide:

SET 1

$.007\overline{).763}$ 109　$.003\overline{)17.862}$ 5,954　$.043\overline{)1.634}$ 38　$.175\overline{)49.525}$ 283　$.526\overline{)77.322}$ 147

SET 2

$.006\overline{)6.0282}$ 1,004.7　$.035\overline{).1715}$ 4.9　$.216\overline{)8.1864}$ 37.9　$.024\overline{)1.2144}$ 50.6　$.382\overline{)8.9388}$ 23.4

SET 3

$.004\overline{).05964}$ 14.91　$.073\overline{).31828}$ 4.36　$.524\overline{).36156}$ .69　$.358\overline{).86278}$ 2.41　$.449\overline{)1.50846}$ 3.36

SET 4

$$\overset{.09}{.009\overline{)\,.00081}} \qquad \overset{.05}{.018\overline{)\,.0009}} \qquad \overset{.07}{.105\overline{)\,.00735}} \qquad \overset{.04}{.096\overline{)\,.00384}} \qquad \overset{.06}{.667\overline{)\,.04002}}$$

SET 5

$$\overset{3.2}{2.548\overline{)\,8.1536}} \qquad \overset{.23}{4.125\overline{)\,.94875}} \qquad \overset{5.7}{6.875\overline{)\,39.1875}} \qquad \overset{1.48}{3.002\overline{)\,4.44296}}$$

$$\overset{24.2}{2.383\overline{)\,57.6686}}$$

SET 6

$$\overset{2,000}{.008\overline{)\,16.000}} \qquad \overset{2,500}{.056\overline{)\,140.000}} \qquad \overset{750}{.144\overline{)\,108.000}} \qquad \overset{56}{.625\overline{)\,35.000}} \qquad \overset{96}{1.375\overline{)\,132.000}}$$

SET 7

$$\overset{270}{.094\overline{)\,25.38}} \qquad \overset{11,270}{.008\overline{)\,90.16}} \qquad \overset{640}{.231\overline{)\,147.84}} \qquad \overset{2,340}{.382\overline{)\,893.88}} \qquad \overset{860}{.292\overline{)\,251.12}}$$

SET 8

$$\overset{87,700}{.005\overline{)\,438.5}} \qquad \overset{9,600}{.072\overline{)\,691.2}} \qquad \overset{8,900}{.043\overline{)\,382.7}} \qquad \overset{5,400}{.337\overline{)\,1819.8}} \qquad \overset{48,300}{.265\overline{)\,12799.5}}$$

SET 9

$$\overset{3,000}{.059\overline{)\,177}} \qquad \overset{9,000}{.108\overline{)\,972}} \qquad \overset{30,000}{.007\overline{)\,210}} \qquad \overset{4,000}{.591\overline{)\,2,364}} \qquad \overset{6,000}{.473\overline{)\,2,838}}$$

SET 10(a)

Find quotient correct to nearest tenth:

$$\overset{291.5}{.542\overline{)\,158}} \qquad \overset{183.3}{.003\overline{)\,.55}} \qquad \overset{229.5}{.061\overline{)\,14}} \qquad \overset{5.2}{1.853\overline{)\,9.647}} \qquad \overset{421.5}{.868\overline{)\,365.9}}$$

SET 10(b)

Find quotient correct to nearest hundredth:

$$\overset{4,166.67}{.006\overline{)\,25}} \qquad \overset{4,330.77}{.013\overline{)\,56.3}} \qquad \overset{27.04}{.314\overline{)\,8.49}} \qquad \overset{3,772.15}{.079\overline{)\,298}} \qquad \overset{1.63}{.592\overline{)\,.964}}$$

SET 10(c)

Find quotient correct to nearest thousandth:

$$\overset{4,714.286}{.007\overline{)\,33}} \qquad \overset{7.643}{.056\overline{)\,.428}} \qquad \overset{86.445}{.723\overline{)\,62.5}} \qquad \overset{149.517}{2.381\overline{)\,356}} \qquad \overset{9.546}{4.007\overline{)\,38.25}}$$

PROBLEM SOLVING

See page 16, *Problem Solving Strategy.*

For problems 1–5, select the letter corresponding to your answer.

1. An airplane flies 1,495.2 kilometers in 2.8 hours. What is its average ground speed?

    a. 534 km/h   **b.** 604 km/h   **c.** 594 km/h   **d.** Answer not given

2. The neighborhood baseball team bought 12 balls at $3.81 each and 9 bats at $5.95 each. If the 9 boys and girls shared the costs equally, how much was each person's share?

   a. $10.93   **b.** $12.15   **c.** $11.03   **d.** $11.15

3. A butcher charged $9.66 for a certain cut of meat at $2.24 a pound. What was the weight of the meat?

   a. 5 lb. 3 oz.   **b.** 4 lb. 11 oz.   **c.** 4 lb. 5 oz.   **d.** Answer not given

4. The fuel consumption of a certain airplane is 214.5 liters per hour. Find the number of hours the airplane can fly if its gas tank holds 1,287 liters.

   a. 5 hours   **b.** 5.5 hours   **c.** 5.75 hours   **d.** 6 hours

5. The telephone rates between 2 zones is $.25 for the first 3 minutes and $.05 for each additional minute. How long was a call for which the charges were $.55?

   a. 6 minutes   **b.** 8 minutes   **c.** 9 minutes   **d.** Answer not given

6. Find amount saved on each can when buying in quantity:

a. 1 dozen cans of peas for $4.50 or $.43 each? $.66

b. 3 cans of corn for $1.00 or $.39 each? $.17

c. 1 case of 24 cans of soup for $6.00 or $.29 each? $.96

7. a. At $.79 per kilogram, how many kilograms of apples can you buy for $3.95? 5 kilograms

   b. At $1.25 per dozen, how many dozen oranges can you buy for $7.50? 6 dozen

   c. At 5 kilograms for $1.49, how many kilograms of potatoes can you buy for $5.96? 20 kilograms

8. a. If 6 dozen pencils cost $4.32, what will one pencil cost? $.06

   b. If 5 chairs cost $32.25, how much will 8 chairs cost? $51.60

   c. If a dozen oranges cost $.90, what will 14 oranges cost? $1.05

9. A student measured a 2-inch metal block by a precision instrument and made the following readings: 1.9962 inches, 2.0008 inches, 2.0012 inches, 1.9993 inches, and 2.0015 inches. Find the average reading and the amount of error. 1.9998 in.; .0002 in.

10. The rates of a local laundry are $5.45 for the first 15 pieces and $.30 for each additional piece. If the laundry charged $7.25, how many pieces were laundered? 21 pieces

11. If the charge for a telephone call between two cities was $2.12 at the rate of $.52 for the first minute and $.32 for each additional minute, how long was the call? 6 min.

12. If the postage for sending a certain package by first class mail was $1.56 at the rate of $.20 for the first ounce and $.17 for each additional ounce, how much did the package weigh? 9 ounces

REFRESH YOUR SKILLS

1. Add:

216,958
133,425
590,643
614,299
271,086
484,137   2,310,548

2. Subtract:

8,304,060
8,293,851
————
10,209

3. Multiply:

3,600
3,600
————
12,960,000

4. Divide:

$$5,280\overline{)2,101,440}\quad 398$$

5. Add:

$1\frac{2}{3}$
$\frac{5}{6}$ $2\frac{1}{2}$

6. Subtract:

$15\frac{7}{12}$
9
—— $6\frac{7}{12}$

7. Multiply:

$\frac{3}{4} \times 2\frac{5}{8}$ $1\frac{31}{32}$

8. Divide:

$32 \div 6\frac{2}{3}$
$4\frac{4}{5}$

9. Add:

$1.4 + .06 + .8$ 2.26

10. Subtract: $600 - 9.08$ 590.92

11. Multiply: $14.4 \times .09$ 1.296

12. Divide: $.02\overline{)1}$ 50

13. Round 48,263,506 to the nearest hundred thousand.   48,300,000

14. Express $\frac{64}{96}$ in lowest terms. $\frac{2}{3}$

15. Change: $\dfrac{11}{12} = \dfrac{?}{60}$ 55

16. What part of 63 is 36? $\frac{4}{7}$

17. $\frac{5}{8}$ of what amount is $14.75? $23.60

18. Write 5.82 *billion* as a complete numeral.   5,820,000,000

19. Round 96.0194 to the nearest hundredth.   96.02

20. a. Is the statement $\dfrac{4}{7} < \dfrac{3}{5}$ true?   Yes

    b. Is the statement .029 > .03 true?   No

UNITED STATES MONEY—REVIEW

1. Add:

$.41 + $.27 $.68 $2 + $.06 $2.06 $8.59 + $.94 $9.53 $6.37 + $15.69 $22.06

2. Subtract:

$.90 − $.71 $.19 $4.85 − $3.49 $1.36 $6 − $.65 $5.35 $8.42 − $.87 $7.55

3. Multiply (find correct to nearest cent):

$80 \times $.05$ $4.00

$18.75
$\times\ 27$ $506.25

$46.92
$.03$ $1.41

$\frac{7}{8}$ of $6.52 $5.71

$91.50
$.04\frac{1}{2}$
————
$4.12

4. Divide:

$8\overline{)\$.96}$ $.75 $.12

$16\overline{)\$12}$ $.75

$.45\overline{)\$10.35}$ 23

$.09\overline{)\$27}$ 300

$1.75\overline{)\$14}$ 8

Multiplying Whole Numbers and Decimals by 10, 100, 1,000, etc.

OBJECTIVE:

I. Aim To multiply whole numbers and decimals by 10, 100, 1,000, etc.

II. Procedure

1. To multiply a whole number by 10, 100, 1,000, etc., annex as many zeros to the right of the given numeral as there are zeros in the given multiplier.
   a. If multiplied by 10, annex one zero.
   b. If multiplied by 100, annex two zeros.
   c. If multiplied by 1,000, annex three zeros, etc.
2. To multiply a decimal by 10, 100, 1,000, etc., first write the figures of the given numeral, then move the decimal point as many places to the right of its original position as there are zeros in the given multiplier.
   a. If multiplied by 10, move decimal point 1 place to the right.
   b. If multiplied by 100, move decimal point 2 places to the right.
   c. If multiplied by 1,000, move decimal point 3 places to the right, etc.

III. Sample Solutions

1. a. $10 \times 7 = 70$ *Answer: 70*
  b. $10 \times .6 = 6$ *Answer: 6*
  c. $10 \times .05 = .5$ *Answer: .5*
  d. $10 \times 2.741 = 27.41$ *Answer: 27.41*
2. a. $100 \times 80 = 8,000$ *Answer: 8,000*
  b. $100 \times .6 = 60$ *Answer: 60*
  c. $100 \times .05 = 5$ *Answer: 5*
  d. $100 \times 2.741 = 274.1$ *Answer: 274.1*
3. a. $1,000 \times 9 = 9,000$ *Answer: 9,000*
  b. $1,000 \times .6 = 600$ *Answer: 600*
  c. $1,000 \times .05 = 50$ *Answer: 50*
  d. $1,000 \times 2.741 = 2,741$ *Answer: 2,741*

DIAGNOSTIC TEST

In examples 1 through 6, multiply the given numbers by 10:

1. 8 ^80^   **2.** 60 ^600^   **3.** .4 ^4^   **4.** .78 ^7.8^   **5.** .06 ^.6^   **6.** 25.324 ^253.24^

In examples 7 through 12, multiply the given numbers by 100:

7. 26 ^2600^   **8.** 500 ^50,000^   **9.** .83 ^83^   **10.** .5 ^50^   **11.** .0987 ^9.87^   **12.** 67.39 ^6,739^

In examples 13 through 18, multiply the given numbers by 1,000:

13. 9   **14.** 420   **15.** .365   **16.** .67   **17.** .8574   **18.** 56.967

9,000   420,000   365   670   857.4   56,967

RELATED PRACTICE EXAMPLES

In sets 1 through 6, multiply the given numbers by 10:

SET 1	SET 2	SET 3	SET 4	SET 5	SET 6
1. 5 50	1. 40 400	1. .3 3	1. .26 2.6	1. .03 .3	1. 5.8 58
2. 9 90	2. 80 800	2. .1 1	2. .57 5.7	2. .09 .9	2. 34.6 346
3. 27 270	3. 100 1,000	3. .5 5	3. .924 9.24	3. .085 .85	3. 96.34 963.4
4. 85 850	4. 150 1,500	4. .9 9	4. .805 8.05	4. .007 .07	4. 49.927 499.27
5. 763 7,630	5. 3,000 30,000	5. .8 8	5. .4326 4.326	5. .0625 .625	5. 540.653 5,406.53

In sets 7 through 12, multiply the given numbers by 100:

SET 7	SET 8	SET 9	SET 10	SET 11	SET 12
1. 7 700	1. 20 2,000	1. .42 42	1. .2 20	1. .721 72.1	1. 8.54 854
2. 51 5,100	2. 60 6,000	2. .33 33	2. .8 80	2. .039 3.9	2. 36.46 3,646
3. 38 3,800	3. 300 30,000	3. .19 19	3. .7 70	3. .5257 52.57	3. 72.03 7,203
4. 423 42,300	4. 590 59,000	4. .67 67	4. .9 90	4. .0416 4.16	4. 5.792 579.2
5. 9,564 956,400	5. 7,400 740,000	5. .95 95	5. .1 10	5. .8547 85.47	5. 25.875 2,587.5

In sets 13 through 18, multiply the given numbers by 1,000:

SET 13	SET 14	SET 15	SET 16	SET 17	SET 18
1. 3 3,000	1. 50 50,000	1. .657 657	1. .35 350	1. .2653 265.3	1. 6.582 6,582
2. 62 62,000	2. 90 90,000	2. .942 942	2. .7 700	2. .5929 592.9	2. 29.37 29,370
3. 597 597,000	3. 200 200,000	3. .485 485	3. .09 90	3. .0357 35.7	3. 81.1 81,100
4. 2,055 2,055,000	4. 780 780,000	4. .076 76	4. .4 400	4. .17425 174.25	4. 176.2563 176,256.3
5. 4,682 4,682,000	5. 5,000 5,000,000	5. .189 189	5. .98 980	5. .00072 .72	5. 250.40575 250,405.75

PRACTICAL APPLICATIONS

1. To change kilowatts (kW) to watts, multiply the number of kilowatts by 1,000. Change to watts:
   a. 8 kW 8,000 b. 40 kW 40,000 c. 6.5 kW 6,500 d. 139 kW 139,000 e. 27.3 kW 27,300

2. To change centimeters (cm) to millimeters, multiply the number of centimeters by 10. Change to millimeters:
   a. 3 cm 30 mm b. 15 cm 150 mm c. 4.2 cm 42 mm d. 87 cm 870 mm e. 18.5 cm 185 mm

3. To change meters (m) to centimeters, multiply the number of meters by 100. Change to centimeters:
   a. 5 m 500 cm b. 20 m 2,000 cm c. 3.7 m 370 cm d. 15.08 m 1,508 cm e. 9.385 m 938.5 cm

4. a. How many dimes are in 28 dollars? 280 dimes
   b. Find the number of pennies in 17 dollars. 1,700 pennies

5. A fuel oil consumer bought 100 liters of heating oil at $.328 per liter. How much did the oil cost? $32.80

Dividing Whole Numbers and Decimals by 10, 100, 1,000, etc.

OBJECTIVE:

I. Aim To divide whole numbers and decimals by 10, 100, 1,000, etc.

II. Procedure

1. First write the numeral for the given whole number or decimal.

2. Then move the decimal point as many places to the left of its original position as there are zeros in the given divisor. In a numeral naming a whole number the decimal point is understood after the units figure.

  a. If divided by 10, move decimal point 1 place to the left.

  b. If divided by 100, move decimal point 2 places to the left.

  c. If divided by 1,000, move decimal point 3 places to the left, etc.

III. Sample Solutions

1. **a.** $2,500 \div 10 = 250$ *Answer:* 250
   b. $.5 \div 10 = .05$ *Answer:* .05
   c. $128.4 \div 10 = 12.84$ *Answer:* 12.84
   d. $67 \div 10 = 6.7$ *Answer:* 6.7

2. **a.** $2,500 \div 100 = 25$ *Answer:* 25
   b. $5 \div 100 = .05$ *Answer:* .05
   c. $128.4 \div 100 = 1.284$ *Answer:* 1.284
   d. $431 \div 100 = 4.31$ *Answer:* 4.31

3. **a.** $2,500 \div 1,000 = 2.5$ *Answer:* 2.5
   b. $.75 \div 1,000 = .00075$ *Answer:* .00075
   c. $128.4 \div 1,000 = .1284$ *Answer:* .1284
   d. $9,678 \div 1,000 = 9.678$ *Answer:* 9.678

4. **a.** $6,400,000 \div 1,000,000 = 6.4$ *Answer:* 6.4
   b. $9,200,000,000 \div 1,000,000,000 = 9.2$ *Answer:* 9.2

DIAGNOSTIC TEST

In examples 1 through 5, divide the given numbers by 10:

1. 80 8	**3.** 6 .6	**5.** 15.683
2. 95 9.5	**4.** .7 .07	1.5683

In examples 6 through 11, divide the given numbers by 100:

6. 400 4	**8.** 92 .92	**10.** .34 .0034
7. 875 8.75	**9.** 8 .08	**11.** 197.2 1.972

In examples 12 through 17, divide the given numbers by 1,000:

12. 65,000 65	**14.** 467 .467	**16.** .675 .000675
13. 2,973 2.973	**15.** 72 .072	**17.** 527.3 .5273

RELATED PRACTICE EXAMPLES

In sets 1 through 5, divide the given numbers by 10:

SET 1	SET 2	SET 3	SET 4	SET 5
1. 20 2	1. 34 3.4	1. 5 .5	1. .2 .02	1. 3.5 .35
2. 50 5	2. 79 7.9	2. 3 .3	2. .9 .09	2. 9.82 .982
3. 600 60	3. 276 27.6	3. 9 .9	3. .32 .032	3. 27.46 2.746
4. 300 30	4. 408 40.8	4. 7 .7	4. .08 .008	4. 39.239
5. 1,000 100	5. 5,426 542.6	5. 4 .4	5. .936 .0936	5. 145.948

4. 3.9239 5. 14.5948

In sets 6 through 11, divide the given numbers by 100:

SET 6	SET 7	SET 8	SET 9	SET 10	SET 11
1. 200 2	1. 382 3.82	1. 59 .59	1. 4 .04	1. .21 .0021	1. 29.5 .295
2. 700 7	2. 829 8.29	2. 32 .32	2. 9 .09	2. .60 .006	2. 128.93 1.2893
3. 3,000 30	3. 4,520 45.2	3. 67 .67	3. 2 .02	3. .8 .008	3. 502.86 5.0286
4. 5,000 50	4. 3,948 39.48	4. 70 .7	4. 3 .03	4. .045 .00045	4. 68.24 .6824
5. 2,700 27	5. 65,726 657.26	5. 85 .85	5. 7 .07	5. .914 .00914	5. 1,500.75 15.0075

In sets 12 through 17, divide the given numbers by 1,000:

SET 12	SET 13	SET 14	SET 15	SET 16	SET 17
1. 8,000 8	1. 3,725 3.725	1. 628 .628	1. 85 .085	1. .925 .000925	1. 284.9 .2849
2. 2,000 2	2. 2,890 2.89	2. 314 .314	2. 93 .093	2. .56 .00056	2. 500.74 .50074
3. 10,000 10	3. 8,575 8.575	3. 200 .2	3. 6 .006	3. .3 .0003	3. 795.82 .79582
4. 28,000 28	4. 15,925	4. 476 .476	4. 8 .008	4. .072 .000072	4. 1,526.1
5. 150,000 150	5. 18,464	5. 957 .957	5. 38 .038	5. .89 .00089	5. 2,963.45

4. 15.925 5. 18.464

4. 1.5261
5. 2.96345

PRACTICAL APPLICATIONS

1. To change from watt-hours (W · h) to kilowatt-hours, divide the number of watt-hours by 1,000. Change to kilowatt-hours: 27.5 kWh
  a. 20,000 W · h 20 kWh   **b.** 135,000 W · h 135 kWh   **c.** 27,500 W · h

2. A bag of sugar, weighing 100 pounds, costs $18.30. What is the cost per pound? 18.3¢

3. Taxes

  a. If the tax rate is $4.80 per hundred dollars, how much must you pay for taxes on a house assessed for $8,900? $427.20

  b. Find the amount of taxes on properties having the following assessed valuations and tax rates:

Assessed valuation	$3,400	$17,500	$51,800	$45,000	$30,200
Tax rate per $100	$5.70	$6.45	$3.60	$4.25	$5.84
	$193.80	$1,128.75	$1,864.80	$1,912.50	$1,763.68

Reducing the rate; $1.35

c. A house is assessed for $6,300 and the tax rate is $2.40 per $100. Which plan will give a greater reduction in taxes, reducing the assessment to $5,700 or reducing the rate to $2.15 per $100? How much greater?

4. Life Insurance

a. What annual premium should a person pay on a $5,000 policy when the rate is $29.18 per $1,000? $145.90

b. Find the annual premium on the following amounts of life insurance at the given annual rates per $1,000:

Face of Policy	$3,000	$2,000	$10,000	$5,000	$12,000
Rate per $1,000	$43.95	$18.40	$33.07	$47.72	$52.81

5. Fire Insurance $131.85 $36.80 $330.70 $238.60 $633.72

a. Mrs. Daniels wishes to insure her house for $34,000. The rate charged by a fire insurance company is $.24 per $100. What is the annual premium? $81.60

b. Find the annual premium on the following amounts of fire insurance at the given annual rates per $100:

Face of Policy	$30,000	$9,000	$25,000	$47,500	$32,800
Rate per $100	$54 $.18	$67.50 $.75	$.46	$.29	$1.05

6. Interest—6% (or 60-day) Method $115.00 $137.75 $344.40

To find the interest on a sum of money borrowed or invested (principal) for 60 days at 6%, divide the principal by 100.

a. Find the interest for 60 days at 6% on the following given principals: (1) $2,000 $20 (2) $580 $5.80 (3) $45 $.45 (4) $1,325   (5) $769 $7.69
$13.25

b. Find the interest on:

(1) $1,000 for 30 days at 6% $5.00   (3) $1,200 for 30 days at 12% $12.00

(2) $500 for 30 days at 3% $1.25   (4) $4,000 for 90 days at 6% $60.00

REFRESH YOUR SKILLS

1. Add: **2.** Subtract: **3.** Multiply: **4.** Divide: 8,009   **5.** Add:

820,142   4,258,003   8,039 $962{\overline{)7,704,658}}$ $4\frac{11}{12}$

963,819   3,167,905   9,670   77,737,130 $3\frac{4}{5}$

610,387   _____   **8.** Divide: $8\frac{43}{60}$

241,299   1,090,098   **6.** Subtract: **7.** Multiply:

872,175   20 $1\frac{7}{8} \times 3\frac{3}{4}$ $7\frac{1}{32}$ $9\frac{1}{7} \div 4\frac{4}{5}$ $1\frac{19}{21}$

_____

3,507,822 $6\frac{7}{10}$

_____

9. Add: $13\frac{3}{10}$ **10.** Subtract: **11.** Multiply: **12.** Divide:

.638 + 2.47 + 92.5 $4.68 − $3   500 × .002 $.08{\overline{)$40}}$

95.608 $1.68   1   500

Changing Common Fractions to Decimals*

OBJECTIVE:

I. Aim To change a common fraction or mixed number to a decimal.

II. Procedure

1. To change a common fraction to a decimal (3 methods):

a. Divide the numerator by the denominator. (See sample solutions 1 and 4.)

b. Or, if the decimal equivalent is known, write the decimal numeral directly. Sometimes it may be necessary first to reduce the given fraction to lowest terms.

c. Or, if the given fraction has as its denominator 10, 100, 1,000, etc., drop the denominator and rewrite the numerator, placing a decimal point in the proper location. (See sample solutions 2 and 3.)

2. To change a mixed number to a mixed decimal:

Change the fraction to a decimal and annex it to the numeral for the whole number. (See sample solution 6.)

III. Sample Solutions

1. Change $\frac{2}{5}$ to a decimal.

$$\frac{2}{5} = 5\overline{)2.0}^{.4}$$

Answer: .4

2. Change $\frac{19}{100}$ to a decimal.

$$\frac{19}{100} = .19$$

Answer: .19

3. Change $\frac{165}{100}$ to a decimal.

$$\frac{165}{100} = 1.65$$

Answer: 1.65

4. Change $\frac{6}{14}$ to a decimal (2 places).

$$\frac{6}{14} = \frac{3}{7} = 7\overline{)3.00}^{.42\frac{6}{7}}$$

Answer: $.42\frac{6}{7}$

5. Change $\frac{5}{4}$ to a decimal.

$$\frac{5}{4} = 4\overline{)5.00}^{1.25}$$

Answer: 1.25

6. Change $1\frac{7}{8}$ to a decimal.

$$\frac{7}{8} = .87\frac{1}{2} \text{ or } .875$$
$$1\frac{7}{8} = 1.87\frac{1}{2} \text{ or } 1.875$$

Answer: $1.87\frac{1}{2}$ or 1.875

*See Exercise 1-61.

Related Resources: *Practical Applications in Mathematics:* Diag. Test and Prac. Ex., p. 50

DIAGNOSTIC TEST

Express the following common fractions or mixed numbers as decimals: (Carry out examples 3, 4, and 6 through 16 to 2 places.)

1. $\frac{9}{10}$ .9   **4.** $\frac{1}{4}$ .25   **7.** $\frac{125}{100}$ 1.25   **10.** $\frac{20}{25}$ .8 or .80   **13.** $1\frac{3}{4}$ 1.75   **16.** $2\frac{7}{16}$ 2.43$\frac{3}{4}$

2. $\frac{1}{2}$ .5   **5.** $\frac{7}{8}$ .87$\frac{1}{2}$ **8.** $\frac{37\frac{1}{2}}{100}$ .37$\frac{1}{2}$ **11.** $\frac{49}{56}$ .87$\frac{1}{2}$ **14.** $\frac{12}{8}$ 1.5 or 1.50   **17.** $\frac{893}{1000}$ .893

3. $\frac{27}{100}$ .27   **6.** $\frac{5}{6}$ .83$\frac{1}{3}$ **9.** $\frac{8}{9}$ .88$\frac{8}{9}$ **12.** $\frac{24}{28}$ .85$\frac{5}{7}$ **15.** $\frac{18}{14}$ 1.28$\frac{4}{7}$ **18.** $\frac{7429}{10000}$ .7429

Basic: Sets 1–13, Misc. Ex. 1–2, Appl. 1–2; *Average:* Sets 1–16, Misc. Ex. 1–4, Appl. 1–4; *Enriched:* Sets 1–18, Misc. Ex. 1–5, Appl. 1–5

RELATED PRACTICE EXAMPLES

Express the following common fractions or mixed numbers as decimals: (Carry out sets 3, 4, and 6 through 16 to 2 decimal places.)

SET 1	SET 2	SET 3	SET 4	SET 5	SET 6
1. $\frac{1}{10}$ .1	**1.** $\frac{4}{5}$ .8	**1.** $\frac{39}{100}$ .39	**1.** $\frac{3}{4}$ .75	**1.** $\frac{3}{8}$ .37$\frac{1}{2}$	**1.** $\frac{1}{3}$ .33$\frac{1}{3}$
2. $\frac{7}{10}$ .7	**2.** $\frac{1}{5}$ .2	**2.** $\frac{54}{100}$ .54	**2.** $\frac{17}{20}$ .85	**2.** $\frac{5}{8}$ .62$\frac{1}{2}$	**2.** $\frac{1}{6}$ .16$\frac{2}{3}$
3. $\frac{2}{10}$ .2	**3.** $\frac{1}{2}$ .5	**3.** $\frac{3}{100}$ .03	**3.** $\frac{14}{25}$ .56	**3.** $\frac{1}{8}$ .12$\frac{1}{2}$	**3.** $\frac{2}{3}$ .66$\frac{2}{3}$
4. $\frac{8}{10}$ .8	**4.** $\frac{2}{5}$ .4	**4.** $\frac{6}{100}$ .06	**4.** $\frac{41}{50}$ .82	**4.** $\frac{9}{16}$ .56$\frac{1}{4}$	**4.** $\frac{1}{12}$ .08$\frac{1}{3}$
5. $\frac{6}{10}$ .6	**5.** $\frac{3}{5}$ .6	**5.** $\frac{91}{100}$ .91	**5.** $\frac{9}{20}$ .45	**5.** $\frac{1}{16}$ .06$\frac{1}{4}$	**5.** $\frac{5}{12}$ .41$\frac{2}{3}$

SET 7	SET 8	SET 9	SET 10	SET 11	SET 12
1. $\frac{115}{100}$ 1.15	**1.** $\frac{33\frac{1}{3}}{100}$ .33$\frac{1}{3}$	**1.** $\frac{4}{7}$ .57$\frac{1}{7}$	**1.** $\frac{18}{36}$ .5 or .50	**1.** $\frac{15}{40}$ .37$\frac{1}{2}$	**1.** $\frac{42}{54}$ .77$\frac{7}{9}$
2. $\frac{175}{100}$ 1.75	**2.** $\frac{62\frac{1}{2}}{100}$ .62$\frac{1}{2}$	**2.** $\frac{3}{11}$ .27$\frac{3}{11}$	**2.** $\frac{21}{28}$ .75	**2.** $\frac{45}{54}$ .83$\frac{1}{3}$	**2.** $\frac{27}{63}$ .42$\frac{6}{7}$
3. $\frac{183}{100}$ 1.83	**3.** $\frac{16\frac{2}{3}}{100}$ .16$\frac{2}{3}$	**3.** $\frac{7}{9}$ .77$\frac{7}{9}$	**3.** $\frac{30}{75}$ .4 or .40	**3.** $\frac{34}{51}$ .66$\frac{2}{3}$	**3.** $\frac{30}{66}$ .45$\frac{5}{11}$
4. $\frac{150}{100}$ 1.5 or 1.50	**4.** $\frac{5\frac{3}{4}}{100}$ .05$\frac{3}{4}$	**4.** $\frac{13}{15}$ .86$\frac{2}{3}$	**4.** $\frac{56}{80}$ .7 or .70	**4.** $\frac{84}{96}$ .87$\frac{1}{2}$	**4.** $\frac{28}{105}$ .26$\frac{2}{3}$
5. $\frac{234}{100}$ 2.34	**5.** $\frac{87\frac{1}{2}}{100}$ .87$\frac{1}{2}$	**5.** $\frac{9}{13}$ .69$\frac{3}{13}$	**5.** $\frac{54}{90}$ .6 or .60	**5.** $\frac{80}{128}$ .62$\frac{1}{2}$	**5.** $\frac{24}{108}$ .22$\frac{2}{9}$

SET 13	SET 14	SET 15	SET 16	SET 17	SET 18
1. $1\frac{1}{2}$ 1.5 or 1.50	**1.** $\frac{8}{5}$ 1.6 or 1.60	**1.** $\frac{14}{9}$ 1.55$\frac{5}{9}$	**1.** $2\frac{3}{8}$ 2.37$\frac{1}{2}$	**1.** $\frac{571}{1000}$ .571	**1.** $\frac{9514}{10000}$ .9514
2. $1\frac{5}{8}$ 1.62$\frac{1}{2}$	**2.** $\frac{14}{8}$ 1.75	**2.** $\frac{10}{7}$ 1.42$\frac{6}{7}$	**2.** $3\frac{11}{12}$ 3.91$\frac{2}{3}$	**2.** $\frac{386}{1000}$ .386	**2.** $\frac{2781}{10000}$ .2781
3. $1\frac{3}{7}$ 1.42$\frac{6}{7}$	**3.** $\frac{12}{9}$ 1.33$\frac{1}{3}$	**3.** $\frac{76}{60}$ 1.26$\frac{2}{3}$	**3.** $\frac{96}{36}$ 2.66$\frac{2}{3}$	**3.** $\frac{49}{1000}$ .049	**3.** $\frac{457}{10000}$ .0457
4. $1\frac{2}{5}$ 1.4 or 1.40	**4.** $\frac{57}{48}$ 1.18$\frac{3}{4}$	**4.** $\frac{65}{35}$ 1.85$\frac{5}{7}$	**4.** $\frac{66}{21}$ 3.14$\frac{2}{7}$	**4.** $\frac{8}{1000}$ .008	**4.** $\frac{26}{10000}$ .0026
5. $1\frac{7}{16}$ 1.43$\frac{3}{4}$	**5.** $\frac{52}{32}$ 1.62$\frac{1}{2}$	**5.** $\frac{50}{45}$ 1.11$\frac{1}{9}$	**5.** $2\frac{5}{6}$ 2.83$\frac{1}{3}$	**5.** $\frac{647}{1000}$ .647	**5.** $\frac{5933}{10000}$ .5933

MISCELLANEOUS EXAMPLES

Express each of the following measurements as a decimal. Use as many decimal places as you need to make the dividend exactly divisible.

1. a. $\frac{3}{8}$ inch .375 **b.** $\frac{1}{8}$ inch .125 **c.** $\frac{3}{4}$ inch .75 **d.** $\frac{5}{8}$ inch .625 **e.** $\frac{7}{8}$ inch .875

2. a. $\frac{5}{16}$ inch .3125 **b.** $\frac{7}{16}$ inch .4375 **c.** $\frac{11}{16}$ inch .6875 **d.** $\frac{3}{16}$ inch .1875 **e.** $\frac{15}{16}$ inch .9375

3. a. $\frac{7}{32}$ inch   **b.** $\frac{25}{32}$ inch   **c.** $\frac{9}{32}$ inch   **d.** $\frac{19}{32}$ inch   **e.** $\frac{13}{32}$ inch

4. a. $\frac{1}{64}$ inch   **b.** $\frac{43}{64}$ inch   **c.** $\frac{17}{64}$ inch   **d.** $\frac{35}{64}$ inch   **e.** $\frac{29}{64}$ inch

5. a. $1\frac{17}{32}$ inches   **b.** $5\frac{9}{16}$ inches   **c.** $2\frac{23}{64}$ inches   **d.** $8\frac{1}{16}$ inches

e. $3\frac{27}{32}$ inches

3.a. .21875 b. .78125 c. .28125 d. .59375 e. .40625
4.a. .015625 b. .671875 c. .265625 d. .546875 e. .453125
5.a. 1.53125 b. 5.5625 c. 2.359375 d. 8.0625 e. 3.84375

PRACTICAL APPLICATIONS

1. Which is longer: $\frac{5}{8}$ in.
a. $\frac{5}{8}$ inch or .619 inch?   **b.** .5347 inch or $\frac{17}{32}$ inch? .5347 in.
2. Which is shorter: .742 in.
a. .742 inch or $\frac{3}{4}$ inch?   **b.** .8129 inch or $\frac{13}{16}$ inch? $\frac{13}{16}$ in.

3. To find the batting average, first find what fractional part of the times at bat each player hit safely, then change the fraction to a decimal correct to three places:

	At Bat	Hits	Average
Todd	32	12	.375
Maria	27	9	.333
José	41	13	.317

4. To find the team standing average, first find what fractional part of the games played is the games won, then change the fraction to a decimal correct to three places:

	Won	Lost	Average
Blues	8	4	.667
Yellows	7	5	.583
Browns	3	9	.250

5. To find the fielding average, first find what fractional part of the total chances (the sum of the put-outs, assists, and errors) are the chances properly handled (the sum of the put-outs and assists). Then change the fraction to a decimal correct to three places:

	Put-Outs	Assists	Errors	Average
Dave	32	37	6	.920
Karen	40	15	5	.917
Felipe	14	49	7	.900
Frank	17	31	8	.857
Rosa	29	27	10	.848

Changing Decimals to Common Fractions* | EXERCISE 1-31

OBJECTIVE:

I. Aim   To change a decimal to a common fraction or a mixed number.

II. Procedure

1. To change a decimal to a common fraction (2 methods):

a. Write a fraction, using the figures of the decimal numeral as the numerator and a power of ten (10, 100, 1,000, etc.) corresponding to the place value of the last figure of the decimal numeral as the denominator. Then simplify and express in lowest terms. (See sample solutions 1, 2, 3, 4, and 6.)

b. Or, if the common fraction equivalent is known, write the common fraction directly.

2. To change a mixed decimal to a mixed number:

a. Change the decimal fraction to a common fraction. Express in lowest terms.

b. Annex this answer to the numeral for the whole number. (See sample solution 5.)

III. Sample Solutions

1. Change .05 to a common fraction.

$$.05 = \tfrac{5}{100} = \tfrac{1}{20}$$

Answer: $\tfrac{1}{20}$

2. Change $.33\tfrac{1}{3}$ to a common fraction.

$$.33\tfrac{1}{3} = \frac{33\tfrac{1}{3}}{100} = \frac{100}{300} = \frac{1}{3}$$

Answer: $\tfrac{1}{3}$

3. Change .625 to a common fraction.

$$.625 = \tfrac{625}{1000} = \tfrac{5}{8}$$

Answer: $\tfrac{5}{8}$

4. Change $.12\tfrac{1}{2}$ to a common fraction.

$$.12\tfrac{1}{2} = \frac{12\tfrac{1}{2}}{100} = 12\tfrac{1}{2} \div 100$$

$$= \frac{\overset{1}{\cancel{25}}}{2} \times \frac{1}{\underset{4}{\cancel{100}}} = \frac{1}{8}$$

Answer: $\tfrac{1}{8}$

5. Change 5.875 to a mixed number.

$$5.875 = 5\tfrac{875}{1000} = 5\tfrac{7}{8}$$

Answer: $5\tfrac{7}{8}$

6. Change .0045 to a common fraction.

$$.0045 = \tfrac{45}{10000} = \tfrac{9}{2000}$$

Answer: $\tfrac{9}{2000}$

---

*See Exercise 1-61.

Related Resources: *Practical Applications in Mathematics:* Diag. Test and Prac. Ex., p. 51 **181**

DIAGNOSTIC TEST

Express the following decimals as common fractions or mixed numbers:

1. .3 $\frac{3}{10}$ 5. .66$\frac{2}{3}$ $\frac{2}{3}$ 9. .672 $\frac{84}{125}$ 13. .0075 $\frac{3}{400}$

2. .25 $\frac{1}{4}$ 6. 1.9 $1\frac{9}{10}$ 10. .028 $\frac{7}{250}$ 14. 7.8125

3. .04 $\frac{1}{25}$ 7. 2.85 $2\frac{17}{20}$ 11. 3.125 $3\frac{1}{8}$ $7\frac{13}{16}$

4. .60 $\frac{3}{5}$ 8. 1.37$\frac{1}{2}$ $1\frac{3}{8}$ 12. .4375 $\frac{7}{16}$

Basic: Sets 1–9, Misc. Ex. 1–2; *Average:* Sets 1–12, Misc. Ex. 1–4; *Enriched:* Sets 1–14, Misc. Ex. 1–6

RELATED PRACTICE EXAMPLES

Express the following decimals as common fractions or mixed numbers:

SET 1	SET 2	SET 3	SET 4	SET 5
1. .6 $\frac{3}{5}$	1. .75 $\frac{3}{4}$	1. .02 $\frac{1}{50}$	1. .40 $\frac{2}{5}$	1. .16$\frac{2}{3}$ $\frac{1}{6}$
2. .2 $\frac{1}{5}$	2. .45 $\frac{9}{20}$	2. .07 $\frac{7}{100}$	2. .70 $\frac{7}{10}$	2. .87$\frac{1}{2}$ $\frac{7}{8}$
3. .5 $\frac{1}{2}$	3. .52 $\frac{13}{25}$	3. .01 $\frac{1}{100}$	3. .90 $\frac{9}{10}$	3. .06$\frac{1}{4}$ $\frac{1}{16}$
4. .1 $\frac{1}{10}$	4. .87 $\frac{87}{100}$	4. .08 $\frac{2}{25}$	4. .10 $\frac{1}{10}$	4. .83$\frac{1}{3}$ $\frac{5}{6}$
5. .9 $\frac{9}{10}$	5. .32 $\frac{8}{25}$	5. .06 $\frac{3}{50}$	5. .80 $\frac{4}{5}$	5. .62$\frac{1}{2}$ $\frac{5}{8}$

SET 6	SET 7	SET 8	SET 9	SET 10
1. 1.2 $1\frac{1}{5}$	1. 1.25 $1\frac{1}{4}$	1. 1.33$\frac{1}{3}$ $1\frac{1}{3}$	1. .125 $\frac{1}{8}$	1. .036 $\frac{9}{250}$
2. 1.5 $1\frac{1}{2}$	2. 2.42 $2\frac{21}{50}$	2. 1.12$\frac{1}{2}$ $1\frac{1}{8}$	2. .875 $\frac{7}{8}$	2. .085 $\frac{17}{200}$
3. 2.8 $2\frac{4}{5}$	3. 2.67 $2\frac{67}{100}$	3. 2.66$\frac{2}{3}$ $2\frac{2}{3}$	3. .946 $\frac{473}{500}$	3. .004 $\frac{1}{250}$
4. 3.4 $3\frac{2}{5}$	4. 1.32 $1\frac{8}{25}$	4. 4.08$\frac{1}{3}$ $4\frac{1}{12}$	4. .384 $\frac{48}{125}$	4. .006 $\frac{3}{500}$
5. 1.7 $1\frac{7}{10}$	5. 3.84 $3\frac{21}{25}$	5. 3.87$\frac{1}{2}$ $3\frac{7}{8}$	5. .192 $\frac{24}{125}$	5. .048 $\frac{6}{125}$

SET 11	SET 12	SET 13	SET 14
1. 1.375 $1\frac{3}{8}$	1. .3125 $\frac{5}{16}$	1. .0025 $\frac{1}{400}$	1. 1.4132 $1\frac{1,033}{2,500}$
2. 1.248 $1\frac{31}{125}$	2. .5625 $\frac{9}{16}$	2. .0054 $\frac{27}{5,000}$	2. 2.0625 $2\frac{1}{16}$
3. 2.964 $2\frac{241}{250}$	3. .0475 $\frac{19}{400}$	3. .0068 $\frac{17}{2,500}$	3. 5.0084 $5\frac{21}{2,500}$
4. 3.755 $3\frac{151}{200}$	4. .9375 $\frac{15}{16}$	4. .0015 $\frac{3}{2,000}$	4. 3.5625 $3\frac{9}{16}$
5. 1.045 $1\frac{9}{200}$	5. .15625 $\frac{5}{32}$	5. .00075 $\frac{3}{4,000}$	5. 6.6875 $6\frac{11}{16}$

MISCELLANEOUS EXAMPLES

Which of the following statements are true?

1. .72 $<$ $\frac{3}{4}$ True   3. $\frac{1}{2}$ $>$ .48   True   5. $\frac{5}{6}$ $<$ .9   True

2. $\frac{5}{8}$ $<$ .61   False   4. .39 $>$ $\frac{2}{5}$ False   6. .375 $>$ $\frac{3}{8}$ False

Table of Equivalents
Percents, Decimals, and Common Fractions

Percent	Decimal	Common Fraction
5%	.05	$\frac{1}{20}$
$6\frac{1}{4}$%	$.06\frac{1}{4}$	$\frac{1}{16}$
$8\frac{1}{3}$%	$.08\frac{1}{3}$	$\frac{1}{12}$
10%	.10 or .1	$\frac{1}{10}$
$12\frac{1}{2}$%	$.12\frac{1}{2}$ or .125	$\frac{1}{8}$
$16\frac{2}{3}$%	$.16\frac{2}{3}$	$\frac{1}{6}$
20%	.20 or .2	$\frac{1}{5}$
25%	.25	$\frac{1}{4}$
30%	.30 or .3	$\frac{3}{10}$
$33\frac{1}{3}$%	$.33\frac{1}{3}$	$\frac{1}{3}$
$37\frac{1}{2}$%	$.37\frac{1}{2}$ or .375	$\frac{3}{8}$
40%	.40 or .4	$\frac{2}{5}$
50%	.50 or .5	$\frac{1}{2}$
60%	.60 or .6	$\frac{3}{5}$
$62\frac{1}{2}$%	$.62\frac{1}{2}$ or. 625	$\frac{5}{8}$
$66\frac{2}{3}$%	$.66\frac{2}{3}$	$\frac{2}{3}$
70%	.70 or .7	$\frac{7}{10}$
75%	.75	$\frac{3}{4}$
80%	.80 or .8	$\frac{4}{5}$
$83\frac{1}{3}$%	$.83\frac{1}{3}$	$\frac{5}{6}$
$87\frac{1}{2}$%	$.87\frac{1}{2}$ or .875	$\frac{7}{8}$
90%	.90 or .9	$\frac{9}{10}$
100%	1.00 or 1	

What Decimal Part One Number Is of Another

OBJECTIVE:

I. Aim   To find what decimal part one number is of another.

II. Procedure

1. Express by a common fraction what part one number is of another. (See Exercise 1-18.)

2. Change this fraction to an equivalent decimal fraction by dividing its numerator by its denominator or use other methods described in Exercise 1-30.

III. Sample Solutions

1. What decimal part of 20 is 9?

$$\frac{9}{20} = 9 \div 20 = 20\overline{)9.00}^{\;.45}$$

Answer: .45

2. 15 is what decimal part of 24?

$$\frac{15}{24} = \frac{5}{8} = 5 \div 8 = 8\overline{)5.000}^{\;.625}$$

Answer: .625

3. What decimal part (2 places) of 3 is 2?

$$\frac{2}{3} = 2 \div 3 = 3\overline{)2.00}^{\;.66\frac{2}{3}}$$

Answer: .66$\frac{2}{3}$

4. What decimal part (to nearest thousandth) of 18 is 7?

$$\frac{7}{18} = 7 \div 18 = 18\overline{)7.0000}^{\;.3888}$$

Answer: .389

DIAGNOSTIC TEST

1. What decimal part of 10 is 3? .3
2. 4 is what decimal part of 5? .8
3. What decimal part of 100 is 21? .21
4. What decimal part of 25 is 17? .68
5. 7 is what decimal part of 8? .875
6. What decimal part (2 places) of 6 is 1? .17
7. What decimal part (to nearest thousandth) of 14 is 9? .643
8. 27 is what decimal part of 36? .75
9. 24 is what decimal part (2 places) of 72? .33
10. What decimal part (to nearest hundredth) of 49 is 21? .43
11. What decimal part of .4 is .02? .05
12. .65 is what decimal part of 7.8? .083
13. What decimal part of $2 is $.60? .3
14. $2.25 is what decimal part of $11.25? .2

Basic: Sets 1–9; *Average:* Sets 1–12; *Enriched:* Sets 1–14

RELATED PRACTICE EXAMPLES

SET 1

1. What decimal part of 10 is 7? .7
2. What decimal part of 2 is 1? .5
3. What decimal part of 5 is 4? .8
4. What decimal part of 10 is 1? .1
5. What decimal part of 5 is 3? .6

SET 2

1. 9 is what decimal part of 10? .9
2. 1 is what decimal part of 5? .2
3. 3 is what decimal part of 10? .3
4. 1 is what decimal part of 2? .5
5. 2 is what decimal part of 5? .4

SET 3

1. What decimal part of 100 is 7? .07
2. What decimal part of 100 is 59? .59
3. 83 is what decimal part of 100? .83
4. 31 is what decimal part of 100? .31
5. What decimal part of 100 is 97? .97

SET 4

1. What decimal part of 4 is 3? .75
2. 14 is what decimal part of 25? .56
3. What decimal part of 20 is 7? .35
4. 13 is what decimal part of 50? .26
5. What decimal part of 25 is 22? .88

SET 5

1. 3 is what decimal part of 8? .375
2. What decimal part of 16 is 11? .6875
3. What decimal part of 8 is 5? .625
4. 9 is what decimal part of 16? .5625
5. 21 is what decimal part of 32? .65625

SET 6
(2 decimal places)

1. What decimal part of 3 is 2? .67
2. 1 is what decimal part of 12? .08
3. 5 is what decimal part of 6? .83
4. What decimal part of 12 is 7? .58
5. 13 is what decimal part of 24? .54

SET 7
(to nearest thousandth)

1. What decimal part of 9 is 4? .444
2. 5 is what decimal part of 7? .714
3. What decimal part of 11 is 8?.727
4. 2 is what decimal part of 15? .133
5. What decimal part of 18 is 13? .722

SET 8

1. 16 is what decimal part of 32? .50
2. What decimal part of 56 is 21? .375
3. 40 is what decimal part of 500? .08
4. 64 is what decimal part of 200? .32
5. What decimal part of 750 is 125?
   .167

SET 9
(2 decimal places)

1. 45 is what decimal part of 54? .83
2. 12 is what decimal part of 36? .33
3. What decimal part of 84 is 35?.42
4. 68 is what decimal part of 96?.71
5. What decimal part of 102 is 85?
   .83

SET 10
(to nearest thousandth)

1. What decimal part of 84 is 24? .286
2. What decimal part of 105 is 40? .381
3. 15 is what decimal part of 27? .556
4. What decimal part of 135 is 36? .267
5. 99 is what decimal part of 121? .818

SET 11

1. What decimal part of .12 is .3? 2.5
2. What decimal part of 0.96 is 0.24? .25
3. 0.39 is what decimal part of 0.65? .6
4. .5 is what decimal part of .75? .667
5. What decimal part of 0.2 is 0.18? .9

SET 12

1. .4 is what decimal part of 1.6? .25
2. 0.03 is what decimal part of 300? .0001
3. What decimal part of 3.65 is 1.46? .4
4. What decimal part of 9.6 is 6? .625
5. .34 is what decimal part of 8.5? .04

SET 13

1. What decimal part of $1 is $.05? .05
2. $7.25 is what decimal part of $25? .29
3. $6 is what decimal part of $7.50? .8
4. What decimal part of $3 is $1.86? .62
5. $.65 is what decimal part of $13? .05

SET 14

1. $.18 is what decimal part of $.90? .2
2. $.86 is what decimal part of $1.29? .667
3. What decimal part of $12.40 is $1.86? .15
4. What decimal part of $27.30 is $16.38? .6
5. $1.11 is what decimal part of $1.48? .75

Finding the Number when a Decimal Part of It Is Known

OBJECTIVE:

I. Aim   To find the number when a decimal part of it is known.

II. Procedure

1. Divide the given number representing the decimal part of the unknown number by the given decimal fraction.
2. Or follow method 2 shown in the sample solutions.
3. Also see solution by equation, Exercise 4-26.

III. Sample Solutions

1. .4 of what number is 12?

Method 1

$$12 \div .4 = .4_\wedge \overline{)12.0_\wedge}^{\;30.}$$

Answer: 30

Method 2

Since .4 of the number = 12
Then .1 of the number = 12 ÷ 4 = 3
10 × .1 of the number = 10 × 3 = 30
    Therefore, the number = 30
Answer: 30

2. 480 is .75 of what number?

$$480 \div .75 = .75_\wedge \overline{)480.00_\wedge}^{\;640.}$$

Answer: 640

DIAGNOSTIC TEST

1. .2 of what number is 14? 70
2. 20 is .5 of what number? 40
3. .8 of what number is 2? 2½
4. 3 is .6 of what number? 5
5. .07 of what number is 56? 800
6. .18 of what number is 27? 150
7. 500 is .625 of what number? 800
8. .0375 of what number is 150? 4,000

9. 11 is 0.8 of what number? 13.75
10. .06 of what number is 100? 1,666⅔
11. 560 is .87½ of what number? 640
12. .33⅓ of what number is 41? 123
13. .24 is .03 of what number? 8
14. .25 of what amount is $1.50? $6.00
15. 324 is 1.08 of what number? 300
16. 1.15 of what amount is $9,200? $8,000

RELATED PRACTICE EXAMPLES

SET 1
1. .3 of what number is 12? 40
2. .5 of what number is 30? 60
3. .9 of what number is 540? 600
4. .8 of what number is 400? 500
5. .7 of what number is 63? 90

SET 2
1. 10 is .2 of what number? 50
2. 96 is .4 of what number? 240
3. 270 is .3 of what number? 900
4. 300 is .6 of what number? 500
5. 83 is .1 of what number? 830

SET 3
1. .8 of what number is 4? 5
2. .4 of what number is 2? 5
3. .2 of what number is 1? 5
4. .9 of what number is 3? 3.33
5. .5 of what number is 5? 10

SET 4
1. 3 is .6 of what number? 5
2. 6 is .8 of what number? 7.5
3. 4 is .4 of what number? 10
4. 2 is .6 of what number? $3\frac{1}{3}$
5. 1 is .1 of what number? 10

SET 5
1. .06 of what number is 42? 700
2. .04 of what number is 19? 475
3. 8 is .01 of what number? 800
4. 72 is .05 of what number? 1,440
5. .09 of what number is 630? 7,000

SET 6
1. .12 of what number is 78? 650
2. .46 of what number is 1,500? 3,260.87
3. 51 is .75 of what number? 68
4. 123 is .82 of what number? 150
5. .60 of what number is 810? 1,350

SET 7
1. 49 is .875 of what number? 56
2. .125 of what number is 152? 1,216
3. 267 is .375 of what number? 712
4. 180 is .045 of what number? 4,000
5. .098 of what number is 588? 6,000

SET 8
1. .3125 of what number is 1,250? 4,000
2. 350 is .4375 of what number? 800
3. .1875 of what number is 375? 2,000
4. .0625 of what number is 2,500? 40,000
5. 4,500 is .0875 of what number?
   $51,428\frac{4}{7}$

SET 9
1. 15 is 0.4 of what number? 37.5
2. 5 is 0.3 of what number? 16.667
3. 0.5 of what number is 28? 56
4. .06 of what number is 16? 266.667
5. 24 is .07 of what number? 342.857

SET 10
1. .75 of what number is 200? $266\frac{2}{3}$
2. .07 of what number is 58? $828\frac{4}{7}$
3. 24 is .625 of what number? 38.4
4. .0525 of what number is 15? $285\frac{5}{7}$
5. 11 is .18 of what number? $61\frac{1}{9}$

SET 11
1. 27 is .$37\frac{1}{2}$ of what number? 72
2. 81 is .$04\frac{1}{2}$ of what number? 1,800
3. .$09\frac{3}{4}$ of what number is 780? 8,000
4. 1,640 is .$10\frac{1}{4}$ of what number? 16,000
5. .$62\frac{1}{2}$ of what number is 900? 1,440

SET 12
1. .$16\frac{2}{3}$ of what number is 107? 642
2. 24 is .$66\frac{2}{3}$ of what number? 36
3. .$83\frac{1}{3}$ of what number is 2,000? 2,400
4. .$08\frac{1}{3}$ of what number is 960? 11,520
5. 458 is .$33\frac{1}{3}$ of what number? 1,374

SET 13

1. .92 is .04 of what number? 23
2. .15 of what number is .9? 6
3. 2.73 is .07 of what number? 39
4. .375 of what number is 5.4? 14.4
5. .09$\frac{1}{4}$ of what number is .74? 8

SET 14

1. .06 of what amount is $210? $3,500
2. .12 of what amount is $.84? $7.00
3. $1.08 is .75 of what amount? $1.44
4. .08$\frac{3}{4}$ of what amount is $700? $8,000
5. $4.75 is .625 of what amount? $7.60

SET 15

1. 500 is 1.25 of what number? 400
2. 312 is 1.04 of what number? 300
$12.30 **3.** 2.5 of what number is $30.75?
4. 1.8 of what number is 9,720? 5,400
5. 33.32 is 1.19 of what number? 28

SET 16

1. 1.75 of what amount is $35? $20
2. 1.06 of what amount is $4.77? $4.50
3. $7.80 is 1.2 of what amount? $6.50
4. 2.3 of what amount is $1,978? $860
5. 1.08$\frac{1}{2}$ of what amount is $6,510? $6,000

ESTIMATING ANSWERS

To estimate a sum, a difference, a product or a quotient, first round the given numbers to convenient place values (see pages 28 and 135) and then perform the operation or operations.

For each of the following select your nearest estimate:

1. .97 + .58 + .89 is approximately:	.200	2	2.5
2. .821 − .517 is approximately:	.03	.3	.003
3. 69 × .97 is approximately:	67	6.7	.67
4. 17 ÷ .3 is approximately:	5.1	57	14
5. 8.51 − .293 is approximately:	.56	8.2	.6
6. .24 × 20.1 is approximately:	50	5	.5
7. .38 + 6.3 + 84 is approximately:	.185	906	91
8. .98 ÷ .04 is approximately:	2.5	25	.25
9. $20.63 + $15.81 + $9.58 is approximately:	$46	$48	$44
10. .06 × $39.85 is approximately:	$24	$.24	$2.40
11. $59.89 ÷ 5 is approximately:	$10	$11	$12
12. $81.05 − $48.99 is approximately:	$32	$33	$34
13. .536 + .694 + .329 is approximately:	.15	15	1.5
14. 42 × $7.90 is approximately:	$500	$400	$300
15. .029 ÷ .6 is approximately:	5	.05	.5
16. 94 − 59.9 is approximately:	30	32	35
17. $11.49 + $16.26 + $17.52 is approximately:	$45	$43	$40
18. $200 − $89.90 is approximately:	$100	$110	$120
19. 3.98 × .007 is approximately:	3	.3	.03
20. $95.96 ÷ $.08 is approximately:	12	120	1,200

CHECK BY CALCULATOR

Use your calculator to check each of the following given answers. Indicate which answers are incorrect. If a calculator is not available, check by computation.

1. Add: b and c are incorrect

a.	b.	c.	d.	e.
29	8,658	49,968	73,893	615,968
496	3,709	38,947	29,529	384,557
5,847	5,266	70,539	83,154	509,664
69	1,975	56,388	40,787	832,375
758	2,763	83,584	78,919	987,976
7,199	22,471	297,426	306,282	3,330,540

2. Subtract: c and e are incorrect

a. 5,629	b. 36,503	c. 694,000	d. 725,137	e. 1,854,006
4,819	8,307	385,638	269,582	860,849
810	28,196	319,362	455,555	983,157

3. Multiply: b and d are incorrect

a. $563 \times 78 = 43,914$
b. $887 \times 645 = 582,115$
c. $609 \times 527 = 320,943$
d. $7,403 \times 9,056 = 67,031,568$

4. Divide: a and c are incorrect

a. $36,421 \div 43 = 837$
b. $10,800 \div 16 = 675$
c. $682,864 \div 728 = 948$
d. $409,236 \div 804 = 509$

5. Add: b and d are incorrect

a. $.905 + .690 + .347 = 1.942$
b. $.8 + .56 = .64$
c. $.836 + 9 = 9.836$
d. $2.67 + 81.5 = 83.72$

6. Subtract: a is incorrect

a. $.582 - .496 = .86$
b. $73.8 - 5 = 68.8$
c. $80 - 69.3 = 10.7$
d. $6.834 - 4.62 = 2.214$

7. Multiply: b and c are incorrect

a. $36 \times .89 = 32.04$
b. $4.5 \times .008 = .0036$
c. $17.9 \times 82.4 = 1,464.96$
d. $.013 \times .004 = .000052$

8. Divide: b and d are incorrect

a. $.84 \div 14 = .06$
b. $60 \div .05 = 120$
c. $9.23 \div .2 = 46.15$
d. $3 \div 40 = .75$

For a Computer Activity see p. T39.

2.a. Seven tenths   b. Six thousandths   c. Four and nine hundredths   d. Eighty thousand five hundred one hundred-thousandths   e. Sixty-two and five thousand four hundred eighty-three ten-thousandths

REVIEW OF UNIT THREE

1. Write each of the following as a decimal numeral:

a. Nine hundredths **b.** Five hundred and forty-three thousandths **c.** Eight and five tenths **d.** Eleven millionths **e.** Seven thousand eight hundred twenty-two ten-thousandths   a. .09   b. 500.043   c. 8.5   d. .000011   e. .7822

2. Write each of the following numerals as a word statement:

   a. .7   **b.** .006   **c.** 4.09   **d.** .80501   **e.** 62.5483

3. Round: **a.** .96 correct to nearest tenth **b.** 8.2915 correct to nearest hundredth **c.** 25.1427 correct to nearest thousandth **d.** $4.279 correct to nearest cent **e.** $70.16⅔ correct to nearest cent   a. 1.0   b. 8.29   c. 25.143   d. $4.28   e. $70.16

4. Add:

a. 8.2	**b.** 8.05	**c.** 8.2 + .19 + 24 32.39
5.9	2.03	**d.** Find the sum of: 49.2, .871, and 6.45 56.521
4.5	5.04	**e.** $6.21 + $58.14 + $.68 + $9.85 + $180.07 $254.95
18.6	1.08	
	16.20	

5. Subtract:

a. 47.512
   39.835 7.677   **b.** .489 − .3 .189   **c.** .8 − .375 .425   **d.** 10 − .04 9.96

e. Subtract $3.25 from $28. $24.75

6. Which is greater: **a.** .8 or .62?   **b.** 1.06 or .305?
    Which is smaller: **c.** .425 or .94?   **d.** 2.9 or .49?

e. Arrange in order of size, greatest first: 1.67, .0167, 0.167, 16.7
                                       16.7, 1.67, 0.167, .0167

7. Multiply:

a. .04	**b.** 90.72	**c.** .018 × .05
.2 .008	.125 11.34	.0009

Find correct to nearest cent:

d. ¾ of $5.26 $3.95   **e.** $28.93
                       .83⅓ $24.11

8. Divide:

a. 78)‾.624 .008   **b.** 1.2)‾11.076 9.23   **c.** 35 ÷ .875 40   **d.** $.04)‾$2 50

e. Find correct to nearest tenth: 39.37)‾582 14.8

9. Multiply by short method:

   a. 100 × 90 9,000   **c.** 1,000 × .4 400   **e.** 1,000 × 28.95
   b. 10 × 3.56 35.6   **d.** 100 × .08 8   28,950

10. Divide by short method:

a. $500 \div 10$ ₅₀   **c.** $.7 \div 100$ .007   **e.** $654 \div 100$ 6.54

b. $240 \div 1,000$ .24   **d.** $93.2 \div 1,000$ .0932

11. Change each of the following fractions to decimals:

a. $\frac{3}{5}$ .6   **b.** $\frac{83}{100}$ .83   **c.** $\frac{17}{25}$ .68   **d.** To 3 decimal places: $\frac{49}{56}$ .875

e. To 4 decimal places: $\frac{13}{16}$ .8125

12. Change each of the following decimals to fractions or mixed numbers:

a. $.4$ $\frac{2}{5}$ **b.** $.25$ $\frac{1}{4}$ **c.** $.375$ $\frac{3}{8}$ **d.** 9.0625 $9\frac{1}{16}$ **e.** $.66\frac{2}{3}$ $\frac{2}{3}$

13. Do as directed:

a. $\frac{4}{5} + .75$ 1.55 or $1\frac{11}{20}$ **c.** $\frac{7}{10} \times .43$ .301   **e.** Find $.12\frac{1}{2}$ of 32

b. $.66\frac{2}{3} - \frac{1}{3}$ .33$\frac{1}{3}$ or $\frac{1}{3}$ **d.** $60 \div .83\frac{1}{3}$ 72   4

14. Do as directed, using short methods:

a. 25×36 900   **c.** $700 \div 50$ 14   **e.** $42 \times 16\frac{2}{3}$

b. $33\frac{1}{3} \times 60$ 2,000   **d.** $150 \div 12\frac{1}{2}$ 12   700

For problems 15 through 17, solve each problem and select the letter in red corresponding to your answer.

15. Find the cost of 940 liters of fuel oil at $.324 per liter.

  a. $305.66   **b.** $303.56   **c.** $298.76   **d.** Answer not given

16. A basketball player scored 871 points in 56 games. Find the number of points averaged per game to the nearest tenth of a point.

  a. 15.4 points   **b.** 15.7 points   **c.** 15.5 points   **d.** 15.6 points

17. The best batter in the American League had a batting average of .349. The leading batter in the National League had an average of .355. Who had the higher average and how much higher?

  a. National League, .060   **c.** National League, .006

  b. National League, .014   **d.** Answer not given

18. The distance by train from St. Louis to Indianapolis is 240 miles, from Indianapolis to Dayton is 109.7 miles, from Dayton to Columbus is 70.7 miles, and from Columbus to Pittsburgh is 109.9 miles. What is the distance by train from St. Louis to Pittsburgh by this route? 530.3 miles

19. If the tax rate is $2.95 per $100, how much must you pay for taxes on a house assessed for $29,800? $879.10

20. A merchant bought 150 pounds of pears for $37.50. He sold 67 pounds at $.49 a pound, 48 pounds at $.39 a pound, and 29 pounds at $.25 a pound. The rest spoiled. How much profit did he make? $21.30

CUMULATIVE PRACTICE

1. Add: 2,958
47,426
8,019
846
25,625 _84,874_

2. Subtract:
1,600,520
549,618
‾‾‾‾‾‾‾
1,050,902

3. Multiply:
9,658
8,692
‾‾‾‾‾‾
83,947,336

4. Divide: _387_
456)176,472

5. Add: $8\frac{3}{4}$
$5\frac{1}{2}$
$2\frac{11}{16}$ $16\frac{15}{16}$

6. Subtract:
$4\frac{1}{5}$
$3\frac{2}{3}$ $\frac{8}{15}$

7. Multiply:
98
$7\frac{3}{4}$ $759\frac{1}{2}$

8. Divide:
$6\frac{7}{8} \div 3\frac{1}{7}$ $2\frac{3}{16}$

9. Add:
$8.3 + .75 + 14$
_23.05_

10. Subtract:
$\$92 - \$3.85$
_$88.15_

11. Multiply:
$2.5 \times .04$
_.1_

12. Divide:
.02)$.001
_.05_

KEYED ACHIEVEMENT TEST

For further practice see Exercises indicated in colored boxes.

1. Add:
82,162
48,279
14,086
53,328
13,497
62,172 [1-4]
273,524

2. Subtract:
415,037
316,287 [1-5]
‾‾‾‾‾‾‾
98,750

3. Multiply:
786
598 [1-6]
‾‾‾‾‾‾
470,028

4. Divide: _796_
6,080)4,839,680 [1-7]

5. Add:
$6\frac{9}{10}$
$11\frac{1}{2}$ $4\frac{3}{5}$ [1-12]

6. Subtract:
$7\frac{3}{8}$
$3\frac{5}{6}$ [1-13]
$3\frac{13}{24}$

7. Multiply:
$6\frac{3}{4} \times 4\frac{1}{6}$ $28\frac{1}{8}$ [1-16]

8. Divide:
$84 \div 2\frac{5}{8}$ 32
[1-17]

$\frac{3}{4}$ **9.** Express $\frac{48}{64}$ in lowest terms. [1-8]

10. What part of $\frac{5}{6}$ 90 is 75? [1-18]

11. Which is greater: $\frac{2}{3}$ $\frac{2}{3}$ or $\frac{7}{12}$? [1-14]

12. Add:
$.96 + .096 + 9.6$ 10.656
[1-23]

13. Subtract: $8.3 - .49$ 7.81
[1-24]

14. Multiply: $2.08 \times .15$ .312
[1-26]

15. Divide: .04)$.006 .15
[1-27]

16. Write as a decimal numeral: Four hundred and twenty-five thousandths [1-21] 400.025

17. Round 29.426 to the nearest hundredth. [1-22] 29.43

18. Change $\frac{7}{20}$ to a decimal. [1-30] .35

19. Change .875 to a common fraction. [1-31] $\frac{7}{8}$

20. Multiply 45.02 by 1,000 using the short method. [1-28] 45,020

COMPETENCY CHECK TEST

Solve each problem and select the letter corresponding to your answer.
The numerals in red boxes indicate Exercises where help may be found.

1. Find the sum of: 897 + 3,568 + 959   1-4
   a. 5,234   b. 4,342   c. 6,434   <u>d.</u> 5,424

2. Subtract: 927 − 696   1-5
   a. 131   b. 141   <u>c.</u> 231   d. Answer not given

3. Multiply: 905 × 580   1-6
   a. 117,650   b. 1,485   c. 45,400   <u>d.</u> 524,900

4. Divide: 79)63,437   1-7
   a. 93   b. 903   <u>c.</u> 803   d. 830

5. Add: $\frac{1}{6} + \frac{3}{4}$ 1-12
   a. $\frac{23}{24}$ <u>b.</u> $\frac{11}{12}$ c. $\frac{13}{64}$ d. $\frac{4}{10}$

6. Subtract: $\frac{7}{10} - \frac{1}{2}$ 1-13
   a. $\frac{6}{8}$ b. $\frac{2}{3}$ <u>c.</u> $\frac{1}{5}$ d. Answer not given

7. Multiply: $4\frac{2}{3} \times \frac{5}{7}$ 1-16
   a. $4\frac{7}{10}$ <u>b.</u> $3\frac{1}{3}$ c. $2\frac{3}{4}$ d. $4\frac{10}{21}$

8. Divide: $12 \div 1\frac{3}{5}$ 1-17
   <u>a.</u> $7\frac{1}{2}$ b. $11\frac{2}{5}$ c. $12\frac{3}{5}$ d. $6\frac{3}{4}$

9. Add: .63 + 2.4   1-23
   a. .87   b. 8.7   c. 8.43   <u>d.</u> 3.03

10. Subtract: .547 − .39   1-24
    a. .508   b. .512   <u>c.</u> .157   d. Answer not given

11. Multiply: .013 × .06   1-26
    a. .078   <u>b.</u> .00078   c. .01306   d. .0078

12. Divide: 9 ÷ .6   1-27
    <u>a.</u> 15   b. 1.5   c. .15   d. 150

13. Which is the smallest: .2 or .019 or .12 or .02?   1-25
    a. .2   <u>b.</u> .019   c. .12   d. .02

14. .08 of what amount is $5.76?   1-33
    a. $.72   b. $7.20   <u>c.</u> $72   d. $720

15. Eva bought a raincoat costing $32.49. How much change should she
    receive from a $50 bill?   1-24
    a. $18.51   b. $28.51   c. $27.51   <u>d.</u> $17.51

"The Milky Way is made up of more than ~~100,000,000,000~~ *100 billion* stars. It is shaped like a disk, about ~~100,000~~ *100 thousand* light-years across and about ~~16,000~~ *16 thousand* light-years thick. The solar system moves all the way around the Milky Way's center about once every ~~200,000,000~~ *200 million* years. The solar system was formed about ~~4,500,000,000~~ *4.5 billion* years ago."

SHORTENED NAMES FOR LARGE WHOLE NUMBERS
Review of Reading Shortened Names

To name a large whole number whose numeral ends in a series of zeros, newspapers and periodicals are now using a numeral followed by the name of the appropriate period of the number scale.

Also the $ symbol is generally used to represent "dollars."

---

28 billion dollars is read "twenty-eight billion dollars" and is the shortened name for 28,000,000,000 dollars.

$497.3 million is read "four hundred ninety-seven point three million dollars" and is the shortened name for $497,300,000.

$3.24 billion is read "three point two four billion dollars" and is the shortened name for $3,240,000,000.

It is *not* read as "three dollars and twenty-four cents billion."

16.05 trillion is read "sixteen point zero five trillion and is the shortened name for 16,050,000,000,000.

REVIEW EXERCISES For ex. 1–5 See
Extended Answers, pp. T73–T74.

Read, or write in words, each shortened number in the following:

1. a. 85 million
   b. $691 billion
   c. 3,487 thousand
   d. 14 trillion

2. a. 1.3 trillion
   b. 16.7 thousand
   c. 79.6 million
   d. $25.8 billion

3. a. 4.95 billion
   b. $23.09 million
   c. $1.42 trillion
   d. 74.38 thousand

4. a. 587.4 million
   b. 914.22 trillion
   c. 609.1 thousand
   d. $445.69 billion

5. a. 8.686 billion
   b. 381.299 million
   c. $8.056 trillion
   d. 736.847 billion

Also see Exercises 1-1, 1-2 and 1-20.

PRACTICAL APPLICATIONS

Read, or write in words, each of the shortened number names in the following statements:

1. a. More than 25 million people own stock in United States companies. Twenty-five million people

b. During a recent year business spent about 37 billion dollars for advertising. Thirty-seven billion dollars

c. Every day 218 million consumers in the United States spend more than 3 billion dollars. Two hundred eighteen million consumers; three billion dollars

d. Recently about 335 billion dollars in United States money was in active use. Three hundred thirty-five billion dollars

e. Congress passed a bill extending up to 4 billion dollars more in loans this year to farmers. Four billion dollars

2. a. Canada announced negotiations for almost $3 billion in standby credits. Three billion dollars

b. The United States' holdings of foreign currencies rose a total of $62 million. Sixty-two million dollars

c. An electric utility requested $190 million increase in rates for the year. One hundred ninety million dollars

d. The U.S. Navy awarded a $55 million contract for sonar equipment. Fifty-five million dollars

e. Current worldwide military expenditures total more than $400 billion a year. Four hundred billion dollars

3. a. Americans imported 2.5 million color television sets during the year. Two point five million sets

b. Sales of cotton for the season totaled 1.12 million bales. 3.b. One point one two million bales

c. Total Canadian newsprint shipments were 2.38 million metric tons. 3.c. Two point three eight million metric tons

d. One of our states has passed a 1.43 billion dollar bill for property tax cuts. One point four three billion dollars

e. In one recent year the personal income increased to 1.375 trillion dollars. One point three seven five trillion dollars

4. a. The Senate may approve a budget deficit of $55.6 billion.

b. Gulf of Mexico leases drew high bids from oil companies totaling $767.4 million. Seven hundred sixty-seven point four million dollars

c. The U.S. trade deficit was an adjusted $2.78 billion in the past month. Two point seven eight million dollars

d. Earnings of a certain corporation rose to $252.3 million from $244.2 million and revenues increased to $3.8 billion from $3.45 billion.

e. The life insurance in force in the United States recently amounted to $1.99 trillion. One point nine nine trillion dollars

4.a. Fifty-five point six billion dollars 4.d. Two hundred fifty-two point three million dollars; Two hundred forty-four point two million dollars; Three point eight billion dollars; Three point four five billion dollars

Writing Complete Numerals for the Shortened Names

OBJECTIVE:

I. Aim   To write the complete numeral for the shortened name of a whole number.

II. Procedure

1. When the numeral prefix in the shortened name is a whole number, just replace the period name by annexing to this number the appropriate number of zeros. (See sample solution 1.)

> For *thousand* use 3 zeros (000).
> For *million* use 6 zeros (000,000).
> For *billion* use 9 zeros (000,000,000).
> For *trillion* use 12 zeros (000,000,000,000).

2. In general, multiply the number named by the numeral prefix of the shortened name by the power of ten (10; 100; 1,000; etc.—see page 254). that is equivalent to the value of the period name (or place value) used. (See sample solutions 2, 3, 4, and 5. Also see Exercise 1-28 for the short method of multiplying by powers of ten.)

III. Sample Solutions

Write the complete numeral for each of the following:

1. 96 *million*   *Answer:* 96,000,000

2. 81.9 *trillion* is 81.9 × 1,000,000,000,000

Answer: 81,900,000,000,000

3. 4.358 *billion* is 4.358 × 1,000,000,000   *Answer:* 4,358,000,000

4. 207.6 *thousand* is 207.6 × 1,000   *Answer:* 207,600

5. 798 *hundred* is 798 × 100   *Answer:* 79,800

6. $53.71 *billion* is $53.71 × 1,000,000,000 *Answer:* $53,710,000,000

DIAGNOSTIC TEST   For ex. 1–20 see Extended Answers, p. T74.

Write the complete numeral for each of the following:

1. 725 million	**8.** 2.06 billion	**15.** $70.04 trillion
2. 51.9 million	**9.** 350.742 billion	**16.** 616 thousand
3. 483.28 million	**10.** $691.18 billion	**17.** 18.947 thousand
4. 6.071 million	**11.** 31 trillion	**18.** $5.09 thousand
5. $34.56 million	**12.** 85.3 trillion	**19.** 1,826 hundred
6. 49 billion	**13.** 408.57 trillion	**20.** 62.3 hundred
7. 98.4 billion	**14.** 9.529 trillion	

Basic: Sets 1–13, Misc. Ex. 1; *Average:* Sets 1–16, Misc. Ex. 1–2; *Enriched:* Sets 1–20, Misc. Ex. 1–3

198 SHORTENED NAMES FOR LARGE WHOLE NUMBERS

RELATED PRACTICE EXAMPLES For sets 1–15 see Extended Answers, p. T74.

Write the complete numeral for each of the following:

SET 1
1. 83 million
2. 9 million
3. 460 million
4. 2,805 million
5. 589 million

SET 2
1. 6.8 million
2. 84.7 million
3. 358.3 million
4. 1,098.5 million
5. 49.4 million

SET 3
1. 17.45 million
2. 5.07 million
3. 227.93 million
4. 1,540.16 million
5. 24.82 million

SET 4
1. 8.562 million
2. 13.468 million
3. 479.704 million
4. 3,872.835 million
5. 93.696 million

SET 5
1. $4.5 million
2. $27.4 million
3. 75.09 million
4. 408.263 million
5. 667.85 million

SET 6
1. 78 billion
2. 5 billion
3. 609 billion
4. 3,570 billion
5. 91 billion

SET 7
1. 9.4 billion
2. 72.8 billion
3. 403.7 billion
4. 2,576.3 billion
5. 80.2 billion

SET 8
1. 12.06 billion
2. 6.95 billion
3. 820.73 billion
4. 3,505.81 billion
5. 49.24 billion

SET 9
1. 7.568 billion
2. 31.047 billion
3. 564.382 billion
4. 1,067.509 billion
5. 81.635 billion

SET 10
1. $8.3 billion
2. $30.7 billion
3. $54.61 billion
4. $935.054 billion
5. $819.37 billion

SET 11
1. 4 trillion
2. 17 trillion
3. 106 trillion
4. 58 trillion
5. 292 trillion

SET 12
1. 3.9 trillion
2. 18.5 trillion
3. 207.8 trillion
4. 83.1 trillion
5. 6.7 trillion

SET 13
1. 5.92 trillion
2. 60.07 trillion
3. 139.78 trillion
4. 17.05 trillion
5. 308.59 trillion

SET 14
1. 8.227 trillion
2. 41.058 trillion
3. 9.464 trillion
4. 250.829 trillion
5. 11.256 trillion

SET 15
1. $9.6 trillion
2. $3.28 trillion
3. $87.92 trillion
4. $40.071 trillion
5. $156.435 trillion

SET 16

1. 9 thousand 9,000
2. 528 thousand 528,000
3. 77 thousand 77,000
4. 3,863 thousand
5. 406 thousand 406,000
4. 3,863,000

SET 17

1. 8.7 thousand 8,700
2. 28.5 thousand 28,500
3. 63.09 thousand 63,090
4. 592.16 thousand 592,160
5. 860.442 thousand
5. 860,442

SET 18

1. $19 thousand $19,000
2. $51.6 thousand $51,600
3. $89.95 thousand $89,950
4. $200.52 thousand $200,520
5. $1,174.49 thousand
5. $1,174,490

SET 19

1. 38 hundred 3,800
2. 952 hundred 95,200
3. 2,433 hundred 243,300
4. $77 hundred $7,700
5. $605 hundred $60,500

SET 20

1. 8.5 hundred 850
2. 63.27 hundred 6,327
3. 107.54 hundred 10,754
4. $83.6 hundred $8,360
5. $4.72 hundred $472

MISCELLANEOUS EXAMPLES

Complete each of the following:

SET 1

1. 6.547 million = __6__ million __547__ thousand
2. 29.38 million = __29__ million __380__ thousand
3. 820.6 million = __820__ million __600__ thousand
4. $52.4 million = __52__ million __400__ thousand dollars
5. $17.075 million = __17__ million __75__ thousand dollars

SET 2

1. 18.697 billion = __18__ billion __697__ million
2. 575.06 billion = __575__ billion __60__ million
3. 7.1 billion = __7__ billion __100__ million
4. $33.85 billion = __33__ billion __850__ million dollars
5. $94.3 billion = __94__ billion __300__ million dollars

SET 3

1. 4.2 trillion = __4__ trillion __200__ billion
2. 1.48 trillion = __1__ trillion __480__ billion
3. 16.952 trillion = __16__ trillion __952__ billion
4. $8.5 trillion = __8__ trillion __500__ billion dollars
5. $21.76 trillion = __21__ trillion __760__ billion dollars

Writing Shortened Names for Large Whole Numbers

OBJECTIVE:

I. Aim To write the shortened name for a large whole number when its numeral ends in a series of zeros.

II. Procedure

1. Replace the required number of zeros by the equivalent period name. (See sample solution 1.)

2. In general, divide the given number by the power of ten (10; 100; 1,000; etc.—see page 254) that is equivalent to the value of the period name (or place value) used. (See sample solutions 2, 3, 4, and 5. Also see Exercise 1-29 for the short method of dividing by powers of ten.)

III. Sample Solutions

Write the shortened number for each of the following:

1. 47,000,000,000 in billions. Divide by 1,000,000,000 *Answer:* 47 billion
2. 8,910,000 in millions. Divide by 1,000,000   *Answer:* 8.91 million
3. 9,832,000,000,000 in trillions.
   Divide by 1,000,000,000,000   *Answer:* 9.832 trillion
4. 566,700 in thousands. Divide by 1,000   *Answer:* 566.7 thousand
5. 7,250 in hundreds. Divide by 100   *Answer:* 72.5 hundreds
6. $68,400,000 in millions of dollars.
   Divide by 1,000,000   *Answer:* $68.4 million

DIAGNOSTIC TEST

Write the shortened number name for each of the following:

In hundreds: **1.** 21,400 214 hundred   **2.** 6,830 68.3 hundred   **3.** 8,745 87.45 hundred

In thousands: **4.** 1,632,000 4. 1,632 thousand   **5.** 750,600 5. 750.6 thousand   **6.** 9,050   **7.** 48,409 6. 9.05 thousand   7. 48.409 thousand

In millions: **8.** 35,000,000 35 million   **9.** 4,700,000 9. 4.7 million

10. 264,010,000 264.01 million **11.** 80,557,000 80.557 million

In billions: **12.** 167,000,000,000 167 billion **13.** 4,200,000,000 4.2 billion

14. 379,460,000,000 14. 379.46 billion   **15.** 50,088,000,000 50.088 billion

In trillions: **16.** 45,000,000,000,000 16. 45 trillion   **17.** 8,100,000,000,000 8.1 trillion

18. 66,090,000,000,000 18. 66.09 trillion   **19.** 724,525,000,000,000 724.525 trillion

20. In thousands of dollars: $27,900 $27.9 thousand

21. In millions of dollars: $838,250,000 $838.25 million

22. In billions of dollars: $96,080,000,000 $96.08 billion

23. In trillions of dollars: $4,202,000,000,000 $4.202 trillion

RELATED PRACTICE EXAMPLES

Write the shortened number name for each of the following:

SET 1
In hundreds:

1. 900 9 hundred
2. 3,200 32 hundred
3. 7,100 71 hundred
4. 12,500 125 hundred
5. 49,300 493 hundred

SET 2
In hundreds:

1. 1,340 13.4 hundred
2. 870 8.7 hundred
3. 4,950 49.5 hundred
4. 20,820 208.2 hundred
5. 1,7060 170.6 hundred

SET 3
In hundreds:

1. 2,568 25.68 hundred
2. 693 6.93 hundred
3. 5,756 57.56 hundred
4. 33,497 334.97 hundred
5. 9,928 99.28 hundred

SET 4
In thousands:

1. 6,000 6 thousand
2. 11,000 11 thousand
3. 1,437,000 1,437 thousand
4. 282,000 282 thousand
5. 5,158,000 5,158 thousand

SET 5
In thousands:

1. 8,500 8.5 thousand
2. 14,400 14.4 thousand
3. 260,800 260.8 thousand
4. 53,700 53.7 thousand
5. 4,839,100 4,839.1 thousand

SET 6
In thousands:

1. 7,130 7.13 thousand
2. 6,520 6.52 thousand
3. 8,040 8.04 thousand
4. 37,290 37.29 thousand
5. 626,750 626.75 thousand

SET 7
In thousands:

1. 2,654 2.654 thousand
2. 9,068 9.068 thousand
3. 13,852 13.852 thousand
4. 824,905 824.905 thousand
5. 1,303,056
   1,303.056 thousand

SET 8
In millions:

1. 8,000,000 8 million
2. 43,000,000 43 million
3. 119,000,000 119 million
4. 2,640,000,000 2,640 million
5. 75,000,000 75 million

SET 9
In millions:

1. 7,300,000 7.3 million
2. 14,600,000 14.6 million
3. 256,800,000 256.8 million
4. 69,200,000 69.2 million
5. 3,108,500,000 3,108.5 million

SET 10
In millions:

1. 11,470,000 11.47 million
2. 6,580,000 6.58 million
3. 73,060,000 73.06 million
4. 408,950,000 408.95 million
5. 1,584,120,000
   1,584.12 million

SET 11
In millions:

1. 28,063,000 28.063 million
2. 4,709,000 4.709 million
3. 85,458,000 85.458 million
4. 396,677,000 396.677 million
5. 54,096,000 54.096 million

SET 12
In billions:

1. 9,000,000,000 9 billion
2. 23,000,000,000 23 billion
3. 102,000,000,000 102 billion
4. 56,000,000,000 56 billion
5. 2,421,000,000,000
   2,421 billion

SET 13
In billions:

1. 7,500,000,000 7.5 billion
2. 15,400,000,000 15.4 bilion
3. 32,700,000,000 32.7 billion
4. 184,300,000,000
5. 1,421,900,000,000
   4. 184.3 billion
   5. 1,421.9 billion

SET 14
In billions:

1. 16,820,000,000 16.82 billion
2. 6,370,000,000 6.37 billion
3. 73,220,000,000 73.22 billion
4. 309,760,000,000
5. 86,940,000,000
   4. 309.76 billion
   5. 86.94 billion

SET 15
In billions:

1. 51,366,000,000 51.366 billion
2. 4,592,000,000 4.592 billion
3. 608,028,000,000 608.028 billion
4. 79,009,000,000
5. 4,312,454,000,000
   4. 79.009 billion
   5. 4,312.454 billion

SET 16
In trillions:

1. 6,000,000,000,000 6 trillion
2. 18,000,000,000,000 18 trillion
3. 263,000,000,000,000 263 trillion
4. 95,000,000,000,000 95 trillion
5. 107,000,000,000,000 107 trillion

SET 17
In trillions:

1. 4,700,000,000,000 4.7 trillion
2. 26,400,000,000,000 26.4 trillion
3. 870,500,000,000,000 870.5 trillion
4. 329,900,000,000,000 329.9 trillion
5. 53,600,000,000,000 53.6 trillion

SET 18
In trillions:

1. 32,080,000,000,000 32.08 trillion
2. 8,450,000,000,000 8.45 trillion
3. 97,530,000,000,000 97.53 trillion
4. 208,410,000,000,000 208.41 trillion
5. 70,250,000,000,000 70.25 trillion

SET 19
In trillions:

1. 17,526,000,000,000 17.526 trillion
2. 2,905,000,000,000 2.905 trillion
3. 56,847,000,000,000 56.847 trillion
4. 290,061,000,000,000 290.061 trillion
5. 388,519,000,000,000 388.519 trillion

SET 20
In thousands of dollars:

1. $63,000 $63 thousand
2. $19,500 $19.5 thousand
3. $420,930 $420.93 thousand
4. $56,078 $56.078 thousand
5. $1,243,610 $1,243.61 thousand

SET 21
In millions of dollars:

1. $84,000,000 $84 million
2. $9,360,000 $9.36 million
3. $450,844,000 $450.844 million
4. $17,525,000 $17.525 million
5. $2,655,007,000 $2,655.007 million

SET 22
In billions of dollars:

1. $58,000,000,000 $58 billion
2. $15,060,000,000 $15.06 billion
3. $8,377,000,000 $8.377 billion
4. $142,552,000,000 $142.522 billion
5. $61,414,000,000 $61.414 billion

SET 23
In trillions of dollars:

1. $6,500,000,000,000 $6.5 trillion
2. $48,320,000,000,000 $48.32 trillion
3. $210,680,000,000,000 $210.68 trillion
4. $36,493,000,000,000 $36.493 trillion
5. $9,057,000,000,000 $9.057 trillion

OPERATIONS WITH SHORTENED NAMES

We add, subtract, multiply, and divide numbers with shortened names just as we do with whole numbers and decimal numbers.

1. Add:

57 million	32 billion	1.4 trillion	47.9 million
166 million	489 billion	6.7 trillion	259.65 million
223 million	58 billion	5.5 trillion	93.78 million
	579 billion	13.6 trillion	401.33 million

2. Add as indicated:

a. 4.6 million + 825 thousand

b. $1.8 billion + $671 million

c. $2.35 trillion + $987 billion

d. $36.83 billion + $56 million

2.a. 5.425 million
  b. $2.471 billion

c. $3.337 trillion
d. $36.886 billion

3. Subtract:

607 billion	56.2 thousand	4.3 million	9.049 trillion
198 billion	7.8 thousand	2.71 million	1.56 trillion
409 billion	48.4 thousand	1.59 million	7.489 trillion

4. Subtract as indicated:

a. 8 billion − 358 million   **c.** $59.12 million − $800 thousand

b. 2.6 trillion − 73 billion   **d.** 64.5 billion − 53 million

5. Multiply: **4.a.** 7.642 billion **b.** 2.527 trillion **c.** $58.32 million **d.** 64.447 billion

a. 5 × $48 billion $240 billion   **c.** 8 × 1.7 trillion 13.6 trillion

b. 12 × $3.09 million $37.08 billion **d.** 25 × $30.64 billion $766 billion

6. Divide:

a. 78 million ÷ 2 39 million   **c.** $13.89 billion ÷ 10 $1.389 billion

b. 69.3 trillion ÷ 5 13.86 trillion   **d.** 826.1 million ÷ 100 million 8.261

7. During a recent year, the United States exported goods worth 129 billion dollars. If the goods imported by the United States were worth 27 billion dollars more than those exported, what was the value of the goods imported? **156 billion dollars**   **8. 33.5 million jobs**

8. Thirty years ago there were about 57 million civilian jobs. Now there are about 90.5 million civilian jobs. Find the amount of increase.

9. Before World War I federal, state, and local governments cost $3 billion. During a recent year they cost $633 billion. How many times as much did government cost recently compared to before World War I? **211**

10. Earnings of a corporation this year doubled from last year. If the earnings last year were $18.2 million, what are the earnings for this year? **$36.4 million**

11. The New York Stock Exchange volume reached 32.85 million shares on Friday. If this is down from 35.47 million shares the day before, find the amount of decrease in volume. **2.62 million shares**

12. Congress proposed to spend $501.3 billion for the year with the government taking in $443.2 billion. What is the deficit under this proposal? **$58.1 billion**

13. The township school budget was $9.79 million last year but is $310,000 more this year. What is the budget this year? **10.1 million dollars**

14. Recently the cost of residential building construction was $6.14 billion, non-residential building construction $3.43 billion, and non-building construction $2.78 billion. What was the total cost of all the construction? **$12.35 billion**

15. The first-quarter earnings of a large corporation rose to $252.5 million from $244.2 million last year and revenues increased to $15.2 billion from $14.2 billion. Find the amounts of increase in both earnings and revenues. **$8.3 million; $1 billion**

204   SHORTENED NAMES FOR LARGE WHOLE NUMBERS

REVIEW OF UNIT FOUR

2.a. Sixty-three point five one million shares; Fifty-two point two eight million shares

1. Read, or write in words, each of the following number names:

a. 657.9 billion

c. $6.73 trillion

b. 71.048 million

d. 903.5 thousand

2. Read, or write in words, each shortened name in the following:

a. Trading volume surged to 63.51 million shares, easily topping the 52.28 million share volume of the previous day.

b. Personal income increased last month by $24.2 billion to a $1.719 trillion annual rate. Twenty-four point two billion dollars; One point seven one nine trillion dollars

c. Total consumer credit expanded by a record $4.07 billion.

d. New homes were started last month at the annual rate of 2.08 million units. 2.c Four point zero seven billion dollars 2.d. Two point zero eight million units

3. Write the complete numeral for each of the following:

a. $18.38 million $18,380,000

c. 220.86 billion 220,860,000,000

b. 257.4 thousand 257,400

d. $8.529 trillion $8,529,000,000,000

4. Write the shortened number name for each of the following:

a. In millions: 67,200,000 67.2 million

b. In billions: 504,730,000,000 504.73 billion

c. In thousands: 1,596,000 1,596 thousand

d. In trillions: 8,331,000,000,000 8.331 trillion

5. a. Add:

948.3 million
26.4 million
397.9 million
_____
1,372.6 million

b. Subtract:

$453.7 billion
233.8 billion
_____
$219.9 billion

c. Multiply: $17.28 trillion

3 × $5.76 trillion

d. Divide: 2.3 million

19.32 million ÷ 8.4

6. a. Add as indicated $10.175 billion

$9.38 billion + $795 million

b. Subtract as indicated: 893 billion

1.4 trillion − 507 billion

7. The current national debt ceiling is $1.08 trillion. If Congress increases the ceiling to $1.131 trillion, what is the amount of increase? $51 billion

8. During a recent year sales of durable goods reached $277.92 billion and sales of nondurable goods were $520.9 billion. What were the total sales? $798.82 billion

9. How many tons below the goal of 1.2 billion tons is this year's coal production of 688.6 million tons? 511.4 million tons

10. Several years ago the sales of a certain corporation were $265.9 million. If sales tripled this year, what is the amount? $797.7 million

COMPETENCY CHECK TEST

Solve each problem and select the letter corresponding to your answer. The numerals in red boxes indicate Exercises where help may be found.

1. Add: 8,416 + 5,267 + 6,329 + 9,755 [1-4]
 a. 39,776 b. 29,677 c. 39,767 d. 29,767

2. Subtract: 40,805 − 30,945 [1-5]
 a. 10,860 b. 9,860 c. 9,850 d. 10,850

3. Multiply: 36 × 68 [1-6]
 a. 504 b. 21,888 c. 2,448 d. Answer not given

4. Divide: 79,550 ÷ 86 [1-7]
 a. 925 b. 840 c. 960 d. 885

5. Find the sum of $5\frac{2}{3}$ and $2\frac{3}{4}$. [1-12]
 a. $7\frac{11}{12}$ b. $8\frac{1}{6}$ c. $8\frac{5}{12}$ d. $8\frac{7}{12}$

6. Subtract: $14\frac{5}{6} - 8$ [1-13]
 a. $5\frac{5}{6}$ b. $5\frac{1}{6}$ c. $6\frac{1}{6}$ d. $6\frac{5}{6}$

7. Multiply: $10 \times 1\frac{3}{5}$ [1-16]
 a. $10\frac{3}{5}$ b. $11\frac{3}{5}$ c. 16 d. 12

8. Divide: $2\frac{5}{8} \div \frac{1}{2}$ [1-17]
 a. $2\frac{3}{8}$ b. $5\frac{1}{4}$ c. $1\frac{5}{16}$ d. $4\frac{3}{8}$

9. Add: 50.8 + .392 [1-23]
 a. 51.192 b. .900 c. 900 d. 50.8392

10. Subtract: .004 − .0003 [1-24]
 a. .001 b. .0043 c. .0037 d. .0001

11. Multiply: .007 × 400 [1-26]
 a. 28 b. 2.8 c. .028 d. .28

12. Divide: 3.5 ÷ .07 [1-27]
 a. .05 b. .5 c. 5 d. 50

13. Write 837.4 million as a complete numeral. [1-34]
 a. 8,374,000 b. 83,740,000,000 c. 837,400,000 d. .8374

14. Write the shortened name for 25,460,000,000 in billions. [1-35]
 a. 2.546 billion c. 254.6 billion
 b. 25.46 billion d. Answer not given

15. The original funding to build the Northeast Corridor railroad system was $1.75 billion. It appears that funding will be increased to $2.5 billion. Find the amount of increase in funding. [1-35]
 a. $1.25 billion b. $750 million c. $75 million d. $1.5 billion

For a Unit Test see p. T44 or *Testing Program*, p. 7.

PERCENT

Earners, consumers, and business people use percent, designated by the symbol "%," extensively in their daily affairs. They experience its use in situations like the following:

Earners may find that a 14% withholding tax, 6.7% social security tax, and perhaps a 3% state or city income tax are deducted from their paychecks. Salespeople may earn 5% commission on what they sell.

Homeowners may receive a 2% discount for paying their real estate taxes in advance, a 3% discount on their gas and electric bills or a 5% discount on their water bills if they pay before the discount period ends. Consumers may buy merchandise at a department store advertising a 40% reduction on certain sales items; they may pay a 12% carrying charge on a new automobile; people may buy jewelry and cosmetics on which they pay a state sales tax of perhaps 1% to 6%. Labels on clothing may indicate the content of cloth like 30% mohair, 70% wool.

In business, a storekeeper may make a 35% profit on sales; a bank may pay $5\frac{1}{2}$% interest on deposits and charge $18\frac{1}{2}$% interest on loans.

Students and teachers also use percent. A student may receive a mark of 83%. A teacher may find the daily class attendance averages 94%.

The Meaning of Percent*

OBJECTIVE:

I. Aim   To find the meaning of percent.

II. Procedure

In our study of fractions we found that $\frac{27}{100}$ is read twenty-seven *hundredths* meaning 27 parts out of 100 equal parts or 27 out of a group of 100. In the study of decimals we found that the decimal .27 also represents 27 *hundredths*. Now we study a third way of representing hundredths. It is called percent and it is indicated by the symbol %. 27% means 27 *hundredths*.

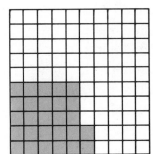

Thus, *hundredths* can be expressed as a percent, a decimal, or a fraction.

It should be noted that 100% of anything is $\frac{100}{100}$ of it or all of it.

A percent may also be considered a ratio (see Exercise 4-27). 27%, which is equivalent to $\frac{27}{100}$, is the ratio of 27 to 100.

III. Sample Solutions

1. Express five hundredths as a percent.   *Answer:* 5%
2. Express $3\frac{1}{2}$ hundredths as a percent.   *Answer:* $3\frac{1}{2}$%
3. Express 19 out of 100 as a percent.   *Answer:* 19%
4. How many hundredths are in 4.9%.   *Answer* 4.9 hundredths
5. Write thirty-seven hundredths as a percent, decimal, and fraction.
          Answer: 37%, .37, and $\frac{37}{100}$

DIAGNOSTIC TEST

1. Express eighteen hundredths as a percent. 18%
2. Express 81 out of 100 as a percent. 81%
3. How many hundredths are in 35%? 35 hundredths
4. Write four hundredths as a percent, decimal, and fraction.
          4%, .04, $\frac{4}{100}$

---

*See page 232.

Basic: Sets 1–2; *Average:* Sets 1–3, Misc. Ex. 1; *Enriched:* Sets 1–4, Misc. Ex. 1–2

208   PERCENT

RELATED PRACTICE EXAMPLES

SET 1

Express each of the following
as a percent:

1. 3 hundredths 3%
2. 9 hundredths 9%
3. 16 hundredths 16%
4. 30 hundredths 30%
5. 74 hundredths 74%
6. 200 hundredths 200%
7. $5\frac{1}{2}$ hundredths $5\frac{1}{2}$%
8. $37\frac{1}{2}$ hundredths $37\frac{1}{2}$%
9. 62.5 hundredths 62.5%
10. $\frac{1}{4}$ hundredth $\frac{1}{4}$%

SET 2

Express each of the following
as a percent:

1. 47 out of 100. 47%
2. 68 out of 100. 68%
3. 7 out of 100. 7%
4. 95 out of 100. 95%
5. 1 out of 100. 1%
6. $4\frac{1}{2}$ out of 100. $4\frac{1}{2}$%
7. $66\frac{2}{3}$ out of 100. $66\frac{2}{3}$%
8. 12.5 out of 100. 12.5%
9. 20 out of 100. 20%
10. $3\frac{3}{4}$ out of 100. $3\frac{3}{4}$%

SET 3

How many hundredths
are in:

1. 4%? 4 hundredths
2. 7%? 7 hundredths
3. 19%? 19 hundredths
4. 45%? 45 hundredths
5. 60%? 60 hundredths
6. 140%? 140 hundredths
7. $87\frac{1}{2}$%? $87\frac{1}{2}$ hundredths
8. $4\frac{3}{4}$%? $4\frac{3}{4}$ hundredths
9. 3.5%? 3.5 hundredths
10. $\frac{1}{2}$%? $\frac{1}{2}$ hundredth

SET 4

Write each of the following
statements as a percent,
decimal, and fraction:

1. Six hundredths 6%, .06, $\frac{6}{100}$
2. Eight hundredths 8%, .08, $\frac{8}{100}$
3. Twenty-three hundredths 23%, .23, $\frac{23}{100}$
4. Sixty-four hundredths 64%, .64, $\frac{64}{100}$
5. Seventy hundredths 70%, .70, $\frac{70}{100}$
6. Twelve and one-half hundredths
7. Two and one-fourth hundredths
8. Three hundred hundredths 300%, 3.00, $\frac{300}{100}$
9. Thirty-two hundredths 32%, .32, $\frac{32}{100}$
10. Three-fourths hundredth $\frac{3}{4}$%, .0075, $\frac{\frac{3}{4}}{100}$

6. $12\frac{1}{2}$%, .125, $\frac{12\frac{1}{2}}{100}$ 7. $2\frac{1}{4}$%, .0225, $\frac{2\frac{1}{4}}{100}$

MISCELLANEOUS EXAMPLES

Explain the meaning of each of the following:

1. Cloth bearing the label: 55% Dacron, 45% wool.
2. Plant food containing 5% nitrogen, 10% phosphoric acid, 5% potash.

1. Cloth content is 55 hundredths Dacron and 45 hundredths wool
2. 5 hundredths of the plant food is nitrogen, 10 hundredths is phosphoric acid, and 5 hundredths is potash

For a Class Activity see p. T33.

Changing Percents to Decimals

OBJECTIVE:

I. Aim To change a percent to a decimal.

II. Procedure

1. Rewrite the given numeral, omitting the percent sign.
2. Move the decimal point two places to the left. In a numeral naming a whole number a decimal point is understood after the units figure.

III. Sample Solutions

1. $5\% = .05$ Answer: .05
2. $84\% = .84$ Answer: .84
3. $30\% = .30$ or $.3$ Answer: .3
4. $125\% = 1.25$ Answer: 1.25
5. $110\% = 1.10$ or 1.1 Answer: 1.10 or 1.1
6. $2\frac{1}{2}\% = .02\frac{1}{2}$ or $.025$ Answer: $.02\frac{1}{2}$ or .025
7. $102\frac{1}{4}\% = 1.02\frac{1}{4}$ or 1.0225 Answer: $1.02\frac{1}{4}$ or 1.0225
8. $37.5\% = .375$ Answer: .375
9. $8.34\% = .0834$ Answer: .0834
10. $.2\%$ or $0.2\% = .002$ Answer: .002
11. $\frac{7}{8}\% = .00\frac{7}{8}$ or $.00875$ Answer: $.00\frac{7}{8}$ or .00875
12. $700\% = 7.00$ or 7 Answer: 7

DIAGNOSTIC TEST

Express each of the following percents as a decimal:

1. 8% .08
2. 53% .53
3. 90% .90 or .9
4. 119% 1.19
5. 160% 1.60 or 1.6

6. 500% 5
7. $12\frac{1}{2}\%$ .12$\frac{1}{2}$ or .125
8. $5\frac{1}{4}\%$ .05$\frac{1}{4}$ or .0525
9. $40\frac{5}{8}\%$ .40$\frac{5}{8}$ or .40625

10. $105\frac{3}{4}\%$
11. 27.5% .275
12. 12.25% .1225
13. 8.6% .086
10. 1.05$\frac{3}{4}$ or 1.0575

14. 4.82% .0482
15. 352.875% 3.52875
16. $.16\%$ .0016
17. $\frac{3}{4}\%$ .00$\frac{3}{4}$ or .0075

Basic: Sets 1–13, Misc. Ex. 1; *Average:* Sets 1–15, Misc. Ex. 1–2; *Enriched:* Sets 1–17, Misc. Ex. 1–3

210   **PERCENT**

RELATED PRACTICE EXAMPLES

Express each of the following percents as a decimal:

SET 1	SET 2(a)	SET 2(b)	SET 3	SET 4
1. 6% .06	**1.** 16% .16	**1.** 27% .27	**1.** 40% .40 or .4	**1.** 134% 1.34
2. 9% .09	**2.** 38% .38	**2.** 81% .81	**2.** 70% .70 or .7	**2.** 157% 1.57
3. 1% .01	**3.** 45% .45	**3.** 99% .99	**3.** 20% .20 or .2	**3.** 148% 1.48
4. 3% .03	**4.** 67% .67	**4.** 13% .13	**4.** 60% .60 or .6	**4.** 106% 1.06
5. 4% .04	**5.** 58% .58	**5.** 42% .42	**5.** 10% .10 or .1	**5.** 175% 1.75

SET 5	SET 6	SET 7	SET 8	SET 9
1. 130% 1.30 or 1.3	**1.** 100% 1	**1.** $37\frac{1}{2}\%$ $.37\frac{1}{2}$ or .375	**1.** $4\frac{1}{2}\%$ $.04\frac{1}{2}$ or .045	**1.** $60\frac{1}{2}\%$ $.60\frac{1}{2}$ or .605
2. 120% 1.20 or 1.2	**2.** 200% 2	**2.** $62\frac{1}{2}\%$ $.62\frac{1}{2}$ or .625	**2.** $6\frac{1}{2}\%$ $.06\frac{1}{2}$ or .065	**2.** $20\frac{1}{2}\%$ $.20\frac{1}{2}$ or .205
3. 180% 1.80 or 1.8	**3.** 300% 3	**3.** $83\frac{1}{3}\%$ $.83\frac{1}{3}$	**3.** $3\frac{3}{4}\%$ $.03\frac{3}{4}$ or .0375	**3.** $10\frac{3}{8}\%$ $.10\frac{3}{8}$ or .10375
4. 150% 1.50 or 1.5	**4.** 600% 6	**4.** $16\frac{2}{3}\%$ $.16\frac{2}{3}$	**4.** $2\frac{7}{8}\%$ $.02\frac{7}{8}$ or .02875	**4.** $50\frac{2}{3}\%$ $.50\frac{2}{3}$
5. 140% 1.40 or 1.4	**5.** 400% 4	**5.** $18\frac{3}{4}\%$ $.18\frac{3}{4}$ or .1875	**5.** $5\frac{4}{5}\%$ $.05\frac{4}{5}$ or .058	**5.** $70\frac{1}{4}\%$ $.70\frac{1}{4}$ or .7025

SET 10	SET 11	SET 12	SET 13	SET 14
$1.00\frac{7}{8}$ or 1.00875 **1.** $100\frac{7}{8}\%$	**1.** 87.5% .875	**1.** 17.75% .1775	**1.** 3.5% .035	**1.** 1.25% .0125
$1.01\frac{1}{4}$ or 1.0125 **2.** $101\frac{1}{4}\%$	**2.** 12.5% .125	**2.** 34.25% .3425	**2.** 2.8% .028	**2.** 3.67% .0367
$1.06\frac{1}{2}$ or 1.065 **3.** $106\frac{1}{2}\%$	**3.** 46.4% .464	**3.** 56.94% .5694	**3.** 4.7% .047	**3.** 5.33% .0533
$1.52\frac{9}{16}$ **4.** $152\frac{9}{16}\%$	**4.** 57.3% .573	**4.** 99.44% .9944	**4.** 9.3% .093	**4.** 2.08% .0208
$1.27\frac{3}{5}$ or 1.276 **5.** $127\frac{3}{5}\%$	**5.** 95.1% .951	**5.** 82.09% .8209	**5.** 6.4% .064	**5.** 7.19% .0719

SET 15	SET 16	SET 17
.26375 **1.** 26.375%	**1.** .7% .007	**1.** $\frac{1}{2}\%$ $.00\frac{1}{2}$ or .005
.31625 **2.** 31.625%	**2.** .02% .0002	**2.** $\frac{1}{4}\%$ $.00\frac{1}{4}$ or .0025
1.28333 **3.** 128.333%	**3.** .9% .009	**3.** $\frac{5}{8}\%$ $.00\frac{5}{8}$ or .00625
4.32125 **4.** 432.125%	**4.** .85% .0085	**4.** $\frac{2}{3}\%$ $.00\frac{2}{3}$
8.95667 **5.** 895.667%	**5.** .49% .0049	**5.** $\frac{7}{16}\%$ $.00\frac{7}{16}$

MISCELLANEOUS EXAMPLES

Express each of the following percents as a decimal:

1. Department store sales increased 8%. .08

2. The increase in the cost of living was 6.5%. .065

3. The market price of a corporation bond remained at $100\frac{5}{8}\%$. $1.00\frac{5}{8}$ or 1.00625

Changing Decimals to Percents

OBJECTIVE:

I. Aim To change a decimal to a percent.

II. Procedure

1. Rewrite the given numeral.

2. Move the decimal point two places to the right. Do not write the decimal point if, after moving it, it is located at the end of the numeral.

3. Write the percent sign, %.

III. Sample Solutions

1. $.07 = 7\%$	*Answer:* 7%
2. $.19 = 19\%$	*Answer:* 19%
3. $1.34 = 134\%$	*Answer:* 134%
4. $.5 = 50\%$	*Answer:* 50%
5. $.06\frac{1}{2} = 6\frac{1}{2}\%$	*Answer:* $6\frac{1}{2}\%$
6. $1.02\frac{3}{4} = 102\frac{3}{4}\%$	*Answer:* $102\frac{3}{4}\%$
7. $.435 = 43.5\%$	*Answer:* 43.5%
8. $.0575 = 5.75\%$	*Answer:* 5.75%
9. $.0036 = .36\%$	*Answer:* .36% or 0.36%
10. $.00\frac{2}{5} = \frac{2}{5}\%$	*Answer:* $\frac{2}{5}\%$
11. 8 or $8.00 = 800\%$	*Answer:* 800%
12. $1.1 = 110\%$	*Answer:* 110%
13. $3.687 = 368.7\%$	*Answer:* 368.7%
14. $.90\frac{1}{4} = 90\frac{1}{4}\%$	*Answer:* $90\frac{1}{4}\%$
15. $.4\frac{1}{2} = .45 = 45\%$	*Answer:* 45%

DIAGNOSTIC TEST

Express each of the following decimals as a percent:

1. .06 6% **5.** 1.4 140% **9.** $1.66\frac{2}{3}$ $166\frac{2}{3}\%$ **13.** 2.875 287.5%

2. .32 32% **6.** $.37\frac{1}{2}$ $37\frac{1}{2}\%$ **10.** .625 62.5% **14.** 2 200%

3. .7 70% **7.** $.04\frac{1}{2}$ $4\frac{1}{2}\%$ **11.** .0467 4.67% **15.** $.00\frac{3}{8}$ $\frac{3}{8}\%$

4. 1.12 112% **8.** $.60\frac{3}{4}$ $60\frac{3}{4}\%$ **12.** $.1\frac{1}{4}$ 12.5% **16.** .0025 .25%

RELATED PRACTICE EXAMPLES

Express each of the following decimals as a percent:

SET 1	SET 2	SET 3	SET 4	SET 5	SET 6
1. .01 1%	**1.** .28 28%	**1.** .6 60%	**1.** 1.39 139%	**1.** 1.2 120%	**1.** .12$\frac{1}{2}$ 12$\frac{1}{2}$%
2. .08 8%	**2.** .75 75%	**2.** .3 30%	**2.** 1.92 192%	**2.** 1.8 180%	**2.** .33$\frac{1}{3}$ 33$\frac{1}{3}$%
3. .04 4%	**3.** .59 59%	**3.** .8 80%	**3.** 1.18 118%	**3.** 1.7 170%	**3.** .83$\frac{5}{6}$ 83$\frac{5}{6}$%
4. .03 3%	**4.** .16 16%	**4.** .1 10%	**4.** 1.50 150%	**4.** 1.3 130%	**4.** .42$\frac{2}{7}$ 42$\frac{2}{7}$%
5. .05 5%	**5.** .93 93%	**5.** .9 90%	**5.** 1.44 144%	**5.** 1.6 160%	**5.** .18$\frac{3}{4}$ 18$\frac{3}{4}$%

SET 7	SET 8	SET 9	SET 10	SET 11	SET 12
1. .01$\frac{1}{2}$ 1$\frac{1}{2}$%	**1.** .10$\frac{1}{2}$ 10$\frac{1}{2}$%	**1.** 1.37$\frac{1}{2}$ 137$\frac{1}{2}$%	**1.** .875 87.5%	**1.** .2625 26.25%	**1.** .2$\frac{1}{4}$ 25%
2. .05$\frac{1}{3}$ 5$\frac{1}{3}$%	**2.** .30$\frac{2}{3}$ 30$\frac{2}{3}$%	**2.** 1.83$\frac{1}{3}$ 183$\frac{1}{3}$%	**2.** .125 12.5%	**2.** .0525 5.25%	**2.** .6$\frac{1}{4}$ 65%
3. .03$\frac{3}{4}$ 3$\frac{3}{4}$%	**3.** .70$\frac{3}{4}$ 70$\frac{3}{4}$%	**3.** 1.00$\frac{1}{2}$ 100$\frac{1}{2}$%	**3.** .347 34.7%	**3.** .0875 8.75%	**3.** .3$\frac{1}{4}$ 32.5%
4. .04$\frac{2}{3}$ 4$\frac{2}{3}$%	**4.** .50$\frac{1}{2}$ 50$\frac{1}{2}$%	**4.** 1.16$\frac{2}{3}$ 116$\frac{2}{3}$%	**4.** .078 7.8%	**4.** .0233 2.33%	**4.** .1$\frac{3}{4}$ 17.5%
5. .06$\frac{5}{6}$ 6$\frac{5}{6}$%	**5.** .40$\frac{7}{8}$ 40$\frac{7}{8}$%	**5.** 1.05$\frac{3}{4}$ 105$\frac{3}{4}$%	**5.** .989 98.9%	**5.** .65125 65.125%	**5.** .8$\frac{7}{8}$ 88.75%

SET 13	SET 14	SET 15	SET 16
1. 1.245 124.5%	**1.** 1 100%	**1.** .00$\frac{1}{4}$ $\frac{1}{4}$%	**1.** .005 .5%
2. 1.375 137.5	**2.** 3 300%	**2.** .00$\frac{1}{2}$ $\frac{1}{2}$%	**2.** .0075 .75%
3. 2.667 266.7%	**3.** 5 500%	**3.** .00$\frac{3}{5}$ $\frac{3}{5}$%	**3.** .0067 .67%
4. 3.7275 372.75%	**4.** 4 400%	**4.** .00$\frac{5}{8}$ $\frac{5}{8}$%	**4.** .008 .8%
5. 2.9625 296.25%	**5.** 6 600%	**5.** .00$\frac{2}{3}$ $\frac{2}{3}$%	**5.** .00875 .875%

PRACTICAL APPLICATIONS

Baseball Averages

1. The pennant-winning team led the league with the standing of .625. What percent of the games played did it win? 62$\frac{1}{2}$%

2. What percent of the number of times at bat must a batter hit safely to average .300? .375? .400? 30%; 37.5%; 40%

3. A pitcher has an average of .800. What percent of the games did she lose? 20%

4. What percent of the games played must a team win to have a standing of .750? 75%

5. On what percent of his chances did a shortstop make an error if his fielding average is .875? 12$\frac{1}{2}$%

OBJECTIVE:

I. Aim To change a percent to a common fraction or a mixed number.

II. Procedure

1. To change a percent to a common fraction (2 methods):

a. Make a fraction by writing the given numeral as the numerator and the numeral 100 as the denominator. Then reduce to lowest terms. (See sample solution 1.)

b. Or, if the common fraction equivalent is known, write the common fraction directly. (See sample solution 2.)

2. If the percent is greater than 100%, the answer is a mixed number. (See sample solution 3.)

III. Sample Solutions

1. $8\% = \frac{8}{100} = \frac{2}{25}$

Answer: $\frac{2}{25}$

2. $40\% = \frac{2}{5}$

Answer: $\frac{2}{5}$

3. $130\% = \frac{130}{100} = 1\frac{3}{10}$

Answer: $1\frac{3}{10}$

DIAGNOSTIC TEST

Express each of the following percents as a common fraction or mixed number:

1. 75% $\frac{3}{4}$ **2.** $16\frac{2}{3}\%$ $\frac{1}{6}$ **3.** 2% $\frac{1}{50}$ **4.** 125% $1\frac{1}{4}$

Basic: Sets 1–3, Appl. 1–3; *Average:* Sets 1–3, Appl. 1–4; *Enriched:* Sets 1–4, Appl. 1–6

RELATED PRACTICE EXAMPLES

Express each of the following percents as a common fraction or mixed number:

SET 1

1. 50% $\frac{1}{2}$ **6.** 60% $\frac{3}{5}$
2. 25% $\frac{1}{4}$ **7.** 30% $\frac{3}{10}$
3. 10% $\frac{1}{10}$ **8.** 40% $\frac{2}{5}$
4. 80% $\frac{4}{5}$ **9.** 70% $\frac{7}{10}$
5. 90% $\frac{9}{10}$ **10.** 20% $\frac{1}{5}$

SET 2

1. $33\frac{1}{3}\%$ $\frac{1}{3}$ **6.** $37\frac{1}{2}\%$ $\frac{3}{8}$
2. $87\frac{1}{2}\%$ $\frac{7}{8}$ **7.** $66\frac{2}{3}\%$ $\frac{2}{3}$
3. $83\frac{1}{3}\%$ $\frac{5}{6}$ **8.** $6\frac{1}{4}\%$ $\frac{1}{16}$
4. $8\frac{1}{3}\%$ $\frac{1}{12}$ **9.** $16\frac{2}{3}\%$ $\frac{1}{6}$
5. $12\frac{1}{2}\%$ $\frac{1}{8}$ **10.** $62\frac{1}{2}\%$ $\frac{5}{8}$

Related Resources: *Practical Applications in Mathematics:* Diag. Test and Prac. Ex., p. 60

213

	SET 3				SET 4		

SET 3

1. 6% $\frac{3}{50}$ **6.** 24% $\frac{6}{25}$

2. 4% $\frac{1}{25}$ **7.** 52% $\frac{13}{25}$

3. 5% $\frac{1}{20}$ **8.** 18% $\frac{9}{50}$

4. 9% $\frac{9}{100}$ **9.** 85% $\frac{17}{20}$

5. 46% $\frac{23}{50}$ **10.** 36% $\frac{9}{25}$

SET 4

1. 110% $1\frac{1}{10}$ **6.** 230% $2\frac{3}{10}$

2. 150% $1\frac{1}{2}$ **7.** 148% $1\frac{12}{25}$

3. $133\frac{1}{3}$% $1\frac{1}{3}$ **8.** 106% $1\frac{3}{50}$

4. $162\frac{1}{2}$% $1\frac{5}{8}$ **9.** 225% $2\frac{1}{4}$

5. 175% $1\frac{3}{4}$ **10.** 180% $1\frac{4}{5}$

PRACTICAL APPLICATIONS

1. 30% of the pupils at the Southeast High School are in the ninth year. What part of the student body is in the ninth year? $\frac{3}{10}$

2. If $87\frac{1}{2}$% of a class received a passing mark in a science test, what part of the class failed in the test? $\frac{1}{8}$

3. What part of a graduating class is planning to go to college if 40% filed applications for entrance? $\frac{2}{5}$

4. If Mrs. Romano pays 25% down on a house, what part of the purchase price is her down payment? $\frac{1}{4}$

5. In a certain area $16\frac{2}{3}$% of all the crops were destroyed by floods. What part of the harvest was saved? $\frac{5}{6}$

6. If the school baseball team won 70% of the games played, what part of the games did it lose? $\frac{3}{10}$

REFRESH YOUR SKILLS

1. Add:

  9,625
  84,517
  862
  54,163
  7,989 157,156

2. Subtract:

  526,809
  348,719
  178,090

3. Multiply:

  784
  697
  546,448

4. Divide: 987

  $2,240\overline{)2,210,880}$

5. Add: $6\frac{2}{3} + 1\frac{7}{12} + 4\frac{5}{6}$ $13\frac{1}{12}$

6. Subtract: $8\frac{1}{4} - 6\frac{7}{10}$ $1\frac{11}{20}$

7. Multiply: $7\frac{5}{6} \times 1\frac{1}{8}$ $8\frac{13}{16}$

8. Divide: $16 \div 3\frac{3}{4}$ $4\frac{4}{15}$

9. Add: $8.9 + .45 + 21$ 30.35

10. Subtract: $8.03 - .7$ 7.33

11. Multiply: $4.06 \times .005$ .0203

12. Divide: $.16\overline{).96}$ 6

13. Round $18.698 to the nearest cent. $18.70

14. Write 407.9 *million* as a complete numeral. 407,900,000

15. .8 of what amount is $560? $700

Changing Common Fractions to Percents

OBJECTIVE:

I. Aim To change common fractions and mixed numbers to percents.

II. Procedure

1. To change a common fraction to a percent (3 methods):

a. (1) Divide the numerator by the denominator, finding the quotient to two decimal places.

(2) Rewrite the quotient, omitting the decimal point.

(3) Then write the percent sign, %. (See sample solutions 1 and 4.)

b. Or, if the percent equivalent of the given fraction is known, write the percent directly. Sometimes it may be necessary to express the given fraction in lowest terms. (See sample solution 2.)

c. Or if possible, raise the given fraction to higher terms or reduce it to lower terms so that its denominator is 100. Then rewrite the numerator, replacing the denominator with the percent sign, %. (See sample solution 3.)

2. To change a mixed number to a percent:

Change the mixed number to a mixed decimal. Then follow step 1a, parts (2) and (3). (See sample solution 7.)

III. Sample Solutions

1. Change $\frac{18}{25}$ to percent.

$$\frac{18}{25} = 25\overline{)18.00} \quad \begin{array}{r} .72 = 72\% \\ \end{array}$$
$$\begin{array}{r} 17\ 5 \\ \hline 50 \\ 50 \\ \hline \end{array}$$

Answer: 72%

2. Change $\frac{24}{40}$ to percent.

$$\frac{24}{40} = \frac{3}{5} = 60\%$$

Answer: 60%

3. Change $\frac{7}{50}$ to percent.

$$\frac{7}{50} = \frac{14}{100} = 14\%$$

Answer: 14%

4. Change $\frac{2}{9}$ to percent.

$$\frac{2}{9} = 9\overline{)2.00} \quad \begin{array}{r} .22\frac{2}{9} = 22\frac{2}{9}\% \\ \end{array}$$

Answer: $22\frac{2}{9}\%$

5. Change $\frac{2}{30}$ to percent.

$$\frac{2}{30} = \frac{1}{15} = 15\overline{)1.00} \quad \begin{array}{r} .06\frac{2}{3} = 6\frac{2}{3}\% \\ \end{array}$$

Answer: $6\frac{2}{3}\%$

6. Change $\frac{11}{8}$ to percent.

$$\frac{11}{8} = 8\overline{)11.00}^{\,1.37\frac{1}{2}} = 137\frac{1}{2}\%$$

Answer: $137\frac{1}{2}\%$

7. Change $1\frac{1}{4}$ to percent.

$$1\frac{1}{4} = 1.25 = 125\%$$

or $\quad 1\frac{1}{4} = 125\%$

since $\quad 1 = 100\%$ and $\frac{1}{4} = 25\%$

Answer: 125%

DIAGNOSTIC TEST

Express each of the following fractions and mixed numbers as a percent:

1. $\frac{3}{4}$ 75% **4.** $\frac{19}{50}$ 38% **7.** $\frac{7}{9}$ $77\frac{7}{9}\%$ **10.** $1\frac{1}{2}$ 150% **13.** $2\frac{2}{3}$ $266\frac{2}{3}\%$

2. $\frac{2}{3}$ $66\frac{2}{3}\%$ **5.** $\frac{8}{400}$ 2% **8.** $\frac{24}{56}$ $42\frac{6}{7}\%$ **11.** $\frac{7}{4}$ 175% **14.** $\frac{12}{5}$ 240%

3. $\frac{3}{100}$ 3% **6.** $\frac{9}{36}$ 25% **9.** $\frac{18}{18}$ 100% **12.** $\frac{72}{64}$ $112\frac{1}{2}\%$

Basic: Sets 1–9, Appl. 1–2; *Average:* Sets 1–12, Appl. 1–4; *Enriched:* Sets 1–14, Appl. 1–5

RELATED PRACTICE EXAMPLES

Express each of the following fractions and mixed numbers as a percent:

SET 1

1. $\frac{1}{4}$ 25%
2. $\frac{2}{5}$ 40%
3. $\frac{1}{2}$ 50%
4. $\frac{7}{10}$ 70%
5. $\frac{4}{5}$ 80%

SET 2

1. $\frac{5}{6}$ $83\frac{1}{3}\%$
2. $\frac{3}{8}$ $37\frac{1}{2}\%$
3. $\frac{1}{3}$ $33\frac{1}{3}\%$
4. $\frac{7}{8}$ $87\frac{1}{2}\%$
5. $\frac{1}{6}$ $16\frac{2}{3}\%$

SET 3

1. $\frac{7}{100}$ 7%
2. $\frac{39}{100}$ 39%
3. $\frac{145}{100}$ 145%
4. $\frac{87\frac{1}{2}}{100}$ $87\frac{1}{2}\%$
5. $\frac{81}{100}$ 81%

SET 4

1. $\frac{27}{50}$ 54%
2. $\frac{4}{25}$ 16%
3. $\frac{13}{20}$ 65%
4. $\frac{31}{50}$ 62%
5. $\frac{16}{25}$ 64%

SET 5

1. $\frac{18}{200}$ 9%
2. $\frac{35}{700}$ 5%
3. $\frac{12}{300}$ 4%
4. $\frac{60}{500}$ 12%
5. $\frac{135}{900}$ 15%

SET 6

1. $\frac{16}{24}$ $66\frac{2}{3}\%$
2. $\frac{39}{65}$ 60%
3. $\frac{40}{48}$ $83\frac{1}{3}\%$
4. $\frac{63}{72}$ $87\frac{1}{2}\%$
5. $\frac{33}{88}$ $37\frac{1}{2}\%$

SET 7

1. $\frac{5}{7}$ $71\frac{3}{7}\%$
2. $\frac{8}{11}$ $72\frac{8}{11}\%$
3. $\frac{4}{9}$ $44\frac{4}{9}\%$
4. $\frac{7}{15}$ $46\frac{2}{3}\%$
5. $\frac{5}{18}$ $27\frac{7}{9}\%$

SET 8

1. $\frac{20}{36}$ $55\frac{5}{9}\%$
2. $\frac{45}{99}$ $45\frac{5}{11}\%$
3. $\frac{42}{49}$ $85\frac{5}{7}\%$
4. $\frac{65}{75}$ $86\frac{2}{3}\%$
5. $\frac{30}{105}$ $28\frac{4}{7}\%$

SET 9

1. $\frac{45}{45}$ 100%
2. $\frac{7}{7}$ 100%
3. $\frac{39}{39}$ 100%
4. $\frac{84}{84}$ 100%
5. $\frac{156}{156}$ 100%

SET 10

1. $1\frac{3}{4}$ 175%
2. $1\frac{2}{5}$ 140%
3. $1\frac{1}{3}$ $133\frac{1}{3}\%$
4. $1\frac{1}{7}$ $157\frac{1}{7}\%$
5. $1\frac{5}{8}$ $162\frac{1}{2}\%$

SET 11

1. $\frac{5}{3}$ $166\frac{2}{3}$%
2. $\frac{13}{8}$ $162\frac{1}{2}$%
3. $\frac{8}{7}$ $114\frac{2}{7}$%
4. $\frac{11}{9}$ $122\frac{2}{9}$%
5. $\frac{9}{5}$ 180%

SET 12

1. $\frac{57}{38}$ 150%
2. $\frac{65}{52}$ 125%
3. $\frac{36}{27}$ $133\frac{1}{3}$%
4. $\frac{70}{49}$ $142\frac{6}{7}$%
5. $\frac{90}{48}$ $187\frac{1}{2}$%

SET 13

1. $2\frac{5}{8}$ $262\frac{1}{2}$%
2. $3\frac{4}{5}$ 380%
3. $4\frac{2}{3}$ $466\frac{2}{3}$%
4. $3\frac{1}{7}$ $314\frac{2}{7}$%
5. $2\frac{3}{4}$ 275%

SET 14

1. $\frac{8}{3}$ $266\frac{2}{3}$%
2. $\frac{14}{4}$ 350%
3. $\frac{17}{6}$ $283\frac{1}{3}$%
4. $\frac{78}{18}$ $433\frac{1}{3}$%
5. $\frac{23}{10}$ 230%

PRACTICAL APPLICATIONS

1. If Maureen received $\frac{3}{5}$ of all votes cast in a homeroom election, what percent of the votes did she get? 60%

2. What percent of the community contributed to the welfare fund if $\frac{5}{6}$ of the community contributed? $83\frac{1}{3}$%

3. If $\frac{7}{10}$ of the student body participated in the interclass athletic games, what percent of the students did not participate? 30%

4. A salesman receives a commission of $\frac{3}{20}$ of the amount of his sales. What percent commission does he receive? 15%

5. A department store advertises a $\frac{1}{4}$ reduction on all merchandise. What is the percent mark-down in this sale? 25%

REFRESH YOUR SKILLS

1. Add:

692,865
734,192
802,371
563,816
814,935
629,683
4,237,862

2. Subtract:

4,502,097
3,968,498
533,599

3. Multiply:

54,008
984
53,143,872

4. Divide: 9,879

$675\overline{)6,668,325}$

5. Add:

$8\frac{3}{4} + 3\frac{17}{32}$ $12\frac{9}{32}$

6. Subtract:

$6\frac{4}{5} - 2\frac{5}{6}$ $3\frac{29}{30}$

7. Multiply:

$3\frac{1}{7} \times 4\frac{2}{3}$ $14\frac{2}{3}$

8. Divide:

$10\frac{1}{2} \div 2\frac{2}{5}$ $4\frac{3}{8}$

9. Add:

$4.26 + $9.83 + $18.97 + $153.62 $186.68

10. Subtract:

$100 - $71.06 $28.94

11. Multiply:

$.62\frac{1}{2} \times $500 $312.50

12. Divide:

$.24\overline{)$18}$ 75

13. Express 80% as a common fraction. $\frac{4}{5}$

14. Express 10.9% as a decimal. .109

15. Express .0613 as a percent. 6.13%

1-41 Finding a Percent of a Number*

OBJECTIVE:

I. Aim   To find a percent of a number.

II. Procedure
1. Change the percent to a decimal or common fraction.
2. Multiply the given number by this decimal or common fraction.

III. Sample Solutions

1. Find 23% of 64.

$$
\begin{array}{r}
64 \\
.23 \\
\hline
1\,92 \\
12\,8 \\
\hline
14.72
\end{array}
$$

Answer: 14.72

2. Find 3% of 18.

$$
\begin{array}{r}
18 \\
.03 \\
\hline
.54
\end{array}
$$

Answer: .54

3. Find 4% of $200.

$$\frac{4}{100} \times \$200 = \$8$$

Answer: $8

4. Find 60% of 175.

$$\frac{3}{5} \times \overset{35}{\cancel{175}} = 105$$

Answer: 105

5. Find 114% of 240.

$$
\begin{array}{r}
240 \\
1.14 \\
\hline
9\,60 \\
24\,0 \\
240 \\
\hline
273.60
\end{array}
$$

Answer: 273.6

6. Find $3\frac{1}{2}$% of 40.

$$
\begin{array}{r}
40 \\
.035 \\
\hline
200 \\
1\,20 \\
\hline
1.400
\end{array}
\quad \text{or} \quad
\begin{array}{r}
40 \\
.03\frac{1}{2} \\
\hline
1\,20 \\
20 \\
\hline
1.40
\end{array}
$$

$$\frac{1}{2} \times \overset{20}{\cancel{40}} = 20$$

Answer: 1.4

7. Find $\frac{3}{4}$% of 650.

$$
\begin{array}{r}
650 \\
.0075 \\
\hline
3250 \\
4\,550 \\
\hline
4.8750
\end{array}
$$

Answer: 4.875

8. Find .4% of 52.

$$
\begin{array}{r}
52 \\
.004 \\
\hline
.208
\end{array}
$$

Answer: .208

9. Find 4% of $12.96 to nearest cent.

$$
\begin{array}{r}
\$12.96 \\
.04 \\
\hline
\$.5184 = \$.52
\end{array}
$$

Answer: $.52

---

*See page 232.

Related Resources: *Practical Applications in Mathematics:* Diag. Test and Prac. Ex., p. 61 **218**

DIAGNOSTIC TEST

Find the following:

1. 18% of 46 8.28   2. 6% of 24 1.44   3. 39% of 6.75 2.6325

Change percents to common fractions in examples 4, 5, and 6.

4. 3% of 5,000 150   5. 25% of 36 9   6. $83\frac{1}{3}$% of 582 485

7. 127% of 743 943.61   8. 140% of 295 413   9. 200% of 75 150

10. $4\frac{1}{2}$% of 624 28.08   11. $50\frac{3}{4}$% of 840 426.3   12. $\frac{1}{4}$% of 300 .75

13. 7.8% of 45 3.51   14. .3% of 160 .48   15. 5% of $29 $1.45

16. 2% of $4.68 (to nearest cent) $.09

Basic: Sets 1–12, Prob. 1–8; **Average:** Sets 1–15(a), Prob. 1–12; **Enriched:** Sets 1–16(b), Prob. 1–14

RELATED PRACTICE EXAMPLES

Find the following:

SET 1	SET 2	SET 3
1. 24% of 52 12.48	1. 2% of 18 .36	1. 2% of 3.6 .072
2. 87% of 41 35.67	2. 4% of 93 3.72	2. 5% of 8.24 .412
3. 63% of 75 47.25	3. 5% of 84 4.2	3. 3% of 9.62 .2886
4. 14% of 80 11.2	4. 9% of 50 4.5	4. 18% of 4.7 .846
5. 92% of 48 44.16	5. 1% of 85 .85	5. 27% of 8.73 2.3571
6. 16% of 240 38.4	6. 3% of 200 6	6. 1% of 15.40 .154
7. 59% of 936 552.24	7. 6% of 325 19.5	7. 6% of 27.75 1.665
8. 45% of 819 368.55	8. 7% of 928 64.96	8. 5% of 295.08 14.754
9. 32% of 2,907 930.24	9. 8% of 1,540 123.2	9. 25% of 75.24 18.81
10. 21% of 4,384 920.64	10. 4% of 8,462 338.48	10. 38% of 251.69 95.6422

Change percents to common fractions in sets 4, 5, and 6.

SET 4	SET 5	SET 6
1. 5% of 200 10	1. 10% of 30 3	1. $12\frac{1}{2}$% of 800 100
2. 4% of 4,000 160	2. 50% of 62 31	2. $37\frac{1}{2}$% of 312 117
3. 8% of 9,000 720	3. 40% of 65 26	3. $62\frac{1}{2}$% of 672 420
4. 7% of 3,000 210	4. 75% of 92 69	4. $83\frac{1}{3}$% of 5,820 4,850
5. 6% of 650 39	5. 30% of 50 15	5. $8\frac{1}{3}$% of 8,280 690
6. 18% of 500 90	6. 60% of 200 120	6. $33\frac{1}{3}$% of 768 256
7. 32% of 2,000 640	7. 90% of 740 666	7. $16\frac{2}{3}$% of 804 134
8. 29% of 10,000 2,900	8. 25% of 328 82	8. $66\frac{2}{3}$% of 987 658
9. 95% of 5,300 5,035	9. 20% of 735 147	9. $87\frac{1}{2}$% of 4,368 3,822
10. 83% of 7,250 6,017.5	10. 80% of 1,225 980	10. $6\frac{1}{4}$% of 9,328 583

SET 7
1. 116% of 28 32.48
2. 105% of 57 59.85
3. 138% of 120 165.6
4. 147% of 825 1,212.75
5. 235% of 1,500 3,525

SET 8
1. 120% of 45 54
2. 150% of 274 411
3. 125% of 848 1,060
4. $133\frac{1}{3}$% of 246 328
5. $137\frac{1}{2}$% of 5,640 7,755

SET 9
1. 100% of 72 72
2. 500% of 400 2,000
3. 200% of 325 650
4. 400% of 2,000 8,000
5. 300% of 3,854 11,562

SET 10
1. $2\frac{1}{2}$% of 30 .75
2. $5\frac{1}{2}$% of 180 9.9
3. $1\frac{1}{2}$% of 2,500 37.5
4. $4\frac{3}{4}$% of 3,000 142.5
5. $5\frac{1}{4}$% of 824 43.26

SET 11
1. $24\frac{1}{2}$% of 20 4.9
2. $142\frac{3}{4}$% of 64 91.36
3. $10\frac{1}{2}$% of 36 3.78
4. $130\frac{1}{2}$% of 245 319.725
5. $98\frac{7}{8}$% of 5,000 4,943.75

SET 12
1. $\frac{1}{2}$% of 48 .24
2. $\frac{1}{3}$% of 930 3.1
3. $\frac{3}{4}$% of 219 1.6425
4. $\frac{5}{8}$% of 600 3.75
5. $\frac{7}{16}$% of 3,500 15.3125

SET 13
1. 2.5% of 36 .9
2. 4.7% of 840 39.48
3. 6.75% of 725 48.9375
4. 5.625% of 192 10.8
5. 15.375% of 2,400 369

SET 14
1. .2% of 8 .016
2. 0.1% of 95 .095
3. 0.25% of 500 1.25
4. .375% of 848 3.18
5. $.66\frac{2}{3}$% of 759 5.06

Find correct to nearest cent:

SET 15(a)
1. 4% of $5 $.20
2. 2% of $94 $1.88
3. 6% of $273 $16.38
4. 1% of $3,000 $30
5. 10% of $50 $5
6. 40% of $240 $96
7. 23% of $19 $4.37
8. 75% of $760 $570
9. $33\frac{1}{3}$% of $6,900 $2,300
10. 41% of $5,643 $2,313.63

SET 15(b)
1. 113% of $42 $47.46
2. 150% of $264 $396
3. $2\frac{1}{2}$% of $300 $7.50
4. $4\frac{1}{2}$% of $740 $33.30
5. $133\frac{1}{3}$% of $600 $800
6. 300% of $493 $1,479
7. .5% of $28 $.14
8. 6.13% of $4,000 $245.20
9. $\frac{3}{4}$% of $2,000 $15
10. $\frac{1}{2}$% of $5,400 $27

SET 16(a)
1. 3% of $.90 $.03
2. 5% of $4.60 $.23
3. 6% of $3.29 $.20
4. 4% of $10.55 $.42
5. 6% of $36.48 $2.19
6. 15% of $8.75 $1.31
7. 34% of $9.23 $3.14
8. 79% of $25.38 $20.05
9. 50% of $94.81 $47.41
10. 25% of $285.96 $71.49

SET 16(b)
1. $87\frac{1}{2}$% of $.56 $.49
2. $33\frac{1}{3}$% of $8.16 $2.72
3. 124% of $6.73 $8.35
4. 160% of $2.25 $3.60
5. $3\frac{1}{2}$% of $9.40 $.33
6. 200% of $35.86 $71.72
7. $137\frac{1}{2}$% of $11.12 $15.29
8. 6.05% of $58 $3.51
9. 0.4% of $18.54 $.07
10. $\frac{1}{4}$% of $32.80 $.08

PROBLEM SOLVING

See page 16, *Problem Solving Strategy.*
For problems 1–5, select the letter corresponding to your answer.

1. How many problems did Joan have right if she received a grade of 85% in a mathematics test of 20 problems?

  a. 15 problems   **c.** 3 problems
  b. 17 problems   **d.** Answer not given

2. Richard received a grade of 60% in a spelling test of 25 words. How many words did he misspell?

  a. 10 words   **b.** 15 words   **c.** 20 words   **d.** 5 words

3. The enrollment in the Weston Junior High School is 850. If the attendance for a certain month was 92%, how many absences were there during the month?

  a. 88 absences   **c.** 68 absences
  b. 782 absences   **d.** Answer not given

4. A house worth $47,900 is insured for 80% of its value. How much would the owner receive if the house were destroyed by fire?

  a. $47,900   **b.** $9,580   **c.** $35,000   **d.** $38,320

5. Mr. Lightfood bought a washing machine for $336. He paid 25% in cash and the balance is to be paid in 12 equal monthly installments. How much must he pay each month?

  a. $84   **b.** $252   **c.** $21   **d.** $3.36

6. How many questions out of 28 may a pupil miss and still get a grade of 75%? 7 questions

7. Arthur's father earns $28,800 a year. He plans to use the following budget: food, 25%; shelter, 20%; transportation, 18%; clothing, 15%; savings, 10%; miscellaneous, 12%. How much does he plan to spend for each item?

8. Mr. Becker bought a house for $39,500 and made a down payment of 20%. What is the amount of the mortgage? $31,600

9. Mrs. Ritter stored her fur coat for the summer and was charged 2% of its value. If the coat is valued at $750, how much did she pay? $15

10. Frank's mother received a 10% increase on a salary of $290 per week. Sometime later her salary was reduced 10%. How does her present salary compare with her original salary before the increase? $2.90 less

11. A certain manganese bronze contains 59% copper. How many pounds of copper are in 300 pounds of manganese bronze? 177 lb.

12. A certain electric motor is 78% efficient. What is the output if the input is 1,400 watts? 1,092 watts

7. Food, $7,200; Shelter, $5,760; Trans., $5,184; Clothing, $4,320; Savings, $2,880; Misc., $3,456

13. The true altitude for a certain air density is 5% less than the indicated altitude. What is the true altitude if the indicated altitude is 4,900 feet? 4,655 ft.

14. a. If the sales tax is 6%, what would the tax be on an article selling for $12.50? $.75

b. Mrs. Gomez purchased a television set at $495.99. If the sales tax on the set is 4%, find the cost of the set including this tax. $515.83

15. Find the sale price of each of the following articles:

a. A camera; regular price, $39.60; reduced 25% $29.70

b. A coat; regular price, $87.50; reduced $33\frac{1}{3}$% $58.33

c. A bicycle; regular price, $125; reduced 16% $105.00

d. A toaster/oven; regular price, $54.98; reduced 20% $43.98

REFRESH YOUR SKILLS

1. Add:

29
6,842
98,386
2,594
73,967
―――
181,818

2. Subtract:

1,000,000
908,147
―――
91,853

3. Multiply:

846
259
―――
219,114

4. Divide:

958
1,728)1,655,424

5. Add: $4\frac{3}{8} + 5\frac{9}{16}$ $9\frac{15}{16}$

6. Subtract: $8\frac{1}{4} - 2\frac{1}{3}$ $5\frac{11}{12}$

7. Multiply: $3\frac{1}{7} \times 4\frac{2}{3}$ $14\frac{2}{3}$

8. Divide: $7\frac{1}{2} \div 6\frac{3}{4}$ $1\frac{1}{9}$

9. Add:

6.03
2.39
3.26
7.32
――
19

10. Subtract:

$9 − $1.32
$7.68

11. Multiply: (nearest cent)

$70.29
$.01\frac{1}{2}$
―――
$1.05

12. Divide: $.09

144)$12.96

13. Round 3,999,506 to the nearest thousand. 4,000,000

14. Express $\frac{48}{144}$ in lowest terms. $\frac{1}{3}$

15. Change .18 to a percent. 18%

16. Change 60% to a fraction. $\frac{3}{5}$

17. Change 56% to a decimal. .56

18. Change $\frac{13}{20}$ to a decimal. .65

19. Find 3% of $56.25 to the nearest cent. $1.69

20. Find $133\frac{1}{3}$% of 924. 1,232

For a Computer Activity see p. T40.

Finding What Percent One Number Is of Another*

OBJECTIVE:

I. Aim   To find what percent one number is of another.

II. Procedure

1. Make a fraction, indicating what fractional part one number is of the other. If possible, reduce the fraction to lowest terms.

2. Change the fraction to a percent, using the percent equivalent if it is known; otherwise change the fraction first to a 2-place decimal by dividing the numerator by the denominator, then change the decimal to a percent. (See sample solution 2 for a short method.)

III. Sample Solutions

1. 27 is what percent of 36?

$$\frac{27}{36} = \frac{3}{4} = 75\%$$

or $\dfrac{27}{36} =$
$$\begin{array}{r} .75 = 75\% \\ 36\overline{)27.00} \\ 25\ 2 \\ \hline 1\ 80 \\ 1\ 80 \\ \hline \end{array}$$

Answer: 75%

2. 9 is _____ percent of 20?

$$\frac{9}{20} = \frac{45}{100} = 45\%$$

Answer: 45%

3. What percent of 26 is 26?

$$\frac{26}{26} = 1 = 100\%$$

Answer: 100%

4. What percent of 48 is 30?

$$\frac{30}{48} = \frac{5}{8} = 62\tfrac{1}{2}\%$$

or $\dfrac{30}{48} =$
$$\begin{array}{r} .62\tfrac{1}{2} = 62\tfrac{1}{2}\% \\ 48\overline{)30.00} \\ 28\ 8 \\ \hline 1\ 20 \\ 96 \\ \hline 24 \end{array}$$

$$\frac{24}{48} = \frac{1}{2}$$

Answer: $62\tfrac{1}{2}\%$

5. 18 is what percent of 16?

$$\frac{18}{16} = \frac{9}{8} = \begin{array}{r} 1.12\tfrac{1}{2} = 112\tfrac{1}{2}\% \\ 8\overline{)9.00} \end{array}$$

or $\frac{9}{8} = 1\tfrac{1}{8} = 112\tfrac{1}{2}\%$

Answer: $112\tfrac{1}{2}\%$

6. $1.50 is what percent of $7.50?

$$\frac{\$1.50}{\$7.50} = \frac{1}{5} = 20\%$$

Answer: 20%

---

*See page 232.

Related Resources: *Practical Applications in Mathematics:* Diag. Test and Prac. Ex., p. 62 **223**

DIAGNOSTIC TEST

Find the following:

1. 4 is what percent of 5? 80%   7. 37 is what percent of 37? 100%
2. What percent of 12 is 6? 50% 8. 8 is ____ % of 4? 200%
3. What percent of 8 is 7? $87\frac{1}{2}$% 9. 25 is what percent of 20? 125%
4. 45 is ____ % of 54? $83\frac{1}{3}$%   10. What percent of 36 is 48? $133\frac{1}{3}$%
5. 2 is what percent of 7? $28\frac{4}{7}$% 11. 561 is what percent of 935? 60%
$55\frac{5}{9}$% 6. What percent of 18 is 10?   12. What percent of $17 is $3.40? 20%
13. $3\frac{1}{2}$ is what percent of $10\frac{1}{2}$? $33\frac{1}{3}$%
14. 9 is ____ % of 100? (Use short method.) 9%
15. 13 is what percent of 25? (Use short method.) 52%
16. What percent of 400 is 16? (Use short method.) 4%

Basic: Sets 1(a)–11, Misc. Ex. 1, Prob. 1–5; *Average:* Sets 1(a)–13, Misc. Ex. 1–2, Prob. 1–8; *Enriched:* Sets 1(a)–16, Misc. Ex. 1–2, Prob. 1–10

RELATED PRACTICE EXAMPLES

Find the following:

SET 1(a)

1. 3 is what percent of 5? 60%
2. 1 is what percent of 4? 25%
3. 3 is ____ % of 4? 75%
4. 1 is ____ % of 5? 20%
5. 9 is ____ % of 10? 90%

SET 1(b)

1. What percent of 4 is 1? 25%
2. What percent of 5 is 2? 40%
3. What percent of 2 is 1? 50%
4. What percent of 10 is 7? 70%
5. What percent of 10 is 3? 30%

SET 2(a)

1. What percent of 36 is 9? 25%
2. What percent of 50 is 15? 30%
3. 4 is ____ % of 40? 10%
4. 39 is ____ % of 52? 75%
5. 42 is what percent of 60? 70%

SET 2(b)

1. 5 is ____ % of 50? 10%
2. 36 is ____ % of 45? 80%
3. 72 is what percent of 96? 75%
4. What percent of 80 is 56? 70%
5. What percent of 90 is 81? 90%

SET 3

1. What percent of 8 is 1? $12\frac{1}{2}$%
2. What percent of 6 is 5? $83\frac{1}{3}$%
3. What percent of 3 is 2? $66\frac{2}{3}$%
4. 3 is ____ % of 8? $37\frac{1}{2}$%
5. 5 is what percent of 8? $62\frac{1}{2}$%

SET 4(a)

1. 9 is ____ % of 27? $33\frac{1}{3}$%
2. 16 is ____ % of 24? $66\frac{2}{3}$%
3. 35 is what percent of 42? $83\frac{1}{3}$%
4. What percent of 40 is 15? $37\frac{1}{2}$%
5. What percent of 72 is 63? $87\frac{1}{2}$%

SET 4(b)

1. 9 is what percent of 54? $16\frac{2}{3}$%
2. 28 is what percent of 32? $87\frac{1}{2}$%
3. What percent of 64 is 24? $37\frac{1}{2}$%
4. What percent of 16 is 10? $62\frac{1}{2}$%
5. 5 is _____ % of 80? $6\frac{1}{4}$%

SET 5

1. 3 is what percent of 7? $42\frac{6}{7}$%
2. 5 is what percent of 12? $41\frac{2}{3}$%
3. What percent of 9 is 7? $77\frac{7}{9}$%
4. What percent of 13 is 8? $61\frac{7}{13}$%
5. 17 is _____ % of 18? $94\frac{4}{9}$%

SET 6(a)

1. What percent of 14 is 4? $28\frac{4}{7}$%
2. What percent of 54 is 30? $55\frac{5}{9}$%
3. 15 is _____ % of 84? $17\frac{6}{7}$%
4. 42 is what percent of 49? $85\frac{5}{7}$%
5. 60 is what percent of 78? $76\frac{12}{13}$%

SET 6(b)

1. 32 is _____ % of 72? $44\frac{4}{9}$%
2. 27 is _____ % of 99? $27\frac{3}{11}$%
3. 70 is what percent of 75? $93\frac{1}{3}$%
4. What percent of 96 is 27? $28\frac{1}{8}$%
5. What percent of 48 is 2? $4\frac{1}{6}$%

SET 7

1. 8 is what percent of 8? 100%
2. 40 is what percent of 40? 100%
3. What percent of 53 is 53? 100%
4. What percent of 82 is 82? 100%
5. 250 is _____ % of 250? 100%

SET 8

1. 2 is _____ % of 1? 200%
2. 8 is _____ % of 2? 400%
3. 60 is what percent of 12? 500%
4. What percent of 10 is 30? 300%
5. What percent of 12 is 72? 600%

SET 9

1. 6 is what percent of 5? 120%
2. 7 is what percent of 4? 175%
3. What percent of 24 is 54? 225%
4. What percent of 50 is 65? 130%
5. 63 is _____ % of 35? 180%

SET 10

1. What percent of 3 is 4? $133\frac{1}{3}$%
2. What percent of 12 is 32? $266\frac{2}{3}$%
3. 90 is _____ % of 48? $187\frac{1}{2}$%
4. 64 is _____ % of 28? $228\frac{4}{7}$%
5. 85 is what percent of 75? $113\frac{1}{3}$%

SET 11

1. 492 is what percent of 656? 75%
2. 516 is what percent of 774? $66\frac{2}{3}$%
3. 645 is _____ % of 1,032? $62\frac{1}{2}$%
4. What percent of 1,225 is 98? 8%
5. 1,014 is _____ % of 2,535? 40%

SET 12(a)

1. What percent of 10 is 2.5? 25%
2. What percent of 7.5 is 3? 40%
3. .54 is _____ % of 9? 6%
4. 4.3 is what percent of 6.45? $66\frac{2}{3}$%
5. 9.6 is what percent of 6.4? 150%

SET 12(b)

1. $.75 is _____ % of $18.75? 4%
2. $1.30 is _____ % of $6.50? 20%
3. $9 is what percent of $6.75? $133\frac{1}{3}$%
4. What percent of $12.25 is $10?
5. What percent of $15 is $3.75? 25%

4. $81\frac{31}{49}$ or 81.63%

SET 13

1. $2\frac{1}{4}$ is what percent of 9? 25%
2. $1\frac{2}{3}$ is what percent of $8\frac{1}{3}$? 20%
3. What percent of $4\frac{2}{3}$ is $3\frac{1}{2}$? 75%
4. What percent of $1\frac{7}{8}$ is $1\frac{1}{4}$? $66\frac{2}{3}$%
5. $4\frac{1}{8}$ is _____ % of $6\frac{7}{8}$? 60%

Use short method:

SET 14	SET 15
1. 15 is ___ % of 100? 15%	**1.** 27 is what percent of 50? 54%
2. 125 is ___ % of 100? 125%	**2.** 7 is what percent of 20? 35%
3. 4.5 is what percent of 100? 4.5%	**3.** What percent of 50 is 34? 68%
4. What percent of 100 is 7? 7%	**4.** What percent of 25 is 21? 84%
5. What percent of 100 is $33\frac{1}{3}$? $33\frac{1}{3}$%	**5.** 2 is ___ % of 25? 8%

SET 16

1. What percent of 200 is 6? 3%   **3.** 120 is ___ % of 1,000? 12%

2. What percent of 900 is 60? $6\frac{2}{3}$%   **4.** 32 is what percent of 800? 4%

5. 480 is what percent of 2,400? 20%

MISCELLANEOUS EXAMPLES

Find each of the following correct to nearest tenth of a percent:

SET 1	SET 2
1. 5 is what % of 9? 55.6%	**1.** What % of 63 is 27? 42.9%
2. What % of 21 is 13? 61.9%	**2.** 30 is what % of 78? 38.5%
3. $.11 is ___ % of $.15? 73.3%	**3.** What % of $42 is $39? 92.9%
4. What % of 16 is 1? 6.3%	**4.** $.60 is ___ % of $1.44? 41.7%
5. $7 is what % of $12? 58.3%	**5.** What % of $171 is $135? 78.9%

PROBLEM SOLVING

See page 16, *Problem Solving Strategy.*

For problems 1–5, select the letter corresponding to your answer.

1. There are 40 pupils in a class. On a certain day 38 pupils were present. What percent of the class attended school?

    a. 38%   **b.** 2%   <u>**c.**</u> 95%   **d.** Answer not given

2. There are 18 girls and 27 boys in a class. What percent are boys?

    a. $66\frac{2}{3}$%   <u>**b.**</u> 60%   **c.** 50%   **d.** 40%

3. Find the rate of reduction allowed when a basketball that regularly sells for $25 was purchased for $20.

    a. 80%   **b.** 60%   **c.** 40%   <u>**d.**</u> 20%

4. The school team won 9 games and lost 6. What percent of the games did the team lose?

    <u>**a.**</u> 40%   **b.** $66\frac{2}{3}$%   **c.** 3%   **d.** 60%

5. Mr. Graham saves $150 each month. If his annual income is $18,000, what percent of his income does he save annually?

    a. 15%   <u>**b.**</u> 10%   **c.** 5%   **d.** Answer not given

6. Charles answered 19 questions correctly and missed 6 questions. What percent of the questions did he answer correctly? 76%

7. Gloria guessed the length of the schoolroom to be 28 feet. When she measured it, she found it to be 32 feet. What was her percent of error on the actual measurement? $12\frac{1}{2}$%

8. To find the efficiency of an electric motor, find what percent the output is of the input. Find the efficiency of a motor if its output is 1,520 watts when the input is 1,600 watts. 95%

9. Find the rate of increase or decrease when:

a. enrollment increased from 250 pupils to 300 pupils; 20% increase

b. lateness dropped from 42 cases to 35 cases; $16\frac{2}{3}$% decrease

c. price changed from 30¢ to 40¢; $33\frac{1}{3}$% increase

d. price changed from 40¢ to 30¢; 25% decrease

e. a certain factory produced 1,605 units last month and 1,284 units the month before. 25% increase

10. Find the rate of reduction allowed on each of the following articles:

a. A lamp; regular price, $125; sale price, $85 32%

b. A desk; regular price, $300; sale price, $250 $16\frac{2}{3}$%

c. A CB radio; regular price, $140; sale price, $99.40 29%

d. A calculator; regular price, $79.50; sale price, $60.42 24%

REFRESH YOUR SKILLS

1. Add:

7,124
83,259
61,435
816
9,527
84,349
246,510

2. Subtract:

815,008
747,598
67,410

3. Multiply:

403
807
325,221

4. Divide: 297

794)235,818

5. Add:

$7\frac{2}{3}$
$4\frac{5}{6}$ $12\frac{1}{2}$

6. Subtract:

$8\frac{3}{4}$
$2\frac{4}{5}$ $5\frac{19}{20}$

7. Multiply:

$4\frac{3}{8} \times 2\frac{3}{10}$ $10\frac{1}{16}$

8. Divide:

$\frac{13}{16} \div 1\frac{1}{12}$ $\frac{3}{4}$

9. Add:

$3.8 + .47 + 25$ 29.27

10. Subtract: $6.825 - .97$ 5.855

11. Multiply: 3.14×80 251.20

12. Divide: 1.15)46 40

13. Find 8% of 60.9 4.872

14. Find $87\frac{1}{2}$% of $526 $460.25

15. What percent of 144 is 64? $44\frac{4}{9}$%

16. What percent of $1.25 is $.50? 40%

For a Class Activity see p. T34.

Finding a Number When a Percent of It Is Known*

OBJECTIVE:

I. Aim To find a number when a percent of it is known.

II. Procedure

1. Change the percent to a decimal or common fraction.
2. a. Divide the given number representing the given percent of the unknown number by this decimal or common fraction.
   b. Or follow the alternate method shown in the sample solutions.
3. Check by finding the given percent of the answer.

III. Sample Solutions

1. 16% of what number is 48?

Method 1

16% = .16
Divide 48 by .16

$$\begin{array}{r} 3\ 00. \\ .16_\wedge \overline{)48.00_\wedge} \end{array}$$

Answer: 300

Method 2

16% of the number = 48
1% of the number = 48 ÷ 16 = 3
100% of the number = 100 × 3 = 300
Therefore the number = 300

Check: 16% of 300 = 48

2. 50% of what number = 40?

Method 1

50% = $\frac{1}{2}$
Divide 40 by $\frac{1}{2}$
40 ÷ $\frac{1}{2}$ = 40 × 2 = 80

Answer: 80

Method 2

50% of the number = 40
$\frac{1}{2}$ of the number = 40
$\frac{2}{2}$ of the number = 2 × 40 = 80
Therefore the number = 80
Check: 50% of 80 = 40

3. 31 is 4% of what number?

4% = .04
Divide 31 by .04

$$\begin{array}{r} 7\ 75. \\ .04_\wedge \overline{)31.00_\wedge} \end{array}$$

Answer: 775

4. $3\frac{1}{4}$% of _____ = 65?

$3\frac{1}{4}$% = .0325
Divide 65 by .0325

$$\begin{array}{r} 2000. \\ .0325_\wedge \overline{)65.0000_\wedge} \end{array}$$

Answer: 2,000

---

*See page 232.

Related Resources: *Practical Applications in Mathematics:* Diag. Test and Prac. Ex., p. 63

5. .2% of what number is 6.4?

.2% = .002

Divide 6.4 by .002

$$.002_\wedge\overline{)6.400_\wedge}$$

3 200.

Answer: 3,200

6. $\frac{3}{8}$% of what number = 15?

$\frac{3}{8}$% = .00375

Divide 15 by .00375

$$.00375_\wedge\overline{)15.00000_\wedge}$$

4000.

15 00

Answer: 4,000

DIAGNOSTIC TEST

Find the missing numbers:

1. 12% of what number is 24? 200
2. 18 is 36% of what number? 50
3. 25% of what number is 6? 24
4. 66$\frac{2}{3}$% of what number is 14? 21
5. 6% of what number is 12? 200
6. 20 is 20% of what number? 100
7. 100% of what number is 70? 70
8. 120% of what number is 108? 90
9. 40% of what number is 12.6? 31.5
10. 2.5% of what number is 2? 80
11. 4$\frac{1}{2}$% of what number is 90? 2,000
12. 187$\frac{1}{2}$% of what number is 105? 56
13. .5% of what number is 4? 800
14. $\frac{3}{4}$% of what number is 27? 3,600
15. 8% of what amount is $180? $2,250

Basic: Sets 1–10, Appl. 1–2; *Average:* Sets 1–12, Appl. 1–4; *Enriched:* Sets 1–15, Appl. 1–5

RELATED PRACTICE EXAMPLES

Find the missing numbers:

SET 1

1. 45% of what number is 90? 200
2. 65% of what number is 260? 400
3. 74% of what number is 370? 500
4. 15% of what number is 18? 120
5. 31% of what number is 279? 900

SET 2

1. 12 is 24% of what number? 50
2. 44 is 55% of what number? 80
3. 9 is 15% of what number? 60
4. 40 is 45% of what number? 88$\frac{8}{9}$
5. 17 is 85% of what number? 20

SET 3

1. 20% of what number is 3? 15
2. 40% of _____ = 48? 120
3. 18 is 60% of what number? 30
4. 91 = 70% of _____ ? 130
5. 75% of what number is 108? 144

SET 4

1. 33$\frac{1}{3}$% of what number is 78? 234
2. 62$\frac{1}{2}$% of _____ = 200? 320
3. 4 is 16$\frac{2}{3}$% of what number? 24
4. 462 = 66$\frac{2}{3}$% of _____ ? 693
5. 83$\frac{1}{3}$% of what number is 290? 348

SET 5

1. 2% of what number is 10? 500
2. 4% of _____ = 26? 650
3. 95 is 6% of what number? 1,583⅓
4. 15 = 1% of _____ ? 1,500
5. 5% of what number is 45? 900

SET 6

1. 50 is 50% of what number? 100
2. 30 = 30% of _____ ? 100
3. 28% of what number is 28? 100
4. 79% of what number is 79? 100
5. 43% of _____ = 43? 100

SET 7

1. 100% of what number is 59? 59
2. 300% of _____ = 240? 80
3. 5 is 100% of what number? 5
4. 36 = 200% of _____ ? 18
5. 100% of what number is 180? 180

SET 8

1. 160% of what number is 72? 45
2. 175% of _____ = 42? 24
3. 513 is 114% of what number? 450
4. 78 = 156% of _____ ? 50
5. 245% of what number is 98? 40

SET 9

1. 42% of what number is 115.5? 275
2. 60% of _____ = 46.8? 78
3. 102.7 is 65% of what number? 158
4. 371.2 = 58% of _____ ? 640
5. 70% of what number is 667.8? 954

SET 10

1. 4.75% of what number is 38? 800
2. 15.4% of _____ = 130.9? 850
3. 8.1 is 8.1% of what number? 100
4. 16.5 = 4.125% of _____ ? 400
5. 26.3% of _____ = 18.41? 70

SET 11

1. $4\frac{3}{4}$% of what number is 19? 400
2. $6\frac{1}{8}$% of _____ = 12.25? 200
3. 18 is $2\frac{1}{2}$% of what number? 720
4. 72 = $28\frac{4}{5}$% of _____ ? 250
5. $18\frac{5}{8}$% of _____ = 931.25? 5,000

SET 12

1. $116\frac{2}{3}$% of what number is 21? 18
2. $287\frac{1}{2}$% of _____ = 230? 80
3. 39 is $162\frac{1}{2}$% of what number? 24
4. 644 = $233\frac{1}{3}$% of _____ ? 276
5. $183\frac{1}{3}$% of _____ = 440? 240

SET 13

1. .4% of what number is 2? 500
2. .875% of _____ = 70? 8,000
3. 2 is .1% of what number? 2,000
4. 12 = $.33\frac{1}{3}$% of _____ ? 3,600
5. .25% of what number is 2.5? 1,000

SET 14

1. $\frac{1}{2}$% of what number is 10? 2,000
2. $\frac{5}{12}$% of _____ = 100? 24,000
3. 9 is $\frac{3}{20}$% of what number? 6,000
4. 60 is $\frac{2}{3}$% of _____ ? 9,000
5. $\frac{1}{4}$% of what number is 6.25? 2,500

SET 15

1. 75% of what amount is $150? $200
2. 6% of what amount is $78? $1,300
3. $.12 is 10% of what amount? $1.20
4. $5,075 is 101.5% of what amount? $5,000
5. $37\frac{1}{2}$% of what amount is $8.13? $21.68

PROBLEM SOLVING

See page 16, *Problem Solving Strategy.*
For problems 1–2, select the letter corresponding to your answer.

1. Betty bought a camera at a 20% reduction sale. If she paid $36 for it, what was the regular price?

    a. $18.20   **b.** $42.50   **c.** $45.00   **d.** Answer not given

2. If 45% of the school are boys and the girls number 858, how many boys are enrolled?

    a. 386 boys   **b.** 702 boys   **c.** 1,560 boys   **d.** Answer not given

3. How much money must be invested at 9% to earn $4,500 per year? $50,000

4. If an ore contains 16% copper, how many metric tons of ore are needed to get 20 metric tons of copper? 125 metric tons

5. Find the regular price of each of the following:

a. A tire, $33 sale price when a 40% reduction is allowed. $55

b. A typewriter, $216 sale price when a 10% reduction is allowed. $240

c. A shirt, $10.60 sale price when a $33\frac{1}{3}$% reduction is allowed. $15.90

d. A clothes dryer, $203.15 sale price when a 15% reduction is allowed. $239

REFRESH YOUR SKILLS

1. Add:
23,845
74,583
96,148
31,599
83,172
   309,347

2. Subtract:
403,070
392,989
   10,081

3. Multiply:
8,004
9,050
72,436,200

4. Divide: 217
365)79,205

5. Add: $4\frac{5}{8}$
    $\frac{3}{4}$
$8\frac{11}{16}$
   $14\frac{1}{16}$

6. Subtract:
$1\frac{1}{6}$
$\frac{1}{2}$
$\frac{2}{3}$

7. Multiply:
$6\frac{2}{3} \times 7\frac{7}{8}$ $52\frac{1}{2}$

8. Divide:
$4\frac{1}{6} \div 2\frac{13}{16}$ $1\frac{13}{27}$

9. Add: $8.29 + $.58 + $46.75 $55.62

10. Subtract: $49 − $.89 $48.11

11. Multiply: 2.8 × .003 .0084

12. Divide: 20).1 .005

13. Find $2\frac{1}{2}$% of $6,000 $150

14. What percent of 120 is 75? .625

15. 2% of what number is 5?

16. 96% of what number is 288? 300

PERCENT–SOLVING BY PROPORTION

Ratio and proportion may be used to solve problems which involve percent. See Exercises 4-27 and 4-28.

We generally think of a percent as another way of writing a fraction. A percent may also be considered a ratio. 75%, which is equivalent to $\frac{75}{100}$, is the ratio of 75 to 100.

The three basic types of percentage problems may be treated as one through the use of proportion.

(1) Find 8% of 25.

8% is the ratio of 8 to 100 or $\frac{8}{100}$.

To find 8% of 25 means to determine the number (n) which compared to 25 is the same as 8 compared to 100.

The proportion $\frac{n}{25} = \frac{8}{100}$ is formed and solved.

$$(1) \quad \frac{n}{25} = \frac{8}{100}$$
$$100\,n = 200$$
$$n = 2$$
Answer: 2

(2) What percent of 25 is 2?

To find what percent of 25 is 2 means to find the number (n) per 100 or the ratio of a number to 100 which has the same ratio as 2 to 25.

The proportion $\frac{n}{100} = \frac{2}{25}$ is formed and solved.

$$(2) \quad \frac{n}{100} = \frac{2}{25}$$
$$25\,n = 200$$
$$n = 8$$
Answer: 8%

(3) 8% of what number is 2?

8% is the ratio of 8 to 100 or $\frac{8}{100}$.

To find the number of which 8% is 2 means to determine the number (n) such that 2 compared to this number is the same as 8 compared to 100.

The proportion $\frac{2}{n} = \frac{8}{100}$ is formed and solved.

$$(3) \quad \frac{2}{n} = \frac{8}{100}$$
$$8\,n = 200$$
$$n = 25$$
Answer: 25

PRACTICE PROBLEMS

1. Write each of the following as a ratio:

a. 6% $\frac{6}{100}$ or $\frac{3}{50}$ **d.** 70% $\frac{70}{100}$ or $\frac{7}{10}$ **g.** 300% $\frac{300}{100}$ or $\frac{3}{1}$ **j.** 2.4% $\frac{2.4}{100}$ or $\frac{3}{125}$ **m.** 57% $\frac{57}{100}$

b. 18% $\frac{18}{100}$ or $\frac{9}{50}$ **e.** 4% $\frac{4}{100}$ or $\frac{1}{25}$ **h.** $62\frac{1}{2}$% $\frac{62\frac{1}{2}}{100}$ or $\frac{5}{8}$ **k.** $\frac{1}{2}$% $\frac{\frac{1}{2}}{100}$ or $\frac{1}{200}$ **n.** $33\frac{1}{3}$% $\frac{33\frac{1}{3}}{100}$ or $\frac{1}{3}$

c. 93% $\frac{93}{100}$ **f.** 120% $\frac{120}{100}$ or $\frac{6}{5}$ **i.** $5\frac{1}{4}$% $\frac{5\frac{1}{4}}{100}$ or $\frac{21}{400}$ **l.** 25% $\frac{25}{100}$ or $\frac{1}{4}$ **o.** 0.7% $\frac{0.7}{100}$ or $\frac{7}{1,000}$

2. Find each of the following:

a. 37% of 16 5.92

b. 8% of 729 58.32

c. 45% of $300 $135

d. 60% of $1,200 $720

e. 3% of $940 $28.20

f. 180% of 685 1,233

g. $16\frac{2}{3}$% of 732 122

h. 5.9% of 28 1.652

i. $4\frac{3}{4}$% of 924 43.89

j. $\frac{1}{4}$% of 600 1.5

k. 200% of 350 700

l. $60\frac{1}{2}$% of 520 314.6

3. Find each of the following:

a. 2 is what percent of 5? 40%

b. 18 is what percent of 54? $33\frac{1}{3}$%

c. What percent of 105 is 70? $66\frac{2}{3}$%

d. What percent of 84 is 63? 75%

e. 49 is what percent of 56? $87\frac{1}{2}$%

f. What percent of 72 is 60? $83\frac{1}{3}$%

g. 6 is what percent of 3? 200%

h. What percent of 8 is 10? 125%

i. What percent of 1.5 is 1.2? 80%

j. 7 is what percent of 11? $63\frac{7}{11}$%

k. What percent of $.45 is $.27? 60%

l. $1.50 is what percent of $9? $16\frac{2}{3}$%

4. Find each of the following:

a. 15% of what number is 3? 20

b. 9% of what number is 72? 800

c. 24 is 75% of what number? 32

d. 7.8 is 4% of what number? 195

e. 100% of what number is 17? 17

f. 48 is 80% of what number? 60

g. 160% of what number is 56? 35

h. 65 is $33\frac{1}{3}$% of what number? 195

i. 140 is $62\frac{1}{2}$% of what number? 224

j. 3.4% of what number is 10.2? 300

k. 19 is $\frac{1}{2}$% of what number? 3,800

l. .7% of what number is 21? 3,000

ESTIMATING ANSWERS

To estimate a product or a quotient, first round the given fractions to nearest whole numbers and the given whole numbers to convenient place values (see pages 28 and 135) and then perform the operation.

For each of the following select your nearest estimate:

1. 6% of 198 is approximately: .12   <u>12</u>   1.2

2. 89% of $7,040 is approximately: $630   <u>$6,200</u> $6,800

3. $5\frac{1}{8}$% of $61 is approximately: $31 $.31   <u>$3.10</u>

4. $9\frac{3}{4}$% of 8,000 is approximately: 75   700   <u>800</u>

5. $7\frac{1}{4}$% of $389 is approximately: $2.80   <u>$28</u> $.28

6. 4 is approximately 50%   or   70%   or   <u>5%</u>   of 79.

7. 9 is approximately 7%   or   8%   or   <u>9%</u>   of 101.

8. 18 is approximately 20%   or   <u>25%</u>   or   30%   of 74.

9. 10 is approximately <u>65%</u>   or   70%   or   75%   of 16.

10. 39 is approximately 50%   or   45%   or   <u>40%</u>   of 98.

CHECK BY CALCULATOR

Use your calculator to check each of the following given answers. Indicate which answers are incorrect. If a calculator is not available, check by computation.

1. Add: b and c are incorrect

a. 693 + 4,856 + 96 + 829 + 9,069 = 15,543
b. 5,826 + 9,658 + 4,178 + 3,907 + 6,455 = 31,024
c. 82,476 + 985 + 7,593 + 6,847 + 91,328 = 188,229
d. 79,327 + 87,419 + 76,678 + 50,724 + 23,586 = 317,734

2. Subtract: b, c, and d are incorrect

a. 8,315 − 5,672 = 2,643
b. 40,681 − 31,584 = 9,197
c. 626,000 − 498,608 = 207,392
d. 8,039,547 − 6,929,759 = 1,109,888

3. Multiply: c and d are incorrect

a. 69 × 82 = 5,658
b. 705 × 980 = 690,900
c. 5,276 × 416 = 2,294,816
d. 8,509 × 6,053 = 51,404,977

4. Divide: a is incorrect

a. 41,993 ÷ 49 = 847
b. 671,550 ÷ 726 = 925
c. 248,472 ÷ 609 = 408
d. 618,576 ÷ 1,578 = 392

5. Add: b and d are incorrect

a. 1.46 + 8.91 + 6.87 = 17.24
b. .914 + 5 = .915
c. .367 + .659 + .348 = 1.374
d. 6.2 + .276 = .896

6. Subtract: b and d are incorrect

a. .74 − .69 = .05
b. 9.435 − 3.15 = 6.385
c. 60.7 − 3.54 = 57.16
d. 8 − 5.6 = 3.6

7. Multiply: a, b, and d are incorrect

a. .4 × .19 = .76
b. .008 × .025 = .002
c. 3,000 × .0002 = .6
d. 3.1416 × 28.03 = 88.05948

8. Divide: b, c, and d are incorrect

a. 60.8 ÷ 16 = 3.8
b. .002 ÷ .04 = .5
c. 65.86 ÷ 8.9 = .74
d. .00054 ÷ .009 = .006

9. Find: b and c are incorrect

a. 5% of 73 = 3.65
b. 175% of 20 = 350
c. 8% of 17.26 = 138.08
d. 4.9% of 52.4 = 2.5676

10. Find each of the following: a and d are incorrect

a. What percent of 96 is 72? $62\frac{1}{2}$%
b. 51 is what percent of 150? 34%
c. 30% of what number is 180? 600
d. 6 is 4% of what number? 15

REVIEW OF UNIT FIVE

1. Write each of the following as a percent, decimal, and fraction:
a. Seven hundredths 7%, .07, $\frac{7}{100}$
b. Fifty-six hundredths 56%, .56, $\frac{56}{100}$
c. 53 out of 100 53%, .53, $\frac{53}{100}$
d. 91 out of 100 91%, .91, $\frac{91}{100}$
e. Thirty-nine and one-third hundredths $39\frac{1}{3}$%, $.39\frac{1}{3}$, $\frac{39\frac{1}{3}}{100}$

2. Express each of the following percents as a decimal:
a. 4% .04   **b.** 59% .59   **c.** 150% 1.50   **d.** $100\frac{3}{4}$% $1.00\frac{3}{4}$ **e.** 6.9% .069

3. Express each of the following decimals as a percent:
a. .03 3%   **b.** .3 30%   **c.** 1.18 118%   **d.** .942 94.2%   **e.** $.66\frac{2}{3}$ $66\frac{2}{3}$%

4. Express each of the following percents as a common fraction or mixed number:
a. 25% $\frac{1}{4}$ **b.** $83\frac{1}{3}$% $\frac{5}{6}$ **c.** 6% $\frac{3}{50}$ **d.** $162\frac{1}{2}$% $1\frac{5}{8}$ **e.** 250% $2\frac{1}{2}$

5. Express each of the following fractions or mixed numbers as a percent:
a. $\frac{3}{5}$ 60%   **b.** $\frac{11}{25}$ 44%   **c.** $\frac{5}{4}$ 125%   **d.** $\frac{71}{71}$ 100%   **e.** $3\frac{2}{3}$ $366\frac{2}{3}$%

6. Find the following:
a. 82% of 350 287   **b.** 6% of $39 $2.34   **c.** 75% of 2,000 1,500
d. $4\frac{1}{2}$% of $92.75 $4.17   **e.** 180% of 560 1,008

7. Find the following:
a. 16 is what percent of 24? $66\frac{2}{3}$
b. What percent of 50 is 29? 58%
c. $.48 is what percent of $3.20 15%
d. 5 is what percent of 2? 250%
e. What percent of 9 is 4? (correct to nearest tenth) 44.4%

8. Find the missing numbers:
a. 28% of what number is 7? 25
b. 72 is 9% of what number? 800
c. 54 is 3.6% of what number? 1,500
d. $2\frac{1}{4}$% of what number is 81? 3,600
e. $137\frac{1}{2}$% of what number is 55? 40

9. Mrs. Chow bought a new car for $8,460. She paid 20% down. If the balance is to be paid in 24 equal monthly installments, how much must she pay each month? $282

10. What percent was a television set reduced if it was marked $450 and sold for $390? $13\frac{1}{3}$%

CUMULATIVE PRACTICE

1. Add:

124,173
790,321
525,808
153,349
630,775
219,683
2,444,109

2. Subtract:

815,927
7,549
808,378

3. Multiply:

5,807
609
3,536,463

4. Divide: 8,094

495)4,006,530

5. Add:

$4\frac{2}{3}$

$5\frac{7}{8}$ $10\frac{13}{24}$

6. Subtract:

20

$6\frac{13}{16}$ $13\frac{3}{16}$

7. Multiply:

$\frac{3}{8} \times 4\frac{9}{16}$ $1\frac{91}{128}$

8. Divide:

$12 \div \frac{3}{4}$ 16

9. Add:

$.06 + $9 $9.06

10. Subtract:

.4 − .39 .01

11. Multiply:

$2.96
.045 $.13

12. Divide:

$.05)$3 60

13. Find 8% of $427 $34.16

14. What percent of 60 is 56? $93\frac{1}{3}$

15. 35% of what number is 14? 40

KEYED ACHIEVEMENT TEST

For additional practice turn to the Exercises indicated by the numerals in the red boxes.

1. Round 93,482,576 to the nearest thousand. 1-3 93,483,000

2. Write as a decimal: Sixty and five thousandths 1-21 60.005

3. Add:

82,197
4,825
399
47,286
8,678 1-4
143,385

4. Subtract:

250,105
176,298 1-5
73,807

5. Multiply:

6,043
7,008 1-6
42,349,344

6. Divide: 306

809)247,554 1-7

7. Add: **8.** Subtract: **9.** Multiply: **10.** Divide:

$6\frac{1}{3}$ $8\frac{1}{2}$ $6\frac{1}{4} \times 3\frac{3}{5}$ 1-16 $2\frac{5}{16} \div \frac{7}{8}$ 1-17

$4\frac{3}{4}$ $2\frac{2}{3}$ 1-13 $22\frac{1}{2}$ $2\frac{9}{14}$

$3\frac{7}{12}$ 1-12 $14\frac{2}{3}$ $5\frac{5}{6}$

11. Add: **12.** Subtract: **13.** Multiply: **14.** Divide:

$4.29 + 97.2 + .687$ $\$6.50 - \$3$ $.24 \times .002$ $.06\overline{)\,.3}$ 5

  102.177 1-23 $\$3.50$ 1-24   .00048 1-26   1-27

15. Change $\frac{3}{8}$ to a decimal fraction. 1-30 .375

16. Change .8 to a common fraction. 1-31 $\frac{4}{5}$

17. Change .069 to a percent. 1-38 6.9%

18. Find 5% of $32.68 1-41 $1.63

19. 45 is what percent of 72? 1-42 $62\frac{1}{2}$%

20. 6% of what number is 270? 1-43 4,500

MEANINGS AND UNDERSTANDINGS

1. What is the answer called when:

a. One number is subtracted from another? remainder or difference

b. Two numbers are multiplied? product

c. One number is divided by another? quotient

d. Two numbers are added? sum

2. Which is equal to 0:

a. Any number, other than zero, divided by itself?

b. Any number subtracted from itself?

3. Is the product *greater than, equal to,* or *smaller than* a given number when the given number is multiplied by: a. equal b. greater c. smaller

a. 1? **b.** A number greater than 1? **c.** A number smaller than 1?

4. Is the quotient *greater than, equal to,* or *smaller than* a given number when the given number is divided by: a. equal b. smaller c. greater

a. 1?   **b.** A number greater than 1?

c. A number smaller than 1 (zero excluded)?

5. Does the value of the fraction change when the numerator and denominator of a given fraction are each:

a. Multiplied by the same number? No

b. Divided by the same number (zero excluded)? No

COMPETENCY CHECK TEST

Solve each problem and select the letter corresponding to your answer.
The numerals in red boxes indicate Exercises where help may be found.

1. Add: 7,798 + 82,596 + 39,495 + 9,389 ⬚1-4
   a. 129,378   <u>b.</u> 139,278   c. 138,168   d. 137,298

2. Subtract: 1,650,000 − 938,074 ⬚1-5
   a. 712,916   b. 728,074   <u>c.</u> 711,926   d. 712,926

3. Multiply: 398 × 207 ⬚1-6
   <u>a.</u> 82,386   b. 10,746   c. 8,286   d. 605

4. Divide: 506)$\overline{15,180}$ ⬚1-7
   a. 3   <u>b.</u> 30   c. 300   d. 3,000

5. Add: $6\frac{3}{4} + 2\frac{5}{8} + 4\frac{1}{2}$ ⬚1-12
   a. $12\frac{1}{2}$ b. $13\frac{3}{8}$ c. $13\frac{11}{16}$ <u>d.</u> $13\frac{7}{8}$

6. Subtract: $12 - 8\frac{7}{10}$ ⬚1-13
   a. $4\frac{3}{10}$ b. $4\frac{7}{10}$ <u>c.</u> $3\frac{3}{10}$ d. $3\frac{7}{10}$

7. Multiply: $4\frac{1}{2} × \frac{5}{6}$ ⬚1-16
   <u>a.</u> $3\frac{3}{4}$ b. $5\frac{1}{4}$ c. $4\frac{5}{12}$ d. $2\frac{7}{8}$

8. Divide: $\frac{2}{3} ÷ 6$ ⬚1-17
   a. 4   b. $\frac{1}{4}$ <u>c.</u> $\frac{1}{9}$ d. 9

9. Find the sum: .8 + .88 + .888 ⬚1-23
   <u>a.</u> 2.568   b. .2568   c. .888888   d. Answer not given

10. Subtract: $57 − $9.43 ⬚1-24
    a. $48.67   <u>b.</u> $47.57   c. $47.67   d. $48.57

11. Multiply: .45 × .004 ⬚1-26
    a. .00018   b. .0180   c. 1.8   <u>d.</u> .0018

12. Divide: $.09)$\overline{\$63}$ ⬚1-27
    a. 7   b. 70   <u>c.</u> 700   d. .07

13. Find 4% of $245 ⬚1-41
    a. $.98   <u>b.</u> $9.80   c. $98   d. $980

14. What percent of 96 is 36? ⬚1-42
    a. 30%   b. $33\frac{1}{3}$%   <u>c.</u> $37\frac{1}{2}$%   d. 40%

15. 8% of what number is 74? ⬚1-43
    a. 592   b. 59.20   c. 5.92   <u>d.</u> 925

16. The school soccer team won 60% of the 25 games played.
    How many games did it lose? ⬚1-41
    a. 5 games   <u>b.</u> 10 games   c. 15 games   d. 20 games

For a Unit Test see p. T44 or *Testing Program*, p. 8.

SQUARES AND SQUARE ROOTS

The square of a number is the product obtained when the given number is multiplied by itself.

The square root of a number is that number which when multiplied by itself produces the given number. It is one of the two equal factors of a product.

The square of 9 is 81 but the square root of 9 is 3.

An irrational number is a number that cannot be expressed as a quotient of two whole numbers (with division by zero excluded). A number that is both a non-terminating and non-repeating decimal like the square root of any positive number (see page 242) other than perfect squares (numbers having an exact square root) is an irrational number.

There is a point on the number line that corresponds to each irrational number. To locate the point corresponding to $\sqrt{2}$ on the number line, construct a square with the side measuring the unit length. By the Rule of Pythagoras (see Exercise 3-27) it can be shown that the length of the diagonal of this square is $\sqrt{2}$. Thus, to locate the points corresponding to $\sqrt{2}$ and $-\sqrt{2}$ (see Exercise 4-4) on the number line, describe an arc using the diagonal as the radius. The points where the arc intersects the number line are the required points.

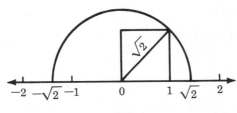

VOCABULARY: squaring a number
square
exponent

OBJECTIVE:

I. Aim   To square a number.

II. Procedure

Multiply the given number by itself.

III. Sample Solutions

1. Square 46

$$
\begin{array}{r}
46 \\
\times\,46 \\
\hline
276 \\
1\,84 \\
\hline
2{,}116
\end{array}
$$

Answer: 2,116

2. Square .03

$$
\begin{array}{r}
.03 \\
\times\,.03 \\
\hline
.0009
\end{array}
$$

Answer: .0009

3. Square $\frac{4}{5}$

$$\frac{4}{5} \times \frac{4}{5} = \frac{16}{25}$$

Answer: $\frac{16}{25}$

IV. Definitions

1. Squaring a number means multiplying the number by itself.

2. The square of a number is the product obtained when the given number is multiplied by itself. The square of 5 is 25 because 25 is the product obtained when 5 is multiplied by 5.

3. A short way of writing 5×5, or 5 squared, is 5^2. The small figure ^{2}, written to the right and a little above the number 5, is called an exponent. The exponent ^{2} indicates the square of a number.

DIAGNOSTIC TEST

Square each of the following numbers:

1. 8 64   **2.** 57 3,249   **3.** .25 .0625   **4.** $\frac{3}{8}$ $\frac{9}{64}$ **5.** $2\frac{1}{4}$ $\frac{81}{16}$ or $5\frac{1}{16}$

6. Find the value of: 14^2 196

RELATED PRACTICE EXAMPLES

Square each of the following numbers:

SET 1	SET 2	SET 3	SET 4	SET 5
1. 4 16	**1.** 16 256	**1.** .2 .04	**1.** $\frac{1}{2}$ $\frac{1}{4}$	**1.** $1\frac{1}{2}$ $\frac{9}{4}$ or $2\frac{1}{4}$
2. 1 1	**2.** 24 576	**2.** .7 .49	**2.** $\frac{5}{6}$ $\frac{25}{36}$	**2.** $1\frac{2}{3}$ $\frac{25}{9}$ or $2\frac{7}{9}$
3. 9 81	**3.** 39 1,521	**3.** .01 .0001	**3.** $\frac{2}{5}$ $\frac{4}{25}$	**3.** $2\frac{1}{8}$ $\frac{289}{64}$ or $4\frac{33}{64}$
4. 7 49	**4.** 50 2,500	**4.** .09 .0081	**4.** $\frac{1}{3}$ $\frac{1}{9}$	**4.** $2\frac{3}{4}$ $\frac{121}{16}$ or $7\frac{9}{16}$
5. 10 100	**5.** 78 6,084	**5.** .75 .5625	**5.** $\frac{7}{8}$ $\frac{49}{64}$	**5.** $3\frac{1}{3}$ $\frac{100}{9}$ or $11\frac{1}{9}$
6. 3 9	**6.** 100 10,000	**6.** .18 .0324	**6.** $\frac{3}{16}$ $\frac{9}{256}$	**6.** $1\frac{1}{4}$ $\frac{25}{16}$ or $1\frac{9}{16}$
7. 6 36	**7.** 156 24,336	**7.** 1.47 2.1609	**7.** $\frac{11}{12}$ $\frac{121}{144}$	**7.** $4\frac{1}{2}$ $\frac{81}{4}$ or $20\frac{1}{4}$
8. 2 4	**8.** 247 61,009	**8.** 2.05 4.2025	**8.** $\frac{5}{7}$ $\frac{25}{49}$	**8.** $1\frac{5}{8}$ $\frac{169}{64}$ or $2\frac{41}{64}$
9. 12 144	**9.** 1,760 3,097,600	**9.** 3.14 9.8596	**9.** $\frac{4}{9}$ $\frac{16}{81}$	**9.** $2\frac{2}{5}$ $\frac{144}{25}$ or $5\frac{19}{25}$
10. 11 121	**10.** 5,280 27,878,400	**10.** 18.73 350.8129	**10.** $\frac{7}{10}$ $\frac{49}{100}$	**10.** $1\frac{1}{6}$ $\frac{49}{36}$ or $1\frac{13}{36}$

SET 6

Find the value of each of the following:

1. 6^2 36   **3.** $(.4)^2$ .16   **5.** $(8.1)^2$ 65.61   **7.** $\left(\frac{4}{5}\right)^2$ .64   **9.** 120^2 14,400

2. 17^2 289   **4.** $(.009)^2$ .000081   **6.** $(2.05)^2$ 4.2025   **8.** $\left(3\frac{1}{2}\right)^2$ 12.25   **10.** $\left(\frac{9}{10}\right)^2$ .81

PRACTICAL APPLICATIONS

1. To find the area of a square when the length of its side is known, square the given side. Find the area of a square if its side measures:*

a. 7 in. 49 sq. in. **b.** 13 mm 169 mm² **c.** $2\frac{1}{2}$ ft. $6\frac{1}{4}$ sq. ft. **d.** 25 m 625 m² **e.** 19 cm 361 cm²

In evaluating formulas it is sometimes necessary to square numbers.

2. The formula for finding the lift of an airplane involves squaring its air speed. Square the following numbers representing air speeds:†

a. 100 m/s 10,000   **b.** 132 ft./sec. 17,424   **c.** 228 m/s 51,984

3. The formula for finding the distance a body falls involves squaring the number of seconds the body falls.

Square the following numbers representing intervals of time:

a. 5 sec. 25   **b.** 12 sec. 144   **c.** 30 sec. 900   **d.** 27 sec. 729   **e.** 45 sec. 2,025

4. The formula for finding the area of a circle involves squaring the radius. Square the following numbers representing radii of circles:

a. 6 mm 36   **b.** 14 cm 196   **c.** $7\frac{1}{2}$ in. $56\frac{1}{4}$ **d.** $\frac{3}{4}$ ft. $\frac{9}{16}$ **e.** 26 m 676

---

*See Exercise 2-1 for meanings of m, cm, and mm.

†See Exercise 2-15 for meaning of m/s.

VOCABULARY: square root

OBJECTIVE:

I. Aim   To find the square root of a number.

II. Procedure

1. Write the numeral under the square root symbol ($\sqrt{}$) and separate the numeral into groups of two figures each, starting at the decimal point and forming the groups, first to the left and then to the right of the decimal point.

   If there is an odd number of figures to the left of the decimal point, there will be one group containing a single figure. However, if there is an odd number of figures to the right of the decimal point, annex a zero so that each group contains two figures.

2. Find the largest square which can be subtracted from the first group at the left. Write it under the first group.

3. Write the square root of this largest square above the first group as the first figure of the square root.

4. Subtract the square number from the first group. Annex the next group to the remainder.

5. Form the trial divisor by multiplying the root already found by 2 and annexing a zero which is not written but is used mentally.

6. Divide the dividend (step 4) by the trial divisor (step 5). Annex the quotient to the root already found; also annex it to the trial divisor to form the complete divisor.

7. Multiply the complete divisor by the new figure of the root.

8. Subtract this product (step 7) from the dividend (step 4).

9. Continue this process until all the groups have been used or the desired number of decimal places has been obtained.

10. Since each figure of the root is placed directly above its corresponding group, the decimal point in the root is placed directly above the decimal point in the numeral for the given number.

11. Check by squaring the root (multiplying the answer by itself) to obtain the given number.

III. Sample Solutions

1. Find the square root of 676.

```
        2  6          Check:
    √6 76               26
     4                  26
  46)2 76              156
    2 76               52
     . . .             676
```
Answer: 26

2. Find the square root of 5,184.

```
        7  2          Check:
    √51 84              72
     49                 72
  142) 2 84            144
      2 84             5 04
      . . .            5,184
```
Answer: 72

3. Find the square root of 214,369.

```
        4  6  3       Check:
    √21 43 69           463
     16                 463
   86) 5 43           1 389
      5 16            27 78
  923)   27 69        185 2
         27 69        214,369
```
Answer: 463

4. Find the square root of 94,249.

```
        3  0  7       Check:
    √9 42 49            307
     9                  307
  607)   42 49        2 149
         42 49        92 10
         . . . .      94,249
```
Answer: 307

5. Find the square root of .2916

```
              .5  4        Check:
          √.29 16            .54
           25               .54
       104) 4 16           216
           4 16            270
   Answer: .54  . . .      .2916
```

6. Find the square root of 7 correct to the nearest hundredth:

```
        2. 6  4  5 = 2.65     Check:
       √7.00 00 00              2.645
        4                      2.645
     46)3 00                  13 225
       2 76                  105 80
     524)  24 00            1 587 0
          20 96             5 290
     5285)  3 04 00         6 996 025
            2 64 25            3 975
                           7.000 000
   Answer: 2.65    39 75
```

IV. Definition

The square root of a number is that number which when multiplied by itself produces the given number. The square root of 36 is 6 because 6 is the number which multiplied by itself equals 36.

DIAGNOSTIC TEST

Find the square root of each of the following numbers. If there is a remainder, find answer correct to nearest hundredth.

1. 64 8
2. 8,100 90
3. .09 .3
4. 529 23

5. 6,084 78
6. 71,289 267
7. 343,396 586
8. 167,281 409

9. 2,673,225 1,635
10. 52,649,536 7,256
11. 36,048,016 6,004
12. 409,600 640

13. .8836 .94
14. 42 6.48
15. 768.4 27.72

Basic: Sets 1–9, Appl. 1–2; *Average:* Sets 1–12, Appl. 1–4; *Enriched:* Sets 1–15, Appl. 1–5

RELATED PRACTICE EXAMPLES

Find the square root of each of the following numbers. If there is a remainder, find answer correct to nearest hundredth:

SET 1	SET 2	SET 3	SET 4
1. 4 2	1. 400 20	1. .04 .2	1. 196 14
2. 49 7	2. 2,500 50	2. .64 .8	2. 625 25
3. 25 5	3. 100 10	3. .81 .9	3. 784 28
4. 9 3	4. 6,400 80	4. .25 .5	4. 576 24
5. 81 9	5. 900 30	5. .01 .1	5. 324 18
6. 16 4	6. 1,600 40	6. .16 .4	6. 961 31
7. 36 6	7. 8,100 90	7. .49 .7	7. 484 22
8. 1 1	8. 3,600 60	8. .0009 .03	8. 289 17
9. 64 8	9. 4,900 70	9. .0036 .06	9. 841 29
10. 100 10	10. 10,000 100	10. .0001 .01	10. 729 27

SET 5	SET 6	SET 7	SET 8
1. 2,601 51	1. 16,384 128	1. 140,625 375	1. 40,401 201
2. 4,356 66	2. 60,516 246	2. 682,276 826	2. 11,881 109
3. 5,929 77	3. 97,344 312	3. 398,161 631	3. 93,636 306
4. 8,836 94	4. 70,225 265	4. 175,561 419	4. 43,264 208
5. 3,481 59	5. 21,904 148	5. 877,969 937	5. 164,836 406
6. 1,369 37	6. 56,169 237	6. 341,056 584	6. 497,025 705
7. 7,396 86	7. 35,721 189	7. 208,849 457	7. 254,016 504
8. 9,801 99	8. 23,716 154	8. 487,204 698	8. 815,409 903
9. 5,625 75	9. 85,849 293	9. 565,504 752	9. 643,204 802
10. 1,444 38	10. 45,369 213	10. 799,236 894	10. 368,449 607

SET 9			
1. 1,628,176	1,276		
2. 3,912,484	1,978		
3. 5,499,025	2,345		
4. 1,108,809	1,053		
5. 6,620,329	2,573		
6. 1,580,049	1,257		
7. 5,692,996	2,386		
8. 9,872,164	3,142		
9. 8,826,841	2,971		
10. 3,418,801	1,849		

SET 9
1. 1,628,176 1,276
2. 3,912,484 1,978
3. 5,499,025 2,345
4. 1,108,809 1,053
5. 6,620,329 2,573
6. 1,580,049 1,257
7. 5,692,996 2,386
8. 9,872,164 3,142
9. 8,826,841 2,971
10. 3,418,801 1,849

SET 10
1. 17,297,281 4,159
2. 46,403,344 6,812
3. 10,349,089 3,217
4. 31,315,216 5,596
5. 19,351,201 4,399
6. 58,491,904 7,648
7. 72,914,521 8,539
8. 92,871,769 9,637
9. 40,640,625 6,375
10. 65,999,376 8,124

SET 11
1. 1,014,049 1,007
2. 4,016,016 2,004
3. 9,030,025 3,005
4. 64,096,036 8,006
5. 16,024,009 4,003
6. 25,090,081 5,009
7. 36,024,004 6,002
8. 81,018,001 9,001
9. 49,112,064 7,008
10. 36,072,036 6,006

SET 12
1. 52,900 230
2. 313,600 560
3. 774,400 880
4. 2,992,900 1,730
5. 67,404,100 8,210
6. 1,040,400 1,020
7. 50,268,100 7,090
8. 2,250,000 1,500
9. 27,040,000 5,200
10. 58,982,400 7,680

SET 13
1. .1296 .36
2. .0324 .18
3. .0841 .29
4. .5476 .74
5. .9409 .97
6. 190.44 13.8
7. 3.0276 1.74
8. 88.5481 9.41
9. 7,157.16 84.6
10. 1,070.5984 32.72

SET 14
1. 2 1.41
2. 8 2.83
3. 15 3.87
4. 39 6.24
5. 128 11.31
6. 346 18.60
7. 983 31.35
8. 1,000 31.62
9. 2,382 48.81
10. 4,976 70.54

SET 15
1. .1 .32
2. .59 .77
3. 147.6 12.15
4. 3.5 1.87
5. 207.59 14.41
6. 6.283 2.51
7. 25.9 5.09
8. .924 .96
9. 94.617 9.73
10. 503.419 22.44

ALTERNATE METHODS OF FINDING THE SQUARE ROOT

Square Root by Estimation, Division, and Average

The approximate square root of a number may also be found by estimation, division, and average as follows:

1. Estimate the square root of the given number.

2. Divide the given number by the estimated square root.

3. Find the average of the resulting quotient and estimated square root.

4. Divide the given number by this average (step 3).

5. Find the average of the divisor used and quotient found in step 4.

6. Continue this process to obtain a greater degree of approximation as the divisor and quotient will eventually approximate each other.

SAMPLE PROBLEMS

Find the square root of 12.
Since 12 is between 9 and
16, its square root will be
between 3 and 4.

1. Use 3.4 as the estimate.
2. Divide 12 by 3.4 to get
   the quotient 3.5.
3. The average of 3.4 and
   3.5 is 3.45.
4. Divide 12 by 3.45 to get
   the quotient 3.48.
5. The average of 3.45 and
   3.48 is 3.465.
6. Divide 12 by 3.465 to
   get the quotient 3.463.
   The average of 3.465
   and 3.463 is 3.464.

$$3.4 \overline{)12.00} = 3.5$$

$$\begin{array}{r} 3.4 \\ + 3.5 \\ \hline 2\overline{)6.9}(3.45 \end{array}$$

$$3.45 \overline{)12.0000} = 3.48$$

$$\begin{array}{r} 3.45 \\ + 3.48 \\ \hline 2\overline{)6.93}(3.465 \end{array}$$

$$3.465 \overline{)12.000000} = 3.463$$

$$\begin{array}{r} 3.465 \\ 3.463 \\ \hline 2\overline{)6.928}(3.464 \end{array}$$

Answer: 3.464

PRACTICE PROBLEMS

Find the square root of each of the following numbers by estimation,
division, and average to 3 decimal places:

a. 6 2.449   **d.** 59 7.681   **g.** 105 10.247   **j.** 92 9.592   **m.** 550 23.452

b. 14 3.742   **e.** 38 6.164   **h.** 86 9.274   **k.** 125 11.180   **n.** 972 31.177

c. 21 4.583   **f.** 75 8.660   **i.** 41 6.403   **l.** 300 17.321   **o.** 697 26.401

Square Root by Use of Table

Square roots of whole numbers 1 to 99 inclusive and of the perfect
squares (squares of whole numbers) given in the table may be found
directly from the table of squares and square roots on page 621.

To find the square root of any whole number from 1 to 99 inclusive,
first locate the given number in the "No." column and then move to the
right to the corresponding "Square Root" column to obtain the required
square root.

To find the square root of a perfect square given in the table, first
locate this number in the "Square" column and then move to the left to
the corresponding "No." column to obtain the required square root.

PRACTICE PROBLEMS

Find the square root of each of the following numbers by use of the table (see page 621):

1. a. 19 4.359   **f.** 29 5.385   **2. a.** 324 18   **f.** 5,476 74
   b. 57 7.550   **g.** 66 8.124   **b.** 1,225 35   **g.** 2,809 53
   c. 95 9.747   **h.** 84 9.165   **c.** 9,216 96   **h.** 7,921 89
   d. 73 8.544   **i.** 33 5.745   **d.** 784 28   **i.** 3,481 59
   e. 48 6.928   **j.** 91 9.539   **e.** 2,209 47   **j.** 3,969 63

PRACTICAL APPLICATIONS

   1. To find the length of the side of a square when its area is known, find the square root of the area of the square.
   Find the length of the side of a square if its area is:*   69.6 yd. or
                                                                    208.7 ft
   a. 25 cm² 5 cm   **b.** 144 sq. in. 12 in.   **c.** 6,889 m² 83 m   **d.** 1 acre

   2. To find the number of seconds it takes a freely falling body to fall a given distance, divide the distance by 16, then find the square root of the quotient.
   Find the number of seconds it takes a freely falling body to fall:
   a. 64 ft. 2 s **b.** 400 ft. 5 s **c.** 4,624 ft. 17 s **d.** 7,056 ft. 21 s **e.** 10,000 ft. 25 s

   3. To find the radius of a circle when its area is known, divide the area of the circle by $\frac{22}{7}$, then find the square root of the quotient.
   Find the radius of a circle when its area is:*
   a. 88 mm² 5,292 mm**b.** 132 sq. ft. 6.481 ft. **c.** 198 cm² 7.937 cm **d.** 440 m² 11.832 m

   4. To find the length of the edge of a cube when the total area of its outside surface is known, divide the total area by 6, then find the square root of the quotient.
   Find the length of the edge of a cube when its total surface area is:*
   a. 294 cm² 7 cm   **b.** 1,536 sq. ft. 16 ft.   **c.** 3,750 m² 25 m   **d.** 8,214 mm² 37 mm

   5. To find the diameter of a circle when its area is known, divide the area of the circle by .7854, then find the square root of the quotient.
   Find the diameter of a circle when its area is:*
   a. 78.54 sq. in.10 in.**b.** 1,963.5 m² 50 m**c.** 314.16 cm²20 cm**d.** 706.86 sq. in. 30 in.

---

*In the following problems, mm² represents square millimeters, cm² represents square centimeters, and m² represents square meters. See Exercise 2-4.

REVIEW OF UNIT SIX

1. Square each of the following numbers:

a. 13 169 **b.** 250 62,500 **c.** .001 .000001 **d.** 4.9 24.01 **e.** $\frac{5}{16}$ .098   **f.** $1\frac{3}{4}$ 3.063

2. Find the value of each of the following:

a. 9^2 81 **b.** $(6.4)^2$ 40.96 **c.** $\left(\frac{7}{12}\right)^2$ .34027 **d.** 58^2 3,364 **e.** $(.23)^2$ .0529 **f.** $\left(5\frac{1}{2}\right)^2$ 30.25

3. Find the square root of each of the following numbers. If there is a remainder, find answer correct to nearest hundredth:

a. 9,409 97   **c.** 49,126,081 7,009   **e.** 62 7.87
b. 85,849 293   **d.** 35,212,356 5,934   **f.** 405.8 20.14

4. Find the square root of each of the following numbers by estimation, division, and average (3 decimal places):

a. 13 3.606   **c.** 76 8.718   **e.** 114 10.677
b. 5 2.236   **d.** 39 6.245   **f.** 200 14.142

5. Find the square root of each of the following numbers by using the table of squares and square roots on page 621:

a. 44 6.63   **c.** 69 8.31   **e.** 841 29
b. 23 4.80   **d.** 8,836 94   **f.** 6,084 78

CUMULATIVE PRACTICE

1. Add:
68,594
53,083
69,418
98,579
43,859 333,533

2. Subtract:
129,306
59,269
70,037

3. Multiply:
8,906
5,007
44,592,342

4. Divide: 96
604)57,984

5. Add:
$8\frac{3}{5} + 9\frac{9}{10} + 7\frac{1}{2}$ 26

6. Subtract:
$11 - 6\frac{5}{8}$ $4\frac{3}{8}$

7. Multiply:
$4\frac{1}{2} \times 2\frac{3}{4}$ 12.375

8. Divide:
$150 \div 3\frac{1}{3}$ 45

9. Add:
$9.42 + .296 + 51.7$ 61.416

10. Subtract:
$70.5 - 6.83$ 63.67

11. Multiply:
$.04 \times .016$ .00064

12. Divide:
$18.3 \div .6$ 30.5

13. Find 6.28% of $8,500. $533.80

14. What percent of $9 is $.81? 9%

15. 6% of what number is 39? 650

16. Find the square root of 413,449. 643

COMPETENCY CHECK TEST

Solve each problem and select the letter corresponding to your answer. The numerals in red boxes indicate Exercises where help may be found.

1. Add: $63,594 + 98,697 + 28,569 + 53,935$ ⬚1-4
   a. 245,695   b. 245,795   c. 244,795   d. Answer not given

2. Subtract: $506,713 - 427,619$ ⬚1-5
   a. 188,194   b. 79,094   c. 189,094   d. 79,194

3. Multiply: 893×576 ⬚1-6
   a. 514,368   b. 415,368   c. 515,458   d. 512,458

4. Divide: $584)\overline{119,720}$ ⬚1-7
   a. 25   b. 250   c. 190   d. 205

5. Add: $5\frac{4}{5} + 3\frac{7}{10}$ ⬚1-12
   a. $8\frac{9}{10}$ b. $9\frac{1}{2}$ c. $9\frac{3}{4}$ d. $9\frac{3}{5}$

6. Subtract: $10\frac{1}{4} - 9\frac{5}{6}$ ⬚1-13
   a. $1\frac{2}{3}$ b. $\frac{7}{8}$ c. $\frac{5}{12}$ d. $1\frac{1}{12}$

7. Multiply: $3\frac{1}{7} \times 4\frac{3}{8}$ ⬚1-16
   a. $13\frac{3}{4}$ b. $12\frac{3}{56}$ c. $12\frac{1}{2}$ d. $13\frac{1}{2}$

8. Divide: $\frac{13}{16} \div 3\frac{1}{4}$ ⬚1-17
   a. $3\frac{1}{2}$ b. $\frac{1}{4}$ c. $\frac{169}{256}$ d. $\frac{3}{8}$

9. Add: $47.9 + 6.45 + 386$ ⬚1-23
   a. 1,510   b. 439.945   c. 440.35   d. Answer not given

10. Subtract: $1.706 - .8$ ⬚1-24
    a. 1.698   b. 1.106   c. .906   d. .894

11. Multiply: $27.4 \times .001$ ⬚1-26
    a. 27.4   b. 2.74   c. .274   d. .0274

12. Divide: $1.25)\overline{.5}$ ⬚1-27
    a. .04   b. .4   c. 4   d. 40

13. Find $7\frac{3}{4}\%$ of $9,000 ⬚1-41
    a. $69.75   b. $697.50   c. $6,975   d. Answer not given

14. $57 is what percent of $75? ⬚1-42
    a. 19%   b. 50%   c. 76%   d. 94%

15. 20% of what amount is $6.85? ⬚1-43
    a. $1.37   b. $13.70   c. $342.50   d. $34.25

16. Find the square root of 67,081. ⬚1-45
    a. 89   b. 301   c. 271   d. 259

17. Find the value of $(10.6)^2$ ⬚1-44
    a. 21.2   b. 112.36   c. 112.6   d. Answer not given

For a Unit Test see p. T45 or *Testing Program*, p. 9.

SUPPLEMENTARY TOPICS

This section contains some topics that are usually studied in a contemporary arithmetic program. However, many of these topics may be selected for study in any basic arithmetic program.

EXERCISE 1-46 Roman System of Numeration

OBJECTIVE: To read and write Roman numerals.

The Romans used the following number symbols in their system of numeration:

Symbol	I	V	X	L	C	D	M
Value	1	5	10	50	100	500	1,000

They formed Roman numerals by writing from left to right as a sum, first the symbol for the greatest possible value, with symbols I, X, or C used as many as three times when necessary, then the symbol for the next smaller value, etc.

When symbols I, X, or C preceded a Roman number symbol of greater value, its value was subtracted from the larger value.

IV = 5 − 1 = 4	XL = 50 − 10 = 40	CD = 500 − 100 = 400
IX = 10 − 1 = 9	XC = 100 − 10 = 90	CM = 1,000 − 100 = 900

The symbols V, L, and D never preceded a Roman number symbol of greater value and were never used in succession. A bar above the symbol indicated that the value of the symbol was multiplied by 1,000.

MDCCXLVIII = ?

Since M = 1,000
     D =   500
    CC =   200
    XL =   40
     V =   5
   III =   3
        1,748

MDCCXLVIII = 1,748

Write the Roman numeral representing 287.

287 = 200 + 50 + 30 + 5 + 2
    = CC + L + XXX + V + II
    = CCLXXXVII

Basic: Prob. 1–2; *Average:* Prob. 1–4; *Enriched:* Prob. 1–5

PRACTICE PROBLEMS

1. What number does each of the following Roman numerals represent?

a. II ₂   VIII ₈   IX ₉   XVII ₁₇   XIV ₁₄
b. XXVI ₂₆ XXIX ₂₉   LXVII ₆₇ XLVIII ₄₈   XCIV ₉₄
c. LXXIII ₇₃ XCIX ₉₉   CLXVI ₁₆₆ CXIV ₁₁₄   CCCXC ₃₉₀
d. CDXI ₄₁₁ CCXXVII ₂₂₇ DCCXLI ₇₄₁ MDCLXIX ₁,₆₆₉ MCCLIV ₁,₂₅₄
e. MCDL ₁,₄₅₀ MCMLXV ₁,₉₆₅ MLXII ₁,₀₆₂ MCMXCIII ₁,₉₉₃ MCDXLIV ₁,₄₄₄

2. Write the Roman numeral for each of the following numbers:

a. 7 VII   4 IV   19 XIX   34 XXXIV 21 XXI   38 XXXVIII 27 XXVII
b. 63 LXIII 76 LXXVI 44 XLIV 97 XCVII 59 LIX   94 XCIV   88 LXXXVIII
c. 104   299   386   847   465   628   910
d. 1492   1970   1984   1992   1241   1776   1999

3. A cornerstone is marked MCMXIX. What date does it represent? 1919

4. What number is represented by the Roman numeral in each of the following:

a. Page VII ₇   **c.** Chapter XLVI ₄₆   **e.** Paragraph LXVIII ₆₈
b. Unit XIV ₁₄   **d.** Item CXC ₁₉₀   **f.** Page CCXXIX ₂₂₉

5. In making an outline, what Roman numeral should you use following:

        IV   XX   XXV   C   XL   L
a. III **b.** XIX **c.** XXIV **d.** XCIX **e.** XXXIX **f.** XLIX

2.c. CIV; CCIC; CCCLXXXVI; DCCCXLVII; CDLXV; DCXXVIII; CMX
2.d. MCDXCII; MCMLXX; MCMLXXXIV; MCMXCII; MCCXLI; MDCCLXXVI; MCMIC

Reading and Writing
Shortened Lists of Numerals

OBJECTIVE: To read and write shortened lists of numerals.

Instead of the "etc." being used with a large group of numbers, in mathematics a shortened list and three dots may be used to indicate the missing numbers.

To indicate an unlimited number of numbers that follow a pattern and continue in this pattern without ending, write the first few numerals and use three dots after the numeral naming the last listed number.

> Write the list of numerals indicating:
> Zero, seven, fourteen, twenty-one, and so on without end.
> *Answer:* 0, 7, 14, 21, . . .

To indicate the missing numbers when there is a limited number of numbers that follow a pattern but too many to list, use three dots before the numeral naming the last number.

> Write the list of numerals indicating:
> One, three, five, seven and so on up to and including twenty-nine.
> *Answer:* 1, 3, 5, 7, . . . , 29

For ex. 1–3 see Extended Answers, p. T74. **PRACTICE PROBLEMS** *Basic:* Prob. 1
Average: Prob. 1–2
Enriched: Prob. 1–3

1. Read, or write in words, each of the following:

a. 0, 1, 2, 3, . . .   **c.** 1, 3, 5, 7, . . .   **e.** 0, 5, 10, 15, . . .
b. 0, 2, 4, 6, . . .   **d.** 2, 3, 5, 7, . . .   **f.** 1, 4, 9, 16, . . .

2. Read, or write in words, each of the following:

a. 0, 1, 2, 3, . . . , 25   **c.** 1, 3, 5, 7, . . . , 61   **e.** 0, 8, 16, 24, . . . , 96
b. 0, 4, 8, 12, . . . , 64 **d.** 0, 10, 20, 30, . . . , 200 **f.** 2, 4, 6, 8, . . . , 50

3. Write the list of numerals indicating:

a. Zero, three, six, nine, and so on without end.
b. Zero, six, twelve, eighteen and so on up to and including seventy-two.
c. Two, three, five, seven and so on up to and including eighty-nine.
d. Zero, twenty, forty, sixty and so on without end.
e. One, two, three, four and so on up to and including one hundred.
f. Three, nine, twenty-seven, eighty-one and so on without end.

Exponents

OBJECTIVE: To read and write
numerals in exponential form.

VOCABULARY: exponent
base
powers of 10

When two equal factors are used in multiplication like 5×5, it may be written in exponential form as 5^2, read "five to the second power" or "the second power of five" or "five squared" or "the square of five."

Three equal factors like $7 \times 7 \times 7$ may be written as 7^3, which is read "seven to the third power" or "seven cubed" or "the cube of seven."

10^4 represents $10 \times 10 \times 10 \times 10$ or the product 10,000 and is read "ten to the fourth power."

2^5 represents $2 \times 2 \times 2 \times 2 \times 2$ or the product 32 and is read "two to the fifth power."

The small numeral written to the upper right (superscript) of the repeated factor is called an exponent. When it names a natural number, it tells how many times the factor is being used in multiplication. The factor that is being repeated is called the base. The number 2^5 uses 2 as the base and 5 as the exponent.

Numbers like $10, 10^2, 10^3, 10^4, 10^5, 10^6$, etc. are called powers of 10

For ex. 1–10 see
Extended Answers, p. T75.

PRACTICE PROBLEMS

Basic: Prob. 1–7
Average: Prob. 1–9
Enriched: Prob. 1–10

1. Read, or write in words, each of the following:

a. 3^4 **b.** 7^8 **c.** 2^{11} **d.** 10^5 **e.** 6^1 **f.** 11^9 **g.** 5^3 **h.** 20^2

2. Write as a numeral:

a. Five to the seventh power **b.** Nine cubed **c.** Fifty squared

3. What is the exponent in: **a.** 4^7? **b.** 10^9? **c.** 5^1? **d.** 3^{12}? **e.** 11^4?

4. What is the base in: **a.** 5^6? **b.** 6^{11}? **c.** 2^5? **d.** 12^9? **e.** 21^{10}?

5. How many times is the base being used as a factor in:

a. 5^8? **b.** 10^{12}? **c.** 3^7? **d.** 2^{15}? **e.** 17^{30}?

6. Use the exponential form to write:

a. $2 \times 2 \times 2 \times 2$ **b.** $6 \times 6 \times 6 \times 6 \times 6 \times 6 \times 6 \times 6 \times 6 \times 6$

7. Express each of the following as a product of a repeated factor:

a. $7^3 = 7 \times 7 \times 7$ **b.** 6^4 **c.** 3^{10} **d.** 8^7 **e.** 12^9 **f.** 9^{12}

8. a. Express 64 as a power of 8. **b.** Express 81 as a power of 3.

9. First express each of the following as a product of a repeated factor, then in exponential form:

a. $36 = 6 \times 6 = 6^2$ **b.** 25 **c.** 8 **d.** 81 **e.** 121 **f.** 625

10. Find the value of each of the following:

a. 9^2 **b.** 4^4 **c.** 2^8 **d.** 3^7 **e.** 8^3 **f.** 5^6

1-49 Powers of Ten

OBJECTIVE: To read and write numerals expressed as powers of 10.

Using 10 as a repeated factor, we may write this table of powers of 10:

$$
\begin{array}{rll}
10 &= 10 &= 10^1 \\
100 &= 10 \times 10 &= 10^2 \\
1,000 &= 10 \times 10 \times 10 &= 10^3 \\
10,000 &= 10 \times 10 \times 10 \times 10 &= 10^4 \\
100,000 &= 10 \times 10 \times 10 \times 10 \times 10 &= 10^5 \\
1,000,000 &= 10 \times 10 \times 10 \times 10 \times 10 \times 10 &= 10^6
\end{array}
$$

Observe in the above table:

1. 10 used one time as a factor may be written with the exponent 1, although the exponent 1 usually is omitted.
2. The number of zeros found after the digit 1 in the product (first column above) corresponds in each case to the exponent of 10 (third column above.)

For example in $100,000 = 10^5$ there are 5 zeros in the product (100,000) and the exponent of 10^5 is 5.

Any numeral having zeros for all digits except the first, such as 30, 600, or 7,000, may be expressed as a product of a digit and a power of ten.

$$
7,000 = 7 \times 1,000 = 7 \times 10^3
$$

Basic: Prob. 1; *Average:* Prob. 1–2; *Enriched:* Prob. 1–3

PRACTICE PROBLEMS

1. Express the following as powers of ten:
   a. 1,000 10^3
   b. 100,000 10^5
   c. 10,000,000 10^7
   d. 1,000,000,000 10^9
   e. 100,000,000,000,000 10^{14}
   f. 10,000,000,000,000,000 10^{16}

2. Express the following as products of a digit and a power of ten:
   a. 50 5×10
   b. 700 7×10^2
   c. 80,000 8×10^4
   d. 6,000 6×10^3
   e. 4,000,000 4×10^6
   f. 200,000 2×10^5
   g. 50,000,000 5×10^7
   h. 3,000,000,000 3×10^9
   i. 700,000,000,000 7×10^{11}
   j. 90,000,000,000 9×10^{10}
   k. 6,000,000,000,000 6×10^{12}
   l. 400,000,000,000,000 4×10^{14}

3. Express the following as products of a digit and a power of ten:
   a. 7 million (7,000,000) 7×10^6
   b. 8 trillion (8,000,000,000,000) 8×10^{12}
   c. 5 billion (5,000,000,000) 5×10^9
   d. 4 quadrillion (4,000,000,000,000,000) 4×10^{15}
   e. 9 sextillion
      (9,000,000,000,000,000,000,000) 9×10^{21}
   f. 3 quintillion
      (3,000,000,000,000,000,000) 3×10^{18}

Writing Decimal Numerals As Polynomials—Expanded Notation

OBJECTIVE: To write decimal numerals as polynomials in expanded notation.

VOCABULARY: polynomial expanded notation

The value of each place in a decimal numeral may be expressed as a power of ten.

The digit 4 in 46 means 4 tens or 4×10.

The digit 7 in 725 means 7 hundreds or $7 \times 100 = 7 \times 10^2$.

The digit 3 in 3,918 means 3 thousands or $3 \times 1,000 = 3 \times 10^3$.

The complete value of each digit in the decimal numeral is equal to the value of the digit itself times its place value.

A decimal numeral may be written as the sum of the products of each digit in the numeral and its place value expressed as a power of ten.

The numeral 5,963

$= 5$ thousands $+ 9$ hundreds $+ 6$ tens $+ 3$ ones

$= (5 \times 1,000) + (9 \times 100) + (6 \times 10) + (3 \times 1)$

$= (5 \times 10^3) + (9 \times 10^2) + (6 \times 10^1) + (3 \times 1)$

Writing a numeral in this expanded form as an indicated sum (polynomial) is sometimes called expanded notation. Observe that the place values are powers of ten arranged in decreasing order from left to right.

We extend this notation to the decimal fraction. The value of each place in a decimal numeral may be expressed as a power of ten.

.1 and $\frac{1}{10}$ are equivalent and $\frac{1}{10}$ may be written as $\frac{1}{10^1}$

.01 and $\frac{1}{100}$ are equivalent and $\frac{1}{100}$ may be written as $\frac{1}{10^2}$

.001 and $\frac{1}{1000}$ are equivalent and $\frac{1}{1000}$ may be written as $\frac{1}{10^3}$

.0001 and $\frac{1}{10,000}$ are equivalent and $\frac{1}{10,000}$ may be written as $\frac{1}{10^4}$

It can be shown that $\frac{1}{10^1} = 10^{-1}, \frac{1}{10^2} = 10^{-2}, \frac{1}{10^3} = 10^{-3}, \frac{1}{10^4} = 10^{-4},$ etc., and $10^0 = 1$.

Thus, the value of each place in a numeral naming a decimal fraction may be expressed as a power of ten with a negative exponent.

The decimal numeral .9247
$= 9$ tenths $+ 2$ hundredths $+ 4$ thousandths $+ 7$ ten-thousandths
$= 9 \times \dfrac{1}{10} + 2 \times \dfrac{1}{100} + 4 \times \dfrac{1}{1,000} + 7 \times \dfrac{1}{10,000}$
$= 9 \times \dfrac{1}{10^1} + 2 \times \dfrac{1}{10^2} + 4 \times \dfrac{1}{10^3} + 7 \times \dfrac{1}{10^4}$
$= (9 \times 10^{-1}) + (2 \times 10^{-2}) + (4 \times 10^{-3}) + (7 \times 10^{-4})$
    Also the numeral 685.379 is written in expanded form as:
$(6 \times 10^2) + (8 \times 10^1) + (5 \times 10^0) + (3 \times 10^{-1}) + (7 \times 10^{-2}) + (9 \times 10^{-3})$

Observe that the place values are powers of ten arranged in a decreasing order from left to right.

Basic: Prob. 1–2; *Average:* Prob. 1–3; *Enriched:* Prob. 1–4

PRACTICE PROBLEMS   For ex. 1–4 see Extended Answers, p. T75.

1. Express each of the following polynomials as a numeral:
  a. $(6 \times 10^4) + (7 \times 10^3) + (2 \times 10^2) + (9 \times 10^1) + (4 \times 1)$
  b. $(3 \times 10^5) + (5 \times 10^4) + (6 \times 10^3) + (8 \times 10^2) + (2 \times 10^1) + (7 \times 1)$

2. Write each of the following numerals in expanded form as a polynomial:

  a. 58   **e.** 37,026   **i.** 46,212,597
  b. 492   **f.** 29,142,358   **j.** 25,384,569,703
  c. 3,564   **g.** 5,688,241   **k.** 6,739,158,245
  d. 892,187   **h.** 823,917,084   **l.** 937,175,486,298

3. Express each of the following polynomials as a numeral:
  a. $(7 \times 10^{-1}) + (3 \times 10^{-2}) + (4 \times 10^{-3})$
  b. $(6 \times 10^{-1}) + (1 \times 10^{-2}) + (7 \times 10^{-3}) + (5 \times 10^{-4})$
  c. $(9 \times 10^1) + (3 \times 10^0) + (6 \times 10^{-1}) + (8 \times 10^{-2})$
  d. $(2 \times 10^2) + (0 \times 10^1) + (7 \times 10^0) + (4 \times 10^{-1}) + (9 \times 10^{-2})$
  e. $(4 \times 10^4) + (8 \times 10^3) + (3 \times 10^2) + (6 \times 10^1) + (8 \times 10^0) +$
    $(0 \times 10^{-1}) + (2 \times 10^{-2}) + (7 \times 10^{-3})$
  f. $(6 \times 10^3) + (4 \times 10^2) + (9 \times 10^1) + (2 \times 10^0) + (5 \times 10^{-1}) +$
    $(0 \times 10^{-2}) + (3 \times 10^{-3}) + (8 \times 10^{-4}) + (1 \times 10^{-5})$

4. Write each of the following numerals in expanded form as a polynomial:

  a. .29   **d.** .68145   **g.** 89.46   **j.** 759.504
  b. .476   **e.** .975263   **h.** 215.83   **k.** 3,146.25
  c. .3509   **f.** 8.2   **i.** 53.927   **l.** 8,372.4917

Number Sentences

OBJECTIVE; To read and
write number sentences.

VOCABULARY: number sentence   inequality
equality   statement

Sentences that deal with numbers are called number sentences or mathematical sentences.

The equality $6 + 9 = 15$ is a number sentence.

It reads "Six plus nine is equal to fifteen."

The symbol $=$ is the equality sign. It means "is equal to" and is the verb in the sentence. The expressions on both sides of the equality sign designate the same number.

When two numbers are compared, one number may be:

(1) equal to the other;

(2) greater than the other; or

(3) less than the other.

In any specific case only one of these three possibilities is true.

When one number is greater than or less than a second number, an inequality exists. An inequality is a number sentence.

Symbols of inequality include: $\neq, >, <, \not>$, and $\not<$. Each symbol is a verb in a number sentence.

Number sentences may be true or they may be false. The number sentence $3 + 7 = 10$ is true but the sentence $8 - 2 = 5$ is false. A sentence that is either true or false is called a statement. A sentence cannot be both true and false at one time.

The symbol $=$ means "is equal to."

$9 - 6 = 3$ is read:

"Nine minus six is equal to three."

The symbol $\neq$ means "is not equal to."

$11 - 5 \neq 8$ is read:

"Eleven minus five is not equal to eight."

The symbol $>$ means "is greater than."

$9 > 5$ is read: "Nine is greater than five."

The symbol $\not>$ means "is not greater than."

$8 \times 3 \not> 30$ is read:

"Eight times three is not greater than thirty."

The symbol $<$ means "is less than."

$11 < 40 \div 2$ is read:

"Eleven is less than forty divided by two."

The symbol $\not<$ means "is not less than."

$7 + 8 \not< 20 - 6$ is read:

"Seven plus eight is not less than twenty minus six."

PRACTICE PROBLEMS For ex. 1 see Extended Answers, pp. T75–T76.

1. Read, or write in words, each of the following:

   a. $34 = 34$
   b. $8 \times 9 = 9 \times 8$
   c. $53 > 49$
   d. $48 - 9 > 37$
   e. $16 < 20$
   f. $4 < 21 - 16$
   g. $17 \neq 10 + 8$
   h. $6 \times 7 \neq 98 \div 2$
   i. $15 \not> 23$
   j. $45 \times 4 \not> 200$
   k. $2 + 3 \not< 3 + 2$
   l. $9^2 \not< 4 + 77$

   m. $15 \times 10 = 10 \times 15$
   n. $6 \times 9 < 49 + 7$
   o. $11 - 4 > 2 \times 3$
   p. $27 \div 3 \neq 3 \div 27$
   q. $37 - 18 \not< 15 - 15$
   r. $16 \times 2 = 64 \div 2$
   s. $63 \div 7 < 63 - 7$
   t. $4 + 0 > 7 \times 0$
   u. $1 \times 1 = 1 \div 1$
   v. $72 \div 9 \not> 2 \times 4$
   w. $.6 - .6 < .6 \div .6$
   x. $\frac{1}{2} + \frac{1}{2} > \frac{1}{2} \times \frac{1}{2}$

2. Write each of the following sentences symbolically:
   a. Eight plus six is equal to fourteen. $8 + 6 = 14$
   b. Seven minus two is greater than zero. $7 - 2 > 0$
   c. Twenty-five is less than four times seven. $25 < 4 \times 7$
   d. Ten times five is greater than eighty divided by two. $10 \times 5 > 80 \div 2$
   e. One minus one is less than one times one. $1 - 1 < 1 \times 1$
   f. Forty is not equal to ninety divided by three. $40 \neq 90 \div 3$
   g. Twenty divided by twenty is not less than twenty minus twenty. $20 \div 20 \not< 20 - 2$
   h. Nine plus eight is not greater than nine times eight. $9 + 8 \not> 9 \times 8$
   i. Five times seven is equal to seven times five. $5 \times 7 = 7 \times 5$
   j. Twelve divided by six is greater than twenty times zero. $12 \div 6 > 20 \times 0$

3. Determine which of these sentences are true and which are false:
   a. $6 + 7 = 7 + 6$ True
   b. $15 \times 8 = 8 \times 15$ True
   c. $25 > 31 - 19$ True
   d. $60 \div 5 = 3^2$ False
   e. $61 > 9 \times 8$ False
   f. $0 < 10 - 4$ True
   g. $12 - 3 \neq 3 + 12$ True
   h. $18 \div 18 \not> 18 - 18$ False
   i. $4 + 8 \not< 0$ True
   j. $2 + 5 + 9 < 9 + 5 + 2$ False
   k. $5^2 > 2^5$ False
   l. $6 \times (3 + 7) \neq (6 \times 3) + (3 + 7)$ True

Even and Odd Numbers

OBJECTIVE: To read and write
even and odd numbers.

VOCABULARY: even number
odd number

Whole numbers may be separated into even and odd numbers.

An even number is a whole number that can be divided exactly (is divisible) by two (2). Zero is considered an even whole number. Numbers whose numerals end with a 0, 2, 4, 6, or 8 are even numbers.

An odd number is a whole number that cannot be divided exactly by two (2). Numbers whose numerals end with a 1, 3, 5, 7, or 9 are odd numbers.

Basic: Prob. 1–7; *Average:* Prob. 1–9; *Enriched:* Prob. 1–10

PRACTICE PROBLEMS

1. Which of the following are <u>even</u> numbers? Odd numbers?

a. <u>10</u>   49   97   <u>284</u>   565   2,183 odd: 49, 97, 565, 2,183

b. 81   <u>36</u>   <u>78</u>   <u>300</u>   <u>902</u>   5,749 odd: 81, 5,749

2. a. Write all the one-digit numerals that name odd whole numbers. 1, 3, 5, 7, 9

b. Write all the one-digit numerals that name even whole numbers. 2, 4, 6, 8

3. a. Write the numerals that name all the even numbers greater than 47 and less than 63.

b. Write the numerals that name all the odd numbers less than 106 and greater than 94.

4. Is the sum an odd number or an even number when we add:

a. Two even numbers? Illustrate. even; 2 + 2 = 4

b. Two odd numbers? Illustrate. even; 3 + 3 = 6

c. An odd number and an even number? Illustrate. odd; 2 + 3 = 5

5. Is the product an odd number or an even number when we multiply:

a. Two even numbers? Illustrate. even; 2 × 2 = 4

b. Two odd numbers? Illustrate. odd; 5 × 3 = 15

c. An odd number and an even number? Illustrate. even; 3 × 2 = 6

6. a. Is there an odd whole number between every pair of even whole numbers? Yes

b. Is there an even whole number between every pair of odd whole numbers? Yes

7. a. Is one more than any whole number an even number? No

b. Is two more than any whole number an even number? No

c. Is two times any whole number an odd number or an even number? Even number

3.a. forty-eight, fifty, fifty-two, fifty-four, fifty-six, fifty-eight, sixty, sixty-two

3.b. ninety-five, ninety-seven, ninety-nine, one hundred one, one hundred three, one hundred five

8. a. Is one more than any even number an odd number or an even number? odd number

b. Is one more than any odd number an odd number or an even number? even number

c. Is two more than any odd number an odd number or an even number? odd number

9. a. Select any even number and square it. Is the square of an even number an odd number or an even number? Why?

b. Select any odd number and square it. Is the square of an odd number an odd number or an even number? Why?

10. Using the odd numbers 1, 3, 5, 7, 9, 11, and 13, find the sum of the:

a. First two odd numbers. 4

b. First three odd numbers. 9

c. First four odd numbers. 16

d. First five odd numbers. 25

e. First six odd numbers. 36

f. All seven odd numbers. 49

The sum is equal to the square of the number of addends.

Study the sums and the number of addends used in each case to find the relationship between the number of addends and their sums.

9.a. Even number; product of multiplication of two even numbers is an even number.
9.b. Odd number; product of multiplication of two odd numbers is an odd number.

REFRESH YOUR SKILLS

1. Add: 8,795
     658
   5,937
  28,556
   5,679
  49,625

2. Subtract:
143,704
52,893
90,811

3. Multiply:
496
879
435,984

4. Divide:
$65\overline{)45,890}$
706

5. Add:
$2\frac{5}{6} + 5\frac{1}{2}$ $8\frac{1}{3}$

6. Subtract:
$10 - 4\frac{7}{10}$ $5\frac{3}{10}$

7. Multiply:
$4\frac{4}{5} \times 1\frac{2}{3}$ 8

8. Divide:
$35 \div 3\frac{1}{8}$ 11.2

9. Add:
$.394 + .19 + .8$ 1.384

10. Subtract:
$.06 - .057$ .003

11. Multiply:
$.008 \times 3.5$ .028

12. Divide:
$2.1 \div .07$ 30

13. Find $87\frac{1}{2}\%$ of $960 $840

14. What percent of 40 is 28? 70%

15. 31 is 5% of what number? 620

16. Square: .06   .0036

17. Find the square root of .9 to the nearest hundredth. .95

18. Write 316.2 *million* as a complete number. 316,200,000

Factors

VOCABULARY: factor   common divisor
product   greatest common factor
common factor

A factor is any one of the numbers used in multiplication to form the answer called the product. In $4 \times 6 = 24$, the numbers 4 and 6 are called factors and the number 24 is called the product.

To determine whether a particular number is a factor of a given whole number, divide it into the given number. If the division is exact (with a zero remainder), the number is a factor.

Any number that is a factor of each of two or more given whole numbers is called a common factor of the numbers. Sometimes it is called the common divisor of the numbers. One (1) is a common factor of any set of numbers.

The greatest common factor of two or more whole numbers is the greatest whole number that will divide all the given numbers exactly.

The factors of 16 are 1, 2, 4, 8, and 16.

The factors of 24 are 1, 2, 3, 4, 6, 8, 12, and 24.

The common factors of 16 and 24 are 1, 2, 4, and 8.

The greatest common factor of 16 and 24 is 8.

Observe that 1 is a common factor of any two or more numbers. Some numbers like 2 and 3 have no common factor except 1; other numbers like 4 and 6 have one common factor (2) other than 1; still other numbers like 16 and 24 have many common factors (2, 4, and 8) other than 1.

Basic: Prob. 1–7; *Average:* Prob. 1–9; *Enriched:* Prob. 1–10

PRACTICE PROBLEMS

1. a. Is 6 a factor of 24? Yes   **c.** Is 9 a factor of 56? No
b. Is 5 a factor of 40? Yes   **d.** Is 18 a factor of 90? Yes
2. What is the second factor of the pair of factors when:
a. 7 is one factor of 21? 3   **d.** 14 is one factor of 98? 7
b. 8 is one factor of 72? 9   **e.** 25 is one factor of 1,000? 40
c. 18 is one factor of 144? 8   **f.** 17 is one factor of 51? 3

For ex. 3–10 see Extended Answers, p. T76.

3. Write all the factors of each of the following numbers:

a. 26	**g.** 48
b. 18	**h.** 84
c. 17	**i.** 250
d. 45	**j.** 72
e. 54	**k.** 400
f. 70	**l.** 132

4. a. What are the factors of 12?
b. What are the factors of 16?
c. What are the common factors of 12 and 16?
5. a. What are the factors of 36?
b. What are the factors of 108?
c. What are the factors of 81?
d. What are the common factors of 36, 108, and 81?

6. For each of the following groups of numbers, first find the factors of each number, then find their common factors:

a. 8 and 12	**d.** 54 and 72	**g.** 24, 60, and 96
b. 4 and 5	**e.** 63 and 84	**h.** 56, 140, and 168
c. 48 and 56	**f.** 225 and 150	**i.** 84, 144, and 360

7. For each of the following groups of numbers write the common factors:

a. 16 and 18	**d.** 45 and 135	**g.** 39, 65, and 91
b. 21 and 9	**e.** 8, 10, and 12	**h.** 252, 588, and 420
c. 32 and 104	**f.** 36, 108, and 156	**i.** 225, 375, and 825

8. a. What are the factors of 126?
b. What are the factors of 162?
c. What are the common factors of 126 and 162?
d. What is the greatest common factor of 126 and 162?
9. a. What are the factors of 48?
b. What are the factors of 80?
c. What are the factors of 112?
d. What are the common factors of 48, 80, and 112?
e. What is the greatest common factor of 48, 80, and 112?

10. Find the greatest common factor of each of the following groups of numbers:

a. 16 and 20	**f.** 120 and 216
b. 3 and 2	**g.** 18, 54, and 198
c. 54 and 144	**h.** 34, 85, and 102
d. 35 and 63	**i.** 250, 375, and 625
e. 150 and 200	

Prime and Composite Numbers

OBJECTIVE: To determine whether a
number is prime or composite.

Whole numbers other than 0 and 1 may be separated into prime and composite numbers.

A prime number is a whole number other than 0 and 1 which is divisible (is divided exactly) only by itself and 1 and by no other whole number. It is any whole number greater than 1 whose only factors are 1 and itself. See Exercise 1-55.

> 23 is a prime number.
> It can be divided exactly only by 23 and by 1.

A composite number is any whole number greater than 1 which is divisible by at least one whole number other than 1 and itself.

> 15 is a composite number.
> It can be divided exactly not only by 15 and by 1
> but also by 3 and 5.

A composite number can be expressed as a product of prime numbers. Each composite number has only one group of prime factors but the factors may be arranged in different orders. A raised dot may be used to indicate multiplication.

> $18 = 2 \cdot 3 \cdot 3$ or $3 \cdot 2 \cdot 3$ or $3 \cdot 3 \cdot 2$

Two prime numbers are called twin primes if one number is two more than the other number.

> 17 and 19 are a pair of twin primes.

Two numbers are said to be relatively prime when they have no common factor other than 1. The numbers do not necessarily have to be prime numbers.

> 16 and 21 are relatively prime to each other because 1 is the only whole number that will divide into both 16 and 21 exactly.

PRACTICE PROBLEMS

1. Which of the following are prime numbers?

     14   <u>23</u>   <u>79</u>   51   85   <u>97</u>   69   117

2. Which of the following are composite numbers?

     <u>81</u>   <u>18</u>   <u>49</u>   2   <u>91</u>   <u>39</u>   53   <u>119</u>

3. Write all the one-digit numerals naming:

a. Prime numbers 2, 3, 5, 7

b. Even prime numbers 2

c. Odd prime numbers 3, 5, 7

4.a. 19, 23, 29, 31

4. Write the group of numerals naming: 4.b. 54, 55, 56, 57, 58, 60, 62, 63, 64, 65, 66, 68, 69

a. All prime numbers greater than 18 and less than 32.

b. All composite numbers less than 70 and greater than 53.

5. a. Are all even numbers composite numbers? No

If not, name an even prime number. 2

b. Are all odd numbers prime numbers? No

If not, name an odd composite number. 9, 15, 21, 25, 27, 33, 35, etc.

6. What twin primes may be found among numbers between 25 and 75? 29 and 31; 41 and 43; 59 and 61; 71 and 73

7. Which of the following pairs of numbers are relatively prime?

a. 9 and 22   Rel. prime   **c.** 54 and 63   Not rel. prime   **e.** 39 and 65   Not rel. prime

b. 31 and 15   Rel. prime   **d.** 27 and 35   Rel. prime   **f.** 57 and 43   Rel. prime

8. Find two prime factors which when multiplied give the product of:

    a. 15   **b.** 26   **c.** 143   **d.** 58   **e.** 178   **f.** 85   **g.** 106

     3×5 2×13 11×13 2×29 2×89 5×17 2×53

9. Find two prime numbers which when added give the sum of:

    a. 8   **b.** 14   **c.** 32   **d.** 76   **e.** 98   **f.** 110   **g.** 62

  3 and 5   3 and 11   3 and 29   19 and 79   3 and 59

            7 and 7   13 and 19   31 and 67   19 and 43

                                     37 and 61   31 and 31

Goldbach, a mathematician, guessed that, "Any even number greater than 4 can be expressed as the sum of two odd prime numbers."

10. a. Select any even number greater than 4 (other than those used Answers will in problem 8) and show that it is the sum of two odd prime numbers. vary.

b. Can you find an even number greater than 4 that is not the sum of two odd prime numbers? No

c. Select an odd number that is the sum of two prime numbers. Illustrate this. Answers will vary.

d. Select an odd number that is the sum of two odd prime numbers. Illustrate this. Cannot be done since sum of any two odd numbers is an even number.

9.d. 3 and 73; 5 and 71, 17 and 59; 23 and 53; 29 and 47
9.f. 3 and 107; 7 and 103; 13 and 97; 31 and 79; 37 and 73; 43 and 67

Factoring a Natural Number— Complete Factorization

OBJECTIVE:

To factor a natural number means to replace the number by its whole-number factors expressed as an indicated product. This indicated product may contain two or more whole-number factors.

A prime number may be expressed only as a product of the given number and 1. The prime number 17 is expressed as 17×1.

To factor 18 means to replace 18 by any one of the following indicated products of:

(1) Two whole-number factors: 1×18 or 2×9 or 3×6
or 18×1 or 9×2 or 6×3

(2) Three whole-number factors: $2 \times 3 \times 3$ or $1 \times 2 \times 9$
or $1 \times 3 \times 6$ arranged in different orders.

(3) More than three whole-number factors since the factor 1 may be repeated:
$18 \times 1 \times 1 \times 1$ or $2 \times 9 \times 1 \times 1 \times 1 \times 1$ or
$3 \times 6 \times 1 \times 1 \times 1 \times 1 \times 1 \times 1$ etc.
However, the factor 1 is usually excluded when there are more than two whole-number factors in the indicated product.

A composite number may be expressed as a product of prime numbers. The number is said to be completely factored only if the factors are all prime numbers (prime factors).

To factor completely, first find two factors of the given number by inspection and then continue factoring any of the factors which are composite numbers until only prime factors result.

$$18 = 6 \times 3 = 2 \times 3 \times 3 \quad \text{or} \quad 18 = 9 \times 2 = 3 \times 3 \times 2$$

Each of these may be arranged as a factor tree:

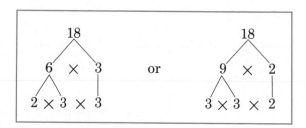

Or we divide the given number and the resulting quotients successively by prime numbers and divide these numbers exactly until a quotient of 1 is obtained. The divisors are the prime factors.

Observe in the model example at the right how the quotients are brought down in each row.

$$
\begin{array}{r}
2\,)\,\overline{18} \\
3\,)\,\overline{9} \\
3\,)\,\overline{3} \\
1
\end{array}
$$

$18 = 2 \times 3 \times 3$

Basic: Prob. 1–2; *Average:* Prob. 1–4; *Enriched:* Prob. 1–5

PRACTICE PROBLEMS   For ex. 1–5 see Extended Answers, pp. T76–T77.

1. Factor each of the following numbers as a product of two whole-number factors in as many ways as possible:

    a. 24   **b.** 19   **c.** 42   **d.** 144   **e.** 108   **f.** 400   **g.** 360

2. Factor each of the following numbers as a product of three whole-number factors in as many ways as possible:

    a. 28   **b.** 30   **c.** 75   **d.** 64   **e.** 120   **f.** 500   **g.** 98

3. Factor each of the following numbers as a product of four whole-number factors in as many ways as possible:

    a. 40   **b.** 72   **c.** 54   **d.** 225   **e.** 160   **f.** 420   **g.** 1,200

4. Factor each of the following numbers as a product of prime numbers:

    a. 14   **b.** 54   **c.** 72   **d.** 135   **e.** 168   **f.** 600   **g.** 960

5. Factor each of the following numbers as a product of prime numbers, using a factor tree:

    a. 20   **b.** 54   **c.** 21   **d.** 48   **e.** 100   **f.** 72   **g.** 150

Number Multiples

**OBJECTIVE: To find the
multiples of a number.**

A multiple of a given whole number is a product of the given number and another whole number factor. A multiple of a given number is divisible by the given number.

> The multiples of 4 are: 0, 4, 8, 12, . . .

Any number which is a multiple of two or more numbers is called the common multiple of the numbers. Numbers may have many common multiples.

The least common multiple (LCM) of two or more numbers is the smallest natural number which is the multiple of all of them. It is the smallest possible natural number that can be divided exactly by all the given numbers.

Zero (0) is excluded when determining the least common multiple, although it is a common multiple of any group of numbers.

> The multiples of 4 are: 0, 4, 8, 12, 16, 20, 24, . . .
> The multiples of 6 are: 0, 6, 12, 18, 24, 30, 36, . . .
> The common multiples of 4 and 6 are: 0, 12, 24, . . .
> The least common multiple of 4 and 6 is 12 since
> 0 is excluded.

The LCM may be found by factoring the given numbers as primes and forming a product of these primes using each the greatest number of times it appears in the factored form of any one number.

> The LCM of 4 and 6 is found as follows:
>
> Since $4 = 2 \cdot 2$
>
> and $6 = 2 \cdot 3$ Then LCM $= \overset{4}{\overline{2 \cdot 2} \cdot 3}$
>
> $\underset{6}{}$
>
> *Answer:* LCM $= 12$

For ex. 1–12 see
Extended Answers, p. T77.

PRACTICE PROBLEMS

1. Which of the following numbers are multiples:

a. Of 4?   (1) 28   (2) 52   (3) 34   (4) 90   (5) 148
b. Of 7?   (1) 54   (2) 84   (3) 67   (4) 91   (5) 119
c. Of 3?   (1) 19   (2) 42   (3) 51   (4) 73   (5) 102

2. Name five different multiples of each of the following numbers:

a. 5   **c.** 6   **e.** 11   **g.** 27   **i.** 100
b. 2   **d.** 9   **f.** 14   **h.** 59   **j.** 225

3. Are all whole numbers multiples of 1?

4. Are all even whole numbers multiples of 2?

5. Write all the multiples of each of the following, listing the first five multiples:

a. 8   **c.** 11   **e.** 40   **g.** 75   **i.** 120
b. 10   **d.** 32   **f.** 17   **h.** 96   **j.** 275

6. What are the factors of 24? Is 24 a multiple of each of its factors?

7. Is 18 a multiple of 6? Is 18 a multiple of 9? Is 18 a common multiple of 6 and 9?

8. What are the first twenty multiples of 8? Of 12? What are the first six common multiples of 8 and 12?

9. Write all the common multiples of each of the following, listing the first four multiples:

a. 3 and 5   **d.** 16 and 18   **g.** 4, 8, and 16
b. 8 and 4   **e.** 25 and 20   **h.** 12, 18, and 30
c. 10 and 12   **f.** 2, 5, and 8   **i.** 14, 21, and 35

10. Is the product of 4 and 10 a common multiple of 4 and 10? Is it the smallest common multiple? If not, what is the smallest natural number that can be divided exactly by both 4 and 10?

11. Find the least common multiple for each of the following:

a. 3 and 8   **e.** 25 and 100   **i.** 105 and 150
b. 6 and 12   **f.** 56 and 42   **j.** 2, 3, and 4
c. 15 and 20   **g.** 84 and 96   **k.** 8, 10, and 12
d. 24 and 16   **h.** 75 and 100   **l.** 15, 40, and 75

12. Find the greatest common factor (GCF) and least common multiple (LCM) of each of the following pairs of numbers. For each pair of numbers compare the product of the GCF and LCM with the product of the two given numbers. What do you find to be true in each case?

a. 6 and 8   **b.** 25 and 10   **c.** 12 and 16   **d.** 24 and 18

Equivalent Fractions

OBJECTIVE: To find equivalent fractions.

Fractions that name the same number are called equivalent fractions. A group of equivalent fractions may be developed from the name of the fractional number in simplest form by expressing it successively in higher terms.

The group of fractions equivalent to $\frac{1}{3}$ is $\frac{1}{3}, \frac{2}{6}, \frac{3}{9}, \frac{4}{12}, \ldots$

The group of fractions equivalent to $\frac{12}{16}$ is $\frac{3}{4}, \frac{6}{8}, \frac{9}{12}, \frac{12}{16}, \ldots$

The three dots at the end of each group above indicate in each case that the numbers that follow the pattern are unlimited in number.

To test whether one fraction is equivalent to another, first express each fraction in lowest terms, then check whether the resulting fractions are the same. If they are, then the given fractions are equivalent.

Are $\frac{4}{16}$ and $\frac{3}{12}$ equivalent fractions?

Yes, $\frac{4}{16} = \frac{3}{12}$ since $\frac{4}{16} = \frac{1}{4}$ and $\frac{3}{12} = \frac{1}{4}$

Or use the *cross product method* where you find the product of the numerator of the first fraction and the denominator of the second fraction and the product of the numerator of the second fraction and the denominator of the first fraction.

When the cross products are equal, the two fractions are equivalent. Also when two fractions are equivalent, the cross products are equal.

Are $\frac{6}{15}$ and $\frac{4}{10}$ equivalent fractions?

$\frac{6}{15} = \frac{4}{10}$ is true if $6 \times 10 = 4 \times 15$.

Since 6×10 and $4 \times 15 = 60$, then $6 \times 10 = 4 \times 15$.

Thus $\frac{6}{15}$ and $\frac{4}{10}$ are equivalent fractions.

PRACTICE PROBLEMS   For ex. 1–2 see Extended Answers, p. T78.

1. Write the group of fractions equivalent to:

a. $\frac{1}{2}$ **c.** $\frac{3}{5}$ **e.** $\frac{5}{8}$ **g.** $\frac{11}{12}$ **i.** $\frac{3}{4}$ **k.** $\frac{9}{13}$ **m.** $\frac{4}{11}$ **o.** $\frac{17}{20}$

b. $\frac{1}{9}$ **d.** $\frac{2}{3}$ **f.** $\frac{13}{16}$ **h.** $\frac{5}{6}$ **j.** $\frac{7}{10}$ **l.** $\frac{8}{15}$ **n.** $\frac{19}{25}$ **p.** $\frac{31}{50}$

2. For each of the following fractions, write the simplest form and three other equivalent fractions:

a. $\frac{7}{42}$ **c.** $\frac{9}{36}$ **e.** $\frac{35}{84}$ **g.** $\frac{48}{54}$ **i.** $\frac{55}{80}$ **k.** $\frac{21}{70}$ **m.** $\frac{170}{200}$ **o.** $\frac{24}{225}$

b. $\frac{24}{30}$ **d.** $\frac{56}{64}$ **f.** $\frac{54}{63}$ **h.** $\frac{77}{132}$ **j.** $\frac{76}{96}$ **l.** $\frac{52}{72}$ **n.** $\frac{145}{250}$ **p.** $\frac{110}{121}$

Test whether each of the following pairs of fractions are equivalent by using the method of:

3. Lowest terms:

a. $\frac{6}{16}$ and $\frac{15}{40}$ Yes   **d.** $\frac{28}{35}$ and $\frac{35}{42}$ No   **g.** $\frac{3}{6}$ and $\frac{10}{18}$ No

b. $\frac{15}{18}$ and $\frac{36}{48}$ No   **e.** $\frac{21}{27}$ and $\frac{24}{33}$ No   **h.** $\frac{10}{24}$ and $\frac{30}{65}$ No

c. $\frac{42}{63}$ and $\frac{16}{24}$ Yes   **f.** $\frac{39}{52}$ and $\frac{51}{68}$ Yes   **i.** $\frac{35}{80}$ and $\frac{21}{48}$ Yes

4. Equal cross products:

a. $\frac{16}{28}$ and $\frac{24}{36}$ No   **d.** $\frac{24}{33}$ and $\frac{18}{26}$ No   **g.** $\frac{48}{75}$ and $\frac{35}{50}$ No

b. $\frac{10}{16}$ and $\frac{15}{24}$ Yes   **e.** $\frac{42}{54}$ and $\frac{28}{35}$ No   **h.** $\frac{68}{80}$ and $\frac{102}{120}$ Yes

c. $\frac{15}{25}$ and $\frac{36}{60}$ Yes   **f.** $\frac{27}{48}$ and $\frac{45}{80}$ Yes   **i.** $\frac{49}{105}$ and $\frac{77}{165}$ Yes

5. Either method:

a. $\frac{14}{35}$ and $\frac{6}{15}$ Yes   **d.** $\frac{24}{32}$ and $\frac{42}{56}$ Yes   **g.** $\frac{10}{25}$ and $\frac{36}{90}$ Yes

b. $\frac{15}{24}$ and $\frac{12}{20}$ No   **e.** $\frac{40}{56}$ and $\frac{49}{63}$ No   **h.** $\frac{45}{54}$ and $\frac{70}{84}$ Yes

c. $\frac{36}{40}$ and $\frac{42}{48}$ No   **f.** $\frac{132}{144}$ and $\frac{77}{84}$ Yes   **i.** $\frac{26}{40}$ and $\frac{36}{54}$ No

---

EXERCISE

1-58 | Comparing Fractions

OBJECTIVE: To compare fractions.

    In Exercise 1-11 we studied how to compare two fractions by expressing the fractions as equivalent fractions with a common denominator. However, we may also use the *cross product method.*

> The symbol $>$ means "is greater than."
> The symbol $<$ means "is less than."

To compare two fractions by the cross product method, find the cross products.

1. If the product of the numerator of the first fraction and the denominator of the second fraction is greater than the product of the numerator of the second fraction and the denominator of the first fraction, then the first fraction is greater than the second fraction.

$\frac{3}{5} > \frac{7}{12}$ is true if $3 \times 12 > 7 \times 5$.

Since $3 \times 12 = 36$ and $7 \times 5 = 35$, then $3 \times 12 > 7 \times 5$ and $\frac{3}{5} > \frac{7}{12}$.

2. If the product of the numerator of the first fraction and the denominator of the second fraction is less than the product of the numerator of the second fraction and the denominator of the first fraction, then the first fraction is less than the second fraction.

$\frac{3}{8} < \frac{2}{5}$ is true if $3 \times 5 < 2 \times 8$.

Since $3 \times 5 = 15$ and $2 \times 8 = 16$, then $3 \times 5 < 2 \times 8$ and $\frac{3}{8} < \frac{2}{5}$.

Summarizing the cross product test:

If a represents the numerator of the first fraction; b the denominator of the first fraction; c the numerator of the second fraction; and d the denominator of the second fraction, then:

(1) Two fractions are equivalent: $\dfrac{a}{b} = \dfrac{c}{d}$ if $a \times d = c \times b$

(2) First fraction is greater: $\dfrac{a}{b} > \dfrac{c}{d}$ if $a \times d > c \times b$

(3) First fraction is smaller: $\dfrac{a}{b} < \dfrac{c}{d}$ if $a \times d < c \times b$

Basic: Prob. 1–2; *Average:* Prob. 1–3; *Enriched:* Prob. 1–4

PRACTICE PROBLEMS

Use the cross product test to determine which of the following statements are true:

1. **a.** $\frac{1}{3} > \frac{1}{5}$ T **b.** $\frac{1}{8} > \frac{1}{6}$ F **c.** $\frac{3}{4} > \frac{5}{8}$ T **d.** $\frac{7}{12} > \frac{9}{16}$ T **e.** $\frac{8}{9} > \frac{11}{13}$ T

2. **a.** $\frac{2}{3} < \frac{3}{5}$ F **b.** $\frac{7}{8} < \frac{9}{10}$ T **c.** $\frac{19}{24} < \frac{7}{9}$ F **d.** $\frac{11}{16} < \frac{16}{25}$ F **e.** $\frac{9}{17} < \frac{3}{5}$ T

3. **a.** $\frac{5}{6} > \frac{3}{4}$ T **b.** $\frac{9}{10} > \frac{15}{16}$ F **c.** $\frac{3}{11} < \frac{5}{18}$ T **d.** $\frac{19}{75} < \frac{6}{25}$ F **e.** $\frac{17}{18} > \frac{19}{20}$ F

4. **a.** $\frac{27}{36} < \frac{35}{42}$ T **b.** $\frac{15}{24} > \frac{21}{30}$ F **c.** $\frac{49}{56} < \frac{60}{72}$ F **d.** $\frac{27}{81} > \frac{18}{48}$ F **e.** $\frac{36}{84} < \frac{28}{63}$ T

OBJECTIVE: To find multiplicative inverses.   VOCABULARY: multiplicative inverse
                                                              reciprocal

If the product of two numbers is one (1), then each factor is called the
multiplicative inverse or reciprocal of the other.

---

8 and $\frac{1}{8}$ are multiplicative inverses of each other
because $8 \times \frac{1}{8} = 1$.
$\frac{4}{3}$ and $\frac{3}{4}$ are multiplicative inverses of each other
because $\frac{4}{3} \times \frac{3}{4} = 1$.

Observe that the numerator of one fraction is the denominator of its
reciprocal, and its denominator is the numerator of its reciprocal. Zero
has no inverse for multiplication.

When we divide 12 by 4, the quotient is 3. $12 \div 4 = 3$

When we multiply 12 by $\frac{1}{4}$, the product is 3. $12 \times \frac{1}{4} = 3$

Dividing a given number by 4 gives the same answer as multiplying
the given number by its multiplicative inverse (or reciprocal) $\frac{1}{4}$.

Thus, to divide a number by another number, you may multiply the
first number by the multiplicative inverse of the divisor. This principle is
used in the division of fractions and mixed numbers. See Exercise 1-17.

$$\tfrac{2}{3} \div \tfrac{3}{4} = \tfrac{2}{3} \times \tfrac{4}{3} = \tfrac{8}{9}$$

Basic: Prob. 1–3; Average: Prob. 1–5; Enriched: Prob. 1–6

PRACTICE PROBLEMS

Write the multiplicative inverse or reciprocal of each of the following:

1. a. 6 $\frac{1}{6}$ **b.** 3 $\frac{1}{3}$ **c.** 1 1   **d.** 9 $\frac{1}{9}$ **e.** 25 $\frac{1}{25}$ **f.** 0 none **g.** 12 $\frac{1}{12}$

2. a. $\frac{1}{4}$ 4   **b.** $\frac{1}{10}$ 10   **c.** $\frac{1}{7}$ 7   **d.** $\frac{1}{2}$ 2   **e.** $\frac{1}{3}$ 3   **f.** $\frac{1}{15}$ 15   **g.** $\frac{1}{20}$ 20

3. a. $\frac{5}{6}$ $\frac{6}{5}$ **b.** $\frac{7}{12}$ $\frac{12}{7}$ **c.** $\frac{3}{8}$ $\frac{8}{3}$ **d.** $\frac{2}{5}$ $\frac{5}{2}$ **e.** $\frac{9}{10}$ $\frac{10}{9}$ **f.** $\frac{11}{16}$ $\frac{16}{11}$ **g.** $\frac{23}{24}$ $\frac{24}{23}$

4. a. $\frac{8}{5}$ $\frac{5}{8}$ **b.** $\frac{5}{4}$ $\frac{4}{5}$ **c.** $\frac{22}{7}$ $\frac{7}{22}$ **d.** $\frac{9}{2}$ $\frac{2}{9}$ **e.** $\frac{13}{6}$ $\frac{6}{13}$ **f.** $\frac{19}{8}$ $\frac{8}{19}$ **g.** $\frac{21}{12}$ $\frac{12}{21}$

5. a. $2\frac{1}{2}$ $\frac{4}{5}$ **b.** $6\frac{2}{3}$ $\frac{3}{20}$ **c.** $1\frac{3}{4}$ $\frac{4}{7}$ **d.** $4\frac{3}{8}$ $\frac{8}{35}$ **e.** $1\frac{5}{12}$ $\frac{12}{17}$ **f.** $6\frac{2}{5}$ $\frac{5}{32}$ **g.** $3\frac{9}{16}$ $\frac{16}{57}$

6. Find the missing numbers:

a. $\frac{3}{8} \times \frac{8}{3} = ?$ 1   **c.** $? \times \frac{1}{9} = 1$ 9   **e.** $? \times \frac{11}{16} = 1$ $\frac{16}{11}$ **g.** $\frac{9}{5} \times ? = 1$ $\frac{5}{9}$

b. $12 \times ? = 1$ $\frac{1}{12}$ **d.** $\frac{5}{6} \times ? = 1$ $\frac{6}{5}$ **f.** $\frac{7}{10} \times \frac{10}{7} = ?$ 1   **h.** $? \times \frac{15}{2} = 1$ $\frac{2}{15}$

OBJECTIVE: To read and write numerals
expressed in scientific notation.

VOCABULARY: scientific notation
significant digit

A brief form of writing numerals for very large or very small numbers is known as scientific notation. A number is expressed in scientific notation when it is greater than 1 but less than 10 multiplied by some power of ten.

9,800,000 is 9.8 million which may be expressed in scientific notation as 9.8×10^6

To write a numeral in scientific notation, we rewrite the significant digits (see Exercise 2-20):
   (1) As a numeral for a whole number if there is only one significant digit, as in 9,000,000.
   (2) As a numeral for a mixed decimal if there are two or more significant digits, as in 9,800,000 or 9,850,000, using the first digit as a numeral for a whole number and for all other digits as the numeral for a decimal fraction.

Then we indicate that this numeral is multiplied by the required power of ten.

The required power of ten may be determined as follows:
   (1) By dividing the whole number or the mixed decimal into the given number and changing the quotient into a power of ten, or
   (2) By counting the number of places the decimal point is being moved.

When the given number is 10 or greater, a positive integer is used for the exponent. (See Exercise 4-1.)

Write 47,600,000 in scientific notation:

$47{,}600{,}000 = 4.76 \times ?$

$$\text{Divide: } 4.76_\wedge \overline{)47{,}600{,}000.00_\wedge} \overset{10{,}000{,}000}{}$$

However, $10{,}000{,}000 = 10^7$

$47{,}600{,}000 = 4.76 \times 10^7$

Answer: 4.76×10^7

or

$47{,}600{,}000 = 4.76 \times 10^7$

The decimal point is moved *7 places* to the left. This numeral names the required exponent.

When the given number is between 0 and 1, a negative integer is used for the exponent. (See Exercise 4-1.)

Write .000053 in scientific notation:

.000053 = 5.3 × ?

$$\overset{\text{.00001}}{\text{Divide: }5.3_{\wedge})\overline{.0_{\wedge}00053}}$$

However, .00001 = 10^{-5}

.000053 = 5.3 × 10^{-5}

Answer: .000053 = 5.3 × 10^{-5}

or   .000053 = 5.3 × 10^{-5}

The decimal point is moved *5 places* to the right.

Basic: Prob. 1; *Average:* Prob. 1–2; *Enriched:* Prob. 1–3

PRACTICE PROBLEMS

For ex. 1–3 see Extended Answers, p. T78.

1. Express each of the following numbers in scientific notation:

a. 50   87   400   914   8,000   3,060

b. 60,000   79,000   520,000   36,000,000   4,875,000

c. 2,000,000,000   308,000,000,000   84,500,000,000,000

d. 960 000 000   510 000 000 000   423 000 000 000 000 000

e. 7 000 000 000 000 000 000   6 700 000 000 000 000 000 000

2. Express each of the following numbers in scientific notation:

a. .4   .9   .3   .34   .02   .88   .69

b. .057   .007   .983   .0542   .0096   .8167   .2005

c. .0018   .00249   .06356   .000784   .0000091

d. .000 036 5   .000 000 79   .000 000 801   .000 000 003 6

e. .000 000 000 043   .000 000 000 000 05   .000 000 000 000 000 682

3. Express the numbers in each of the following in scientific notation:

a. The sun at any second develops 500,000,000,000,000,000,000,000 horsepower.

b. The star Alpha Hercules is 3 860 000 000 kilometers in diameter.

c. The weight of the earth is about 6,600,000,000,000,000,000,000 tons.

d. The distance from the sun to the planet Jupiter is about 779 000 000 kilometers.

e. (1) 49 quadrillion (49,000,000,000,000,000)
(2) 528 quintillion (528,000,000,000,000,000,000)

OBJECTIVE: To write common
fractions as repeating
decimals.

Repeating Decimals

VOCABULARY: repeating decimal   terminating decimal

When the fraction $\frac{1}{3}$ is changed to a decimal, the 1 is divided by the 3. However, the division is not exact. The remainder at each step is the same (1). Examination of the quotient will show that the digit 3 repeats and will keep repeating endlessly as the division is extended.

When the fraction $\frac{8}{11}$ is changed to a decimal, this division also is not exact. The remainder at each second step is the same (8). Examination of the quotient will show that the pair of digits 72 repeats and will keep repeating endlessly as the division is extended.

Decimals like $.33\overline{3}$ . . . (or $.\overline{3}$ . . . or simply $.\overline{3}$) and $.72\overline{72}$ . . . (or $.\overline{72}$ . . . or simply $.\overline{72}$) which have a digit or a group of digits repeating endlessly are called repeating decimals.

The bar indicates the repeating sequence and the dots indicate that the sequence repeats endlessly.

When $\frac{1}{4}$ is changed to a decimal, the quotient is exactly .25 with a remainder 0. The decimal form .25 is called a terminating decimal. However, since .25 could be written in the repeating form $.25\overline{0}$ . . . or $.250\overline{0}$ . . . , terminating decimals may also be considered repeating decimals.

Summarizing, observe that if a remainder of 0 occurs in the division, the quotient is a terminating decimal. If, however, after a series of divisions a remainder other than 0 repeats, then the sequence of digits obtained in the quotient between occurrences of this remainder will repeat endlessly.

```
      .111
  3)1.000
     9
    ──
     10
      9
    ──
     10
      9
    ──
      1
```

```
      .7272
  11)8.0000
     7 7
     ───
      30
      22
     ──
       80
       77
      ──
       30
       22
      ──
        8
```

```
      .25
  4)1.00
     8
    ──
     20
     20
    ──
      0
```

Basic: Prob. 1–18; *Average:* Prob. 1–22; *Enriched:* Prob. 1–25

PRACTICE PROBLEMS

Express each of the following common fractions as a numeral naming a repeating decimal. Indicate the repeating sequence by a horizontal bar.

1. $\frac{2}{3}$ $.\overline{6}$
2. $\frac{7}{8}$ $.8750\overline{0}$
3. $\frac{1}{6}$ $.1\overline{6}$
4. $\frac{3}{7}$ $.\overline{428571}$
5. $\frac{5}{9}$ $.\overline{5}$

6. $\frac{11}{16}$ $.68750\overline{0}$
7. $\frac{5}{12}$ $.41\overline{6}$
8. $\frac{13}{33}$ $.\overline{39}$
9. $\frac{7}{11}$ $.\overline{63}$
10. $\frac{9}{13}$ $.\overline{692307}$

11. $\frac{4}{5}$ $.80$
12. $\frac{11}{18}$ $.6\overline{1}$
13. $\frac{7}{15}$ $.4\overline{6}$
14. $\frac{19}{24}$ $.791\overline{6}$
15. $\frac{8}{27}$ $.\overline{296}$

16. $\frac{20}{21}$ $.9\overline{52380}$
17. $\frac{9}{11}$ $.\overline{81}$
18. $\frac{3}{4}$ $.750\overline{0}$
19. $\frac{14}{15}$ $.93\overline{3}$
20. $\frac{2}{9}$ $.\overline{2}$

21. $\frac{11}{12}$ $.916\overline{6}$
22. $\frac{23}{32}$ $.718750\overline{0}$
23. $\frac{17}{30}$ $.56\overline{6}$
24. $\frac{28}{33}$ $.\overline{84}$
25. $\frac{5}{14}$ $.3\overline{571428}$

Expressing a Repeating Decimal as a Common Fraction

OBJECTIVE: To express a repeating decimal as a common fraction.

To change a repeating decimal to a common fraction:

1. Multiply the given repeating decimal by some power of 10 so that there is a whole number to the left of the decimal point and the repeating sequence begins to the right of the decimal point.

   If the repeating sequence has 1 digit, multiply by 10.

   If the repeating sequence has 2 digits, multiply by 100, etc.

2. Subtract the given number from this product to get a whole number for the difference. (See sample solutions (1) and (2).)

3. Solve the resulting equation (see Exercise 4-21) to find the required common fraction.

4. When a repeating decimal begins with digits other than those used in the repeating sequence, another step is necessary as shown in sample solution (3).

(1)	(2)	(3)
$.77\overline{7} \ldots = ?$	$.2424\overline{24} \ldots = ?$	$.833\overline{3} \ldots = ?$
Let $n = .777 \ldots$	Let $n = .242424 \ldots$	Let $n = .83333 \ldots$
$10n = 7.777 \ldots$	$100n = 24.242424 \ldots$	$100n = 83.3333 \ldots$
$10n - n = 7$	$100n - n = 24$	and $10n = 8.3333 \ldots$
$9n = 7$	$99n = 24$	then $90n = 75$
$n = \frac{7}{9}$	$n = \frac{24}{99}$	$n = \frac{75}{90}$
	$n = \frac{8}{33}$	$n = \frac{5}{6}$
Answer: $\frac{7}{9}$	*Answer:* $\frac{8}{33}$	*Answer:* $\frac{5}{6}$

Basic: Prob. 1–9; *Average:* Prob. 1–12; *Enriched:* Prob. 1–15

PRACTICE PROBLEMS

Using the equation method, express each of the following repeating decimals as a common fraction.

1. $.44\overline{4} \ldots$ $\frac{4}{9}$
2. $.\overline{5}$ $\frac{5}{9}$
3. $.5454\overline{54} \ldots$ $\frac{6}{11}$
4. $.\overline{63}$ $\frac{7}{11}$
5. $.\overline{148}$ $\frac{4}{27}$
6. $.\overline{481}$ $\frac{13}{27}$
7. $.9\overline{4}$ $\frac{17}{18}$
8. $.58\overline{3}$ $\frac{7}{12}$

9. $.79\overline{16}$ $\frac{19}{24}$
10. $.6\overline{0}$ $\frac{3}{5}$
11. $.375\overline{0}$ $\frac{3}{8}$
12. $.\overline{15}$ $\frac{5}{33}$
13. $.15\overline{0}$ $\frac{3}{20}$
14. $.52\overline{7}$ $\frac{19}{36}$
15. $.\overline{428571}$ $\frac{3}{7}$

6.a. Eight plus six is equal to six plus eight
6.b. Twenty minus nine is less than four times three

REVIEW OF UNIT SEVEN

6.c. Fifteen is greater than ninety divided by ten
6.d. Twenty-eight minus three is not less than five times five

1. What year does MCMXLIV represent? 1944

2. Read, or write in words, each of the following:

a. 0, 11, 22, 33, . . . zero, eleven, twenty-two, thirty-three and so on endlessly

b. 5, 10, 15, 20, . . . , 500 five, ten, fifteen, twenty and so on up to and including five hundred

3. a. Find the value of 3^5. 243

b. Express 64 as a power of 2. 8^2

4. Express 80,000,000,000 as a product of a digit and a power of ten. 8.0×10^{10}

5. Write 725.96 in expanded form as a polynomial. $(7 \times 10^2) + (2 \times 10^1) + (5 \times 10^0)$ $+ (9 \times 10^{-1}) + (6 \times 10^{-2})$

6. Read, or write in words:

a. $8 + 6 = 6 + 8$ **c.** $15 > 90 \div 10$
b. $20 - 9 < 4 \times 3$ **d.** $28 - 3 \nless 5 \times 5$

7. Determine which of the following sentences are true and which are false:

a. $53 - 18 < 19 + 16$ False   **c.** $150 > 100 \div \frac{1}{2}$ True
b. $29 + 7 = 6^2$ True   **d.** $2 - 2 \ngtr 2 \div 2$ True

8. a. Is 289 an odd number? Yes

b. Is 50 an even number? Yes

9. What is the greatest common factor of 72, 108, and 216? 216

10. a. Is 123 a prime number? No

b. Is 57 a composite number? Yes

11. Factor 196 as a product of prime numbers. $2 \times 2 \times 7 \times 7$

12. Find the least common multiple of 12, 20, and 24. 120

13. Are $\frac{20}{35}$ and $\frac{28}{49}$ equivalent fractions? Yes   14. $\frac{5}{8}, \frac{10}{16}, \frac{20}{32}, \frac{40}{64}$

14. Write the simplest form and 3 other fractions equivalent to $\frac{45}{72}$.

15. Determine by the cross product test which of the following statements are true:

a. $\frac{11}{12} < \frac{9}{10}$ False
b. $\frac{7}{20} > \frac{5}{16}$ True

16. Write the multiplicative inverse or reciprocal of each of the following:

a. 4 $\frac{1}{4}$ **b.** $\frac{1}{6}$ 6   **c.** $\frac{7}{10}$ $\frac{10}{7}$ **d.** $\frac{11}{2}$ $\frac{2}{11}$ **e.** $3\frac{1}{3}$ $\frac{3}{10}$

17. Express 290,000,000 in scientific notation. 2.9×10^8

18. Express .00000658 in scientific notation. 6.58×10^{-6}

19. Express $\frac{4}{9}$ as a repeating decimal. .4̄

20. Express $.\overline{45}$ as a numeral naming a common fraction. $\frac{5}{11}$

For a Unit Test see p. T45 or *Testing Program*, p. 10.

1. Eighty-five million, two hundred sixty-seven thousand, four hundred twenty-eight

1-1. Read the numeral 85,267,428 or write it in words.

1-2. Write the numeral naming:
Ninety-two million, four hundred six thousand, fifty 92,406,050

1-3. Round 581,560 to the nearest thousand. 582,000

1-4. Add: 472 + 5,973 + 15,682 + 33 + 280 22,440

1-5. Take 638 from 4,000. 3,362   **1-6.** Multiply: $8 \times 14 \times 1,728$ 193,536

1-7. Divide: $48\overline{)33,744}$ 703   **1-8.** Express $\frac{28}{48}$ in lowest terms. $\frac{7}{12}$

1-9. Change $\frac{9}{2}$ to a mixed number. $4\frac{1}{2}$ **1-10.** Change $\frac{7}{12}$ to 60ths. $\frac{35}{60}$

1-11. Find the lowest common denominator of the fractions: $\frac{3}{8}$ and $\frac{11}{16}$ 16

1-12. Add: $9\frac{1}{2} + 3\frac{2}{5}$ $12\frac{9}{10}$ **1-13.** Subtract: $4\frac{1}{4}$

1-14. Which is smaller: $\frac{5}{8}$ or $\frac{5}{6}$? $\frac{5}{8}$ $2\frac{1}{3}$

1-15. Change $6\frac{2}{3}$ to an improper fraction. $\frac{20}{3}$ $1\frac{11}{12}$

1-16. Multiply: $3\frac{7}{8} \times 2$ $7\frac{3}{4}$ **1-17.** Divide: $\frac{3}{4} \div 6$ .125

1-18. What part of 32 is 20? $\frac{5}{8}$ **1-19.** $\frac{9}{20}$ of what number is 36? 80

1-20. Read the numeral .0626 or write it in words. Six hundred twenty-six ten-thousandths

1-21. Write as a decimal numeral: Fifteen and seven tenths 15.7

1-22. Find correct to nearest hundredth: 26.8539 26.85

1-23. Add: $ 6.57   **1-24.** Subtract: 54.3 − 8.28 46.02
   .36   **1-25.** Which is greater: .46 or .462? .462
  12.40   **1-26.** Multiply: 3.14×15 47.1
   .82 $20.15   **1-27.** Divide: $.06\overline{)\$7.44}$ 124

1-28. Multiply by short method: 10×8.69 86.9

1-29. Divide by short method: $1.95 \div 100$ .0195

1-30. Change $\frac{11}{16}$ to a decimal (2 places). .69

1-31. Change .65 to a common fraction. $\frac{13}{20}$

1-32. What decimal part of 20 is 13? .65

1-33. .06 of what number is 18? 300

1-34. Write 14.6 *million* completely as a numeral. 14,600,000

1-35. Write the shortened name for 63,700,000,000 in billions. 63.7 billion

1-36. Write sixty hundredths as a percent, as a decimal, and as a common fraction. 60%, .60, $\frac{60}{100}$ or $\frac{3}{5}$

.07 **1-37.** Express 7% as a decimal.   **1-38.** Express .49 as a percent. 49%

1-39. Express 20% as a common fraction. $\frac{1}{5}$

12% **1-40.** Express $\frac{3}{25}$ as a percent.   **1-41.** Find 35% of 560. 196

$81\frac{1}{4}$% **1-42.** 52 is what percent of 64?   **1-43.** 75% of what number is 60? 80

1-44. Square 140. 19,600   **1-45.** Find the square root of 290,521. 539

1. Seven million, fifteen thousand, six hundred seventy-three

1-1. Read the numeral 7,015,673 or write it in words.

1-2. Write the complete numeral naming: 816 *million* 816,000,000

1-3. Round 496,280,074 to the nearest million. 496,000,000

1-4. Add: 15,976
2,734
38,592
21,085
6,588 84,975

1-5. Subtract: 178,346
77,482 100,864

1-6. Multiply 3,600 by 807. 2,905,200

1-7. Divide: 4,843,630 ÷ 605 8,006

1-8. Express $\frac{63}{105}$ in lowest terms. $\frac{3}{5}$ **1-9.** Change $\frac{21}{12}$ to a mixed number. $1\frac{3}{4}$

1-10. Raise to higher terms: $\frac{2}{3} = \frac{?}{27}$ 18

1-11. Find the lowest common denominator of the fractions: $\frac{2}{3}$, $\frac{3}{4}$, and $\frac{11}{12}$ 12

1-12. Add: $\frac{3}{4}$
$2\frac{5}{16}$
$4\frac{19}{32}$ $7\frac{21}{32}$

1-13. Subtract $3\frac{3}{8}$ from 4. $\frac{5}{8}$

1-14. Which is smaller: $\frac{9}{10}$ or $\frac{11}{12}$? $\frac{9}{10}$

1-15. Change $4\frac{5}{16}$ to an improper fraction. $\frac{69}{16}$

1-16. Multiply: $2\frac{1}{3} \times 3\frac{1}{7}$ $7\frac{1}{3}$

1-17. Divide: $\frac{15}{16} \div 1\frac{1}{2}$ .625 or $\frac{5}{8}$

1-18. 60 is what part of 100? $\frac{3}{5}$

1-19. $\frac{1}{6}$ of what number is 17? 102

1-20. Read the numeral 740.835 or write it in words.

1-21. Write as a decimal numeral: Five hundred and six thousandths 500.006

1-22. Find correct to nearest cent: $4.2046 $4.20

1-23. Add: 1.93 + .541 + 43.8 46.271

1-24. Subtract: $452.46
189.57 $262.89

1-25. Which is greater: .92 or .919? .92

1-26. Multiply 6.42 by .25 1.605

1-27. Divide: .04)‾.00012 .003

1-28. Multiply by short method: 100 × 250 25,000

1-29. Divide by short method: 18.53 ÷ 1,000 .01853

1-30. Change $\frac{3}{7}$ to a decimal (2 places). .43

1-31. Change $.37\frac{1}{2}$ to a common fraction. $\frac{3}{8}$

.375 **1-32.** What decimal part of 64 is 24? **1-33.** .375 of what number is 900? 2,400

1-34. Write 532.7 *billion* completely as a numeral. 532,700,000,000

1-35. Write the shortened name for 594,200,000 in millions. 594.2 million

1-36. Write ninety-four hundredths as a percent, as a decimal, and as a common fraction. 94%, .94, $\frac{94}{100}$

1-37. Express 4.5% as a decimal. .045 **1-38.** Express 1.05 as a percent. 105%

$1\frac{3}{4}$ **1-39.** Express 175% as a mixed number. **1-40.** Express $\frac{24}{300}$ as a percent. 8%

1-41. Find 140% of $30. $42.00 **1-42.** What percent of $7.00 is $1.75? 25%

1-43. $6\frac{1}{2}$% of what number is 26? 400 **1-44.** Square $1\frac{3}{5}$. $\frac{64}{25}$ or $2\frac{14}{25}$

1-45. Find the square root of 6 correct to nearest hundredth. 2.45

20. Seven hundred forty and eight hundred thirty-five thousandths

1. Five billion, nine hundred sixty-five million, four hundred eight thousand

1-1. Read the numeral 5,965,408,000 or write it in words.

1-2. Write the numeral naming:
Six billion, four hundred million, ten thousand 6,400,010,000

1-3. Round 147,492,350 to the nearest hundred thousand. 147,500,000

1-4. Add: 93,574
69,898
74,379
58,486
96,798
―――
393,135

1-5. From 823,849 subtract 812,958. 10,891

1-6. Find the product of 6,009 and 7,908. 47,519,172

1-7. Divide 72,576 by 1,728. 42

1-8. Express $\frac{1400}{3600}$ in lowest terms. $\frac{7}{18}$

1-9. Change $\frac{51}{16}$ to a mixed number. $3\frac{3}{16}$

1-10. Raise to higher terms: $\frac{13}{16} = \frac{?}{64}$ 52

1-11. Find the lowest common denominator of the fractions: $\frac{3}{5}$, $\frac{1}{4}$, and $\frac{2}{3}$ 60

1-12. Add: $4\frac{1}{8} + \frac{7}{10}$ $4\frac{33}{40}$

1-13. Subtract: $8\frac{3}{16} - \frac{7}{8}$ $7\frac{5}{16}$

1-14. Is the statement $\frac{9}{32} < \frac{5}{16}$ true? Yes

1-15. Change $3\frac{7}{12}$ to an improper fraction. $\frac{43}{12}$

1-16. Multiply: $1\frac{1}{2} \times 2\frac{1}{4} \times 1\frac{1}{3}$ $4\frac{1}{2}$

1-17. Divide: $(3\frac{3}{4} \div 2\frac{1}{2}) \div \frac{3}{4}$ 2

1-18. What part of 184 is 115? $\frac{5}{8}$

1-19. $\frac{9}{16}$ of what number is 45? 80

1-20. Read the numeral 583.29 *million* or write it in words.

1-21. Write as a decimal: Four hundred seventy-one thousandths .471

1-22. Find correct to nearest ten-thousandth: .00385 .0039

$24.70 **1-23.** Add: $8.24 + $5.62 + $.76 + $10.08

1-24. Subtract .02 from .2 .18

1-25. Is the statement .201 > .21 true? No

$10.94 **1-26.** Find $1\frac{3}{4}$ of $6.25 (to nearest cent).

1-27. Divide 27 by 1.5 18

1-28. Multiply by short method: 1,000 × 48.21 48,210

1-29. Divide by short method: 37.2 ÷ 10 3.72

1-30. Change $\frac{25}{30}$ to a decimal (2 places). .83

1-31. Change .016 to a common fraction. $\frac{2}{125}$

.7 **1-32.** $3.50 is what decimal part of $5? **1-33.** 78 is .12 of what number? 650

1-34. Write 17.594 *billion* completely as a numeral. 17,594,000,000

1-35. Write the shortened name for 4,129,000,000,000 in trillions. 4.129 trillion

1-36. Write eighteen and one-half hundredths as a percent, as a decimal, and as a common fraction. $18\frac{1}{2}$%, .185, $\frac{18\frac{1}{2}}{100}$

1-37. Express 150% as a decimal. 1.5 **1-38.** Express .046 as a percent. 4.6%

$\frac{2}{25}$ **1-39.** Express 8% as a common fraction. **1-40.** Express $\frac{8}{18}$ as a percent. $44\frac{4}{9}$%

$1.25 **1-41.** Find 8.3% of $15 (nearest cent). **1-42.** 12 is what percent of 6? 200%

1-43. 125% of what number is 70? 56 **1-44.** Square 3.5 12.25

1-45. Find the square root of 0.9 correct to nearest hundredth. .95

20. Five hundred eighty-three million, two hundred ninety thousand

Basic Skills—Achievement Test

1. Four trillion, three hundred six billion, nine million, eighty-five thousand

1-1. Read the numeral 4,306,009,085,000 or write it in words.

1-2. Write the complete numeral naming: 59 *billion* 59,000,000,000

1-3. Round 6,382,957,429 to the nearest hundred million. 6,400,000,000

1-4. Add: 462,741
893,826
652,158
894,923
579,109
——————
3,482,757

1-5. Subtract 37,989 from 5,000,000. 4,962,011

1-6. Multiply: 7,963 × 5,897 46,957,811

1-7. Divide 79,183,450 by 9,806. 8,075

1-8. Express $\frac{39}{169}$ in lowest terms. $\frac{3}{13}$

1-9. Change $\frac{100}{24}$ to a mixed number. $4\frac{1}{6}$

1-10. Raise to higher terms: $\frac{13}{15} = \frac{?}{120}$ 104

1-11. Find the lowest common denominator of the fractions: $\frac{5}{8}$, $\frac{7}{12}$, and $\frac{9}{20}$ 120

1-12. Add: $5\frac{5}{6} + \frac{7}{8} + 4\frac{11}{12}$ $11\frac{15}{24}$ or $11\frac{5}{8}$

1-13. Subtract: $9\frac{2}{5} - 6\frac{3}{4}$ $2\frac{13}{20}$

1-14. Which is greater: $\frac{7}{15}$ or $\frac{12}{25}$? $\frac{12}{25}$

1-15. Change $7\frac{9}{10}$ to an improper fraction. $\frac{79}{10}$

1-16. Multiply: $11\frac{3}{16} \times 4\frac{2}{3}$ $52\frac{5}{24}$

1-17. Divide: $6\frac{1}{4} \div (\frac{5}{8} \div 3\frac{1}{3})$ $33\frac{1}{3}$

1-18. 21 is what part of 98? $\frac{3}{14}$

1-19. 91 is $\frac{13}{18}$ of what number? 126

1-20. Read the numeral 47.592 billion or write it in words.

1-21. Write as a decimal numeral: Fifty and nine hundredths 50.09

1-22. Find correct to the nearest millionth: .0015867 .001587

1-23. Add: .06 + 6.0 + .60 6.66

1-24. Subtract: .803 − .79 .013

1-25. Which is smaller: .9 or .885? .885

1-26. Multiply: .004 × .05 .0002

1-27. Divide: .012)‾.6 50

1-28. Multiply by short method: 1,000 × 86.05 86,050

1-29. Divide by short method: 63.49 ÷ 1,000 .06349

1-30. Change $1\frac{3}{8}$ to a decimal (2 places). 1.38

1-31. Change 2.75 to a mixed number. $2\frac{3}{4}$

1-32. What decimal part (2 places) of 96 is 80? .83

1-33. $330 is $.08\frac{1}{4}$ of what amount? $4000

1-34. Write 2.83 *trillion* completely as a numeral. 2,830,000,000,000

1-35. Write the shortened name for 199,560,000,000 in billions. 199.56 billion

1-36. Write thirty-one hundredths as a percent, as a decimal, and as a common fraction. 31%, .31, $\frac{31}{100}$

1-37. Express 7.4% as a decimal. .074

1-38. Express $.00\frac{7}{8}$ as a percent. $\frac{7}{8}$%

$2\frac{1}{2}$ **1-39.** Express 250% as a mixed number.

1-40. Express $\frac{36}{40}$ as a percent. 90%

1-41. Find $\frac{1}{4}$% of $660. $1.65

1-42. What percent of 60 is 27? 45%

1-43. 8% of what amount is $12? $150

1-44. Square 1.73 2.9929

1-45. Find the square root of 4.2849 2.07

20. Forty-seven billion, five hundred ninety-two million

For an additional Basic Skills-Achievement Test see p. T46 or *Testing Program* p. 11.

Basic Units of Measures

2

The numeral in color following each problem indicates the Exercise where explanatory material may be found.

For ex. 1–20 see Extended Answers, p. T78.

1. Change: **a.** 42 cm to mm   **b.** 690 m to km   2-1

2. Change: **a.** 1.8 kg to g   **b.** 5,500 mg to g   2-2

3. How many: **a.** cL are in 7 L?   **b.** L are in 665 ml?   2-3

4. Change: **a.** 49 km^2 to hectares   **b.** 800 m^2 to cm^2 2-4

5. Change: **a.** 8.3 m^3 to cm^3 **b.** 9,250 cm^3 to dm^3 2-5

     c. 6.75 L of water occupies _____ cm^3 and weighs _____ kg.

6. How many feet are in 264 inches?   2-6

7. What part of a pound is 12 ounces?   2-7

8. Find the number of pints in 9 gallons.   2-8

9. How many bushels are in 96 pecks?   2-9

10. Change 108 square inches to square feet.   2-10

11. How many cubic feet are in 7 cubic yards?   2-11

12. What part of a year is 10 months?   2-13

13. Subtract 103°42′ from 180°.   2-14

14. Change 72 km/h to m/min.   2-15

15. Is a temperature of 15°C colder than 20°F?   2-16

16. Express in A.M. or P.M. time: **a.** 2105   **b.** 0030   2-17

17. At 11 A.M. in Atlanta, what time is it in Seattle?   2-18

18. **a.** Find the greatest possible error and the relative error in   2-19
each of the following measurements:
(1) 80 cm   (2) 25,000 kg   (3) 7.4 L   (4) $4\frac{2}{3}$ hr.
   b. Which is more precise: 0.060 mm or 0.06 mm?
Which is more accurate?
   c. Arrange the following measurements in order of precision, most precise first.
Arrange in order of accuracy, most accurate first:
(1) 69 cm   (2) 40 m   (3) 8.2 mm   (4) 750,000 km

19. How many significant digits are in each of the following?
What are they?   2-20
   a. 9,008   **c.** 89,026,571   **e.** 3.8×10^5
   b. .0040   **d.** 67.139   **f.** 5,630,000

20. Compute the following approximate numbers as indicated: 2-21
   a. 6.7 + 3.41   **c.** $5.9 \times .213$
   b. 47.732 − 9.8   **d.** $36\overline{)8.67}$

UNITS OF MEASURE

Many years ago there were no standards of measure. Early humans used their fingers, feet, and arms to measure length or distance. The width across the open hand at the base of the fingers, called the *palm,* the breadth of a finger, called a *digit,* the greatest stretch of the open hand, called the *span,* and the length of the forearm from the elbow to the end of the middle finger, called the *cubit,* were some of the units used. However, since these measurements varied depending upon the size of the person, they were unsatisfactory.

Today we use metric and customary units of measure which are standard. See pages 621–623 for the tables of measure. Probably in the near future we shall be using in the United States a modernized metric system, the International System of Units, generally known as SI. It should be noted that sometimes the metric unit meter is spelled as *metre* and liter as *litre.*

Changing to a Smaller Unit of Measure

To change a given number of units of one denomination to units of smaller denomination, we find the number of units of the smaller denomination that is equivalent to one unit of the larger denomination. This number is sometimes called the conversion factor. Then we multiply the given number of units of the larger denomination by this conversion factor.

Since each conversion factor in the metric system is some power of ten, short methods of computation may be used. See Exercise 1-28.

On the next page are several examples of changing to a smaller unit of measure and references to Exercises where you will find more practice.

2-1 Metric—Length
23 centimeters to millimeters:
$10 \times 23 = 230$
Answer: 230 millimeters

2-2 Metric—Weight
6 kilograms to grams:
$1,000 \times 6 = 6,000$
Answer: 6,000 grams

2-3 Metric—Capacity
4 liters to centiliters:
$100 \times 4 = 400$
Answer: 400 centiliters

2-4 Metric—Area
9 km^2 to m^2:
$1,000,000 \times 9 = 9,000,000$
Answer: 9,000,000

2-5 Metric—Volume
7 dm^3 to cm^3:
$1,000 \times 7 = 7,000$
Answer: 7,000 cm^3

2-6 Customary—Length
5 feet to inches:
$12 \times 5 = 60$
Answer: 60 inches

2-7 Customary—Weight
8 pounds to ounces:
$16 \times 8 = 128$
Answer: 128 ounces

2-8 Customary—Liquid
13 gallons to quarts:
$4 \times 13 = 52$
Answer: 52 quarts

2-9 Customary—Dry
6 bushels to pecks:
$4 \times 6 = 24$
Answer: 24 pecks

2-10 Customary—Area
17 sq. yd. to sq. ft.:
$9 \times 17 = 153$
Answer: 153 sq. ft.

2-11 Customary—Volume
3 cu. ft. to cu. in.:
$1,728 \times 3 = 5,184$
Answer: 5,184 cu. in.

2-13 Time
12 hours to minutes:
$60 \times 12 = 720$
Answer: 720 minutes

2-14 Angles and Arcs
29 degrees to minutes:
$60 \times 29 = 1,740$
Answer: 1,740 minutes

Changing to a Larger Unit of Measure

To change a given number of units of one denomination to units of a larger denomination, we find the number of units of the smaller denomination that is equivalent to one of the larger denomination. Then we divide the given number of units of the smaller denomination by this conversion factor.

Since each conversion factor in the metric system is some power of ten, short methods of computation may be used. See Exercise 1-29.

Below are several examples of changing to a larger unit of measure and references to Exercises where you will find additional practice.

2-1 Metric—Length
9,000 meters to kilometers:
$9,000 \div 1,000 = 9$
Answer: 9 kilometers

2-2 Metric—Weight
58 milligrams to centigrams:
$58 \div 10 = 5.8$
Answer: 5.8 centigrams

2-3 Metric—Capacity
725 centiliters to liters:
$725 \div 100 = 7.25$
Answer: 7.25 liters

2-4 Metric—Area
670 mm^2 to cm^2:
$670 \div 100 = 6.7$
Answer: 6.7 cm^2

2-5 Metric—Volume
490,000 dm^3 to m^3:
$490,000 \div 1,000 = 490$
Answer: 490 m^3

2-6 Customary—Length
21 feet to yards:
$21 \div 3 = 7$
Answer: 7 yards

2-7 Customary—Weight
16,000 lb. to short tons:
$16,000 \div 2,000 = 8$
Answer: 8 short tons

2-8 Customary—Liquid
10 pints to quarts:
$10 \div 2 = 5$
Answer: 5 quarts

2-9 Customary—Dry
24 pecks to bushels:
$24 \div 4 = 6$
Answer: 6 bushels

2-10 Customary—Area
288 sq. in. to sq. ft.:
$288 \div 144 = 2$
Answer: 2 sq. ft.

2-11 Customary—Volume
189 cu. ft. to cu. yd.:
$189 \div 27 = 7$
Answer: 7 cu. yd.

2-13 Time
48 months to years:
$48 \div 12 = 4$
Answer: 4 years

2-14 Angles and Arcs
540 minutes to degrees:
$540 \div 60 = 9$
Answer: 9 degrees

WOMEN'S FREESTYLE RACES
50 METERS
100 METERS
400 METERS

METRIC SYSTEM OF MEASURE

Our monetary system is a decimal system in which the dollar is the basic unit. In the metric system, the meter (m) is the basic unit of length, the gram (g) is the basic unit of weight or mass, and the liter (L) is the basic unit of capacity (dry and liquid measures). Other metric units of length, weight, and capacity are named by adding the following prefixes to the basic unit of measure:

Prefix	Symbol		Value
kilo-	k	meaning	thousand (1,000)
hecto-	h	meaning	hundred (100)
deka-	da	meaning	ten (10)
deci-	d	meaning	one-tenth (.1 or $\frac{1}{10}$)
centi-	c	meaning	one-hundredth (.01 or $\frac{1}{100}$)
milli-	m	meaning	one-thousandth (.001 or $\frac{1}{1000}$)

The units of measure chart on the next page shows the relationship among units. Note the similarity of the metric system of measures to our decimal numeration system and to our monetary system.

It should be noted that just as 3 dollars 8 dimes 4 cents may be written as a single numeral, 3.84 *dollars* or $3.84; so 3 meters 8 decimeters 4 centimeters may also be written as a single numeral, 3.84 *meters*.

Abbreviations or symbols are written without the period after the last letter. The same symbol is used for both one or more quantities. Thus, "cm" is the symbol for *centimeter* or *centimeters*.

UNITS OF MEASURE

	1,000	100	10	1	.1	.01	.001
Decimal Place Value	thousands	hundreds	tens	ones	tenths	hundredths	thousandths
United States Money	$1,000 bill	$100 bill	$10 bill	dollar	dime	cent	mill
Metric Length	kilometer km	hectometer hm	dekameter dam	meter m	decimeter dm	centimeter cm	millimeter mm
Metric Weight	kilogram kg	hectogram hg	dekagram dag	gram g	decigram dg	centigram cg	milligram mg
Metric Capacity	kiloliter kL	hectoliter hL	dekaliter daL	liter L	deciliter dL	centiliter cL	milliliter mL

The above chart reveals that in the metric system ten (10) of any unit of measure is equivalent to one (1) unit of the next larger size.

> The prefix "micro" means one-millionth.
> The prefix "mega" means one million.
> A *micrometer* is one-millionth of a meter.
> A *microgram* is one-millionth of a gram.
> A *microliter* is one-millionth of a liter.
> A *megameter* is one million (1 000 000) meters.
> A *megagram* is one million (1 000 000) grams.
> A *megaliter* is one million (1 000 000) liters.

Measure of Length—Metric

OBJECTIVE: To use metric measures of length.

Let us examine the following section of a metric ruler:

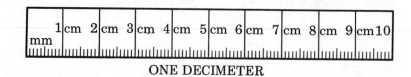

ONE DECIMETER

Each of the smallest subdivisions shown here indicates a measure of 1 millimeter (mm). Observe that ten (10) of these millimeter subdivisions form the next larger subdivision, called a centimeter (cm), and that ten (10) of the centimeter subdivisions form the next larger subdivision, called a decimeter (dm). The meter stick measuring one meter (m), the basic metric unit measuring length, has markings that show 10 decimeter divisions, 100 centimeter divisions, and 1,000 millimeter divisions.

To change a given number of units of one metric denomination to units of another metric denomination, follow the procedures that are explained on pages 285–287. For short methods of computation, see Exercises 1-28 and 1-29.

Basic: Prob. 1–18; *Average:* Prob. 1–22; *Enriched:* Prob. 1–25

PRACTICE PROBLEMS

1.a. kilometer   b. centimeter   c. meter   d. millimeter   e. decimeter   f. dekameter   g. hectometer

1. What metric unit of length does each of these symbols represent?
  a. km   **b.** cm   **c.** m   **d.** mm   **e.** dm   **f.** dam   **g.** hm

2. Use this section of the metric ruler to answer the following questions:

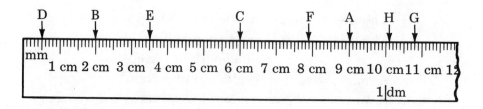

2.a. 2 cm; 6 cm; 3.5 cm; 10.1 cm; 9 cm; $\frac{1}{2}$ cm; 10.8 cm; 7.9 cm

a. What measurement is indicated by the point labeled:
  B?   C?   E?   H?   A?   D?   G?   F?

2.b. 4 cm; 3 cm; 1.8 cm; 5.5 cm; 2.2 cm

b. How far from point B is point C? From point D is point E? From point A is point G? From point E is point A? From point F is point H?

c. How many centimeters is it from point B to point A? From point C to point G? From point E to point F? 7 cm; 4.8 cm; 4.4 cm

d. How many millimeters is it from point E to point H? From point D to point F? From point F to point G? 66 mm; 74 mm; 29 mm

e. How many millimeters are in the length from 0 to each of the following markings:

8 cm? 80 mm 6 cm 2 mm? 62 mm 3 cm 4 mm? 34 mm 1 dm 1 cm 6 mm? 116 mm

f. In each of the following find the sum of the measurements by locating the mark for the first measurement on the ruler and adding on to this the second measurement. Simplify each sum as indicated:

8 mm + 6 mm = __14__ mm = __1__ cm __4__ mm

27 mm + 54 mm = __81__ mm = __8__ cm __1__ mm

2 cm 5 mm + 4 cm 2 mm = __6__ cm __7__ mm

5 cm 9 mm + 3 cm 7 mm = __8__ cm __16__ mm = __9__ cm __6__ mm

3 cm 6 mm + 2 cm 4 mm = __5__ cm __10__ mm = __6__ cm

1 dm 2 mm + 3 cm 5 mm = __1__ dm __3__ cm __7__ mm

3 cm 8 mm + 6 cm 2 mm = __9__ cm __10__ mm = __1__ dm

g. In each of the following locate the marking for the first measurement on the ruler and take away from this the second measurement to find your answer:

14 mm − 7 mm = __7__ mm

9 cm 6 mm − 4 cm 5 mm = __5__ cm __1__ mm

6 cm − 9 mm = __5__ cm __1__ mm

4 cm 2 mm − 6 mm = __3__ cm __6__ mm

7 cm 3 mm − 6 cm 8 mm = __0__ cm __5__ mm

5 cm − 3 cm 4 mm = __1__ cm __6__ mm

1 dm − 4 cm 2 mm = __5__ cm __8__ mm

h. Find the 108 mm mark on the metric ruler. Subtract from it a measurement of 4.3 cm. What measurement does the mark you reach indicate? 6.5 cm

3. Complete each of the following:

a. __10__ mm = 1 cm **e.** __10__ dam = 1 hm **i.** $\frac{1}{10}$ m = 1 dm

b. __10__ cm = 1 dm **f.** __10__ hm = 1 km **j.** $\frac{1}{10}$ dam = 1 m

c. __10__ dm = 1 m **g.** $\frac{1}{10}$ cm = 1 mm **k.** $\frac{1}{10}$ hm = 1 dam

d. __10__ m = 1 dam **h.** $\frac{1}{10}$ dm = 1 cm **l.** $\frac{1}{10}$ km = 1 hm

4. Complete each of the following:

a. _1,000_ mm = 1 m

b. _100_ cm = 1 m

c. _10_ dm = 1 m

d. _10_ mm = 1 cm

e. _100_ mm = 1 dm

f. _1,000_ m = 1 km

g. $\frac{1}{1,000}$ km = 1 m

h. $\frac{1}{100}$ m = 1 cm

i. $\frac{1}{1,000}$ m = 1 mm

j. $\frac{1}{10}$ m = 1 dm

k. 1 m = _10_ dm = _100_ cm = _1,000_ mm

l. 1 km = _10_ hm = _100_ dam = _1,000_ m

m. 1 km = _1,000_ m = _10,000_ dm = _100,000_ cm = _1,000,000_ mm

n. 1 mm = $\frac{1}{1,000}$ m = $\frac{1}{1,000,000}$ km

5. Express each of the following in meters:

a. 7 km 5 hm 8 dam 3 m = _7,583_ m

b. 8 m 2 dm 6 cm 5 mm = _8.265_ m

c. 6 m 4 cm 9 mm = _6.049_ m

d. 3 km 5 m 1 cm 6 mm = _____ m 3,005.016 m

e. 9 km 4 hm 2 dam 7 m 3 dm 7 cm 5 mm = _____ m 9,427.375 m

6. Change each of the following to <u>millimeters</u>:

34 cm; 340   28 m; 28,000   4 km; 4,000,000 8.5 cm; 85   7.6 m 7,600

7. Change each of the following to <u>centimeters</u>:

4.7 dm; 47   6.4 m; 640   851 mm; 85.1   8 km; 800,000 9.3 m 930

8. Change each of the following to <u>decimeters</u>:

439 m; 4,390   760 cm; 76   6,875 mm; 68.75 16.7 m; 167   2.6 km 26,000

9. Change each of the following to <u>meters</u>:

a. 78 dam; 780   4,206 cm; 42.06 3.9 hm; 390   60.4 dm; 6.04   569 mm .569

b. 0.57 km; 570 21.4 dam; 214   5 cm; .05   12,600 mm; 12.6 9.625 km 9,625

10. Change each of the following to <u>dekameters</u>:

46 m; 4.6   53 hm; 530   7.2 km; 720   38.9 m; 3.89   3,625 cm 3.625

11. Change each of the following to <u>hectometers</u>:

60 km; 600   16.4 m; .164   400 dam; 40   225.38 m; 2.2538   2.6 km 26

12. Change each of the following to <u>kilometers</u>:

81 hm; 8.1   5,328 m; 5.328 794,000 cm; 7.94 362.3 dam; 3.623 9,270,000 mm 9.27

13. Find the missing equivalent measurements:

km	hm	dam	m	dm	cm	mm
a. ? .7	? 7	? 70	700	? 7,000	? 70,000	? 700,000
b. 3	? 30	? 300	? 3,000	? 30,000	? 300,000	? 3,000,000
c. ? .05	? .5	? 5	? 50	? 500	5,000	? 50,000
d. ? 6.8	? 68	? 680	? 6,800	? 68,000	? 680,000	6,800,000

14. Complete each of the following:

a. 7 cm 4 mm = __74__ mm   **d.** 3 m 7 mm = __3,007__ mm

b. 5 m 9 cm = __509__ cm   **e.** 1 cm 8 mm = __18__ mm

c. 8 km 620 m = __8,620__ m   **f.** 6 m 39 cm = __639__ cm

15. Complete each of the following:

a. 4 m 8 dm 2 cm = __482__ cm   **c.** 1 km 9 m 3 cm = __100,903__ cm

b. 2 m 6 cm 5 mm = __2,065__ mm **d.** 7 m 5 cm 4 mm = __7,054__ mm

16. Complete each of the following:

a. 3 cm 8 mm = __3.8__ cm   **d.** 2 cm 9 mm = __2.9__ cm

b. 5 km 2 m = __5.002__ km   **e.** 4 km 218 m = __4.218__ km

c. 5 m 62 mm = __5.062__ m   **f.** 6 m 7 cm = __6.07__ m

17. Complete each of the following:

a. 6 m 9 cm 4 mm = __6.094__ m **c.** 8 km 6 m 9 cm = __8.00609__ km

b. 4 km 17 m 52 cm = __4.01752__ km **d.** 5 m 2 cm 6 mm = __5.026__ m

18. Complete each of the following:

a. 6 m 1 cm 8 mm = __601.8__ cm **c.** 3 cm 7 mm = __.037__ m

b. 7 km 4 m 3 cm = __7,004.03__ m **d.** 92 m 6 cm = __.09206__ km

19. If each curtain panel requires 125 cm of fabric and you need 8 panels, how many meters of fabric should you buy? 10 meters

20. Sound travels in water at a speed of 1,450 meters per second. How many kilometers does it travel in 15 seconds? 21.75 km

21. How much longer is a metal strip that measures 1.2 meters than one that measures 894 millimeters? .306 m or 306 mm

22. Which measurement is greater:

a. 9,000 mm or 7 m?   **d.** 800 mm or 6 m?

b. 4 km or 3,000 m?   **e.** 5.1 cm or 72 mm?

c. 4,000 cm or 12 m?   **f.** 4.5 m or 450 cm?
                                      equal

23. Which measurement is smaller:

a. 7 m or 9 km?   **d.** 73 mm or 6.9 cm?

b. 40 cm or 3 m?   **e.** 2.7 km or 2,994 m?

c. 5 km or 260,000 mm?   **f.** 8.49 m or 849 mm?

24. Arrange the following measurements in order of size (longest first): 6,800 m; 6.7 km; 526,000 cm; 692,400 mm

6,800 m;   526,000 cm;   6.7 km;   692,400 mm

25. Arrange the following measurements in order of size (shortest first): 84.5 mm; 845 m; 8.45 km; 8,450,000 cm

8.45 km;   84.5 mm;   845 m;   8,450,000 cm

OBJECTIVE: To use metric measures
of mass or weight.

VOC: mass
kilogram
weight

Technically, the kilogram is a unit used to measure the mass of an object. However, in everyday use the word "weight" almost always means "mass." Therefore, for our purposes we will continue to speak of the kilogram as a unit of weight

Since the weight of one (1) gram is so small, the kilogram is generally considered the practical basic unit of weight. Note also that the weight of 1,000 kilograms is equivalent to one (1) metric ton (t).

To change a given number of units of one metric denomination to units of another metric denomination, follow the procedures explained on pages 285–287. For short methods of computation, see Exercises 1-28 and 1-29.

Basic: Prob. 1–15; *Average:* Prob. 1–19; *Enriched:* Prob. 1–22

PRACTICE PROBLEMS

1. What metric unit of weight does each of these symbols represent?

a. mg   **b.** kg   **c.** dg   **d.** g   **e.** cg   **f.** hg   **g.** dag
   milligram   kilogram   decigram   gram   centigram   hectogram   dekagram

2. Complete each of the following:

a. __10__ mg = 1 cg   **e.** __10__ dag = 1 hg   **i.** $\frac{1}{10}$ g = 1 dg

b. __10__ cg = 1 dg   **f.** __10__ hg = 1 kg   **j.** $\frac{1}{10}$ dag = 1 g

c. __10__ dg = 1 g   **g.** $\frac{1}{10}$ cg = 1 mg   **k.** $\frac{1}{10}$ hg = 1 dag

d. __10__ g = 1 dag   **h.** $\frac{1}{10}$ dg = 1 cg   **l.** $\frac{1}{10}$ kg = 1 hg

3. Complete each of the following:

a. __1,000__ mg = 1 g   **j.** __100__ dag = 1 kg

b. __100__ cg = 1 g   **k.** __10,000__ dg = 1 kg

c. $\frac{1}{100}$ g = 1 cg   **l.** __100,000__ cg = 1 kg

d. $\frac{1}{1,000}$ g = 1 mg   **m.** _____ mg = 1 kg 1,000,000

e. __1,000__ g = 1 kg   **n.** $\frac{1}{10}$ kg = 1 hg

f. $\frac{1}{1,000}$ kg = 1 g   **o.** $\frac{1}{100}$ kg = 1 dag

g. 1 kg = __1,000__ g =1,000,000mg   **p.** $\frac{1}{10,000}$ kg = 1 dg

h. 1 mg = $\frac{1}{1,000}$ g = $\frac{1}{1,000,000}$ kg   **q.** $\frac{1}{100,000}$ kg = 1 cg

i. __10__ hg = 1 kg   **r.** $\frac{1}{1,000,000}$ kg = 1 mg

4. Change each of the following to milligrams:

9 cg; 90   16 g; 16,000   4.7 cg; 47   2.36 kg; 2,360,000 0.841 g 841

5. Change each of the following to centigrams:

39 mg; 3.9   5.74 dg; 57.4   62 g; 6,200   1.4 kg; 140,000   28.9 g 2,890

6. Change each of the following to decigrams:

25 g; 250   78 cg; 7.8   3.1 kg; 31,000   4,500 mg; 45   9.4 g 94

7. Change each of the following to grams:

60 dag; 600   530 dg; 53   9 kg; 9,000   25.8 cg; .258   750 mg .750

8. Change each of the following to dekagrams:

3,000 g; 300   68 kg; 6,800   76.7 cg; .0767   5.63 hg; 56.3   9,100 mg .91

9. Change each of the following to hectograms:

533 dag; 53.3   7,000 g; 70   40 kg; 400   451.9 g; 4.519   6.5 kg 65

10. Change each of the following to kilograms:

7,864 g; 7.864 85 hg; 8.5   961,000 cg; 9.61   423.6 dag; 4.236 7,400,000 mg 7.4

11. Change each of the following to metric tons:

3,000 kg; 3   570 kg; .57   6,300,000 g; 6.3   19,000 hg; 1.9   82,600 kg 82.6

12. Find the missing equivalent weights:

	kg	g	cg	mg
a.	? 5	5,000	? 500,000	? 5,000,000
b.	7.3	? 7,300	? 730,000	? 7,300,000
c.	? .029	? 29	2,900	? 29,000
d.	? .8	? 800	? 80,000	800,000
e.	? .456	456	? 45,600	? 456,000

13. Complete each of the following:

a. 5 cg 6 mg = __56__ mg

b. 8 kg 130 g = __8,130__ g

c. 7 g 25 cg = __725__ cg

d. 2 g 84 mg = __2,084__ mg

e. 3 cg 9 mg = __39__ mg

f. 6 kg 370 g = __6,370__ g

14. Complete each of the following:

a. 4 g 8 dg 5 cg = __485__ cg

b. 3 g 6 cg 9 mg = __3,069__ mg

c. 7 kg 2 g 8 cg = __700,208__ cg

d. 9 kg 1 g 5 mg = __9,001,005__ mg

15. Complete each of the following:

a. 5 cg 2 mg = __5.2__ cg

b. 15 kg 260 g = __15.260__ kg

c. 9 g 4 cg = __9.04__ g

d. 7 g 1 mg = __7.001__ g

e. 8 kg 7 g = __8.007__ kg

f. 19 g 58 mg = __19.058__ g

16. Complete each of the following:

a. 8 g 5 cg 2 mg = __8.052__ g

b. 5 kg 9 g 4 cg = __5.00904__ kg

c. 6 kg 7 g 1 mg = __6.007001__ kg

d. 3 g 6 cg 6 mg = __3.066__ g

17. Complete each of the following:

a. 9 g 1 cg 6 mg = <u>901.6</u> cg   **c.** 5 cg 3 mg = <u>.053</u> g

b. 2 kg 8 g 4 cg = _____ g 2,008.04   **d.** 650 g 47 cg = _____ kg .65047

18. If your new diet will reduce your weight an average of 150 grams per day, how long will it take to reduce 6 kilograms? 40 days

19. Which weight is heavier:

a. 9 kg or 9,000 g? equal   **d.** 7,500 cg or <u>8 kg</u>?

b. 56 mg or <u>6.1 cg</u>?   **e.** 16 mg or <u>5.2 g</u>?

c. <u>19 g</u> or 470 mg?   **f.** 600 g or <u>4.9 kg</u>?

20. Which weight is lighter:

a. 9 kg or <u>3,800 g</u>?   **d.** <u>8,100 g</u> or 6.7 kg?

b. <u>44 cg</u> or 3.8 g?   **e.** 4.32 g or <u>5,075 mg</u>?

c. <u>29 mg</u> or 2.9 cg? equal   **f.** 8.85 kg or <u>7,990 g</u>?

21. Arrange in order of weight (lightest first):

6,500 g; ④   5.9 kg; ③   625,000 mg; ①   67,000 cg ②

22. Arrange in order of weight (heaviest first):

40.75 cg; ④   4,080 g; ①   4.06 kg; ②   4,010,000 mg ③

EXERCISE

2-3 | Measure of Capacity—Metric

OBJECTIVE: To use metric measures of capacity.

The units that measure capacity also measure volume. See Exercise 2-5 on page 300 where units of cubic measure are studied.

To change a given number of units of one metric denomination to units of another denomination, follow the procedures explained on pages 285–287. For short methods of computation, see Exercises 1-28 and 1-29.

Basic: Prob. 1–13; *Average:* Prob. 1–16; *Enriched:* Prob. 1–20

PRACTICE PROBLEMS

1. What metric unit of capacity does each symbol represent?

a. kL   **b.** L   **c.** mL   **d.** cL   **e.** hL   **f.** dL   **g.** daL

2. Complete each of the following:

a. <u>10</u> mL = 1 cL   **e.** <u>10</u> daL = 1 hL   **i.** <u>$\frac{1}{10}$</u> L = 1 dL

b. <u>10</u> cL = 1 dL   **f.** <u>10</u> hL = 1 kL   **j.** <u>$\frac{1}{10}$</u> daL = 1 L

c. <u>10</u> dL = 1 L   **g.** <u>$\frac{1}{10}$</u> cL = 1 mL   **k.** <u>$\frac{1}{10}$</u> hL = 1 daL

d. <u>10</u> L = 1 daL   **h.** <u>$\frac{1}{10}$</u> dL = 1 cL   **l.** <u>$\frac{1}{10}$</u> kL = 1 hL

1.a. kiloliter b. liter c. milliliter d. centiliter e. hectoliter f. deciliter g. dekaliter

3. Complete each of the following:

a. __1,000__ mL = 1 L

b. __100__ cL = 1 L

c. __1,000__ L = 1 kL

d. $\frac{1}{1,000}$ kL = 1 L

e. $\frac{1}{100}$ L = 1 cL

f. $\frac{1}{1,000}$ L = 1 mL

g. 1 kL = __1,000__ L = __1,000,000__ mL

h. 1 mL = $\frac{1}{1,000}$ L = $\frac{1}{1,000,000}$ kL

4. Change each of the following to <u>milliliters</u>:

7 cL; 70   4.6 L; 4,600   53 dL; 5,300   0.375 L; 375   80.4 cL 804

5. Change each of the following to <u>centiliters</u>:

85 L; 8,500   9 dL; 90   18 mL; 1.8   6.7 L; 670   44.2 mL 4.42

6. Change each of the following to <u>deciliters</u>:

31 L; 310   68 cL; 6.8   2.25 L; 22.5   520 mL; 5.2   767 cL 76.7

7. Change each of the following to <u>liters</u>:

9 dL; .9   60 hL; 6,000   21.6 kL; 21,600   400 cL; 4   8,405 mL 8.405

8. Change each of the following to <u>dekaliters</u>:

29 hL; 290   108 L; 10.8   4.7 hL; 47   16.4 L; 1.64   332 dL 3.32

9. Change each of the following to <u>hectoliters</u>:

7.9 kL; 79   5.2 daL; .52   6,008 L; 60.08   21.57 kL; 215.7   825.1 L 8.251

10. Change each of the following to <u>kiloliters</u>:

6,582 L; 6.582   43 hL; 4.3   875 L; .875   79.5 hL; 7.95   200 daL 2

11. Find the missing equivalent capacities:

	L	dL	cL	mL
a.	7	? 70	? 700	? 7,000
b.	? 64.5	645	? 6,450	? 64,500
c.	? .9	? 9	? 90	900
d.	3.8	? 38	? 380	? 3,800
e.	? 5.452	? 54.52	545.2	? 5,452

12. Complete each of the following:

a. 7 cL 4 mL = __74__ mL

b. 1 L 6 cL = __106__ cL

c. 5 L 8 mL = __5,008__ mL

d. 8 kL 600 L = __8,600__ L

e. 4 L 91 mL = __4,091__ mL

f. 13 L 25 cL = __1,325__ cL

13. Complete each of the following:

a. 6 L 1 cL 9 mL = __6,019__ mL

b. 1 L 57 cL 8 mL = __1,578__ mL

14. Complete each of the following:

a. 3 cL 7 mL = __3.7__ cL

b. 5 L 5 cL = __5.05__ L

c. 4 kL 46 L = __4.046__ kL

d. 8 L 2 mL = __8.002__ L

e. 9 L 54 cL = __9.54__ L

f. 6 cL 9 mL = __6.9__ cL

15. Complete each of the following:

   a. 8 L 3 cL 7 mL = <u>8.037</u> L   **b.** 3 L 6 cL 8 mL = <u>3.068</u> L

16. Complete each of the following:

   a. 4 L 1 cL 4 mL = <u>401.4</u> cL   **b.** 5 L 9 cL 6 mL = <u>509.6</u> cL

17. Frank mixed 15 cL of a weed-killer solution with 485 cL of water. How many liters of the mixture did he have? 5 liters

18. Which capacity is smaller:

   a. 40 cL or <u>65 mL</u>?   **c.** <u>54.8 mL</u> or 1 L?   **e.** <u>65 dL</u> or 65 daL?

   b. <u>6.8 L</u> or 2,000 cL?   **d.** 2.6 cL or 3 <u>mL</u>?   **f.** <u>8.1 L</u> or 9,700 mL?

19. Arrange the following capacities in order of size (largest first):

63.8 cL;   637 mL;   0.639 L;   6.39 dL

20. Arrange the following capacities in order of size (smallest first):

19 L;   6,235 mL;   89.4 dL;   622 cL

19. 0.639 L and 6.39 dL are equal, 63.8 cL, 637 mL
20. 622 cL, 6,235 mL, 89.4 dL, 19 L

For a Class Activity see p. T33.

EXERCISE

2-4 | Measure of Area—Metric

OBJECTIVE: To use metric measures of area.

   We have found in linear measure that ten (10) of any metric unit is equivalent to one (1) of the next higher metric unit. However, in the square measure one hundred (100) of any metric unit is equivalent to one (1) of the next higher unit. As we see in the following table, in the metric measure of area the exponent ^{2} is used to represent the word "square."

   100 square millimeters (mm^2) = 1 square centimeter (cm^2)
   100 square centimeters (cm^2) = 1 square decimeter (dm^2)
   100 square decimeters (dm^2) = 1 square meter (m^2)
   100 square meters (m^2)   = 1 square dekameter (dam^2)
   100 square dekameters (dam^2) = 1 square hectometer (hm^2)
   100 square hectometers (hm^2) = 1 square kilometer (km^2)

1 cm

Area = 1 cm^2

Related Resources: *Practical Applications in
Mathematics:* Diag. Test and Prac. Ex., p. 81

Special names are sometimes used. *Centare* may be used instead of square meter, *are* instead of square dekameter, and *hectare* instead of square hectometer. The hectare is widely used. Thus,

$$100 \text{ centares} = 1 \text{ are (a)}$$
$$100 \text{ ares} \quad = 1 \text{ hectare (ha)}$$
$$100 \text{ hectares} = 1 \text{ square kilometer}$$

To change from one metric unit of square measure to another metric unit of square measure, follow the procedures explained on pages 285–287.

Basic: Prob. 1–7; *Average:* Prob. 1–9; *Enriched:* Prob. 1–10

PRACTICE PROBLEMS

1.a. square centimeter   b. square kilometer · c. square meter   d. square millimeter

1. What unit of metric square measure does each of these symbols represent? 1.e. hectare   f. square decimeter   g. square dekameter

    a. cm^2 **b.** km^2 **c.** m^2 **d.** mm^2 **e.** ha   **f.** dm^2 **g.** dam^2

2. a. How many square centimeters are in 1 square meter? 10,000 cm²

    b. How many square millimeters are in 1 square meter? 1,000,000 mm²

    c. How many square meters are in 1 square kilometer? 1,000,000 m²

    d. How many square meters are in 1 hectare? 10,000 m²

    e. How many hectares are in 1 square kilometer? 100 ha

3. Change each of the following to square millimeters:

    6 cm^2; 600 2.15 m^2; 2,150,000 1.7 dm^2; 17,000 41.9 cm^2; 4,190 0.85 m^2 850,000

4. Change each of the following to square centimeters:

    37 dm^2; 3,700 8.4 m^2; 84,000 700 mm^2; 7 0.66 m^2; 6,600 $5,300 \text{ mm}^2$ 53

5. Change each of the following to square decimeters:

    819 cm^2; 8.19 30 m^2; 3,000 $600,000 \text{ mm}^2$; 60 7.5 m^2; 750 240.7 cm^2 2.407

6. Change each of the following to square meters (or centares):

45,000,000 45 km^2; 19 dm^2; .19 $20,000 \text{ cm}^2$; 2 $809,000 \text{ mm}^2$; .809 0.259 km^2 259,000

7. Change each of the following to square dekameters (or ares):

    $8,000 \text{ m}^2$; 80 31 hectares;3,100 2.77 m^2;.0277 6.8 hectares; 680 $12,300 \text{ m}^2$ 123

8. Change each of the following to hectares (or square hectometers):

    $75,000 \text{ m}^2$; 7.5 78.2 dam^2; .782 14 km^2; 1,400 387.5 m^2; .03875 4.3 km^2 430

9. Change each of the following to square kilometers:

    500 hectares; 5 72.6 hectares; .726 $3,825,000 \text{ m}^2$; 3.825 $609,220 \text{ m}^2$ .60922

10. A farm contains 40 hectares of land. How many square meters does it measure? What part of a square kilometer is it?
    400,000 m²   .4 km²

2-5 | Measure of Volume—Metric

OBJECTIVE: To use metric measures of volume.

The volume, also called capacity, is generally the number of units of cubic measure contained in a given space. However, the units of capacity (see Exercise 2-3 on page 296) also are used to measure volume. Refer to the relationship given below between the cubic decimeter and the liter.

In linear measure ten (10) of any metric unit is equivalent to one (1) of the next higher metric unit. In the square measure one hundred (100) of any metric unit is equivalent to one (1) of the next higher unit.

Volume = 1 cm³

However, in the cubic measure one thousand (1,000) of any metric unit is equivalent to one (1) of the next higher unit. As shown below, in the metric measure of volume the exponent ³ is used to represent the word "cubic."

1,000 cubic millimeters (mm³) = 1 cubic centimeter (cm³)
1,000 cubic centimeters (cm³) = 1 cubic decimeter (dm³)
1,000 cubic decimeters (dm³) = 1 cubic meter (m³)

To change from one metric unit of cubic measure to another unit of cubic measure, follow the procedures explained on pages 285–287.

Also note the following relationships:

The volume of one cubic decimeter has the same capacity as one (1) liter. Since 1 cubic decimeter is equivalent to 1,000 cubic centimeters (cm³) and 1 liter is equivalent to 1,000 milliliters (mL), then the volume of 1 cubic centimeter (cm³) has the same capacity as 1 milliliter (mL).

Also a gram is the weight of one (1) cubic centimeter (or 1 milliliter) of water at a temperature of 4 degrees Celsius, and a kilogram is the weight of 1,000 cubic centimeters (or 1 liter) of water at a temperature of 4 degrees Celsius.

MISCELLANEOUS EQUIVALENTS

---

1 liter = 1 cubic decimeter (dm³) = 1,000 cubic centimeters (cm³)
1 milliliter (mL) = 1 cubic centimeter (cm³)
1 liter of water weighs 1 kilogram (kg)
1 milliliter or cubic centimeter of water weighs 1 gram (g)

1.a. cubic meter b. cubic millimeter c. cubic kilometer d. cubic centimeter e. cubic decimeter f. cubic hectometer g. cubic dekameter

PRACTICE PROBLEMS

1. What unit of metric cubic measure does each of the following
symbols represent?

  a. m^3 **b.** mm^3 **c.** km^3 **d.** cm^3 **e.** dm^3 **f.** hm^3 **g.** dam^3

2. a. How many cubic millimeters are in 1 cubic decimeter? 1,000,000 mm³

  b. How many cubic millimeters are in 1 cubic meter? 1,000,000,000 mm³

  c. How many cubic centimeters are in 1 cubic meter? 1,000,000 cm³

For ex. 3–6
see Extended
Answers, p. T78.

3. Change each of the following to cubic millimeters:
8 cm^3;   15 m^3;   6.1 dm^3;   25.8 cm^3;   53.6 m^3

4. Change each of the following to cubic centimeters:
27 m^3;   4.95 dm^3;   6,226 mm^3;   8.5 m^3;   740 mm^3

5. Change each of the following to cubic decimeters:
63 m^3;   2,930 cm^3;   17.9 m^3;   73,250 mm^3;   86.4 cm^3

6. Change each of the following to cubic meters:
17,000 dm^3;   36,800 cm^3;   9,400,000 mm^3;   575 dm^3;   600,000 cm^3

7. a. 21 milliliters of liquid will fill how many cubic centimeters? 21 cm³

  b. 9 liters of liquid will fill how many cubic centimeters of space? 9,000 cm³

  c. A space of 430 cubic centimeters will hold how many milliliters of
liquid? 430 mL

  d. A space of 18.6 cubic decimeters will hold how many liters of
liquid? 18.6 L

  e. A space of 7,900 cubic centimeters will hold how many liters of
liquid? 7.9 L

  f. 5 liters of water weigh approximately how many kilograms? 5 kg

  g. 67 milliliters of water weigh how many grams? 67 g

  h. 82 centiliters of water weigh how many grams? 820 g

  i. 9.1 liters of water weigh how many kilograms? 9.1 kg

  j. 0.615 liter of water weighs how many grams? 615 g

  k. How many liters of water are in a container if the water weighs 4
kilograms? 4 L

  l. How many milliliters of water weigh 75 grams? 75 mL

  m. How many liters of water are in a container if the water weighs
545 grams? .545 L

  n. How many cubic decimeters do 26 kilograms of water occupy? 26 dm³

  o. Water weighing 19 grams occupies a space of how many cubic
centimeters? 19 cm³

  p. Water weighing 8.8 kilograms fills a space of how many cubic
centimeters? 8,800 cm³

8. How much space is occupied by each of the following capacities?

a. In cubic centimeters (cm^3):

7 mL; 7   43.6 mL; 43.6 61 cL; 610   4 dL; 400   0.75 L 750

b. In cubic decimeters (dm^3):

6 L; 6   84 dL; 8.4   257 cL; 2.57 14.2 L; 14.2   8.25 L 8.25

9. Find the capacity that will fill each of the following volumes:

a. In milliliters (mL):

13 cm^3; 13 2.5 cm^3; 2.5 800 mm^3; .8 7 dm^3; 7,000   0.09 m^3 90,000

b. In centiliters (cL):

8 cm^3; .8   3.18 m^3; ^{318,000} 2.6 dm^3; 260 5.07 cm^3; .507 9,100 mm^3 .91

c. In liters (L):

7 dm^3; 7   28.3 dm^3; 28.3 5.9 m^3; 5.900 6,000 cm^3; 6 43,000 mm^3 .043

10. Find the weight of each of the following volumes or capacities of water:

a. In kilograms (kg):

5 dm^3; 5   79 L; 79   196 cm^3; .196 4.81 L; 4.81   6,300 cm^3 6.3

b. In grams (g):

11 cm^3; 11   0.225 L; 225 8.7 mL; 8.7   6.08 cL; 60.8   5,600 mm^3 5.6

11. Find the volume or capacity occupied by each of the following weights of water:

a. In cubic centimeters (cm^3):

6 kg; 6,000   30 g; 30   0.6 kg; 600   950 mg; .950   8.74 g 8.74

b. In liters (L):

15 kg; 15   4,100 g; 4.1   520 g; .520   0.004 kg; .004   7,090 mg .00709

c. In milliliters (mL):

7 g; 7   138.51 g; 138.51 1.8 kg; 1,800   6,270 mg; 6.270 9,500 cg 95

d. In centiliters (cL):

26 g; 2.6   5.7 kg; 570   615 g; 61.5   17,400 cg; 17.4 9,000 mg .9

12. A tank holds 800 liters of oil. How many cubic decimeters of space are in the tank? 800 dm³

13. It takes 190,000 liters of water to fill a swimming pool. What is the weight of the water when the pool is nine-tenths full? 171,000 kg

14. How many liters of water will fill an aquarium if it occupies a space of 6,600 cubic centimeters? What is the weight of the water when the aquarium is full? 6.6 L; 6.6 kg

15. A water tank has a volume of 7.54 cubic meters. How many liters of water can the tank hold? What is the weight of the water when the tank is full? 7,540 L; 7,540 kg

For Class Activities see pp. T33 and T34.

For a Test on the Metric System of Measure see p. T46 or *Testing Program*, p. 12.

CUSTOMARY SYSTEM OF MEASURE

Since the complete changeover to the metric system is a gradual one, we will find the customary units of measure still being used. Consequently the following sections deal with the customary units of measure for possible use in our everyday affairs.

We find that the conversion factors used in the customary system consist of many different numbers such as 12; 3; 36; 5,280; 1,760; 16; 2,000; etc. When we change from a small customary unit of measure to a larger unit, we sometimes get complicated answers. For example:

> 95 inches changed to yards is $2\frac{23}{36}$ yards.
>
> 27 ounces changed to pounds is $1\frac{11}{16}$ pounds.

However, we have seen that the conversion factors used in the metric system are all some power of ten. When we change from any unit of measure to another unit in the metric system, the computation is quick and easy because we work with the decimal system. For example:

> 58.7 centimeters = 587 millimeters
>
> 349 centiliters = 3.49 liters
>
> 16.5 kilograms = 16,500 grams

In the following Exercises 2-6 through 2-11, to change a given number of units of one customary denomination to units of another denomination, follow the procedures that are explained on pages 285–287. See pages 621–623 for tables of measure.

Fundamental Operations with Denominate Numbers

Any numbers which are expressed in terms of units of measure are called denominate numbers

Addition

To add denominate numbers, arrange like units in columns, then add each column. Where the sum of any column is greater than the number of units that make the next larger unit, it is simplified as illustrated in the model.

Add:

6 ft. 7 in.
1 ft. 9 in.
7 ft. 16 in. = 8 ft. 4 in.

Answer: 8 ft. 4 in.

Subtraction

To subtract denominate numbers, arrange like units under each other, then subtract, starting from the right. When the number of units in the subtrahend is greater than the number of corresponding units in the minuend, take one of the next larger units in the minuend and change it to an equivalent number of smaller units to permit subtraction as shown in the model.

Subtract:

7 hr. 10 min. = 6 hr. 70 min.
2 hr. 25 min. = 2 hr. 25 min.
              4 hr. 45 min.

Answer: 4 hr. 45 min.

Multiplication

To multiply denominate numbers, multiply each unit by the multiplier. Where the product of any column is greater than the number of units that make the next larger unit, it is simplified as shown in the model.

Multiply:

5 yd. 11 in.
       4
20 yd. 44 in. = 21 yd. 8 in.

Answer: 21 yd. 8 in.

Division

To divide denominate numbers, divide each unit by the divisor. If the unit is not exactly divisible, change the remainder to the next smaller unit and combine with the given number of the smaller unit to form the next partial dividend.

Divide:

```
     3 lb. 5 oz.
5)16 lb. 9 oz.
  15 lb.
    1 lb. 9 oz. = 25 oz.
                  25 oz.
```

Answer: 3 lb. 5 oz.

Measure of Length—Customary

OBJECTIVE: To use customary measures of length.

VOC: denominate number

PRACTICE PROBLEMS

1. Find the number of <u>inches</u> in:

a. 8 ft.; 96   6.25 ft.; 75 $3\frac{1}{2}$ ft.; 42   50 ft. 600

b. 9 yd; 324   1.75 yd.; 63 $4\frac{2}{3}$ yd.; 168   16 yd. 576

c. 5 ft. 7 in.; 67   9 ft. 3 in.; 111   1 yd. 11 in.; 47   5 yd. 21 in. 201

2. Find the number of <u>feet</u> in:

a. 6 yd.; 18   1.4 yd.; 4.2 $\frac{3}{8}$ yd.; 1.125 $15\frac{1}{2}$ yd. 46.5

b. 8 mi.; 42,240   6.5 mi.; 34,320 $23\frac{1}{4}$ mi.; 122,760   39 mi. 205,920

c. 4 rd.; 66   3.25 rd.; 53.625 $5\frac{3}{4}$ rd.; 94.875   26 rd. 429

d. 5 yd. 2 ft.; 17   47 yd. 1 ft.; 142   11 mi. 1,000 ft.; 59,080   3 rd. 10 ft. 59.5

e. 24 in.; 2   192 in.; 16   57 in.; 4.75   588 in. 49

3. Find the number of <u>yards</u> in:

a. 7 mi.; 12,320   13.1 mi.; 23,056 $4\frac{1}{2}$ mi.; 7,920   38 mi. 66,880

b. 9 rd.; 49.5   4.6 rd.; 25.3 $6\frac{1}{4}$ rd.; 34.375   25 rd. 137.50

c. 4 mi. 200 yd.; 7,240   15 mi. 800 yd.; 27,200   3 rd. 5 yd.; 21.5   13 rd. 3 yd. 74.5

d. 180 in.; 5   756 in.; 21   50.4 in.; 1.4   576 in. 16

e. 21 ft.; 7   57 ft.; 19   89 ft.; $29\frac{2}{3}$ 7.5 ft. 2.5

4. Find the number of <u>rods</u> in:

a. 10 mi.; 3,200   4.2 mi.; 1,344 $9\frac{3}{4}$ mi.; 3,120   39 mi. 12,480

b. 99 ft.; 6 $8\frac{1}{4}$ ft.; $\frac{1}{2}$ 198 yd.; 36 $41\frac{1}{4}$ yd. 7.5

5. Find the number of <u>miles</u> in:

a. 15,840 ft.; 3   36,960 ft.; 7   3,300 ft.; $\frac{5}{8}$ 132,000 ft. 25

b. 8,800 yd.; 5   70,400 yd.; 40   4,840 yd.; $2\frac{3}{4}$ 6,000 yd. 3.41

6. a. What part of a <u>foot</u> is:

4 in.; $\frac{1}{3}$ 9 in.; $\frac{3}{4}$ 6 in.; $\frac{1}{2}$ 1 in.; $\frac{1}{12}$ 3 in.; $\frac{1}{4}$ 11 in.; $\frac{11}{12}$ 10 in.? $\frac{5}{6}$

b. What part of a <u>yard</u> is:

21 in.; $\frac{7}{12}$ 30 in.; $\frac{5}{6}$ 24 in.; $\frac{2}{3}$ 7 in.; $\frac{7}{36}$ 1 ft.; $\frac{1}{3}$ 2 ft.; $\frac{2}{3}$ $1\frac{1}{2}$ ft.? $\frac{1}{2}$

c. What part of a <u>mile</u> is:

440 yd.; $\frac{1}{4}$ 2,640 ft.; $\frac{1}{2}$ 1,100 yd.; $\frac{5}{8}$ 280 rd.; $\frac{7}{8}$ 1,320 yd.; $\frac{3}{4}$ 1,980 ft.? $\frac{3}{8}$

7. Add and simplify:

2 ft. 9 in.	3 yd. 27 in.	4 mi. 700 yd.	5 yd. 1 ft. 10 in.
11 ft. 2 in.	5 yd. 19 in.	1 mi. 850 yd.	12 yd. 2 ft. 8 in.
		3 mi. 320 yd.	6 yd. 1 ft. 6 in.
13 ft. 11 in.	9 yd. 10 in.	9 mi. 110 yd.	25 yd.

8. Subtract:

9 yd. 7 in.	10 ft.	4 mi. 1,200 ft.	8 yd. 1 ft. 3 in.
4 yd. 5 in.	7 ft. 8 in.	3,700 ft.	5 yd. 2 ft. 9 in.

5 yd. 2 in.   2 ft. 4 in.   3 mi. 2,780 ft.   2 yd. 1 ft. 6 in.

9. Multiply and simplify:

5 ft. 2 in.	4 mi. 300 yd.	2 yd. 9 in.	4 yd. 1 ft. 11 in.
4	9	8	6

20 ft. 8 in.   37 mi. 940 yd.   18 yd.   27 yd. 2 ft. 6 in.

10. Divide:

4)12 ft. 8 in.   7)17 yd. 4 in.   5)6 mi. 100 ft.   3)7 yd. 2 ft. 6 in.
   3 ft. 2 in.   2 yd. 16 in.   1 mi. 1,076 ft.   2 yd. 1 ft. 10 in.

11. Change:

a. To <u>feet</u>:

36,480 6 naut. mi.;   2.8 naut. mi.; $4\frac{1}{4}$ naut. mi.;   40 naut. mi.
                         17,024   25,840   243,200

b. To yards:

$22,293\frac{1}{3}$ 11 naut. mi.;   8.75 naut. mi.; $\frac{7}{8}$ naut. mi.;   145 naut. mi.
c. To <u>statute miles</u>: $17,733\frac{1}{3}$ $1,773\frac{1}{3}$ $293,866\frac{2}{3}$

7 naut. mi.;   9.3 naut. mi.; $36\frac{1}{2}$ naut. mi.;   145 naut. mi.
   8.05   10.695   41.975   166.75

12. Change to <u>nautical miles</u>:

a. 30,400 ft.; 5   18,240 ft.; 3   10,000 ft.; 1.6   152,000 ft. 25

b. 42,560 yd.; 21   4,000 yd.; 1.97   11,500 yd.; 5.67   54,720 yd. 27

c. 5 stat. mi.; 4.34 61.25 stat. mi.; 53.19 $28\frac{3}{4}$ stat. mi.; 24.97 130 stat. mi. 112.89

In problems 13 through 15, solve each problem and select the letter in red corresponding to your answer.

13. If the floor boards are 2 inches wide, how many will be required to cover a floor 17 feet wide?

a. 34 boards   c. 84 boards
<u>b.</u> 102 boards   d. Answer not given

14. How many yards of ribbon are needed to make 30 school officer badges if it takes 9 inches of ribbon to make one badge?

a. $27\frac{1}{2}$ yards   c. 12 yards
<u>b.</u> $7\frac{1}{2}$ yards   d. $6\frac{2}{3}$ yards

15. How many miles high is Mount Everest if its elevation is 29,028 feet? Find answer correct to nearest tenth.

<u>a.</u> 5.5 miles   c. 54.9 miles
b. 5.8 miles   d. Answer not given

Measure of Weight—Customary

OBJECTIVE: To use customary measures of weight.
Basic: Prob. 1-9; *Average:* Prob. 1-12; *Enriched:* Prob. 1-15

PRACTICE PROBLEMS

1. Find the number of <u>ounces</u> in:

a. 6 lb.; 96   2.2 lb.; $35\frac{1}{5}$ $5\frac{3}{4}$ lb.; 92   37 lb. 592

b. 1 lb. 2 oz.; 18   4 lb. 8 oz.; 72   9 lb. 7 oz.; 151   18 lb. 13 oz. 301

2. Find the number of <u>pounds</u> in:

a. 7 s.t.; 14,000   8.75 s.t.; 17,500 $4\frac{1}{2}$ s.t.; 9,000   148 s.t. 296,000

b. 59 l.t.; 132,160   9.5 l.t.; 21,280 $17\frac{3}{4}$ l.t.; 39,760   31 l.t. 69,440

c. 6 s.t. 400 lb.; 12,400   9 l.t. 500 lb.; 20,660   20 s.t. 1,200 lb. 41,200

d. 64 oz.; 4   176 oz.; 11   26 oz.; $1\frac{5}{8}$ 368 oz. 23

3. Find the number of <u>short tons</u> in:

6,000 lb.; 3   24,000 lb.; 12   7,000 lb. $3\frac{1}{2}$

4. Find the number of <u>long tons</u> in:

6,720 lb.; 3   17,920 lb.; 8   4,000 lb. $1\frac{11}{14}$

5. a. What part of a <u>pound</u> is:

12 oz.; $\frac{3}{4}$ 8 oz.; $\frac{1}{2}$ 14 oz.; $\frac{7}{8}$ 5 oz.? $\frac{5}{16}$

b. What part of a <u>short ton</u> is:

1,200 lb.; $\frac{3}{5}$ 500 lb.; $\frac{1}{4}$ 750 lb.; $\frac{3}{8}$ 1,800 lb.? $\frac{9}{10}$

c. What part of a <u>long ton</u> is:

560 lb.; $\frac{1}{4}$ 1,120 lb.; $\frac{1}{2}$ 840 lb.; $\frac{3}{8}$ 1,960 lb.? $\frac{7}{8}$

6. Add and simplify:

		4 lb. 10 oz.	2 l.t. 1,200 lb.
3 lb. 6 oz.	2 s.t. 1,000 lb.	7 lb. 9 oz.	1 l.t. 850 lb.
1 lb. 8 oz.	5 s.t. 1,200 lb.	6 lb. 13 oz.	5 l.t. 1,000 lb.
4 lb. 14 oz.	8 s.t. 200 lb.	19 lb.	9 l.t. 810 lb.

7. Subtract:

6 lb. 13 oz.	9 lb. 5 oz.	4 s.t. 700 lb.	5 lb.
3 lb. 4 oz.	8 lb. 14 oz.	2 s.t. 1,900 lb.	3 lb. 6 oz.
3 lb. 9 oz.	7 oz.	1 s.t. 800 lb.	1 lb. 10 oz.

8. Multiply and simplify:

3 lb. 2 oz.	7 st. 400 lb.	1 lb. 8 oz.	6 lb. 4 oz.
6 18 lb. 12 oz.	5	9	8
	36 s.t.	13.5 lb.	50 lb.

9. Divide:

5)10 lb. 15 oz.   4)5 l.t. 160 lb.   6)14 lb. 4 oz.   8)10 s.t. 200 lb.
2 lb. 3 oz.   1 l.t. 600 lb.   2 lb. 6 oz.   1 s.t. 525 lb.

Related Resources: *Practical Applications in Mathematics:* Diag. Test and Prac. Ex., p. 76

In problems 10 through 15, solve each problem and select the letter in red corresponding to your answer.

10. What is the cost of a chicken weighing 5 lb. 12 oz. at $.96 per pound?

   a. $5.24   **b.** $5.48   <u>**c.**</u> $5.52   **d.** Answer not given

11. At $2.88 per pound, how many ounces of cake should you get for $1.80?

   a. 9 oz.   <u>**b.**</u> 10 oz.   **c.** 11 oz.   **d.** 12 oz.

12. How many 3-oz. boxes of cookies can be made from a 9-lb. box?

   a. 36   **b.** 40   **c.** 44   <u>**d.**</u> 48

13. How much is saved by buying a 1-lb. box of cereal at $1.67 instead of two 8-oz. boxes at $.89 each?

   a. $.08   **b.** $.09   **c.** $.10   <u>**d.**</u> $.11

14. Find the difference in price per ounce when a 12-oz. jar of jelly sells for $.78 and a 2-lb. jar sells for $1.92.

   a. 1¢   **b.** $\frac{3}{4}$¢   <u>**c.**</u> $\frac{1}{2}$¢   **d.** $\frac{1}{4}$¢

15. If the price of steel increases $10 per ton (2,000 lb.), what is the increase per pound?

   a. $.50   **b.** $.05   <u>**c.**</u> $.005   **d.** Answer not given

---

EXERCISE
2-8 | Liquid Measure—Customary

OBJECTIVE: To use customary measures of liquids.
Basic: Prob. 1–8; *Average:* Prob. 1–10; *Enriched:* Prob. 1–12

PRACTICE PROBLEMS

1. Find the number of <u>ounces</u> in:

   a. 9 pt.; 144   4.75 pt.; 76 $5\frac{1}{2}$ pt.; 88   17 pt. 272

   b. 5 qt.; 160   1.1 qt.; 35.2 $\frac{4}{5}$ qt.; 25.6   24 qt. 768

   c. 7 gal.; 896   .25 gal.; 32 $6\frac{3}{4}$ gal.; 864   39 gal. 4,992

   d. 2 pt. 8 oz.; 40   5 qt. 12 oz.; 172   1 gal. 48 oz. 176

2. Find the number of <u>pints</u> in:

   a. 3 qt.; 6   5.5 qt.; 11 $4\frac{1}{4}$ qt.; $8\frac{1}{2}$ 18 qt. 36

   b. 6 gal.; 48   16.8 gal.; 134.4 $7\frac{1}{2}$ gal.; 60   40 gal. 320

   c. 5 qt. 1 pt.; 11   7 qt. 3 pt.; 17   9 qt. 2 pt. 20

   d. 48 oz.; 3   72 oz.; 4.5   256 oz.; 16   352 oz. 22

Related Resources: *Practical Applications in Mathematics:* Diag. Test and Prac. Ex., p. 75

3. Find the number of quarts in:

a. 8 gal.; 32   10.25 gal.; 41 $17\frac{1}{2}$ gal.; 70   100 gal. 400

b. 6 gal. 3 qt.; 27   7 gal. 1 qt.; 29   1 gal. 2 qt. 6

c. 14 pt.; 7   6.5 pt.; $3\frac{1}{4}$ 9 pt.; 4.5   27 pt. 13.5

d. 96 oz.; 3   320 oz.; 10   192 oz.; 6   700 oz. $21\frac{7}{8}$

4. Find the number of gallons in:

a. 20 qt.; 5   76 qt.; 19   6 qt.; $1\frac{1}{2}$ 35 qt. $8\frac{3}{4}$

b. 24 pt.; 3   56 pt.; 7   18 pt.; $2\frac{1}{4}$ 31 pt. $3\frac{7}{8}$

c. 256 oz.; 2   768 oz.; 6   192 oz.; $1\frac{1}{2}$ 96 oz. $\frac{3}{4}$

5. a. What part of a pint is:

8 oz.; $\frac{1}{2}$ 12 oz.; $\frac{3}{4}$ 7 oz.; $\frac{7}{16}$ 4 oz.; $\frac{1}{4}$ 6 oz.; $\frac{3}{8}$ 10 oz.? $\frac{5}{8}$

b. What part of a quart is:

1 pt.; $\frac{1}{2}$ $1\frac{1}{2}$ pt.; $\frac{3}{4}$ 16 oz.; $\frac{1}{2}$ $\frac{1}{2}$ pt.; $\frac{1}{4}$ 24 oz.; $\frac{3}{4}$ 4 oz.? $\frac{1}{8}$

c. What part of a gallon is:

2 qt.; $\frac{1}{2}$ 3 qt.; $\frac{3}{4}$ 32 oz.; $\frac{1}{4}$ 1 pt.; $\frac{1}{8}$ 96 oz.; $\frac{3}{4}$ $3\frac{1}{2}$ qt.? $\frac{7}{8}$

6. Add and simplify:

		3 gal. 2 qt.	2 qt. 1 pt. 11 oz.
5 gal. 2 qt.	2 qt. 14 oz.	6 gal. 3 qt.	1 qt. 1 pt. 6 oz.
3 gal. 1 qt.	1 qt. 7 oz.	2 gal. 3 qt.	3 qt. 1 pt. 7 oz.
8 gal. 3 qt.	3 qt. 1 pt. 5 oz.	13 gal.	8 qt. 8 oz.

7. Subtract:

		4 gal.	3 qt. 1 pt. 7 oz.
5 qt. 1 pt.	1 pt. 8 oz.	2 gal. 2 qt.	1 qt. 1 pt. 13 oz.
2 qt. 1 pt. 3 qt.	14 oz. 10 oz.	1 gal. 2 qt.	1 qt. 1 pt. 10 oz.

8. Multiply and simplify:

2 qt. 1 pt.	3 qt. 8 oz.	1 pt. 7 oz.	5 gal. 3 qt.
5 3 gal. 1 pt.	8	4	6
	6 gal. 2 qt.	2 qt. 1 pt. 12 oz.	34 gal. 2 qt.

9. Divide:

2)8 gal. 2 qt.   3)9 pt. 12 oz.   6)8 qt. 2 oz.   9)20 gal. 1 qt.

4 gal. 1 qt.   3 pt. 4 oz.   1 qt. 11 oz.   2 gal. 1 qt.

10. How many ounces of water must be added to:

a. A 6-oz. can of frozen lemon concentrate to make 1 quart of lemonade? 26 oz.

b. A 12-oz. can of frozen grapefruit concentrate to make 3 pints of grapefruit juice? 36 oz.

c. A 6-oz. can of frozen grape concentrate to make $1\frac{1}{2}$ pints of grape juice? 18 oz.

d. A 27-oz. can of frozen orange concentrate to make 1 gallon of orange juice? 101 oz.

In problems 11 and 12, solve each problem and select the letter in red corresponding to your answer.

11. How much is saved by buying one gallon of paint for $13.99 instead of four 1-qt. cans at $4.50 each?

    a. $5.01   b. $4.51   c. $4.01   d. Answer not given

12. Find the difference in price per liquid ounce when a 1-pt. 14-oz. bottle of beverage sells for $.48 and a 10-oz. bottle sells for $.25.

    a. 1¢   b. 0.9¢   c. 0.8¢   d. 1.2¢

EXERCISE

2-9 Dry Measure—Customary

OBJECTIVE: To use customary measures of dry quantities.
Basic: Prob. 1–5; Average: Prob. 1–7; Enriched: Prob. 1–9

PRACTICE PROBLEMS

1. Find the number of pints in:

4 qt.; 8   30 qt.; 60 $3\frac{3}{4}$ qt.; 7.5   9 qt. 1 pt. 19

2. Find the number of quarts in:

a. 2 pk; 16   15 pk.; 120 $2\frac{1}{4}$ pt.; $1\frac{1}{8}$ 4 pk. 2 qt. 34

b. 12 pt.; 6   56 pt.; 28   5 pt.; $2\frac{1}{2}$ 1 pt. $\frac{1}{2}$

3. Find the number of pecks in:

a. 9 bu.; 36   28 bu.; 112 $\frac{3}{4}$ bu.; 3 $7\frac{1}{2}$ bu. 30

b. 2 bu. 1 pk.; 9   17 bu. 3 pk.; 71   40 qt.; 5   22 qt. $2\frac{3}{4}$

4. Find the number of bushels in:

36 pk.; 9   92 pk.; 23   23 pk.; $5\frac{3}{4}$ 108 qt. $3\frac{3}{8}$

5. a. What part of a bushel is: 2 pk.; $\frac{1}{2}$ 1 pk.; $\frac{1}{4}$ 3 pk.; $\frac{3}{4}$ 4 qt.? $\frac{1}{8}$

b. What part of a peck is: 6 qt.; $\frac{3}{4}$ 4 qt.; $\frac{1}{2}$ 5 qt.; $\frac{5}{8}$ 6 pt.? $\frac{3}{8}$

6. Add and simplify:

                                    4 bu. 2 pk.   2 pk. 5 qt.

  7 bu. 1 pk.   6 pk. 1 qt.   2 bu. 3 pk.   6 pk. 1 qt.

  4 bu. 2 pk.   2 pk. 5 qt.   1 bu. 3 pk.   3 pk. 4 qt.

11 bu. 3 pk.   8 pk. 6 qt.   9 bu.   12 pk. 2 qt.

7. Subtract:

  12 pk. 7 qt.   5 bu.   7 pk. 1 qt.   12 bu. 2 pk.

   8 pk. 2 qt.   2 bu. 3 pk.   6 qt.   8 bu. 3 pk.

4 pk. 5 qt.   2 bu. 1 pk.   6 pk. 3 qt.   3 bu. 3 pk.

8. Multiply and simplify:

  5 pk. 2 qt.   6 bu. 3 pk.

       3   6

3 bu. 3 pk. 6 qt.   40 bu. 2 pk.

9. Divide:

4)12 pk. 4 qt.   5)13 bu. 3 pk.

  3 pk. 1 qt.   2 bu. 3 pk.

Related Resources: *Practical Applications in Mathematics:* Diag. Test and Prac. Ex., pp. 75–76

Measure of Area—Customary

OBJECTIVE: To use customary measures of area.
Basic: Prob. 1–8; *Average:* Prob. 1–10; *Enriched:* Prob. 1–12

PRACTICE PROBLEMS

Change:

1. To square inches:

16 sq. ft.; 2,304   5 sq. yd.; 6,480   14.75 sq. ft.; 2,124 $\frac{2}{3}$ sq. yd. 864

2. To square feet:

1,584 sq. in.; 11   12 sq. yd.; 108   40 sq. rd.; 10,890   5 acres 217,800

3. To square yards:

45 sq. ft.; 5   9,072 sq. in.; 7   7 acres; 33,880   3.25 sq. mi. 10,067,200

4. To square rods:

23 acres; 3,680   847 sq. yd.; 28   1,452 sq. yd.; 48   16.875 acres 2,700

5. To acres:

800 sq. rd.; 5   3.6 sq. mi.; 2,304   24,200 sq. yd.; 5   13,068 sq. ft. $\frac{3}{10}$

6. To square miles:

10,880 acres; 17   18,585,600 sq. yd.; 6   14,720 acres 23

7. What part of a square foot is:

72 sq. in.; $\frac{1}{2}$ 126 sq. in.; $\frac{7}{8}$ 81 sq. in.? $\frac{9}{16}$

8. What part of a square yard is:

8 sq. ft.; $\frac{8}{9}$ 6 sq. ft.; $\frac{2}{3}$ 648 sq. in.? $\frac{1}{2}$

9. What part of a square mile is:

240 acres; $\frac{3}{8}$ 61,952 sq. yd.; $\frac{1}{50}$ 278,784 sq. ft.? $\frac{1}{100}$

10. What part of an acre is:

120 sq. rd.; $\frac{3}{4}$ 1,210 sq. yd.; $\frac{1}{4}$ 27,225 sq. ft.? $\frac{5}{8}$

In problems 11 and 12, solve each problem and select the letter in red corresponding to your answer.

11. A kitchen has 108 sq. ft. of floor space. How many square yards of linoleum are needed to cover the entire floor?

a. 10 sq. yd.   b. 16 sq. yd.   c. 12 sq. yd.   d. 18 sq. yd.

12. The air pressure at sea level is 14.7 lb. per sq. in. Find the pressure on 2 sq. ft. of surface at sea level.

a. 3,654 lb.   b. 4,222 lb.   c. 7.33 lb.   d. Answer not given

2-11 Measure of Volume—Customary

OBJECTIVE: To use customary measures of volume.
Basic: Prob. 1–8; *Average:* Prob. 1–10; *Enriched:* Prob. 1–12

PRACTICE PROBLEMS

Change:

1. To <u>cubic inches</u>:

38 cu. ft.; 65,664 $\frac{3}{4}$ cu. yd.; 34,992   7.25 cu. ft. 12,528

2. To <u>cubic inches</u>:

17 cu. yd.; 793,152 $2\frac{1}{2}$ cu. ft.; 4,320   9.375 cu. yd. 437,400

3. To <u>cubic feet</u>:

40 cu. yd.; 1,080   6.5 cu. yd.; $175\frac{1}{2}$ 5,184 cu. in. 3

4. To <u>cubic feet</u>:

29,376 cu. in.; 17   432 cu. in.; $\frac{1}{4}$ $7\frac{2}{3}$ cu. yd. 207

5. To <u>cubic yards</u>:

135 cu. ft.; 5   93,312 cu. in.; 2   63 cu. ft. $2\frac{1}{3}$

6. To <u>cubic yards</u>:

233,280 cu. in.; 5   688.5 cu. ft.; 25.5   793,152 cu. in. 17

7. What <u>part of a cubic foot</u> is:

432 cu. in.; $\frac{1}{4}$ 756 cu. in.; $\frac{7}{16}$ 1,440 cu. in.? $\frac{5}{6}$

8. What <u>part of a cubic yard</u> is:

9 cu. ft.; $\frac{1}{3}$ $6\frac{3}{4}$ cu. ft.; $\frac{1}{4}$ 21 cu. ft.? $\frac{7}{9}$

9. What <u>part of a cubic foot</u> is:

576 cu. in.; $\frac{1}{3}$ 108 cu. in.; $\frac{1}{16}$ 1,152 cu. in.? $\frac{2}{3}$

10. What <u>part of a cubic yard</u> is:

15 cu. ft.; $\frac{5}{9}$ 6 cu. ft.; $\frac{2}{9}$ $13\frac{1}{2}$ cu. ft.? $\frac{1}{2}$

In problems 11 and 12, solve each problem and select the letter in red corresponding to your answer.

11. Lead weighs .41 lb. per cu. in. Find the weight of one cubic foot of lead.

     a. 59.04 lb.   b. 4.92 lb.   c. 4,100 lb.   <u>d.</u> 708.48 lb.

12. A certain truck can hold 2 cu. yd. of dirt. If a cubic foot of dirt weighs 100 lb., what is the weight of a truck load of dirt?

     a. 50 lb.   b. 2,700 lb.   <u>c.</u> 5,400 lb.   d. Answer not given

Volume, Capacity, and Weight Relationships—Customary

OBJECTIVE: To use the customary relationships among volume, capacity, and weight.
Basic: Prob. 1–9; *Average:* Prob. 1–12; *Enriched:* Prob. 1–13

PRACTICE PROBLEMS

1. Find the capacity in gallons equal to a volume of 3,234 cu. in.
14 gal.

2. Find the volume in cubic inches equal to a capacity of 20 gallons.
4,620 cu. in.

3. Find the capacity in gallons equal to a volume of 52 cu. ft.
390 gal.

4. Find the volume in cubic feet equal to a capacity of 195 gallons.
26 cu. ft.

5. What is the weight in pounds of 48 cu. ft. of sea water?
3,072 lb.

6. What volume in cubic feet do 8,320 lb. of sea water occupy?
130 cu. ft.

7. What is the weight in pounds of 250 cu. ft. of fresh water?
15,625 lb.

8. What volume in cubic feet do 4,750 lb. of fresh water occupy?
76 cu. ft.

9. Find the capacity in bushels equal to a volume of 100 cu. ft.
80 bu.

10. Find the volume in cubic feet equal to a capacity of 92 bushels.
115 cu. ft.

11. Find the equivalent:

a. Capacity in gallons: 1,848 cu. in.; $_8$ 21 cu. ft.; $_{157\frac{1}{2}}$ 4.9 cu. ft. $_{36\frac{3}{4}}$

b. Capacity in bushels: 80 cu. ft.; $_{64}$ $12\frac{1}{2}$ cu. ft.; $_{10}$ 35 cu. yd. $_{756}$

c. Volume in cubic feet: 45 gal.; $_6$ 900 gal.; $_{120}$ 206 bu. $_{257\frac{1}{2}}$

d. Volume in cubic inches: 7 gal.; $_{1,617}$ 15.8 gal.; $_{3,649\frac{4}{5}}$ $4\frac{1}{2}$ gal. $_{1,039\frac{1}{2}}$

e. Volume in cubic feet:
4,800 lb. of sea water; $_{75}$ 6,000 lb. of fresh water $_{96}$

f. Weight in pounds:
75 cu. ft. of fresh water;   120 cu. ft. of sea water
$4,687\frac{1}{2}$ 7,680

12. A submarine displaces 1,400 tons (1 long ton = 2,240 lb.) of sea water when submerged. How many cubic feet of sea water does it displace? 49,000 cu. ft.

13. The normal flow of water over Niagara Falls is 500,000 tons (1 short ton = 2,000 lb.) of fresh water a minute. Find the rate of flow in gallons per minute. 120,000,000 gal. per. min.

Related Resources: *Practical Applications in Mathematics:* Diag. Test and Prac. Ex., pp. 79–80 **313** For a Test on the Customary Units of Measure see p. T47 or *Testing Program,* p. 13.

OTHER BASIC MEASURES

OBJECTIVE: To use measures of time.

See pages 285–287 for directions to change units of time and page 304 for directions to compute with these units. The table of measure is found on page 623.

Basic: Prob. 1–18; *Average:* Prob. 1–22; *Enriched:* Prob. 1–25

PRACTICE PROBLEMS

1. Find the number of <u>seconds</u> in:

a. 24 min.; 1,440   8.3 min.; 498 $16\frac{1}{4}$ min.; 975   52 min. 3,120

b. 10 hr.; 36,000   6.25 hr.; 22,500 $1\frac{1}{2}$ hr.; 5,400   17 hr. 61,200

c. 5 min. 20 sec.; 320   14 min. 32 sec.; 872   56 min. 15 sec.; 3,375

2 hr. 15 min. 6 sec. 8,106

2. Find the number of <u>minutes</u> in:

a. 5 hr.; 300   1.8 hr.; 108 $\frac{3}{4}$ hr.; 45   21 hr. 1,260

b. 1 hr. 30 min.; 90   4 hr. 29 min.; 269   13 hr. 52 min.; 832   23 hr. 8 min. 1,388

c. 540 sec.; 9   1,800 sec.; 30   225 sec.; $3\frac{3}{4}$ 390 sec. $6\frac{1}{2}$

3. Find the number of <u>hours</u> in:

a. 4 days; 96 $8\frac{1}{6}$ days; 196 $3\frac{5}{8}$ days; 87   30 days 720

b. 1 day 6 hr.; 30   5 days 13 hr.; 133   11 days 9 hr.; 273   8 days 21 hr. 213

c. 480 min.; 8   3,180 min.; 53   100 min.; $1\frac{2}{3}$ 10,800 sec. 3

4. Find the number of <u>days</u> in:

a. 8 wk.; 56   4.5 wk.; $31\frac{1}{2}$ $6\frac{3}{7}$ wk.; 45   26 wk. 182

*__**b.** 2 yr.; 730   3.4 yr.; 1,241 $1\frac{4}{5}$ yr.; 657   12 yr. 4,383

c. 4 wk. 6 days; 34   20 wk. 3 days; 143   3 yr. 200 days;   1 yr. 95 days 460

d. 48 hr.; 2   192 hr.; 8   30 hr.; $\begin{smallmatrix}1,295\\1\frac{1}{4}\end{smallmatrix}$ 76 hr. $3\frac{1}{6}$

---

*Where 4 or more years are involved, use 366 days for every fourth year.

Related Resources: *Practical Applications in Mathematics:* Diag. Test and Prac. Ex., p. 77

5. Find the number of <u>months</u> in:

a. 7 yr.; 84   5.75 yr.; 69 $2\frac{1}{2}$ yr.; 30   10 yr. 120

b. 4 yr. 4 mo.; 52   1 yr. 6 mo.; 18   3 yr. 10 mo.; 46   12 yr. 5 mo. 149

6. Find the number of <u>weeks</u> in:

3 yr.; 156 $5\frac{3}{4}$ yr.; 299   2 yr. 8 wk.; 112   1 yr. 39 wk. 91

7. Find the number of <u>years</u> in:

a. 84 mo.; 7   35 mo.; $2\frac{11}{12}$ 156 mo.; 13   300 mo. 25

b. 156 wk.; 3   260 wk.; 5   39 wk.; $\frac{3}{4}$ 208 wk. 4

c. 1,095 days; 3   219 days; $\frac{3}{5}$ 511 days; $1\frac{2}{5}$ 2,555 days 7

8. a. What <u>part of a minute</u> is: 45 sec.; $\frac{3}{4}$ 18 sec.; $\frac{3}{10}$ 5 sec.; $\frac{1}{12}$ 36 sec.? $\frac{3}{5}$

b. What <u>part of an hour</u> is: 40 min.; $\frac{2}{3}$ 15 min.; $\frac{1}{4}$ 30 min.; $\frac{1}{2}$ 54 sec.? $\frac{3}{200}$

c. What <u>part of a day</u> is: 4 hr.; $\frac{1}{6}$ 16 hr.; $\frac{2}{3}$ 20 hr.; $\frac{5}{6}$ 15 hr.? $\frac{5}{8}$

d. What <u>part of a week</u> is: 2 days; $\frac{2}{7}$ 6 days; $\frac{6}{7}$ $3\frac{1}{2}$ days; $\frac{1}{2}$ 5 days? $\frac{5}{7}$

e. What <u>part of a year</u> is: 9 mo.; $\frac{3}{4}$ 13 wk.; $\frac{1}{4}$ 146 days; $\frac{2}{5}$ 4 mo.? $\frac{1}{3}$

9. Add and simplify:

                           5 yr. 6 mo.   8 hr. 19 min. 43 sec.

17 hr. 22 min.   1 wk. 5 days   4 yr. 9 mo.   6 hr. 15 min. 29 sec.

23 hr.   5 hr. 38 min.   7 wk. 4 days   9 yr. 8 mo.   5 hr. 24 min. 48 sec.

                   9 wk. 2 days   19 yr. 11 mo.   20 hr.

10. Subtract:

4 days 8 hr.   10 yr. 4 mo.   8 min.   9 hr. 28 min. 35 sec.

2 days 15 hr.   7 mo.   5 min. 49 sec.   6 hr. 40 min. 51 sec.

1 day 17 hr.   9 yr. 9 mo.   2 min. 11 sec.   2 hr. 47 min. 44 sec.

11. Multiply and simplify:

5 yr. 3 mo.   2 days 7 hr.   3 hr. 20 min.   1 hr. 42 min. 10 sec.

15 yr. 9 mo.   3   5   6   15

                  11 days 11 hr.   20 hr.   25 hr. 32 min. 30 sec.

12. Divide: 4)12 yr. 8 mo. 3 yr. 2 mo.   6)10 days 2 hr. 1 day 16 hr. 20 min.

3)21 hr. 34 min. 7 hr. 11 min. 20 sec. 15)8 hr. 17 min. 30 sec. 33 min. 20 sec.

13. Which is greater:

a. A <u>half-hour</u> or 25 minutes?

b. 8 months or <u>three-quarters of a year?</u>

14. Which is less:

a. A third of a minute or <u>16 seconds?</u>   **b.** <u>10 hours</u> or a half-day?

15. Write as A.M. or P.M. time:

a. 9 o'clock at night. 9 P.M.

b. 11 o'clock in the morning. 11 A.M.

c. Half past two in the afternoon. 2:30 P.M.

d. Quarter after six in the morning. 6:15 A.M.

e. Twenty-five minutes to ten at night. 9:35 P.M.

16. On a clock, how long does it take the minute hand to move from the figure:

a. 4 to 5? 5 min.   **b.** 2 to 7? 25 min.   **c.** 1 to 9? 40 min.   **d.** 8 to 3? 35 min.

17. Find the length of time:

a. From 2 A.M. to 6 P.M. the same day. 16 hr.

b. From 10:45 A.M. to 7:15 P.M. the same day. $8\frac{1}{2}$ hr.

c. From 12:26 P.M. one day to 3:10 A.M. the following day. 14 hr. 44 min.

18. Look at your calendar. On what date this month does: Answers will vary.

a. the second Monday fall?   **c.** the third Saturday fall?

b. the fourth Thursday fall?   **d.** the first Wednesday fall?

19. How many days are in the month of:

a. May? 31   **c.** April? 30   **e.** November? 30   **g.** February? 28

b. October? 31   **d.** January? 31   **f.** June? 30   **h.** December? 31

20. Find the exact number of days from:

a. March 15 to May 15. 60   **d.** June 12 of one year to January 8

b. April 3 to October 16. 195   of the following year. 208

c. July 16 to December 9. 145

21. How old will Lisa be on her next birthday if she was born on June 3, 1972? Answers will vary.

In problems 22 through 25, solve each problem and select the letter in red corresponding to your answer.

22. How far can an airplane fly in 4 hr. 48 min. if its average ground speed is 900 kilometers per hour?

a. 3,900 km   c. 4,320 km

b. 4,560 km   d. Answer not given

23. School starts at 8:45 A.M. How long is the school day if the last class period ends at 3:20 P.M.?

a. 6 hr. 25 min.   c. 6 hr. 35 min.

b. 5 hr. 45 min.   d. 5 hr. 55 min.

24. Miss Gomez parked her car at the airport. When she entered the parking lot, her ticket read 30 NOV 7:35 A.M. How long was her car in the parking lot if she picked it up on December 2 at 6:10 P.M.?

a. 3 days 4 hr. 35 min.   c. 2 days 1 hr. 25 min.

b. 2 days 10 hr. 35 min.   d. Answer not given

25. A bus leaves the station every 17 minutes for the center city. If you missed the 10:56 A.M. bus, what time is the next scheduled bus?

a. 11:12 A.M.   b. 11:10 A.M.   c. 11:11 A.M.   d. 11:13 A.M.

Measure of Angles and Arcs

**OBJECTIVE: To use measures
of angles and arcs.**

See pages 285–287 for directions to change units measuring angles and arcs and page 304 for directions to compute with these units. The table of measure is found on page 623.

Basic: Prob. 1–8; *Average:* Prob. 1–10; *Enriched:* Prob. 1–12

PRACTICE PROBLEMS

Change:

1. To minutes: $45°$; 2,700' $32°6'$; 1,926' $2,940''$; 49' $13°$ 780'

2. To <u>seconds</u>: $38'$; 2,280 $5°$; 18,000 $12' 45''$; 765 $2° 18' 50''$ 8,330

3. To degrees: $180'$; 3° $4,320'$; 72° $10,800''$; 3° $36,000''$ 10°

4. Add and simplify:

$14° 26'$	$63° 27' 40''$	$44° 21' 19''$	$73° 35' 45''$
$52° 19'$	$39° 18' 20''$	$27° 53' 24''$	$32° 18' 53''$
		$102° 39' 43''$	$65° \ 5' 22''$
$\overline{66° 45'}$	$\overline{102° 46'}$	$\overline{174° 54' 26''}$	$\overline{171°}$

5. Subtract:

$120° 48'$	$90°$	$78° \ 6' 37''$	$180°$
$\ 67° 19'$	$38° 45'$	$56° 15' 20''$	$95° 24' 31''$
$\overline{53° 29'}$	$\overline{51° 15'}$	$\overline{21° 51' 17''}$	$\overline{84° 35' 29''}$

6. Multiply and simplify:

$24° 13'$	$12° 32'$	$28° 15' 24''$	$13° 49' 30''$
$\ \ \ \ 4$	$\ \ \ \ 6$	$\ \ \ \ \ \ \ 5$	$\ \ \ \ \ \ \ 15$
$\overline{96° 52'}$	$\overline{75° 12'}$	$\overline{141° 17'}$	$\overline{207° 22' 30''}$

7. Divide: 35'   7° 23' 27''   8° 34' 46''

$6\overline{)42° 18'}$ 7°3' $15\overline{)8° 45'}$ $4\overline{)29° 33' 48''}$ $15\overline{)128° 41' 30''}$

8. An angle measuring $112°$ is to be divided into 4 equal angles. What is the measure of each angle? 28°

9. Find the sum of measures of two angles when one angle measures $27°$ and the other measures $145°$. 172°

10. Angle C is 3 times the size of angle D. Angle D measures $19° 35'$. What is the measure of angle C? 58° 45'

11. One angle measures $50°$. A second angle measures $37° 54'$. How much greater is the measure of the first angle? 12°6'

12. A circle, measuring 360 arc degrees, is divided into 5 equal arcs. What is the measure of each arc? 72°

2-15 Rates of Speed

OBJECTIVE: To change given units to required units.

We change the given units to the required units (in the metric model: kilometer to 1,000 meters and hour to 3,600 seconds; in the customary model: mile to 5,280 feet and hour to 3,600 seconds), then we perform the necessary arithmetical operations. A knot is one (1) nautical mile per hour. A nautical mile is approximately equivalent to 6,080 ft. or 1.15 statute miles or 1.85 km. A kilometer is approximately equal to 0.54 nautical mile. See tables of measure on pages 622–623.

Note that in the metric system rates of speed are abbreviated in the following ways: meters per second (m/s); kilometers per hour (km/h); centimeters per second (cm/s); millimeters per minute (mm/min); and so on.

Metric	**Customary**
Change 54 km/h to m/s:	Change 60 m.p.h. to ft. per sec.:
$54 \text{ km/h} = \dfrac{54 \text{ km}}{1 \text{ h}}$	$60 \text{ m.p.h.} = \dfrac{60 \text{ mi.}}{1 \text{ hr.}}$
$\dfrac{54 \times 1,000}{3,600} = 15 \text{ m/s}$	$\dfrac{60 \times 5,280}{3,600} = 88 \text{ ft. per sec.}$
Answer: 15 m/s	*Answer:* 88 ft. per sec.

Basic: Prob. 1–7; *Average:* Prob. 1–9; *Enriched:* Prob. 1–10

PRACTICE PROBLEMS

Change, finding answers to the nearest hundredth wherever necessary:

1. To <u>meters per second</u>: 400 cm/s; 4 150 km/h; $\overset{41.67}{}$ 480 mm/min .01
2. To <u>meters per minute</u>: 75 mm/s; 4.5 360 cm/min; 3.6 60 km/h 1,000
3. To <u>centimeters per second</u>: 90 km/h; 85 m/min; 540 mm/s
4. To <u>kilometers per hour</u>: 30 m/s; 108 720 cm/s; 25.92 115 knots 212.75
5. To <u>knots</u>: 70 km/h; 37.84 600 m/s; 1,167.57 250 km/h 135.14
6. To <u>feet per second</u>: 195 stat. m.p.h.; 400 stat. m.p.h.; 29 knots
7. To <u>statute miles per hour</u>: 352 ft. per sec.; 240 90 knots; 103.6 1,800 ft. per min. 20.45
8. To <u>nautical miles per hour</u>: 16.4 knots; 16.4 50 stat. m.p.h.; 43.4 100 ft. per sec. 59.21
9. To <u>knots</u>: 69.2 naut. m.p.h.; 86 stat. m.p.h.; 132 ft. per sec.
10. To <u>feet per minute</u>: 88 ft. per sec.; 60 stat. m.p.h.; 145 knots
   5,280   5,280 $14,693\frac{1}{3}$

Related Resources: *Practical Applications in Mathematics:* Diag. Test and Prac. Ex. p. 82

318

3. 2,500; 141.67; 54
6. 286; $586\frac{2}{3}$; 48.98
9. 69.2; 74.68; 78.16

For a Class Activity see p. T34.

Temperature 2-16

OBJECTIVE: To change from one unit of measure of temperature to another among the three scales: Celsius, Fahrenheit, and Kelvin.

VOC: Celsius scale   degree
Fahrenheit scale   kelvin
Kelvin scale

The three thermometer scales that are used to measure temperature are the Celsius, the Fahrenheit, and the Kelvin. The Celsius scale (formerly called centigrade) is based on 100 divisions, each called a degree. On the Celsius scale the freezing point of water is indicated as 0° and the boiling point as 100°. On the Fahrenheit scale the freezing point and the boiling point of water are indicated as 32° and 212° respectively for a 180° interval. It now appears that the Celsius scale will soon replace the Fahrenheit scale in the United States.

The Kelvin temperature scale, used in SI measurement, is related to the Celsius scale. One degree Celsius is exactly equal to one *kelvin* (the name used to mean degree Kelvin). The reading of a specific temperature on the Kelvin scale is approximately 273 kelvins more than its reading on the Celsius scale.

To change a Celsius temperature reading to a kelvin reading, add 273° to the Celsius reading or use the formula:

$$K = C + 273$$

To change a kelvin temperature reading to a Celsius reading, subtract 273° from the kelvin reading or use the formula:

$$C = K - 273$$

To change a Celsius temperature reading to a Fahrenheit reading, add 32° to nine-fifths of the Celsius temperature reading or use the formula:

$$F = \tfrac{9}{5}C + 32$$

To change a Fahrenheit temperature reading to a Celsius reading, subtract 32° from the Fahrenheit temperature reading and take five-ninths of this answer or use the formula:

$$C = \tfrac{5}{9}(F - 32)$$

PRACTICE PROBLEMS

1. Change each of the following Celsius temperature readings to a corresponding kelvin reading:

    a. 50°C 323 K **b.** 96°C 369 K **c.** 8°C 281 K **d.** 0°C 273 K **e.** 160°C 433 K

2. Change each of the following kelvin temperature readings to a corresponding Celsius reading:

    a. 280 K 7°C **b.** 346 K 73°C **c.** 405 K 132°C **d.** 318 K 45°C **e.** 371 K 98°C

3. Change each of the following Fahrenheit readings to a corresponding Celsius temperature reading:

    a. 86°F 30°C **b.** 32°F 0°C **c.** 113°F 45°C **d.** 212°F 100°C **e.** 60°F 15.56°C

4. Change each of the following Celsius temperature readings to a corresponding Fahrenheit temperature reading:

    a. 60°C 140°F **b.** 0°C 32°F **c.** 100°C 212°F **d.** 85°C 185°F **e.** 18°C 64.4°F

5. If the classroom thermostat is set at 68°F, what is the corresponding temperature reading on the Celsius scale? 20°C

For a Computer Activity see p. T40.

EXERCISE

2-17 Twenty-Four Hour Clock

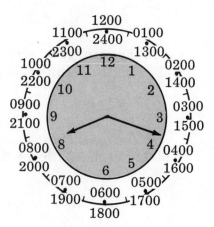

OBJECTIVE:

To tell time on a 24-hour clock, a 4-digit numeral is used, the first two digits indicating the hour and the last two the minutes. 1 P.M. is thought of as the 13th hour, 2 P.M. the 14th hour, etc. Thus 1200 is added to the given digits when expressing time from 1 P.M. to 12 midnight. The numeral 12 representing the hour is replaced by two zeros when expressing the time from midnight to 1 A.M.

6:42 A.M. is expressed as 0642	0500 indicates 5:00 A.M.
7:13 P.M. is expressed as 1913	1425 indicates 2:25 P.M.
3:00 P.M. is expressed as 1500	1018 indicates 10:18 A.M.
12:47 A.M. is expressed as 0047	2131 indicates 9:31 P.M.

Related Resources: *Practical Applications in Mathematics:* Problems, p. 211

PRACTICE PROBLEMS

1. Using the 24-hour system, express each of the following:

a. 11:26 A.M. 1126 **f.** 2:00 A.M. 0200 **k.** 10:37 P.M. 2237 **p.** 12:37 P.M. 1237

b. 10:09 A.M. 1009 **g.** 7:00 P.M. 1900 **l.** 11:02 P.M. 2302 **q.** 1:26 A.M. 0126

c. 9:45 A.M. 0945 **h.** 10:00 P.M. 2200 **m.** 12:18 A.M. 0018 **r.** 5:35 P.M. 1735

d. 3:38 A.M. 0338 **i.** 4:29 P.M. 1629 **n.** 12:54 A.M. 0054 **s.** 11:40 A.M. 1140

e. 6:00 A.M. 0600 **j.** 9:14 P.M. 2114 **o.** 12:23 P.M. 1223 **t.** 8:25 P.M. 2025

2. Express each of the following in A.M. or P.M. time:

a. 0648 6:48 A.M. **f.** 0400 4:00 A.M. **k.** 2342 11:42 P.M. **p.** 1258 12:58 P.M.

b. 0905 9:05 A.M. **g.** 1300 1:00 P.M. **l.** 2230 10:30 P.M. **q.** 0539 5:39 A.M.

c. 1023 10:23 A.M. **h.** 2100 9:00 P.M. **m.** 0025 12:25 P.M. **r.** 2206 10:06 P.M.

d. 1152 11:52 A.M. **i.** 1952 7:52 P.M. **n.** 0040 12:40 P.M. **s.** 0007 12:07 A.M.

e. 1000 10:00 A.M. **j.** 2315 11:15 P.M. **o.** 1233 12:33 P.M. **t.** 2350 11:50 P.M.

3. Find the difference in time between:

a. 0415 and 0842 4 hr. 27 min.   **f.** 0153 and 0300 1 hr. 7 min.

b. 0031 and 1058 10 hr. 27 min.   **g.** 0641 and 1135 4 hr. 54 min.

c. 0500 and 1625 11 hr. 25 min.   **h.** 1028 and 1902 8 hr. 34 min.

d. 1307 and 2219 9 hr. 12 min.   **i.** 0919 and 2115 11 hr. 56 min.

e. 0922 and 2050 11 hr. 28 min.   **j.** 0058 and 2347 22 hr. 49 min.

4. An airplane took off from Seattle at 0950 and arrived in San Diego at 1315. How long did it take to make the trip? 3 hr. 25 min.

5. A ship left port at 0530 and arrived at its port of destination at 1305 the next day. Find the time it took the ship to reach its destination. 31 hr. 35 min.

EXERCISE

Time Zones | **2-18**

OBJECTIVE: To use the time zones.

There are four standard time belts in the United States, excluding Alaska and Hawaii: Eastern (EST), Central (CST), Mountain (MST), and Pacific (PST). Central time is one hour earlier than Eastern time, Mountain time is one hour earlier than Central time, and Pacific time is one hour earlier than Mountain time. The meridians at 75°, 90°, 105°, and 120° west longitude are used to determine the time in these zones.

Alaska has four time zones determined by the meridians at 120°, 135°, 150°, and 165° west longitude. Hawaii uses the time zone of the meridian at 150° west longitude. Parts of Canada east of Maine are in another time zone, the Atlantic Standard Time Zone, which is one hour later than Eastern Standard Time.

Related Resources: *Practical Applications in Mathematics:* Problems, p. 211

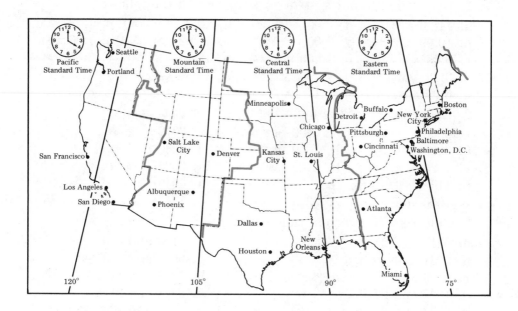

PRACTICE PROBLEMS

1. In what time zone is your city located? Answers will vary.

2. If it is 11 A.M. in the Mountain Time Zone, what time is it in:

a. Pacific Zone? 10:00 A.M. **b.** Eastern Zone? 1:00 P.M. **c.** Central Zone? 12:00 noon

3. When it is 8 P.M. in Dallas, what time is it in:

a. Phoenix? 7 P.M.   **e.** Atlanta? 9 P.M.   **i.** Detroit? 9 P.M.

b. Washington, D.C.? 9 P.M. **f.** Kansas City? 8 P.M.   **j.** New York? 9 P.M.

c. St. Louis? 8 P.M.   **g.** Denver? 7 P.M.   **k.** Houston? 8 P.M.

d. San Francisco? 6 P.M.   **h.** Seattle? 6 P.M.   **l.** Your city? varies

4. If a nationwide program is telecast at 10 P.M. from Los Angeles, what time is it seen in:

12:00 midnight

a. Pittsburgh? 1:00 A.M.   **c.** Minneapolis?   **e.** Salt Lake City? 11:00 P.M.

b. Albuquerque? 11:00 A.M.**d.** Miami? 1:00 A.M.   **f.** Baltimore? 1:00 A.M.

5. How long did it take an airplane to fly from Boston to New Orleans if it left Boston at 9:30 P.M. (EST) and arrived in New Orleans at 12:08 A.M. (CST) the next day? 3 hr. 38 min.

6. If, exactly at noon, you place a telephone call in Buffalo for San Francisco, what time is it in San Francisco? 9:00 A.M.

For a Test on Other Basic Measures see p. T47 or *Testing Program,* p. 14.

MEASUREMENT

Precision and Accuracy in Measurement

EXERCISE
2-19

VOC.: precision relative error accuracy tolerance

Measurement is never exact, it is approximate. The same length measured on each of the following scales shows a measurement of:

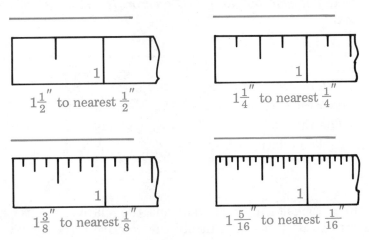

$1\frac{1}{2}''$ to nearest $\frac{1}{2}''$

$1\frac{1}{4}''$ to nearest $\frac{1}{4}''$

$1\frac{3}{8}''$ to nearest $\frac{1}{8}''$

$1\frac{5}{16}''$ to nearest $\frac{1}{16}''$

Observe that $1\frac{1}{2}''$ is precise to the nearest $\frac{1}{2}''$; $1\frac{1}{4}''$ is precise to the nearest $\frac{1}{4}''$; $1\frac{3}{8}''$ is precise to the nearest $\frac{1}{8}''$; $1\frac{5}{16}''$ is precise to the nearest $\frac{1}{16}''$. *The smaller the unit, the more precise is the measurement.* The measurement $1\frac{5}{16}''$ is the most precise of the above measurements.

Precision is the closeness to the true measurement. It is determined by the smallest unit of measure used to make the measurement.

The precision of a measurement named by a numeral for a whole number ending in zeros is indicated by an underlined zero as shown below. (Also see page 326.)

16,0$\underline{00}$ kilometers is precise to the nearest kilometer.
16,0$\underline{0}$0 kilometers is precise to the nearest 10 kilometers.
16,$\underline{0}$00 miles is precise to the nearest 100 miles.
16,000 kilometers is precise to the nearest 1,000 kilometers.
16,000.0 miles is precise to the nearest tenth of a mile.

In measuring the length $1\frac{5}{16}''$ on page 323, the unit used was $\frac{1}{16}$ inch. The real length is between $1\frac{9}{32}''$ and $1\frac{11}{32}''$. The greatest possible error between the measurement $1\frac{5}{16}''$ and the true measurement can only be $\frac{1}{32}''$ since *the greatest possible error is one-half of the unit used to measure*.

The relative error is the ratio of the greatest possible error to the measurement. The accuracy of the measurement is determined by the relative error. *The smaller the relative error, the more accurate is the measurement.*

The measurement $2\frac{3}{8}''$ has the greatest possible error of $\frac{1}{16}''$. $\frac{1}{16}$ to $2\frac{3}{8}$
The ratio of $\frac{1}{16}''$ to $2\frac{3}{8}''$ is $\frac{1}{16} \div 2\frac{3}{8}$
$\frac{1}{38}$. The relative error is $\frac{1}{38}$. $\frac{1}{16} \times \frac{8}{19} = \frac{1}{38}$

Sometimes, to make quick comparisons, the relative error is expressed as a percent. To do this we multiply the relative error by 100 (or see Exercise 1-40).

The allowance for error in measurement is called tolerance. A required measurement of 7.3 inches with a tolerance of .05 inch allowed in both directions would be indicated as $7.3'' \pm .05''$ and would represent any measurement from 7.25 inches to 7.35 inches inclusive.

PRACTICE PROBLEMS

1. To what unit of measure are these measurements precise?

a. 0.09 cm .1 mm   **e.** 16.92 kg dam   **i.** $7\frac{1}{2}$ pt. $\frac{1}{2}$ pt.   **m.** 0.2 mg .1 mg

b. 35.6 km hm   **f.** 8,006 L liter   **j.** $8\frac{3}{4}$ lb. $\frac{1}{4}$ lb.   **n.** 5 kg 7 g gram

c. 0.0005 g .1 mg   **g.** $2\frac{13}{16}$ in. $\frac{1}{16}$ in.   **k.** 6.8 sec. $\frac{1}{10}$ sec. **o.** 10 hr. 14 min. min.

.001 mm **d.** 0.030 mm   **h.** $9\frac{3}{5}$ hr. $\frac{1}{5}$ hr.   **l.** 9.54 m cm   **p.** 11 ft. $8\frac{1}{4}$ in. $\frac{1}{4}$ in.

2. Find the precision (unit of measure) of each measurement:

100 tons centiliter
a. 649 m meter   **c.** 50,0<u>0</u>0 tons   **e.** 1,834 cL 10,000 mi.**g.** 34,000 km 1,000 km

kilogram **b.** 8,5<u>0</u>0 kg   **d.** 720 sec. 10 sec.   **f.** 2,9<u>0</u>0,000 mi.   **h.** 470,0<u>0</u>0 ft. 10 ft.

3. Which measurement in each of the following is more precise?

a. 9.6 m or 18.0 m? equally precise   **d.** $8\frac{3}{4}$ mi. or $5\frac{2}{3}$ yd.?

b. $7\frac{7}{16}$ in. or $3\frac{17}{32}$ in.?   **e.** $2\frac{5}{6}$ hr. or <u>10 hr. 8 min.</u>?

c. <u>3.9 g</u> or 12.15 kg?   **f.** $5\frac{1}{2}$ ft. or <u>3 ft. 7 in.</u>?

For ex. 4–10
see Extended
Answers,
pp. T78–T79.
4. Find the greatest possible error in each of these measurements:

a. 65 kg   **f.** $9\frac{1}{4}$ hr.   **k.** 760 m   **p.** 6 m 9 cm

b. 4.6 L   **g.** $5\frac{7}{8}$ lb.   **l.** 40,<u>0</u>00 tons   **q.** 2 ft. 11 in.

c. 2.057 cm   **h.** 867 gal.   **m.** 270,0<u>0</u>0 km   **r.** 1 lb. 4 oz.

d. 0.08 mm   **i.** 9,526.8 ft.   **n.** 5,8<u>0</u>0,000 mi.   **s.** 2 hr. 19 min.

e. 473 km   **j.** $1\frac{25}{32}$ in.   **o.** 0.039 mg   **t.** 2,06<u>0</u>,000 mm

5. Find the relative error in each measurement in problem 4.

6. Find the relative error as a percent for each measurement:

a. 0.9 m   **b.** 50 km   **c.** $4\frac{2}{3}$ yd.   **d.** $9\frac{3}{8}$ in.   **e.** 12 mm

7. Which measurement in each of the following is more accurate:

a. 28.1 cm or 6.52 cm?   **d.** 75,0<u>0</u>0 km or 0.9 mm?

b. 30 mm or 0.003 mm?   **e.** 58 ft. or 5.8 ft.?

c. 0.06 m or 250 km?   **f.** $3\frac{1}{2}$ in. or $10\frac{7}{8}$ mi.?

8. Which is more precise: 65,000 m or 0.004 mm? Which is more accurate?

9. Arrange the following measurements in order of precision, most precise first; and also in order of accuracy, most accurate first:

a. 0.09 mm   90 mm   0.9 mm   9 mm   0.009 mm

b. 63.4 ft.   500 mi.   9.375 yd.   7.26 in.   83.0 mi.

10. What range of measurements does each of these represent?

a. 8.16 mm $\pm$ .05 mm   **d.** 3.07″ $\pm$ .075″

b. $2\frac{3}{4}$″ $\pm \frac{1}{8}$″   **e.** 2 mm $\pm$ .001 mm

c. 4.8 cm $\pm$ .025 cm   **f.** $9\frac{1}{2}$″ $\pm \frac{3}{32}$″

2-20 Significant Digits

Digits are significant in an approximate number when they indicate the precision which is determined by the value of the place of the last significant digit on the right.

In 64.75, the significant digit 5 indicates precision to the nearest hundredth.

In 83,0<u>0</u>0, the underlined 0 indicates precision to the nearest ten.

The digits described below in items (1) to (4) inclusive are significant; those described in items (5) and (6) are not significant except as noted in (5).

(1) All non-zero digits are significant.

(2) Zeros located between non-zero digits are significant.

(3) When a decimal or mixed decimal ends in zeros, these zeros are significant.

(4) When numbers are expressed in scientific notation, all the digits in the first factor are significant.

(5) When a number ends in zeros, these zeros are not significant unless they are specified as being significant or indicated as significant by a line drawn under them.

(6) In a decimal fraction (when the number is between 0 and 1) the zeros immediately following the decimal point are not significant.

---

(1) 684 has 3 significant digits (6, 8, 4).

(2) 5,006 has 4 significant digits (5, 0, 0, 6).

(3) 86.190 has 5 significant digits (8, 6, 1, 9, 0).

(4) In 7.32×10^9, digits 7, 3, and 2 are significant.

(5) 51,000 has 2 significant digits (5, 1).
51,0<u>0</u>0 has 3 significant digits (5, 1, 0).
51,00<u>0</u> has 5 significant digits (5, 1, 0, 0, 0).
4,030 has 3 significant digits (4, 0, 3).

(6) .00296 has 3 significant digits (2, 9, 6) but 2.0096 has 5 significant digits (2, 0, 0, 9, 6). .007<u>0</u> has 2 significant digits (7, 0).

PRACTICE PROBLEMS

Determine the number of significant digits in each of the following. What are they in each case?

1. 28 two; 2, 8

2. 600 one; 6

3. 4,271 four; 4, 2, 7, 1

4. 37,000 two; 3, 7

5. 9 one; 9

6. .08 one; 8

7. .0095 two; 9, 5

8. .013 two; 1, 3

9. .0130 three; 1, 3, 0

10. 9,050,000 three; 9, 0, 5

11. 23.05 four; 2, 3, 0, 5

12. .0001000 four; 1, 0, 0, 0

13. 87,000 four; 8, 7, 0, 0

14. 350,000 three; 3, 5, 0

15. 6.3×10^5 two; 6, 3

16. 8.49×10^7 three; 8, 4, 9

17. 2,500 two; 2, 5

18. .0007 one; 7

19. 6,810 three; 6, 8, 1

20. 51.0 three; 5, 1, 0

21. 3.1416

22. 7,060

23. 95,280

24. 16,003

25. 40,000

26. .000000015

27. 9.00000027

28. 8,259,078

29. 4.0060

30. 203,000

21. five; 3, 1, 4, 1, 6 22. three; 7, 0, 6 23. four; 9, 5, 2, 8 24. five; 1, 6, 0, 0, 3 25. three; 4, 0, 0 26. two; 1, 5 27. nine; 9, 0, 0, 0, 0, 0, 0, 2, 7 28. seven; 8, 2, 5, 9, 0, 7, 8 29. five; 4, 0, 0, 6, 0 30. six; 2, 0, 3, 0, 0, 0

EXERCISE
Approximate Numbers 2-21

OBJECTIVE: To compute with approximate numbers.

Numbers used to count are exact. Numbers used to measure are approximate. An approximate number is a number that is almost equal to the true number.

Since measurement is approximate, we may be required to compute with approximate numbers arising through measurement. The result of a computation with approximate numbers cannot be more accurate than the least accurate approximate number involved in the computation. There are several methods of computing approximate numbers, one of which is shown here and on the next page.

Addition

To add approximate numbers, add as you usually do, then round the sum using the unit of the least precise addend.

Add:
1.7
2.564
1.42
$5.684 = 5.7$
Answer: 5.7

Subtraction

To subtract approximate numbers, subtract as you usually do, then round the difference using the unit of the less precise of the given numbers.

Multiplication

To multiply approximate numbers, multiply as you usually do, then round the product so that it contains the same number of significant digits as the factor having the smaller number of significant digits.

Division

To divide approximate numbers, divide as you usually do, then round the quotient so that it contains the same number of significant digits as there are in either the dividend or divisor, whichever is less.

Subtract:

8.543
3.82
———
$4.723 = 4.72$

Answer: 4.72

Multiply:

3.49
2.5
———
1745
698
———
$8.725 = 8.7$

Answer: 8.7

Divide:

$$\begin{array}{r} 4.08 = 4.1 \\ 1.6_\wedge \overline{)6.5_\wedge 34} \\ 6\ 4 \\ \hline 1\ 34 \\ 1\ 28 \\ \hline 6 \end{array}$$

Answer: 4.1

Basic: Prob. 1–2; *Average:* Prob. 1–3; *Enriched:* Prob. 1–4

PRACTICE PROBLEMS

Compute the following approximate numbers as indicated:

1. Add:

a. $6.8 + 3.44$ 10.2

b. $8.14 + 7.275 + 3.6$ 19.0

c. $19 + .84 + 9.3$ 29

2. Subtract:

a. $93.174 - 8.32$ 84.85

b. $27.6 - .039$ 27.6

c. $58 - 1.8$ 56

3. Multiply:

a. 5.9×6.7 40

b. $48 \times .903$ 43

c. 6.1504×13.85 85.18

4. Divide:

a. $12\overline{)8.29}$ .69

b. $9\overline{).022}$ .002

c. $3.8\overline{)954.15}$ 250

REVIEW OF PART 2

The numeral in color following each problem indicates the Exercise
where explanatory material may be found.

1. Change: **a.** 63.8 m to cm 6,380 cm   **c.** 52 km to m 52,000 m   2-1
   b. 419 mm to cm 41.9 cm   **d.** 2,965 cm to m 29.65 m

2. Change: **a.** 12.4 kg to g 12,400 g   **c.** 9.2 cg to mg 92 mg   2-2
   b. 23.78 g to cg 2,378 g   **d.** 35 mg to g .035 g

3. Complete: **a.** 8.5 cL = ___85___ mL   **c.** 12 L = ___1,200___ cL   2-3
   b. 759 mL = ___.759___ L   **d.** 46 dL = ___4.6___ L

4. Complete: **a.** 26.45 cm² = ___2,645___ mm²   2-4
   b. 9,100 m² = ___.91___ hectares
   c. 568 cm² = ___.0568___ m²
   d. 31.7 km² = ___3,170___ hectares

5. a. Change: (1) 14.95 cm³ to dm³   (3) 83.9 m³ to dm³   83,900 dm³   2-5
   (1) .01495 dm³ (2) 6.3 cm³ to mm³ 6,300 mm³ (4) 2.8 m³ to cm³ 2,800,000 cm³

   b. 33.4 liters of water occupies a space of ___33,400___ cm³ and
   weighs ___33.4___ kg.

   c. How many liters of water will fill a pool if it occupies a
   space of 600,000 dm³? 600,000 L

6. How many yards are in $6\frac{2}{3}$ miles? 11,733$\frac{1}{3}$ yd.   2-6

7. Change $2\frac{5}{8}$ pounds to ounces. 42 oz.   2-7

8. How many liquid ounces are in 3 quarts? 96 oz.   2-8

9. Change $10\frac{1}{2}$ bushels to pecks. 42 pecks   2-9

10. What part of a square foot is 48 square inches? $\frac{1}{3}$ sq. ft.   2-10

11. Change 297 cubic feet to cubic yards. 11 cu. yd.   2-11

12. What part of a minute is 45 seconds? $\frac{3}{4}$ min.   2-13

13. Multiply 62° 20′ by 15. Simplify your answer. 935°   2-14

14. Change 54 km/h to m/s. 15 m/s   2-15

15. Is a temperature of 50°F warmer than 30°C? No   7:15 A.M.   2-16

16. Express in A.M. or P.M. time: **a.** 0150 1:50 A.M. **b.** 1915   2-17

17. If it is midnight in Cincinnati, what time is it in Baltimore?   2-18
In Denver? In Portland, Oregon? 1 A.M.; 11 P.M.; 10 P.M.

18. a. Find the greatest possible error and the relative error in   2-19
each of the following measurements:

   (1) 4.6 cm   (2) 19 g   (3) 240 L   (4) $1\frac{5}{6}$ hr.   (5) $6\frac{5}{8}$ in.

   b. Which is more precise: 16,000 km or 0.05 mm? Which is
   more accurate? 0.05 mm; 16,000 km

18.a. (1) .05 cm; 1.09% (2) $\frac{1}{2}$ g; 2.63% (3) 5 L; $2\frac{1}{2}$% (4) $\frac{1}{12}$ hr.; $4\frac{6}{11}$% (5) $\frac{1}{16}$ in.; .94%

c. Arrange the following measurements in order of precision, most precise first. Precision: 0.007 mm; 0.07 mm; 0.7 mm; 7 mm; 70 mm
Arrange in order of accuracy, most accurate first: Accuracy: all the same
0.07 mm   70 mm   0.7 mm   7 mm   0.007 mm

19. How many significant digits are in each of the following?   2-20
    a. 9,050 three   **c.** 8.9×10^8 two   **e.** .000080 two
    b. 68,517,209 eight   **d.** 3,400,000 two   **f.** 4.007 four

20. Compute the approximate numbers as indicated: 2-21
    a. $4.66 + 2.5 + 9.425$ 16.6   **c.** $6.09 \times .278$ 1.69
    b. $83.49 - 6.513$ 76.98   **d.** $5.6\overline{)7.382}$ 1.3

COMPETENCY CHECK TEST

Solve each problem and select the letter corresponding to your answer. The numerals in red boxes indicate Exercises where help may be found.

1. Which measurement is the shortest? 2-1
    a. 6.47 m   <u>b.</u> 38.9 cm   c. 0.006 km   d. 2,725 mm

2. Which weight is the heaviest? 2-2
    a. 0.002 kg   b. 97.8 mg   c. 2,500 cg   <u>d.</u> 395 g

3. Which capacity is the largest? 2-3
    <u>a.</u> 500 cL   b. 4.5 L   c. 37.6 dL   d. 1,890 mL

4. What measurement does point A indicate on the ruler? 2-1

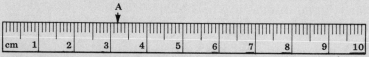

    <u>a.</u> 3.2 cm   b. 35 mm   c. 37 mm   d. 3.8 cm

5. What part of an hour is 40 minutes? 2-13
    a. $\frac{1}{2}$ hour   b. $\frac{7}{10}$ hour   c. $\frac{4}{5}$ hour   <u>d.</u> $\frac{2}{3}$ hour

6. Change $2\frac{1}{2}$ yards to inches. 2-6
    <u>a.</u> 90 in.   b. 25 in.   c. 30 in.   d. Answer not given

7. 10 ounces is what part of a pound? 2-7
    a. $\frac{3}{4}$ lb.   <u>b.</u> $\frac{5}{8}$ lb.   c. $\frac{2}{5}$ lb.   d. $\frac{1}{2}$ lb.

8. How many square yards are in 54 square feet? 2-10
    a. 18 sq. yd.   b. 162 sq. yd.   <u>c.</u> 6 sq. yd.   d. 9 sq. yd.

9. If it is noon by Pacific Time, what time is it by Eastern Time? 2-18
    a. 9 A.M.   b. 11 A.M.   c. 2 P.M.   <u>d.</u> 3 P.M.

10. How many significant digits are there in .000560? 2-20
    a. 2   b. 5   <u>c.</u> 3   d. 6

For a Test on Measurement see p. T48 or *Testing Program,* p. 16.

Facts and Skills in Geometry

3

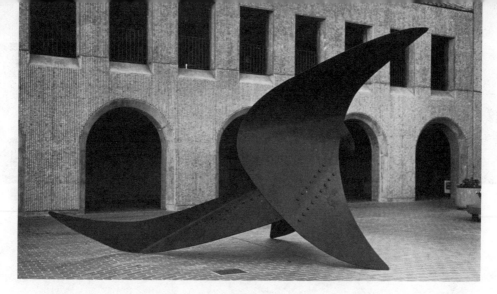

Geometry is the study of points, lines, planes, and space, of measurement and construction of geometric figures, and of geometric facts and relationships.

The word "geometry" means "earth measure." The Egyptians developed many geometric ideas because the flooding of the Nile River destroyed land boundaries which had to be restored.

Euclid, a Greek mathematician who lived in the third century B.C., collected and organized the geometric principles known at that time and wrote a book on geometry.

1. Line segment BC, line NR, ray AT, line CF, line segment SY, ray PQ

INVENTORY TEST

The numeral in color following each problem indicates the Exercise where explanatory material may be found.

2.a. Line segment MS or line segment SM; $\overline{MS}$ or $\overline{SM}$

1. Read, or write in words, each of the following: 3-1
$\overline{BC}, \overleftrightarrow{NR}, \overrightarrow{AT}, \overleftrightarrow{CF}, \overline{SY}, \overrightarrow{PQ}$

2. Name each of the following and express them symbolically: 3-1

2.b. Ray ED, $\overrightarrow{ED}$ 2.c. Line CJ or line JC; $\overleftrightarrow{CJ}$ or $\overleftrightarrow{JC}$

a. M ●———● S   b. ←——— D ●———● E ———→   c. ●C   J

3. a. Name the point where $\overleftrightarrow{AB}$ and $\overleftrightarrow{CD}$ cross. E   3-1

b. Are points C, E, and B collinear? No

c. Are points A, E, and B collinear? Yes

d. Are points B and C collinear? Why?

e. Are $\overleftrightarrow{AB}$ and $\overleftrightarrow{CD}$ concurrent? Yes

3.d. Yes, because a straight line can be drawn through any two points and points that lie on the same straight line are collinear.

332

4.–6. Constructions

4. Draw a representation of: 3-1

  a. a curved line   **b.** a straight line   **c.** a broken line

5. Draw a representation of a line in: 3-1

  a. a slanting position   **b.** a vertical position
           c. a horizontal position

6. Draw a representation of a pair of: 3-1

  a. intersecting lines   **b.** parallel lines
       c. perpendicular lines

7. If the distance from R to A is 50 miles, what is the distance   3-3
from A to B?

R   A   B

125 miles

8. a. With a protractor draw an angle of 125°. S   3-8
  b. Name the angle at the right: Angle *SRT*   3-7
           or *TRS*
  c. Draw any obtuse angle.
  d. Draw a right angle.
  e. Draw any acute angle. R   T 9.–18. Constructions

9. Draw a circle with a diameter of 6.8 centimeters.   3-4

10. Draw a regular hexagon, in which each side measures 23   3-20
millimeters.

11. Construct a triangle with sides measuring $1\frac{5}{8}$ in., $2\frac{1}{8}$ in., and   3-19
$1\frac{3}{4}$ in.

12. Construct a triangle with sides measuring 5 cm and 4.5 cm   3-19
and an included angle of 75°.

13. With the scale of $\frac{1}{4}$ in. = 1 ft. construct a triangle with   3-19
angles measuring 30° and 60° and an included side meas-
uring 9 ft.

14. Draw any line segment. Use compasses to bisect this seg-   3-23
ment.

15. Draw any angle. Use compasses to bisect this angle.   3-25

16. Draw any line. Use compasses to construct a perpendicular
to this line:

  a. At a point on the line.   3-21

  b. From a point outside the line.   3-22

17. Draw with a protractor an angle of 55°. Then construct   3-24
with compasses an angle equal to it. Check with protractor.

18. Draw any line. Locate a point outside this line. Through   3-26
this point construct a line parallel to the first line.

19. What is the complement of an angle measuring 69°? 21°   3-11

20. What is the supplement of an angle measuring 108°? 72°   3-11

21. If $m\angle 2 = 49°$ and $m\angle 3 = 75°$, find the measure of $\angle 1$.   3-11
Of $\angle 4$. $m\angle 1 = 56°; m\angle 4 = 105°$

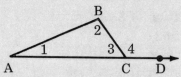

22. If $m\angle 3 = 67°$, what is the measure of $\angle 1$? $\angle 2$? $\angle 4$?   3-11
Ex. 22   Ex. 23

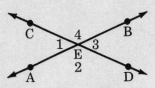   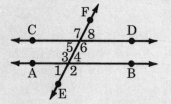

$m\angle 1 = 67°; m\angle 2 = 113°; m\angle 4 = 113°$

23. Parallel lines AB and CD are cut by the transversal EF. If   3-12
$m\angle 6 = 95°$, what is the measure of $\angle 1$? $\angle 2$? $\angle 3$? $\angle 4$?
$\angle 5$? $\angle 7$? $\angle 8$?

$m\angle 1 = 85°; m\angle 2 = 95°; m\angle 3 = 95°; m\angle 4 = 85°; m\angle 5 = 85°; m\angle 7 = 95°; m\angle 8 = 85°$
Find the perimeter of:

24. A rectangle 138 meters long and 57 meters wide. 390 m   3-5

25. A square whose side measured 106 millimeters. 424 mm   3-5

26. A triangle with sides measuring $8\frac{9}{16}$ in., $7\frac{5}{8}$ in., and $6\frac{3}{4}$ in. $22\frac{15}{16}$ in. 3-5

Find the circumference of:

27. A circle whose radius is 49 feet. 308 ft.   3-6

28. A circle whose diameter is 64 centimeters. 200.96 cm   3-6

Find the area of:

29. A rectangle 293 meters long and 185 meters wide. 54,205 m²   3-30

30. A square whose side measures 78 kilometers. 6,084 km²   3-30

31. A parallelogram with an altitude of 98 feet and a base of   3-30
107 feet. 10,486 sq. ft.

                                                           $47\frac{1}{4}$ sq. in.
32. A triangle whose altitude is $10\frac{1}{2}$ inches and base is 9 inches.   3-30

33. A trapezoid with bases of 87 meters and 41 meters and a   3-30
height of 62 meters. 3,968 m²

34. A circle whose radius is 3.8 kilometers. 45.34 km²   3-30

35. A circle whose diameter is 56 inches. 2,464 sq. in.   3-30

Find the volume of:

36. A rectangular solid 109 centimeters long, 97 centimeters wide, and 63 centimeters high. 666,099 cm³   3-31

37. A cube whose side measures 2 feet 8 inches. $18\frac{26}{27}$ cu. ft. or 32,768 cu. in.   3-31

38. A right circular cylinder with a radius of 8 centimeters and a height of 10 centimeters. 2,009.6 cm³   3-31

39. A sphere whose diameter is 4.5 meters. 47.69 m³   3-31

40. A right circular cone 10 inches in diameter and 9 inches high. 235.5 cu. in.   3-31

41. A square pyramid 16 yards on each side of the base and 13 yards high. $1,109\frac{1}{3}$ cu. yd.   3-31

Find the total area of the outside surface of:

42. A rectangular solid $7\frac{1}{2}$ inches long, 6 inches wide, and $2\frac{3}{4}$ inches high. $164\frac{1}{4}$ sq. in.   3-30

43. A cube whose side measures 9.3 meters. 518.94 m²   3-30

44. A right circular cylinder 20 centimeters in diameter and 34 centimeters high. 2,763.2 cm²   3-30

45. A sphere whose diameter is 84 millimeters. 22,176 mm²   3-30

46. Find the base of a right triangle if the hypotenuse is 146 meters and the altitude is 96 meters. 110 m   3-27

47. Find the hypotenuse of a right triangle if the altitude is 112 feet and the base is 384 feet. 400 ft.   3-27

48. Find the altitude of a right triangle if the base is 45 centimeters and the hypotenuse is 117 centimeters. 108 cm   3-27

49. Find the height of a flagpole that casts a shadow of 450 feet at a time when a girl, $5\frac{1}{2}$ feet tall, casts a shadow of 33 feet. 75 ft.   3-28

50. The letters in the following problems represent parts of right triangle ABC with angle C, the right angle; altitude a, the side opposite angle A; and base b, the side opposite angle B.   3-29

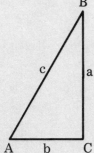

    a. Find side a when angle $A = 89°$ and side $b = 250$ centimeters. 14,322.5 cm

    b. Find side a when angle $A = 71°$ and side $c = 800$ yards. 756.4 yd.

    c. Find side b when angle $A = 47°$ and side $c = 500$ meters. 341 m

    d. How high is the top of a TV tower if, at a horizontal distance of 400 feet from its base, the angle of elevation is 73°? 1,308.36 ft.

APPLICATIONS USING
UNITS OF LENGTH

EXERCISE

3-1 Points and Lines

VOCABULARY: point
line
half-line
line segment
interval
ray
intersecting lines

parallel lines
perpendicular lines

OBJECTIVE: (1) To name points, lines, line segments, and rays, and (2) to identify kinds of lines, positions of lines, and intersecting, parallel, and perpendicular lines.

A geometric point is an exact location in space. It has no size nor can it be seen. Any dot we generally use to indicate it is only a representation of a geometric point.

A geometric line is a collection of points. The pencil or chalk lines we draw are only representations of geometric lines. A line may be extended indefinitely in both directions because it is endless; it has an infinite number of points

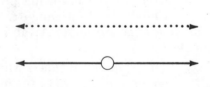

but no endpoints. A definite part of a line has length but no width or thickness. We cannot see a geometric line.

A point separates a line into two half-lines. Each half-line extends indefinitely in one direction only and does not include the point that separates the line into two half-lines.

Related Resources: *Practical Applications in Mathematics:* Problems, pp. 88–89 **336**

I. Naming Points, Lines, Line Segments, and Rays

A capital letter is used to label and name a point.

> ·A   is "point *A*."

A line is represented as: ⟷ The arrowheads are used to show that a line is endless in both directions.

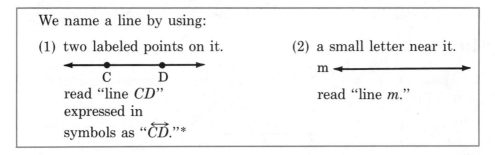

A definite part of a line including both of its endpoints is called a line segment or segment. It consists of two endpoints and all the points between. We name a line segment by its endpoints.

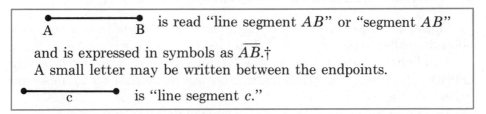

A definite part of a line excluding its endpoints is called an interval
A half-line which includes one endpoint is called a ray. This endpoint is the one that separates the line into two half-lines. To name a ray we use the letter first which names the endpoint and then the letter which names one other point on the ray.

> ●━━●━━▶   is named "ray *MN*," expressed in symbols as $\overrightarrow{MN}$.
> M   N
> But ◀━━●   is named "ray *NM*," expressed in symbols as $\overrightarrow{NM}$.
>   M   N

---

*Also read "line *DC*" and expressed as $\overleftrightarrow{DC}$.
†Also read "line segment *BA*" and expressed as $\overline{BA}$.

II. Kinds of Lines

Lines may be straight, curved, or broken. Usually a straight line is simply called a line. All these lines are sometimes called curves.

Straight Line   Curved Line   Broken Line

III. Position of Lines

Lines may be in vertical, horizontal, or slanting (sometimes called oblique) positions.

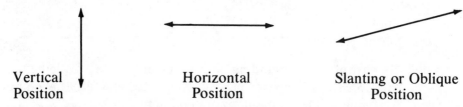

Vertical   Horizontal   Slanting or Oblique
Position   Position   Position

IV. Intersecting, Parallel, and Perpendicular Lines

Lines that meet are intersecting lines. Lines that have a common point are called concurrent lines. Intersecting lines are concurrent lines.

Two lines in the same plane (see Exercise 3-26) that do not meet are called parallel lines. Two lines not in the same plane that do not meet are called skew lines.

Two intersecting lines or rays or segments or a line and a ray or a line and a segment or a ray and a segment that form a right angle are said to be perpendicular to each other. (See Exercises 3-21 and 3-22.)

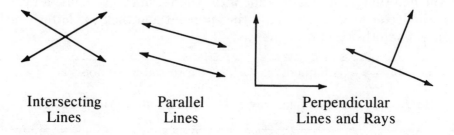

Intersecting   Parallel   Perpendicular
Lines   Lines   Lines and Rays

PRACTICE PROBLEMS

1. Name the endpoints of the following line segment:

B   C   *B, C*

2. Name the following lines:

   A   G   t   R   S   p

line *AG* or *GA*   line *t*   line *RS* or *SR*   line *P*

3. Name the following line segments:

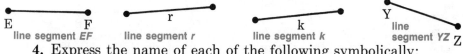

E   F   r   k   Y

line segment *EF*   line segment *r*   line segment *k*   line segment *YZ*   Z

4. Express the name of each of the following symbolically:

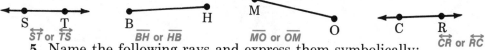

  S   T   B   H   M   C   R

$\overleftrightarrow{ST}$ or $\overleftrightarrow{TS}$ $\overline{BH}$ or $\overline{HB}$ $\overline{MO}$ or $\overline{OM}$ O $\overleftrightarrow{CR}$ or $\overleftrightarrow{RC}$

5. Name the following rays and express them symbolically:

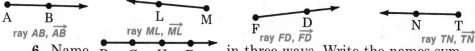

A   B   L   M   F   D   N   T

ray *AB*, $\overrightarrow{AB}$ ray *ML*, $\overrightarrow{ML}$ ray *FD*, $\overrightarrow{FD}$ ray *TN*, $\overrightarrow{TN}$

6. Name  in three ways. Write the names symbolically.   ray *BG*, ray *BH*, ray *BD*; $\overrightarrow{BG}$, $\overrightarrow{BH}$, $\overrightarrow{BD}$

B   G   H   D

7. Read, or write in words, each of the following:

$\overleftrightarrow{CN}$, $\overrightarrow{SV}$, $\overline{KM}$, $\overrightarrow{BR}$, $\overleftrightarrow{GT}$, $\overline{DS}$, $\overleftrightarrow{NJ}$, $\overrightarrow{FL}$, $\overline{ME}$, $\overleftrightarrow{AD}$

line *CN*; ray *SV*; line segment *KM*; ray *BR*; line *GT*; line segment *DS*; line *NJ*; ray *FL*; line segment *ME*; line *AD*

8. Name the point where $\overleftrightarrow{DE}$ and $\overleftrightarrow{FG}$ cross.   *H*

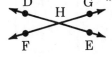

D   H   G

F   E

9. $\overline{RT}$ is divided into 2 parts by point *S*. Name the two segments.

line segment *RS*, line segment *ST*

R   S   T

10. Points *M* and *N* separate $\overline{KL}$ into three parts. Name the three segments.

line segment *KM*, line segment *MN*, line segment *NL*

K   M   N   L

11. Point *D* bisects $\overline{BC}$. Name the two equal segments.

line segment *BD*, line segment *DC*

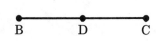

B   D   C

12.a. line segment *FG*, line segment *GH*, line segment *HF*

12. Name the line segments in each of the following figures:

12.b. line segment *a*, line segment *b*, line segment *c*

a. G   **b.**   **c.**   P

F   H   b   L   O

                                                   M   N

12.c. line segment *ML*, line segment *LP*, line segment *PO*, line segment *ON*, line segment *NM*

13. $\overline{DE}$ is the sum of what segments?

line segment *DF*, line segment *FG*, line segment *GE*

D   F   G   E

14. $\overline{AB}$ is the difference of what segments?

line segment *AC*, line segment *BC*

A   B   C

15. What kind of line is shown by each of the following figures:
a? b? c? d? e? f? g? h? i? j?

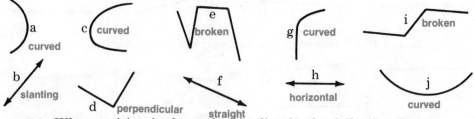

a
curved

c curved

e
broken

g curved

i
broken

b
slanting

d perpendicular

f
straight

h
horizontal

j
curved

16. What position is shown by the line in the following figures:
a? b? c? d? e? f? g? h? i? j?

16.a. horizontal

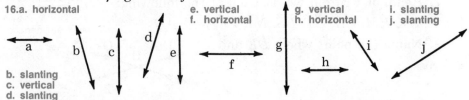

e. vertical
f. horizontal

g. vertical
h. horizontal

i. slanting
j. slanting

a

b   c   d   e   f

g   h   i   j

b. slanting
c. vertical
d. slanting

17. Which of the following are intersecting lines or rays?
Which are parallel lines? Which are perpendicular lines or rays?

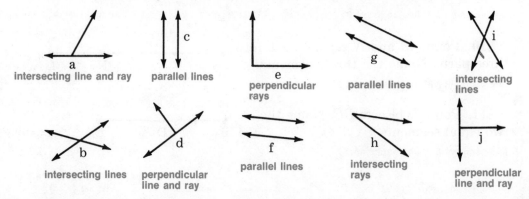

a
intersecting line and ray

c
parallel lines

e
perpendicular
rays

g
parallel lines

i
intersecting
lines

b
intersecting lines

d
perpendicular
line and ray

f
parallel lines

h
intersecting
rays

j
perpendicular
line and ray

18. Which line is parallel to line *p*?
line *n*

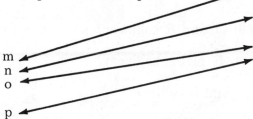

19. a. Which ray is perpendicular to $\overleftrightarrow{AB}$? $\overrightarrow{CE}$

b. Which ray is perpendicular to $\overrightarrow{CD}$? $\overrightarrow{CF}$

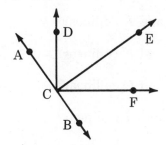

20. At what point do $\overrightarrow{PQ}$ and $\overleftrightarrow{ST}$ intersect? R

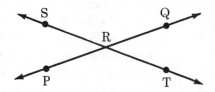

21. Name the point where each of the following pairs of segments intersect.

a. $\overline{CD}$ and $\overline{BD}$ D **c.** $\overline{BD}$ and $\overline{CB}$ B

b. $\overline{BC}$ and $\overline{CD}$ C

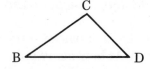

22. Name the point where each of the following pairs of segments intersect.

a. $\overline{ML}$ and $\overline{LO}$ L **c.** $\overline{LO}$ and $\overline{ON}$ O

b. $\overline{MN}$ and $\overline{NO}$ **d.** $\overline{LM}$ and $\overline{MN}$ M

23. Points that lie on the same straight line are called collinear points

a. Are points *P*, *Q*, and *R* collinear? Yes

b. Are points *A*, *B*, and *C* collinear? No

Are points *A* and *B* collinear? Yes

Are points *B* and *C* collinear? Yes

Are points *A* and *C* collinear? Yes

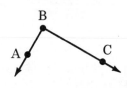

c. Are points *E*, *O*, and *F* collinear? Yes

Are points *G*, *O*, and *F* collinear? No

Are points *H*, *O*, and *E* collinear? No

Are points *G*, *O*, and *H* collinear? Yes

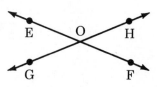

24. Yes; an infinite number; yes; yes

24. Label a point on your paper as *D*. Draw a line through point *D*. Draw a different line through point *D*. Draw a third line through point *D*. Can more than one straight line be drawn through a point? How many lines can be drawn through a point? Can we say that *an infinite number of straight lines can be drawn through a point?* Since these lines have a common point, can we call them concurrent lines?

25. Label two points on your paper as *M* and *N*. Draw a straight line through points *M* and *N*. Draw another line through points *M* and *N*. How many straight lines can be drawn through two points? Can we say that *two points determine a straight line and that one and only one straight line can pass through any two points?* Are any two points collinear? One; yes; yes

26. One; one; yes

26. Draw a pair of intersecting lines. Draw another pair of intersecting lines. At how many points can two straight lines intersect? How many points in common does a pair of intersecting lines have? Can we say that *two straight lines can intersect in only one point?*

27. Which of the three kinds of lines is the shortest path between points *R* and *S*? Can we say that *the shortest path between two points is along a straight line?*

The straight line; yes

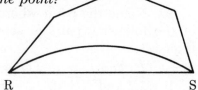

Drawing and Measuring Line Segments

VOCABULARY: straightedge
ruler
compass

A straightedge is used to draw line segments. The ruler, which is a straightedge with calibrated measurements, is used both to draw and to measure line segments of varying lengths.

Sometimes an instrument called a compass is used along with the ruler to measure line segments or to draw line segments of specified lengths. Since the compass is used to draw circles and arcs of circles, the distance between the metal point and the pencil point of the compass corresponds to the radius of the circle or the radius of the arc of the circle that may be drawn.

Generally we measure the length of a line segment by using only the ruler. However when we use both a compass and a ruler to do this, we open the compass so that the metal point and the pencil point fit exactly on the endpoints of the line segment. We then transfer the compass to a ruler to determine the measurement between the metal point and the pencil point.

We reverse this operation to draw a line segment of a given length by using both a compass and a ruler. We first set the compass so that the distance between the metal point and pencil point corresponds to the given measurement on a ruler. We then apply this compass setting to a light pencil working line, using the metal point as one endpoint and the pencil point as the second endpoint.

Basic: Prob. 1–4; *Average:* Prob. 1–6; *Enriched:* Prob. 1–7

PRACTICE PROBLEMS

1. Using a customary ruler, draw line segments having the following dimensions: construction

a. $4\frac{1}{2}$ in.   **b.** $2\frac{3}{4}$ in.   **c.** $3\frac{5}{8}$ in.   **d.** $4\frac{13}{16}$ in.

2. Write your estimate, then measure the length of each of the following line segments, using the ruler only for segments in a, b, and c, and a compass and ruler for segments in d and e.

a. ────────────────────────── $2\frac{1}{2}$ in.

b. ─────────── $\frac{7}{8}$ in.

c. ──────────────────────────────── $3\frac{1}{4}$ in.

d. ─────────────────── $1\frac{9}{16}$ in.

e. ────────────────────── $2\frac{3}{8}$ in.

Related Resources: *Practical Applications in Mathematics:* Problems, pp. 90–91

3. Using a metric ruler, draw line segments having the following dimensions: construction

a. 6 cm **b.** 49 mm **c.** 13 cm **d.** 108 mm

4. Draw a line segment 15 centimeters long. Using a compass, lay off in succession on this segment lengths of 2.8 centimeters, 4.9 centimeters, and 3.6 centimeters. How long is the remaining segment? 3.7 cm

5. The symbol $m\overline{BR}$ is read "the measure of line segment BR" and represents the length of the segment.

5.a. $m\overline{BR}$ = 44 mm; $m\overline{AD}$ = 3.5 cm;
$m\overline{ST}$ = 42 mm; $m\overline{NH}$ = 35 mm

a. Find $m\overline{BR}$; $m\overline{AD}$; $m\overline{ST}$; $m\overline{NH}$.

b. Does $m\overline{AD} = m\overline{NH}$? Yes

c. Does $m\overline{ST} = m\overline{BR}$? No

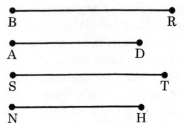

6. Draw any line segment EF. Label a point not on $\overline{EF}$ as A. Draw a line segment AG equal in length to $\overline{EF}$.

7. a. Draw a line segment twice as long as $\overline{BR}$ in problem 5.
b. Draw a line segment three times as long as $\overline{AD}$ in problem 5.
6–7.b. Constructions

REFRESH YOUR SKILLS

1. Add:
9,586
4,835
2,978
4,769
22,168

2. Subtract:
506,950
425,869
81,081

3. Multiply:
496
857
425,072

4. Divide: 780
903)704,340

5. Add:
$3\frac{1}{6} + 2\frac{1}{3}$ $5\frac{1}{2}$

6. Subtract:
$4\frac{2}{5} - \frac{1}{2}$ $3\frac{9}{10}$

7. Multiply:
$9\frac{3}{8} \times 3\frac{1}{5}$ 30

8. Divide:
$\frac{3}{16} \div 3\frac{3}{4}$ $\frac{1}{20}$

9. Add:
$.926 + .47 + .3$ 1.696

10. Subtract:
$.02 - .001$ .019

11. Multiply:
$.06 \times .005$ .0003

12. Divide:
$.4)\overline{1}$ 2.5

13. Find $8\frac{1}{2}\%$ of $12,000. $1,020

14. 6 is what percent of 4? 150%

15. $.90 is 15% of what amount? $6.00

16. Find the value of 25^2. 625

17. Find the square root of 600.25 24.5

18. Write 327.5 *million* as a complete numeral. 327,500,000

EXERCISE

Scale | **3-3**

The scale shows the relationship between the dimensions of a drawing, plan, or map and the actual dimensions.

A scale like 1 inch = 8 feet means that 1 scale inch represents 8 actual feet. This scale may also be written as $\frac{1}{8}$ inch = 1 foot or as the representative fraction $\frac{1}{96}$ or as the ratio 1:96. These all indicate that each scale inch represents 96 actual inches or 8 feet.

The scale ratio 1:4,000,000 indicates that 1 scale millimeter represents 4,000,000 actual millimeters which equal 4,000 meters or 4 kilometers. Therefore, the ratio 1:4,000,000 could also be written as 1 mm = 4 km.

On maps we usually find a scale of miles like that shown at the right. This scale indicates that each scale inch represents 80 actual miles and is equivalent to the scale 1 inch = 80 miles.

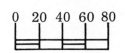

(1) *To find the actual distance when the scale and scale distance are known,* multiply the scale distance by the scale value of a unit (cm in the following):

If the scale is 1 cm = 30 km, what actual distance is represented by 4.7 cm?

Scale: 1 cm = 30 km   Scale distance 4.7 cm = ? km

$$4.7 \times 30 = 141.0$$

Answer: 141 km

---

(2) *To find the scale distance when the scale and the actual distance are known,* divide the actual distance by the scale value of a unit (inch in the following):

If the scale is 1 inch = 16 feet, how many inches represent 78 feet?

Scale: 1 inch = 16 feet   ? scale distance = 78 feet

$$78 \div 16 = 4\tfrac{7}{8}$$

Answer: $4\tfrac{7}{8}$ inches

---

(3) *To find the scale when the actual and scale distances are known,* divide the actual distance by the scale distance:

Find the scale when the actual distance is 60 kilometers and the scale distance is 12 millimeters.

$$60 \div 12 = 5$$

Each scale millimeter represents 5 actual kilometers or 5,000,000 actual millimeters.

Thus, the scale is 1 mm = 5 km which may be written as the ratio 1:5,000,000.

Answer: 1 mm = 5 km or 1:5,000,000

Basic: Prob. 1–23; *Average:* Prob. 1–26; *Enriched:* Prob. 1–30

PRACTICE PROBLEMS

1. If the scale is 1 mm = 15 m, what actual distance is represented by:
    105 m   195 m   975 m   615 m   1,590 m
  a. 7 mm   **b.** 13 mm   **c.** 65 mm   **d.** 41 mm   **e.** 106 mm

2. If the scale is 1 cm = 30 km, what actual distance is represented by:
    180 km   690 km   225 km   18 km   513 km
  a. 6 cm   **b.** 23 cm   **c.** 8.5 cm   **d.** 0.6 cm   **e.** 17.1 cm

3. If the scale is 1 inch = 48 miles, what actual distance is represented by:
                                                  93 mi.
  a. 5 in. 240 mi. **b.** $6\frac{1}{2}$ in. 312 mi. **c.** $2\frac{3}{4}$ in. 132 mi. **d.** $3\frac{5}{8}$ in. 174 mi. **e.** $1\frac{15}{16}$ in.

4. If the scale is $\frac{1}{4}$ inch = 1 foot, what actual distance is represented by:
                                                  22.25 ft.
  a. 3 in. 12 ft. **b.** $4\frac{1}{2}$ in. 18 ft. **c.** $1\frac{5}{8}$ in. 6.5 ft. **d.** $7\frac{3}{4}$ in. 31 ft. **e.** $5\frac{9}{16}$ in.

5. If the scale is $\frac{1}{24}$, what actual distance is represented by: $15\frac{3}{8}$ ft.
  a. 2 in. 4 ft. **b.** $6\frac{1}{2}$ in. 13 ft. **c.** $9\frac{1}{4}$ in. $18\frac{1}{2}$ ft. **d.** $3\frac{7}{8}$ in. $7\frac{3}{4}$ ft. **e.** $7\frac{11}{16}$ in.

6. The scale ratio 1:600 means:
  1 scale cm = __600__ actual cm = __6__ actual m

7. The scale ratio 1:5,000,000 means:
  1 scale mm = __5,000,000__ actual mm = __5,000__ m = __5__ km

8. The scale ratio 1:1,250,000 means:
  1 scale mm = __1,250,000__ actual mm = __1,250__ m = __1.25__ km

9. The scale 1 mm = 75 m means:
  1 scale mm = __75,000__ actual mm = scale ratio 1: __75,000__

10. The scale 1 cm = 20 km means:

1 scale cm = ^{2,000,000} actual cm = scale ratio 1: ^{100,000}

11. Write the scale ratio for each of the following scales:

a. 1 mm = 4 km 1:4,000,000   **c.** 1 cm = 25 m 1:2,500

b. 1 cm = 45 km 1:4,500,000   **d.** 1 mm = 90 m 1:90,000

12. Write in another way the scale for each of the following scale ratios: Answers will vary.

     a. 1:3,000,000   **b.** 1:800   **c.** 1:100,000   **d.** 1:2,500,000

13. If the scale is 1:600, what actual distance is represented by:

a. 4 m   **b.** 27 cm   **c.** 9.3 cm   **d.** 63 mm   **e.** 154 mm

14. If the scale is 1:2,500,000, what actual distance is represented by:

a. 18 mm 45 km **b.** 7 mm 17.5 km **c.** 69 mm $^{172.5 \text{ km}}$ **d.** 1 cm 25 km **e.** 15.6 cm 390 km

15. If the scale is 1 inch = 64 miles, how many inches represent:

a. 192 mi. 3 in. **b.** 160 mi. 2$\frac{1}{2}$ in. **c.** 304 mi. 4$\frac{3}{4}$ in. **d.** 104 mi. 1$\frac{5}{8}$ in. **e.** 372 mi. 5$\frac{13}{16}$ in.

16. If the scale is $\frac{1}{8}$ inch = 1 foot, how many inches represent:

a. 48 ft. 6 in.   **b.** 60 ft. 7.5 in.   **c.** 126 ft. 15.75 in. **d.** 29 ft. 3.625 in. **e.** 167 ft. 20.875 in.

17. If the scale is $\frac{1}{48}$, how many inches represent:

a. 20 ft. 5 in.   **b.** 54 ft. 13$\frac{1}{2}$ in.   **c.** 3 ft. $\frac{3}{4}$ in.   **d.** 117 ft. 29$\frac{1}{4}$ in.   **e.** 30$\frac{1}{2}$ ft. 7$\frac{5}{8}$ in.

18. If the scale is 1 cm = 25 m, how many centimeters represent:

a. 75 m 3 cm   **b.** 5 m $\frac{1}{5}$ cm   **c.** 325 m 13 cm   **d.** 110 m 4.4 cm   **e.** 42.5 m 1.7 cm

19. If the scale is 1:2,000,000, how many millimeters represent:

a. 8 km 4 mm   **b.** 30 km 15 mm   **c.** 116 km 58 mm   **d.** 98 km 49 mm   **e.** 47 km 23$\frac{1}{2}$ mm

20. Find the scale when: 20.a. 1:60,000,000 b. 1:192

a. The scale length of 7 mm represents an actual distance of 420 km.

b. The scale length of 6$\frac{5}{8}$ in. represents an actual distance of 106 ft.

c. The actual distance of 870 km is represented by the scale length of 5.8 cm. 1:15,000,000

d. The actual distance of 750 mi. is represented by the scale length of 4$\frac{11}{16}$ in. 1:10,137,600

e. The scale length of 10.7 cm represents an actual distance of 192.6 m. 1:1,800

f. The scale length of 3$\frac{7}{8}$ in. represents an actual distance of 186 ft. 1:576

21. Using the scale 1 cm = 100 km, draw line segments representing: 21.–23. constructions

     a. 300 km   **b.** 60 km   **c.** 140 km   **d.** 290 km   **e.** 450 km

22. Using the scale $\frac{1}{8}$ inch = 1 foot, draw line segments representing:

     a. 24 ft.   **b.** 10 ft.   **c.** 19 ft.   **d.** 35 ft.   **e.** 61$\frac{1}{2}$ ft.

13.a. 2,400 m or 2.4 km b. 16,200 cm or 162 m c. 5,580 cm or 55.8 m d. 37,800 mm or 37.8 m
e. 92,400 mm or 92.4 m

23. Using the scale 1:1,500,000, draw line segments representing:

a. 6 km   **b.** 90 km   **c.** 48 km   **d.** 105 km   **e.** 210 km

24. What are the actual dimensions of a warehouse floor if plans drawn to the scale of 1:200 shows dimensions of 12 cm by 9 cm? 24 m by 18 m

25. Draw a floor plan of a schoolroom 33 feet long and 18 feet wide, using the scale 1 inch = 12 feet. Construction

26. Using the scale of *miles* 0 12 24 36 48 find the distance represented by each of the following line segments:

a. _____ 72 mi.   **c.** _____ 90 mi.

b. _____ 36 mi.   **d.** _____ 108 mi.

e. _____ 159 mi.

27. Using the scale of *kilometers* 0 10 20 30 40 50 find the distance represented by each of the following line segments:

a. _____ 60 km   **c.** _____ 37 km

b. _____ 25 km   **d.** _____ 74 km

e. _____ 93 km

28. Find the scales used to draw the following line segments representing the given distances:

a. _____ 150 mi. _____
Scale: 1 inch = ? 80 mi.

b. _____ 41 ft. _____
Scale: 1 inch = ? 16 ft.

c. _____ 54 m _____
Scale: 1:? 900

d. _____ 180 km _____
Scale: 1:? 4,000,000

29. If the distance from A to B is 56 miles, what is the distance from G to H? _____   _____
   A   B   G   112 mi.   H

30. If the distance from D to E is 70 kilometers, what is the distance from M to N?

_____   _____
D   E   M   130 km   N

For a Class Activity see p. T34.

Geometric Figures

VOCABULARY: plane   simple closed figure
            coplanar points   triangle
            coplanar lines   quadrilateral
            space   circle
            angle   solid figure

Geometric figures consist of collections of points. Plane geometric figures are figures with all their points in the same plane.

I. Planes and Space

A geometric plane or flat surface is a set of points. We cannot see a geometric plane. It is endless and extends beyond any line boundaries we use to represent it. When the plane is limited, it has length and width which can be measured but no height or thickness. A wall, floor, and desk top are common representations of a limited plane.

A line separates a plane into two half-planes.

To name a plane, use letters which name three points not on the same line belonging to the plane or by two capital letters at opposite outside corners or by one capital letter in an interior corner.

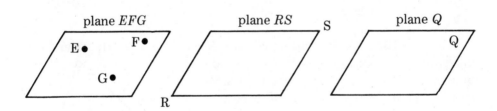

plane *EFG*   plane *RS*   plane *Q*

Points that lie in the same plane are called coplanar points. Lines that are in the same plane are called coplanar lines.

Space is the infinite set of all points. Its length, width, and height are endless. A limited space can be measured. A plane separates space into two half-spaces.

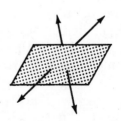

II. Angles

An angle is the figure formed by two different rays having the same endpoint.

See Exercise 3-7 for the complete study of angles.

III. Simple Closed Plane Figures or Simple Closed Curves

Geometric figures consist of collections of points. Plane geometric figures are figures with all of their points in the same plane. A simple closed plane figure (or curve) begins at a point and returns to this point without crossing itself. It divides a plane into three collections of points, those in the interior, those in the exterior, and those on the figure.

A polygon is a simple closed plane figure made up of line segments (called sides). It is the union of three or more line segments. Each pair of intersecting sides meets in a point called a vertex. In a polygon the union of two adjacent sides (line segments) forms an angle, since these line segments are parts of rays. A polygon with all sides of equal length and all angles of equal measure is called a regular polygon. It should be noted that the collection of points contained in the sides is the figure and not the region enclosed by the figure. A polygon is named by reading the letters at the vertices. A line segment connecting two nonadjacent vertices of a polygon is called a diagonal. Some common polygons are: *triangle,* 3 sides; *quadrilateral,* 4 sides; *pentagon,* 5 sides; *hexagon,* 6 sides; *octagon,* 8 sides; *decagon,* 10 sides; and *dodecagon,* 12 sides.

IV. Triangles

When all three sides of a triangle are equal in length, the triangle is called an equilateral triangle; when two sides are equal, an isosceles triangle; when no sides are equal, a scalene triangle. When all three angles of a triangle are equal in size, the triangle is called an equiangular triangle; a triangle with a right angle, a right triangle; with an obtuse angle, an obtuse triangle; and with three acute angles, an acute triangle. The altitude of a triangle is the perpendicular segment from any vertex of a triangle to the opposite side or extension of that side. The median of a triangle is the line segment connecting any vertex of a triangle to the midpoint of the opposite side.

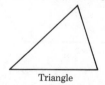

Triangle

In a right triangle the side opposite the right angle is called the hypotenuse. The other two sides or legs are the altitude and base of the triangle. The base is generally the side on which the triangle rests. In an isosceles triangle the angle formed by the two equal sides is called the vertex angle. It is opposite to the base. The angles opposite the equal sides are called the base angles.

Congruent triangles are triangles which have exactly the same shape and the same size. (See Exercise 3-13.)

Similar triangles are triangles which have the same shape but differ in size. (See Exercise 3-14.)

V. Quadrilaterals

The following properties describe special quadrilaterals:

The rectangle has two pairs of opposite sides which are equal and parallel and four angles which are right angles.

The square has four equal sides with the opposite sides parallel and four angles which are right angles.

The parallelogram has two pairs of opposite sides which are parallel and equal.

The trapezoid has only one pair of opposite sides that are parallel.

The square is a special rectangle, and the rectangle and square are special parallelograms.

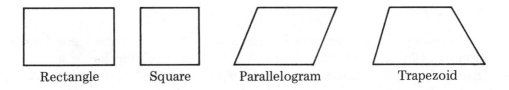

Rectangle   Square   Parallelogram   Trapezoid

VI. Circles

A circle is the collection of points in a plane which are equidistant from a fixed point in the plane called the center. It is a simple closed curve. The radius of a circle is a line segment which has one endpoint at the center of the circle and the other endpoint on the circle. The diameter of a circle is a line segment which has both of its endpoints on the circle but passes through the center. A chord of a circle is a line segment which has both of its endpoints on the circle. An arc is a part of the circle. If the endpoints of an arc are the endpoints of a diameter, the arc is a semicircle. An angle whose vertex is at the center of a circle is called a central angle. The circumference is the distance around the circle. A tangent to a circle is a line that has one and only one point in common with the circle.

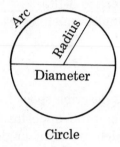

Circle

Concentric circles are circles in the same plane which have the same center but different radii.

Concentric
Circles

VII. Solid or Space Figures

A closed geometric figure consisting of four or more polygons and their interiors, all in different planes, is called a polyhedron. The polygons and their interiors are called faces. These faces intersect in line segments called edges. These edges intersect in points called vertices.

Common polyhedra are the rectangular solid (right rectangular prism), the cube, and the pyramid.

(1)   (2)   (3)

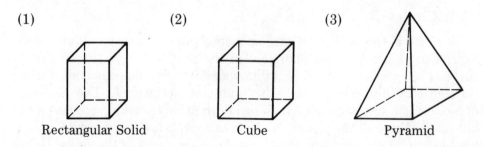

Rectangular Solid   Cube   Pyramid

(1) The rectangular solid has six rectangular faces.
(2) The cube has six squares for its faces. All the edges are equal in length.
(3) The pyramid has any polygon as its base and triangular faces that meet in a common vertex.

Other common solid geometric figures are the cylinder, sphere, and cone.

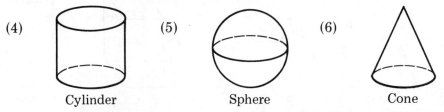

(4) Cylinder   (5) Sphere   (6) Cone

(4) The cylinder has two equal and parallel circles as bases and a lateral curved surface.
(5) The sphere has a curved surface on which every point is the same distance from the center within.
(6) The cone has a circle for the base and a curved surface that comes to a point called the vertex.

VIII. Euler Formula

The Euler formula expresses the relationship of the faces, edges and vertices of a polyhedron. The formula $F + V - E = 2$ tells us that "the number of faces plus the number of vertices minus the number of edges is equal to two."

Basic: Prob. 1–8; *Average:* Prob. 1–10; *Enriched:* Prob. 1–11

PRACTICE PROBLEMS

1. Name each of the following planes:

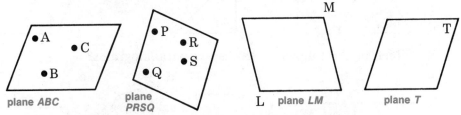

plane *ABC*   plane *PRSQ*   L   plane *LM*   plane *T*

2. What is the name of each of the following figures?

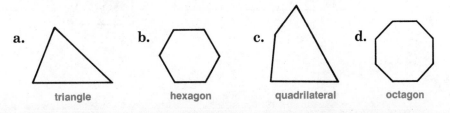

a. triangle   b. hexagon   c. quadrilateral   d. octagon

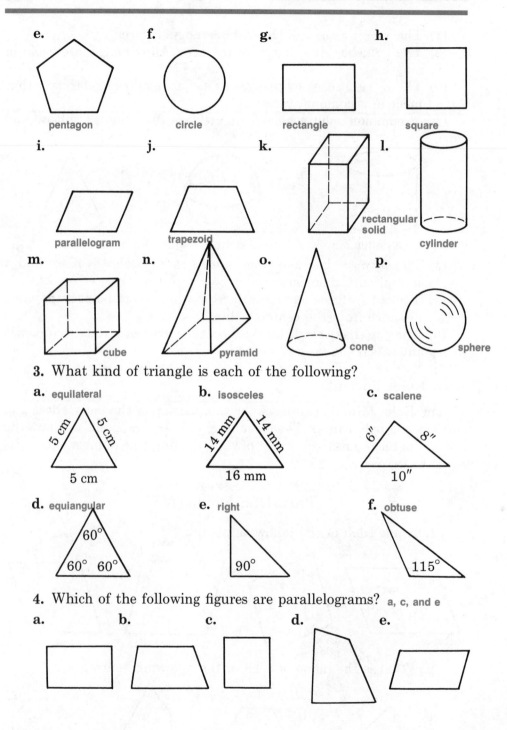

e. pentagon

f. circle

g. rectangle

h. square

i. parallelogram

j. trapezoid

k. rectangular solid

l. cylinder

m. cube

n. pyramid

o. cone

p. sphere

3. What kind of triangle is each of the following?

a. equilateral

5 cm 5 cm 5 cm

b. isosceles

14 mm 14 mm 16 mm

c. scalene

6″ 8″ 10″

d. equiangular

60° 60° 60°

e. right

90°

f. obtuse

115°

4. Which of the following figures are parallelograms? a, c, and e

a. **b.** **c.** **d.** **e.**

5. How many diagonals can be drawn from any one vertex in each of the following figures?

a.

b.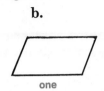

c.

d.

e.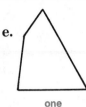

three one two five one

6. a. How many faces (F) does a rectangular solid have? How many vertices (V)? How many edges (E)? a. 6 faces; 8 vertices; 12 edges, yes

Does $F + V - E = 2$?   b. 5 faces; 5 vertices; 8 edges, yes

b. How many faces (F) does a pyramid with a square base have? How many vertices (V)? How many edges (E)? Does $F + V - E = 2$?

7. Select 3 points on a straight line. Can more than one plane pass through these points?

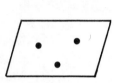

Select 3 points not on a line. Can more than one plane pass through these 3 points? Can we say that *through 3 points not on the same straight line one and only one plane can pass?* Points in the same plane are called coplanar points. Yes; no; yes

8. How many points does a geometric plane have? How many lines can be drawn through these points? How many lines does a geometric plane contain? Can we say that *a geometric plane contains an infinite number of points and lines?* Lines in the same plane are sometimes called coplanar lines. Can we say that *two intersecting lines determine one and only one plane?* An infinite number; an infinite number; an infinite number; yes; yes

9. Draw two different planes that intersect as shown. What geometric figure is their intersection? Can we say that *when two different planes intersect, their intersection is a straight line?* Planes that do not intersect are called parallel planes They have no point in common. Straight line; yes

10. Draw a plane and a line that is not in the plane but which intersects the plane. Can we say that their intersection is one and only one point? A plane and a line not in the plane are parallel when they have no point in common. Yes

11. A line perpendicular to a plane is a line that is perpendicular to every line in the plane that passes through its foot (the point of intersection). Can we say that *a line that is perpendicular to each of two intersecting lines at their point of intersection is perpendicular to the plane in which these lines lie?* Yes

Problem Solving Techniques for Geometric Problems

1. *Read* the problem carefully to find:
   a. The facts, dimensions or values that are given.
   b. The fact, dimension or value that is to be determined.

2. *Decide* which formula or formulas are needed to solve the problem by selecting:
   The formula that relates the variable representing the value to be found with the variables representing the given values.

3. *Solve* by arranging the solution in three columns.
   a. In left column
      (1) Draw the geometric figure described in the problem.
      (2) Mark the given dimensions on the figure.
      (3) Below the figure write the variables with corresponding given values and the variable representing the unknown value.
   b. In center column, evaluate the formula (see page 485)
      (1) Write the formula that relates the variables.
      (2) Substitute the known given values for the corresponding variables in the formula.
      (3) Perform the necessary operations (see step c below).
      (4) Write answer.
      (5) When necessary, solve the resulting equation. See page 501.
   c. In right column
      Perform the necessary arithmetic operations.

4. *Rewrite* answer.

For sample solution see completely worked-out solution on page 357.

Perimeter

EXERCISE 3-5

VOC.: perimeter of a rectangle
perimeter of a square
perimeter of a triangle

OBJECTIVE: To determine the perimeter
of some geometric figures.

The distance around a polygon is called the perimeter. It is the sum of the lengths of the sides of the polygon.

I. Rectangle

The perimeter of a rectangle is equal to twice the length added to twice the width.

$$\text{Formula: } p = 2\,l + 2\,w$$

Find the perimeter of a rectangle
26 meters long and 17 meters wide.

26 m
17 m

$l = 26$ m
$w = 17$ m
$p = ?$

$$p = 2\,l + 2\,w$$
$$p = 2 \times 26 + 2 \times 17$$
$$p = 52 + 34$$
$$p = 86 \text{ meters}$$

$$\begin{array}{cc} 26 & 17 \\ \times\,2 & \times\,2 \\ \hline 52 & 34 \end{array}$$

$$\begin{array}{r} 52 \\ + 34 \\ \hline 86 \end{array}$$

Answer: 86 meters

Basic: Prob. 1–7; *Average:* Prob. 1–9; *Enriched:* Prob. 1–10

PRACTICE PROBLEMS

1. What is the perimeter of a rectangle if its length is 47 millimeters and width is 21 millimeters? 136 mm

2. Find the perimeters of rectangles having the following dimensions:

a. 50 cm; 224 ft.; 446 km; 1,108 yd.; 1,518 m
b. 26.5 m; 20.5 km; $24\frac{3}{8}$ in.; $18\frac{1}{3}$ yd.; 8 ft. 10 in.

a.	Length	16 cm	63 ft.	125 km	208 yd.	432 m
	Width	9 cm	49 ft.	98 km	346 yd.	327 m

b.	Length	3.5 m	8.375 km	$7\frac{3}{8}$ in.	$6\frac{2}{3}$ yd.	2 ft. 7 in.
	Width	9.75 m	1.875 km	$4\frac{13}{16}$ in.	$2\frac{1}{2}$ yd.	1 ft. 10 in.

3. How many meters of fencing are required to enclose a rectangular garden 59 m long and 39 m wide? If each 2-meter section costs $11.99, how much will the fencing cost? 196 m; $1,175.02

4. How many meters of fringe are needed for a border on a bedspread 182 cm by 268 cm? 9 m

5. Fred wishes to make a frame for his class picture. The picture measures 25 in. by $12\frac{1}{2}$ in. If he allows $2\frac{1}{4}$ inches extra for each corner, how many feet of molding will he need? 7 ft.

6. How many feet of baseboard are needed for a room 20 ft. 6 in. long and 16 ft. wide if 5 ft. must be deducted for doorways? 68 ft.

7. At 11¢ per foot, what is the total cost of weatherstripping for 5 window frames, each measuring 32 in. by 66 in.; 3 window frames, each measuring 24 in. by 58 in.; and 2 doorways, each measuring 43 in. by 78 in.? $17.93

8. Which lot will take more fencing to enclose it, lot A, 135 m long and 93 m wide, or lot B, 124 m long and 101 m wide? How much more fencing? Lot A; 6 m

9. How many times must you walk around the school grounds, 325 ft. long and 203 ft. wide, to walk a mile? 5 times

10. How many feet of chrome edging do you need to finish a table top 54 inches long and 30 inches wide? 14 ft.

For a Computer Activity see p. T40.

II. Square

The perimeter of a square is equal to 4 times the length of its side.

Formula: $p = 4s$

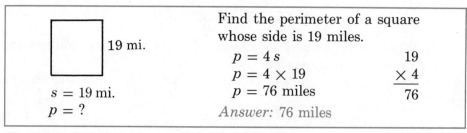

	Find the perimeter of a square whose side is 19 miles.	
19 mi.	$p = 4s$	19
$s = 19$ mi.	$p = 4 \times 19$	$\times\ 4$
$p = ?$	$p = 76$ miles	76
	Answer: 76 miles	

Basic: Prob. 1–2; *Average:* Prob. 1–3; *Enriched:* Prob. 1–4

PRACTICE PROBLEMS

1. What is the perimeter of a square whose side is 8 kilometers? 32 km

2. Find the perimeters of squares whose sides measure:

a. 25 cm	**c.** 880 yd.	**e.** 5,280 ft.	**g.** 17.5 cm	**i.** $20\frac{5}{8}$ in.
b. 11 km	**d.** 76 mm	**f.** 0.25 m	**h.** $8\frac{3}{4}$ ft.	**j.** 2 ft. 9 in.
a. 100 cm	c. 3,520 yd.	e. 21,120 ft.	g. 70 cm	i. $82\frac{1}{2}$ in.
b. 44 km	d. 304 mm	f. 1 m	h. 35 ft.	j. 11 ft.

3. If the distance between bases is 90 feet, how many yards does a batter run when he hits a home run? 120 yards

4. Find the cost of the wire needed to make a fence of 5 strands around a square lot 60 m by 60 m if a 400-meter spool of wire costs $34.69.

$104.07

III. Triangle

The perimeter of a triangle is equal to the sum of its sides.

Formula: $p = a + b + c$

Since the sides of an equilateral triangle are equal, its perimeter is equal to 3 times the length of its side.

Formula: $p = 3s$

Basic: Prob. 1–3; *Average:* 1–5; *Enriched:* 1–6

PRACTICE PROBLEMS

1. Find the perimeters of triangles with sides measuring:

a. 18 m, 9 m, 15 m 42 m

b. 21 cm, 17 cm, 29 cm 67 cm

c. 6.23 km, 4.7 km, 3.59 km 14.52 km

d. $4\frac{1}{4}$ mi., $7\frac{5}{8}$ mi., $6\frac{1}{2}$ mi. $18\frac{3}{8}$ mi.

e. 2 ft. 8 in., 1 ft. 9 in., 1 yd. 7 ft. 5 in. or 2 yd. 1 ft. 5 in.

2. Find the perimeters of equilateral triangles with sides measuring:

a. 63 cm   **b.** 14 km   **c.** 4.75 m   **d.** $16\frac{5}{8}$ in.   **e.** 2 ft. 4 in.

3. Find the perimeter of an isosceles triangle if each of the equal sides is 15 centimeters and the third side is 8 centimeters. 38 cm

4. How many meters of hedge are needed to enclose a triangular lot with sides measuring 196 meters, 209 meters, and 187 meters? 592 m

5. How many bricks, each 8 inches long, are needed to enclose a triangular vegetable patch with sides measuring 11 ft. 4 in., 8 ft. 11 in., and 7 ft. 9 in.? At $.20 per brick, how much will they cost? 42 bricks (with one

6. Find the perimeter of each of the following triangles: broken); $8.40

a. 16.6 m   **b.** 33 ft.   **c.** 315 cm

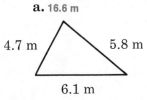

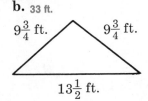

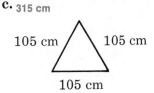

2.a. 189 cm   b. 42 km   c. 14.25 m   d. $49\frac{7}{8}$ in.   e. 7 ft.

3-6 Circumference of a Circle

OBJECTIVE: To determine the
circumference of a circle.

VOC.: circumference
diameter
radius
pi

The distance around a circle is called circumference.
The parts of the circle are related as follows:

a. The diameter is twice the radius. Formula: $d = 2\,r$

b. The radius is one-half the diameter. Formula: $r = \dfrac{d}{2}$

c. The circumference of a circle is equal to pi (π) times the diameter.
Formula: $c = \pi d$ where $\pi = 3\frac{1}{7}$ or $\frac{22}{7}$ or 3.14. For greater accuracy 3.1416 is
used.

d. The circumference of a circle is equal to 2 times pi (π) times the
radius. Formula: $c = 2\,\pi r$

e. The diameter of a circle is equal to the circumference divided by
pi (π). Formula: $d = \dfrac{c}{\pi}$

---

8 cm

Find the circumference of a circle
when its diameter is 8 centimeters.

$d = 8$ cm
$\pi = 3.14$
$c = ?$

$c = \pi d$
$c = 3.14 \times 8$
$c = 25.12$ cm

Answer: 25.12 centimeters

$$
\begin{array}{r}
3.14 \\
\times\, 8 \\
\hline
25.12
\end{array}
$$

---

21 in.

Find the circumference of a circle
when its radius is 21 inches.

$r = 21$ in.
$\pi = \frac{22}{7}$
$c = ?$

$c = 2\,\pi r$
$c = 2 \times \frac{22}{7} \times 21$
$c = 132$ in.

Answer: 132 inches

$$\frac{2}{1} \times \frac{22}{7} \times \frac{\overset{3}{\cancel{21}}}{1} = 132$$

PRACTICE PROBLEMS

1. How long is the diameter if the radius is: d. $21\frac{3}{8}$ in. or 1 ft. $9\frac{3}{8}$ in.

   a. 7 cm? 14 cm **b.** 23 m? 46 m **c.** 6.5 km? 13 km **d.** $10\frac{11}{16}$ in.?   **e.** 2 ft. 3 in.? $4\frac{1}{2}$ ft.

2. How long is the radius if the diameter is: $5\frac{1}{3}$ ft.   4 yd. 1 ft.

   a. 38 m? 19 m **b.** 5 mm? 2.5 mm **c.** 8.9 km? 4.45 km **d.** $10\frac{2}{3}$ ft.?   **e.** 8 yd. 2 ft.?

188.4 m **3.** What is the circumference of a circle whose diameter is 60 meters?

4. Find the circumference of a circle having a diameter of:

   a. 5 mm 15.7 mm or $15\frac{5}{7}$ mm*   **f.** 1.8 m 5.652 m

   b. 35 cm 109.9 cm or 110 cm*   **g.** 6.3 km 19.782 km

   c. 260 m 816.4 m or $817\frac{1}{7}$ m*   **h.** $\frac{3}{4}$ in. $2\frac{5}{14}$ in.*

   d. 49 mi. 153.86 mi. or 154 mi.*   **i.** $3\frac{1}{2}$ ft. 11 ft.*

   e. 440 yd. 1,381.6 yd. or $1,382\frac{6}{7}$ yd.*   **j.** 1 ft. 11 in. 72.22 in.

163.28 m **5.** What is the circumference of a circle whose radius is 26 meters?

6. Find the circumference of a circle having a radius of:

   a. 90 m 565.2 m or $565\frac{5}{7}$ m*   **f.** 0.25 m 1.57 m

   b. 7 mm 43.96 mm or 44 mm*   **g.** 8.4 km 52.752 km

   c. 1,000 yd. 6,280 yd. or $6,285\frac{5}{7}$ yd.*   **h.** $5\frac{1}{2}$ mi. 34.54 mi. or $34\frac{4}{7}$ mi.*

   d. 56 km 351.68 km or 352 km*   **i.** $4\frac{3}{8}$ in. $27\frac{1}{2}$ in.*

   e. 382 ft. 2,398.96 ft. or $2,401\frac{1}{7}$ ft.*   **j.** 2 ft. 4 in. 175.84 in. or 176 in.*

7. What is the diameter of a circle whose circumference is 286 centimeters? 91 cm*

8. Find the diameter of a circle when its circumference is:

   a. 176 m   **b.** 198 cm   **c.** 330 yd.   **d.** 40 mm   **e.** $6\frac{7}{8}$ in.

9. If the diameter of a circular table is 42 inches, what is the circumference of the table? 132 in.*

10. What distance in feet does the tip of a propeller travel in one revolution if its length (diameter) is 7 feet? 21.98 ft. or 22 ft.*

11. How long a metal bar do you need to make a basketball hoop with a diameter of 48 centimeters? 150.72 cm

12. How much farther do you ride in one turn of a merry-go-round if you sit in the outside lane, 21 ft. from the center, than if you sit in the inside lane, 14 ft. from the center? 44 ft.*

13. If the diameter of each wheel is 28 inches, how far does a bicycle go when the wheels revolve once? How many times do the wheels revolve in a distance of 1 mile? 88 in.*; 720 times each

14. Find the circumferences of the earth, moon, and sun if their respective diameters are 7,900 mi., 2,200 mi., and 864,000 mi.

15. The circumference of a tree is 176 cm. Find its diameter. 56 cm

16. What should the diameter of a circular track be if the mile is to be run in 20 laps around the track? 84 ft.*

8.a. 56 m* b. 63 cm* c. 105 yd.* d. 12.7 mm* e. $2\frac{3}{16}$ in.*

14. earth: 24,806 mi.; moon: 6,908 mi.; sun: 2, 712,960 mi.

*$\pi = 3\frac{1}{7}$ was used to determine answers marked with asterisk. Otherwise 3.14 was used.

REVIEW

1. line *RT*; line segment *OM*; ray *DG*

1. Read, or write in words, each of the following: $\overleftrightarrow{RT}$; $\overline{OM}$; $\overrightarrow{DG}$.

2. Name each of the following:
line *CR* or *RC*; ray *NB*

C R   B   N

3. Name the endpoints in the above drawings: None; *N*

4. Name the line segments in the following triangle.

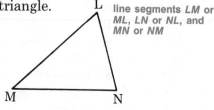

line segments *LM* or
ML, *LN* or *NL*, and
MN or *NM*

5. At what point do $\overline{MN}$ and $\overline{LN}$ meet? at *N*

6. Are points *M*, *L*, and *N* collinear? No

7. Are $\overline{LM}$ and $\overline{LN}$ concurrent? Yes

8–9. Constructions

8. Draw a representation of a pair of:
a. perpendicular lines
b. parallel lines
c. intersecting lines

9. Draw a representation of a line in a: **a.** slanting position **b.** horizontal position **c.** vertical position

10. If the scale is 1 cm = 100 km, what actual distance is represented by the scale distance of 5.7 centimeters? 570 km

11. What is the perimeter of a rectangle 135 meters long and 92 meters wide? 454 m

12. Find the circumference of a circle whose diameter is 84 centimeters. 264 cm*

13. What is the perimeter of a square whose side measures $7\frac{5}{8}$ inches? $30\frac{1}{2}$ in.

14. What is the circumference of a circle whose radius is 120 feet? 753.6 ft.

15. Find the perimeter of a triangle whose sides measure 2.7 meters, 4.2 meters, and 3.5 meters. 10.4 m

16. How many yards of linoleum border are needed for a kitchen floor 12 feet by 10 feet 6 inches? 15 yd.

17. Which is longer, the perimeter of a bridge table measuring 76 centimeters on a side or the circumference of a circular table measuring 90 centimeters in diameter? bridge table

18. Allowing $\frac{1}{2}$ inch for the overlap, how long should a paper label be in order to go around a metal container having a diameter of 7 inches? 22.48 in. or $22\frac{1}{2}$ in.*

19. How many lines can be drawn through a point? Through two points? An infinite number; one

20. a. When two lines intersect, what is their intersection? a point
b. When two planes intersect, what is their intersection? a straight line
c. How many points determine a line? two points
d. How many points determine a plane? three points not on a straight line

For a Test on Applications Using Units of Length see p. T49 or *Testing Program*, p. 17.

APPLICATIONS USING
ANGLES AND ARCS

OBJECTIVE: To name angles
and to identify kinds of
angles.

Facts About Angles

VOC.: angle   right angle
       vertex   acute angle
       sides   obtuse angle

I. Meaning of an Angle

An angle is the figure formed by two different rays having the same
endpoint. This common endpoint is called the vertex of the angle and the
two rays are called the sides of the angle.

An angle may be considered as the rotation
of a ray about a fixed endpoint, the angle being
formed as the ray turns from one position to
another. An angle is sometimes used to show
direction.

The symbol $\angle$ designates the word "angle."

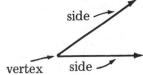

II. Naming Angles

Angles are identified or named in
the following ways:

(1) By reading the capital letter at
the vertex.

(2) By reading the inside letter or
numeral.

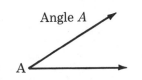

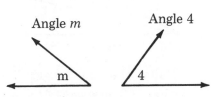

(3) By reading the three letters associated with the vertex and one
point on each of the sides. The
middle letter always indicates
the vertex.

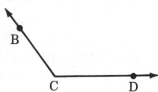

Angle *BCD* or Angle *DCB*

363

Related Resources: *Practical Applications in
Mathematics:* Problems, pp. 94–95

PRACTICE PROBLEMS

1. Name the sides and vertex of the following angle:
  Side *ED* and side *DF*; vertex *D*

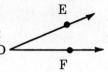

2. What is the symbol used to denote the word "angle"?
  ∠

3. Name the following angle in three ways:
  Angle *O*, angle *TOP*, angle *POT*

4. Name the following angle in two ways:
  Angle *F*, angle *x*

5. Name the following angle in two ways:
  Angle *Y*, angle *2*

6. Name each of the following angles:

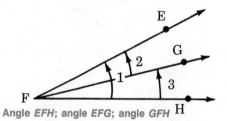

Angle *e*   Angle *BOC*   Angle *N*   Angle *3*   Angle *DEF*   Angle *RTS*   Angle *LAK*

7. Name ∠1, ∠2, and ∠3 in the figure below, using three letters.

8. Name each angle of the triangle below in four ways.

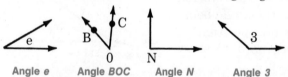

Angle *EFH*; angle *EFG*; angle *GFH*

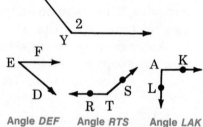

Angles *K, MKL, LKM,* and *1*; angles *M, KML, LMK,* and *2*; angles *L, MLK, KLM,* and *3*

9. Name the four angles formed when $\overleftrightarrow{AB}$ intersects $\overleftrightarrow{MN}$ at point *T*, using 3 letters for each angle.
  Angle *ATN*, angle *NTB*, angle *BTM*, and angle *MTA*

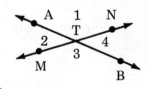

10. In the following triangle, find the angle opposite to:
  a. Side *PR*   **b.** Side *RQ*   **c.** Side *PQ*
      ∠*Q*   ∠*P*   ∠*R*

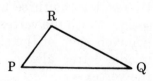

III. Kinds of Angles

The degree, indicated by the symbol (°) is the unit of measure of angles and arcs. A degree is $\frac{1}{360}$ part of the entire angular measure about a point in a plane. If a circle is divided into 360 equal parts and lines are drawn from the center to these points of divison, 360 equal central angles are formed each measuring 1 degree. Each of the corresponding 360 equal arcs also measures 1 degree.

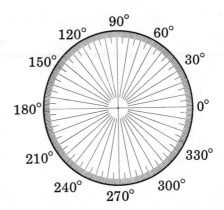

The degree (°) is divided into 60 equal parts called minutes (').
The minute (') is divided into 60 equal parts called seconds (").

> To use these symbols:
>
> 360 degrees is written 360°,
> 60 minutes is written 60',
> 60 seconds is written 60".

When a ray turns from one position to another about its fixed endpoint, one complete rotation is equal to 360°.

A right angle is one-fourth of a complete rotation; it is an angle whose measure is 90°.

Right

An acute angle is an angle whose measure is greater than 0° but less than 90°.

Acute

An obtuse angle is an angle whose measure is greater than 90° but less than 180°.

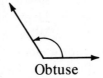

Obtuse

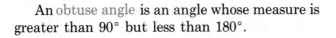

A straight angle is one half of a complete rotation; it is an angle whose measure is 180°. The two rays that form a straight angle extend in opposite directions along a straight line that passes through the vertex.

Straight

A reflex angle is an angle whose measure is greater than 180° but less than 360°.

Reflex

Basic: Prob. 1-7; *Average:* Prob. 1-9; *Enriched:* Prob. 1-10

PRACTICE PROBLEMS

1. What is the unit called that is used to measure angles? What symbol designates it? °   degrees

2. One complete rotation measures __360__°.

3. A right angle measures __90__° and is __$\frac{1}{4}$__ of a rotation.

4. The measure of an acute angle is greater than __0__° and less than __90__°.

5. The measure of an obtuse angle is greater than __90__° and less than __180__°.

6. A straight angle is __$\frac{1}{2}$__ of a rotation and measures __180__°.

7. Which of the following are measures of an acute angle?
<u>67°</u>   98°   <u>89°</u>   155°   90°

8. Which of the following are measures of an obtuse angle?
39°   <u>126°</u>   <u>94°</u>   88°   191°

9. Indicate by writing corresponding letters which of the following angles are:

a. right angles **b.** acute angles **c.** obtuse angles **d.** straight angles

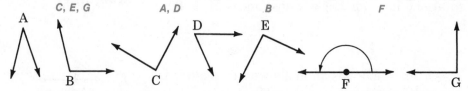

C, E, G   A, D   B   F

10. a. Draw any acute angle.
b. Draw any obtuse angle.
c. What kind of angle is formed by the hands of a clock at 3 o'clock? Describe only the smaller of the two angles formed. A right angle

Measuring and Drawing Angles

**OBJECTIVE: To measure
and draw angles.**

VOC.: protractor

I. Measuring Angles

Measuring an angle means to determine how many units of angular measure are contained in it. The protractor is an instrument used to measure an angle.

The size of an angle does not depend on the length of its sides. The symbol "$m \angle ABC$" is read "the measure of angle ABC."

To measure an angle, place the straight edge of the protractor on one side of the angle with its center mark at the vertex of the angle. Read the number of degrees at the point where the other side of the angle cuts the protractor, using the scale which has its zero on one side of the angle.

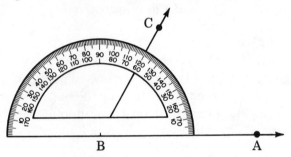

This angle measures 60°, written either as $m \angle ABC = 60°$ or briefly as $\angle ABC = 60°$.

Basic: Prob. 1–2; Average: Prob. 1–4; Enriched: Prob. 1–5

PRACTICE PROBLEMS

1. Estimate the size of each of the following angles. Then measure each angle with a protractor.

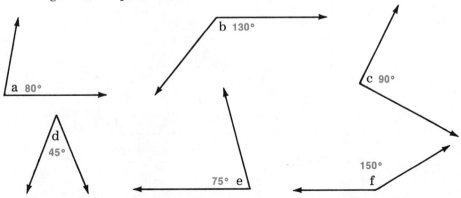

2. Write the amount of error and percent of error of each estimate you made in Problem 1. Answers will vary.

3. a. Which of the following angles is smaller? angle s

b. The size of an angle does not depend upon the length of the sides .

4. How many degrees are in the angle formed by the hands of a clock:

    a. at 5 o'clock? 150°   **c.** at 2 o'clock? 60°

    b. at 3 o'clock? 90°   **d.** at 6 o'clock? 180°

5. Through how many degrees does the minute hand of a clock turn in:

    a. 30 minutes? 180°   **c.** 15 minutes? 90°   **e.** 40 minutes? 240°

    b. 10 minutes? 60°   **d.** 25 minutes? 150°   **f.** 1 hour? 360°

II. Drawing Angles

To draw an angle with a given measure, draw a ray to represent one side of the angle. Sometimes this ray (or line segment which is part of a ray) is already drawn. Place the protractor so that its straight edge falls on this ray, and its center mark is on the endpoint which becomes the vertex of the angle. This vertex may also be any point on the line.

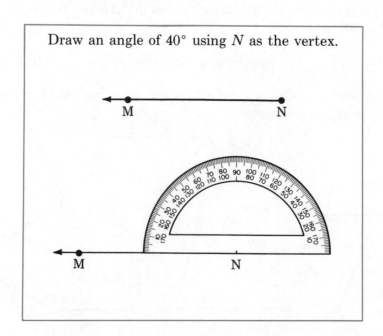

Draw an angle of 40° using N as the vertex.

Counting on the scale which has its zero on the ray, locate the required number of degrees and indicate its position by a dot. Remove the protractor and then draw a ray from the vertex through this dot.

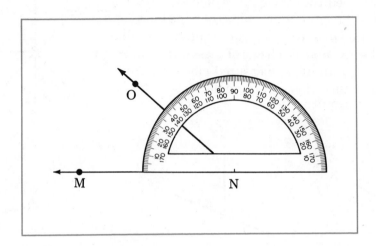

Basic: Prob. 1; *Average:* Prob. 1–2; *Enriched:* Prob. 1–3

1.–3. Constructions

PRACTICE PROBLEMS

1. For each of the following first draw a ray with the endpoint either on the left or on the right as required, then with a protractor draw the angle of the given measure.

   a. Left endpoint as the vertex:

      30°, 75°, 83°, 120°, 270°, 145°, 100°, 58°, 225°, 305°

   b. Right endpoint as the vertex:

      80°, 45°, 9°, 165°, 330°, 25°, 124°, 67°, 96°, 200°

2. With a protractor measure each of the following angles, then draw an angle equal to it.

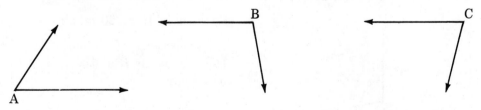

3. With a protractor draw: **a.** a right angle   **b.** a straight angle

OBJECTIVE: To use angles
in navigation applications.

VOC.: course   air speed
      bearing   ground speed
      azimuth   vector

An angle is often used to show the direction (usually called the course) in which an airplane flies or the direction from which the wind blows or the direction of one object from another (usually called the bearing). Each one of these directions is indicated by an angle measured clockwise from the north direction. This angle is sometimes called an azimuth.

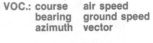

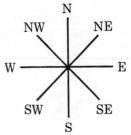

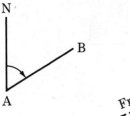

Course: 60°

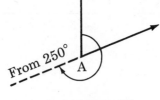

Wind from 250°

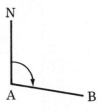

Bearing of B
100° from A

Winds decrease or increase the speed developed by an airplane in flight through the air. The speed of the plane in still air is called air speed. The actual speed of the plane measured by land markings is called ground speed. The direction, represented by an angle, in which the plane points is called the heading.

A vector is an arrow which represents a speed or force and at the same time indicates a direction. This vector is drawn to scale usually as one side of the angle which represents the direction measured clockwise from the north.

Find the position of an airplane at the end of 1 hour when flying on a course of 100° with a ground speed of 416 m.p.h.

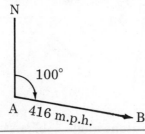

Scale: 1 inch = 320 m.p.h.

Airplane takes off at point A.
Reaches point B at the end of 1 hour.

PRACTICE PROBLEMS

1. If you face north, then turn clockwise to face west, how many degrees do you turn? 270°

2. If you face northwest, then turn clockwise to face southeast, how many degees do you turn? 180°

3. If you face east, then turn clockwise to face southwest, how many degrees do you turn? 135°

In each of the following constructions, first draw a vertical line to indicate the north direction.

4. Draw the following course angles: 4.–9. Constructions

a. 80° **b.** 145° **c.** 210° **d.** 300° **e.** 25° **f.** 160° **g.** 64° **h.** 237° **i.** 350°

5. Draw angles indicating the following directions *from* which winds are blowing:

a. 50° **b.** east (90°) **c.** west (270°) **d.** south (180°) **e.** 130° **f.** 200° **g.** 335° **h.** southwest (225°) **i.** 95° **j.** northwest (315°)

6. Draw angles indicating the following bearings of objects from a given point:

a. 35° **b.** 160° **c.** 240° **d.** 312° **e.** 23° **f.** 98° **g.** 187° **h.** 205° **i.** 341°

7. Using appropriate scales, make scale drawings showing the positions of airplanes at the end of one hour flying:

a. On a course of 130° with a ground speed of 850 km/h

b. With a heading of 70° and an air speed of 360 m.p.h.

c. On a course of 48° with a ground speed of 560 km/h

d. With a heading of 310° and an air speed of 780 km/h

e. On a course of 263° with a ground speed of 460 m.p.h.

f. With a heading of 97° and an air speed of 640 km/h

8. Draw vectors representing the following wind velocities:

a. 50 m.p.h. wind from 45° **b.** 40 km/h wind from 200° **c.** 25 km/h north wind (0°) **d.** 15 m.p.h. wind from 135° **e.** 45 m.p.h. wind from 325° **f.** 60 km/h wind from 225°

9. Make a scale drawing showing the position of an airplane at the end of 3 hours when flying on a course of 60° with a ground speed of 750 km/h.

10. An airplane, flying due east (90°) at an air speed of 280 m.p.h., was blown off its couse by a 35 m.p.h. south wind (180°). Make a scale drawing showing the position of the airplane at the end of one hour. How far was the airplane from its starting point? What was its ground speed? What course (or track) was it actually flying? 282 mi.; 282 m.p.h.; 83°

3-10 Sum of Angles of Polygons

OBJECTIVE: To add the angles of a polygon.

I. Triangle

Draw any triangle. Measure its three angles. What is the sum of the measure of these three angles? Draw a second triangle. Find the sum of the measures of its three angles. Find the sum of the measures of the three angles of a third triangle.

Do you see that *the sum of the measures of the angles of any triangle is 180°?*

Basic: Prob. 1–4; *Average:* Prob. 1–6; *Enriched:* Prob. 1–7

PRACTICE PROBLEMS

1. In each of the following, find the measure of the third angle of a triangle when the other two angles measure:

a. 29° and 87° 64°

b. 63° and 72° 45°

c. 106° and 51° 23°

d. 90° and 45° 45°

e. 18° and 121° 41°

f. 133° and 6° 41°

2. What is the measure of each angle of an equiangular triangle? 60°

3. If each of the equal angles of an isosceles triangle measures 59°, find the measure of the third angle. 62°

4. If the vertex angle of an isosceles triangle measures 70°, find the measure of each of the other two angles. 55°

5. Why can there be only one right or one obtuse angle in a triangle? What kind of angle must each of the other two angles be?

6. In a right triangle if one of the acute angles measures 34°, what is the measure of the other acute angle? 56°

7. If two angles of one triangle are equal respectively to two angles of another triangle, why are the third angles equal?

5. Since the sum of the measures of the three angles of any triangle is 180°, the measure of any two angles cannot equal or exceed 180°; acute

II. Quadrilateral

Draw a parallelogram and a trapezoid. Measure the four angles in each. What is the sum of the measures of the angles of the parallelogram? Of the trapezoid?

Draw any other quadrilateral. Find the sum of the measures of the angles of the quadrilateral.

Do you see that *the sum of the measures of the angles of any quadrilateral is 360°?*

7. The third angle in each triangle equals 180° minus the sum of the other two angles. Since the sum of the two given angles in each triangle is the same, the third angles are equal.

PRACTICE PROBLEMS

1. What is the sum of the measures of the angles of a rectangle? 360° Of a square? 360°

2. In each of the following, find the measure of the fourth angle of a quadrilateral when the other three angles measure:

a. 45°, 85°, and 97° 133° **b.** 119°, 90°, and 90° 61° **c.** 76°, 102°, and 161° 21°

3. The opposite angles of a parallelogram are equal. If one angle measures 65°, find the measures of the other three angles. 65°; 115°; 115°

4. Three angles of a trapezoid measure 90°, 90°, and 108°. What is the measure of the fourth angle? 72°

III. Other Polygons

Draw a pentagon, a hexagon, an octagon, and a decagon. What is the sum of the measures of the angles of each of these geometric figures? Check whether your sum of measures of the angles of each of these figures matches the angular measure determined by substituting the number of sides in the polygon for n in the expression $180(n - 2)$ and performing the required operations. Is this also true for the triangle and the quadrilateral?

Do you see that *the sum of measures of the angles of a polygon of n sides is 180(n − 2)?*

REFRESH YOUR SKILLS

1. Add: 9,657
14,986
789
7,494
$\overline{32,926}$

2. Subtract:
400,000
69,058
$\overline{330,942}$

3. Multiply:
490
807
$395,\overline{430}$

4. Divide:
$56\overline{)41,328}$
738

5. Add:
$6\frac{2}{3} + 1\frac{3}{5}$ $8\frac{4}{15}$

6. Subtract:
$\frac{5}{6} - \frac{3}{8}$ $\frac{11}{24}$

7. Multiply:
$5 \times 4\frac{7}{10}$ $23\frac{1}{2}$

8. Divide:
$37\frac{1}{2} \div 3\frac{1}{8}$ 12

9. Add:
$7.58 + .2$ 7.78

10. Subtract:
$9.4 - .51$ 8.89

11. Multiply:
$15 \times .04$ $\frac{3}{5}$

12. Divide:
$\$25\overline{)\$10}$ .4

13. Find 6.13% of $8,450. $517.99

14. What percent of 300 is 42? 14%

15. 125% of what number is 70? 56

16. Square $2\frac{2}{3}$ $7\frac{1}{9}$

17. Find the square root of .9216 .96

18. Write 9.48 *billion* as a complete numeral. 9,480,000,000

3-11 Pairs of Angles

OBJECTIVE: To name and identify pairs of angles.

VOC.: complementary angle
supplementary angle
exterior angle
vertical angles
opposite angles

I. Complementary Angles

Complementary angles are two angles whose sum of measures is 90°.

Basic: Prob. 1–3; *Average:* Prob. 1–4; *Enriched:* Prob. 1–5

PRACTICE PROBLEMS

1. Which of the following pairs of angles are complementary?

a. $m \angle E = 50°$, $m \angle G = 40°$
b. $m \angle R = 37°$, $m \angle T = 53°$
c. $m \angle 6 = 22°$, $m \angle 3 = 68°$
d. $m \angle a = 39°$, $m \angle b = 41°$
e. $m \angle D = 80°$, $m \angle H = 100°$
f. $m \angle 4 = 76°$, $m \angle 5 = 14°$

2. Find the measure of the angle that is the complement of each of the following angles.

a. $m \angle 2 = 21°$ 69°   **c.** $m \angle B = 6°$ 84°   **e.** $m \angle x = 43° \ 18'$ 46°42'
b. $m \angle L = 87°$ 3°   **d.** $m \angle 8 = 33°$ 57°   **f.** $m \angle P = 68° \ 51' \ 15''$ 21°8'45''

3. Angle ABC is a right angle.

a. If $m \angle 2 = 26°$, find the measure of $\angle 1$. 64°
b. If $m \angle 1 = 53°$, find the measure of $\angle 2$. 37°
c. If $m \angle 2 = 14°$, find the measure of $\angle 1$. 76°
d. If $m \angle 1 = 67°$, find the measure of $\angle 2$. 23°
e. If $m \angle 2 = 42°30'$, find the measure of $\angle 1$.
47°30'

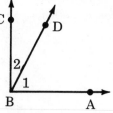

4. In the figure $\overrightarrow{FB} \perp \overrightarrow{FD}$ and $\overrightarrow{FC} \perp \overleftrightarrow{AE}$.

a. If $m \angle 4 = 55°$, find the measure of $\angle 1$. Of $\angle 2$. Of $\angle 3$. 35°; 55°; 35°

b. If $m \angle 1 = 61°$, find the measure of $\angle 2$. Of $\angle 3$. Of $\angle 4$. 29°; 61°; 29°

c. If $m \angle 3 = 74°$, find the measure of $\angle 1$. Of $\angle 2$. Of $\angle 4$. 74°; 16°; 16°

d. If $m \angle 2 = 19°$, find the measure of $\angle 3$. Of $\angle 1$. Of $\angle 4$. 71°; 71°; 19°

e. If $m \angle 4 = 25° \ 45'$, find the measure of $\angle 3$. Of $\angle 2$. Of $\angle 1$. 64°15'; 25°45'; 64°15'

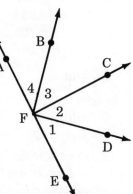

5. An acute angle is an angle greater than 0° but less than 90°. An acute angle subtracted from a right angle must result in an angle that is greater than 0° but less than 90°, or another acute angle. **5.** Why is the complement of an acute angle also an acute angle?

Related Resources: *Practical Applications in Mathematics:* Problems, p. 107

II. Supplementary Angles

Supplementary angles are two angles whose sum of measures is 180°.

Basic: Prob. 1–2; *Average:* Prob. 1–3; *Enriched:* Prob. 1–4

PRACTICE PROBLEMS

1. Which of the following pairs of angles are supplementary?

a. $m\angle A = 134°, m\angle N = 46°$
b. $m\angle 3 = 85°, m\angle 4 = 105°$
c. $m\angle B = 79°, m\angle D = 91°$
d. $m\angle c = 127°, m\angle e = 53°$
e. $m\angle T = 32°, m\angle F = 148°$
f. $m\angle G = 63°, m\angle K = 27°$

2. Find the measure of the angle that is the supplement of each of the following angles.

a. $m\angle H = 44°$ 136°
b. $m\angle a = 117°$ 63°
c. $m\angle 5 = 89°$ 91°

d. $m\angle R = 8°$ 172°
e. $m\angle Q = 96° \ 20'$ 83°40'
f. $m\angle E = 170° \ 42' \ 59''$ 9°17'1''

3. Measure $\angle CDF$ and $\angle EDF$. Is the sum of their measures 180°? Do you see that *when one straight line meets another, the adjacent angles, which have the same vertex and a common side, are supplementary?* Yes; yes

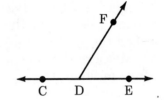

a. What is the measure of $\angle CDF$ when $m\angle EDF = 32°$? 148°
b. Find the measure of $\angle EDF$ when $m\angle CDF = 116°$. 64°
c. Find the measure of $\angle CDF$ when $m\angle EDF = 90°$. 90°
d. What is the measure of $\angle EDF$ when $m\angle CDF = 145°$? 35°
e. Find the measure of angle CDF when $m\angle EDF = 76° \ 30'$. 103°30'

4. Is the supplement of an obtuse angle also an obtuse angle? Explain you answer. No, an acute angle. Since an obtuse angle is greater than 90°, its supplement must be less than 90°.

III. Opposite or Vertical Angles

Draw two intersecting lines, forming four angles. Measure a pair of angles that are directly opposite to each other. Are their measures equal? Measure the other pair of opposite angles. Are their measures equal?

Do you see that, *when two straight lines intersect, the opposite (vertical angles) are equal?*

PRACTICE PROBLEMS

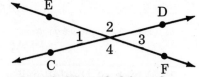

1. In the drawing at the right, what angle is opposite to $\angle 1$? Does $\angle 1 = \angle 3$? What angle is opposite to $\angle 4$? Does $\angle 4 = \angle 2$? $\angle 3$; yes; $\angle 2$; yes

2. If $m\angle 3 = 68°$, what is the measure of $\angle 1$? Of $\angle 2$? Of $\angle 4$?

3. If $m\angle 4 = 110°$, what is the measure of $\angle 2$? Of $\angle 1$? Of $\angle 3$?

4. If $m\angle 1 = 17°$, what is the measure of $\angle 4$? Of $\angle 2$? Of $\angle 3$?

5. If $m\angle 2 = 145°$, what is the measure of $\angle 3$? Of $\angle 4$? Of $\angle 1$?

2. 68°; 112°; 112°   3. 110°; 70°; 70°   4. 163°; 163°; 17°   5. 35°; 145°; 35°

IV. Exterior Angle of a Triangle

Draw a triangle and extend one side like the drawing at the right. Label the angles as shown in the drawing.

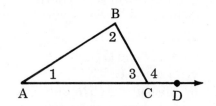

The angle formed by a side of the triangle and the adjacent side extended (see $\angle 4$ in the drawing) is called the exterior angle.

Measure the three angles of the triangle and the indicated exterior angle in your drawing.

What is the sum of the measures of $\angle 1$, $\angle 2$, and $\angle 3$? What is the sum of the measures of $\angle 4$ and $\angle 3$?

Why does the measure of $\angle 4$ equal the sum of the measures of $\angle 1$ and $\angle 2$?

Do you see that *the measure of an exterior angle of a triangle is equal to the sum of the measures of the opposite two interior angles?*

Basic: Prob. 1–4; *Average:* Prob. 1–5; *Enriched:* Prob. 1–6

PRACTICE PROBLEMS

111°; 69° **1.** If $m\angle 1 = 42°$ and $m\angle 2 = 69°$, find the measure of $\angle 4$. Of $\angle 3$.

52°; 127° **2.** If $m\angle 3 = 53°$ and $m\angle 1 = 75°$, find the measure of $\angle 2$. Of $\angle 4$.

61°; 99° **3.** If $m\angle 2 = 38°$ and $m\angle 3 = 81°$, find the measure of $\angle 1$. Of $\angle 4$.

84°; 70° **4.** If $m\angle 4 = 110°$ and $m\angle 1 = 26°$, find the measure of $\angle 2$. Of $\angle 3$.

49°; 86° **5.** If $m\angle 2 = 45°$ and $m\angle 4 = 131°$, find the measure of $\angle 3$. Of $\angle 1$.

6. Can an exterior angle of a triangle have the same measure as one of the angles of the triangle? As one of the opposite interior angles of the triangle? Explain your answers. For ex. 6 see Extended Answers, p. T79.

Parallel Lines and Angle Relationships

OBJECTIVE: To determine the angle relationships when parallel lines are cut by another line.

VOC.: alternate-interior angles
corresponding angles
transversal

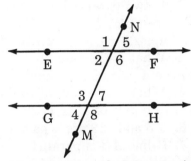

In the figure at the right, $\overleftrightarrow{EF}$ and $\overleftrightarrow{GH}$ are parallel and cut by $\overleftrightarrow{MN}$; $\angle 2$ and $\angle 7$ form a pair of alternate-interior angles and $\angle 3$ and $\angle 6$ form another pair of alternate-interior angles. These angles are between the parallel lines and the related angles fall on alternate sides.

$\angle 1$ and $\angle 3$, $\angle 2$ and $\angle 4$, $\angle 5$ and $\angle 7$, and $\angle 6$ and $\angle 8$ are pairs of corresponding angles. Each pair of angles is in a corresponding position. Observe that when two parallel lines are cut by a third line (called a transversal), both the corresponding angles are equal and the alternate-interior angles are equal.

Also, when two lines are cut by a transversal making a pair of corresponding angles or a pair of alternate-interior angles equal, the lines are parallel.

Basic: Prob. 1–6; *Average:* Prob. 1–8; *Enriched:* Prob. 1–9

PRACTICE PROBLEMS

Use the figure below for all problems. Complete each statement as required in problems 1 through 6.

1. If lines AB and CD are cut by a third line EF,
  a. the line EF is called a _____ . transversal
  b. $\angle 6$ and $\angle 3$ are _____ angles. alternate-interior angles
  c. $\angle 5$ and $\angle 4$ are _____ angles. alternate-interior angles
  d. $\angle 5$ and $\angle 1$ are _____ angles.
  e. $\angle 8$ and $\angle 4$ are _____ angles.

d. corresponding angles

e. corresponding angles

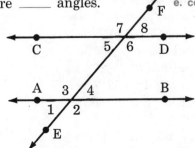

2. If two parallel lines are cut by a transversal, the alternate-interior angles are equal . Therefore, $\angle 3 = $ $\angle 6$, $\angle 5 = $ $\angle 4$.

3. If two parallel lines are cut by a transversal, the corresponding angles are equal .

Therefore, $\angle 2 = $ $\angle 6$, $\angle 7 = $ $\angle 3$, $\angle 1 = $ $\angle 5$, $\angle 8 = $ $\angle 4$.

4. What is the sum of the measures of $\angle 5$ and $\angle 7$? 180°

They are a pair of _____ angles. supplementary

5. What is the sum of the measures of $\angle 1$ and $\angle 2$? 180°

They are a pair of _____ angles. supplementary

6. $\angle 5$ and $\angle 8$ are a pair of _____ angles. vertical or opposite

For ex. 7–8
see Extended
Answers, p. T79.

7. If line AB is parallel to line CD,

a. Show that $\angle 2 = \angle 7$.

b. Show that $\angle 1$ and $\angle 6$ are supplementary angles.

c. Show that $\angle 8 = \angle 1$.

d. Show that $\angle 3$ and $\angle 5$ are supplementary angles.

e. Show that $\angle 8$ and $\angle 2$ are supplementary angles.

8. If line AB is parallel to line CD, what is the measure of:

a. $\angle 6$ if $m\angle 2 = 130°$?

b. $\angle 7$ if $m\angle 6 = 106°$?

c. $\angle 5$ if $m\angle 4 = 75°$?

d. $\angle 2$ if $m\angle 7 = 93°$?

e. $\angle 3$ if $m\angle 8 = 47°$?

f. $\angle 8$ if $m\angle 1 = 64°$?

g. $\angle 2$ if $m\angle 5 = 83°$?

h. $\angle 4$ if $m\angle 6 = 115°$?

i. all other angles when $m\angle 8 = 48°$?

j. all other angles when $m\angle 3 = 125°$?

9. If two lines are cut by a transversal making a pair of corresponding angles or a pair of alternate-interior angles equal, the lines are _____. parallel

10. Are lines AB and CD parallel when:

a. $m\angle 1 = 70°$ and $m\angle 5 = 70°$? Yes

b. $m\angle 7 = 120°$ and $m\angle 3 = 120°$? Yes

c. $m\angle 7 = 135°$ and $m\angle 4 = 45°$? Yes

d. $m\angle 2 = 108°$ and $m\angle 8 = 82°$? No

e. $m\angle 8 = 52°$ and $m\angle 1 = 52°$? Yes

f. $m\angle 4 = 81°$ and $m\angle 6 = 99°$? Yes

g. $m\angle 1 = 89°$ and $m\angle 6 = 89°$? No

h. $m\angle 5 = 69°$ and $m\angle 2 = 111°$? Yes

i. $m\angle 3 = 94°$ and $m\angle 7 = 86°$? No

Congruent Triangles

**OBJECTIVE: To identify
congruent triangles.**

VOC.: congruent triangles

Congruent triangles are triangles which have exactly the same shape and the same size. The corresponding sides are equal in length and the corresponding angles are equal in size. The symbol ≅ means "is congruent to."

Two triangles are congruent when any of the following combinations of three parts are known:

(1) Three sides of one triangle are equal to three sides of the second triangle.

(2) Two sides and an included angle of one triangle are equal respectively to two sides and an included angle of the other.

(3) Two angles and an included side of one triangle are equal respectively to two angles and an included side of the other.

Basic: Prob. 1–4; *Average:* Prob. 1–6; *Enriched:* Prob. 1–7

PRACTICE PROBLEMS For ex. 1–3 see Extended
Answers, p. T79.

Select in each of the following groups two triangles which are congruent and state the reason why.

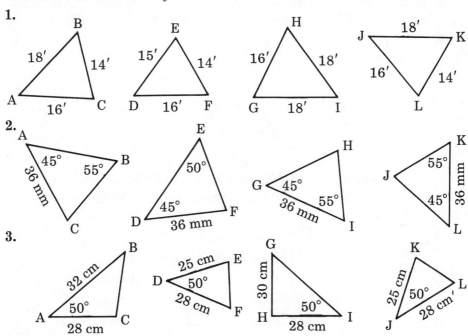

1.

2.

3.

Related Resources: *Practical Applications in
Mathematics:* Problems, p. 109

Find the indicated missing parts in the following triangles:

4.

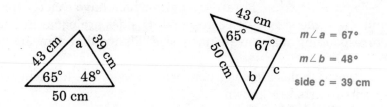

m∠a = 67°

m∠b = 48°

side c = 39 cm

5.

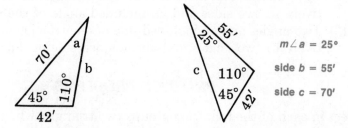

m∠a = 25°

side b = 55′

side c = 70′

6. In triangle *ABC*, side *AB* = side *BC* and *BD* bisects ∠*ABC* making ∠*ABD* = ∠*DBC*. Prove that triangle *ABD* and triangle *DBC* are congruent.

For ex. 6–7
see Extended
Answers, p. T79.

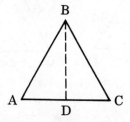

7. In parallelogram *EFGH*, side *EF* = side *HG* and side *EH* = side *FG*. Prove that the diagonal *EG* divides the parallelogram into two congruent triangles.

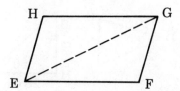

Similar Triangles

**OBJECTIVE: To identify
similar triangles.**

Similar triangles are triangles which have the same shape but differ in size. Two triangles are similar when any of the following conditions are known:

(1) Two angles of one triangle are equal to two angles of the other triangle.
(2) The ratios of the corresponding sides are equal.
(3) Two sides of one triangle are proportional (equal ratios) to two corresponding sides of the other triangle and the included angles are equal.

Basic: Prob. 1–4; *Average:* Prob. 1–6; *Enriched:* Prob. 1–7

PRACTICE PROBLEMS

Select in each of the following groups two triangles which are similar and state the reason why.

1.

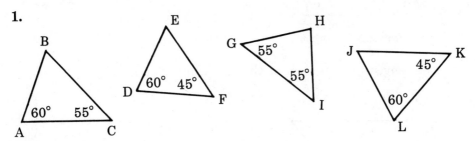

Triangles *DEF* and *JKL* are similar because two angles of triangle *DEF* are equal to two angles of triangle *JKL*.

2.

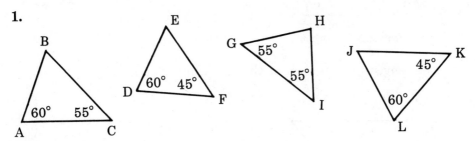

Triangles *DEF* and *GHI* are similar because the ratios of the corresponding sides are equal.

Related Resources: *Practical Applications in Mathematics:* Problems, p. 110

3.

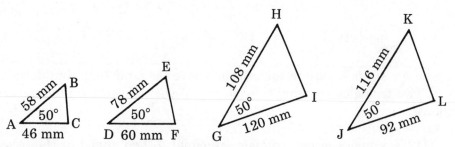

Triangles *ABC* and *JKL* are similar because two sides of triangle *ABC* are proportional to two corresponding sides of triangle *JKL* and the included angles are equal.

Find the lengths of *a* and *b* in each of the following figures:

4.

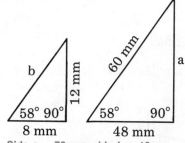

Side *a* = 72 mm; side *b* = 10 mm

5.

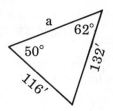

Side *a* = 144′; side *b* = 33′

Find the length of *a* in each of the following figures:

6.

a

225 m   210 m

42 m   45 m

55 m

Side *a* = 275 m

7.

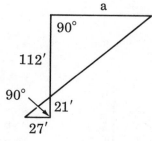

Side *a* = 144′

For a Class Activity see p. T34.

Longitude, Latitude, and Nautical Mile

OBJECTIVE: To find latitudes, longitudes, and nautical miles.

The position of any point on the earth's surface is determined by the intersection of its meridian of longitude and its parallel of latitude. Meridians of longitude are imaginary circles which pass through the North Pole and South Pole. Parallels of latitude are imaginary circles which are parallel to the equator.

The prime meridian from which longitude is calculated is the meridian that passes through Greenwich near London, England. West longitude extends from this prime meridian (0° longitude) westward halfway around the earth to the International Date Line (180° longitude). East longitude extends eastward from the prime meridian to the International Date Line.

The equator is 0° latitude. North latitude is measured north of the equator and south latitude is measured south of the equator. The North Pole is 90° north latitude and the South Pole is 90° south latitude.

North latitude is indicated by the letter N, south latitude by S, east longitude by E, and west longitude by W.

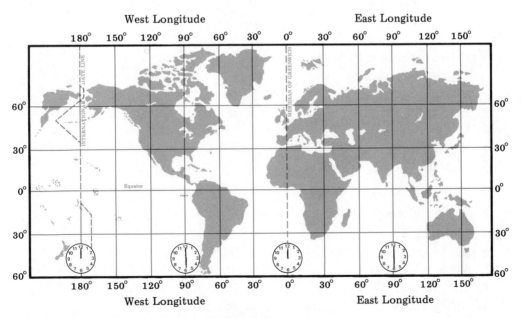

MAP OF THE WORLD

Since one minute of arc of latitude equals one nautical mile, *to find the distance between two points on the earth's surface located on the same meridian,* find the difference in latitudes, convert to minutes of arc, and then to nautical miles.

Since one minute of arc of longitude measured at the equator only also equals one nautical mile, *to find the distance between two points on the equator,* find the difference in longitude, convert to minutes of arc, and then to nautical miles.

How many nautical miles apart are two ships if one is located 24° N. 36° W. and the other 17° N. 36° W.?

Both ships are located on 36° W. Therefore,

| 24° N. | | 7° = 420′ |
| 17° N. | | 420′ = 420 nautical miles |

7° difference in latitudes

Answer: 420 nautical miles

Basic: Sets 1–4; *Average:* Sets 1–6; *Enriched:* Sets 1–7

PRACTICE PROBLEMS

Find the difference in latitudes:

SET 1

1. 57° N. and 12° N. 45°
2. 29° S. and 63° S. 34°
3. 43° N. and 8° S. 51°
4. 38° S. and 51° N. 89°
5. 0° and 21° S. 21°

SET 2

1. 16° 25′ N. and 42° 40′ N. 26°15′
2. 58° 16′ S. and 23° 27′ S. 34°49′
3. 27° 32′ N. and 10° 51′ S. 38°23′
4. 39° 6′ S. and 4° 30′ N. 43°36′
5. 42° 18′ 25″ N. and 15° 14′ 49″ S.
   57°33′14″

Find the difference in longitudes:

SET 3

1. 36° W. and 93° W. 57°
2. 58° E. and 14° E. 44°
3. 119° W. and 25° E. 144°
4. 45° E. and 0° 45°
5. 7° E. and 62° W. 69°

SET 4

1. 93° 48′ W. and 67° 29′ W. 26°19′
2. 6° 17′ E. and 31° 14′ E. 24°57′
3. 135° 5′ W. and 43° 52′ E. 178°57′
4. 28° 40′ E. and 104° 36′ W. 133°16′
5. 72° 57′ 26″ W. and 64° 2′ 51″ E.
   137°0′17″

SET 5

Find how far apart in nautical miles ships A and B are if:

Position of Ship A	Position of Ship B
1. 46° N., 140° W.	18° N., 140° W. 1,680 naut. mi.
2. 62° S., 52° W.	49° S., 52° W. 780 naut. mi.
3. 29° S., 46° W.	7° N., 46° W. 2,160 naut. mi.
4. 0°, 27° 12′ W.	0°, 42° 54′ W. 942 naut. mi.
5. 26° 43′ N., 176° 39′ E.	18° 26′ S., 176° 39′ E. 2,709 naut. mi.

SET 6

Find how far apart airports X and Y are if:

Location of Airport X	Location of Airport Y
1. 56° N., 94° W.	48° N., 94° W. 480 naut. mi.
2. 37° 43′ N., 83° 25′ W.	33° 29′ N., 83° 25′ W. 254 naut. mi.
3. 46° 32′ N., 118° 43′ W.	48° 6′ N., 118° 43′ W. 94 naut. mi.
4. 28° 17′ N., 105° 39′ W.	35° 58′ N., 105° 39′ W. 461 naut. mi.
5. 54° 27′ 45″ N., 110° 8′ 15″ W.	63° 14′ 30″ N., 110° 8′ 15″ W. 526.75 naut. mi.

SET 7

1. How many nautical miles away is a Coast Guard cutter located 47° 38′ N., 25° 30′ W. from a boat in distress located 49° 28′ N., 25° 30′ W.? How long will it take the cutter, traveling at 20 knots, to reach the boat in distress? 110 naut. mi.; 5½ hr.

2. A ship left a United States port located 33° N., 80° W. and sailed along the 80th meridian to a Cuban port located 23° N., 80° W. How many nautical miles did the ship sail? 600 naut. mi.

3. An airplane flew from a city located 39° N., 93° W. to a city located 45° N., 93° W. How many nautical miles apart are the two cities? How many statute miles apart? If it took the airplane 1 hour 48 minutes to make the flight, what was its ground speed in knots? In statute miles per hour? 360 naut. mi.; 414 stat. mi.; 200 knots; 230 stat. m.p.h.

4. A ship steamed east along the equator from 168° 20′ W. to 178° 45′ E. before turning south. How many nautical miles did the ship steam along the equator? 775 naut. mi.

5. How long should it take an airplane to fly at an average speed of 460 statute miles per hour from Pittsburgh, located 40° 27′ N., 80° W. to Miami, located 25° 47′ N., 80° W.? 2 hr. 12 min.

3-16 Longitude and Time

OBJECTIVE: To express arcs of longitude as units of time and to express units of time as arcs of longitude.

I. Time Zones*

The earth rotates from the west to the east about its axis, making one complete revolution each day. It takes the earth 24 hours to pass through 360° of longitude or one hour for 15° of longitude or 4 minutes for one degree of longitude.

The sun or solar time is the same at any given instant for all places located on any one meridian. As the sun crosses each meridian, it becomes noon sun or solar time for all the places located on this meridian. To avoid the confusion that would arise if towns, just a few miles apart, used their own local solar time, the earth is divided into time zones each about 15° longitude in width. All places within each zone use the sun time of approximately its central meridian. The difference in time of adjacent zones is one hour, earlier to the west, later to the east.

II. Longitude Expressed in Units of Time

In navigation the longitude is sometimes expressed in units of time. One hour of time is equivalent to 15° of longitude, one minute of time to 15′ of longitude, one second of time to 15″ of longitude, 1° of longitude to 4 minutes of time, and 1′ of longitude to 4 seconds of time.

Both world time and longitude are calculated from the meridian of Greenwich. Thus, the longitude of any place is equal to the difference between its local sun time and Greenwich time.

To change units measuring longitude to units of time, divide the units measuring longitude by 15. Simplify the answer where possible.

To change units of time to units measuring longitude, multiply the units of time by 15. Simplify the answer where possible.

Change 44° 15′ of arc of longitude to units of time.

$$
\begin{array}{r}
2 \text{ hr. } 57 \text{ min.} \\
15)\overline{44°15'} \\
30° \\
\hline
14°15' = 855' \\
855'
\end{array}
$$

Answer: 2 hr. 57 min.

Change 3 hr. 41 min. 18 sec. to units measuring longitude.

$$
\begin{array}{r}
3 \text{ hr. } \quad 41 \text{ min.} \quad 18 \text{ sec.} \\
15 \\
\hline
45° \quad 615' \quad 270'' = 45° \ 619' \ 30'' \\
\quad\quad = 55° \quad 19' \quad 30''
\end{array}
$$

Answer: 55° 19′ 30″

---

*Also see Exercise 2-18.

Related Resources: *Practical Applications in Mathematics:* Problems, p. 212

To find the time at point B when the longitude and time at point A and the longitude at point B are known, convert the difference in longitude to units of time. If point B is east of point A, add the difference in time to the time at point A. If point B is west of point A, subtract the difference in time.

If it is 0814 at 63° 45′ W. longitude, what time is it at 46° 30′ W. longitude?

63° 45′	1 hr. 9 min.
46° 30′	15)17° 15′
17° 15′ difference in longitude	15°
	2° 15′ = 135′
	135′

0814 time at 63° 45′ W.

109

0923 time at 46° 30′ W.

Answer: It is 0923 at 46° 30′ W.

To find the longitude at point B when the longitude and time at point A and the time at point B are known, convert the difference in time to units of longitude. Then use the longitude at point A and this difference in longitude to obtain the longitude at point B.

Basic: Sets 1–12; *Average:* Sets 1–14; *Enriched:* Sets 1–16

PRACTICE PROBLEMS

Express the following arcs of longitude as units of time:

SET 1	SET 2	SET 3	SET 4
1. 45° 3 hr.	**1.** 30′ 2 min.	**1.** 60° 45′ 1 hr. 3 min.	**1.** 20° 1 hr. 20 min.
2. 75° 5 hr.	**2.** 45′ 3 min.	**2.** 135° 30′ 9 hr. 2 min.	**2.** 175° 11 hr. 40 min.
3. 180° 12 hr.	**3.** 15″ 1 sec.	**3.** 45′ 15″ 3 min. 1 sec.	**3.** 38′ 2 min. 32 sec.
4. 105° 7 hr.	**4.** 30″ 2 sec.	**4.** 15° 30′ 45″	**4.** 52′ 3 min. 28 sec.
5. 300° 20 hr.	**5.** 15′ 1 min.	**5.** 120° 15′ 30″	**5.** 84° 5 hr. 36 min.

4. 1 hr. 2 min. 3 sec.
5. 8 hr. 1 min. 2 sec.

SET 5	SET 6	SET 7	SET 8
1. 8° 32 min.	**1.** 17° 45′ 1 hr. 11 min.	**1.** 49° 26′ 3 hr. 17 min. 44 sec.	**1.** 25° 18′ 30″
2. 14° 56 min.	**2.** 31° 15′ 2 hr. 5 min.	**2.** 28° 20′ 1 hr. 53 min. 20 sec.	**2.** 106° 37′ 15″
3. 5′ 20 sec.	**3.** 55′ 30″ 3 min. 42 sec.	**3.** 6° 51′ 27 min. 24 sec.	**3.** 3° 42′ 45″
4. 11′ 44 sec.	**4.** 7° 15′ 29 min.	**4.** 1° 43′ 6 min. 52 sec.	**4.** 9° 20′ 15″
5. 2° 8 min.	**5.** 26′ 45″ 1 min. 47 sec.	**5.** 2° 39′ 10 min. 36 sec.	**5.** 16° 58′ 30″

Set 8: 1. 1 hr. 41 min. 14 sec.   2. 7 hr. 6 min. 36 sec.   3. 14 min. 51 sec.   4. 37 min. 21 sec.
5. 1 hr. 7 min. 54 sec.

Express the following units of time as arcs of longitude:

SET 9	SET 10	SET 11	SET 12	SET 13
1. 4 hr.	**1.** 30 min.	**1.** 15 sec.	**1.** 2 hr. 15 min.	**1.** 3 hr. 45 min. 30 sec.
2. 7 hr.	**2.** 45 min.	**2.** 40 sec.	**2.** 1 hr. 24 min.	**2.** 2 hr. 12 min. 48 sec.
3. 2 hr.	**3.** 20 min.	**3.** 12 sec.	**3.** 4 hr. 50 min.	**3.** 1 hr. 29 min. 10 sec.
4. 23 hr.	**4.** 36 min.	**4.** 30 sec.	**4.** 15 hr. 31 min.	**4.** 8 hr. 3 min. 54 sec.
5. 10 hr.	**5.** 15 min.	**5.** 49 sec.	**5.** 9 hr. 6 min.	**5.** 5 hr. 37 min. 25 sec.

SET 14

1. If it is 1200 at Greenwich, what time is it at these longitudes?
a. 30° W.   **b.** 21° E.   **c.** 135° E.   **d.** 75° 45′ W.   **e.** 64° 15′ E.

2. If it is 1455 at Greenwich, what time is it at these longitudes?
a. 45° E.   **b.** 120° W.   **c.** 60° 30′ W.   **d.** 49° 20′ E.   **e.** 122° 37′ W.

3. If it is 0732 at 50° W. longitude, what time is it at 72° W. longitude?

4. What time is it at 81° 45′ W. longitude if it is 1826 at 94° 30′ W. longitude?

5. If it is 0907 at 5° 18′ W. longitude, what time is it at 13° 45′ E. longitude?

SET 15

1. If it is 1200 at Greenwich, at what longitude is the time 1100? 1500? 0600? 1720? 0348?

2. If it is 1040 at Greenwich, at what longitude is the time 1120? 0856? 0200? 1812? 0736?

3. If it is 2330 at 150° E. longitude, at what longitude is the time 1445?

4. At what longitude is the time 0050 if it is 1150 at 10° 52′ W. longitude?

5. If it is 1324 at 81° 31′ W. longitude, at what longitude is the time 1248?

SET 16

1. Find the difference between the standard time and the solar time at each of the following cities:

Cleveland, 81° 34′ W.; Boston, 71° 04′ W.; Chicago, 87° 37′ W.; Los Angeles, 118° 14′ W.; Salt Lake City, 111° 53′ W.; Des Moines, 93° 41′ W.; San Francisco, 122° 26′ W.

2. If the sun time at Philadelphia, 75° 10′ W., is 1700, what is the sun time at each of the following cities:

Louisville, 85° 46′ W.; Washington, D.C., 77° 04′ W.; Seattle, 122° 19′ W.; Denver, 104° 57′ W.; El Paso, 106° 29′ W.; New York City, 73° 58′ W.; Kansas City, 94° 35′ W.

The Magnetic Compass

OBJECTIVE: To find the magnetic course, the compass course, or the true course.

When navigating an airplane by chart and compass, called dead reckoning, the navigator frequently determines the compass course from the true course and the true course from the compass course.

The course of an airplane is the direction in which it flies over the earth's surface and is expressed as an angle. It is called the true course when it is measured clockwise from the true north (North Pole). However, since the compass needle points to the magnetic north, the pilot must correct his true course reading. This correction is called magnetic variation or variation and the corrected course reading is called the magnetic course. This variation is designated as west variation when the magnetic north is west of the true north and as east variation when the magnetic north is east of the true north. The variation for any locality may be found on aeronautical maps.

Since the metal parts of the airplane affect the compass, the magnetic course reading must be corrected. This corrected course reading is called the compass course. The correction caused by the magnetism of the plane is called deviation and is designated east or west.

The following angular equivalents in degrees may be used for the given directions:

North (N.) = 0°   South (S.) = 180°
Northeast (N.E.) = 45°   Southwest (S.W.) = 225°
East (E.) = 90°   .   West (W.) = 270°
Southeast (S.E.) = 135°   Northwest (N.W.) = 315°

To find the magnetic course from the true course, add west variation but subtract east variation.

To find the compass course from the magnetic course, add west deviation but subtract east deviation.

Find the compass course if the true course is 125°; variation, 7° W.; and deviation, 2° W. (See page 390.)

125° true course
+ 7° west variation
───
132° magnetic course
+ 2° west deviation
───
134° compass course

Answer: Compass course is 134°

To find the compass course from the true course, first change true course to magnetic course, then to compass course. (See page 389.)

To find the magnetic course from the compass course, subtract west deviation but add east deviation.

> Find the true course if the compass course is 54°; variation 11° W.; and deviation, 5° E.
>
> $\quad$ 54° compass course
> $\underline{+\ 5°\text{ east deviation}}$
> $\quad$ 59° magnetic course
> $\underline{-\ 11°\text{ west variation}}$
> $\quad$ 48° true course
>
> *Answer:* True course is 48°

To find the true course from the magnetic course, subtract west variation but add east variation.

To find the true course from the compass course, first change compass course to magnetic course, then to true course.

Basic: Sets 1–3; *Average:* Sets 1–4; *Enriched:* Sets 1–5

PRACTICE PROBLEMS

Find the compass course:

SET 1

True Course	Variation	Deviation		True Course	Variation	Deviation	
1. 238°	9° W.	5° W.	252°	1. 302°	14° E.	1° E.	287°
2. N.E.	3° E.	6° E.	36°	2. 224°	8° E.	6° W.	222°
3. 85°	7° W.	3° E.	89°	3. W.	2° W.	4° E.	268°
4. 130°	4° E.	5° W.	131°	4. 113°	0°	0°	113°
5. 69°	0°	4° W.	73°	5. 51°	5° W.	3° W.	59°

(SET 2 headers: True Course, Variation, Deviation)

Find the true course:

SET 3

Compass Course	Deviation	Variation		Compass Course	Deviation	Variation	
1. 246°	1° W.	5° W.	240°	1. 80°	6° W.	5° E.	79°
2. 327°	2° W.	8° E.	333°	2. 349°	2° E.	16° E.	7°
3. 90°	5° E.	12° W.	83°	3. 228°	5° W.	19° W.	204°
4. 50°	6° E.	13° W.	43°	4. 215°	3° E.	0°	218°
5. 17°	0°	7° E.	24°	5. 106°	1° E.	9° W.	98°

(SET 4 headers: Compass Course, Deviation, Variation)

SET 5

1. The true course from airport A to airport B is 102°. If the variation is 5° W. and the deviation is 2° E., what compass course should be steered? 105°

2. At a certain point in flight a navigator's compass course is 267°. 277° If the variation is 6° E. and deviation 4° E., what is the true course?

3. What compass course should a pilot steer if the true course is 61°, variation 7° E., and deviation 5° W.? 59°

4. A pilot is steering a compass course of 175°. If the variation is 12° W. and deviation is 3° E., what is the true course? 166°

5. Determine the compass course of an airplane using the following data: true course, 340°; variation, 9° W.; deviation, 1° W. 350°

6. What compass course should be steered if the true course is 233°, variation 10° W., and deviation 2° W.? 245°

7. A plane is steered along a compass course of 198°. If the variation is 11° W. and deviation is 3° E., what is its true course? 190°

8. If the true course is 87°, magnetic course 94°, and compass course 91°, what is the variation and deviation? variation = 7°W.; deviation = 3°E.

9. Compute the compass course if the true course is 126°, variation 6° E., and deviation 2° E. 118°

10. What is the true course if the compass course is 56°, variation 8° E., and deviation 1° W.? 63°

REFRESH YOUR SKILLS

1. Add:
638
1,259
17
387
52,495
3,798
———
58,594

2. Subtract:
100,000
96,024
——— 3,976

3. Multiply:
6,080
705
——— 4,286,400

4. Divide: 98
$5{,}280\overline{)517{,}440}$

5. Add:
$6\frac{7}{8}$
$9\frac{1}{4}$
$1\frac{5}{16}$ $17\frac{7}{16}$

6. Subtract:
$3\frac{11}{16}$
$\frac{5}{6}$
$2\frac{41}{48}$

7. Multiply:
$4\frac{1}{2} \times 1\frac{3}{8}$ $6\frac{3}{16}$

8. Divide:
$\frac{5}{12} \div 3\frac{1}{3}$ $\frac{1}{8}$

9. Add: $.96 + $4.28 + $19.09 + $6.57 + $.46 $31.36

10. Subtract: $9,150 − $79.25 $9,070.75

11. Multiply: 48 × $23.94 $1,149.12

12. Divide: $.15$\overline{)$9.}$ 60

13. Find 140% of $37.08 $51.91

14. $.54 is what percent of $.72 75%

15. 20% of what number is 9? 45

REVIEW

1. Name the following angle in three ways:
Angle *LMN*, angle *NML*, angle *M*

2. Which of the following are measures of an acute angle?
a. 98°   **b.** 90°   **c.** 89°

3. Measure the following angle: 70°

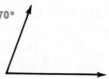

4.–5. Constructions
4. Draw an angle measuring 110°.

5. Draw: **a.** an acute angle   **b.** an obtuse angle   **c.** a right angle

6. Find the measure of the third angle of a triangle when the other two angles measure 67° and 59°. 54°

7. What is the complement of an angle measuring 41°? 49°

8. Find the supplement of an angle measuring 16° 30′. 163°30′

9. In the following figure find $m\angle 3 = 55°$ $m\angle 2 = 93°$ the measures of $\angle 3$ and $\angle 2$ when $m\angle 4 = 125°$ and $m\angle 1 = 32°$.

10. $\overleftrightarrow{CD}$ and $\overleftrightarrow{EF}$ intersect forming $\angle$s 1, 2, 3, and 4. If $m\angle 2 = 105°$, what is the measure of $\angle 1$? $\angle 3$? $\angle 4$? $m\angle 1 = 75°$; $m\angle 3 = 75°$; $m\angle 4 = 105°$

11. Parallel lines *MN* and *OP* are cut by transversal *QR*. If $m\angle 5 = 79°$, what is the measure of $\angle 4$? $\angle 6$? $\angle 3$? $\angle 2$? $\angle 8$? $\angle 1$? $\angle 7$?

12. How many nautical miles apart are two ships if one ship is located at 18° N. 29° W. and the other ship is at 2° S. 29° W.? 1,200 naut. mi.

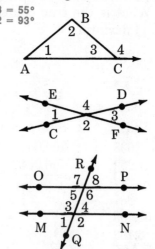

13. What time is it at 60° W. longitude when it is 1200 at Greenwich? 0800

14. What compass course should a pilot steer if the true course is 86°, variation is 9° W., and deviation is 2° E.? 93°

15. Make a scale drawing, using the scale 1 cm = 100 km/h, showing the position of an airplane at the end of one hour when flying on a course of 285° with a ground speed of 650 km/h. Construction

11. $m\angle 4 = 79°$; $m\angle 6 = 101°$; $m\angle 3 = 101°$; $m\angle 2 = 101°$; $m\angle 8 = 79°$; $m\angle 1 = 79°$; $m\angle 7 = 101°$

For a Test on Applications Using Angles and Arcs see p. T50 or *Testing Program*, p. 19.

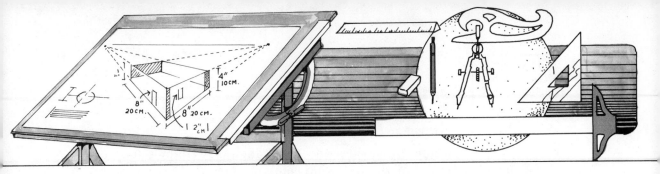

CONSTRUCTIONS

Copying a Line Segment

OBJECTIVE:

To copy a line segment means to draw a line segment equal in length to the given line segment.

(1) *Using a ruler,* first measure the given line segment, then draw another line segment of the same length.

(2) *Using a compass,* first draw a pencil working line. Then select a point on this line where the copy of the given line segment is to begin, indicating it by a dot. Open the compass, placing the metal point on one endpoint of the given line segment and the pencil point on the other endpoint. Transfer this fixed compass setting to the working line, placing the metal point on the marked dot.

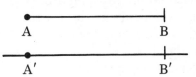

This point forms one endpoint of the copy of the given line segment. Draw an arc, cutting the working line. This intersection forms the second endpoint of the copy of the given line segment. Then draw the segment of the working line between these endpoints heavier to indicate the required line segment. *Basic:* Prob. 1; *Average:* Prob. 1–2; *Enriched:* Prob. 1–3

1.–3. Constructions

PRACTICE PROBLEMS

1. Make a copy of each of the following line segments, using a ruler:

_____

2. Make a copy of each of the following line segments, using a compass and a straightedge. Check with ruler.

_____   _____

3. Draw any line segment and label the endpoints D and E. Label a point not on $\overline{DE}$ as F. Using a compass and a straightedge, draw a line segment FG equal in length to $\overline{DE}$.

OBJECTIVE: To construct triangles.

A triangle contains three sides and three angles. A triangle may be constructed when any of the following combinations of three parts are known:

I. Three sides.
II. Two sides and an included angle.
III. Two angles and an included side.

These are discussed in this Exercise.

I. Three Sides

To construct a triangle whose sides measure 4 cm, 3.6 cm, and 2.9 cm, first use a compass to lay off a line segment 4 cm long. With one of the endpoints as the center and setting the compass so that the radius is 3.6 cm, draw an arc. With the other endpoint as the center and a radius of 2.9 cm, draw an arc crossing the first arc. From this point of intersection draw line segments to the endpoints of the base line to form the required triangle.

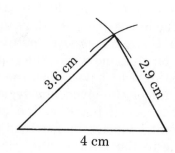

PRACTICE PROBLEMS

1. Construct triangles having sides that measure: Construction
a. 39 mm, 43 mm, 27 mm
b. $2\frac{3}{4}$ in., $2\frac{1}{4}$ in., 3 in.
c. 5.8 cm, 4.6 cm, 6.3 cm

2. Construct an equilateral triangle whose sides are each 3.1 cm long. Measure the three angles. Are their measures equal? Yes

3. Construct an isosceles triangle whose base is $2\frac{3}{16}$ in. long and each of whose two equal sides is $1\frac{5}{8}$ in. long.

4. Using the scale 1 inch = 8 feet, construct triangles with sides:
a. 16 ft., 22 ft., 9 ft.
b. 30 ft., 24 ft., 15 ft.
c. 12 ft., 12 ft., 12 ft.
d. 18 ft., 10 ft., 10 ft.

5. Using the scale 1:2,000,000, construct triangles with sides:

a. 40 km, 70 km, 58 km   **c.** 60 km, 54 km, 72 km

b. 30 km, 42 km, 38 km   **d.** 26 km, 36 km, 48 km

6. Construct a triangle with sides equal to these line segments:

_____   _____   _____

7. Construct a scalene triangle. Check whether:

a. Of any two sides, the side opposite the greater angle is greater.

b. Of any two angles, the angle opposite the greater side is greater.

II. Two Sides and an Included Angle

To construct a triangle with sides measuring $1\frac{3}{4}$ in. and $1\frac{3}{8}$ in. and an included angle of 90°, draw a line segment $1\frac{3}{4}$ in. long. Using the left endpoint as the vertex, draw an angle of 90°. Along the ray just drawn measure $1\frac{3}{8}$ in. from the vertex. Then draw a line segment connecting endpoints to form the required triangle.

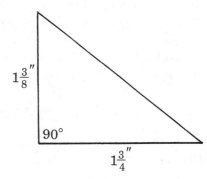

Basic: Prob. 1–3; *Average:* Prob. 1–4; *Enriched:* Prob. 1–5

PRACTICE PROBLEMS

1.–3. Constructions

1. Construct triangles having the following sides and included angles:

a. 61 mm, 48 mm, 55°

b. 5.7 cm, 4.6 cm, 70°

c. $2\frac{7}{8}$ in., $2\frac{7}{8}$ in., 90°

2. Construct a right triangle having:

a. A base of 6.5 cm and an altitude of 5.9 cm

b. A base of $2\frac{15}{16}$ in. and an altitude of $3\frac{5}{8}$ in.

3. Construct a triangle having two sides and an included angle equal to these two line segments and angle. (See Exercises 3-2, 3-18, and 3-24.)

_____   _____

Construction **4.** Construct an isosceles triangle in which the equal sides each measure $3\frac{1}{8}$ in. and the vertex angle formed by these sides measures 63°. Measure the angles opposite the equal sides. Are their measures equal? Yes

Construction **5.** Using the scale 1:500, construct a triangle in which two sides and the included angle measure:

     a. 15 m, 10 m, and 50°   **c.** 18 m, 27 m, and 80°
     b. 20 m, 35 m, and 75°   **d.** 31 m, 24 m, and 110°

III. Two Angles and an Included Side

To construct a triangle with angles measuring 60° and 48° and the included side measuring 4 cm, first draw a line segment 4 cm long. Using the left endpoint as the vertex, draw an angle of 60°. Using the right endpoint as the vertex, draw an angle of 48°. Extend the sides until they meet to form the required triangle.

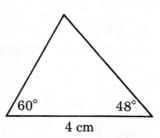

Basic: Prob. 1–4; *Average:* Prob. 1–6; *Enriched:* Prob. 1–7

PRACTICE PROBLEMS

1. Construct triangles having the following angles and included sides that measure: Construction

     a. 90°, 45°, 5 cm   **b.** 105°, 35°, $3\frac{1}{4}$ in.   **c.** 70°, 30°, 58 mm

2. Construct a triangle having two equal angles each measuring 50° and the included side measuring 46 mm. Measure the sides opposite the equal angles. Are their measures equal? Construction; yes

Construction **3.** Construct a triangle having two equal angles each measuring 60° and the included side measuring $2\frac{11}{16}$ in. Measure the third angle. Measure the other two sides. Is the triangle equiangular? Equilateral? Yes; yes

4. Construct a triangle having two angles and an included side equal to these angles and line segment. (See Exercises 3-2, 3-18, and 3-24.)

4.–5. Constructions

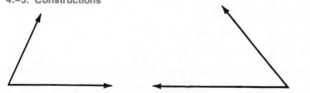

5. Using the scale $\frac{1}{4}$ in. = 1 ft., construct a triangle in which two angles and an included side measure:

     a. 60°, 50°, 16 ft.   **c.** 95°, 55°, 22 ft.
     b. 45°, 45°, 24 ft.   **d.** 70°, 65°, 19 ft.

6. Construct right triangles with angles and included sides measuring:

Constructions **a.** 90°, 30°, 53 mm   **c.** 90°, 30°, 6.7 cm
　　　　　　 b. 90°, 30°, $2\frac{3}{8}$ in.   **d.** 90°, 65°, 72 mm

e. Also select your own measurement for the included side and, using the measures of 90° and 30° for the angles, construct a right triangle.

f. In each of the above triangles measure the hypotenuse and the side opposite the 30° angle. Then compare these measurements. In each case is the hypotenuse twice as long as the side opposite the 30° angle? Yes

7. Construct two triangles, each containing two angles measuring 75° and 60°. The included side in the first triangle measures 3 inches and in the second triangle it measures $2\frac{5}{8}$ inches. Are the triangles congruent? Are they similar? (See Exercises 3-4, 3-13, and 3-14.) Construction; no; yes

Constructing Regular Polygons and Other Polygons

EXERCISE 3-20

OBJECTIVE:
To construct regular polygons and other polygons.

VOC.: regular polygon   inscribed polygon

A regular polygon is a polygon that is both equilateral (all of its sides are of equal length) and equiangular (all of its angles are of equal size). An inscribed polygon in a circle is a polygon whose vertices are points on the circle.

Although there are other ways to construct some of the regular polygons, in general we use a method that is based on the geometric fact that equal central angles of a circle intercept equal arcs and equal chords. A regular polygon constructed in this way is inscribed within a circle.

Therefore, *to construct a regular polygon,* first draw a circle. Then divide this circle into the same number of equal arcs as there are sides in the required polygon by drawing a corresponding number of equal central angles. Draw line segments (chords) connecting the points of division to form the polygon.

To draw a regular hexagon, which is a polygon of six equal sides, first determine the measure of each of the six equal central angles by dividing 360° by 6. This measure is 60°. Draw six central angles each measuring 60°, with its sides (radii) intercepting the circle dividing it into six equal arcs. Then draw line segments to connect the points of division to form the regular hexagon.

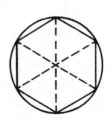

A regular hexagon and an equilateral triangle may also be drawn by the following special method:

Divide the circle into six equal arcs by using the radius of the circle as the radius of the arc. Then connect these points of division to get a regular hexagon.

An equilateral triangle may be constructed by drawing line segments to connect alternate points of division after the circle is divided into six equal arcs.

A square may be constructed by drawing two diameters of a circle perpendicular to each other to divide the circle into four equal arcs. Then connect these points of division to get a square.

Rectangles and parallelograms of specific measurement may be constructed provided enough of these measurements are given so that basic constructions may be used.

Basic: Prob. 1–3; *Average:* Prob. 1–4; *Enriched:* Prob. 1–5

PRACTICE PROBLEMS

1.–5. Constructions

1. Construct each of the following regular polygons:

   a. Pentagon—5 sides   **d.** Dodecagon—12 sides
   b. Octagon—8 sides   **e.** Equilateral triangle
   c. Decagon—10 sides   **f.** Square

2. Construct a rectangle:

   a. 64 mm long and 41 mm wide   **c.** 7.6 cm long and 5.3 cm wide
   b. $3\frac{1}{2}$ in. long and $2\frac{3}{8}$ in. wide   **d.** $4\frac{1}{8}$ in. long and $3\frac{9}{16}$ in. wide

3. Construct a square whose side measures:

   a. 5 cm   **b.** $3\frac{1}{4}$ inches   **c.** 68 mm   **d.** $\frac{7}{8}$ inch

4. Construct a parallelogram with:

   a. A base 71 mm long, a side 52 mm long, and an included angle of 60°

   b. A base $4\frac{3}{8}$ inches long, a side $3\frac{1}{4}$ inches long, and an included angle of 45°

   c. A base 5 inches long, a side $3\frac{13}{16}$ inches long, and an included angle of 80°

   d. A base 7 cm long, a side 6.2 cm long, and an included angle of 120°

5. Draw a regular hexagon, each side measuring:

   a. 4 cm   **b.** 54 mm   **c.** $2\frac{7}{8}$ inches

In each case draw a circle whose radius has the same measure as each required side.

Constructing a Perpendicular to a Line At or Through a Point on the Line

OBJECTIVE:
To construct a perpendicular to a given line at or through a given point on the given line.

VOCABULARY: perpendicular lines

Two lines (or rays or segments) that meet to form right angles are called perpendicular lines (or rays or segments). Each line is said to be perpendicular to the other. (See page 338.)

The symbol ⊥ means "is perpendicular to."

(1) *Using a protractor,* draw a 90° angle with the given point on the line as the vertex. The ray drawn to form the angle is perpendicular to the given line.

(2) *Using a compass* (see figure): To draw a line (or ray or segment) perpendicular to $\overleftrightarrow{AB}$ at C, use point C as the center and with any radius draw an arc cutting $\overleftrightarrow{AB}$ at D and E. With D and E as centers and with a radius greater than $\overline{CD}$, draw arcs crossing at F. Draw $\overrightarrow{CF}$ which is perpendicular to $\overleftrightarrow{AB}$ at point C. Or draw the line FC passing through point C.

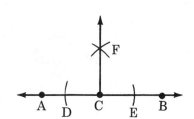

Basic: Prob. 1–3; *Average:* Prob. 1–5; *Enriched:* Prob. 1–6

PRACTICE PROBLEMS

1.–4. Constructions

1. Draw any line. Select a point on this line. Construct a line perpendicular to the line you have drawn at your selected point.

2. Construct a rectangle 43 mm long and 35 mm wide.

3. Construct a square with each side $2\frac{1}{2}$ inches long.

4. Make a plan of a room 33 ft. by 21 ft., using the scale 1 in. = 8 ft.

5. Draw a circle. Draw any diameter of this circle. At the center construct another diameter perpendicular to the first diameter, dividing the circle into four arcs. Are these four arcs equal in length? Use a compass to check. Construction; yes

6. Draw a circle. Select any point on this circle and label it B. Construct a line perpendicular to the radius at point B. Construction

A tangent to a circle at point B is a line that has one and only one point in common with the circle. Observe that a line that is perpendicular to the radius at a point on the circle is tangent to the circle.

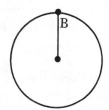

Related Resources: *Practical Applications in Mathematics:* Problems, p. 104

Constructing a Perpendicular to a Line From or Through a Point Not on the Line

OBJECTIVE: To construct a perpendicular to a given line from or through a given point not on the given line.

Using a compass (see figure): To draw a ray from point C perpendicular to $\overleftrightarrow{AB}$, use point C as the center and draw an arc cutting $\overleftrightarrow{AB}$ at D and E. With D and E as centers and a radius of more than one-half the distance from D to E, draw arcs crossing at F. Draw $\overrightarrow{CF}$ which is perpendicular to $\overleftrightarrow{AB}$. Or draw the line FC passing through point C.

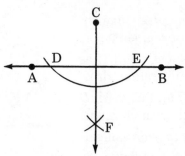

Basic: Prob. 1–3; *Average:* Prob. 1–4; *Enriched:* Prob. 1–5

PRACTICE PROBLEMS

1. Draw any line. Select a point not on this line. Construct a perpendicular to the line you have drawn from your selected point Construction

2. Construct an equilateral triangle with each side 6.2 cm long. From each vertex construct a perpendicular to the opposite side. Is each angle bisected? Is each side bisected? Construction; yes; yes

3. Draw any acute triangle. From each vertex construct a perpendicular to the opposite side. What do these perpendicular line segments represent in a triangle? Are they concurrent? Construction; altitude; yes

4. Draw any right triangle. Construct the altitude to each side. Are they concurrent? If so, what is the common point? Construction; yes; the vertex of the right angle

Construction **5.** Draw a circle. Draw any chord in this circle except the diameter. **a.** Construct a perpendicular from the center of the circle to this chord. Check whether the chord is bisected. **b.** Extend the perpendicular line segment so that it intersects the circle. Check with a compass whether the arc corresponding to the chord is also bisected. Observe that a radius of a circle that is perpendicular to a chord bisects the chord and its corresponding arc.

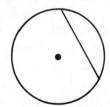

Related Resources: *Practical Applications in Mathematics:* Problems, p. 104

Bisecting a Line Segment

OBJECTIVE:

To bisect a line segment means to divide it into two equal parts. The point that separates the segment into equal parts is called the midpoint.

(1) *Using a ruler,* first measure the line segment, then mark off half the measurement.

(2) *Using a compass* (see figure): To bisect $\overline{AB}$, set the compass so that the radius is more than half the length of $\overline{AB}$. With A and B as centers draw arcs which cross above and below the segment at C and D. Then draw $\overleftrightarrow{CD}$ bisecting $\overline{AB}$ at E. Observe that $\overleftrightarrow{CD}$ is also perpendicular to $\overline{AB}$. Thus a line like $\overleftrightarrow{CD}$, which both bisects a line segment and is perpendicular to it, is called the perpendicular bisector.

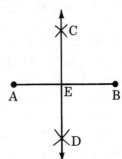

Basic: Prob. 1–3; *Average:* Prob. 1–4; *Enriched:* Prob. 1–5

PRACTICE PROBLEMS

1.–2. Constructions

1. Draw line segments which measure: **a.** 52 mm **b.** $4\frac{3}{8}''$ **c.** 4.6 cm Bisect each line segment, using a compass. Check with a ruler.

2. Copy, then using a compass, divide the following segment into four equal parts. Check with a ruler.

---

3. Draw any line segment. Bisect it, using a compass. Then use a protractor to check whether each of the angles formed is a right angle. What do we call a line that bisects and is perpendicular to a segment?

4. Draw any triangle. Bisect each side by constructing the perpendicular bisector of that side. Are these perpendicular bisectors concurrent? Do they meet in a point equidistant from the vertices of the triangle? Check by measuring. Using this common point as the center and the distance from this point to any vertex of the triangle as the radius, draw a circle through the three vertices of the triangle. Construction; yes; yes

When each side of the triangle is a chord of the circle or each vertex is a point on the circle, we say that *the circle is circumscribed about the triangle* or that *the triangle is inscribed in a circle.*

5. Draw any triangle. Find the midpoint of each side by constructing the perpendicular bisector of that side. Draw the median from each vertex to the midpoint of the opposite side. Are these medians concurrent? Along each median, is the distance from the common point to the vertex twice the distance from the common point to the opposite side?

3. Construction; perpendicular bisector
5. Construction; yes; yes

3-24 Copying a Given Angle

OBJECTIVE:

To copy a given angle means to construct an angle equal in size to the given angle.

(1) *Using a protractor,* measure the given angle and draw another angle of the same size.

(2) *Using a compass* (see figure): To construct an angle at point C on $\overleftrightarrow{AB}$ equal to $\angle MNO$, take point N as center and draw an arc cutting side MN at P and side NO at Q. With the same radius and point C as center, draw an arc cutting $\overleftrightarrow{AB}$ at D. With a radius equal to PQ and point D as center, draw an arc crossing the first arc at E. Draw CE. $m \angle BCE$ is equal to $m \angle MNO$.

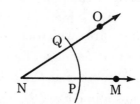

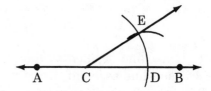

Basic: Prob. 1–3; *Average:* Prob. 1–4; *Enriched:* Prob. 1–5

PRACTICE PROBLEMS

1.–5. Constructions

1. Draw angles having the following measures, using a protractor. For each angle construct with a compass an angle of equal size. Check your copy of the angle with a protractor.

    a. 40°   **d.** 21°

    b. 150°   **e.** 163°

    c. 65°   **f.** 90°

2. Draw an acute angle. Construct with a compass an angle of equal size. Check both angles with a protractor.

3. Draw any obtuse angle. Construct with a compass an angle of equal size. Check both angles with a protractor.

4. Draw a right angle, using a protractor. Construct with a compass an angle of equal size. Check with a protractor.

5. Draw any angle. Construct with a compass an angle having the same measure. Check both angles with a protractor.

Bisecting an Angle

OBJECTIVE:

To bisect an angle means to divide it into two equal angles.

(1) *Using a protractor,* measure the given angle and mark off one-half the measurement. Then draw a ray from the vertex.

(2) *Using a compass* (see figure): To bisect $\angle ABC$ with B as the center and any radius, draw an arc cutting side AB at D and side BC at E. With D and E as centers and a radius of more than half the distance from D to E, draw arcs crossing at F. Then draw $\overrightarrow{BF}$ bisecting $\angle ABC$.

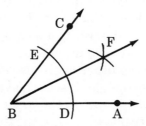

Basic: Prob. 1–6; *Average:* Prob. 1–8; *Enriched:* Prob. 1–9

PRACTICE PROBLEMS

1.–4. Constructions

1. Draw angles having the following measures. Bisect each angle using a compass. Check with a protractor.

    a. 70°   **b.** 130°   **c.** 54°   **d.** 145°   **e.** 43°   **f.** 137°

2. Draw any angle. Bisect it by using a compass only. Check by measuring your angle and each bisected angle.

3. Copy, then bisect each of the following angles:

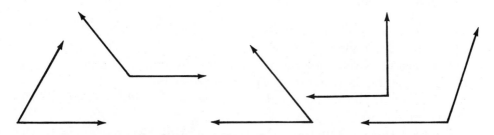

4. Draw any angle. Using a compass only, divide the angle into four equal angles. Check with protractor.

Related Resources: *Practical Applications in Mathematics:* Prob. 5–10, p. 103

5. Draw any triangle. Bisect each angle. Do the bisectors meet at a common point? Are these bisectors concurrent? From this common point draw a perpendicular to any side. With this common point as center and with the perpendicular distance from the common point to any side as the radius, draw a circle. Observe that each side of the triangle is tangent to the circle. Construction; yes; yes

When each side of the triangle is tangent to the circle, we say that *the triangle is circumscribed about the circle* or that *the circle is inscribed in the triangle.*

Construction **6.** Construct with a compass an angle measure: **a.** 45°   **b.** 135°

7. Draw any equilateral triangle. Select an angle and bisect it. Construct from the vertex of this selected angle the altitude (line segment that is perpendicular to the opposite side). Does this perpendicular line bisect this opposite side? From this same vertex also draw the median to the opposite side. In an equilateral triangle, are all three of these lines (angle bisector, altitude, and median) one and the same line? If this is so, will it be true if each of the other angles is selected? Check. Construction; yes; yes

8. Draw any isosceles triangle. Bisect the vertex angle (angle formed by the two equal sides and opposite to the base). Then construct from the vertex of this angle the altitude (line segment that is perpendicular to the base). Does this perpendicular line bisect the base of the isosceles triangle? From this same vertex also draw the median to the base. In an isosceles triangle, are all three of these lines (bisector of the vertex angle, altitude drawn to the base, and the median drawn to the base) one and the same line? Construction; yes; yes

In an isosceles triangle the angles opposite the equal sides are called base angles. Check whether the bisector of a base angle, the altitude and the median both drawn from the vertex of the same base angle are all one and the same line.

9. Draw any scalene triangle and do the following: Construction

a. Select an angle and bisect it.

b. Construct from the vertex of this bisected angle the line segment that is perpendicular (altitude) to the opposite side.

c. Construct the perpendicular bisector of the side opposite to the bisected angle.

d. Draw the median from the vertex of this bisected angle to the opposite side. Are any of these four lines (angle bisector, altitude, perpendicular bisector, and median) the same line? If so, which?
No

Constructing a Line Parallel to a Line Through a Point Not on the Line

OBJECTIVE: To construct a
line parallel to a given
line through a given
point not on the given line.

VOCABULARY: parallel lines

Lines in the same plane which do not meet are called parallel lines
The symbol ‖ means "is parallel to."

To construct a line parallel to
$\overleftrightarrow{AB}$ through point C (see figure),
draw any line $\overleftrightarrow{DE}$ through C meet-
ing $\overleftrightarrow{AB}$ at F.

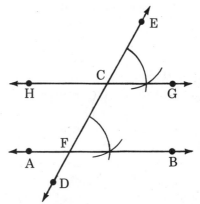

(1) *Using a protractor,* measure
$\angle BFC$ and draw, at point C on $\overleftrightarrow{DE}$,
$\angle GCE$ equal to the corresponding
$\angle BFC$. Then extend $\overleftrightarrow{GC}$ through H.
$\overleftrightarrow{HG}$ is parallel to $\overleftrightarrow{AB}$.

(2) *Using a compass,* construct, at point C on $\overleftrightarrow{DE}$, $\angle GCE$ equal to
the corresponding $\angle BFC$ by following the procedure explained previously
in Exercise 3-24. Then extend $\overleftrightarrow{GC}$ through H. $\overleftrightarrow{HG}$ will then be parallel to
$\overleftrightarrow{AB}$.

Basic: Prob. 1; *Average:* Prob. 1–2; *Enriched:* Prob. 1–3

PRACTICE PROBLEMS

1.–2. Constructions
1. Draw any line. Select a point that is not on this line. Through this
point construct a line parallel to the line you have drawn.
2. Construct a parallelogram with a base 75 mm long, a side 58 mm
long, and an included angle of 60°.
3. Draw any acute triangle. Bisect one of the sides. Draw a line
segment from this midpoint, parallel to one of the other sides, until it
intersects the third side. Does this line segment bisect the third side?
Check by measuring. Also check whether this line segment is one-half as
long as the side to which it is parallel. Draw other triangles and check
whether the above findings are true no matter which side is used first or
which side is used as the parallel side. Construction; yes

Related Resources: *Practical Applications in*
Mathematics: Prob. 3–4, p. 105

REVIEW

1.–10. Constructions

1. Make a copy of the following line segment using a compass and a straightedge.

_____

2. Construct a triangle with sides measuring 45 mm, 53 mm, and 39 mm.

3. Construct a triangle with two angles measuring 95° and 40° and an included side $2\frac{3}{4}$ inches long.

4. Construct a triangle with two sides measuring 6.4 cm and 5.8 cm and an included angle of 70°.

5. Draw a regular hexagon, each side measuring 37 mm.

6. Draw a line segment which measures 6.2 cm. Bisect it using a compass. Check with a ruler.

7. Draw an angle measuring 68°. Bisect it using a compass. Check with a protractor.

8. Draw any line. Using a compass, construct a perpendicular to it:
a. At a point on the line.   **b.** From a point not on the line.

9. Draw with a protractor an angle of 75°. Then construct with a compass an angle equal to it. Check your copy of the angle with a protractor.

10. Draw any line. Select a point outside this line. Through this point construct a line parallel to the line you have drawn.

REFRESH YOUR SKILLS

1. Add:

48,296
38,453
91,027
52,834
69,521
_____
300,131

2. Subtract:

1,000,000
842,056
_____
157,944

3. Multiply:

1,728
950
_____
1,641,600

4. Divide:

696
$659\overline{)458,664}$

5. Add: $4\frac{3}{16} + 2\frac{7}{10}$ $6\frac{71}{80}$

6. Subtract: $9\frac{1}{6} - \frac{2}{3}$ $8\frac{1}{2}$

7. Multiply: $8\frac{2}{5} \times 3\frac{1}{7}$ $26\frac{2}{5}$

8. Divide: $4 \div \frac{1}{3}$ 12

9. Add: $.6 + .87$ 1.47

10. Subtract: $4.5 - .23$ 4.27

11. Multiply: $8.02 \times .5$ 4.01

12. Divide: $1.2\overline{)}.9$ .75

13. Find $2\frac{1}{2}\%$ of $8,250. $206.25

14. What percent of $.75 is $.36?   48%

15. 96% of what number is 24?   25

For a Test on Constructions see p. T51 or *Testing Program*, p. 21.

INDIRECT MEASUREMENT

Not all distances or lengths can be measured directly. The distances of ships from the shore, the heights of clouds and mountains, the distances across rivers and lakes, and the distances from the earth to the sun and moon are measured indirectly.

OBJECTIVE: To find the measure of the third side of a right triangle given the measure of the other two sides, using the rule of Pythagoras.

EXERCISE
3-27

Rule of Pythagoras*

VOCABULARY: hypotenuse
altitude
base

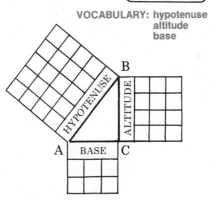

An important mathematical principle used in finding distances by indirect means is the hypotenuse rule, sometimes called the rule of Pythagoras. It expresses the relationship of the sides of a right triangle. A right triangle is a triangle having a right angle. The side opposite the right angle is called the hypotenuse. The other two sides or legs are the altitude and base of the triangle.

The area of the square drawn on the hypotenuse illustrated in the diagram is 25 square units and is equal to the sum of the areas of the squares drawn on the altitude and base (16 square units and 9 square units respectively). This is the rule of Pythagoras. It is usually stated as: *the square of the hypotenuse is equal to the sum of the squares of the other two sides*. As a formula this relationship is expressed:

$$h^2 = a^2 + b^2$$

h representing the hypotenuse, a, the altitude, and b, the base.

---

*Pythagoras, a Greek mathematician, lived about 550 B.C.

If any two sides of a right triangle are known, the third side may be found by the Pythagorean relation expressed in one of the following simplified forms:

$$h = \sqrt{a^2 + b^2} \qquad a = \sqrt{h^2 - b^2} \qquad b = \sqrt{h^2 - a^2}$$

In each case the square root is used because to determine the length of the side of a square when its area is known, the square root of the area must be found.*

Find the hypotenuse of a right triangle if the altitude is 4 millimeters and the base is 3 millimeters.

$$h = \sqrt{a^2 + b^2}$$
$$h = \sqrt{4^2 + 3^2}$$
$$h = \sqrt{16 + 9}$$
$$h = \sqrt{25}$$
$$h = 5 \text{ mm}$$

$$4^2 = 4 \times 4 = 16$$
$$3^2 = 3 \times 3 = 9$$

$$\begin{array}{c} 16 \\ + 9 \\ \hline 25 \end{array} \qquad \begin{array}{c} 5 \\ \sqrt{25} \end{array}$$

Answer: 5 millimeters

$a = 4 \text{ mm}$
$b = 3 \text{ mm}$
$h = ?$

(diagram: right triangle with hypotenuse h, altitude 4 mm, base 3 mm)

Find the altitude of a right triangle if the hypotenuse is 20 centimeters and the base is 12 centimeters.

$$a = \sqrt{h^2 - b^2}$$
$$a = \sqrt{20^2 - 12^2}$$
$$a = \sqrt{400 - 144}$$
$$a = \sqrt{256}$$
$$a = 16 \text{ cm}$$

$$20^2 = 20 \times 20 = 400$$
$$12^2 = 12 \times 12 = 144$$

$$\begin{array}{c} 400 \\ - 144 \\ \hline 256 \end{array} \qquad \begin{array}{c} 16 \\ \sqrt{256} \\ \begin{array}{r} 1 \\ 26\overline{)156} \\ 156 \\ \hline \end{array} \end{array}$$

$h = 20 \text{ cm}$
$b = 12 \text{ cm}$
$a = ?$

(diagram: right triangle with hypotenuse 20 cm, altitude a, base 12 cm)

Answer: 16 centimeters

---

*See Exercise 1-45.

PRACTICE PROBLEMS

1. Find the hypotenuse of each right triangle with the following dimensions:

Altitude	12 m	8 cm	60 km	33 ft.	180 m	136 yd.
Base	9 m	15 cm	25 km	56 ft.	112 m	255 yd.
	15 m	17 cm	65 km	65 ft.	212 m	289 yd.

2. Find the altitude of each right triangle with the following dimensions:

Hypotenuse	13 mm	35 cm	89 ft.	53 km	219 ft.	325 m
Base	5 mm	28 cm	80 ft.	45 km	144 ft.	165 m
	12 mm	21 cm	39 ft.	28 km	165 ft.	280 m

3. Find the base of each right triangle with the following dimensions:

Hypotenuse	25 cm	87 m	73 mm	91 yd.	153 ft.	477 km
Altitude	24 cm	63 m	48 mm	84 yd.	72 ft.	252 km
	7 cm	60 m	55 mm	35 yd.	135 ft.	405 km

4. By rule of Pythagoras and by actual measurement:

a. Find the diagonal (line segment joining opposite corners) of a rectangle 8 cm long and 6 cm wide. 10 cm

b. Find the diagonal of a square whose side measures 7 m. 9.9 m

5. What is the shortest distance from first base to third base if the distance between bases is 90 ft.? 127.3 ft.

6. How high up on a wall does a 25-foot ladder reach if the foot of the ladder is 7 feet from the wall? 24 ft.

7. A child lets out 60 m of string in flying a kite. The distance from a point directly under the kite to where the child stands is 36 m. If the child holds the string 1.5 m from the ground, how high is the kite? Disregard any sag. 49.5 m

8. An airplane, flying 252 kilometers from town A due west to town B, drifts off its course in a straight line and is 39 kilometers due south of town B. What distance did the airplane actually fly? 255 km

9. Two poles, 34 ft. and 48 ft. high respectively, are 75 ft. apart. What is the distance from the top of one pole to the top of the second pole? 76.3 ft.

10. The escalator between the first and second floors of a store measures 28 ft. vertically and 25 ft. horizontally. How many feet are you carried when traveling from the first to the second floor? 37.5 ft.

VOCABULARY: similar triangles
congruent triangles
is congruent to

OBJECTIVE: To use similar triangles to find the measure of a side of one of the triangles given the measures of its corresponding side of the other triangle and another pair of corresponding sides.

The relationship between the corresponding sides of similar triangles presents another method of measuring lengths and distances indirectly.

Similar triangles have the same shape but differ in size. Their corresponding angles are equal. Also, the ratios of their corresponding sides are equal. See Exercise 3-14.

Two triangles are said to be similar when any one of the following conditions is known:

(1) Two angles of one triangle are equal to two angles of the other.

(2) The ratios of all the corresponding sides are equal.

(3) Two sides of one triangle are proportional (equal ratios) to two corresponding sides of the other and the included angles are equal.

Triangles that have the same shape and the same size are called congruent triangles. The corresponding sides are equal in length and the corresponding angles are equal in size. The symbol ≅ means "is congruent to." See Exercise 3-13.

To measure indirectly a distance or length by means of similar triangles, use the given data to draw two similar triangles. Since the corresponding sides of similar triangles have equal ratios (or are proportional), form and solve a proportion, using three known sides and the required distance as the fourth side. (See Exercise 4-28.)

A tree casts a shadow (*AB*) of 27 ft. while a 4-foot post (*EF*) nearby casts a shadow (*DE*) of 3 ft. What is the tree's height (*BC*)?

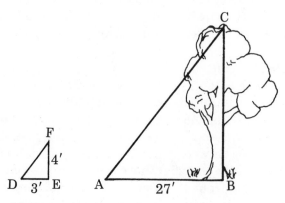

Triangles *ABC* and *DEF* are similar, therefore the corresponding sides have the same ratio. The side representing the shadow of the tree (*AB*) corresponds to the side representing the shadow of the post (*DE*). The side representing the height of the tree (*BC*) corresponds to the side representing the height of the post (*EF*).

The ratio of the shadow of the tree to the shadow of the post is 27 to 3 or 9 to 1.

Therefore the height of the tree must be 9 times the height of the post or 9 × 4 ft. = 36 ft. *Answer:* The tree is 36 feet high.

Basic: Prob. 1–3; *Average:* Prob. 1–5; *Enriched:* Prob. 1–6

PRACTICE PROBLEMS

1. In Figure 1, triangles *ABE* and *CDE* are similar. Which side in triangle *ABE* corresponds to side *CE*? to side *DE*? to side *AB*? Find the length (*AB*) of the lake.
EB; EA; CD; 85 m

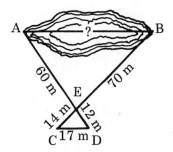

Figure 1

2. Triangles *ABC* and *CDE* are similar in Fig. 2. Which side in triangle *ABC* corresponds to side *CE*? to side *DE*? What is the distance (*AB*) across the stream?
AC; AB; 100 m

3. A flagpole casts a shadow 15 m long. At the same time a 2-meter pole casts a shadow of 3 m. How high is the flagpole?
10 m

4. Find the height of a building that casts a shadow of 26 ft. at a time when a student, 5 feet tall, casts a shadow of 2 ft.
65 ft.

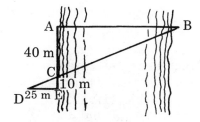

Figure 2

5. Using Figure 2, find the distance across the stream if *DE* is 32 m, *CE* is 9 m, and *AC* is 45 m.
160 m

6. What is the height of a TV antenna tower that casts a shadow of 12 m when the adjacent building, 20 m high, casts a shadow of 2.5 m?
96 m

VOCABULARY: tangent (tan)
sine (sin)
cosine (cos)

OBJECTIVE:
To solve problems using trigonometric ratios.

In trigonometry (meaning triangle measure) the relationships between the sides and the angles of the right triangle are used to determine certain parts of the triangle when the other parts are known. The ratios of the sides of the right triangle are related to the acute angles as follows:

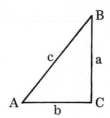

The ratio of the side opposite an acute angle to the adjacent side is called the tangent of the angle (abbreviated tan). The expression tan A means tangent of angle A.

In right triangle ABC, $\tan A = \dfrac{a}{b}$ and $\tan B = \dfrac{b}{a}$.

The ratio of the side opposite an acute angle to the hypotenuse is called the sine of the angle (abbreviated sin).

In the right triangle ABC, $\sin A = \dfrac{a}{c}$ and $\sin B = \dfrac{b}{c}$.

The ratio of the adjacent side of an acute angle to the hypotenuse is called the cosine of the angle (abbreviated cos).

In the right triangle ABC, $\cos A = \dfrac{b}{c}$ and $\cos B = \dfrac{a}{c}$.

Angles are measured both vertically and horizontally. An angle measured vertically between the horizontal line and the observer's line of sight to an object is called the angle of elevation when the object is above the observer and the angle of depression when the object is below the observer.

To solve problems using trigonometric ratios, use these three steps:

a. Draw a right triangle if one is not given and place the given dimensions on it.

b. Select the proper formula and substitute the given values, using the table of trigonometric values (page 417) when necessary.

c. Then solve the resulting equation.

In problems where the complement of an angle (see page 374) is simpler to use for computation than the given angle, it may be determined by either of the following formulas:

$$A = 90° - B \quad \text{or} \quad B = 90° - A.$$

In right triangle ABC, find side a when angle $A = 53°$ and side $b = 100$ feet.

$b = 100$ ft.
$A = 53°$
$a = ?$

$$\tan A = \frac{a}{b}$$

$$\tan 53° = \frac{a}{100}$$

$$1.3270 = \frac{a}{100}$$

$$a = 132.7 \text{ ft.}$$

Answer: 132.7 feet

In right triangle ABC, find side b (to nearest tenth) when angle $B = 26°$ and side $c = 45$ meters.

$c = 45$ m
$B = 26°$
$b = ?$

$$\sin B = \frac{b}{c}$$

$$\sin 26° = \frac{b}{45}$$

$$.4384 = \frac{b}{45}$$

$$19.728 = b$$

$$b = 19.7 \text{ m}$$

Answer: 19.7 meters

In right triangle ABC, find side a when angle $B = 77°$ and side $c = 80$ centimeters.

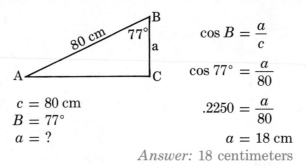

$$\cos B = \frac{a}{c}$$

$$\cos 77° = \frac{a}{80}$$

$c = 80$ cm
$B = 77°$
$a = ?$

$$.2250 = \frac{a}{80}$$

$$a = 18 \text{ cm}$$

Answer: 18 centimeters

Basic: Prob. 1–7; *Average:* Prob. 1–9; *Enriched:* Prob. 1–10

PRACTICE PROBLEMS

The letters in the following problems represent parts of the right triangle ABC with angle C, the right angle; altitude a, the side opposite angle A; and base b, the side opposite angle B. See the drawing on page 412.

1. Find the value of:

a. $\tan 43°$; .9325 $\tan 81°$; 6.3138 $\tan 63° \ 30'$; 2.0065 $\tan 37° \ 45'$ .7744

b. $\sin 79°$; .9816 $\sin 7°$; .1219 $\sin 41° \ 30'$; .6626 $\sin 24° \ 20'$ .4120

c. $\cos 18°$; .9511 $\cos 56°$; .5592 $\cos 6° \ 15'$; .9940 $\cos 71° \ 40'$ .3145

2. Find angle A when:

a. $\tan A = .3839$; 21° $\tan A = 3.0777$; 72° $\tan A = .2586$; 14°30'
$\tan A = 1.6483$ 58°45'

b. $\sin A = .9063$; 65° $\sin A = .3420$; 20° $\sin A = .7489$; 48°30'
$\sin A = .9940$ 83°45'

c. $\cos A = .8988$; 26° $\cos A = .6293$; 51° $\cos A = .5983$; 53°15'
$\cos A = .6626$ 48°30'

3. Find angle B when:

a. $\tan B = 19.0811$; 87° $\tan B = .9004$; 42° $\tan B = 1.2649$; 51°40'
$\tan B = .6129$ 31°30'

b. $\sin B = .6561$; 41° $\sin B = .9986$; 87° $\sin B = .5075$; 30°30'
$\sin B = .9469$ 71°15'

c. $\cos B = .1045$; 84° $\cos B = .9998$; 1° $\cos B = .4423$; 63°45'
$\cos B = .9984$ 3°12'

4. Use the tangent ratio to find:

a. Side a if angle $A = 30°$ and side $b = 50$ m 28.87 m
b. Side b if angle $B = 65°$ and side $a = 80$ cm 171.56 cm
c. Side b if angle $A = 22°$ and side $a = 101$ ft. 250 ft.
d. Side a if angle $B = 70°$ and side $b = 140$ km 50.96 km
e. Angle A if side $a = 202$ m and side $b = 500$ m 22°
f. Angle B if side $a = 1,000$ ft. and side $b = 1,804$ ft. 61°

5. Find the indicated part in each of the following right triangles to nearest tenth:

a. Find side a: 392.5 m   **b.** Find side b: 303.1 km   **c.** Find $\angle A$: 20°

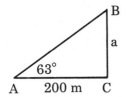 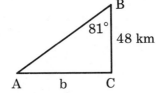 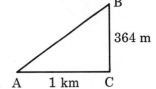

6. Use the sine ratio to find:

a. Side a if angle $A = 60°$ and side $c = 75$ m 64.95 m
b. Side b if angle $B = 15°$ and side $c = 40$ in. 10.352 in.
c. Side b if angle $A = 86°$ and side $c = 3.49$ km .244 km
d. Side a if angle $B = 32°$ and side $c = 80$ m 67.84 m
e. Side c if angle $A = 52°$ and side $a = 394$ ft. 500 ft.
f. Side c if angle $B = 20°$ and side $b = 855$ cm 2,500 cm
g. Angle A if side $a = 170.5$ m and side $c = 250$ m 43°
h. Angle B if side $b = 106$ yd. and side $c = 125$ yd. 58°

7. Find the indicated part in each of the following right triangles to nearest tenth:

a. Find side a: 21.1 ft.   **b.** Find side b: 116.7 km   **c.** Find $\angle B$: 31°

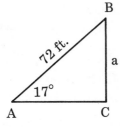 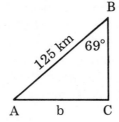 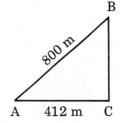

8. Use the cosine ratio to find:

a. Side b if angle $A = 46°$ and side $c = 100$ m ₆₉.₄₇ m 69.47 m

b. Side a if angle $B = 62°$ and side $c = 250$ yd. 117.375 yd.

c. Side c if angle $A = 70°$ and side $b = 17.1$ km 50 km

d. Side c if angle $B = 59°$ and side $a = 515$ cm 1,000 cm

e. Angle A if side $b = 23$ mm and side $c = 46$ mm 60°

f. Angle B if side $a = 309$ mi. and side $c = 1,000$ mi. 72°

9. Find the indicated part in each of the following right triangles to nearest tenth:

a. Find side b: 223.7 yd.   **b.** Find side c: 250 cm   **c.** Find $\angle A$: 30°

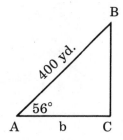
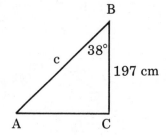
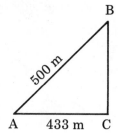

10. Find:

a. Side a if angle $A = 61°$ and side $b = 25$ cm 45.1 cm

b. Side a if angle $A = 40°$ and side $c = 50$ m 32.14 m

c. Side b if angle $A = 66°$ and side $c = 175$ ft. 71.1725 ft.

d. Side b if angle $B = 49°$ and side $a = 1,000$ yd. 1,150.4 yd.

e. Side b if angle $B = 13°$ and side $c = 80$ km 18 km

f. Side a if angle $B = 27°$ and side $c = 100$ m 89.1 m

g. Side b if angle $A = 30°$ and side $a = 46$ mm 79.6766 mm

h. Angle A if side $a = 451$ ft. and side $b = 250$ ft. 61°

i. Angle B if side $a = 511.5$ mi. and side $c = 750$ mi. 47°

j. Angle A if side $a = 53.46$ m and side $c = 60$ m 63°

11. The light from a searchlight is seen on a cloud at a horizontal distance of 2,800 feet from the searchlight. What is the height of the cloud if the angle of elevation from the searchlight is 65°? 6,004.6 ft.

12. From a cliff 300 meters above the sea, the angle of depression of a boat is 87°. How far is the boat from the foot of the cliff? 5,724.33 m

13. How high is a kite when 160 feet of string is let out and the string makes an angle of 50° with the ground? 122.56 ft.

14. The altitude of a right triangle is 220 kilometers and the angle opposite the base is 77°. How long is the hypotenuse? $977\frac{7}{9}$ km

15. What is the elevation of a road if the road rises 364 feet in a horizontal distance of 1,000 feet? 20°

Table of Trigonometric Values

Angle	Sine	Cosine	Tangent	Angle	Sine	Cosine	Tangent
0°	.0000	1.0000	.0000	46°	.7193	.6947	1.0355
1°	.0175	.9998	.0175	47°	.7314	.6820	1.0724
2°	.0349	.9994	.0349	48°	.7431	.6691	1.1106
3°	.0523	.9986	.0524	49°	.7547	.6561	1.1504
4°	.0698	.9976	.0699	50°	.7660	.6428	1.1918
5°	.0872	.9962	.0875	51°	.7771	.6293	1.2349
6°	.1045	.9945	.1051	52°	.7880	.6157	1.2799
7°	.1219	.9925	.1228	53°	.7986	.6018	1.3270
8°	.1392	.9903	.1405	54°	.8090	.5878	1.3764
9°	.1564	.9877	.1584	55°	.8192	.5736	1.4281
10°	.1736	.9848	.1763	56°	.8290	.5592	1.4826
11°	.1908	.9816	.1944	57°	.8387	.5446	1.5399
12°	.2079	.9781	.2126	58°	.8480	.5299	1.6003
13°	.2250	.9744	.2309	59°	.8572	.5150	1.6643
14°	.2419	.9703	.2493	60°	.8660	.5000	1.7321
15°	.2588	.9659	.2679	61°	.8746	.4848	1.8040
16°	.2756	.9613	.2867	62°	.8829	.4695	1.8807
17°	.2924	.9563	.3057	63°	.8910	.4540	1.9626
18°	.3090	.9511	.3249	64°	.8988	.4384	2.0503
19°	.3256	.9455	.3443	65°	.9063	.4226	2.1445
20°	.3420	.9397	.3640	66°	.9135	.4067	2.2460
21°	.3584	.9336	.3839	67°	.9205	.3907	2.3559
22°	.3746	.9272	.4040	68°	.9272	.3746	2.4751
23°	.3907	.9205	.4245	69°	.9336	.3584	2.6051
24°	.4067	.9135	.4452	70°	.9397	.3420	2.7475
25°	.4226	.9063	.4663	71°	.9455	.3256	2.9042
26°	.4384	.8988	.4877	72°	.9511	.3090	3.0777
27°	.4540	.8910	.5095	73°	.9563	.2924	3.2709
28°	.4695	.8829	.5317	74°	.9613	.2756	3.4874
29°	.4848	.8746	.5543	75°	.9659	.2588	3.7321
30°	.5000	.8660	.5774	76°	.9703	.2419	4.0108
31°	.5150	.8572	.6009	77°	.9744	.2250	4.3315
32°	.5299	.8480	.6249	78°	.9781	.2079	4.7046
33°	.5446	.8387	.6494	79°	.9816	.1908	5.1446
34°	.5592	.8290	.6745	80°	.9848	.1736	5.6713
35°	.5736	.8192	.7002	81°	.9877	.1564	6.3138
36°	.5878	.8090	.7265	82°	.9903	.1392	7.1154
37°	.6018	.7986	.7536	83°	.9925	.1219	8.1443
38°	.6157	.7880	.7813	84°	.9945	.1045	9.5144
39°	.6293	.7771	.8098	85°	.9962	.0872	11.4301
40°	.6428	.7660	.8391	86°	.9976	.0698	14.3007
41°	.6561	.7547	.8693	87°	.9986	.0523	19.0811
42°	.6691	.7431	.9004	88°	.9994	.0349	28.6363
43°	.6820	.7314	.9325	89°	.9998	.0175	57.2900
44°	.6947	.7193	.9657	90°	1.0000	.0000	
45°	.7071	.7071	1.0000				

5. If the umbrella is fitted into the suitcase diagonally, it will form the hypotenuse of a right triangle with the sides of the suitcase. It should fit by .017 of an inch.

418   GEOMETRY—INDIRECT MEASUREMENT

REVIEW

1. Find the hypotenuse of a right triangle if the altitude is 76 meters and the base is 57 meters. 95 m

2. What is the base of a right triangle if the hypotenuse is 377 feet and the altitude is 145 feet? 348 ft.

3. Find the altitude of a right triangle if the base is 66 centimeters and the hypotenuse is 130 centimeters. 112 cm

4. Find the height of a building that casts a shadow of 40 feet at the time a 7-foot fence casts a shadow of 5 feet. 56 ft.

5. Show by the rule of Pythagoras that a 30-inch umbrella can be fitted into a suitcase having inside dimensions of 26 inches by 15 inches.

Find the indicated parts of the following right triangles:

6. Find side a:
357.54 m

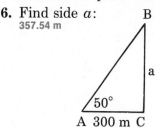

7. Find side b:
606.2 yd.

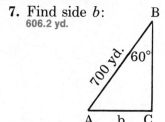

8. Find side b:
122.88 mm

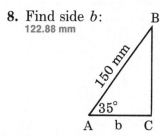

9. Find side a:
128.67 km

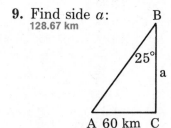

10. Find side b:
84 m

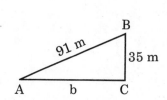

11. Find angle B:
47°

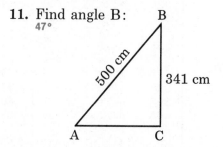

435.67 ft. **12.** The angle of elevation from a ship to the top of a lighthouse, 150 feet high, is 19°. How far from the foot of the lighthouse is the ship?

For a Test on Indirect Measurement see p. T52 or *Testing Program*, p. 23.

VOCABULARY: area
area of a rectangle
area of a square
area of a parallelogram
area of a triangle
area of a trapezoid
area of a circle
total area of a
  rectangular solid

total area of
  a cube
lateral area of
  a cylinder
total area of a
  cylinder
area of the
  surface of
  a sphere

APPLICATIONS MEASURING AREA AND VOLUME

EXERCISE **3-30**

Measuring Area

OBJECTIVE: To measure area.

The area of any surface is the number of units of square measure contained in the surface. When computing the area of the interior of a geometric figure, *express all linear units in the same denomination.*

I. Rectangle

The area of a rectangle is equal to the length times the width.

Formula: $A = lw$

Or, the area is equal to the altitude times the base.

Formula: $A = ab$

---

23 cm

16 cm

$l = 23$ cm
$w = 16$ cm
$A = ?$

Find the area of a rectangle 23 centimeters long and 16 centimeters wide.

$A = lw$
$A = 23 \times 16$
$A = 368$ cm^2

Answer: 368 cm^2

$$\begin{array}{r} 23 \\ \times\,16 \\ \hline 138 \\ 23 \\ \hline 368 \end{array}$$

Related Resources: *Practical Applications in*
Mathematics: Prob., pp. 118–124, 129–130 **419**

PRACTICE PROBLEMS

1. What is the area of a rectangle if its length is 14 meters and its width is 9 meters? 126 m²

2. Find the areas of rectangles having the following dimensions:

a.

Length	23 mm	125 cm	6.8 m	$1\frac{1}{3}$ yd.	2 yd.
Width	17 mm	95 cm	1.625 m	$4\frac{1}{2}$ yd.	20 in.

391 mm²   11,875 cm²   11.05 m²   6 sq. yd.   1,440 sq. in. or 10 sq. ft. or $1\frac{1}{9}$ sq. yd.

b.

Altitude	12 in.	150 m	0.4 km	$1\frac{7}{8}$ in.	3 ft. 7 in.
Base	8 in.	200 m	3.5 km	$2\frac{3}{4}$ in.	1 ft. 3 in.

96 sq. in.   30,000 m²   1.4 km² $5\frac{5}{32}$ sq. in.   645 sq. in. or 4.48 sq. ft.

3. Find the cost of each of the following:

$247.50 **a.** Cementing a driveway 3 m by 5 m at $16.50 per square meter.

b. Covering a floor 4 m by 4.5 m with linoleum at $12.75 per square meter. $229.50

c. Resilvering a mirror 36 in. by 42 in. at $9.25 per square foot. $97.13

d. Sodding a lawn 10 m by 6.2 m at $3.50 per square meter. $217.00

e. Refinishing a hardwood floor 16 ft. 4 in. by 14 ft. 8 in. at $.40 per square foot. $95.82

4. A schoolroom is 27 ft. long and 20 ft. wide. It has 6 windows, each measuring 3 ft. by 9 ft. There are also 5 blackboards, each 5 ft. by 3 ft.; 6 blackboards, each $2\frac{1}{2}$ ft. by 5 ft.; and one bulletin board 6 ft. by $3\frac{1}{2}$ ft. What is the ratio of the window area to the floor area? Find the ratio of the total blackboard area to the area of the bulletin board. $\frac{3}{10}$; $7\frac{1}{7}$

$81,250 **5.** Find the value of a field 325 m by 200 m at $12,500 per hectare.

6. Find the area of a rectangular wing of an airplane if the span (length) is 47 ft. 6 in. and chord (width) is 8 ft. 3 in. $391\frac{7}{8}$ sq. ft.

7. What is the cost per square meter of 1.25-meter width material costing $6.75 per running meter? $5.40

8. How many 4 cm by 6 cm tickets can be cut from 1.2 m by 1.5 m stock? 750 tickets

9. How many acres (nearest tenth) are contained in a football field 120 yd. long and 160 ft. wide? 1.3 acres

10. A field measuring 48 rd. by 40 rd. yields 420 bu. of corn. Find the average yield per acre. 35 bu.

For a Computer Activity see p. T40.

II. Square

The area of a square is equal to the length of its side times itself or the side squared. Formula: $A = s^2$

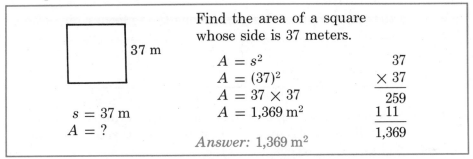

Find the area of a square whose side is 37 meters.

$$A = s^2$$
$$A = (37)^2$$
$$A = 37 \times 37$$
$$A = 1,369 \text{ m}^2$$

Answer: 1,369 m²

```
    37
  × 37
   259
  1 11
 1,369
```

Basic: Prob. 1–2; *Average:* Prob. 1–4; *Enriched:* Prob. 1–5

PRACTICE PROBLEMS

1. What is the area of a square whose side is 23 feet? **529 sq. ft.**

2. Find the areas of squares whose sides measure:

a. 10 cm **b.** 42 mm **c.** 32 in. **d.** 6,080 ft. **e.** 1,760 m
f. 1.61 km **g.** 39.37 in. **h.** $18\frac{1}{2}$ ft. **i.** $4\frac{3}{8}$ in. **j.** 3 ft. 8 in.

3. What is the cross sectional area of a square beam 14.5 cm on a side? **210.25 cm²**

4. At $19.95 per square meter how much will a broadloom rug 5 m by 5 m cost? **$498.75**

5. The base of the Great Pyramid is 746 feet square. How many acres (to nearest hundredth) does it cover? **12.78 acres**

III. Parallelogram

The area of a parallelogram is equal to the altitude times the base.

$$\text{Formula: } A = ab$$

Or, the area is equal to the base times the height.

$$\text{Formula: } A = bh$$

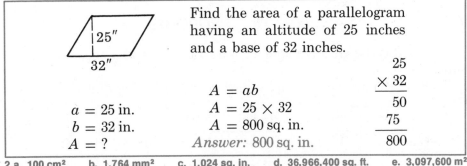

Find the area of a parallelogram having an altitude of 25 inches and a base of 32 inches.

$a = 25$ in.
$b = 32$ in.
$A = ?$

$$A = ab$$
$$A = 25 \times 32$$
$$A = 800 \text{ sq. in.}$$

Answer: 800 sq. in.

```
    25
  × 32
    50
    75
   800
```

2.a. 100 cm² **b.** 1,764 mm² **c.** 1,024 sq. in. **d.** 36,966,400 sq. ft. **e.** 3,097,600 m²
f. 2.5921 km² **g.** 1,550 sq. in. **h.** 342.25 sq. ft. **i.** $19\frac{9}{64}$ sq. in. **j.** 1,936 sq. in. or $13\frac{4}{9}$ sq. ft.

PRACTICE PROBLEMS

1. What is the area of a parallelogram if its altitude is 6 meters and its base is 8 meters? 48 m²

2. Find the areas of parallelograms having the following dimensions:

Altitude	26 in.	75 cm	8.3 km	$1\frac{1}{4}$ in.	3 yd. 2 ft.
Base	14 in.	98 cm	4.7 km	$\frac{1}{2}$ in.	4 yd. 1 ft.

364 sq. in.   7,350 cm²   39.01 km² $\frac{5}{8}$ sq. in.   143 sq. ft.

3. Find the cost of seeding a lawn shaped like a parallelogram with a base of 6 m and altitude of 5 m. One kilogram of grass seed covers 30 m² and costs $7.49. $7.49

IV. Triangle

The area of a triangle is equal to one half the altitude times the base.

Formula: $A = \frac{1}{2}ab$ which may be written as $A = \dfrac{ab}{2}$. Or, the area is equal to one half the base times the height.

Formula: $A = \frac{1}{2}bh$

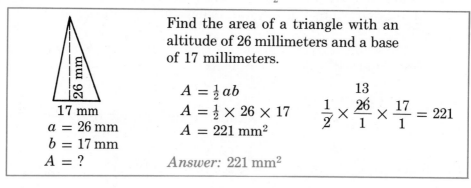

Find the area of a triangle with an altitude of 26 millimeters and a base of 17 millimeters.

$A = \frac{1}{2}ab$
$A = \frac{1}{2} \times 26 \times 17$
$A = 221$ mm²

$\dfrac{1}{2} \times \dfrac{\overset{13}{\cancel{26}}}{1} \times \dfrac{17}{1} = 221$

Answer: 221 mm²

26 mm
17 mm
$a = 26$ mm
$b = 17$ mm
$A = ?$

PRACTICE PROBLEMS

1. What is the area of a triangle if its altitude is 10 inches and base is 8 inches? 40 sq. in.

2. Find the areas of triangles having the following dimensions:

a.

Altitude	18 cm	13 m	5 km	27 ft.	1 ft. 4 in.
Base	12 cm	10 m	7 km	$16\frac{1}{2}$ ft.	2 ft.

108 cm²   65 m² $17\frac{1}{2}$ km² $222\frac{3}{4}$ sq. ft. $1\frac{1}{3}$ sq. ft.

b.

Base	8 km	4.8 m	$1\frac{7}{8}$ in.	$9\frac{1}{3}$ yd.	4 yd. 7 in.
Height	11 km	3.4 m	$2\frac{1}{4}$ in.	$6\frac{5}{8}$ yd.	1 ft. 11 in.

44 km²   8.16 m² $2\frac{7}{64}$ sq. in. $30\frac{11}{12}$ sq. yd.   1,736.5 sq. in. or 1.34 sq. yd.

3. How many square feet of surface does one side of a triangular sail expose if it has a base of 10 ft. and a height of 12 ft. 6 in.? $62\frac{1}{2}$ sq. ft.

V. Trapezoid

The area of a trapezoid is equal to the height times the average of the two parallel sides (bases).

$$\text{Formula: } A = h \times \frac{b_1 + b_2}{2}$$

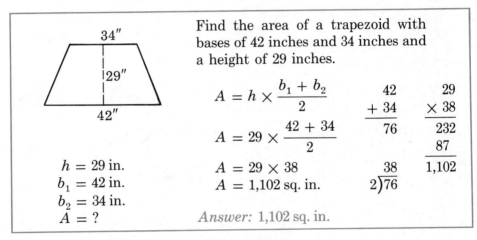

Find the area of a trapezoid with bases of 42 inches and 34 inches and a height of 29 inches.

$$A = h \times \frac{b_1 + b_2}{2}$$

$$A = 29 \times \frac{42 + 34}{2}$$

$$A = 29 \times 38$$
$$A = 1,102 \text{ sq. in.}$$

$h = 29$ in.
$b_1 = 42$ in.
$b_2 = 34$ in.
$A = ?$

```
   42          29
 + 34        × 38
 ----        ----
   76         232
               87
   38        1,102
2)76
```

Answer: 1,102 sq. in.

Basic: Prob. 1; *Average:* Prob. 1–2; *Enriched:* Prob. 1–3

PRACTICE PROBLEMS

1. What is the area of a trapezoid if the height is 7 inches and the parallel sides are 8 inches and 14 inches? 77 sq. in.

2. Find the areas of trapezoids having the following dimensions:

Height	8 cm	5 mm	18 m	6 ft.	10 in.
Upper Base	4 cm	9 mm	29 m	$11\frac{3}{4}$ ft.	1 ft.
Lower Base	10 cm	13 mm	36 m	$14\frac{1}{2}$ ft.	1 ft. 4 in.

56 cm²   55 mm²   585 m² $78\frac{3}{4}$ sq. ft.   140 sq. in.

3. A section of a tapered airplane wing has the shape of a trapezoid. If the two parallel sides (chords), measuring $3\frac{1}{2}$ ft. and $5\frac{1}{2}$ ft., are 18 ft. apart, find the area of the section. 81 sq. ft.

VI. Circle

The area of a circle is equal to pi (π) times the radius squared.

$$\text{Formula: } A = \pi r^2$$

Any one of the following values of π may be used: $3\frac{1}{7}$ or $\frac{22}{7}$ or 3.14 or, for greater accuracy, 3.1416.

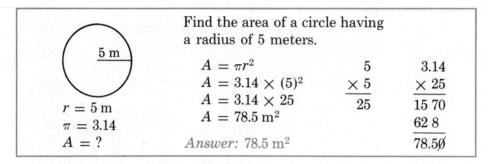

Find the area of a circle having a radius of 5 meters.

$A = \pi r^2$

$A = 3.14 \times (5)^2$

$A = 3.14 \times 25$

$A = 78.5 \text{ m}^2$

$r = 5 \text{ m}$
$\pi = 3.14$
$A = ?$

Answer: 78.5 m²

$$\begin{array}{r} 5 \\ \times\, 5 \\ \hline 25 \end{array} \qquad \begin{array}{r} 3.14 \\ \times\, 25 \\ \hline 15\,70 \\ 62\,8 \\ \hline 78.5\emptyset \end{array}$$

Or, the area of a circle is equal to one fourth times pi (π) times the diameter squared. Formula: $A = \frac{1}{4}\pi d^2$

Sometimes the formula $A = .7854\,d^2$ is also used.

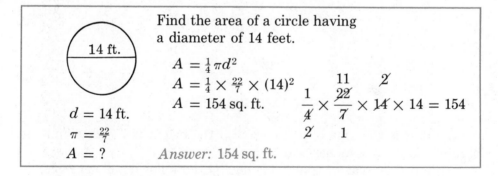

Find the area of a circle having a diameter of 14 feet.

$A = \frac{1}{4}\pi d^2$

$A = \frac{1}{4} \times \frac{22}{7} \times (14)^2$

$A = 154 \text{ sq. ft.}$

$d = 14 \text{ ft.}$
$\pi = \frac{22}{7}$
$A = ?$

$$\frac{1}{\underset{2}{\cancel{4}}} \times \frac{\overset{11}{\cancel{22}}}{7} \times \overset{2}{\cancel{14}} \times 14 = 154$$

Answer: 154 sq. ft.

Basic: Prob. 1–10; *Average:* Prob. 1–12; *Enriched:* Prob. 1–15

PRACTICE PROBLEMS

1. What is the area of a circle whose radius is 6 centimeters? 113.04 cm²

2. Find the areas of circles having the following radii:

a. 13 m   **c.** 52 in   **e.** 100 ft.   **g.** 1.375 mm   **i.** $\frac{1}{2}$ mi.

b. 28 cm   **d.** 91 mm   **f.** 4.5 km   **h.** $5\frac{2}{3}$ yd.   **j.** 7 ft. 4 in.

3. What is the area of a circle whose diameter is 24 feet? 452.16 sq. ft.

2.a. 530.66 m²   b. 2,461.76 cm² or 2,464 cm²*   c. 8,490.56 sq. in.   d. 26,002.34 mm² or 26,026 mm²*   e. 31,400 sq. ft.   f. 63.585 km²   g. 5.94 mm²   h. 100.92 sq. yd.*   i. .785 sq. mi.   j. 169.02 sq. ft.*

*$\pi = 3\frac{1}{7}$ was used to determine answers marked with asterisk. Otherwise, 3.14 was used.

4. Find the areas of circles having the following diameters:

a. 2 cm   **c.** 19 yd.   **e.** 220 yd.   **g.** 16.1 m   **i.** $8\frac{1}{6}$ ft.

b. 84 m   **d.** 63 mm   **f.** 0.75 km   **h.** $\frac{5}{8}$ in.   **j.** 5 yd. 1 ft.

5. If a forest ranger can see from his tower for a distance of 42 km in all directions, how many square kilometers can he watch? **5,538.96 km²**

6. The dial of one of the world's largest clocks has a diameter of 50 ft. What is the area of the dial? **1,962.5 sq. ft.**

7. A revolving sprinkler sprays a lawn for a distance of 7 m. How many square meters does the sprinkler water in 1 revolution? **153.86 m²**

8. If a station can televise programs for a distance of 91 km, over what area may the programs be received? **26,002.34 km²**

9. Which is larger: the area of a circle 6 cm in diameter or the area of a square whose side is 6 cm? How much larger? **Area of square; 7.74 cm²**

10. What is the area of one side of a washer if its diameter is $\frac{3}{4}$ in. and the diameter of the hole is $\frac{1}{4}$ in.? **.3925 sq. in.**

11. Find the area of the head of the piston if the bore (diameter) of the piston is $3\frac{1}{2}$ in. **$9\frac{5}{8}$ sq. in.***

12. A steel tube has a wall thickness of 0.125 cm and an outside diameter of 2 cm. What is the cross-sectional area of the metal? **.74 cm²**

13. A circular flower bed 12 ft. in diameter has a circular walk 3 ft. wide paved around it. At $1.90 per sq. ft., how much did the walk cost? **$268.47**

14. What is the area of one side of a penny? nickel? quarter? **.44 sq. in.; .52 sq. in.; .69 sq. in.**

15. Find the minimum number of 5-centimeter pipes necessary to draw off the water supplied by an 8-centimeter pipe under full pressure. **3 pipes**

VII. Total Area of a Rectangular Solid

The total area of the outside surface of a rectangular solid is equal to the sum of twice the product of the length and width, twice the product of the length and height and twice the product of the width and height.
Formula: $A = 2\,lw + 2\,lh + 2\,wh$

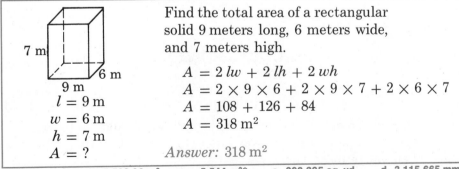

Find the total area of a rectangular solid 9 meters long, 6 meters wide, and 7 meters high.

7 m
6 m
9 m
$l = 9$ m
$w = 6$ m
$h = 7$ m
$A = ?$

$A = 2\,lw + 2\,lh + 2\,wh$
$A = 2 \times 9 \times 6 + 2 \times 9 \times 7 + 2 \times 6 \times 7$
$A = 108 + 126 + 84$
$A = 318$ m²

Answer: 318 m^2

4.a. 3.14 cm²   b. 5,538.96 m²   or   5,544 m²*   c. 283.385 sq. yd.   d. 3,115.665 mm²
e. 37,994 sq. yd.   f. .44 km²   g. 203.48 m²   h. .31 sq. in.*   i. 52.4 sq. ft.*   j. 200.96 sq. ft.

PRACTICE PROBLEMS

1. Find the total areas of rectangular solids with the following dimensions:

Length	23 cm	61 mm	8.1 m	$9\frac{1}{2}$ in.	5 ft. 4 in.
Width	14 cm	37 mm	2.7 m	8 in.	4 ft. 3 in.
Height	19 cm	25 mm	5.9 m	$10\frac{3}{8}$ in.	7 ft. 6 in.

2,050 cm²   9,414 mm²   171.18 m² $515\frac{1}{8}$ sq. in.   27,228 sq. in. or $189\frac{1}{12}$ sq. ft.

2. How many square meters of plywood will be needed to make a packing box with the dimensions 2.6 m by 2.5 m by 1.8 m? **31.36 m²**

3. A room is 18 ft. long, 15 ft. wide, and 9 ft. high. Find the total area of the walls and ceiling, allowing a deduction of 64 sq. ft. for the windows and doorway. How many gallons of paint are needed to cover the walls and ceiling with two coats if a gallon will cover 400 sq. ft. one coat? At $14.69 per gallon, how much will the paint cost? **800 sq. ft.; 4 gal.; $58.76**

VIII. Total Area of a Cube

The total area of the outside surface of a cube is equal to 6 times the area of one of its faces.

Formula: $A = 6\,s^2$ or $A = 6\,e^2$

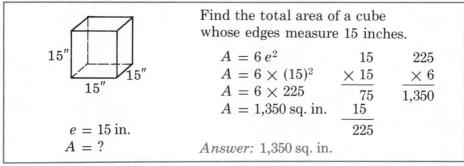

Find the total area of a cube
whose edges measure 15 inches.

$A = 6\,e^2$	15	225
$A = 6 \times (15)^2$	$\times\ 15$	$\times\ 6$
$A = 6 \times 225$	75	1,350
$A = 1{,}350$ sq. in.	15	
	225	

$e = 15$ in.
$A = ?$

Answer: 1,350 sq. in.

PRACTICE PROBLEMS

1. Find the total areas of cubes whose edges measure:

a. 37 mm   **b.** 20 cm   **c.** 9.2 m   **d.** $3\frac{7}{8}$ in.   **e.** 1 ft. 11 in.

2. A carton 1.5 m by 1.5 m by 1.5 m is made of corrugated paper. How many square meters of paper were used to make it if 10% extra was allowed for waste in cutting? **14.85 m²**

1.a. 8,214 mm²   b. 2,400 cm²   c. 507.84 m²   d. $90\frac{3}{32}$ sq. in.   e. 3,174 sq. in. or $22\frac{1}{24}$ sq. ft.

IX. Lateral Area and Total Area of a Right Circular Cylinder

The lateral area or area of the outside curved surface of a cylinder is equal to pi (π) times the diameter times the height.

Formula: $A = \pi dh$

Sometimes the formula $A = 2\,\pi rh$ is used.

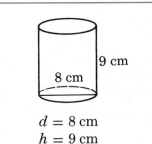

$d = 8$ cm
$h = 9$ cm
$\pi = 3.14$
$A = ?$

Find the lateral area of a cylinder having a diameter of 8 centimeters and a height of 9 centimeters.

$$A = \pi dh$$
$$A = 3.14 \times 8 \times 9$$
$$A = 226.08 \text{ cm}^2$$

Answer: 226.08 cm²

The total area of a cylinder equals the lateral area added to the area of the two bases.

Formula: $A = 2\,\pi rh + 2\,\pi r^2$

or $\qquad A = \pi dh + \frac{1}{2}\pi d^2$

or $\qquad A = 2\,\pi r(h + r)$

$r = 21$ ft.
$h = 30$ ft.
$\pi = \frac{22}{7}$
$A = ?$

Find the total area of a cylinder when its radius is 21 feet and its height is 30 feet.

$$A = 2\,\pi rh + 2\,\pi r^2$$
$$A = 2 \times \tfrac{22}{7} \times 21 \times 30 + 2 \times \tfrac{22}{7} \times (21)^2$$
$$A = 3{,}960 + 2{,}772$$
$$A = 6{,}732 \text{ sq. ft.}$$

Answer: 6,732 sq. ft.

PRACTICE PROBLEMS

1. Find the lateral areas of cylinders with the following dimensions:

a.

Diameter	4 m	60 cm	28 mm	$3\frac{1}{2}$ in.	2 ft. 9 in.
Height	10 m	45 cm	32 mm	$5\frac{1}{4}$ in.	4 ft.

126.6 m²   8,478 cm²   2,816 mm²* $57\frac{3}{4}$ sq. in.* $34\frac{4}{7}$ sq. ft.*

b.

Radius	3 mm	14 m	42 cm	$2\frac{3}{4}$ in.	8 ft. 6 in.
Height	5 mm	20 m	18 cm	8 in.	5 ft. 4 in.

94.2 mm²   1,760 m²*   4,752 cm²* $138\frac{2}{7}$ sq. in.*   284.95 sq. ft.* or
                                                    40,995.84 sq. in.

2. How many square inches of paper are needed to make a label on a can 4 inches in diameter and 5 inches high? 62.8 sq. in.

3. Find the total areas of cylinders with the following dimensions:

a.

Radius	2 cm	7 m	30 mm	$4\frac{3}{8}$ in.	1 ft. 6 in.
Height	4 cm	12 m	25 mm	20 in.	3 ft.

75.36 cm²   836 m²*   10,362 mm²   670.31 sq. in.*   42.39 sq. ft.

b.

Diameter	5 m	4 mm	35 cm	$2\frac{1}{2}$ in.	2 ft. 5 in.
Height	8 m	2 mm	40 cm	7 in.	1 ft. 8 in.

164.85 m²   50.24 mm²   6,325 cm²*   64.82 sq. in.*   3,141.57 sq. in.

4. How much asbestos paper covering is needed to enclose the curved surface and the two ends of a hot water storage tank 35 centimeters in diameter and 1.5 meters high? 18,408.25 cm² or 1.840825 m²

X. Sphere

The area of the surface of a sphere is equal to 4 times pi (π) times the radius squared. Formula: $A = 4\pi r^2$

Sometimes the formula $A = \pi d^2$ is used.

PRACTICE PROBLEMS

1.a. 1,808.64 m²   b. 9,856 mm²*   c. 498.96 cm²*   d. $1,164\frac{5}{8}$ sq. in.*   e. 46,735.76 sq. in.

1. Find the surface areas of spheres having the following radii:

a. 12 m   **b.** 28 mm   **c.** 6.3 cm   **d.** $9\frac{5}{8}$ in.   **e.** 5 ft. 1 in.

2. Find the surface areas of spheres having the following diameters:

a. 16 mm   **b.** 70 cm   **c.** 14.8 m   **d.** $6\frac{3}{4}$ yd.   **e.** 4 ft. 6 in.

3. What is the area of the earth's surface if its diameter is 7,900 miles?

2.a. 803.84 mm²   b. 15,386 cm²   c. 687.79 m²   d. 143.2 sq. yd.*   e. 63.585 sq. ft.
3. 195,967,400 sq. mi.

*$\pi = 3\frac{1}{7}$ was used to determine answers marked with asterisk. Otherwise, 3.14 was used.

VOC.: volume   volume of a rectangular solid
     volume of a   volume of a sphere
     cube   volume of a right circular cone
     volume of a   volume of a
     cylinder   pyramid

EXERCISE
3-31

Measuring Volume

OBJECTIVE: To measure volume.

The volume, also called cubical contents or capacity, is the number of units of cubic measure contained in a given space. When computing the volume of a geometric figure, *express all linear units in the same denomination.*

I. Rectangular Solid

The volume of a rectangular solid is equal to the length times the width times the height. Formula: $V = lwh$

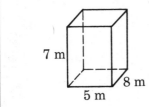

7 m

8 m

5 m

$l = 8\,\text{m}$
$w = 5\,\text{m}$
$h = 7\,\text{m}$
$V = ?$

Find the volume of a rectangular solid 8 meters long, 5 meters wide and 7 meters high.

$$V = lwh$$
$$V = 8 \times 5 \times 7$$
$$V = 280\,\text{m}^3$$

$$\begin{array}{r} 8 \\ \times 5 \\ \hline 40 \end{array} \qquad \begin{array}{r} 40 \\ \times 7 \\ \hline 280 \end{array}$$

Answer: 280 m³

Basic: Prob. 1–4; Average: Prob. 1–6; Enriched: Prob. 1–7

PRACTICE PROBLEMS

1. What is the volume of a rectangular solid if it is 7 inches long, 4 inches wide, and 9 inches high? **252 cu. in.**

2. Find the volumes of rectangular solids having the following dimensions:

a.

	144 mm³	1,080 m³	4,284 cu. in.	16,848 m³	47,120 cm³
Length	8 mm	12 m	17 in.	36 m	62 cm
Width	3 mm	9 m	18 in.	36 m	40 cm
Height	6 mm	10 m	14 in.	13 m	19 cm

b.

	39.6 m³	$95\frac{25}{64}$ cu. in.	165 cu. ft.	198.9 cu. ft. or 343,710 cu. in.	2 cu. ft.
Length	3.2 m	$5\frac{1}{2}$ in.	$8\frac{1}{4}$ ft.	7 ft. 6 in.	2 ft.
Width	4.5 m	$3\frac{3}{4}$ in.	6 ft.	4 ft. 9 in.	1 ft. 4 in.
Height	2.75 m	$4\frac{5}{8}$ in.	$3\frac{1}{3}$ ft.	5 ft. 7 in.	9 in.

3. A schoolroom is 12 m long, 7 m wide, and 4.5 m high. How many cubic meters of air space does the room contain? Allowing 6 m³ of air space per pupil, what is the maximum number of pupils that should be assigned to the room? **378 m³; 63 pupils**

4. How many cubic yards of dirt must be removed when digging for a foundation of a building if the excavation is 94 ft. long, 36 ft. wide, and 25 ft. deep? **3,133⅓ cu. yd.**

5. Find the weight of a steel bar 12 ft. long, 2 in. wide, and ½ in. thick if steel weighs 490 lb. per cu. ft. **40⅚ lb.**

6. How many cartons, 2 ft. by 1 ft. 6 in. by 1 ft., can be stored to a height of 9 ft. in a space 10 ft. by 12 ft.? **360 cartons**

7. What is the capacity of a freight car in cubic meters if the inside dimensions are 12.4 m by 3 m by 3.1 m? **115.32 m³**

II. Cube

The volume of a cube is equal to the length of the edge times itself times itself or the edge or side cubed.

Formula: $V = e^3$ or $V = s^3$

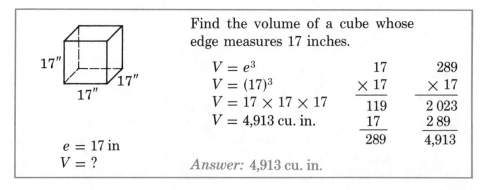

Find the volume of a cube whose edge measures 17 inches.

$$V = e^3$$
$$V = (17)^3$$
$$V = 17 \times 17 \times 17$$
$$V = 4,913 \text{ cu. in.}$$

17	289
× 17	× 17
119	2 023
17	2 89
289	4,913

$e = 17$ in
$V = ?$

Answer: 4,913 cu. in.

Basic: Prob. 1–2; *Average:* Prob. 1–3; *Enriched:* Prob. 1–4

PRACTICE PROBLEMS

1. What is the volume of a cube whose edge is 25 centimeters? **15,625 cm³**

2. Find the volumes of cubes whose edges measure:

a. 9 m	**c.** 11 ft.	**e.** 3.1 m	**g.** 1.09 m	**i.** 5 yd. 2 ft.
b. 14 mm	**d.** 27 cm	**f.** 4.5 cm	**h.** 4¾ ft.	**j.** 1 ft. 10 in.

3. How many cu. ft. of space are in a bin 6 ft. by 6 ft. by 6 ft.? **216 cu. ft.**

4. The volume of an 8-centimeter cube is how many times as large as the volume of a 2-centimeter cube? **64 times as large**

2.a. 729 m³   b. 2,744 mm³   c. 1,331 cu. ft.   d. 19,683 cm³   e. 29.791 m³   f. 91.125 cm³
g. 1.30 m³   h. 107 11/64 cu. ft.   i. 4,913 cu. ft.   j. 10,648 cu. in.

III. Right Circular Cylinder

The volume of a cylinder is equal to pi (π)* times the square of the radius of the base times the height. Formula: $V = \pi r^2 h$

When the diameter is known, the following formula may be used: $V = \frac{1}{4}\pi d^2 h$

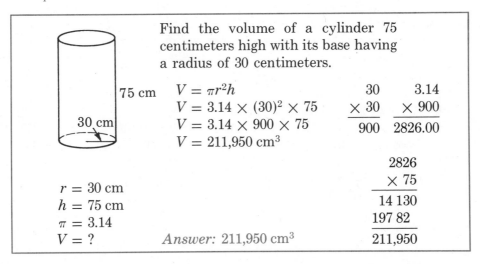

Find the volume of a cylinder 75 centimeters high with its base having a radius of 30 centimeters.

$$V = \pi r^2 h$$
$$V = 3.14 \times (30)^2 \times 75$$
$$V = 3.14 \times 900 \times 75$$
$$V = 211{,}950 \text{ cm}^3$$

$$\begin{array}{r} 30 \\ \times\, 30 \\ \hline 900 \end{array} \qquad \begin{array}{r} 3.14 \\ \times\, 900 \\ \hline 2826.00 \end{array}$$

$$\begin{array}{r} 2826 \\ \times\, 75 \\ \hline 14\,130 \\ 197\,82 \\ \hline 211{,}950 \end{array}$$

$r = 30$ cm
$h = 75$ cm
$\pi = 3.14$
$V = ?$

Answer: 211,950 cm³

Basic: Prob. 1–7; *Average:* Prob. 1–9; *Enriched:* Prob. 1–10

PRACTICE PROBLEMS

1. What is the volume of a cylinder if the radius of its base is 3 meters and the height is 6 meters? **169.56 m³**

2. What is the volume of a cylinder if the diameter of its base is 10 feet and the height is 16 feet? **1,256 cu. ft.**

3. Find the volumes of cylinders having the following dimensions:

a.

Radius	5 mm	14 m	19 ft.	$2\frac{5}{8}$ in.	1 ft. 6 in.
Height	8 mm	10 m	25 ft.	$7\frac{1}{2}$ in.	9 ft. 2 in.

628 mm³   6,160 m³*   28,338.5 cu. ft.   162.27 cu. in.   64.76 cu. ft.

b.

Diameter	4 cm	56 mm	12 m	$6\frac{1}{8}$ in.	1 ft. 2 in.
Height	6 cm	20 mm	28 m	9 in.	2 ft. 3 in.

75.36 cm³   49,280 mm³*   3,168 m³* $265\frac{37}{128}$ cu. in.*   4,158 cu. in. or $2\frac{13}{32}$ cu. ft.*

*$\pi = 3\frac{1}{7}$ or $\frac{22}{7}$ or 3.14. For greater accuracy 3.1416 may be used.

*$\pi = 3\frac{1}{7}$ was used to determine answers marked with asterisk. Otherwise, 3.14 was used.

4. Which container holds more, one 3 cm in diameter and 4 cm high or one 4 cm in diameter and 3 cm high? How many cubic centimeters more? **Second container; $9\frac{3}{7}$ cm³ more***

5. How many cubic yards of dirt must be dug to make a well 4 ft. 6 in. in diameter and 42 ft. deep? **$24\frac{3}{4}$ cu. yd.***

6. What is the weight of a round steel rod 8 ft. long and $\frac{3}{4}$ in. in diameter? A cubic foot of steel weighs 490 lb. **12 lb. $\frac{1}{2}$ oz.***

7. Find the displacement of a piston (volume of an engine cylinder) whose bore (diameter) is 4 in. and whose stroke (height piston moves) is 5 in. **62.8 cu. in.**

8. A silo has an inside diameter of 4.7 m. It is 9.4 m high. How many cubic meters of silage will it hold? **163 m³**

9. How many cubic meters of water does a tank hold if its diameter is 45 cm and height is 2 m? Find its capacity in liters. What is the weight of the water when the tank is full? **.317925 m³; 317.925 L; 317.925 kg**

10. How many gallons of oil does a tank car hold if it is 10 ft. in diameter and 42 ft. long? 1 cu. ft. $= 7\frac{1}{2}$ gal. **24,727.5 gal.**

IV. Sphere

The volume of a sphere is equal to $\frac{4}{3}$ times pi (π) times the cube of the radius. Formula: $V = \frac{4}{3}\pi r^3$

Sometimes the formula $V = \dfrac{\pi d^3}{6}$ is used.

Basic: Prob. 1; *Average:* Prob. 1–2; *Enriched:* Prob. 1–3

PRACTICE PROBLEMS

1. Find the volumes of spheres having the following radii:

a. 30 m   **b.** 56 ft.   **c.** 10.4 cm   **d.** $6\frac{5}{16}$ in.   **e.** 2 ft. 8 in.

2. Find the volumes of spheres having the following diameters:

a. 26 mm   **b.** 84 cm   **c.** 11.3 m   **d.** $12\frac{1}{2}$ in.   **e.** 4 ft. 5 in.

3. How many cubic feet of air does a basketball contain if its diameter is 10 in.? **.3 cu. ft.**

V. Right Circular Cone

The volume of a right circular cone is equal to $\frac{1}{3}$ times pi (π) times the square of the radius of the base times the height.

Formula: $V = \frac{1}{3}\pi r^2 h$

1.a. 113,142$\frac{6}{7}$ m³* b. 735,914$\frac{2}{3}$ cu. ft.* c. 4,709.43 cm³ d. 1,054.06 cu. in.*
e. 137,188.69 cu. in. or 79.39 cu. ft.   2.a. 9,206.48 mm³* b. 310,464 cm³* c. 755.12 m³
d. 1,023.07 cu. in.* e. 77,912.3 cu. in. or 45.088 cu. ft.

*$\pi = 3\frac{1}{7}$ was used to determine answers marked with asterisk. Otherwise, 3.14 was used.

PRACTICE PROBLEMS

1. Find the volumes of cones having the following dimensions:

a.

Radius	8 cm	14 mm	6.5 m	4 ft. 3 in.
Height	23 cm	12 mm	9 m	6 ft.

1,540.69 cm³   2,464 mm³*   398.36 m³*   113.43 cu. ft.

b.

Diameter	25 m	42 cm	7 ft.	1 ft. 5 in.
Height	15 m	58 cm	$4\frac{1}{2}$ ft.	2 ft.

2,453.13 m³   26,796 cm³* $57\frac{3}{4}$ cu. ft.*   1,814.92 cu. in.

2. A conical pile of sand is 2.5 m in diameter and 9 m high. How many cubic meters of sand are in the pile? 14.72 m³

3. How many bushels of grain are in a conical pile 16 ft. in diameter and 5 ft. high? 267.95 bu.

VI. Pyramid

The volume of a pyramid is equal to $\frac{1}{3}$ times the area of the base times the height.

Formula: $V = \frac{1}{3} Bh$

PRACTICE PROBLEMS

1. Find the volume of a pyramid when the side of its square base is 40 m and the height is 27 m. 14,400 m³

2. What is the volume of a pyramid when its rectangular base is 23 m long and 18 m wide and the height is 35 m? 4,830 m³

3. How many cu. ft. of space are inside a tent in the shape of a square pyramid 15 ft. on each side of the base and 16 ft. high? 1,200 cu. ft.

REVIEW

Find the area of:

1. A rectangle 65 meters long and 43 meters wide. 2,795 m²

2. A circle whose radius is 37 centimeters. 4,298.66 cm²

3. A square whose side measures 21 feet. 441 sq. ft.

4. A parallelogram with an altitude of 159 mm and a base of 250 mm. 39,750 mm²

5. A circle whose diameter is $8\frac{3}{4}$ inches. 60.1 sq. in.

6. A triangle with a base of 60 km and a height of 82 km. 2,460 km²

7. A trapezoid with bases of 101 ft. and 53 ft. and a height of 64 ft.
4,928 sq. ft.

*$\pi = 3\frac{1}{7}$ was used to determine answers marked with asterisk. Otherwise, 3.14 was used.

Find the total area of:

8. A cube whose edge measures 19 centimeters. 2,166 cm²

3,530 sq. yd. **9.** A rectangular solid 31 yd. long, 27 yd. wide, and 16 yd. high.

10. A right circular cylinder 63 mm in diameter and 75 mm high.

11. A sphere whose radius is 6 meters. 452.16 m²

12. How many 9″ by 9″ square tiles are needed for a floor measuring 18′ long and 12′ wide? 384 tiles

13. The cooking area of a circular grill having a diameter of 60 centimeters is how many times as large as the area of a grill having a diameter of 40 centimeters? $2\frac{1}{4}$ times larger

10. 21,067.83 mm² or 21,087 mm²*

Find the volume of:

14. A cube whose edge measures 38 millimeters. 54,872 mm³

5,024 cu. ft. **15.** A cylinder 16 feet high with the radius of its base 10 feet.

46,440 cm³ **16.** A rectangular solid 60 cm long, 43 cm wide, and 18 cm high.

17. A sphere whose diameter is 9 meters. 381.51 m³

18. A right circular cone 40 ft. high with the diameter of its base 28 ft.

19. A square pyramid 12.5 m on each side and 15 m high.

20. A circular wading pool is 96 inches in diameter and 14 inches deep. How many cubic feet of water will it hold? 58.61 cu. ft. or $58\frac{2}{3}$ cu. ft.*

18. 8,205.87 cu. ft. or $8,213\frac{1}{3}$ cu. ft.*

REFRESH YOUR SKILLS

1. Add:

```
       513
    79,884
     8,935
   484,629
     9,356
    97,803
```
681,120

2. Subtract:

```
   614,904
   523,994
```
90,910

3. Multiply:

```
    4,968
      759
```
3,770,712

4. Divide: 994

$$806\overline{)801,164}$$

5. Add:

$2\frac{7}{8}$
$1\frac{3}{4}$
$5\frac{13}{16}$ $10\frac{7}{16}$

6. Subtract:

$8\frac{1}{6}$
$2\frac{4}{5}$
$5\frac{11}{30}$

7. Multiply

$\frac{3}{8} \times 5\frac{1}{4}$ $1\frac{31}{32}$

8. Divide:

$2\frac{11}{12} \div 1\frac{1}{8}$
$2\frac{16}{27}$

9. Add:

.406 + 9.25 + 63.8
73.456

10. Subtract:

$60 − $2.75
$57.25

11. Multiply:

3.1416
.25
.7854

12. Divide:

$.25\overline{)$10}$
40

For a Test on Applications Measuring Area and Volume see p. T52 or *Testing Program*, p. 24.

*$\pi = 3\frac{1}{7}$ was used to determine answers marked with asterisk. Otherwise, 3.14 was used.

REVIEW OF PART 3

The numeral in color following each problem indicates the Exercise where explanatory material may be found.

1. a. Read, or write in words, each of the following: $\overline{BR}$, $\overleftrightarrow{MY}$, $\quad$ 3-1
$\overrightarrow{LT}$. Line segment *BR*; line *MY*, ray *LT*

b. Name each of the following: $\qquad\qquad$ 3-1

(1) line *CN* $\qquad$ (2) ray *ZG* $\qquad$ (3) line segment *RM*

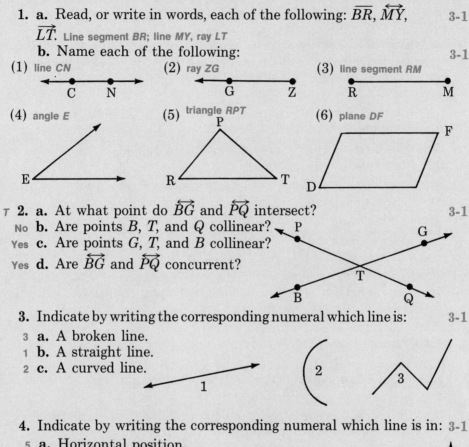

(4) angle *E* $\qquad$ (5) triangle *RPT* $\qquad$ (6) plane *DF*

T **2. a.** At what point do $\overleftrightarrow{BG}$ and $\overrightarrow{PQ}$ intersect? $\qquad$ 3-1
No **b.** Are points *B*, *T*, and *Q* collinear?
Yes **c.** Are points *G*, *T*, and *B* collinear?
Yes **d.** Are $\overleftrightarrow{BG}$ and $\overrightarrow{PQ}$ concurrent?

3. Indicate by writing the corresponding numeral which line is: $\qquad$ 3-1

3 **a.** A broken line.
1 **b.** A straight line.
2 **c.** A curved line.

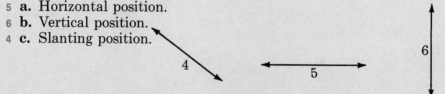

4. Indicate by writing the corresponding numeral which line is in: 3-1

5 **a.** Horizontal position.
6 **b.** Vertical position.
4 **c.** Slanting position.

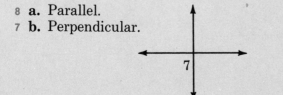

5. Indicate by writing the numeral which pairs of lines are: $\qquad$ 3-1

8 **a.** Parallel.
7 **b.** Perpendicular.

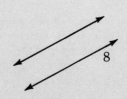

6. If the distance from E to F is 70 kilometers, what is the distance from L to H? 96 km   3-3

7.–17. Constructions

7. **a.** With a protractor draw an angle measuring 115°.   3-8
   b. Draw any acute angle.
   c. Draw any obtuse angle.
   d. Draw a right angle.

8. Construct a triangle with sides measuring 56 mm, 47 mm, and 63 mm.   3-19

9. Construct a triangle with angles measuring 39° and 101° and an included side measuring $3\frac{1}{8}$ in.   3-19

10. Construct a triangle with sides measuring 7.2 cm and 5.1 cm and an included angle of 51°.   3-19

11. Draw a regular hexagon, each side measuring 29 millimeters.   3-20

12. Draw a line segment 46 mm long. Using a compass, bisect this segment. Check with a ruler.   3-23

13. Draw an angle measuring 62°. Using a compass, bisect this angle. Check with a protractor.   3-25

14. Draw any line. Using a compass, construct a perpendicular to this line at a point on the line.   3-21

15. Draw any line. Using a compass, construct a perpendicular to this line from a point not on the line.   3-22

16. Draw with a protractor an angle of 74°. Then construct with a compass an angle equal to it. Check with a protractor.   3-24

17. Draw any line. Locate a point outside this line. Through this point construct a line parallel to the first line.   3-26

18. What is the complement of an angle of 61°? 29°   3-11

19. What is the supplement of an angle of 97°? 83°   3-11

20. What is the sum of the measures of angles 1, 2, and 3 of triangle PQR? 180°   3-10

21. If $m\angle 2 = 47°$ and $m\angle 4 = 121°$, find the measure of $\angle 1$ and $\angle 3$ of triangle PQR.   3-11

$m\angle 1 = 74°$; $m\angle 3 = 59°$

22. $\overleftrightarrow{KL}$ and $\overleftrightarrow{MN}$ intersect at G. If $m \angle 2 = 104°$, what is the measure of $\angle 1$? $\angle 3$? $\angle 4$? 3-11
$m \angle 1 = 76°; m \angle 3 = 76°; m \angle 4 = 104°$

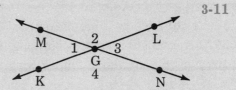

23. Parallel lines AB and CD are cut by transversal EF. If $m \angle 8 = 56°$, what is the measure of $\angle 5$? $\angle 6$? $\angle 7$? $\angle 3$? $\angle 4$? $\angle 1$? $\angle 2$? 3-12
$m \angle 5 = 56°; m \angle 6 = 124°; m \angle 7 = 124°;$
$m \angle 3 = 124°; m \angle 4 = 56°; m \angle 1 = 56°;$
$m \angle 2 = 124°$

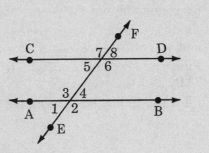

Find the perimeter of:

24. A rectangle 41 m long and 17 m wide. 116 m   3-5

25. A triangle with sides measuring $4\frac{3}{4}$ in., $3\frac{5}{8}$ in., and $5\frac{11}{16}$ in.   3-5

26. A square whose side measures 26 cm.   14.0625 in. or $14\frac{1}{16}$ in.   3-5
104 cm

Find the circumference of:

27. A circle whose diameter is 28 mm. 87.92 mm   3-6

28. A circle whose radius is 80 ft. 502.4 ft.   3-6

Find the area of:

29. A square whose side measures 47 ft. 2,209 sq. ft.   3-30

30. A rectangle 51 cm long and 18 cm wide. 918 cm²   3-30

31. A circle whose radius is 9 in. 254.34 sq. in.   3-30

32. A triangle whose altitude is 39 cm and base is 28 cm. 546 cm²   3-30

33. A parallelogram with an altitude of 24 m and a base of 49 m.   3-30
1,176 m²

34. A circle whose diameter is 56 in. 2,461.76 sq. in.   3-30

35. A trapezoid with bases of 43 mm and 37 mm and a height of   3-30
25 mm. 1,000 mm²

Find the total area of:

36. A cube whose edge measures $4\frac{1}{2}$ ft. $121\frac{1}{2}$ cu. ft.   3-30

37. A right circular cylinder 54 cm in diameter and 75 cm in   3-30
height. 17,295.12 cm²

Find the volume of:

38. A cube whose edge measures 79 mm. 493,039 mm³   3-31

58.374 m³ **39.** A rectangular solid 6.9 m long, 1.8 m wide, and 4.7 m high.   3-31

40. A right circular cylinder with a diameter of 70 ft. and a   3-31
height of 46 ft. 176,939 cu. ft.

41. A sphere whose diameter is 84 cm. 310,181.76 cm³   3-31

42. Find the hypotenuse of a right triangle if the altitude is   3-27
63 m and the base is 60 m. 87 m

43. Find the base of a right triangle if the hypotenuse is 97 cm   3-27
and the altitude is 72 cm. 65 cm

44. Find the altitude of a right triangle if the hypotenuse is   3-27
73 ft. and the base is 55 ft. 48 ft.

45. Find the indicated missing parts: 3-13

a = 49 mm
b = 64 mm

46. Find the indicated missing parts: 3-14

a = 21 m
b = 30 m

47. Find the height of a TV antenna tower that casts a shadow   3-28
6 m long when a nearby light pole, 2.4 m high, casts a
shadow 0.4 m long. 36 m

Find the indicated parts of the following right triangles with   3-29
answers to the nearest tenth. See table on page 417.

48. Find side *a*:   **49.** Find side *b*:   **50.** Find side *b*:
214.2 m   246.28 ft.   33.2 cm

COMPETENCY CHECK TEST

Solve each problem and select the letter corresponding to your answer. The numerals in red boxes indicate Exercises where help may be found.

1. Which line is in a horizontal position? ⬚3-1

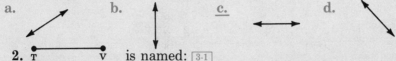

a.   b.   <u>c.</u>   d.

2. T •————• V is named: ⬚3-1

a. line TV   b. ray TV   <u>c.</u> line segment TV   d. plane TV

3. Which pair of lines are parallel? ⬚3-1

a.   <u>b.</u>   c.   d.

4. Which figure is a square? ⬚3-4

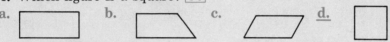

a.   b.   c.   <u>d.</u>

5. If the scale is 1:2,500,000, what is the actual distance represented by 31 mm? ⬚3-3

<u>a.</u> 77.5 km   b. 6.75 km   c. 795 km   d. Answer not given

6. Which triangle is isosceles? ⬚3-4

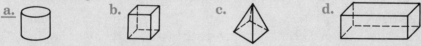

7. An angle measuring 48° is a(an): ⬚3-7

a. right angle   <u>b.</u> acute angle   c. obtuse angle   d. straight angle

8. Which figure illustrates a cylinder? ⬚3-4

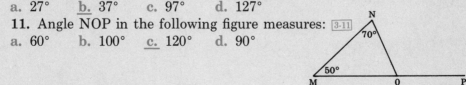

<u>a.</u>   b.   c.   d.

9. The measure of the third angle of a triangle, when the other two angles measure 74° and 39°, is: ⬚3-10

a. 113°   b. 16°   c. 90°   <u>d.</u> 67°

10. The complement of an angle measuring 53° is: ⬚3-11

a. 27°   <u>b.</u> 37°   c. 97°   d. 127°

11. Angle NOP in the following figure measures: ⬚3-11

a. 60°   b. 100°   <u>c.</u> 120°   d. 90°

12. If the scale is 1 cm = 50 m, a distance of 220 m is represented by: ⬚3-3

a. 4.5 cm   b. 45 cm   c. 44 cm   <u>d.</u> 4.4 cm

Continued on page 440.

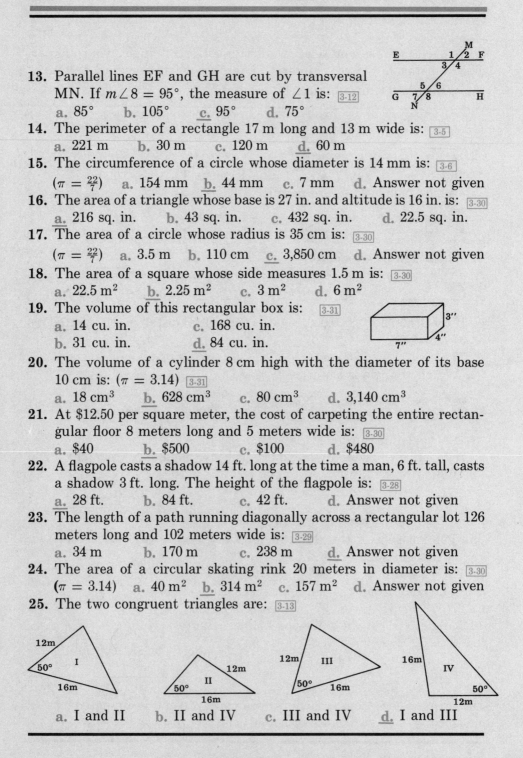

13. Parallel lines EF and GH are cut by transversal MN. If $m\angle 8 = 95°$, the measure of $\angle 1$ is: ⟦3-12⟧

    a. 85°   **b.** 105°   **c.** 95°   **d.** 75°

14. The perimeter of a rectangle 17 m long and 13 m wide is: ⟦3-5⟧

    a. 221 m   **b.** 30 m   **c.** 120 m   **d.** 60 m

15. The circumference of a circle whose diameter is 14 mm is: ⟦3-6⟧

    $(\pi = \frac{22}{7})$ **a.** 154 mm   **b.** 44 mm   **c.** 7 mm   **d.** Answer not given

16. The area of a triangle whose base is 27 in. and altitude is 16 in. is: ⟦3-30⟧

    a. 216 sq. in.   **b.** 43 sq. in.   **c.** 432 sq. in.   **d.** 22.5 sq. in.

17. The area of a circle whose radius is 35 cm is: ⟦3-30⟧

    $(\pi = \frac{22}{7})$ **a.** 3.5 m   **b.** 110 cm   **c.** 3,850 cm   **d.** Answer not given

18. The area of a square whose side measures 1.5 m is: ⟦3-30⟧

    a. 22.5 m²   **b.** 2.25 m²   **c.** 3 m²   **d.** 6 m²

19. The volume of this rectangular box is: ⟦3-31⟧

    a. 14 cu. in.   **c.** 168 cu. in.

    b. 31 cu. in.   **d.** 84 cu. in.

20. The volume of a cylinder 8 cm high with the diameter of its base 10 cm is: $(\pi = 3.14)$ ⟦3-31⟧

    a. 18 cm³   **b.** 628 cm³   **c.** 80 cm³   **d.** 3,140 cm³

21. At $12.50 per square meter, the cost of carpeting the entire rectangular floor 8 meters long and 5 meters wide is: ⟦3-30⟧

    a. $40   **b.** $500   **c.** $100   **d.** $480

22. A flagpole casts a shadow 14 ft. long at the time a man, 6 ft. tall, casts a shadow 3 ft. long. The height of the flagpole is: ⟦3-28⟧

    a. 28 ft.   **b.** 84 ft.   **c.** 42 ft.   **d.** Answer not given

23. The length of a path running diagonally across a rectangular lot 126 meters long and 102 meters wide is: ⟦3-29⟧

    a. 34 m   **b.** 170 m   **c.** 238 m   **d.** Answer not given

24. The area of a circular skating rink 20 meters in diameter is: ⟦3-30⟧

    $(\pi = 3.14)$ **a.** 40 m²   **b.** 314 m²   **c.** 157 m²   **d.** Answer not given

25. The two congruent triangles are: ⟦3-13⟧

    a. I and II   **b.** II and IV   **c.** III and IV   **d.** I and III

Elements of Algebra

4

The algebra in this section of the book includes:

1. The study of positive and negative numbers, thus extending the number system.

2. The reading, writing, and evaluation of algebraic expressions and sentences including formulas, thus developing the basic language and using the symbolism of algebra.

3. The solution and graphing of open sentences in one variable including both equations and inequalities, thus developing a very essential tool for solving problems. The use of the ratio and proportion concepts for solving problems is also stressed.

INVENTORY TEST

The numeral in color following each problem indicates the Exercise where explanatory material may be found.

1. Which of the following numerals name <u>positive</u> numbers? **86, 1.7**
Which name <u>negative</u> numbers? $-\frac{2}{3}, -.625, -5\frac{1}{2}, -300$ **4-1**

$$-\frac{2}{3} \qquad +86 \qquad -.625 \qquad -5\frac{1}{2} \qquad +1.7 \qquad -300$$

2. Which of the following numerals name integers? Which name rational numbers? Which name irrational numbers? Which name real numbers? **4-2**

$$-.3 \qquad +8\frac{3}{4} \qquad -69 \qquad -\sqrt{57} \qquad -\frac{7}{10} \qquad +\frac{36}{6}$$

3. Find the value of: **a.** $|+63|$ **63** **b.** $|-\frac{4}{5}|$ $\frac{4}{5}$ **4-3**

4. Draw the graph of -3, -1, 0, and 2 on the number line. **4-4**

5. Write the coordinates of which the following is the graph: **4-4**
$-5, -2, 0, 3, 6$

2. Integers: -69; $+\frac{36}{6}$

Rational numbers:
$-.3$; $8\frac{3}{4}$; -69; $-\frac{7}{10}$; $+\frac{36}{6}$

$$-6 \quad -5 \quad -4 \quad -3 \quad -2 \quad -1 \quad 0 \quad 1 \quad 2 \quad 3 \quad 4 \quad 5 \quad 6$$

Irrational numbers: $-\sqrt{57}$

442

Real numbers: $-.3$; $8\frac{3}{4}$; -69; $-\sqrt{57}$; $-\frac{7}{10}$; $+\frac{36}{6}$

4.
$$-4 \quad -3 \quad -2 \quad -1 \quad 0 \quad 1 \quad 2 \quad 3 \quad 4$$

6. If -5 kilograms means 5 kilograms underweight, what does
$+3$ kilograms mean? **3 kg overweight**   4-5

7. Which of the following are <u>true</u>?   4-6
    a. $-3 > -1$ <u>**b.**</u> $+2 \not< -2$ **c.** $0 < -4$ <u>**d.**</u> $+4 > +6$

8. Using the number line as the scale, draw the vector and **(See opposite page.)**
write the numeral represented by this vector that illus-
trates a movement from: **For ex. 8 see Extended Answers, p. T80.**   4-7
    a. 0 to -4 **b.** -3 to $+5$ **c.** $+1$ to -6 **d.** $+7$ to $+2$

9. a. What is the opposite of -18? **+18**   4-8
    b. What is the additive inverse of $+23$? **-23**
    c. Write symbolically: The opposite of negative ten is
    positive ten. **$-(-10) = +10$**
    d. Which has the greater opposite number: $\underline{+2}$ or $+5$?

10. Add: 4-9
    a. $\begin{array}{r} -5 \\ +9 \\ \hline \end{array}$ **4**   **b.** $\begin{array}{r} +6 \\ +7 \\ \hline \end{array}$ **13**   **c.** $\begin{array}{r} -12 \\ +8 \\ \hline \end{array}$ **-4**   **d.** $\begin{array}{r} -9 \\ -7 \\ \hline \end{array}$ **-16**   **e.** $\begin{array}{r} -3 \\ +3 \\ \hline \end{array}$ **0**
    f. $(-7) + (+4) + (-9) + (+11)$ **-1**
    g. Simplify: $-4 + 6 - 3 - 7 + 5$ **-3**

11. Subtract: 4-10
    a. $\begin{array}{r} +6 \\ -8 \\ \hline \end{array}$ **14**   **b.** $\begin{array}{r} +9 \\ +11 \\ \hline \end{array}$ **-2**   **c.** $\begin{array}{r} -5 \\ -4 \\ \hline \end{array}$ **-1**   **d.** $\begin{array}{r} -14 \\ +5 \\ \hline \end{array}$ **-19**   **e.** $\begin{array}{r} 0 \\ -4 \\ \hline \end{array}$ **4**
    f. $(+5) - (-5)$ **10**   **g.** Take 6 from 2. **-4**

12. Multiply: 4-11
    a. $\begin{array}{r} +9 \\ -7 \\ \hline \end{array}$ **-63**   **b.** $\begin{array}{r} -4 \\ +8 \\ \hline \end{array}$ **-32**   **c.** $\begin{array}{r} -10 \\ -10 \\ \hline \end{array}$ **100**   **d.** $\begin{array}{r} +7 \\ +3 \\ \hline \end{array}$ **21**   **e.** $\begin{array}{r} 0 \\ -5 \\ \hline \end{array}$ **0**
    f. $(-1)(-5)(-2)(+7)$ **-70**   **g.** Find the value of $(-2)^4$. **16**

13. Divide: 4-12
    a. $\dfrac{-40}{-8}$ **5** **b.** $\dfrac{-72}{+9}$ **-8** **c.** $\dfrac{+24}{-1}$ **-24** **d.** $\dfrac{+60}{+15}$ **4** **e.** $\dfrac{-16}{+16}$ **-1**

    f. $(-56) \div (-7)$ **8**   **g.** Simplify: $\dfrac{6(-2) + 5(-3)}{-3(4 - 7)}$ **-3**

14. Write as an algebraic expression: 4-14
$l + w$ **a.** The sum of l and w.
10 − 4   **b.** The difference between ten and four.
$\frac{6}{r}$ **c.** The quotient of six divided by r.
9a   **d.** The product of a and nine.
6s²   **e.** Six times the square of the side (s).

15. Write each of the following as a sentence symbolically: 4-15
   a. Some number y decreased by five is equal to twelve. $y - 5 = 12$
   b. Each number x increased by two is greater than eight. $x + 2 > 8$
   c. Ten times each number n is less than or equal to fourteen. $10 \times n \leq 14$
   d. Each number s is greater than negative three and less than positive nine. $-3 < s < 9$

16. Express as a formula: 4-16

Centripetal force (F) equals the product of the weight of the body (w) and the square of the velocity (v) divided by the product of the acceleration of gravity (g) and the radius of the circle (r). $F = \frac{wv^2}{gr}$

17. Find the value of: 4-17

   a. $m + 9n$ when $m = 8$ and $n = 3$. 35
   b. $a + b(a - b)$ when $a = 6$ and $b = 2$. 14
   c. $2x^2 - 3xy - y^2$ when $x = 4$ and $y = -3$. 59
   d. $\dfrac{r^2 - s^2}{(r - s)^2}$ when $r = 7$ and $s = 5$. 6

18. Find the value of E when $I = 5$, $r = 16$, and $R = 20$.   4-18
Formula: $E = Ir + IR$ 180

19. a. Which of these equations have 4 as the solution? 2, 3, 5   4-19

   (1) $n - 3 = 7$ (3) $n = 4$ (5) $6n - 7 = 17$

   (2) $n + 9 = 13$ (4) $\dfrac{n}{8} = 2$ (6) $9n + n = 20$

   b. Which of the above equations are equivalent? 2, 3, 5
   c. Which equation of the equivalent equations is in simplest form? 3

20. What operation with what number do you use on both sides   4-20
of each of the following equations to get an equivalent equation in simplest form? Also write this equivalent equation.

   a. $7x = 56$ Divide by 7; $x = 8$ **c.** $y - 11 = 6$ Add 11; $y = 17$

   b. $m + 8 = 14$ Subtract 8, $m = 6$ **d.** $\dfrac{a}{15} = 9$ Multiply by 15; $a = 135$

21. Solve and check, using all the real numbers: 4-21
   a. $x + 6 = 13$ $x = 7$ **d.** $b - 7 = 4$ $b = 11$ **f.** $6n + 17 = 35$ $n = 3$

   b. $-7y = 63$ $y = -9$ **e.** $\dfrac{a}{9} = 18$ $a = 162$ **g.** $5x - x = 10$ $x = 2.5$

   c. $0 = 8r - 24$ $r = 3$ **h.** $4m = 76$ $m = 19$

22. Find the solution of:

    a. $n - 5 = 7$ when the replacements for the variable are all the integers. $n = 12$

    b. $3x + 11 = 2$ when the replacements for the variable are all the integers. $x = -3$

    c. $4b - 3 = 21$ when the replacements for the variable are all the prime numbers. no solution   23. a. $+10$ or -10

23. Find the solutions of: **a.** $|n| = 10$ **b.** $|y| + 6 = 15$ b. $+9$ or -9

24. Find the value of t when $v = 217$, $V = 25$, and $g = 32$. Formula: $v = V + gt$ $t = 6$

25. On a number line draw the graph of each of the following equations when the replacements for the variable are all the real numbers:

    a. $x + 4 = 9$ **b.** $-7c = 28$ **c.** $5y - 3 = 7$

26. Solve by equation method:

    Pedro bought a football at a 40% reduction sale paying $10.65 for it. What was the regular price? $17.75

27. a. Find the ratio of:

    (1) 96 to 16   (2) 30 to 75   (3) 12 minutes to 2 hours
        6 to 1   2 to 5   1 to 10

    b. Solve and check: $\dfrac{n}{54} = \dfrac{13}{18}$ $n = 39$

For ex. 28–30 see
Extended Answers, p. T80.

28. Find the solutions of each of the following inequalities when the replacements for the variable are all the real numbers:

    a. $8x > 136$ **d.** $4a + 13 \geq 1$ **g.** $-7y \not< -21$

    b. $n + 3 < 11$ **e.** $6a - 7a < 0$ **h.** $2n - 5n > -15$

    c. $t - 5 \not> -7$ **f.** $\dfrac{d}{3} \neq -1$ **i.** $\dfrac{3}{5}x \leq 42$

29. Find the solutions of:

    a. $3d + 6 > 21$ when the replacements for the variable are all the integers.

    b. $-5y \leq -60$ when the replacements for the variable are all the multiples of 3.

    c. $2x - 4 < -12$ when the replacements for the variable are -4, -3, -2, -1, 0, 1, 2, 3, and 4.

30. On a number line draw the graph of each of the following inequalities when the replacements for the variable are all the real numbers:

    a. $n < 5$ **c.** $s - 3 \geq 1$ **e.** $2y - 8y \not> 18$

    b. $8a > -16$ **d.** $5b + 2 \neq 7$ **f.** $-1 < x < 2$

25.a. -2 -1 0 1 2 3 4 5 6   b. -5 -4 -3 -2 -1 0 1 2 3   c. -5 -4 -3 -2 -1 0 1 2 3

POSITIVE AND NEGATIVE NUMBERS
Extension of the Number System

Stock quotation changes, usually found in the business section of a newspaper, consist of numerals containing "+" or "−" signs. Weather stations in the larger cities report daily the high and low temperature readings. In the winter some of the numerals indicating these reported temperature readings contain "−" signs.

These signs preceding the digits in the numerals do not mean addition or subtraction. They are used to represent quantities which are opposites.

In arithmetic the numerals naming the whole numbers, the rational numbers, and the irrational numbers contained no sign. Now we are enlarging our number system to include new numbers which are named by numerals that contain either the symbol "+" or the symbol "−". They are called positive numbers and negative numbers, respectively.

EXERCISE

4-1 Integers

OBJECTIVE: To read and name integers.

VOC.: positive number
negative number
integer

Positive and Negative Numbers

A number named by a numeral containing a prefixed "+" sign is a positive number.

$$+8, \ +\frac{3}{5}, \text{ and } +\sqrt{11} \text{ are } positive \ numbers.$$

A number named by a numeral containing a prefixed "−" sign is a negative number.

$$-4, \ -\frac{5}{8}, \text{ and } -\sqrt{7} \text{ are } negative \ numbers.$$

Related Resources: *Practical Applications in Mathematics:* Problems, p. 147

446

On page 18 we found that numerals may be arranged in a definite order on the number line so that they correspond one-to-one with points on the line.

Thus far the number line has been restricted to points which are associated with the whole numbers and fractions of arithmetic and are located to the right of the point labeled 0.

Now let us extend the number line to the left of the point labeled 0. Using the interval between 0 and 1 as the unit measure, we locate equally spaced points to the left of 0. The first new point is labeled -1, read "negative one"; the second point -2, read "negative two"; the third point -3, read "negative three"; etc.

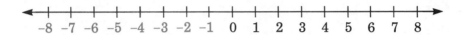

The numbers corresponding to the points to the left of 0 are *negative numbers* and their numerals are identified by the centered $(-)$ or raised dash symbols that precede the digits.

> -4 or $^-4$ names the number *negative four.*

The numbers corresponding to the points to the right of 0 are *positive numbers* and their numerals may contain no sign or centered $(+)$ or raised plus symbols that precede the digits.

> $+5$ or $^+5$ or just 5 names the number *positive five.*

Then the number line would look like this:

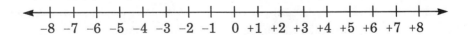

Positive and negative numbers are sometimes called signed numbers or directed numbers. 0 is neither a positive nor a negative number. All numbers greater than 0 are positive numbers and their numerals may be written without a sign as we have seen in arithmetic.

Opposites

Examination of the number line below will show that the points that correspond to +5 and −5 fall on opposite sides of the point marked 0 but are the same distance from 0.

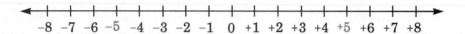

A pair of numbers, one positive and the other negative, such as +5 and −5, which have the same absolute value (see Exercise 4-3) are called opposites.

−5 is the opposite of +5 and +5 is the opposite of −5. Each is the opposite of the other.

The opposite of zero is zero.

Integers

The group of numbers which consists of all the whole numbers and their opposites is called the integers.
Both ..., −2, −1, 0, 1, 2, ... and ..., −2, −1, 0, +1, +2, ...
describe the integers.

Observe that the three dots are used at both ends since the integers continue in both directions without ending. Thus,
1, 2, 3, ... describes the positive integers (or the natural numbers);
0, 1, 2, 3, ... describes the non-negative integers (or whole numbers);
..., −3, −2, −1 describes the negative integers; and
..., −3, −2, −1, 0 describes the non-positive integers.

The even integers consist of the integers that are divisible by two (2). The even integers are described in this way: ..., −4, −2, 0, 2, 4, ... The odd integers consist of all the integers that are not divisible by two (2). The odd integers are described in this way: ..., −3, −1, 1, 3, ...

PRACTICE PROBLEMS

g. positive sixty-three
h. negative one hundred forty-seven

1. Read, or write in words, each of the following: i. negative seventy-four

a. −6 negative six   **d.** +15 positive fifteen   **g.** +63

b. +10 positive ten   **e.** −33 negative thirty-three **h.** −147

c. −7 negative seven   **f.** +50 positive fifty   **i.** −74

2. Write symbolically:

a. Negative twelve −12   **e.** Positive twenty-eight +28

b. Positive nine +9   **f.** Negative seventy −70

c. Positive sixteen +16   **g.** Positive forty-one +41

d. Negative four −4   **h.** Negative ninety-five −95

3. What number is the opposite of each of the following?

a. +7 −7   **d.** −19 +19   **g.** −100 +100

b. −3 +3   **e.** +46 −46   **h.** +79 −79

c. 0 no opposite **f.** −22 +22   **i.** −61 +61

4. a. Does −8 name a whole number? An integer? No; yes

b. Does 17 name a whole number? An integer? Yes; yes

5. List all the numbers described by the following:

a. −11, −10, −9, ... , 3 −11, −10, −9, −8, −7, −6, −5, −4, −3, −2, −1, 0, 1, 2, 3

b. −9, −7, −5, ... , 9 −9, −7, −5, −3, −1, 1, 3, 5, 7, 9

c. −12, −10, −8, ... , 10 −12, −10, −8, −6, −4, −2, 0, 2, 4, 6, 8, 10

```
EXERCISE
 4-2
```

Rational Numbers; Real Numbers

OBJECTIVE: To read and name rational numbers and real numbers.

VOC.: rational numbers
irrational numbers
real numbers

On page 75 we found that numerals naming the fractional numbers of arithmetic may be arranged in a definite order on the number line to the right of the point labeled 0. There is a point on the number line corresponding to each fractional number.

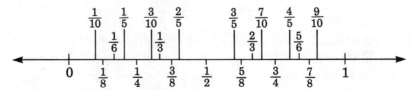

Related Resources: *Practical Applications in Mathematics:* Problems, p. 146

Examination of the number line below will show that there are fractional subdivisions to the left of zero matching those to the right of zero with each point of division corresponding to a fractional number. Since there are an infinite number of positive and negative fractional numbers, only a few are indicated. However, there is a point on the number line corresponding to each positive and negative fractional number.

The numeral $+\dfrac{2}{5}$ or $^{+}\left(\dfrac{2}{5}\right)$ or just $\dfrac{2}{5}$ names the

number *positive two-fifths*.

The numeral $-\dfrac{1}{4}$ or $^{-}\left(\dfrac{1}{4}\right)$ names the number

negative one-fourth.

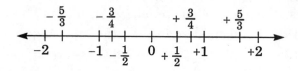

Observe in numerals with raised signs like $^{-}\frac{1}{3}$ or $^{+}\frac{5}{8}$ where parentheses are not used the $^{-}$ or $^{+}$ symbol belongs to the entire fraction although it may appear that the symbol belongs to the numerator alone.

Each positive fractional number has an opposite negative fractional number and each negative fractional number has an opposite positive fractional number.

$-\dfrac{3}{4}$ is the opposite of $+\dfrac{3}{4}$.

$+2\dfrac{3}{8}$ is the opposite of $-2\dfrac{3}{8}$.

All the fractional numbers and their opposites are called the rational numbers. This includes the integers, since each integer may be named in fraction form. For example $-7 = \frac{-7}{1}$, $+2 = \frac{+2}{1}$, and $0 = \frac{0}{1} = \frac{0}{2} = \frac{0}{3}$, etc. A rational number may be described as a number named by a numeral that expresses a quotient of two integers with division by zero excluded.

An irrational number is a number that is both a non-terminating and a non-repeating decimal. See pages 239 and 275. The square root of any positive number other than a perfect square (a number having an exact square root) is an irrational number. $\sqrt{43}$ and $-\sqrt{7}$ are irrational numbers. An irrational number cannot be named by a quotient of two integers with division by zero excluded. All the positive and negative irrational numbers are called the irrational numbers.

All the rational numbers and all the irrational numbers are real numbers. They include all the integers, all the positive and negative fractional numbers, and all the positive and negative irrational numbers. There are an infinite number of real numbers.

Basic: Prob. 1–3; *Average:* Prob. 1–4; *Enriched:* Prob. 1–5

PRACTICE PROBLEMS For ex. 1–5 see Extended Answers, pp. T80–T81.

1. Read, or write in words, each of the following:

   a. $+\frac{5}{6}$ **c.** $-6\frac{1}{3}$ **e.** -17.3 **g.** $-\frac{11}{16}$

   b. $-\frac{13}{4}$ **d.** $+.89$ **f.** $+2\frac{2}{5}$ **h.** $-.004$

2. Write symbolically:

   a. Negative one fifth   **e.** Negative nine and forty-
   b. Positive seven twelfths   eight hundredths
   c. Positive nine hundredths   **f.** Positive one hundred
   d. Negative four and three tenths   sixty-one thousandths

3. What number is the opposite of each of the following?

   a. $+\frac{5}{8}$ **c.** $+.07$ **e.** $-1\frac{3}{5}$ **g.** $+6\frac{4}{7}$

   b. $-\frac{9}{4}$ **d.** -2.75 **f.** $+.102$ **h.** -58.3

4. **a.** Is $-\frac{3}{5}$ an integer? A rational number? A real number?
   b. Is $+7\frac{2}{3}$ an integer? A rational number? A real number?
   c. Is $+.45$ an integer? A rational number? A real number?
   d. Is -8.9 an integer? A rational number? A real number?
   e. Is -6 an integer? A rational number? A real number?
   f. Is $-\sqrt{31}$ an integer? A rational number? A real number?

5. Which of the following numerals name integers? Which name rational numbers? Which name irrational numbers? Which name real numbers?

   a. $-17;$ $+\frac{3}{4};$ $-.05;$ $-\sqrt{29};$ $\frac{16}{2};$ $-7\frac{1}{3}$

   b. $-\frac{4}{7};$ $+100;$ $-5.08;$ $-2\frac{3}{4};$ $+\sqrt{13};$ $-\frac{18}{6}$

   c. $-.375;$ $-\frac{5}{8};$ $-\sqrt{16};$ $-69;$ $+\frac{28}{4};$ $+3\frac{5}{6}$

   d. $-\sqrt{5};$ $-\frac{15}{16};$ $+97;$ $+.18;$ $-4\frac{7}{12};$ $+\frac{21}{4}$

OBJECTIVE: To read and name absolute value.   VOC.: absolute value

The absolute value of any number is the value of the corresponding arithmetic number which has no sign.

> The absolute value of −4 is 4.
> The absolute value of +4 is 4.
> The absolute value of 0 is 0.

A pair of vertical bars | | is the symbol used to designate absolute value.

> |−4| is read "the absolute value of negative four."

Basic: Prob. 1–4; *Average:* Prob. 1–6; *Enriched:* Prob. 1–7

PRACTICE PROBLEMS

1. Read, or write in words, each of the following:
   a. $|+11|$ **b.** $|-3|$ **c.** $|+35|$ **d.** $|-\frac{7}{10}|$ **e.** $|-.09|$ **f.** $|+5\frac{2}{3}|$

2. Write symbolically, using the absolute value symbol:
   a. The absolute value of negative thirteen. $|-13|$
   b. The absolute value of positive eight fifteenths. $|\frac{8}{15}|$
   c. The absolute value of positive three and nine tenths. $|3\frac{9}{10}|$
   d. The absolute value of negative fourteen thousandths. $|\frac{-14}{1,000}|$

3. Find the absolute value of each of the following:

   a. $|-7|$ 7   **d.** $|0|$ 0   **g.** $|-2\frac{1}{2}|$ $2\frac{1}{2}$ **j.** $|-.82|$ .82

   b. $|+3|$ 3   **e.** $|-.8|$ .8   **h.** $|-27|$ 27   **k.** $|-\frac{9}{4}|$ $\frac{9}{4}$

   c. $|-6|$ 6   **f.** $|+\frac{2}{3}|$ $\frac{2}{3}$ **i.** $|+19|$ 19   **l.** $|-1.75|$ 1.75

   In each of the following, first find the absolute value of each number as required; then apply the necessary operation to obtain the answer.

4. **a.** $|-4| + |-8|$ 12   **c.** $|+2\frac{4}{5}| + |-4\frac{1}{4}|$ $7\frac{1}{20}$
   b. $|-\frac{3}{4}| + |-\frac{3}{4}|$ $1\frac{1}{2}$ **d.** $|-.91| + |+1.8|$ 2.71

5. **a.** $|+13| - |-7|$ 6   **c.** $|-1\frac{1}{2}| + |-3\frac{2}{3}|$ $5\frac{1}{6}$
   b. $|-\frac{4}{5}| - |+\frac{7}{10}|$ $\frac{1}{10}$ **d.** $|-2.5| - |+.15|$ 2.35

6. **a.** $|-9| \times |+3|$ 27   **c.** $|-3\frac{2}{5}| \times |+20|$ 68
   b. $|-\frac{2}{3}| \times |-\frac{3}{8}|$ $\frac{1}{4}$ **d.** $|-6.08| \times |-.07|$ .4256

7. **a.** $|-54| \div |-9|$ 6   **c.** $|+3\frac{5}{6}| \div |-1\frac{7}{12}|$ $2\frac{8}{19}$
   b. $|-\frac{5}{8}| \div |+\frac{2}{5}|$ 1.5625 or $1\frac{9}{16}$ **d.** $|-6.3| \div |-.09|$ 70

1.a. The absolute value of positive eleven   **452**   b. The absolute value of negative three
   c. The absolute value of positive thirty-five   d. The absolute value of negative seven tenths.
   e. The absolute value of negative nine hundredths   f. The absolute value of positive five and two-thirds

The Real Number Line;
Graph of a Group of Numbers

EXERCISE
4-4

OBJECTIVE: To draw a graph of a number or a group of numbers on the real number line.

VOC.: real number line   graph   graph of a group   coordinate   of numbers

The real number line is the complete collection of points which corresponds to all the real numbers. The real number line is endless in both directions and only a part of it is shown at any one time. There are an infinite number of points on the real number line. However, there is one and only one point that corresponds to each real number and one and only one real number that corresponds to each point on the real number line.

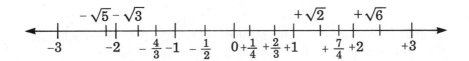

Usually the real number line is labeled only with the numerals naming the integers.

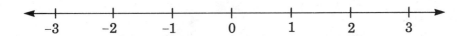

Each point on the number line is called the graph of the number to which it corresponds and each number is called the coordinate of the related point on the line. Capital letters are generally used to identify particular points.

Thus, the graph of a number is a point on the number line whose coordinate is the number.

Point B is the graph of -2.

-2 is the coordinate of point B.

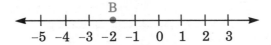

453

The graph of a group of numbers is made up of the points on the number line whose coordinates are the numbers.

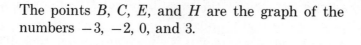

The points *B*, *C*, *E*, and *H* are the graph of the numbers −3, −2, 0, and 3.

To draw the graph of a group of numbers, first draw an appropriate number line and locate the point or points whose coordinate or coordinates are listed in the given numbers. Use heavy solid or colored dots to indicate these points.

The graph of −2, 1, and 4 is:

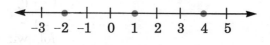

Sometimes a restricted number line is required. To construct this kind of number line, draw a straight line and, using a convenient unit of measure, locate the required points of division. These points of division are then properly labeled.

The number line that shows the points of division associated with the numbers −4, −3, −2, −1, 0, 1, and 2 is:

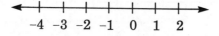

PRACTICE PROBLEMS   For ex. 1–3 see Extended Answers, p. T81.

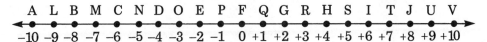

A L B M C N D O E P F Q G R H S I T J U V

-10 -9 -8 -7 -6 -5 -4 -3 -2 -1 0 $+1$ $+2$ $+3$ $+4$ $+5$ $+6$ $+7$ $+8$ $+9$ $+10$

1. a. What number corresponds to each of the following points?

E   G   M   S   F   L   J   R   C   V

b. What letter labels the point corresponding to each of the following numbers?

-3 $+9$ -8 -4 0 $+6$ -5 $+1$ -10 $+7$

2. Draw the graph of each of the following groups of numbers on a number line:

a. -3

b. $-5, +2$

c. $-7, -4, 0, +1$

d. $-4, -2, 0, 2, 4$

e. $-6, -3, -1, 3, 5, 7$

f. $4, -2, +1$

g. $+4, -6, +1, -1, +7, -5, -3$

h. $-8, 4, 0, -6, -2, 5, 9, -1$

i. $-5, -4, -3, \ldots, 4$

j. $-9, -8, -7, \ldots, 2$

3. Draw the graph of each of the following on a number line:

a. all integers less than $+1$ and greater than -1.

b. all integers greater than -4 and less than 0.

c. all integers less than $+5$ and greater than -5.

4. Write the coordinates of which each of the following is the graph:

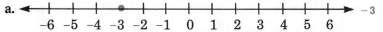

a.

-6 -5 -4 -3 -2 -1 0 1 2 3 4 5 6 -3

b.

-6 -5 -4 -3 -2 -1 0 1 2 3 4 5 6 $-4, -1, 2, 5$

c.

-6 -5 -4 -3 -2 -1 0 1 2 3 4 5 6 $-5, 3$

d.

-6 -5 -4 -3 -2 -1 0 1 2 3 4 5 6 $-3, -1, 5$

e.

-9 -8 -7 -6 -5 -4 -3 -2 -1 0 1 2 3 $-9, -7, -5, -1, 1, 2$

f.

-8 -7 -6 -5 -4 -3 -2 -1 0 1 2 3 4 5 6 7 8

g.

-9 -8 -7 -6 -5 -4 -3 -2 -1 0 1 2 3 4 5 6 7 8 9

f. $-8, -6, -4, -2, 0, 2, 4, 6, 8$

g. $-9, -8, -2, -1, 0, 6, 7, 9$

4-5 Opposite Meanings

OBJECTIVE: To read and interpret opposite meanings.

Positive and negative numbers are used in science, statistics, weather reports, stock reports, sports, and many other fields to express opposite meanings or directions. For example:

If $+1\frac{1}{4}$ points indicates a $1\frac{1}{4}$-point *gain* in a stock, -2 points would indicate a 2-point *loss* is a stock.

Basic: Prob. 1–10; *Average:* Prob. 1–12; *Enriched:* Prob. 1–15

PRACTICE PROBLEMS

1. If $+250$ meters represents 250 meters above sea level, what does -60 meters represent? 60 meters below sea level

2. If -30 kilograms represents a downward force of 30 kilograms, what does $+75$ kilograms represent? an upward force of 75 kilograms

3. If an increase of 6% in the cost of living is indicated by $+6\%$, how can a 4% decrease in the cost of living be indicated? -4%

4. If 85 degrees west longitude is indicated by $-85°$, how can 54 degrees east longitude be indicated? $+54°$

5. If $+18$ m.p.h. indicates a tail wind of 18 m.p.h., what does -23 m.p.h. represent? a head wind of 23 m.p.h.

6. If a deficiency of 47 millimeters of rainfall is indicated by -47 millimeters, how can an excess of 39 millimeters be indicated? $+39$ millimeters

72° above zero **7.** If $-14°$ means 14 degrees below zero, what does $+72°$ mean?

8. If \$70 deposited in the bank is represented by $+\$70$, how can \$25 withdrawn from the bank be represented? $-\$25$

9. If an inventory shortage of 120 items is represented by -120 items, how can 58 items over be represented? $+58$ items

-17 amperes **10.** If a charge of 15 amperes of electricity is indicated by $+15$ amperes, how can a discharge of 17 amperes of electricity be indicated?

11. If $+6$ pounds means 6 pounds overweight, what does -9 pounds mean? 9 pounds underweight

15° N. latitude **12.** If $-10°$ means 10 degrees south latitude, what does $+15°$ mean?

13. If 8 yards lost in a football game is indicated by -8 yards, how can 5 yards gained be indicated? $+5$ yards

14. If 4 degrees above normal temperature is represented by $+4°$, how can 7 degrees below normal temperature be represented? $-7°$

15. If 25 meters to the left is indicated by -25 meters, how can 18 meters to the right be indicated? $+18$ meters

Related Resources: *Practical Applications in Mathematics:* Problems 1–10, p. 147 **456**

Comparing Integers

OBJECTIVE: To compare integers.

The number line may be drawn either horizontally or vertically.

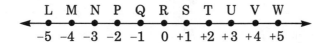

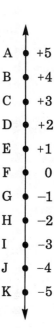

On the horizontal number line, of two numbers the number corresponding to the point on the line *farther to the right* is the greater number.

On the vertical number line, the numbers *above* zero are positive and those *below* zero are negative. Any number corresponding to a point is greater than any number corresponding to a point located below it.

Basic: Prob. 1–7; *Average:* Prob. 1–9; *Enriched:* Prob. 1–10

PRACTICE PROBLEMS

1. On the horizontal number line, which point corresponds to the greater number:

a. Point *T* or point *M*?

b. Point *P* or point *V*?

c. Point *Q* or point *N*?

d. Point *U* or point *L*?

2. On the vertical number line, which point corresponds to the greater number:

a. Point *C* or point *H*?

b. Point *G* or point *B*?

c. Point *F* or point *J*?

d. Point *K* or point *D*?

3. Which number is greater?

a. +9 or +6

b. −8 or −5

c. 0 or +3

d. −7 or 0

e. +8 or −8

f. −1 or +4

g. −2 or +2

h. +5 or −10

4. Which number is smaller?

a. $\underline{+3}$ or $+7$

b. -9 or $\underline{-10}$

c. $+4$ or $\underline{0}$

d. 0 or $\underline{-6}$

e. $\underline{-8}$ or $+5$

f. $+2$ or $\underline{-1}$

g. -4 or $\underline{-7}$

h. $\underline{-9}$ or $+11$

5. Which of the following sentences are true?

a. $+3 > -4$ True

b. $0 < -5$ False

c. $+2 < +8$ True

d. $-12 > +10$ False

e. $-6 < -3$ True

f. $-7 < -7$ False

g. $+6 > +1$ True

h. $+5 < -4$ False

i. $0 > +4$ False

j. $-5 > -2$ False

k. $+8 \not< +2$ True

l. $-1 \not> +1$ True

m. $-9 < +9$ True

n. $-2 \not> 0$ True

o. $+15 \not< -21$ True

p. $-3 < -7$ False

6. Rewrite each of the following and insert between the two numerals the symbol $=$, $<$, or $>$ which will make the sentence true:

a. -8 _____ -1 $-8 < -1$

b. $+6$ _____ $+2$ $+6 > +2$

c. $+9$ _____ -4 $+9 > -4$

d. -7 _____ -9 $-7 > -9$

e. $+11$ _____ -14 $+11 > -14$

f. -5 _____ -5 $-5 = -5$

g. -3 _____ 0 $-3 < 0$

h. -12 _____ -2 $-12 < -2$

i. 0 _____ -1 $0 > -1$

7.a. $-9, -6, -4, -3, -2, 0, +4, +5, +7, +10$
b. $-10, -7, -5, -3, -1, 0, +2, +8, +10, +11$
8.a. $+9, +8, +4, +3, 0, -1, -3, -7, -8, -12$
b. $+10, +8, +5, +1, 0, -1, -4, -6, -9, -11$

7. Name the following numbers in order of size (smallest first):

a. $+4, -3, +7, -9, -2, +10, -4, 0, -6, +5$

b. $-1, +2, 0, -3, +8, -5, -7, +11, -10, +10$

8. Name the following numbers in order of size (greatest first):

a. $-3, +8, -7, +4, -8, 0, +3, -12, +9, -1$

b. $+5, -6, +1, -1, 0, +10, -9, -11, +8, -4$

9. Which is greater:

a. The absolute value of $\underline{-8}$ or the absolute value of $+3$?

b. The absolute value of $+9$ or the absolute value of -9? both are equal to 9

10. Which has the greater opposite number?

a. $+7$ or $\underline{+4}$

b. -3 or $\underline{-8}$

c. $\underline{-1}$ or $+6$

d. $+5$ or $\underline{-10}$

Opposite Directions; Vectors

**OBJECTIVE: To represent opposite directions
and to draw and interpret vectors.**

VOC.: vector

Positive and negative numbers, used as directed numbers, indicate movements in opposite directions. A movement to the right of a particular point is generally considered as moving in a positive direction from the point and a movement to the left of a point as moving in a negative direction from the point.

The sign in the numeral naming a directed number indicates the direction, and the absolute value of the directed number represents the magnitude (distance in units) of the movement.

> For example:
>
> +6 means moving 6 units to the right.
> −7 means moving 7 units to the left.

An arrow that represents a directed line segment is called a vector. It is used to picture the size and the direction of the movement that is indicated by a signed number. The absolute value of the signed number indicates the length of the arrow and the sign indicates the direction.

> +5 is represented by the following vector:
>
> +5
>
> -2 -1 0 1 2 3 4 5 6
>
> −3 is represented by the following vector:
>
> −3
>
> -5 -4 -3 -2 -1 0 1 2 3

To determine the numeral that describes the movement from one point to another point on the number line, draw the vector between the points with the arrowhead pointing in the direction of the movement. Use the number of units of length in the vector as the absolute value of the numeral. Select the sign according to the direction of the vector. If it points to the right, the sign is positive ($+$), if it points to the left, the sign is negative ($-$).

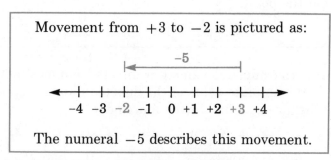

Movement from $+3$ to -2 is pictured as:

The numeral -5 describes this movement.

Basic: Prob. 1–4; Average: Prob. 1–5; Enriched: Prob. 1–6

PRACTICE PROBLEMS

1. A movement of how many units of distance and in what direction is represented by each of the following numbers?

a. -7	**c.** -2	**e.** $+1$	**g.** $-2\frac{1}{4}$
b. $+9$	**d.** $+12$	**f.** -4.25	**h.** $+\frac{2}{3}$

2. Represent each of the following movements by a numeral naming a signed number:

a. Moving 4 units to the left. -4 **c.** Moving 7.5 units to the right. $+7.5$
b. Moving 8 units to the right. $+8$ **d.** Moving $12\frac{1}{4}$ units to the left. -12.25

3. Write the numeral that is represented by each of the following vectors?

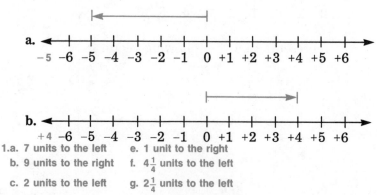

1.a. 7 units to the left   e. 1 unit to the right
  b. 9 units to the right   f. $4\frac{1}{4}$ units to the left
  c. 2 units to the left   g. $2\frac{1}{4}$ units to the left
  d. 12 units to the right   h. $\frac{2}{3}$ of a unit to the right

c.
+9

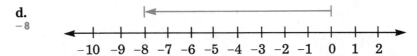

d.
−8

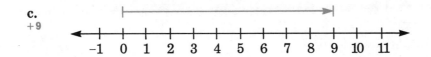

4. Using the number line as a scale, draw a vector representing: **Answers will vary.**

a. −6   **c.** −9   **e.** $+2\frac{1}{2}$ **g.** −11

b. +7   **d.** +3   **f.** −3.5   **h.** $+4\frac{3}{4}$

5. Write the numeral that is represented by each of the following vectors:

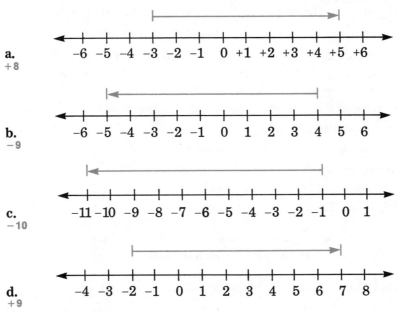

a.
+8

b.
−9

c.
−10

d.
+9

6. Use the number line as a scale to draw a vector that illustrates each of the following movements. Write the numeral that is represented by each of these vectors. For ex. 6 see Extended Answers, pp. T81–T82.

a. From 0 to +4   **g.** From −1 to −6   **m.** From −1 to −9
b. From 0 to −8   **h.** From −7 to −2   **n.** From −10 to −5
c. From −6 to 0   **i.** From +4 to −4   **o.** From +2 to −4
d. From +7 to 0   **j.** From −9 to +10   **p.** From −11 to +11
e. From +2 to +8   **k.** From +11 to −7   **q.** From +9 to −2
f. From +10 to +3   **l.** From +3 to +12   **r.** From +7 to +1

Additive Inverse of a Number; More About Opposites

OBJECTIVE:
1) To determine the addition inverse of a number.
2) To read and write expressions involving opposites.

VOC.: opposite
additive inverse

Sometimes the opposite of a number is called the additive inverse of the number.

When the sum of two numbers is zero, each addend is said to be the additive inverse of the other addend.

We shall see in Exercises 4–9 and 4–10 that the sum of a number and its opposite is zero.

-4 is the opposite of or the additive inverse of $+4$ (or 4).
$+4$ (or 4) is the opposite of or the additive inverse of -4.

A centered dash symbol is used in algebra to indicate the opposite of a number.

The expression $-(4)$ and sometimes -4 are read as:
  "The opposite of four."
The sentence $-(-4) = +4$ is read as:
  "The opposite of negative four is positive four."
The sentence $-(4) = -4$ is read as:
  "The opposite of four is negative four."

Basic: Prob. 1–4d; *Average:* Prob. 1–5; *Enriched:* Prob. 1–6

PRACTICE PROBLEMS

1. What number is the additive inverse of each of the following numbers?

a. -7 $+7$ **c.** 0 0 **e.** $+3.08$ -3.08 **g.** $+\frac{5}{6}$ $-\frac{5}{6}$

b. $+26$ -26 **d.** $-\frac{5}{8}$ $+\frac{5}{8}$ **f.** $-4\frac{1}{2}$ $+4\frac{1}{2}$ **h.** $-\frac{11}{3}$ $+\frac{11}{3}$

2. What number is the opposite of each of the following numbers?

a. $+10$ -10 **c.** $-\frac{7}{16}$ $+\frac{7}{16}$ **e.** 0 0 **g.** $-3\frac{3}{4}$ $+3\frac{3}{4}$

b. -6 $+6$ **d.** $-.85$ $+.85$ **f.** $-\frac{16}{5}$ $+\frac{16}{5}$ **h.** $+35$ -35

3. Read, or write in words, each of the following: For ex. 3 see Extended Answers, p. T82.

a. -9 **c.** $-(-\frac{5}{6})$ **e.** $-(-8) = +8$ **g.** $-(15) = -15$

b. $-(+2)$ **d.** $-(-11)$ **f.** $-(+10) = -10$ **h.** $-(-6.4) = 6.4$

4. Write symbolically, using the "opposite" symbol:
a. The opposite of thirty-six. $-(36)$
b. The opposite of positive eight. $-(+8)$
c. The opposite of negative five. $-(-5)$
d. The opposite of positive nineteen. $-(+19)$
e. The opposite of negative twenty-three. $-(-23)$
f. The opposite of forty is negative forty. $-(40) = -40$
g. The opposite of negative one is positive one. $-(-1) = +1$
h. The opposite of positive fifty-six is negative fifty-six. $-(+56) = -56$
i. The opposite of negative sixty-seven is positive sixty-seven. $-(-67) = +67$
j. The opposite of the opposite of ninety-five is ninety-five. $-(-95) = 95$
5. a. Find the opposite of the opposite of: -7; $+6$; $+\frac{3}{8}$; $-.9$; $-3\frac{1}{2}$
b. What is always true about the opposite of the opposite of a number? The opposite of the opposite of a number is equal to the original number
6. Find the value of each of the following:

a. $-(+9) = ?$ -9 **d.** $-(+6.3) = ?$ -6.3 **g.** $-\left(\frac{11}{4}\right) = ?$ $-\frac{11}{4}$
b. $-(-13) = ?$ $+13$ **e.** $-(-3\frac{2}{5}) = ?$ $+3\frac{2}{5}$ **h.** $-(-.08) = ?$ $+.08$
c. $-\left(-\frac{5}{8}\right) = ?$ $+\frac{5}{8}$ **f.** $-(54) = ?$ -54 **i.** $-(+1\frac{7}{10}) = ?$ $-1\frac{7}{10}$

5a. -7; $+6$; $+\frac{3}{8}$; $-.9$; $-3\frac{1}{2}$

REFRESH YOUR SKILLS

1. Add:
3,589
6,974
2,698
5,783
───
19,044

2. Subtract:
506,300
94,802
───
411,498

3. Multiply:
3,060
409
───
1,251,540

4. Divide:
$175\overline{)53,375}$
305

5. Add:
$7\frac{7}{10} + 3\frac{1}{2}$ $11\frac{1}{5}$

6. Subtract:
$16\frac{1}{3} - \frac{3}{4}$ $15\frac{7}{12}$

7. Multiply:
$\frac{2}{5} \times 4\frac{3}{8}$ $1\frac{3}{4}$

8. Divide:
$14 \div 1\frac{3}{4}$ 8

9. Add:
$.9 + .99 + .999$ 2.889

10. Subtract: $.02 - .007$ $.013$

11. Multiply: $2.5 \times .4$ 1

12. Divide:
$.08\overline{).8}$ 10

13. Find 20% of $15 $3

14. What percent of 400 is 36? 9%

15. 9% of what number is 11? $122.\overline{2}$

16. Find the square root of 81,796. 286

For a Test on Positive and Negative Numbers see p. T53 or *Testing Program*, p. 25.

COMPUTATION WITH
POSITIVE AND NEGATIVE NUMBERS

Arrows or vectors may be used to add or subtract positive and negative numbers on the number line. The absolute value of a signed number indicates the length of the vector and the sign of the given number indicates the direction of the vector. On the horizontal number line, moving to the right is the positive direction and to the left is the negative direction.

<table>
<tr><td>EXERCISE
4-9</td><td>## Addition of Integers
and Rational Numbers</td></tr>
</table>

OBJECTIVE: To add integers and rational numbers.

To add on a number line, draw a vector for the first addend, starting at the point labeled 0; then draw the vector representing the second addend, starting from the point reached by the first vector. If there are more than two addends, continue in this way for each addend. The coordinate of the final point reached is the sum.

When two addends are both positive numbers, the final direction is positive. When two addends are both negative numbers, the final direction is negative. When one addend is a positive number and the other is a negative number, the number containing the greater absolute value determines the final direction.

Related Resources: *Practical Applications in Mathematics:* Problems, p. 147

Using Vectors

Add +2 and +3:

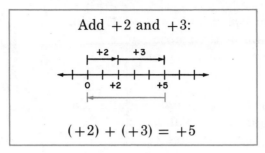

$$(+2) + (+3) = +5$$

Add −4 and −3:

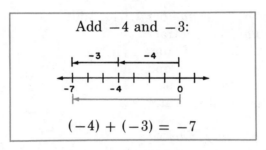

$$(-4) + (-3) = -7$$

Add −6 and +4:

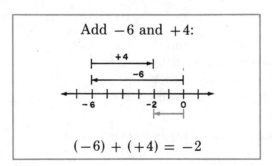

$$(-6) + (+4) = -2$$

Add −3 and +7:

$$(-3) + (+7) = +4$$

Algebraically

Thus algebraically,

1. *To add two positive numbers,* find the sum of their absolute values and prefix the numeral with a positive sign.

2. *To add two negative numbers,* find the sum of their absolute values and prefix the numeral with a negative sign.

3. *To add a positive number and a negative number,* subtract the smaller absolute value from the greater absolute value and prefix the difference with the sign of the number having the greater absolute value.

4. *The sum of any number and its opposite is zero.*

5. *To add three or more signed numbers,* add algebraically first the positive numbers, then the negative numbers, and finally the two answers.

1. Add: $+2$
$+3$
$\overline{+5}$

2. Add: -4
-3
$\overline{-7}$

3. Add: $\quad -6 \qquad -3 \qquad +9$
$\quad\quad +4 \qquad +7 \qquad -6$
$\quad\quad \overline{-2} \qquad \overline{+4} \qquad \overline{+3}$

4. Add: $\quad +7 \qquad -5$
$\quad\quad -7 \qquad +5$
$\quad\quad \overline{0} \qquad \overline{0}$

5. $(+2) + (-9) + (+6) + (-3)$
$\quad\quad = (+8) + (-12)$
$\quad\quad = -4$

To simplify an algebraic expression such as $6 - 9 + 4 - 3$, think of it as $(+6) + (-9) + (+4) + (-3)$ where the given signs are part of the numerals for the signed numbers but the operation is considered to be addition. See page 468.

$$6 - 9 + 4 - 3 = (+6) + (-9) + (+4) + (-3) = -2$$

Basic: Prob. 1–2; *Average:* Prob. 1–4; *Enriched:* Prob. 1–5

PRACTICE PROBLEMS

1. Add on the number line, using vectors:

$+5$	-4	$+7$	$+2$	-8	-1	$+4$	0
$+3$	-1	-3	-6	$+2$	$+5$	-4	-9
$\overline{+8}$	$\overline{-5}$	$\overline{+4}$	$\overline{-4}$	$\overline{-6}$	$\overline{+4}$	$\overline{0}$	$\overline{-9}$

2. Add:

a.

$$
\begin{array}{l}
+3 \\
\underline{+6}
\end{array}_{+9}
\quad
\begin{array}{l}
+9 \\
\underline{+7}
\end{array}_{+16}
\quad
\begin{array}{l}
+24 \\
\underline{+19}
\end{array}_{+43}
\quad
\begin{array}{l}
+8 \\
\underline{+15}
\end{array}_{+23}
\quad
\begin{array}{l}
-5 \\
\underline{-4}
\end{array}_{-9}
\quad
\begin{array}{l}
-9 \\
\underline{-14}
\end{array}_{-23}
\quad
\begin{array}{l}
-27 \\
\underline{-35}
\end{array}_{-62}
\quad
\begin{array}{l}
-83 \\
\underline{-49}
\end{array}_{-132}
$$

b.

$$
\begin{array}{l}
+5 \\
\underline{-2}
\end{array}_{+3}
\quad
\begin{array}{l}
+11 \\
\underline{-3}
\end{array}_{+8}
\quad
\begin{array}{l}
+25 \\
\underline{-14}
\end{array}_{+11}
\quad
\begin{array}{l}
+91 \\
\underline{-37}
\end{array}_{+54}
\quad
\begin{array}{l}
-9 \\
\underline{+1}
\end{array}_{-8}
\quad
\begin{array}{l}
-8 \\
\underline{+6}
\end{array}_{-2}
\quad
\begin{array}{l}
-23 \\
\underline{+12}
\end{array}_{-11}
\quad
\begin{array}{l}
-54 \\
\underline{+39}
\end{array}_{-15}
$$

c.

$$
\begin{array}{l}
+1 \\
\underline{-8}
\end{array}_{-7}
\quad
\begin{array}{l}
+3 \\
\underline{-10}
\end{array}_{-7}
\quad
\begin{array}{l}
+12 \\
\underline{-36}
\end{array}_{-24}
\quad
\begin{array}{l}
+20 \\
\underline{-41}
\end{array}_{-21}
\quad
\begin{array}{l}
-6 \\
\underline{+8}
\end{array}_{+2}
\quad
\begin{array}{l}
-2 \\
\underline{+15}
\end{array}_{+13}
\quad
\begin{array}{l}
-18 \\
\underline{+27}
\end{array}_{+9}
\quad
\begin{array}{l}
-45 \\
\underline{+68}
\end{array}_{+23}
$$

d.

$$
\begin{array}{l}
+6 \\
\underline{-6}
\end{array}_{0}
\quad
\begin{array}{l}
-12 \\
\underline{+12}
\end{array}_{0}
\quad
\begin{array}{l}
0 \\
\underline{+7}
\end{array}_{+7}
\quad
\begin{array}{l}
-5 \\
\underline{0}
\end{array}_{-5}
\quad
\begin{array}{l}
+\frac{3}{4} \\
\underline{-\frac{1}{4}}
\end{array}_{+\frac{1}{2}}
\quad
\begin{array}{l}
-\frac{7}{8} \\
\underline{-\frac{2}{3}}
\end{array}_{-1\frac{13}{24}}
\quad
\begin{array}{l}
-.4 \\
\underline{+1.5}
\end{array}_{+1.1}
\quad
\begin{array}{l}
-5.28 \\
\underline{-2.64}
\end{array}_{-7.92}
$$

e.

$$
\begin{array}{l}
+8 \\
+5 \\
\underline{+7}
\end{array}_{+20}
\quad
\begin{array}{l}
-3 \\
-8 \\
\underline{-9}
\end{array}_{-20}
\quad
\begin{array}{l}
-3 \\
+6 \\
\underline{-5}
\end{array}_{-2}
\quad
\begin{array}{l}
+9 \\
-15 \\
\underline{+3}
\end{array}_{-3}
\quad
\begin{array}{l}
+2 \\
+1 \\
+4 \\
\underline{+6}
\end{array}_{+13}
\quad
\begin{array}{l}
-3 \\
-1 \\
-5 \\
\underline{-2}
\end{array}_{-11}
\quad
\begin{array}{l}
-4 \\
+7 \\
-8 \\
\underline{+5}
\end{array}_{0}
\quad
\begin{array}{l}
-6 \\
-9 \\
+7 \\
\underline{-2}
\end{array}_{-10}
$$

3. Add as indicated:

a. $(-3) + (-4)$ ₋₇
b. $(+8) + (-15)$ ₋₇
c. $(-1) + (+11)$ ₊₁₀
d. $(+5) + (-14) + (+7)$ ₋₂
e. $(-3) + (-9) + (-5)$ ₋₁₇
f. $[(+8) + (-11)] + (-4)$ ₋₇
g. $(-10) + [(-8) + (-2)]$ ₋₂₀
h. $(+7) + (-5) + (-10) + (+9)$ ₊₁
i. $(-7) + (+4) + (-1) + (+6)$ ₊₂
j. $(-12) + (+9) + (-6) + (+9)$ ₀

4. Simplify:

a. $2 + 6$ 8
b. $4 - 8$ -4
c. $-6 + 7$ +1
d. $-5 - 3$ -8
e. $18 - 9 + 5$ 14
f. $9 - 7 + 2$ 4

g. $5 - 6 - 8$ -9
h. $-3 + 8 - 7$ -2
i. $14 - 5 + 3 - 11$ 1
j. $-9 - 6 + 7 - 2$ -10
k. $6 - 8 + 2 - 4 + 5$ +1
l. $-2 - 3 + 9 - 1 + 3$ +6

5. Simplify by performing the indicated operations and finding the absolute values:

a. $|7 + 4|$ 11
b. $|9 - 11|$ 2
c. $|-4 - 5|$ 9
d. $|-8 + 2|$ 6
e. $|-10| + |+7|$ 17

f. $|(-6) + (-2)|$ 8
g. $-|14 - 5|$ -9
h. $-|16 - 24|$ -8
i. $-9 + |-8|$ -1
j. $-|-(-4) + (-6)|$ -2

Subtraction of Integers and Rational Numbers

OBJECTIVE: To subtract integers and rational numbers.

Used as the inverse operation of addition, the operation of subtraction requires that we find the addend which when added to the given addend (subtrahend) will equal the given sum (minuend).

$(+3) - (-2) = ?$ is thought of as $(-2) + ? = (+3)$

Thus, *to subtract on the number line,* draw the vectors for the given addend and the sum. Find the missing addend by calculating the distance and determining the direction (or representing them by a vector) from the point associated with the given addend (subtrahend) to the point associated with the sum (minuend).

Using Vectors

$$(+3) - (-2) = ?$$

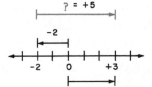

The distance from -2 to $+3$ is 5 units. The direction is to the right or the positive direction.

$$(+3) - (-2) = +5$$

This distance in units is the absolute value of the missing addend and the direction indicates the sign of the addend. Counting to the right is the positive direction and to the left is the negative direction.

In addition we learned that $(+3) + (+2) = +5$
In the box above we find that $(+3) - (-2) = +5$

A comparison of $(+3) - (-2) = +5$
and $(+3) + (+2) = +5$
shows that *subtracting* -2 from $+3$ gives the same answer as *adding* $+2$ to $+3$. That is, *subtracting a number gives the same answer as adding its opposite.*

Thus algebraically,

To subtract a number (positive, negative, or zero) *from another number* (positive, negative, or zero), add to the minuend the additive inverse (or opposite) of the number to be subtracted.

Subtract:	Add:	Subtract:	Add:	Subtract:	Add:
+2	+2	+2	+2	−2	−2
+8 ⟶	−8	−8 ⟶	+8	+8 ⟶	−8
	−6		+10		−10
Answer: −6		*Answer:* +10		*Answer:* −10	
Subtract:	Add:	Subtract:	Add:	Subtract:	Add:
−2	−2	−8	−8	0	0
−8 ⟶	+8	−2 ⟶	+2	+8 ⟶	−8
	+6		−6		−8
Answer: +6		*Answer:* −6		*Answer:* −8	

We may also subtract by changing the sign of the given subtrahend mentally and adding this number to the minuend. This method, generally used in traditional mathematics, is another way of finding the additive inverse and adding it to the minuend.

Basic: Prob. 1–2; *Average:* Prob. 1–4; *Enriched:* Prob. 1–5

PRACTICE PROBLEMS

1. Subtract on the number line, using vectors:

+6	+2	−5	−3	+4	−2	0	−4
+3 ₊₃	+9 ₋₇	−1 ₋₄	−7 ₊₄	−6 ₊₁₀	+5 ₋₇	−8 ₊₈	−4 ₀

2. Subtract:

a.

9	+11	+15	+17	5	6	+2	+28
3 ₆	+6 ₊₅	+9 ₊₆	+8 ₊₉	7 ₋₂	15 ₋₉	+14 ₋₁₂	+45 ₋₁₇

b.

−6	−10	−8	−12	−5	−7	−6	−51
−4 ₋₂	−5 ₋₅	−3 ₋₅	−7 ₋₅	−8 ₊₃	−9 ₊₂	−13 ₊₇	−70 ₊₁₉

c.

+7	+9	+6	+18	−12	−6	−5	−16
−2 ₊₉	−11 ₊₂₀	−5 ₊₁₁	−27 ₊₄₅	+3 ₋₁₅	+19 ₋₂₅	+1 ₋₆	+25 ₋₄₁

d.

+6	−5	0	+8	+11	−.6	$-\frac{7}{8}$	$-2\frac{1}{2}$
+6 ₀	−5 ₀	−1 ₊₁	0 ₊₈	−11 ₊₂₂	−.7 ₊.₁	$+\frac{3}{4}$ $-1\frac{5}{8}$	$-1\frac{2}{3}$ $-\frac{5}{6}$

e.

+5	−6	−13	+20	−3	−14	+.9	0
+9 ₋₄	−2 ₋₄	+7 ₋₂₀	−15 ₊₃₅	0 ₋₃	+14 ₋₂₈	−1.5 ₊₂.₄	−17 ₊₁₇

3. Subtract as indicated:

a. $(-8) - (+3)$ −11   **f.** $(8 + 7) - (6 - 13)$ +22
b. $(+6) - (-11)$ +17   **g.** $[(-2) + (-7)] - (-9)$ 0
c. $(2) - (-2)$ +4   **h.** $(+4) - [(-5) - (+6)]$ +15
d. $(4 - 3) - (6 - 10)$ +5   **i.** $[(-5) - (+8)] - (-12)$ −1
e. $(2 - 6) - (7 - 8)$ −3   **j.** $(-6) - [(-10) + (+3)]$ +1

4. a. From -5 subtract 5. −10   **d.** Subtract 10 from 3. −7
b. Take 4 from 1. −3   **e.** Take -9 from $+9$. +18
c. From -3 take -7. +4   **f.** From 0 subtract -8. +8

5. Simplify by performing the indicated operations and finding the absolute values:

a. $|-4| - |-9|$ −5   **d.** $|-3 - 8| - 6$ 5
b. $|6 - 2| - |-5|$ −1   **e.** $8 - |5 + 9|$ −6
c. $|1 - 3| - |4 - 7|$ −1   **f.** $|(0 - 2) - (4 - 2)|$ 4

EXERCISE
4-11
Multiplication of Integers and Rational Numbers

OBJECTIVE: To multiply integers   VOC.: Multiplicative inverse
and rational numbers.

The expression three times four may be written as 3×4, $3 \cdot 4$, $(3) \times (4)$, $3(4)$, $(3)\ 4$, or $(3)(4)$. Observe that the $\times$ symbol or raised dot is used or the factors may be written next to each other with parentheses without any multiplication symbol.

Since a numeral without a sign represents a positive number and multiplication may be thought of as repeated addition, then

$$3 \times 4 \text{ or } (+3)(+4) = (+4) + (+4) + (+4) = +12$$

The product of two positive numbers is a positive number.

Similarly, $(+3)(-4) = (-4) + (-4) + (-4) = -12$
and since $(+4)(-3) = -12$ and $(-3)(+4) = (+4)(-3)$ by the commutative property of multiplication,

then
$$(-3)(+4) = -12$$

The product of a positive number and a negative number is a negative number.

Related Resources: *Practical Applications in Mathematics:* Problems, p. 148

To determine what $(-3)(-4)$ equals, examine the following step-by-step development.

1. $(-3) \cdot 0 = 0$ — The product of any number and zero is zero.

2. But $[(+4) + (-4)] = 0$ — The sum of a number and its additive inverse is zero.

3. Then
$(-3) \cdot [(+4) + (-4)] = 0$ — By substituting $[(+4) + (-4)]$ for 0 in step (1).

4. $(-3)(+4) + (-3)(-4) = 0$ — Multiplication is distributive over addition.

5. $-12 + (-3)(-4) = 0$ — Since $(-3)(+4) = -12$

6. Therefore $(-3)(-4) = +12$ — In order for the sentence $-12 + (-3)(-4) = 0$ to be true, the product of $(-3)(-4)$ must be the additive inverse of -12, which is $+12$.

The product of two negative numbers is a positive number.

Summarizing:

1. *To multiply two positive numbers or two negative numbers,* find the product of their absolute values and prefix the numeral for this product with a positive sign.

2. *To multiply a positive number by a negative number or a negative number by a positive number,* find the product of their absolute values and prefix the numeral for this product with a negative sign.

1. Multiply:	
$+5$	-5
$+6$	-6
$+30$	$+30$

2. Multiply:	
$+5$	-5
-6	$+6$
-30	-30

3. Observe that an odd number of negative factors produces a negative number as a product while an even number of negative factors produces a positive number as a product provided zero is not a factor.

3. $(-2)(+1)(-4)(-2) = -16$
$(-1)(-5)(-2)(-1) = +10$

If the product of two numbers is one (1), then each factor is said to be the multiplicative inverse or the reciprocal of the other.

The multiplicative inverse of $\frac{2}{3}$ is $\frac{3}{2}$. Of $-\frac{3}{2}$ is $-\frac{2}{3}$.

PRACTICE PROBLEMS

1. Multiply:

a. $\begin{matrix} +5 \\ +9 \end{matrix}$ +45 $\begin{matrix} +8 \\ +6 \end{matrix}$ +48 $\begin{matrix} +7 \\ +7 \end{matrix}$ +49 $\begin{matrix} +6 \\ +4 \end{matrix}$ +24 $\begin{matrix} -2 \\ -3 \end{matrix}$ +6 $\begin{matrix} -9 \\ -4 \end{matrix}$ +36 $\begin{matrix} -6 \\ -7 \end{matrix}$ +42 $\begin{matrix} -8 \\ -5 \end{matrix}$ +40

b. $\begin{matrix} -7 \\ +8 \end{matrix}$ −56 $\begin{matrix} -2 \\ +9 \end{matrix}$ −18 $\begin{matrix} -4 \\ +12 \end{matrix}$ −48 $\begin{matrix} -13 \\ +5 \end{matrix}$ −65 $\begin{matrix} +6 \\ -6 \end{matrix}$ −36 $\begin{matrix} +12 \\ -8 \end{matrix}$ −96 $\begin{matrix} +4 \\ -5 \end{matrix}$ −20 $\begin{matrix} +6 \\ -9 \end{matrix}$ −54

c. $\begin{matrix} 0 \\ +3 \end{matrix}$ 0 $\begin{matrix} -5 \\ 0 \end{matrix}$ 0 $\begin{matrix} 0 \\ -8 \end{matrix}$ 0 $\begin{matrix} +4 \\ 0 \end{matrix}$ 0 $\begin{matrix} -.4 \\ -.2 \end{matrix}$ +.08 $\begin{matrix} +2.5 \\ -1.4 \end{matrix}$ −3.5 $\begin{matrix} -3\frac{1}{2} \\ +2\frac{1}{4} \end{matrix}$ −7$\frac{7}{8}$ $\begin{matrix} -12 \\ -1\frac{2}{3} \end{matrix}$ +20

2. Multiply as indicated:

a. $(-3)(-9)$ +27   **d.** $-8(-7-5)$ +96   **g.** $-5(3-8)-(1-4)$ +28

b. $(+4)(-7)$ −28   **e.** $(+6)(-2)(-4)$ +48   **h.** $[(+10)(-4)] \times (-3)$ +120

c. $11(3-14)$ −121   **f.** $(-2) \times [(-6)(-7)]$ −84   **i.** $(-2)(+3)(-1)(-4)$ −24

3. Find the value of each of the following:

a. $(-3)^2$ +9   **c.** $(-5)^3$ −125   **e.** $(-4)^5$ −1,024   **g.** $(-3)^4$ +81

b. $(+6)^2$ +36   **d.** $(+2)^4$ +16   **f.** $(-1)^6$ +1   **h.** $(-1)^5 \cdot (-3)^3$ +27

4. Find the value of each of the following:

a. $|(-1)(-5)|$ 5   **c.** $-10 \times |-6|$ −60   **e.** $(-6) \times |(-2)(-9)(-1)|$ −108

b. $|-4| \times |-7|$ 28   **d.** $|12-3| \times |6-8|$ 18   **f.** $|-5|^2 \times |(-4)(-11)|$ 1,100

5. Name the multiplicative inverse of each of the following:

a. -6 $\frac{-1}{6}$ **b.** $+14$ $\frac{1}{14}$ **c.** $-\frac{2}{5}$ $\frac{-5}{2}$ **d.** $-\frac{17}{4}$ $\frac{-4}{17}$ **e.** $+\frac{5}{6}$ $\frac{6}{5}$ **f.** $-3\frac{1}{2}$ $\frac{-2}{7}$

EXERCISE 4-12

Division of Integers and Rational Numbers

OBJECTIVE: To divide integers and rational numbers.

1. Since $(+3) \times (+4) = +12$,
then $(+12) \div (+3) = +4$ or $\dfrac{+12}{+3} = +4$

2. Since $(-3) \times (+4) = -12$,
then $(-12) \div (-3) = +4$ or $\dfrac{-12}{-3} = +4$

When the dividend and the divisor are both positive numbers or both negative numbers, the quotient is a positive number.

3. Since $(+3) \times (-4) = -12$,
then $(-12) \div (+3) = -4$ or $\dfrac{-12}{+3} = -4$

4. Since $(-3) \times (-4) = +12$,
then $(+12) \div (-3) = -4$ or $\dfrac{+12}{-3} = -4$

Related Resources: *Practical Applications in Mathematics:* Div. Prob., p. 148

When the dividend is a positive number and the divisor is a negative number or vice-versa, the quotient is a negative number.

Summarizing:

To divide a positive number by a positive number or a negative number by negative number, divide their absolute values and prefix the numeral for this quotient with a positive sign.

To divide a positive number by a negative number or a negative number by a positive number, divide their absolute values and prefix the numeral for this quotient with a negative sign.

Basic: Prob. 1–2; *Average:* Prob. 1–4; *Enriched:* Prob. 1–5

PRACTICE PROBLEMS

1. Divide:

a. $\dfrac{+30}{+5}$ $+6$ $\dfrac{+16}{+4}$ $+4$ $\dfrac{+54}{+9}$ $+6$ $\dfrac{+60}{+10}$ $+6$ $\dfrac{-8}{-2}$ $+4$ $\dfrac{-15}{-5}$ $+3$ $\dfrac{-90}{-18}$ $+5$ $\dfrac{-72}{-12}$ $+6$

b. $\dfrac{+56}{-7}$ -8 $\dfrac{+18}{-3}$ -6 $\dfrac{+45}{-5}$ -9 $\dfrac{+36}{-6}$ -6 $\dfrac{-48}{+8}$ -6 $\dfrac{-63}{+7}$ -9 $\dfrac{-80}{+10}$ -8 $\dfrac{-144}{+16}$ -9

c. $\dfrac{-9}{-9}$ $+1$ $\dfrac{+7}{-7}$ -1 $\dfrac{-15}{-1}$ $+15$ $\dfrac{-40}{+1}$ -40 $\dfrac{0}{+6}$ 0 $\dfrac{0}{-3}$ 0 $\dfrac{0}{+15}$ 0 $\dfrac{0}{-20}$ 0

2. Divide as indicated:

a. $(+36) \div (-9)$ -4

b. $(-20) \div (-4)$ $+5$

c. $(-42) \div (+7)$ -6

d. $(+75) \div (+5)$ $+15$

e. $(0) \div (-7)$ 0

f. $(-8\frac{1}{2}) \div (-\frac{3}{4})$ $11\frac{1}{3}$

g. $(+.24) \div (-.3)$ $-.8$

h. $(-40) \div (-40)$ $+1$

i. $(+19) \div (-1)$ -19

j. $(-100) \div (+25)$ -4

3. Divide as indicated:

a. $+2\overline{)-14}$ -7

b. $-8\overline{)-64}$ $+8$

c. $-3\overline{)-27}$ $+9$

d. $-5\overline{)+50}$ -10

e. $.4\overline{)-3.2}$ -8

f. $-7\overline{)+21}$ -3

4. Simplify by performing the indicated operations:

a. $\dfrac{8 - 17}{3}$ -3

b. $\dfrac{-10 - 4}{-7}$ $+2$

c. $\dfrac{12 - 3(4)}{-2}$ 0

d. $\dfrac{8(-1) + 7(-6)}{-5(5)}$ $+2$

e. $\dfrac{(-4)^3 - (-2)^4}{(-10)^2}$ $\frac{-4}{5}$

f. $\dfrac{6(4 - 7) - 5(6 - 12)}{-3(1 - 5)}$ $+1$

5. Find the value of each of the following:

a. $\dfrac{|-16|}{8}$ 2

b. $\dfrac{|-24|}{|-2|}$ 12

c. $\left|\dfrac{-30}{-5}\right|$ 6

d. $\dfrac{|-8| \times |-6|}{|-9| - |-5|}$ 12

e. $\dfrac{|(-12)(+7)|}{|8 - 9| - |1 - 5|}$ -28

f. $\dfrac{|-3|^3 - |1|^4}{|(-6) + (-7)|}$ 2

For a Test on Computation with Positive and Negative Numbers see p. T54 or *Testing Program,* p. 27.

LANGUAGE OF ALGEBRA

In algebra we continue to use the operational symbols of arithmetic. The signs of operation may assume any one of several meanings.

The addition symbol or plus symbol + means:
sum, add, more than, increased by, exceeded by.

The subtraction symbol or minus symbol − means:
difference, subtract, take away, less, less than, decreased by, diminished by.

The multiplication symbol × or the raised dot · means:
product, multiply, times.

The division symbol ÷ or $\overline{)}$ means:
quotient, divide.

The fraction bar, as in $\dfrac{c}{d}$, is generally used in algebra to indicate

division. $\dfrac{c}{d}$ is sometimes read "c over d.".

Symbols—Review

OBJECTIVE: To read or write symbols.

VOC.: constant
variable

Other symbols used in both algebra and arithemetic include exponents, the square root symbol, parentheses, and the verbs of mathematical sentences such as $=$, $\neq$, $>$, $\not>$, $<$, $\not<$, $\geq$, $\not\geq$, $\leq$, and $\not\leq$.

An exponent is the small numeral written to the upper right (superscript) of the base. When it names a natural number, it tells how many times the factor is being used in multiplication.

The square root symbol $\sqrt{}$ written over the numeral indicates the square root of the corresponding number. The square root of a number is that number which when multiplied by itself produces the given number.

Parentheses ( ) are generally used to group together two or more numerals so that they are treated as a single quantity. Sometimes parentheses are used to set off a numeral from another so that the meaning will not be misunderstood.

A numeral is sometimes called a constant because it names a definite number.

A variable is a letter (small letter, capital letter, letter with a subscript such as b , read "b sub one," or letter with prime marks such as S , read "S prime") or a frame (such as $\square$, $\square$, $\bigcirc$, and $\triangle$) or a blank which holds a place open for a number. Sometimes the first letter of a key word is used as the variable. A variable may represent any number, but under certain conditions it represents a specific number or numbers.

Basic: Prob. 1; *Average:* Prob. 1–2; *Enriched:* Prob. 1–2

PRACTICE PROBLEMS

For ex. 1 see Extended Answers, p. T82.

1. Read, or write in words, each of the following:

a. $(-5) + (+7)$ e. $\sqrt{65}$ i. $(-15)(-4)$ m. $18 - (20 - 7)$

b. 36×28 f. $12 - 30$ j. 6^8 n. $9 \times (12 \times 5)$

c. 53^2 g. $24 \cdot 19$ k. $|-13|$ o. $-(-20)$

d. $\frac{42}{6}$ h. 17^3 l. $7(21 + 9)$ p. $|(-6) + (-3)|$

2. What symbol represents the variable in each of the following?

a. $\square + 8 = 15$ $\square$ d. $4\,y = 36$ y g. $70 = a + 25$ a

b. $14 - ? = 6$?   e. $23 = n - 7$ n h. $3 = 12\,x$ x

c. $x + 9 = 11$ x f. $\dfrac{b}{5} = 12$ b i. $\triangle - 10 = 10$ $\triangle$

Mathematical Expressions or Mathematical Phrases

OBJECTIVE: To read and write mathematical phrases and mathematical expressions.

VOC.: numerical coefficient

A numerical expression or numerical phrase consists of a single numeral with or without operational symbols like 18 or 3^4 or two or more numerals with operational symbols like $7 + 5$; $28 - 14$; 5×43; $63 \div 7$; $8 \times (3 + 11)$; etc.

An algebraic expression or algebraic phrase may be a numerical expression as described above or an expression containing one or more variables joined by operational symbols like b; $a - 5x$; $3y^2 - 7x + 9$; etc.

Both numerical expressions and algebraic expressions are mathematical expressions.

In an algebraic expression no multiplication symbol is necessary when the factors are two letters (variables) or a numeral and a letter. In the latter case the numeral always precedes the letter or the variable. This numeral is called the numerical coefficient of the variable.

---

b times d may be expressed as $b \times d$, $b \cdot d$, bd (preferred), $(b)(d)$, $b(d)$, or $(b)d$.

In $5x^3$, the 5 is the numerical coefficient of x^3.

To write algebraic expressions, write numerals, variables, and operational symbols as required in the proper order.

Write each of the following as an algebraic expression:

1. The sum of five and seven.   *Answer:* $5 + 7$

2. The difference between x and eight.   *Answer:* $x - 8$

3. The product of n and four.   *Answer:* $4n$

4. The area (A) divided by the length (l).   *Answer:* $\dfrac{A}{l}$

5. The square of the velocity (v).   *Answer:* v^2

6. The square root of the radius (r).   *Answer:* $\sqrt{r}$

7. Ten times the sum of m and six.   *Answer:* $10(m + 6)$

Basic: Prob. 1–2; *Average:* Prob. 1–3; *Enriched:* Prob. 1–4

PRACTICE PROBLEMS

1. Write each of the following as an algebraic expression:

a. Six added to four. $4 + 6$ or $6 + 4$

b. Twelve times nineteen. 12×19 or $12 \cdot 19$

c. From ten subtract two. 10 − 2
d. Fifteen divided by three. $\frac{15}{3}$ or 15 ÷ 3
e. The square of nine. 9²
f. The square root of twenty. $\sqrt{20}$
g. The sum of eight and three. 8 + 3
h. The difference between six and two. 6 − 2
i. The product of nine and five. 9 × 5 or 9 · 5
j. The quotient of twelve divided by four. $\frac{12}{4}$ or 12 ÷ 4
k. The sum of b and x. b + x
l. a times y. ay
m. The product of g and eight. 8g
n. The difference between c and g. c − g
o. The quotient of d divided by r. $\frac{d}{r}$ or d ÷ r
p. Two times the sum of l and w. 2(l + w)
q. The cube of r. r³
r. The square root of s. $\sqrt{s}$
s. The product of b and h. bh
t. Three times the difference between t and nine. 3(t − 9)

2.a. Five times n b. nineteen minus seven c. y plus twenty-five d. m divided by x e. c times d squared times x cubed f. Six times a minus nine times b g. The quantity m plus n times the quantity m minus n. h. Three times x squared minus the quantity four times x times y plus the quantity seven times y squared

2. Read, or write in words, each of the following:

a. $5n$ c. $y + 25$ e. cd^2x^3 g. $(m + n)(m - n)$

b. $19 - 7$ d. $\dfrac{m}{x}$ f. $6a - 9b$ h. $3x^2 - 4xy + 7y^2$

3. Write each of the following as an algebraic expression:

a. Twice the radius (r). 2r
b. Three times the side (s). 3s
c. The principal (p) plus the interest (i). p + i
d. 90° decreased by angle B. 90° − B
e. The sum of angles A, B, and C. A + B + C
f. The circumference (c) divided by pi (π). c ÷ π or $\frac{c}{\pi}$
g. The length (l) times the width (w) times the height (h). lwh
h. The profit (p) added to the cost (c). c + p
i. One half the product of the altitude (a) and base (b). $\frac{1}{2}$ ab
j. Angle B subtracted from 180°. 180° − B
k. The square of the side (s). s²
l. The base (b) multiplied by the height (h) divided by two. bh ÷ 2 or $\frac{bh}{2}$
m. The sum of twice the length (l) and twice the width (w). 2l + 2w
n. The cube of the edge (e). e³
o. The quotient of the interest (i) for one year divided by the principal (p). i ÷ p.or $\frac{i}{p}$
p. The sum of bases b_1 and b_2 divided by two. $\frac{b_1 + b_2}{2}$

$\frac{4}{3}\pi r^3$ **q.** Four thirds the product of pi (π) and the cube of the radius (r).

$a^2 + b^2$ **r.** The square of the altitude (a) added to the square of the base (b).

s. Twice pi (π) times the radius (r) times the sum of the height (h) and radius. $2\pi r(h + r)$

t. The square root of the difference between the square of the hypotenuse (h) and the square of the base (b). $\sqrt{h^2 - b^2}$

4. Write an algebraic expression for each of the following:

a. Marilyn has n cents. She spends 95 cents for lunch. How many cents does she have left? $n - 95$

b. Scott is x years old. How old will he be in fifteen years? $x + 15$

c. Steve has d dollars in the bank. If he deposits y dollars, how many dollars will he then have in the bank? $d + y$

$\frac{m}{r}$ **d.** How long does it take a train averaging r m.p.h. to travel m miles?

e. How many notebooks can be bought for x cents if one notebook costs y cents? $\frac{x}{y}$

EXERCISE

4-15 | Mathematical Sentences

VOC.: mathematical sentence
inequality equation
open sentence inequation
disjunction conjunction
compound sentence

OBJECTIVE: To read and write mathematical sentences.

Sentences such as:

$7 + 9 = 16$	read as "Seven plus nine is equal to sixteen"
or $x - 4 = 9$	read as "Some number x decreased by four is equal to nine"
or $8\,a > 20$	read as "Eight times each number a is greater than twenty"
or $n + 10 < 14$	read as "Each number n increased by ten is less than fourteen"

are mathematical sentences.

Open Sentences: Equations and Inequalities

An open sentence is a mathematical sentence that contains a variable. $x - 4 = 9$, $8\,a > 20$, and $n + 10 < 14$ are open sentences.

An equation is an open sentence that has the equality sign "$=$" as its verb. The sentence $x - 4 = 9$ is an equation.

If a sentence uses $\neq$, $>$, $<$, $\not>$, $\not<$, $\geq$, $\leq$, $\not\geq$, or $\not\leq$ as its verb, it is called an inequation or inequality. The sentences $8\,a > 20$ and $n + 10 < 14$ are inequalities.

An open sentence is neither true nor false. It is only after a number is substituted for the variable that the open sentence becomes a statement and the sentence can be determined to be true or false.

Related Resources: *Practical Applications in Mathematics:* **Problems, pp. 136–137**

Compound Sentences

Sentences like $6n \leq 30$, $8 < x < 12$, and $20 \geq b \geq 15$ are compound sentences with connectives "or" or "and."

The sentence "$6n \leq 30$" is the shortened form of "$6n < 30$ or $6n = 30$," a disjunction, and is read "Six times each number n is less than or equal to thirty."

The sentence $8 < x < 12$ is the shortened form of "$8 < x$ and $x < 12$," a conjunction, and is read "Eight is less than each number x and each number x is less than twelve" or "Each number x is greater than eight and less than twelve."

The sentence $20 \geq b \geq 15$ is read "Twenty is greater than or equal to each number b which is greater than or equal to fifteen" or "Each number b is less than or equal to twenty and greater than or equal to fifteen."

Basic: Prob. 1; *Average:* Prob. 1–2; *Enriched:* Prob. 1–2

PRACTICE PROBLEMS   For ex. 1–2 see Extended Answers, p. T83.

1. Read, or write in words, each of the following:

a. $n < 16$

b. $x > 29$

c. $3y \neq 17$

d. $b + 6 > 25$

e. $a - 5 < 53$

f. $7t \not< -14$

g. $9m + 4 = 15$

h. $8c - 9 \not> c + 7$

i. $2n + 6 < 3n - 8$

j. $\dfrac{b}{5} > 20$

k. $15x - 4 \not> 18 - 7x$

l. $h \geq -6$

m. $19w \leq 38$

n. $5d + 7 \leq 12$

o. $n - 16 \geq 4n + 9$

p. $-3 < n < 8$

q. $18 > x > 0$

r. $4 \leq b \leq 27$

s. $12 \geq 7r \geq -6$

t. $2 < a \leq 45$

2. Write each of the following as an open sentence symbolically:

a. Some number x increased by ten is equal to forty-one.

b. Each number n decreased by four is less than eighteen.

c. Four times each number t is greater than twelve.

d. Nine times each number y plus two is not equal to fifty.

e. Each number a increased by five is not less than nine.

f. Each number m divided by seven is not greater than fourteen.

g. Seven times each number x is greater than or equal to thirty.

h. Each number c plus nine is less than or equal to ten.

i. Each number t is less than negative one and greater than negative five.

j. Twelve times each number y is greater than or equal to four and less than or equal to fifteen.

OBJECTIVE: To read and write formulas.

A special kind of equation, called a formula, is a mathematical rule expressing the relationship of two or more quantities by means of numerals, variables, and operating symbols. The formula contains algebraic expressions.

To express mathematical and scientific principles as formulas, write numerals, operating symbols, and letters (variables) representing the given quantities in the required order to show the relationship between quantities. A quantity may be represented by the first letter of a key word.

Write as a formula:

The circumference (c) of a circle is equal to pi (π) times the diameter (d).

Answer: $c = \pi d$

To translate a formula to a word statement, write a word rule stating the relationship expressed by the formula.

Translate:

$$a = \frac{360°}{n}$$ where a = central angle of a regular polygon and n = the number of sides.

Answer: A central angle of a regular polygon equals 360° divided by the number of sides.

PRACTICE PROBLEMS

Basic: Prob. 1; *Average:* Prob. 1–2; *Enriched:* Prob. 1–2

1. Express each of the following as a formula:

a. The perimeter of a square (p) is equal to four times the length of the side (s). $p = 4s$

b. The area of a rectangle (A) is equal to the length (l) multiplied by the width (w). $A = lw$

c. The circumference of a circle (c) is equal to twice pi (π) times the radius (r). $c = 2\pi r$

d. The interest (i) is equal to the principal (p) times the rate of interest per year (r) times the time in years (t). $i = prt$

e. The radius of a circle (r) is equal to the diameter (d) divided by two. $r = \frac{d}{2}$

f. The total amount (A) is equal to the sum of the principal (p) and the interest (i). $A = p + i$

g. The area of a parallelogram (A) is equal to the product of the base (b) and height (h). $A = bh$

h. The volume of a cube (V) is equal to the cube of the edge (e). $V = e^3$

i. The area of a circle (A) is equal to pi (π) times the square of the radius (r). $A = \pi r^2$

j. The area of a triangle (A) is equal to one half the product of the altitude (a) and base (b). $A = \frac{1}{2}ab$

k. The sum of two complementary angles $(A$ and $B)$ equals $90°$. $A + B = 90°$

l. The net price (n) is equal to the list price (l) less the discount (d). $n = l - d$

m. The volume of a cylinder (V) is equal to one fourth pi (π) times the square of a diameter (d) times the height (h). $V = \frac{1}{4}\pi d^2 h$

n. The perimeter of an isosceles triangle (p) is equal to the base (b) added to twice the length of the equal side (e). $p = b + 2e$

o. The rate of commission (r) is equal to the commission (c) divided by the sales (s). $r = \frac{c}{s}$

p. The perimeter of a rectangle (p) is equal to twice the sum of the length (l) and width (w). $p = 2(l + w)$

q. The area of a trapezoid (A) is equal to the height (h) times the sum of the two parallel bases, b_1 and b_2, divided by two. $A = h\frac{b_1 + b_2}{2}$

r. The area of the surface of a rectangular solid (A) is equal to the sum of twice the product of the length (l) and width (w), twice the product of the length and height (h), and twice the product of the width and height. $A = 2lw + 2lh + 2wh$

s. The hypotenuse of a right triangle (h) is equal to the square root of the sum of the square of the altitude (a) and the square of the base (b). $h = \sqrt{a^2 + b^2}$

t. The total area of a cylinder (A) is equal to twice pi (π) times the radius (r) times the sum of the height (h) and the radius. $A = 2\pi r(h + r)$

For ex. 2 see
Extended
Answers, p. T83. **2.** Express each of the following formulas as a word statement:

a. $p = 3s$ where p = perimeter of an equilateral triangle and s = length of triangle side.

b. $E = IR$ where E = electromotive force in volts, I = current in amperes, and R = resistance in ohms.

c. $t = \dfrac{d}{r}$ where t = time of travel, d = distance traveled, and r = average rate of speed.

d. $F = 1.8\,C + 32°$ where F = Fahrenheit temperature reading and C = Celsius temperature reading.

e. $d = \frac{1}{2}gt^2$ where d = distance a freely falling body drops, g = acceleration of gravity, t = time of falling.

OBJECTIVE: To evaluate algebraic expressions.

The value of an algebraic expression is dependent on the numerical values assigned to each variable of the expression. If these values change, the value of the expression usually changes.

To evaluate an algebraic expression,
(1) copy the expression,
(2) substitute the given numerical value for each variable,
(3) perform the necessary operations indicated in the expression.

Observe in the following models how a fractional expression, an expression containing parentheses, and an expression containing exponents are evaluated.

---

1. Find the value of $2\,d + x$ when $d = 7$ and $x = 9$.

$$2\,d + x$$
$$= (2 \cdot 7) + 9$$
$$= 14 + 9$$
$$= 23$$

Answer: 23

2. Find the value of $\dfrac{A}{b}$ when $A = 28$ and $b = 4$.

$$\frac{A}{b}$$
$$= \frac{28}{4}$$
$$= 7$$

Answer: 7

3. Find the value of $2(l + w)$ when $l = 17$ and $w = 14$.

$$2(l + w)$$
$$= 2(17 + 14)$$
$$= 2(31)$$
$$= 62$$

Answer: 62

4. Find the value of $c^2 + d^2$ when $c = 8$ and $d = 3$.

$$c^2 + d^2$$
$$= (8)^2 + (3)^2$$
$$= 64 + 9$$
$$= 73$$

Answer: 73

Basic: Prob. 1–8; *Average:* Prob. 1–10; *Enriched:* Prob. 1–12

PRACTICE PROBLEMS

Find the value of each of the following algebraic expressions:

1. When $a = 12$ and $b = 6$:

a. $a + b$ 18

c. ab 72

e. a^2 144

g. $4(a - b)$ 24

b. $a - b$ 6

d. $\dfrac{a}{b}$ 2

f. $7\,a$ 84

h. $a^2 - b^2$ 108

Related Resources: *Practical Applications in Mathematics:* Problems, p. 138

2. When $x = 4$ and $y = 3$:

a. $x + 6y$ 22   **c.** $4x - 8y$ −8   **e.** $9x^2y$ 432   **g.** $(x + y)^2$ 49

b. $5xy$ 60   **d.** $x(x - y)$ 4   **f.** $x^2 - y^2$ 7   **h.** $\dfrac{x + y}{2x - y}$ $1\frac{2}{5}$

3. When $m = -8$, $n = -2$, and $x = 4$:

a. $3mnx$ 192   **c.** $(m + n)(m - n)$ +60 **e.** $m - n(m + n)$ −28

b. $10mn - 7nx$ 216 **d.** $\dfrac{(n - x)^2}{n^2 - x^2}$ −3   **f.** $5m^2 + 2mx - 3x^2$ 208

4. a. $4s$ when $s = 6$ 24

  b. $2r$ when $r = 12$ 24

  c. bh when $b = 10$ and $h = 7$ 70

  d. πd when $\pi = 3.14$ and $d = 26$ 81.64

  e. ab when $a = 45$ and $b = 51$ 2,295

  f. lwh when $l = 11$, $w = 7$, and $h = 15$ 1,155

  g. $2\pi r$ when $\pi = \frac{22}{7}$ and $r = 21$ 132

  h. prt when $p = 400$, $r = .06$, and $t = 8$ 192

  i. πdh when $\pi = \frac{22}{7}$, $d = 63$, and $h = 20$ 3,960

  j. $2\pi rh$ when $\pi = 3.14$, $r = 48$, and $h = 85$ 25,622.4

5. a. $p + i$ when $p = 500$ and $i = 120$ 620

  b. $b + 2e$ when $b = 32$ and $e = 29$ 90

  c. $b_1 + b_2$ when $b_1 = 47$ and $b_2 = 35$ 82

  d. $2l + 2w$ when $l = 63$ and $w = 54$ 234

  e. $2lw + 2lh + 2wh$ when $l = 18$, $w = 14$, and $h = 17$ 1,592

6. a. $90 - B$ when $B = 72$ 18

  b. $A - p$ when $A = 214$ and $p = 185$ 29

  c. $l - d$ when $l = 67$ and $d = 29$ 38

  d. $180 - A$ when $A = 104$ 76

  e. $A - i$ when $A = 843$ and $i = 196$ 647

7. a. $\dfrac{d}{2}$ when $d = 46$ 23

  b. $\dfrac{ab}{2}$ when $a = 25$ and $b = 30$ 375

  c. $\dfrac{c}{\pi}$ when $c = 12.56$ and $\pi = 3.14$ 4

  d. $\dfrac{Bh}{3}$ when $B = 216$ and $h = 17$ 1,224

  e. $\dfrac{b_1 + b_2}{2}$ when $b_1 = 52$ and $b_2 = 38$ 45

8. a. a^2 when $a = 6$ ₃₆ 36

　b. s^2 when $s = 27$ 729

　c. b^2 when $b = 105$ 11,025

　d. h^2 when $h = 300$ 90,000

　e. πr^2 when $\pi = \frac{22}{7}$ and $r = 28$ 2,464

　f. $.7854\, d^2$ when $d = 50$ 1,963.5

　g. $6\, s^2$ when $s = 41$ 10,086

　h. e^3 when $e = 19$ 6,859

　i. $\frac{4}{3}\pi r^3$ when $\pi = 3.14$ and $r = 20$ $33,493\frac{1}{3}$

　j. $\pi r^2 h$ when $\pi = \frac{22}{7}$, $r = 70$, and $h = 45$ 693,000

9. a. $\frac{1}{2}ab$ when $a = 72$ and $b = 66$ 2,376

　b. $\frac{1}{3}Bh$ when $B = 135$ and $h = 20$ 900

　c. $\frac{1}{4}\pi d^2$ when $\pi = 3.14$ and $d = 55$ 2,374.625

　d. $\frac{4}{3}\pi r^3$ when $\pi = \frac{22}{7}$ and $r = 42$ 310,464

　e. $\frac{1}{3}\pi r^2 h$ when $\pi = 3.14$, $r = 69$, and $h = 80$ 398,654.4

10. a. $a^2 + b^2$ when $a = 11$ and $b = 9$ 202

　b. $h^2 - a^2$ when $h = 36$ and $a = 27$ 567

　c. $V + gt$ when $V = 150$, $g = 32$, and $t = 7$ 374

　d. $p + prt$ when $p = 300$, $r = .06$, and $t = 8$ 444

　e. $2\pi rh + 2\pi r^2$ when $\pi = 3.14$, $r = 20$, and $h = 75$ 11,932

　f. $\pi dh + \frac{1}{2}\pi d^2$ when $\pi = \frac{22}{7}$, $d = 14$, and $h = 39$ 2,024

11. a. $2(l + w)$ when $l = 37$ and $w = 28$ 130

　b. $2\pi r(h + r)$ when $\pi = \frac{22}{7}$, $r = 7$, and $h = 13$ 880

　c. $I(r + R)$ when $I = 8$, $r = 3$, and $R = 10$ 104

　d. $(n - 1)d$ when $n = 10$ and $d = 6$ 54

　e. $h\left(\dfrac{b_1 + b_2}{2}\right)$ when $h = 34$, $b_1 = 27$, and $b_2 = 19$ 782

　f. $\frac{5}{9}(F - 32)$ when $F = -13$ −25

12. a. $\sqrt{A}$ when $A = 49$ 7

　b. $\sqrt{\dfrac{V}{\pi h}}$ when $V = 1,100$, $\pi = \frac{22}{7}$, and $h = 14$ 5

　c. $\sqrt{a^2 + b^2}$ when $a = 18$ and $b = 24$ 30

　d. $\sqrt{h^2 - a^2}$ when $h = 78$ and $a = 72$ 30

　e. $\sqrt{h^2 - b^2}$ when $h = 145$ and $b = 116$ 87

Evaluating Formulas

OBJECTIVE: To evaluate formulas.

To determine the required value, (1) copy the formula,
(2) substitute the given values for the variables,
and (3) perform the necessary operations.

Find the value of p when $s = 16$, using the formula $p = 4\,s$. $$p = 4\,s$$ $$p = 4 \cdot 16$$ $$p = 64$$ *Answer: p = 64*	Find the value of I when $W = 100$ and $E = 25$, using $I = \dfrac{W}{E}$. $\qquad I = \dfrac{W}{E}$ $\qquad\qquad I = \dfrac{100}{25}$ *Answer: I = 4* $I = 4$

Basic: Prob. 1–15; *Average:* Prob. 1–17; *Enriched:* Prob. 1–20

PRACTICE PROBLEMS

Find the value of:

1. d when $r = 85$. Formula: $d = 2\,r$ **170**
2. A when $b = 32$ and $h = 49$. Formula: $A = bh$ **1,568**
3. c when $\pi = \frac{22}{7}$ and $d = 84$. Formula: $c = \pi d$ **264**
4. A when $p = 620$ and $i = 53$. Formula: $A = p + i$ **673**
5. p when $l = 73$ and $w = 28$. Formula: $p = 2\,l + 2\,w$ **202**
6. A when $B = 47$. Formula: $A = 90 - B$ **43**
7. i when $p = 650$, $r = .04$, and $t = 9$. Formula: $i = prt$ **234**
8. V when $l = 41$, $w = 27$, and $h = 30$. Formula: $V = lwh$ **33,210**
9. A when $\pi = 3.14$, $r = 48$, and $h = 75$. Formula: $A = 2\,\pi rh$ **22,608**
10. A when $s = 50$. Formula: $A = s^2$ **2,500**
11. V when $e = 8$. Formula: $V = e^3$ **512**
12. V when $\pi = 3.14$, $r = 40$, and $h = 65$. Formula: $V = \pi r^2 h$ **326,560**
13. p when $l = 427$ and $w = 393$. Formula: $p = 2\,l + 2\,w$ **1,640**
14. B when $A = 75$ and $C = 48$. Formula: $B = 180 - (A + C)$ **57**
15. l when $a = 6$, $n = 12$, and $d = 5$. Formula: $l = a + (n - 1)d$ **61**
16. I when $E = 220$ and $R = 11$. Formula: $I = \dfrac{E}{R}$ **20**
17. P when $F = 90$, $d = 10$, and $t = 3$. Formula: $P = \dfrac{FD}{t}$ **300**
18. v when $V = 27$, $g = 32$, and $t = 5$. Formula: $v = V + gt$ **187**
19. A when $p = 150$, $r = .08$, and $t = 4$. Formula: $A = p(1 + rt)$ **198**
20. I when $E = 220$, $n = 4$, $R = 8$, and $r = 3$. Formula: $I = \dfrac{nE}{R + nr}$ **44**

Related Resources: *Practical Applications in Mathematics:* Problems, p. 139   **485**   For a Test on the Language of Algebra see p. T55 or *Testing Program*, p. 29.

OPEN SENTENCES IN ONE VARIABLE

VOC.: solve an equation
in one variable
satisfy the equa-
tion

root
solution
checking an
equation
members (sides)
of equation
equivalent
equation

EXERCISE 4-19 Introduction to the Solution of Equations in One Variable

OBJECTIVE: To find the number which, when
substituted for the variable, makes the sentence true.

To solve an equation in one variable means to find the number represented by the variable which, when substituted for the variable, will make the sentence true.

Any number that makes the sentence true is said to satisfy the equation and to be the root or the solution of the open sentence.

The equation $n + 4 = 20$ indicates that some number n plus four equals twenty. When 16 is substituted for the variable n, the resulting sentence $16 + 4 = 20$ is a true sentence. Thus 16 is said to be the root of the equation $n + 4 = 20$.

Checking an equation is testing whether some number is a solution to this equation by substituting the number for the variable in the equation. If the resulting sentence is true, the number is a solution; if not, the number is not a solution.

---

16 is a solution to $n + 4 = 20$ because $16 + 4 = 20$.
12 is not a solution because $12 + 4 \neq 20$.

486

The expressions at the left and at the right of the equality sign in an equation are called members or sides of the equation.

> In the equation $n + 4 = 20$, the expression $n + 4$ is the left member and the expression 20 is the right member.

Equations which have exactly the same solutions are called equivalent equations.

> The equations $n + 4 = 20$ and $n = 16$ are equivalent equations because they both have the same solution, 16.

An equation is considered to be in its simplest form when one member contains only the variable itself and the other member is a constant term.

> The equation $n = 16$ is an equation in the simplest form.

Basic: Prob. 1–10; *Average:* Prob. 1–12; *Enriched:* Prob. 1–15

PRACTICE PROBLEMS

1. Which of the following numbers will make the sentence $x + 4 = 8$ true?

  a. $x = 12$ **b.** $x = 2$ **c.** $x = -4$ **d.** $x = \frac{1}{2}$ **e.** $x = 4$

2. Which of the following numbers will make the sentence $n - 2 = 10$ true?

  a. $n = 8$ **b.** $n = 5$ **c.** $n = 12$ **d.** $n = -8$ **e.** $n = -\frac{1}{5}$

3. Which of the following numbers will make the sentence $9\,r = 63$ true?

  a. $r = 72$ **b.** $r = 54$ **c.** $r = 0$ **d.** $r = 7$ **e.** $r = -6$

4. Which of the following numbers will make the sentence $\dfrac{b}{3} = 18$ true?

  a. $b = 6$ **b.** $b = \frac{1}{6}$ **c.** $b = 15$ **d.** $b = 54$ **e.** $b = 21$

5. Which of the following numbers is the root of the equation $m + 6 = 6$?

    a. $m = 12$ **b.** $m = -12$ **c.** $m = 0$ **d.** $m = 1$ **e.** $m = 6$

6. Which of the following numbers is the root of the equation $8s = 24$?

    a. $s = 16$ **b.** $s = 32$ **c.** $s = -\frac{1}{3}$ **d.** $s = 0$ **e.** $s = 3$

7. Which of the following numbers is the root of the equation $d - 9 = 3$?

    a. $d = 6$ **b.** $d = 3$ **c.** $d = -6$ **d.** $d = 12$ **e.** $d = -12$

8. Which of the following numbers is the root of the equation $\frac{a}{8} = 4$?

    a. $= 2$ **b.** $a = \frac{1}{2}$ **c.** $a = 32$ **d.** $a = 12$ **e.** $a = 4$

9. Which of the following is a solution to the equation $12y = 6$?

    a. 6   **b.** 2   **c.** 18   **d.** $\frac{1}{2}$ **e.** -2

10. Which of the following is a solution to the equation $n + 7 = 21$?

    a. 3   **b.** 28   **c.** $\frac{1}{3}$ **d.** 14   **e.** -3

11. Which of the following is a solution to the equation $\frac{x}{12} = 8$?

    a. 20   **b.** 4   **c.** $\frac{2}{3}$ **d.** 96   **e.** $1\frac{1}{2}$

12. Which of the following is a solution to the equation $w - 12 = 12$?

    a. -24 **b.** 0   **c.** 1   **d.** 24   **e.** 144

13. a. Write the left member of the equation $7x + 9 = 37$. 7x + 9
    b. Write the right member of the equation $5n - 3 = 42$. 42

14. Which of the following equations have 6 as a root? b., c., d., e.
    Which are equivalent equations? b., c., d., e.

    a. $y - 2 = 8$ **b.** $\frac{y}{3} = 2$ **c.** $9y = 54$ **d.** $y = 6$ **e.** $y + 17 = 23$

15. Which of the following equations have 7 as a root? a., c., d.
    Which are equivalent equations? a., c., d.
    Which one of the equivalent equations is in the simplest form? c.

    a. $9x = 63$ **b.** $x - 4 = 11$ **c.** $x = 7$ **d.** $\frac{x}{7} = 1$ **e.** $x + 15 = 8$

Inverse Operations; Properties of Equality

OBJECTIVE: To do inverse operations and to use the properties of equality to obtain an equation in simplest form.

On page 5 we studied that *inverse operations undo each other*. That is: subtraction undoes addition, addition undoes subtraction, division undoes multiplication, and multiplication undoes division.

We found that:

(1) When we first add a number to the given number n and then subtract this number from the sum, we return to the given number n. That is: $(n + 1) - 1 = n$; $(n + 2) - 2 = n$; $(n + 3) - 3 = n$.

(2) When we first subtract a number from the given number n and then add this number to the answer, we return to the given number n. That is: $(n - 1) + 1 = n$; $(n - 2) + 2 = n$; $(n - 3) + 3 = n$.

(3) When we first multiply the given number n by a number and then we divide the product by this number, we return to the given number n.

That is: $(n \times 2) \div 2 = n$ or $\dfrac{2\,n}{2} = n$;

$(n \times 3) \div 3 = n$ or $\dfrac{3\,n}{3} = n$;

$(n \times 4) \div 4 = n$ or $\dfrac{4\,n}{4} = n$; etc.

(4) When we first divide the given number n by a number and then we multiply the quotient by this number, we return to the given number n.

That is: $(n \div 2) \times 2 = n$ or $\dfrac{n}{2} \times 2 = n$;

$(n \div 3) \times 3 = n$ or $\dfrac{n}{3} \times 3 = n$;

$(n \div 4) \times 4 = n$ or $\dfrac{n}{4} \times 4 = n$; etc.

The *properties of equality* which state that the results are equal when equals are increased or decreased or multiplied or divided by equals, with division by zero excluded, are called axioms.

These axioms allow us in each of the following cases to obtain an equation in the simplest form.

(1) When we subtract the same number from both sides of a given equation:

> For example: Given: $n + 2 = 12$
>
> Subtracting 2 from each side: $(n + 2) - 2 = 12 - 2$
>
> We get: $n = 10$

(2) When we add the same number to both sides of a given equation.

> For example: Given: $n - 2 = 12$
>
> Adding 2 to each side: $(n - 2) + 2 = 12 + 2$
>
> We get: $n = 14$

(3) When we divide both sides of given equation by the same non-zero number.

> For example: Given: $2n = 12$
>
> Dividing each side by 2: $\dfrac{2n}{2} = \dfrac{12}{2}$
>
> We get: $n = 6$

(4) When we multiply both sides of a given equation by the same non-zero number.

> For example: Given: $\dfrac{n}{2} = 12$
>
> Multiplying each side by 2: $2 \cdot \dfrac{n}{2} = 2 \cdot 12$
>
> We get: $n = 24$

PRACTICE PROBLEMS

Find the missing number and, where required, the missing operation.

1. **a.** $(11 + 8) - \boxed{8} = 11$ **b.** $(16 + 22)\,\underset{+}{?}\,\boxed{22} = 16$ **c.** $(n + 5)\,\underset{+}{?}\,\boxed{5} = n$

2. **a.** $(9 \times 7) \div \boxed{7} = 9$ **b.** $(54 \times 31)\,\underset{\div}{?}\,\boxed{31} = 54$ **c.** $(h \times 6)\,\underset{\div}{?}\,\boxed{6} = h$

3. **a.** $(26 - 5) + \boxed{5} = 26$ **b.** $(40 - 18)\,\underset{+}{?}\,\boxed{18} = 40$ **c.** $(x - 12)\,\underset{+}{?}\,\boxed{12} = x$

4. **a.** $(3 \times 8) \div \boxed{3} = 8$ **b.** $(23 \times 43)\,\underset{\div}{?}\,\boxed{23} = 43$ **c.** $(7\,m)\,\underset{\div}{?}\,\boxed{7} = m$

5. **a.** $(36 \div 4) \times \boxed{4} = 36$ **b.** $(19 \div 6)\,\underset{\times}{?}\,\boxed{6} = 19$ **c.** $(y \div 8)\,\underset{\times}{?}\,\boxed{8} = y$

6. **a.** $\frac{7}{12} \times \boxed{12} = 7$ **b.** $\frac{13}{5}\,\overset{\times}{?}\,\boxed{5} = 13$ **c.** $\frac{n}{4}\,\overset{\times}{?}\,\boxed{4} = n$

7. **a.** $(r - 10)\,\underset{+}{?}\,\boxed{10} = r$ **c.** $(w + 25)\,\underset{-}{?}\,\boxed{25} = w$ **e.** $(11x)\,\underset{\div}{?}\,\boxed{11} = x$

   b. $(16\,n)\,\underset{\div}{?}\,\boxed{16} = n$ **d.** $\left(\dfrac{b}{21}\right)\,\underset{\times}{?}\,\boxed{21} = b$ **f.** $(t - 37)\,\underset{+}{?}\,\boxed{37} = t$

8. In each of the following, what number do you subtract from both sides of the given equation to get an equivalent equation in simplest form? Also write this equivalent equation. **b.** 12; x = 15 **c.** 9; y = 24 **d.** 24; r = 48
a. 3; n = 16 **a.** $n + 3 = 19$ **b.** $x + 12 = 27$ **c.** $y + 9 = 33$ **d.** $r + 24 = 72$

9. In each of the following, what number do you add to both sides of the given equation to get an equivalent equation in simplest form? Also write this equivalent equation. **b.** 19; s = 25 **c.** 28; h = 56 **d.** 15; x = 15
a. 4; w = 15 **a.** $w - 4 = 11$ **b.** $s - 19 = 6$ **c.** $h - 28 = 28$ **d.** $x - 15 = 0$

10. In each of the following, by what number do you divide both sides of the given equation to get an equivalent equation in simplest form? Also write this equivalent equation. **b.** 15; y = 6 **c.** 20; b = 1.75 **d.** 12; P = $\frac{7}{12}$
a. 9; m = 8 **a.** $9\,m = 72$ **b.** $15\,y = 90$ **c.** $20\,b = 35$ **d.** $12\,p = 7$

11. In each of the following, by what number do you multiply both sides of the given equation to get an equivalent equation in simplest form? Also write this equivalent equation.

a. $\dfrac{t}{3} = 6$ **b.** $\dfrac{m}{8} = 2$ **c.** $\dfrac{x}{14} = 9$ **d.** $\dfrac{a}{10} = 25$
      3; t = 18   8; m = 16   14; x = 126   10; a = 250

12. What operation with what number do you use on both sides of the given equation to get an equivalent equation in simplest form? Also write this equivalent equation.

add 5; x = 18   multiply by 8; n = 56 divide by 8; h = 7   add 110; d = 150
a. $x - 5 = 13$ **e.** $\dfrac{n}{8} = 7$ **i.** $8\,h = 56$ **m.** $d - 110 = 40$

divide by 6; a = 7   subtract 9; b = −9 subtract 45; x = 63   subtract 17; a = 58
b. $6\,a = 42$ **f.** $b + 9 = 0$ **j.** $x + 45 = 108$ **n.** $75 = a + 17$

subtract 11; m = 10 multiply by 6; t = 72 multiply by 10; n = 40 add 28; x = 70
c. $m + 11 = 21$ **g.** $\dfrac{t}{6} = 12$ **k.** $4 = \dfrac{n}{10}$ **o.** $42 = x - 28$

divide by 12; y = $\frac{3}{4}$ add 31; s = 62   divide by 8; n = $\frac{1}{4}$ divide by 18; g = 8
d. $12\,y = 9$ **h.** $s - 31 = 31$ **l.** $2 = 8\,n$ **p.** $18\,g = 144$

Solving Equations in One Variable

OBJECTIVE: To solve equations in one variable.

To solve an equation in one variable, transform the given equation to the simplest equivalent equation which has only the variable itself as one member and a constant naming a number as the other member. To make this transformation use the properties of equality (axioms) with inverse operations or with the additive inverse or multiplicative inverse as required.

An equation may be compared to balanced scales. To keep the equation in balance, any change on one side of the equality sign must be balanced by an equal change on the other side of the equality sign.

Observe in the models on the next several pages the following four basic types of equations:

Type I	$n + 4 = 20$
Type II	$n - 4 = 20$
Type III	$4n = 20$
Type IV	$\dfrac{n}{4} = 20$

These equations are solved both by inverse operations and by the additive inverse or the multiplicative inverse.

When the inverse operation method is used, *the operation indicated in the given equation by the variable and its connected constant is undone* by performing the inverse operation on both sides of the equation.

To check whether the number found is a solution, substitute this number for the variable in the given equation. If the resulting sentence is true, then the number is a solution. A check mark is sometimes used to indicate that the resulting sentence is true. In this section the solutions will be only one number.

Basic Type I

Solve and check: $n + 4 = 20$

Solution by inverse operation:

The indicated operation of $n + 4$ is addition. To find the root subtract 4 from each member.

$$n + 4 = 20$$
$$n + 4 - 4 = 20 - 4$$
$$n = 16$$

The root is 16

Answer: 16

Solution by additive inverse:

The additive inverse of 4
is -4. Add the additive
inverse to both members,
using the addition axiom.

$$n + 4 = 20$$
$$n + 4 + (-4) = 20 + (-4)$$
$$n = 16$$

Answer: 16

Check:
$$n + 4 = 20$$
$$16 + 4 = 20$$
$$20 = 20 \checkmark$$

Observe that the equation $n + 4 = 20$ may also appear as $4 + n = 20$ or $20 = n + 4$ or $20 = 4 + n$. These are forms of the same equation. In each case the constant 4 must be eliminated in order to get an equivalent equation in the simplest form.

This is $n = 16$ when solving $n + 4 = 20$ or $4 + n = 20$, and $16 = n$ when solving $20 = n + 4$ or $20 = 4 + n$. The equation $16 = n$ may be rewritten as $n = 16$.

Basic: Prob. 1–22; *Average:* Prob. 1–26; *Enriched:* Prob. 1–30

PRACTICE PROBLEMS

Solve and check:

1. $c + 6 = 14$ $c = 8$
2. $b + 18 = 31$ $b = 13$
3. $i + 37 = 63$ $i = 26$
4. $48 + w = 75$ $w = 27$
5. $29 + p = 106$ $p = 77$
6. $54 + n = 162$ $n = 108$
7. $67 = a + 28$ $a = 39$
8. $50 = m + 33$ $m = 17$
9. $105 = y + 79$ $y = 26$
10. $87 = 40 + x$ $x = 47$
11. $36 = 19 + d$ $d = 17$
12. $210 = 96 + r$ $r = 114$
13. $n + 80 = 80$ $n = 0$
14. $c + \$.09 = \$.23$ $c = \$.14$
15. $\$50 = i + \$42.50$ $i = \$7.50$

16. $t + \frac{3}{4} = 4$ $t = 3\frac{1}{4}$
17. $1\frac{1}{2} + x = 3\frac{5}{8}$ $x = 2\frac{1}{8}$
18. $6.5 = .48 + n$ $n = 6.02$
19. $a + \$9 = \$10.50$ $a = \$1.50$
20. $8\frac{1}{3} = 4\frac{1}{4} + d$ $d = 4\frac{1}{12}$
21. $b + 9 = 4$ $b = -5$
22. $3 + y = 1$ $y = -2$
23. $a + 5 = -3$ $a = -8$
24. $-9 = r + 10$ $r = -19$
25. $g + 15 = 0$ $g = -15$
26. $62 = 62 + x$ $x = 0$
27. $n + 7 = 3.8$ $n = -3.2$
28. $25 + a = 97$ $a = 72$
29. $7 = s + .4$ $s = 6.6$
30. $\frac{3}{8} = \frac{1}{4} + x$ $x = \frac{1}{8}$

Basic Type II

Solve and check: $n - 4 = 20$

Solution by inverse
operation: $n - 4 = 20$
The indicated operation $n - 4 + 4 = 20 + 4$
of $n - 4$ is subtraction. $n = 24$
To find the root add   The root is 24.
4 to both members.   *Answer: 24*

Solution by additive
inverse: $n - 4 = 20$
The additive inverse of $n + (-4) = 20$
-4 is $+4$. Add the $n + (-4) + (+4) = 20 + (+4)$
additive inverse to both $n = 24$
members, using the
addition axiom.   *Answer: 24*

Check:
$n - 4 = 20$
$24 - 4 = 20$
$20 = 20 \checkmark$

Observe that the equation $n - 4 = 20$ may also appear as $20 = n - 4$ but not as $4 - n = 20$ or $20 = 4 - n$. When solving $n - 4 = 20$ or $20 = n - 4$, the constant 4 must be eliminated in each case in order to get an equivalent equation in the simplest form.

This is $n = 24$ when solving $n - 4 = 20$, or $24 = n$ when solving $20 = n - 4$. The equation $24 = n$ may be rewritten as $n = 24$.

Basic: Prob. 1–22; Average: Prob. 1–26; Enriched: Prob. 1–30

PRACTICE PROBLEMS

Solve and check:

1. $c - 7 = 21$ $c = 28$ **5.** $16 = b - 47$ $b = 63$ **9.** $c - 40 = 40$ $c = 80$

2. $l - 18 = 5$ $l = 23$ **6.** $28 = y - 16$ $y = 44$ **10.** $52 = m - 52$ $m = 104$

3. $A - 42 = 124$ $A = 166$ **7.** $t - 9 = 0$ $t = 9$ **11.** $s - \$51 = \$120$ $s = \$171$

4. $35 = n - 19$ $n = 54$ **8.** $0 = y - 26$ $y = 26$ **12.** $9 = b - 2\frac{1}{2}$ $b = 11\frac{1}{2}$

13. $y - \$4.75 = \$3.25$ $_{y \,=\, \$8}$ **19.** $x - 18 = -10$ $_{x \,=\, 8}$ **25.** $51 = t - 51$ $_{t \,=\, 102}$

14. $d - \frac{7}{8} = 1\frac{1}{4}$ $_{d \,=\, 2\frac{1}{8}}$ **20.** $b - 6 = -6$ $_{b \,=\, 0}$ **26.** $w - 34 = -2$ $_{w \,=\, 32}$

15. $9.6 = n - 7$ $_{n \,=\, 16.6}$ **21.** $x - 2\frac{1}{2} = 2\frac{1}{2}$ $_{x \,=\, 5}$ **27.** $a - 29 = 76$ $_{a \,=\, 105}$

16. $a - \$.08 = \$.75$ $_{a \,=\, \$.83}$ **22.** $18 = n - 1.3$ $_{n \,=\, 19.3}$ **28.** $\frac{1}{2} = d - 1\frac{3}{4}$ $_{d \,=\, 2\frac{1}{4}}$

17. $c - 4 = -9$ $_{c \,=\, -5}$ **23.** $0 = x - 75$ $_{x \,=\, 75}$ **29.** $x - \$.75 = \$.35$ $_{x \,=\, \$1.10}$

18. $r - 23 = -45$ $_{r \,=\, -22}$ **24.** $-5 = r - 5$ $_{r \,=\, 0}$ **30.** $y - 1 = -9$ $_{y \,=\, -8}$

Basic Type III

Solve and check: $4\,n = 20$

Solution by inverse operation:
The indicated operation of $4\,n$ is multiplication. To find the root divide both members by 4.

$$4\,n = 20$$
$$\frac{4\,n}{4} = \frac{20}{4}$$
$$n = 5$$

The root is 5.
Answer: 5

Solution by multiplicative inverse:
The multiplicative inverse of 4 is $\frac{1}{4}$. Multiply both members by $\frac{1}{4}$, using the multiplication axiom.

$$4\,n = 20$$
$$\tfrac{1}{4} \cdot 4\,n = \tfrac{1}{4} \cdot 20$$
$$n = 5$$

Answer: 5

Check:
$$4\,n = 20$$
$$4 \cdot 5 = 20$$
$$20 = 20\,\surd$$

Observe that the equation $4\,n = 20$ may also appear as $20 = 4\,n$. When solving $4\,n = 20$ or $20 = 4\,n$, the numerical coefficient 4 must be eliminated in each case in order to get an equivalent equation in simplest form.

This is $n = 5$ when solving $4\,n = 20$, or $5 = n$ when solving $20 = 4\,n$. The equation $5 = n$ may be rewritten as $n = 5$.

PRACTICE PROBLEMS

Solve and check:

1. $5a = 45$ $a = 9$
2. $9c = 54$ $c = 6$
3. $14n = 42$ $n = 3$
4. $91 = 7y$ $y = 13$
5. $60 = 12x$ $x = 5$
6. $105 = 15b$ $b = 7$
7. $2r = 18$ $r = 9$
8. $48 = 4s$ $s = 12$
9. $10w = 10$ $w = 1$
10. $19 = 19h$ $h = 1$
11. $8y = 0$ $y = 0$
12. $0 = 16x$ $x = 0$
13. $5c = 8$ $c = 1\frac{3}{5}$
14. $18w = 27$ $w = 1\frac{1}{2}$
15. $4y = 1$ $y = \frac{1}{4}$

16. $12d = 10$ $d = \frac{5}{6}$
17. $23 = 9n$ $n = 2\frac{5}{9}$
18. $4 = 5b$ $b = \frac{4}{5}$
19. $54 = 8r$ $r = 6\frac{3}{4}$
20. $12 = 16x$ $x = \frac{3}{4}$
21. $7t = 2.8$ $t = .4$
22. $.4n = 36$ $n = 90$
23. $10c = \$.90$ $c = \$.09$
24. $\$.05a = \$1.45$ $a = 29$
25. $\frac{1}{4}c = 20$ $c = 80$
26. $\frac{7}{8}n = 42$ $n = 48$
27. $24 = \frac{2}{3}n$ $n = 36$
28. $3\frac{1}{2}t = 14$ $t = 4$
29. $.06p = \$240$ $p = \$4,000$
30. $\frac{9}{16}r = 72$ $r = 128$

31. $8x = -32$ $x = -4$
32. $-5m = 40$ $m = -8$
33. $-7y = -56$ $y = 8$
34. $-x = -10$ $x = 10$
35. $-64 = -4c$ $c = 16$
36. $-y = 19$ $y = -19$
37. $44 = -18b$ $b = -2\frac{4}{9}$
38. $-20s = -5$ $s = \frac{1}{4}$
39. $10c = -50$ $c = -5$
40. $10 = -50c$ $c = -\frac{1}{5}$
41. $12m = 300$ $m = 25$
42. $140 = 15y$ $y = 9\frac{1}{3}$
43. $3.14d = 12.56$ $d = 4$
44. $2\frac{1}{4}n = 13\frac{1}{2}$ $n = 6$
45. $15 = \frac{3}{8}h$ $h = 40$

Basic Type IV

Solve and check: $\dfrac{n}{4} = 20$

Solution by inverse operation:
The indicated operation of $\dfrac{n}{4}$ is division. To find the root multiply both members by 4.

$$\frac{n}{4} = 20$$

$$4 \cdot \frac{n}{4} = 4 \cdot 20$$

$$n = 80$$

The root is 80.
Answer: 80

Solution by multiplicative inverse:
Since $\dfrac{n}{4} = \dfrac{1}{4}n$, use the multiplicative inverse of $\dfrac{1}{4}$, which is 4.
Multiply both members by 4, using the multiplication axiom.

$$\frac{n}{4} = 20$$

$$4 \cdot \frac{1}{4}n = 4 \cdot 20$$

$$n = 80$$

Answer: 80

Check:

$$\frac{n}{4} = 20 \qquad\qquad \frac{80}{4} = 20 \qquad\qquad 20 = 20 \checkmark$$

Observe that the equation $\frac{n}{4} = 20$ may also appear as $20 = \frac{n}{4}$. When solving $\frac{n}{4} = 20$ or $20 = \frac{n}{4}$, the denominator 4 must be eliminated in each case in order to get an equivalent equation in simplest form.

This is $n = 80$ when solving $\frac{n}{4} = 20$, or $80 = n$ when solving $20 = \frac{n}{4}$. The equation $80 = n$ may be rewritten as $n = 80$.

Basic: Prob. 1–17; *Average:* Prob. 1–21; *Enriched:* Prob. 1–24

PRACTICE PROBLEMS

Solve and check:

1. $\frac{d}{2} = 19$ *d* = 38

2. $\frac{s}{4} = 24$ *s* = 96

3. $\frac{b}{7} = 3$ *b* = 21

4. $\frac{n}{5} = 0$ *n* = 0

5. $\frac{t}{8} = 8$ *t* = 64

6. $12 = \frac{b}{3}$ *b* = 36

7. $45 = \frac{m}{5}$ *m* = 225

8. $1 = \frac{s}{9}$ *s* = 9

9. $\frac{x}{3} = 9$ *x* = 27

10. $\frac{c}{3.14} = 100$ *c* = 314

11. $\frac{b}{5} = 1.4$ *b* = 7

12. $\frac{1}{2}a = 17$ *a* = 34

13. $\frac{3}{4}c = \$.36$ *c* = $.48

14. $0 = \frac{g}{8}$ *g* = 0

15. $\frac{w}{5} = 30$ *w* = 150

16. $\frac{k}{12} = 3$ *k* = 36

17. $\frac{n}{4} = -5$ *n* = −20

18. $\frac{r}{9} = -7$ *r* = −63

19. $\frac{y}{-3} = 12$ *y* = −36

20. $\frac{a}{-6} = -18$ *a* = 108

21. $-8 = \frac{d}{-11}$ *d* = 88

22. $\frac{x}{6} = 1$ *x* = 6

23. $-5 = \frac{z}{7}$ *z* = −35

24. $\frac{b}{16} = 16$ *b* = 256

Solution by More Than One Axiom

The solution of more difficult equations may involve the use of more than one axiom. When a member of a given equation consists of two or more terms containing the variable, first combine these terms.

Solve and check: $7x + 2 = 72$

$$7x + 2 = 72$$
$$7x + 2 - 2 = 72 - 2$$
$$7x = 70$$
$$\frac{7x}{7} = \frac{70}{7}$$
$$x = 10$$

Answer: $x = 10$

Check:
$$7x + 2 = 72$$
$$(7 \cdot 10) + 2 = 72$$
$$70 + 2 = 72$$
$$72 = 72 \checkmark$$

Solve and check: $7x + 2x = 72$

$$7x + 2x = 72$$
$$9x = 72$$
$$\frac{9x}{9} = \frac{72}{9}$$
$$x = 8$$

Answer: $x = 8$

Check:
$$7x + 2x = 72$$
$$(7 \cdot 8) + (2 \cdot 8) = 72$$
$$56 + 16 = 72$$
$$72 = 72 \checkmark$$

Basic: Prob. 1–22; *Average:* Prob. 1–26; *Enriched:* Prob. 1–30

PRACTICE PROBLEMS

Solve and check:

1. $2b + 18 = 46$ b = 14
2. $3 + 5n = 62$ n = 11⅘
3. $51 = 9y + 6$ y = 5
4. $120 = 76 + 2w$ w = 22
h = 10 5. $22h + 154 = 374$
c = 45 6. $1.8c + 32 = 113$
7. $7t - 11 = 38$ t = 7
8. $0 = 10n - 30$ n = 3
9. $6a - 7 = 29$ a = 6
10. $37 = 14b - 19$ b = 4

11. $n + n = 50$ n = 25
12. $9x + 3x = 84$ x = 7
13. $p + .06p = 689$
14. $7y - y = 18$ y = 3
15. $a - .05a = 760$
16. $6b = 91 - 15$
17. $82 = 5a + 9a$
18. $1.8c + 32 = 50$
19. $l - .35l = 195$
20. $\frac{3}{5}b = 21$ b = 35

21. $\frac{5}{8}d = \$.75$ d = $1.20
22. $t + \frac{1}{4}t = 6\frac{1}{2}$ t = 5⅕
23. $\frac{7}{8}n + 18 = 46$ n = 32
24. $\frac{5}{9}b - 13 = 27$ b = 72
25. $8a + 43 = 11$ a = −4
26. $6y - 4y = -28$ y = −14
27. $2 - 3b = 8$ b = −2
28. $5n - 9n = -56$ n = 14
29. $\$2.76 = \frac{3}{4}a$ a = $3.68
30. $c + .25c = \$15.50$ c = $12.40

13. p = 650 15. a = 800 16. b = 12⅔ 17. a = 5⁶⁄₇ 18. c = 10 19. l = 300

Solving Equations When the Replacements for the Variable are Restricted

OBJECTIVE: To solve equations when the replacements for the variable are restricted.

The solution of an equation depends not only on the equation but also on how restricted the replacements for the variable are. The same equation with different replacements for the variable may have different solutions.

For the equation $5n = 15$

When the replacements are all the whole numbers, the solution is 3.

When the replacements are all the prime numbers, the solution is 3.

When the replacements are all the even numbers, there is no solution.

For the equation $5n = 2$

When the replacements are all the real numbers, the solution is $\frac{2}{5}$.

When the replacements are all the whole numbers, there is no solution.

In the latter case, since the solution is a fraction and only whole numbers may be used as the replacements for the variable, there is no solution.

When the replacements for the variables are not given in this book, use all the real numbers as the replacements.

Basic: Prob. 1–8; *Average:* Prob. 1–10; *Enriched:* Prob. 1–11

PRACTICE PROBLEMS

Find the solution of each of the following:

1. When the replacements for the variable are all the whole numbers:

a. $8c = 56$ **b.** $\dfrac{n}{3} = 4$ **c.** $8 + y = 5$ **d.** $0 = t - 7$

$c = 7$ $n = 12$ no solution $t = 7$

2. When the replacements for the variable are all the prime numbers:

a. $m + 1 = 6$
$m = 5$

b. $5a + 3a = 48$
no solution

c. $5t - 1 = 9$
$t = 2$

d. $\dfrac{b}{7} = 5$
no solution

3. When the replacements for the variable are all the integers:

a. $6b = 9$
no solution

b. $\dfrac{c}{4} = -7$
$c = -28$

c. $2n - n = 8$
$n = 8$

d. $4a + 9 = 1$
$a = -2$

4. When the replacements for the variable are all the non-positive integers:

a. $x + 25 = 14$
$x = -11$

b. $-5s = 30$
$s = -6$

c. $\dfrac{a}{-4} = -2$
no solution

d. $7n + 8 = 8$
$n = 0$

5. When the replacements for the variable are all the rational numbers:

a. $-12d = 24$
$d = -2$

b. $x + 1\frac{1}{2} = 5$
$x = 3\frac{1}{2}$

c. $.04p = 78$
$p = 1,950$

d. $16y - y = 10$
$y = \frac{2}{3}$

6. When the replacements for the variable are all the multiples of 3:

a. $t - 8 = 4$
$t = 12$

b. $\dfrac{m}{2} = 12$
$m = 24$

c. $6r + 5 = 53$
no solution

d. $8y + 2y = 270$
$y = 27$

7. When the replacements for the variable are all the one-digit, odd whole numbers:

a. $y + 1 = 7$
no solution

b. $9h + 3h = 60$
$h = 5$

c. $-7x = 21$
no solution

d. $\dfrac{c}{3} = 3$
$c = 9$

8. When the replacements for the variable are all the even integers greater than -11 and less than 17:

a. $\dfrac{s}{5} = 2$
$s = 10$

b. $-2x = 40$
no solution

c. $7m - 3m = 32$
$m = 8$

d. $w + 11 = 9$
$w = -2$

9. When the replacements for the variable are 0, 1, 2, 3, 4, 5, 6, 7, 8, 9, and 10:

a. $n + 5 = 11$
$n = 6$

b. $a - 4 = 9$
no solution

c. $9x = 24$
no solution

d. $\dfrac{b}{2} = 6$
no solution

10. When the replacements for the variable are $-4, -3, -2, -1, 0, 1, 2, 3,$ and 4:

a. $r - 1 = 1$
$r = 2$

b. $9y = -45$
no solution

c. $\dfrac{w}{-5} = -1$
no solution

d. $3n + 6 = -3$
$n = -3$

11. When the replacements for the variable are all the negative integers greater than -11:

a. $x + 15 = 4$
no solution

b. $\dfrac{n}{-3} = -3$
no solution

c. $8x = -24$
$x = -3$

d. $5y - 7y = 14$
$y = -7$

Solving Equations with Absolute Values

OBJECTIVE: To solve equations with absolute values.

Since $|+4| = 4$ and $|-4| = 4$, replacement of the variable in the equation $|x| = 4$ by either $+4$ or -4 will make the sentence true.

Therefore the roots of $|x| = 4$ are $+4$ and -4, or its solutions are $+4$ and -4.

The equality axioms are used to solve equations such as the one shown below.

$$|x| + 5 = 9$$
$$|x| + 5 - 5 = 9 - 5$$
$$|x| = 4$$
$$x = +4 \text{ or } -4$$

Answers: $+4$ and -4

Basic: Prob. 1–7; *Average:* Prob. 1–9; *Enriched:* Prob. 1–10

PRACTICE PROBLEMS

Find the solutions of each of the following equations:

1. $|r| = 7$ $r = +7$ or -7

2. $|n| = 15$ $n = +15$ or -15

3. $|x| - 2 = 9$ $x = 11$ or -11

4. $|y| + 1 = 6$ $y = +5$ or -5

5. $9 \times |t| = 108$ $t = +12$ or -12

6. $|d| - 15 = 7$ $d = +22$ or -22

7. $6 \times |a| = 84$ $a = +14$ or -14

8. $|s| + 11 = 29$ $s = +18$ or -18

9. $\dfrac{|b|}{4} = 5$ $b = +20$ or -20

10. $\dfrac{|y|}{8} = 6$ $y = +48$ or -48

Using Formulas

OBJECTIVE: To use formulas.

To determine the value of any variable in a formula when the values of the other variables are known, (1) copy the formula, (2) substitute the given values for the variables, and (3) perform the necessary operations.

Then solve the resulting equation for the value of the required variable.

Find the value of w when $A = 54$ and $l = 9$, using the formula $A = lw$.

$$A = lw$$
$$54 = 9w$$
$$\frac{54}{9} = \frac{9w}{9}$$
$$6 = w$$
$$w = 6$$

Answer: $w = 6$

PRACTICE PROBLEMS

Find the value of:

1. s when $p = 18$. Formula: $p = 3\,s$ 6
2. r when $d = 56$. Formula: $d = 2\,r$ 28
3. w when $A = 96$ and $l = 6$. Formula: $A = lw$ 16
4. d when $c = 15.7$ and $\pi = 3.14$. Formula: $c = \pi d$ 5
5. t when $i = 140$, $p = 400$, and $r = 5\%$. Formula: $i = prt$ 7
6. h when $V = 6{,}280$, $\pi = 3.14$, and $r = 10$. Formula: $V = \pi r^2 h$ 20
7. b when $A = 72$ and $h = 18$. Formula: $A = bh$ 4
8. l when $A = 777$ and $w = 21$. Formula: $A = lw$ 37
9. p when $i = 108$, $r = 4\%$, and $t = 9$. Formula: $i = prt$ 300
10. w when $V = 1{,}536$, $l = 16$, and $h = 12$. Formula: $V = lwh$ 8
11. r when $A = 660$, $\pi = \frac{22}{7}$, and $h = 21$. Formula: $A = 2\,\pi rh$ 5
12. p when $A = 428$ and $i = 43$. Formula: $A = p + i$ 385
13. b when $p = 91$ and $e = 29$. Formula: $p = b + 2\,e$ 33
14. A when $C = 42$. Formula: $C = A - 273$ 315
15. A when $L = 3{,}500$ and $C = 9{,}300$. Formula: $L = A - C$ 12,800
16. l when $p = 162$ and $w = 34$. Formula: $p = 2\,l + 2\,w$ 47
17. n when $S = 900$. Formula: $S = 180\,n - 360$ 7
18. V when $v = 323$, $g = 32$, and $t = 9$. Formula: $v = V + gt$ 35
19. h when $A = 572$, $\pi = \frac{22}{7}$ and $d = 14$. Formula: $A = \pi dh + \frac{1}{2}\pi d^2$ 6
20. r when $A = 640$, $p = 500$, $t = 7$. Formula: $A = p + prt$ .04 or $\frac{1}{25}$
21. V when $P = 75$, $V' = 38$, $P' = 25$. Formula: $PV = P'V'$ $12\frac{2}{3}$
22. s when $m = 8$. Formula: $m = \dfrac{s}{60}$ 480
23. c when $d = 15$ and $\pi = 3.14$. Formula: $d = \dfrac{c}{\pi}$ 47.1
24. h when $V = 56$ and $B = 24$. Formula: $V = \dfrac{Bh}{3}$ 7
25. a when $A = 144$ and $b = 18$. Formula: $A = \dfrac{ab}{2}$ 16
26. d when $l = 51$, $a = 7$, and $n = 12$. Formula: $l = a + (n - 1)d$ 4
27. C when $F = 14$. Formula: $F = 1.8\,C + 32$ –10
28. t when $v = 219$, $V = 27$, and $g = 32$. Formula: $v = V + gt$ 6
29. p when $A = 560$, $r = .08$, and $t = 5$. Formula: $A = p + prt$ 400
30. I when $E = 70$, $r = 5$, and $R = 9$. Formula: $E = Ir + IR$ 5

Graphing an Equation in One Variable on the Number Line

OBJECTIVE: To graph an equation in one variable on the number line.

The graph of an equation in one variable on a number line is the collection of all points on the number line whose coordinates are the numbers forming the solutions to the equation. Also see Exercise 4-4.

The graph of an equation is the graph of its solutions.

To draw the graph of an equation in one variable on the number line, (1) solve the given equation to find its solutions, (2) locate the point or points on the number line whose coordinate or coordinates are numbers forming all the solutions to the given equation, and (3) indicate these points by solid or colored dots.

Where the solution is one number, the graph will consist of one point. Where there are two solutions, the graph will consist of two points; see illustration **4.** Observe in illustration **5** that the solution of the given equation is an infinite collection of numbers which is shown as the entire number line.

---

1. The solution of the equation $n + 2 = 8$ is 6. The graph of $n + 2 = 8$ is the point whose coordinate is 6.

2. The solution of the equation $7n + 6 = 41$ is 5. The graph of $7n + 6 = 41$ is the point whose coordinate is 5.

3. There is no solution of the equation $x = x + 3$. We have no points to draw the graph.

4. The solutions of $|x| = 2$ are $+2$ and -2. The graph of $|x| = 2$ is the points whose coordinates are $+2$ and -2.

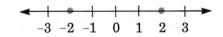

---

5. The solutions of $3x + 12 = 3(x + 4)$ are all the real numbers since the equation (identity) is satisfied by every real number. The graph of $3x + 12 = 3(x + 4)$ is the entire number line indicated by a heavy or colored line with an arrowhead in each direction to show that it is endless.

$$\longleftarrow \underset{-4\ -3\ -2\ -1\ \ \ 0\ \ \ 1\ \ \ 2\ \ \ 3\ \ \ 4}{+\ +\ +\ +\ +\ +\ +\ +\ +} \longrightarrow$$

6. The graph 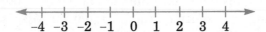 is a picture of
the equation $x = 3$ or any equivalent equation such as $4x = 12$. $x - 1 = 2$, $x + 2 = 5$, etc.

Basic: Prob. 1; *Average:* Prob. 1–2

PRACTICE PROBLEMS

For ex. 1 see Extended Answers, p. T83.

1. For each of the following equations draw an appropriate number line, then graph its solution. The replacements for the variable are all the real numbers.

a. $x = 5$ **e.** $8\,a = 32$ **i.** $5\,b - 6 = 29$
b. $b = -3$ **f.** $3\,y + 8 = 5$ **j.** $|x| = 3$
c. $x + 4 = 7$ **g.** $6\,n - n = 15$ **k.** $z = z + 2$
d. $a - 5 = 2$ **h.** $\dfrac{b}{3} = 2$ **l.** $5(x + 2) = 5\,x + 10$

2. Write a corresponding equation which is pictured by each of the following graphs:

a. $\longleftarrow \underset{-2\ -1\ \ \ 0\ \ \ 1\ \ \ 2\ \ \ 3\ \ \ 4}{+\ +\ +\ +\ \bullet\ +\ +} \longrightarrow$ $x = 2$ or any equivalent equation

b. $\longleftarrow \underset{-4\ -3\ -2\ -1\ \ \ 0\ \ \ 1\ \ \ 2}{+\ \bullet\ +\ +\ +\ +\ +} \longrightarrow$ $x = -3$ or any equivalent equation

c. $\longleftarrow \underset{-3\ -2\ -1\ \ \ 0\ \ \ 1\ \ \ 2\ \ \ 3}{+\ +\ +\ \bullet\ +\ +\ +} \longrightarrow$ $x = 0$ or any equivalent equation

d. $\longleftarrow \underset{-3\ -2\ -1\ \ \ 0\ \ \ 1\ \ \ 2\ \ \ 3}{+\ +\ \bullet\ +\ \bullet\ +\ +} \longrightarrow$ $|x| = 1$ or any equivalent equation

Fraction, Decimal, and Percent Number Relationships— Problem Solving by Equation

OBJECTIVE: To solve fraction, decimal, and percent number relationships by equation.

The equation may be used to find:

a. A fractional part or decimal part or percent of a number.

b. What fractional part or decimal part or percent one number is of another.

c. A number when a fractional part or decimal part or percent of it is known.

See the corresponding rows **a, b** and **c** in the box below.

Problem Solving—Equation Method

1. Read each problem carefully to find the facts which are related to the missing number.
2. Represent this unknown number by a variable, usually a letter.
3. Form an equation by translating two equal facts, with at least one containing the unknown, into algebraic expressions and writing one expression equal to the other.
4. Solve the equation.
5. Check answer directly with the facts of the given problem.

a.

Find $\frac{3}{4}$ of 20.

$$\frac{3}{4} \times 20 = n$$
$$15 = n$$

Answer: 15

Find .75 of 20.

$$.75 \times 20 = n$$
$$15 = n$$

Answer: 15

Find 75% of 20.

$$75\% \times 20 = n$$
$$.75 \times 20 = n$$
$$15 = n$$
$$n = 15$$

Answer: 15

b.

What fractional part of 20 is 15?

$$n \times 20 = 15$$
$$20\,n = 15$$
$$\frac{20\,n}{20} = \frac{15}{20}$$
$$n = \frac{3}{4}$$

Answer: $\frac{3}{4}$

What decimal part of 20 is 15?

$$n \times 20 = 15$$
$$20\,n = 15$$
$$\frac{20\,n}{20} = \frac{15}{20}$$
$$n = \frac{3}{4}$$
$$n = .75$$

Answer: .75

What percent of 20 is 15?

$$n\% \times 20 = 15$$
$$\frac{n}{100} \times 20 = 15$$
$$\frac{n}{5} = 15$$
$$n = 75$$

Answer: 75%

c.

$\frac{3}{4}$ of what number is 15?	.75 of what number is 15?	75% of what number is 15?
$\frac{3}{4} \times n = 15$	$.75 \times n = 15$	$75\% \times n = 15$
$\frac{3}{4} n = 15$	$.75\,n = 15$	$.75\,n = 15$
$\frac{4}{3} \cdot \frac{3}{4} n = \frac{4}{3} \cdot 15$	$\dfrac{.75\,n}{.75} = \dfrac{15}{.75}$	$\dfrac{.75\,n}{.75} = \dfrac{15}{.75}$
$n = 20$	$n = 20$	$n = 20$
Answer: 20	*Answer:* 20	*Answer:* 20

Basic: Prob. 1–7; *Average:* Prob. 1–9; *Enriched:* Prob. 1–10

PRACTICE PROBLEMS

By the equation method find the required number in each of these:

1. a. $\frac{1}{3}$ of 54 18
 b. $\frac{3}{8}$ of 72 27
 c. $\frac{2}{5}$ of 140 56
 d. $\frac{5}{6}$ of 258 215
 e. $\frac{7}{10}$ of 530 371

2. a. .4 of 7 2.8
 b. .75 of 92 69
 c. .36 of 250 90
 d. .625 of 496 310
 e. .19$\frac{3}{4}$ of 1,000 197.5

3. a. 8% of 50 4
 b. 60% of 18 10.8
 c. 25% of 392 98
 d. 140% of 475 665
 e. 87$\frac{1}{2}$% of 280 245

4. Express in fractions:
 a. What part of 25 is 15? $\frac{3}{5}$
 b. What part of 63 is 36? $\frac{4}{7}$
 c. 27 is what part of 81? $\frac{1}{3}$
 d. 45 is what part of 54? $\frac{5}{6}$

5. Express in decimals:
 a. What part of 10 is 3? .3
 b. What part of 25 is 17? .68
 c. 12 is what part of 16? .75
 d. 49 is what part of 56? .875

6. a. What percent of 50 is 9? 18%
 b. 30 is what percent of 48? 62$\frac{1}{2}$%
 c. 28 is what percent of 40? 70%
 d. What percent of 300 is 57? 19%

7. a. $\frac{1}{3}$ of what number is 29? 87
 b. $\frac{4}{5}$ of what number is 56? 70
 c. 63 is $\frac{7}{12}$ of what number? 108
 d. 1$\frac{3}{4}$ times what number is 84? 48

8. a. .06 of what number is 300? 5,000
 b. .375 of what number is 96? 256
 c. 54 is .9 of what number? 60
 d. 1.04 of what number is 364? 350

9. a. 25% of what number is 18? 72
 b. 8% of what number is 52? 650
 c. 37 is 10% of what number? 370
 d. 477 is 1.06% of what number? 45,000

10. a. If the school basketball team won 14 games, or $\frac{2}{3}$ of the games played, how many games were played? 21 games

 b. José received .54 of all votes cast in the election for school president. If he received 405 votes, how many students voted? 750 students

 c. If 817 students, or 95% of the school enrollment, were promoted to the next grade, how many students were enrolled in the school? 860 students

 d. There are 18 boys in a certain mathematics class. If this represents $\frac{3}{5}$ of the class enrollment, how many girls are in the class? 12 girls

VOCABULARY: ratio
equivalent ratios

OBJECTIVE: To find a ratio.

Two numbers may be compared by subtraction or by division. When we compare 12 to 4, we find by subtraction that 12 is eight more than 4 and by division that 12 is three times as large as 4. When we compare 4 with 12, we find by subtraction that 4 is eight less than 12 and by division that 4 is one-third as large as 12.

The answer we obtain when we compare two quantities *by division* is called the ratio of two quantities.

When 12 is compared to 4, 12 is divided by 4 as $\dfrac{12}{4}$ which equals $\dfrac{3}{1}$.

The ratio is $\dfrac{3}{1}$ or 3 to 1. This tells us that 12 is three times as large as 4.

When 4 is compared to 12, 4 is divided by 12 as $\dfrac{4}{12}$ which equals $\dfrac{1}{3}$.

The ratio is $\dfrac{1}{3}$ or 1 to 3. This tells us that 4 is one-third as large as 12.

A ratio has two terms, the number that is being compared (found in the numerator) and the number to which the first number is being compared (found in the denominator). The ratio in fraction form is usually expressed in lowest terms.

The ratio of 3 meters to 8 meters is $\dfrac{3}{8}$.

The fraction $\dfrac{3}{8}$ is an indicated division meaning $3 \div 8$.

Usually the ratio is expressed as a common fraction but it may also be expressed as a decimal fraction or a percent.

The ratio $\dfrac{3}{8}$ is sometimes written as 3:8. The colon may be used instead of the fraction bar. This ratio in either form is read 3 to 8 (three to eight).

Related Resources: *Practical Applications in Mathematics: Problems*, p. 84

If the quantities compared are denominate numbers, they must first be expressed in the same units. The ratio is an abstract number; it contains no unit of measurement.

The ratio of 8 millimeters to 2 centimeters is the ratio of 8 millimeters to 20 millimeters. Since $\dfrac{8}{20} = \dfrac{2}{5}$, the ratio is $\dfrac{2}{5}$ or 2 to 5, read "two to five."

A ratio may be used to express a rate. The average rate of speed of an automobile when it travels 400 kilometers in 5 hours may be expressed by the ratio $\dfrac{400}{5}$ or 80 km/h. If oranges sell at the rate of 6 for 59¢, the ratio expressing this rate is $\dfrac{6}{59}$ and is read "6 for 59."

When a ratio is used to express a rate, the two quantities have different names like kilometers and hours, oranges and cents, etc.

Ratios such as $\dfrac{2}{4}, \dfrac{6}{12}, \dfrac{9}{18}, \dfrac{10}{20}$, although written with different number pairs, express the same comparison or rate. In simplest form they are the same, the ratio $\dfrac{1}{2}$. These ratios are called equivalent ratios

Basic: Prob. 1–10; *Average:* Prob. 1–12; *Enriched:* Prob. 1–15

PRACTICE PROBLEMS

1. Use the fraction bar to express each of the following ratios:
   a. Ratio of five to eleven $\frac{5}{11}$ c. Ratio of sixteen to one $\frac{16}{1}$
   b. Ratio of nine to two $\frac{9}{2}$ d. Ratio of n to twelve $\frac{n}{12}$

2. Use the colon to express each of the following ratios:
   a. Ratio of six to nineteen 6:19   c. Ratio of one to eight 1:8
   b. Ratio of forty to seven 40:7   d. Ratio of x to twenty x:20

Find the ratio of:

3. a. 5 to 7 $\frac{5}{7}$ b. 8 to 15 $\frac{8}{15}$ c. 1 to 6 $\frac{1}{6}$ d. 4 to 11 $\frac{4}{11}$

4. a. 2 to 10 $\frac{1}{5}$ b. 6 to 8 $\frac{3}{4}$ c. 30 to 75 $\frac{2}{5}$ d. 72 to 96 $\frac{3}{4}$

5. a. 8 to 2 $\frac{4}{1}$ or 4   b. 32 to 8 $\frac{4}{1}$ or 4   c. 60 to 12 $\frac{5}{1}$ or 5   d. 75 to 25 $\frac{3}{1}$ or 3

6. a. 11 to 4   b. 7 to 3   c. 15 to 2   d. 20 to 9

7. a. 10 to 8 $\frac{5}{4}$ b. 16 to 6 $\frac{8}{3}$ c. 25 to 15 $\frac{5}{3}$ d. 70 to 21 $\frac{10}{3}$

8. a. n to 12 $\frac{n}{12}$ b. x to 100 $\frac{x}{100}$ c. 9 to n $\frac{9}{n}$ d. 25 to y $\frac{25}{y}$

9. a. 2 to 5 $\frac{2}{5}$ e. 5 to 4 $\frac{5}{4}$ i. 16 to 20 $\frac{4}{5}$ m. 1,000 to 25 $\frac{40}{1}$ or 40
   b. 4 to 24 $\frac{1}{6}$ f. 72 to 56 $\frac{9}{7}$ j. 25 to 85 $\frac{5}{17}$ n. 49 to 56 $\frac{7}{8}$
   c. 18 to 6 $\frac{3}{1}$ or 3 g. 27 to 81 $\frac{1}{3}$ k. 100 to 30 $\frac{10}{3}$ o. 84 to 28 $\frac{3}{1}$ or 3
   d. 14 to 21 $\frac{2}{3}$ h. 54 to 18 $\frac{3}{1}$ or 3 l. 49 to 63 $\frac{7}{9}$ p. 39 to 65 $\frac{3}{5}$

6.a. $\frac{11}{4}$ or $2\frac{3}{4}$ b. $\frac{7}{3}$ or $2\frac{1}{3}$ c. $\frac{15}{2}$ or $7\frac{1}{2}$ d. $\frac{20}{9}$ or $2\frac{2}{9}$

10. Find the ratio of:

a. 6 cm to 15 cm $\frac{2}{5}$ **i.** 18 things to 3 dozen $\frac{1}{2}$

b. 48 hr. to 20 hr. $\frac{48}{20}$ or $2\frac{2}{5}$ **j.** 45 sec. to 2 min. $\frac{3}{8}$

c. 18 mm to 4 mm $\frac{18}{4}$ or $4\frac{1}{2}$ **k.** A dime to a half dollar $\frac{1}{5}$

d. 3 things to 1 dozen $\frac{1}{4}$ **l.** 2 ft. to 5 yd. $\frac{2}{15}$

e. 4 in. to 10 in. $\frac{2}{5}$ **m.** 50 cL to 3 L $\frac{1}{6}$

f. 40 min. to 1 hr. $\frac{2}{3}$ **n.** 3 quarters to 4 nickels $\frac{75}{20}$ or $3\frac{3}{4}$

g. A dollar to a nickel $\frac{100}{5}$ or 20   **o.** 5 m to 250 mm $\frac{5,000}{250}$ or 20

h. 1 kg to 400 g $\frac{1,000}{400}$ or $2\frac{1}{2}$ **p.** 18 mo. to 6 yr. $\frac{1}{4}$

11. Express each of the following rates as a ratio:

    a. 264 kilometers in 3 hours $\frac{88}{1}$ or 88

    b. 5 apples for 57¢ $\frac{5}{57}$

    c. $25 for 2 shirts $\frac{25}{2}$

    d. 3,600 liters in 20 minutes $\frac{180}{1}$ or 180

    e. 384 miles on 24 gallons $\frac{16}{1}$ or 16

    f. 400 meters in 50 seconds $\frac{8}{1}$ or 8

    g. 9,000 revolutions in 5 min. $\frac{1,800}{1}$ or 1,800

    h. 4 tablets for 99¢ $\frac{4}{99}$

    i. $175 for 30 m² of linoleum $\frac{35}{6}$

    j. 800 feet in 16 seconds $\frac{50}{1}$ or 50

12. Which of the following are pairs of equivalent ratios?

a. $\frac{6}{8}$ and $\frac{4}{6}$ **d.** $\frac{15}{100}$ and $\frac{6}{40}$ **g.** $\frac{35}{42}$ and $\frac{15}{18}$

b. $\frac{4}{10}$ and $\frac{18}{45}$ **e.** $\frac{12}{15}$ and $\frac{28}{35}$ **h.** $\frac{27}{45}$ and $\frac{45}{72}$

c. $\frac{16}{24}$ and $\frac{10}{15}$ **f.** $\frac{20}{40}$ and $\frac{9}{24}$ **i.** $\frac{16}{28}$ and $\frac{36}{63}$

13. There are 18 girls and 12 boys in our class. What is the ratio of:

    a. boys to girls? $\frac{2}{3}$

    b. girls to boys? $\frac{3}{2}$

    c. girls to entire class? $\frac{3}{5}$

    d. boys to entire class? $\frac{2}{5}$

14. The aspect ratio is the ratio of the length of an airplane wing to its width. What is the aspect ratio of a wing 32 m long and 4 m wide? $\frac{32}{4}$

15. The compression ratio is the ratio of the total volume of a cylinder of an engine to its clearance volume. Find the compression ratio of a cylinder of an engine if its total volume is 1,200 cm³ and its clearance volume is 25 cm³. $\frac{48}{1}$ or 48

4-28 Proportions

OBJECTIVE: To solve proportions

A proportion is a mathematical sentence which states that two ratios are equivalent.

Using the equivalent ratios $\frac{6}{9}$ and $\frac{8}{12}$ we may write the proportion $\frac{6}{9} = \frac{8}{12}$. This proportion may also be expressed as $6:9 = 8:12$. The proportion in both forms is read "6 is to 9 as 8 is to 12" or "6 compared to 9 is the same as 8 compared to 12."

There are four terms in a proportion as shown:

$$\text{first} \rightarrow \frac{6}{9} = \frac{8}{12} \begin{array}{l} \leftarrow \text{third} \\ \leftarrow \text{fourth} \end{array}$$

$$\begin{array}{cc} \text{1st} \;\; \text{2nd} & \text{3rd} \;\; \text{4th} \\ \downarrow \;\;\; \downarrow & \downarrow \;\;\; \downarrow \\ 6 : 9 & 8 : 12 \end{array}$$

In the proportion $6:9 = 8:12$, the first term (6) and the fourth term (12) are called the extremes; the second term (9) and the third term (8) are called the means. Observe that the product of the extremes (6×12) is equal to the product of the means (9×8). In the form $\frac{6}{9} = \frac{8}{12}$ observe that the cross products are equal ($6 \times 12 = 9 \times 8$). These products are equal *only* when the ratios are equivalent. We can check whether two ratios are equivalent by either expressing each ratio in lowest terms or by determining whether the cross products are equal.

If any three of the four terms of the proportion are known quantities, the fourth may be determined. Use the idea that the product of the extremes is equal to the product of the means to transform the given proportion to a simpler equation.

Thus find the cross products by multiplying the numerator of the first fraction by the denominator of the second fraction and the numerator of the second fraction by the denominator of the first fraction. Write one product equal to the other, then solve the resulting equation.

Related Resources: *Practical Applications in Mathematics*: Problems, p. 153

Solve and check: $\dfrac{n}{32} = \dfrac{3}{4}$

Solution: *Check:*

$\dfrac{n}{32} = \dfrac{3}{4}$ $\dfrac{24}{32} = \dfrac{3}{4}$

$4 \times n = 32 \times 3$ $\dfrac{3}{4} = \dfrac{3}{4}$

$4\,n = 96$

$\dfrac{4\,n}{4} = \dfrac{96}{4}$

$n = 24$ *Answer: 24*

Often in the solution of problems there are four possible ways to write the proportion. *Corresponding terms must be matched.*

For the problem "Traveling at the rate of 225 kilometers in 3 hours, how far can you go in 8 hours?" the proportion may be written:

1. $\dfrac{n \text{ km}}{8 \text{ h}} = \dfrac{225 \text{ km}}{3 \text{ h}}$ which becomes $\dfrac{n}{8} = \dfrac{225}{3}$

2. $\dfrac{8 \text{ h}}{n \text{ km}} = \dfrac{3 \text{ h}}{225 \text{ km}}$ which becomes $\dfrac{8}{n} = \dfrac{3}{225}$

3. $\dfrac{n \text{ km}}{225 \text{ km}} = \dfrac{8 \text{ h}}{3 \text{ h}}$ which becomes $\dfrac{n}{225} = \dfrac{8}{3}$

4. $\dfrac{225 \text{ km}}{n \text{ km}} = \dfrac{3 \text{ h}}{8 \text{ h}}$ which becomes $\dfrac{225}{n} = \dfrac{3}{8}$

Note that the cross products in each of the above proportions will be the same: $3\,n = 1{,}800$ *Solution:* $3\,n = 1{,}800$

$n = 600$ kilometers

Answer: 600 kilometers

Basic: Prob. 1–12; *Average:* Prob. 1–14; *Enriched:* Prob. 1–17

PRACTICE PROBLEMS

Write each of the following as a proportion:

1. a. 26 compared to 13 is the same as 6 compared to 3. $\frac{26}{13} = \frac{6}{3}$

   b. 4 is to 12 as 5 is to 15. $\frac{4}{12} = \frac{5}{15}$

   c. 45 compared to 80 is the same as 18 compared to 32. $\frac{45}{80} = \frac{18}{32}$

   d. 56 is to 14 as 76 is to 19. $\frac{56}{14} = \frac{76}{19}$

   e. 150 compared to 100 is the same as 54 compared to 36. $\frac{150}{100} = \frac{54}{36}$

2. a. Some number x compares to 30 as 9 compares to 54. $\frac{x}{30} = \frac{9}{54}$

   b. Some number n is to 18 as 48 is to 72. $\frac{n}{18} = \frac{48}{72}$

   c. 24 is to y as 6 is to 11. $\frac{24}{y} = \frac{6}{11}$

   d. 35 compared to 500 is the same as n compared to 60. $\frac{35}{500} = \frac{n}{60}$

   e. 42 is to 63 as 12 is to x. $\frac{42}{63} = \frac{12}{x}$

3. Which of the following are true proportions?

   a. $\frac{9}{15} = \frac{4}{10}$ **b.** $\frac{2}{3} = \frac{3}{2}$ **c.** $\frac{15}{90} = \frac{2}{12}$ **d.** $18:14 = 27:21$

                                                      e. $8:20 = 30:100$

Solve and check:

4. $\frac{n}{15} = \frac{4}{5}$ $n = 12$ $\frac{t}{32} = \frac{3}{4}$ $t = 24$ $\frac{x}{9} = \frac{49}{63}$ $x = 7$ $\frac{n}{21} = \frac{5}{6}$ $n = 17.5$ $\frac{b}{5} = \frac{2}{3}$ $b = 3\frac{1}{3}$

5. $\frac{90}{54} = \frac{a}{3}$ $a = 5$ $\frac{12}{21} = \frac{y}{14}$ $y = 8$ $\frac{5}{6} = \frac{c}{78}$ $c = 65$ $\frac{18}{63} = \frac{r}{84}$ $r = 24$ $\frac{1}{6} = \frac{t}{5}$ $t = .833$ or $\frac{5}{6}$

6. $\frac{16}{x} = \frac{2}{7}$ $x = 56$ $\frac{24}{n} = \frac{108}{18}$ $n = 4$ $\frac{9}{n} = \frac{3}{38}$ $n = 114$ $\frac{4}{a} = \frac{10}{25}$ $a = 10$ $\frac{3}{b} = \frac{13}{6}$ $b = 1\frac{5}{13}$

7. $\frac{5}{8} = \frac{15}{a}$ $a = 24$ $\frac{6}{7} = \frac{54}{s}$ $s = 63$ $\frac{105}{35} = \frac{27}{n}$ $n = 9$ $\frac{81}{180} = \frac{9}{x}$ $x = 20$ $\frac{8}{15} = \frac{7}{d}$ $d = 13\frac{1}{8}$ or 13.125

8. $\frac{1.5}{.3} = \frac{a}{8}$ $\frac{.04}{.12} = \frac{.6}{n}$ $\frac{x}{.01} = \frac{.2}{.16}$ $\frac{.005}{c} = \frac{1.4}{.28}$ $\frac{17}{24} = \frac{17}{a}$

9. $\frac{y}{3\frac{1}{2}} = \frac{8}{7}$ $\frac{\frac{1}{2}}{10} = \frac{2\frac{1}{2}}{x}$ $\frac{24}{b} = \frac{2\frac{1}{3}}{5\frac{5}{6}}$ $\frac{n}{2\frac{1}{4}} = \frac{15}{\frac{3}{4}}$ $\frac{21}{x} = \frac{\frac{7}{10}}{\frac{3}{5}}$

10. At the rate of 3 items for 16¢, how many items can you buy for 80¢? 15 items

11. A recipe calls for 4 cups of flour to 6 tablespoons of shortening. How many tablespoons of shortening are needed for 6 cups of flour? 9

12. A tree casts a shadow of 15 meters when a 2-meter post nearby casts a shadow of 3 meters. Find the height of the tree. 10 m

13. If a scale distance of 3.5 centimeters on a map represents an actual distance of 175 kilometers, what actual distance does a scale distance of 5.7 centimeters represent? 285 km

14. Find the value of V when $V' = 9$, $T = 24$, and $T' = 6$. $V = 36$ Formula: $\dfrac{V}{V'} = \dfrac{T}{T'}$

15. Find the value of R_3 when $R_1 = 6$, $R_2 = 8$, and $R_4 = 32$. $R_3 = 24$ Formula: $\dfrac{R_1}{R_2} = \dfrac{R_3}{R_4}$

16. Find the value of W when $F = 60$, $h = 12$, and $d = 9$. $W = 45$ Formula: $\dfrac{F}{W} = \dfrac{h}{d}$

17. Find the value of P when $V = 45$, $V' = 15$, and $P' = 57$. $P = 19$ Formula: $\dfrac{V}{V'} = \dfrac{P'}{P}$

8. $a = 40$; $n = 1.8$; $x = .0125$; $c = .001$; $a = 24$
9. $y = 4$; $x = 50$; $b = 60$; $n = 45$; $x = 18$

OBJECTIVE: To solve inequalities in one variable.

To solve an inequality in one variable means to find all the numbers each of which, when substituted for the variable, will make the inequality a true sentence.

Study the solutions of the basic types of inequalities in their simplest forms, given below. Generally solutions are listed where convenient, especially when the replacements for the variable are restricted. See illustrations **1, 2, 4,** and **6.**

---

1. When the replacements for the variable are all the natural numbers, the solutions of $n < 4$ are the natural numbers less than 4.

Therefore, $n = 1, 2,$ or 3.

2. When the replacements for the variable are all the whole numbers, the solutions of $x > 11$ are the whole numbers greater than 11.

Therefore, $x = 12, 13, 14, \ldots$

3. When the replacements for the variable are all the real numbers, the solutions of $y \neq 6$ are all the real numbers except 6.

Therefore, $y =$ every real number except 6.

4. When the replacements for the variable are all the integers, the solutions of $b \geq -3$ are -3 and all integers greater than -3.

Therefore, $b = -3, -2, -1, 0, 1, 2, \ldots$

5. The solutions of $c \leq 5$ do not exist when the replacements for the variable are 6, 7, 8, 9, and 10.

Therefore, c has no solution.

6. When the replacements for the variable are all the integers, the solutions of $x \not> 2$ are the same as the solutions of $x \leq 2$ which are 2 and all the integers less than 2.

Therefore $x = \ldots, -2, -1, 0, 1, 2.$

7. When the replacements for the variable are all the real numbers, the solutions of $n \not< 4$ are the same as the solutions of $n > 4$ which are 4 and all the real numbers greater than 4.

Therefore, $n = 4$ or any real number greater than 4.

Inequalities, like equations, are solved by using transformations based on the axioms and on the additive and multiplicative inverses to get an equivalent inequality in each case which has the same solution as the given inequality. This simple equivalent inequality is expressed with only the variable itself as one member of the inequality and a numeral for the other member.

The axioms used to solve inequalities are separated into the following three parts.

I. When the same number is added to or subtracted from both members of an inequality, another inequality of the *same order* results.

Solve $n + 2 < 6$ when the replacements for the variable are all the whole numbers. $$n + 2 < 6$$ $$n + 2 - 2 < 6 - 2$$ $$n < 4$$ $$n = 0, 1, 2, \text{ or } 3$$ *Answer:* 0, 1, 2, or 3	Solve $x - 2 > 3$ when the replacements for the variable are all the real numbers. $$x - 2 > 3$$ $$x - 2 + 2 > 3 + 2$$ $$x > 5$$ $$x = \text{all real numbers}$$ $$\text{greater than 5}$$ *Answer:* all real numbers greater than 5

II. When both members of an inequality are either multiplied or divided by the same positive number, another inequality of the *same order* results.

Solve $3c < 27$ when the replacements for the variable are all the natural numbers. $$3c < 27$$ $$\frac{3c}{3} < \frac{27}{3}$$ $$c < 9$$ $$c = 1, 2, 3, \ldots, 8$$ *Answer:* 1, 2, 3, . . . , 8	Solve $4y + 7 \geq 7$ when the replacements for the variable are $-2, -1, 0, 1,$ and 2. $$4y + 7 \geq 7$$ $$4y + 7 - 7 \geq 7 - 7$$ $$4y \geq 0$$ $$y \geq 0$$ $$y = 0, 1, \text{ or } 2$$ *Answer:* 0, 1, or 2

III. When both members of an inequality are either multiplied or divided by the same negative number, an inequality of the *reverse order* results.

Solve $-5x > 30$ when the replacements for the variable are all the integers. $$-5x > 30$$ $$\frac{-5x}{-5} < \frac{30}{-5}$$ $$x < -6$$ $$x = \ldots, -9, -8, -7$$ *Answer:* $\ldots, -9, -8, -7$	Solve $-2b < -6$ when the replacements for the variable are all the integers. $$-2b < -6$$ $$\frac{-2b}{-2} > \frac{-6}{-2}$$ $$b > 3$$ $$b = 4, 5, 6, \ldots$$ *Answer:* $4, 5, 6, \ldots$

Basic: Prob. 1–8, 1–8, 1–7; *Average:* Prob. 1–10, 1–10, 1–10; *Enriched:* Prob. 1–11, 1–11, 1–10

PRACTICE PROBLEMS

Type I For ex. 1–11 see Extended Answers, pp. T83–T84.

Find the solutions of each of the following inequalities when the replacements for the variable are all the real numbers:

1. **a.** $n + 7 > 10$ **b.** $a + 6 > 6$ **c.** $23 + d > 12$
2. **a.** $c - 5 > 18$ **b.** $x - 3 > -4$ **c.** $r - \frac{1}{2} > 3\frac{1}{2}$
3. **a.** $b + 1 < 7$ **b.** $g + 9 < 2$ **c.** $m + .5 < 2.4$
4. **a.** $d - 9 < 6$ **b.** $t - 10 < 18$ **c.** $h - 3 < -5$
5. **a.** $m + 8 \geq 12$ **b.** $f + 17 \geq 0$ **c.** $w + 8 \not< -8$
6. **a.** $x - 11 \geq 4$ **b.** $k - 3.5 \geq 8.7$ **c.** $a - 7 \not< -4$
7. **a.** $y + 2 \leq 9$ **b.** $e + \frac{1}{4} \not> \frac{3}{4}$ **c.** $v + 16 \leq 13$
8. **a.** $r - 4 \leq 10$ **b.** $c - 6 \not> -8$ **c.** $x - 9 \leq 9$
9. **a.** $s + 9 \neq 17$ **b.** $n + .3 \neq 1.6$ **c.** $b + 12 \neq 1$
10. **a.** $w - 11 \neq 6$ **b.** $y - \frac{7}{8} \neq 1\frac{1}{2}$ **c.** $r - 2 \neq -2$
11. **a.** $x - 5 > 4$ **d.** $a - 9 \leq 11$ **g.** $r + 4 > 0$
    b. $y + 8 < 3$ **e.** $t + 3 \neq 3$ **h.** $z - 12 \not< 7$
    c. $n + 7 \geq 1$ **f.** $b - 5 < -9$ **i.** $w + 6 \leq -4$

Type II For ex. 1–11 see Extended Answers, pp. T84–T85.

Find the solutions of each of the following inequalities when the replacements for the variable are all the real numbers:

1. **a.** $7x > 14$ **b.** $9y > -72$ **c.** $8a > 2$

2. **a.** $\frac{m}{4} > 2$ **b.** $\frac{c}{10} > 7$ **c.** $\frac{r}{9} > -6$

3. a. $6\,z < 30$　　　**b.** $12\,s < -96$　　　**c.** $16\,b < 24$

4. a. $\dfrac{h}{8} < 3$　　　**b.** $\dfrac{p}{15} < -1$　　　**c.** $\dfrac{w}{7} < 5$

5. a. $10\,n \geq 50$　　　**b.** $18\,a \geq 12$　　　**c.** $\frac{3}{4}\,x \not< 24$

6. a. $\dfrac{b}{30} \geq 0$　　　**b.** $\dfrac{r}{4} \geq -6$　　　**c.** $\dfrac{z}{5} \not< 3$

7. a. $8\,s \leq 72$　　　**b.** $48\,h \not> 96$　　　**c.** $.05\,n \leq 20$

8. a. $\dfrac{y}{6} \leq 7$　　　**b.** $\dfrac{t}{14} \not> 2$　　　**c.** $\dfrac{b}{3} \leq -10$

9. a. $9\,c \neq 144$　　　**b.** $\frac{5}{8}\,x \neq 80$　　　**c.** $24\,f \neq 0$

10. a. $\dfrac{n}{12} \neq 5$　　　**b.** $\dfrac{d}{6} \neq -2$　　　**c.** $\dfrac{y}{10} \neq 9$

11. a. $15\,b < 60$　　　**d.** $9\,y \leq 39$　　　**g.** $\dfrac{n}{12} > 0$

　　b. $\dfrac{a}{8} > -2$　　　**e.** $6\,m \neq -48$　　　**h.** $\dfrac{z}{5} \not< 15$

　　c. $20\,x \geq -15$　　　**f.** $\dfrac{t}{7} < 7$　　　**i.** $\frac{7}{16}\,x \not> 56$

Type III For ex. 1–10 see Extended Answers, p. T85.

Find the solutions of each of the following inequalities when the replacements for the variable are all the real numbers:

1. a. $-5\,d > 30$　　　**b.** $-7\,c > 98$　　　**c.** $-.4\,h > 1.2$
2. a. $-4\,b > -28$　　　**b.** $-3\,m > -40$　　　**c.** $-\frac{3}{8}\,z > -15$
3. a. $-9\,x < 81$　　　**b.** $-16\,r < 96$　　　**c.** $-8\,a < 36$
4. a. $-2\,t < -16$　　　**b.** $-17\,y < -51$　　　**c.** $-1.5\,m < -9$
5. a. $-3\,y \geq 21$　　　**b.** $-10\,r \geq 0$　　　**c.** $-x \not< 2$
6. a. $-7\,n \geq -63$　　　**b.** $-\frac{4}{9}\,g \geq -72$　　　**c.** $-4\,t \not< -17$
7. a. $-9\,s \leq 180$　　　**b.** $-3\,y \not> 0$　　　**c.** $-18\,x \leq 90$
8. a. $-6\,z \leq -216$　　　**b.** $-7\,n \not> 63$　　　**c.** $-\frac{2}{3}\,y \leq -60$

9. a. $\dfrac{n}{-3} < 12$　　　**d.** $\dfrac{c}{-10} \geq 9$　　　**g.** $\dfrac{s}{-12} \not< -5$

　　b. $\dfrac{b}{-6} > -2$　　　**e.** $\dfrac{a}{-3} < -7$　　　**h.** $\dfrac{m}{-9} > 4$

　　c. $\dfrac{y}{-8} \leq 6$　　　**f.** $\dfrac{x}{-5} > 8$　　　**i.** $\dfrac{d}{-4} \not> -1$

10. a. $-7\,m < 56$　　　**d.** $-10\,a \geq 45$　　　**g.** $-15\,y > 105$
　　b. $-12\,z > -132$　　　**e.** $-3\,r < 0$　　　**h.** $-2\,n \not< -1$

　　c. $-4\,x \leq 92$　　　**f.** $\dfrac{b}{-7} > -8$　　　**i.** $-13\,g \not> -91$

Miscellaneous For ex. 1–2 see Extended Answers, pp. T85–T86.

Find the solutions of each of the following inequalities when the replacements for the variable are all the real numbers:

1. **a.** $2x + 7 < 15$ **e.** $8n + 3 \neq -61$ **i.** $9t + 6 < -57$
   b. $5a - 3 > -28$ **f.** $3y - 7 < 20$ **j.** $14z - 8 \not> 69$
   c. $6m + 1 \leq 13$ **g.** $10b + 5 > 0$ **k.** $7w + 15 > 113$
   d. $7s - 5 \geq 9$ **h.** $17r - 9 \not< 42$ **l.** $4x - 11 \neq -9$

2. **a.** $3x + 2x > 15$ **g.** $12b + 13b < -100$
   b. $9c - 3c < -72$ **h.** $4x - 8x \not< 36$
   c. $5d - 7d \geq 14$ **i.** $n - 2n \not> -5$
   d. $y + y \neq -30$ **j.** $10r + 3r \neq 65$
   e. $8m - 3m \leq 0$ **k.** $12w - 5w > -84$
   f. $6t - 11t > -85$ **l.** $7a - 15a < 144$

Restricted Replacements For ex. 1–9 see Extended Answers, p. T86.

Find the solutions of each of the following inequalities:

1. When replacements for the variable are 0, 1, 2, 3, 4, 5, 6, 7, 8, 9:

$x + 5 > 8$ $n - 6 < 1$ $2b \leq 6$ $-3t \not< -12$

2. When replacements for the variable are $-3, -2, -1, 0, 1, 2, 3$:

$b - 2 \neq 0$ $5n + 3 < 13$ $\dfrac{d}{6} > -1$ $2a - 5 \leq -1$

3. When replacements for the variable are all the prime numbers:

$12d - 4d < 56$ $m - 7 \not> 16$ $6c - c \neq 55$ $-2n > -8$

4. When the replacements for the variable are all the integers:

$d + 5 < -2$ $-7s \geq 35$ $\dfrac{y}{-4} \not> -5$ $4a - 7 \neq -15$

5. When the replacements are all the even whole numbers:

$6b < 45$ $2a - 5 > 9$ $8a - 2a \not> 18$ $\dfrac{d}{9} \not< 4$

6. When the replacements are all the non-positive integers:

$m - 1 > -6$ $-5x < -20$ $4t - 9 \not> -13$ $11y - y \geq -10$

7. When the replacements for the variable are all the integers greater than -3 and less than 4:

$-7r < -21$ $b - 8 > -4$ $6x - 12 \leq 0$ $9m - 4m \neq 10$

8. When the replacements are all the one-digit prime numbers:

$b - 2 > 1$ $10c < 20$ $\dfrac{b}{5} \neq 1$ $12x + 7 \geq 79$

9. When the replacements for the variable are all the multiples of 4 which are greater than 8 and less than 30:

$-5g < -60$ $2t - 13 > 15$ $4a - 16a \not< 144$ $\dfrac{n}{8} \not> 3$

Graphing on the Number Line an Inequality in One Variable

**OBJECTIVE: To graph on the number line
an inequality in one variable.**

The graph of an inequality in one variable is the collection of all points on the number line whose coordinates are the solutions of the given inequality.

We shall see below that the basic graph on a number line of an inequality in one variable in simplest form may be a half-line, a ray, a line segment, or an interval.

A line is a collection of points. A point on the line separates the line into two half-lines. The half-line extends indefinitely in one direction only and does not include the endpoint separating it from the other half-line. An open dot indicates the exclusion of this endpoint and the arrowhead indicates the direction and that the half-line is endless.

A ray is a half-line which includes one endpoint. A heavy solid or colored dot is used to indicate the inclusion of this endpoint.

A definite part of a line which includes its endpoints is called a line segment or segment. Heavy solid or colored dots are used to indicate the inclusion of the two endpoints.

A definite part of a line which excludes its endpoints is called an interval. Open dots are used to indicate that the two endpoints are excluded from the graph.

Study each of the following basic graphs of inequalities in simplest form. The real numbers are used as the replacements for the variables.

The graph of all real numbers greater than a given number is a half-line extending to the right along the number line.

Graph of $x > 2$

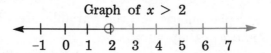

The graph of all real numbers less than a given number is a half-line extending to the left along the number line.

Graph of $x < 2$

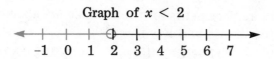

The graph of all real numbers greater than or equal to a given number is a ray extending to the right along the number line. A ray is a half-line which includes one endpoint. A solid dot indicates the inclusion of this endpoint.

Graph of $x \geq 2$

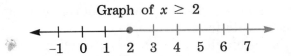

The graph of all real numbers less than or equal to a given number is a ray extending to the left along the number line.

Graph of $x \leq 2$

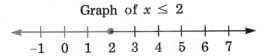

The graph of all real numbers excluding one number is the entire real number line except the point corresponding to the excluded number—or two half-lines.

Graph of $x \neq 1$

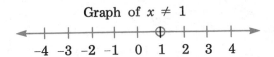

The graph of all real numbers between two given numbers and including the two given numbers is a line segment which is a definite part of a line including the two endpoints.

Graph of $-3 \leq x \leq 2$

The graph of all real numbers between the two given numbers but not including the two given numbers is an interval which is a definite part of a line excluding the two endpoints.

Graph of $-3 < x < 2$

To draw on a number line the graph of an inequality in one variable not in simplest form, first find the solutions of the inequality and then draw the graph of these solutions. Or generally when the given inequality is simplified, a basic type may be identified and its graph easily drawn.

> Draw the graph of $4x - 9 < 3$ when all the real numbers are used as replacements for the variable.
> Solving the inequality $4x - 9 < 3$, we find $x < 3$.
> The graph of $4x - 9 < 3$ is the graph of $x < 3$.
>
>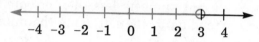
> -4 -3 -2 -1 0 1 2 3 4

Basic: Prob. 1; *Average:* Prob. 1–2; *Enriched:* Prob. 1–3

PRACTICE PROBLEMS   For ex. 1–3 see Extended Answers, p. T86.

1. For each of the following inequalities draw an appropriate number line, then graph its solutions. The replacements for the variable are all the real numbers.

a. $x > 1$ **f.** $d - 3 \geq 1$ **k.** $-7y \not< -7$

b. $y < -3$ **g.** $\dfrac{x}{2} \leq 2$ **l.** $10d - 6 < 24$

c. $a \geq -2$
d. $4c > 20$ **h.** $2x - 1 \neq 1$ **m.** $2b - 5b > -12$
                          i. $-9x > 18$ **n.** $4r + 9 < -7$

e. $a + 5 < 7$ **j.** $n + 2n \not> -12$ **o.** $\dfrac{m}{3} \not< -1$

2. Draw the graph for each of the following on a number line:
a. $-3 < x < 4$ **c.** $-1 \leq x < 3$ **e.** $2 \geq x \geq -5$
b. $0 \leq x \leq 5$ **d.** $6 > x > 2$ **f.** $-1 > x \geq -4$

3. Write a corresponding inequality which is pictured by each of the following graphs:

a.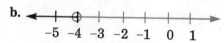
-3 -2 -1 0 1 2 3

d.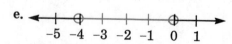
 1 2 3 4 5 6 7

b.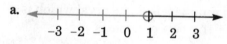
-5 -4 -3 -2 -1 0 1

e.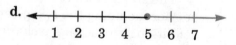
-5 -4 -3 -2 -1 0 1

c.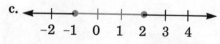
-2 -1 0 1 2 3 4

f.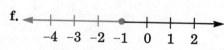
-4 -3 -2 -1 0 1 2

For a Test on Open Sentences in One Variable see p. T56 or *Testing Program,* p. 31.

REVIEW OF PART 4

The numeral in color following each problem indicates the Exercise where explanatory material may be found.

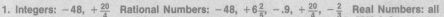
1. Integers: -48, $+\frac{20}{4}$ Rational Numbers: -48, $+6\frac{2}{5}$, $-.9$, $+\frac{20}{4}$, $-\frac{2}{3}$ Real Numbers: all

1. Which of the following numerals name integers? Which name rational numbers? Which name real numbers?   4-1
4-2

$$-48 \qquad +6\frac{2}{5} \qquad -.9 \qquad +\frac{20}{4} \qquad -\sqrt{31} \qquad -\frac{2}{3}$$

2. **a.** Draw a graph of -4, -3, 0, and 3 on a number line.   4-4
   b. Write the coordinates of which the following is the graph: $-6, -1, 0, 2, 4$

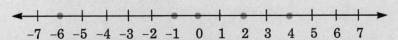

$$\begin{array}{ccccccccccccccc} -7 & -6 & -5 & -4 & -3 & -2 & -1 & 0 & 1 & 2 & 3 & 4 & 5 & 6 & 7 \end{array}$$

3. Which of the following are true?   4-6
   a. $-3 < -1$ **b.** $-10 > +6$ **c.** $+2 \not< -5$ **d.** $-4 \not> 0$

4. **a.** What is the opposite of -20? $+20$ 4-8
   b. What is the additive inverse of $+14$? -14
   c. Find the value of: $|-50|$ 50 4-3

5. Add: **a.** $+\ 8$ **b.** $-\ 4$ **c.** $(-6) + (+11) + (-7)$ -2 4-9
   $\underline{-10}$ $_{-2}$ $\underline{-5}$ $_{-9}$
   d. Simplify: $-6 + 8 - 3 - 2 + 7$ 4

6. Subtract: **a.** $-\ 2$ **b.** $-\ 6$ **c.** $(-9) - (-12)$ $+3$ 4-10
   $\underline{-9}$ $_{+7}$ $\underline{+6}$ $_{-12}$
   d. Take 9 from 5. -4

7. Multiply: **a.** $-\ 8$ **b.** $-\ 9$ **c.** $(-3)(+4)(-2)$ 24 4-11
   $\underline{+3}$ $_{-24}$ $\underline{-7}$ $_{63}$
   d. Find the value of: $(-3)^5$ -243

8. Divide: **a.** $\dfrac{-90}{-15}$ 6 **b.** $\dfrac{+32}{-8}$ -4 **c.** $(-54) \div (+9)$ -6 4-12

   d. Simplify: $\dfrac{6(-7+5) - 4(-1-1)}{-2(9-13)}$ $\frac{-1}{2}$

9. Express as a formula: 4-16

   The distance (d) a freely falling body drops is one half the product of the acceleration due to gravity (g) and the square of the time of falling (t). $d = \frac{1}{2}(gt^2)$

2.a.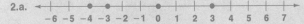
$$\begin{array}{cccccccccccccc} -6 & -5 & -4 & -3 & -2 & -1 & 0 & 1 & 2 & 3 & 4 & 5 & 6 & 7 \end{array}$$

10. Find the value of: 4-17

    a. $x - 4y$ when $x = 5$ and $y = 2$. -3
    b. $6r(r - s)$ when $r = 3$ and $s = -4$. 126
    c. $3m^2 + 2mn - n^2$ when $m = 4$ and $n = -2$. 28
    d. $\dfrac{b + c(b - c)}{(b + c)(b - c)}$ when $b = 10$ and $c = 5$. $\frac{7}{15}$

11. Find the value of A when $\pi = \frac{22}{7}$, $r = 14$, and $h = 10$, using   4-18
    the formula: $A = 2\pi r(h + r)$ A = 2, 112

12. Solve and check, using all the real numbers: 4-21

    a. $x - 6 = 11$ 17   **d.** $\dfrac{c}{14} = 10$ 140

    b. $n + 9 = 7$ -2   **e.** $8b - 19 = 13$ 4

    c. $-12y = 84$ -7   **f.** $7x + 8x = 75$ 5

13. **a.** Find the solution of $2n + 15 = 3$ when the replace-   4-22
       ments for the variable are all the even integers. n = -6
    b. Find the solutions of: $|x| - 5 = 0$ x = +5 or -5   4-23

14. Find the value of d when $l = 61$, $a = 7$, and $n = 9$, using   4-24
    the formula: $l = a + (n - 1)d$ d = 6.75

For ex. 15–20
see Extended
Answers,
pp. T86–T87.
15. On a number line draw the graph of each of the following   4-25
    equations when the replacements for the variable are all
    the real numbers:

    a. $n + 6 = 3$ **b.** $-2y = -8$ **c.** $5x - 7 = 8$

16. Solve by the equation method: 4-26

    If the school baseball team won 15 games, or $\frac{5}{8}$ of the
    games played, how many games were lost?

17. **a.** Find the ratio of 27 to 36.   4-27
    b. Solve and check: $\dfrac{x}{100} = \dfrac{18}{75}$ 4-28

18. Find the solutions of each of the following inequalities   4-29
    when the replacements for the variable are all the real
    numbers:
    a. $5n < 60$ **c.** $b + 2 \not< -3$ **e.** $-7y \not< 14$
    b. $b - 4 > 9$ **d.** $3x - 4 \neq -10$ **f.** $\dfrac{n}{6} \geq -5$

19. Find the solutions of $5n - 9 < 46$ when the replacements   4-29
    for the variable are all the prime numbers.

20. On a number line draw the graph of each of the following   4-30
    inequalities when the replacements for the variable are all
    the real numbers:

    a. $x > 4$ **c.** $r + 6 \leq 5$ **e.** $5m - 9m \not< 12$
    b. $2n < -10$ **d.** $3y - 5 \neq 4$ **f.** $4 > n > -1$

COMPETENCY CHECK TEST

Solve each problem and select the letter corresponding to your answer. The numerals in red boxes indicate Exercises where help may be found.

1. Which of the following names a negative number? `4-1`

    a. $+9$ b. 0 <u>c.</u> $-\frac{3}{4}$ d. $+4\frac{1}{6}$

2. Which of the following names an integer? `4-1`

    a. $-\sqrt{53}$ b. $+7\frac{2}{3}$ c. $-.89$ <u>d.</u> -3

3. Which number is greatest? `4-6`

    a. -6 b. 0 <u>c.</u> $+2$ d. -14

4. Which has the greatest opposite number? `4-6`

    <u>a.</u> -10 b. $+14$ c. 0 d. -2

5. The value of $|-16|$ is: `4-3`

    a. 4 b. -16 <u>c.</u> 16 d. -4

6. The coordinates of which this is the graph are: `4-4`

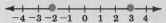

    a. $-3, 2$ b. $-2, 2$ <u>c.</u> $-2, 3$ d. Answer not given

7. The sum of $+2$ and -7 is: `4-9`

    a. -9 <u>b.</u> -5 c. $+9$ d. $+5$

8. The product of -8 and -3 is: `4-10`

    <u>a.</u> $+24$ b. -11 c. -24 d. -5

9. When -4 is subtracted from -2, the answer is: `4-11`

    a. -2 b. -6 <u>c.</u> $+2$ d. $+6$

10. When -18 is divided by $+6$, the quotient is: `4-12`

    a. -12 <u>b.</u> -3 c. $+3$ d. $-\frac{1}{3}$

11. The value of $-(-8)$ is: `4-8`

    a. -8 <u>b.</u> $+8$ c. 0 d. Answer not given

12. "Twice the length of the equal side (e) subtracted from the perimeter (p)" may be written as the algebraic expression: `4-14`

    a. $2e - p$ <u>b.</u> $p - 2e$ c. $e - 2p$ d. Answer not given

13. The sentence "The average (A) of two numbers m and n equals the sum of these two numbers divided by two" expressed as a formula is: `4-16`

    a. $A = 2(m + n)$ b. $A = \dfrac{2}{m + n}$ <u>c.</u> $A = \dfrac{m + n}{2}$

                         d. Answer not given   Continued on page 524.

14. The value $x^2 - 5y$ when $x = -6$ and $y = 2$ is: [4-17]

     a. -12 b. -46 c. -22 <u>d.</u> 26

15. The value of $\dfrac{r - s}{r + s}$ when $r = 10$ and $s = 5$ is: [4-17]

     a. 3   b. $\frac{1}{2}$ c. 2   <u>d.</u> $\frac{1}{3}$

16. The value of v when $V = 45$, $g = 32$, and $t = 5$, using the formula $v = V + gt$ is: [4-18]

     <u>a.</u> 205   b. 385   c. 82   d. Answer not given

17. Which of these equations has -3 as its solution? [4-19]

     a. $y - 6 = 9$ b. $-y = -3$ <u>c.</u> $y + 5 = 2$ d. $4y - 9 = 3$

18. Which of these equations is equivalent to $x = 4$? [4-19]

     a. $8x = 2$ <u>b.</u> $5x - 3 = 17$ c. $2x + 7x = 54$ d. $3x + 15 = -3$

19. The solution of $2n + 11 = 25$ is: [4-19] [4-21]

     a. 18   <u>b.</u> 7   c. 14   d. Answer not given

20. The value of C when $F = 68$, using the formula $F = 1.8C + 32$ is: [4-24]

     a. 2   b. 33.8   c. 36   <u>d.</u> 20

21. The solution of $\dfrac{5}{6} = \dfrac{n}{30}$ is: [4-28]

     a. 20   <u>b.</u> 25   c. 36   d. Answer not given

22. The amount of money that must be invested at the annual rate of 8% to earn \$10,000 per year is: [4-26]

     <u>a.</u> \$125,000   b. \$100,000   c. \$250,000   d. \$175,000

23. The regular price of a TV set that sells for \$423.30 at a 15% reduction sale is: [4-26]

     a. \$449   <u>b.</u> \$498   c. \$525.75   d. \$438.30

24. The solution of $n + 2 > 8$ when the replacements for the variable are all the one-digit prime numbers is: [4-29]

     a. 3 and 5   b. 7 and 9   c. 2   <u>d.</u> 7

25. The following is the graph of the inequality: [4-30]

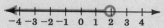

     a. $x > 2$ b. $x + 2 > 0$ <u>c.</u> $x < 2$ d. $x \geq 2$

Graphs, Statistics, and Probability

5

The numeral in color following each problem indicates the Exercise where explanatory material may be found.

1. Construct a bar graph showing the elevations of the following mountain peaks in the United States: **5-1**

Mount Rainier, 4,395 meters; Mount Washington, 1,918 meters; Mount McKinley, 6,198 meters; Mount Hood, 3,427 meters; and Guadalupe Peak, 2,669 meters.

2. Construct a line graph showing the average number of days per month that Seattle and New Orleans have precipitation. Use a line of one color for Seattle and a line of another color for New Orleans. **5-2**

	Jan.	Feb.	Mar.	Apr.	May	June	July	Aug.	Sept.	Oct.	Nov.	Dec.
Seat.	20	16	18	14	10	10	5	6	9	14	18	21
N. O.	10	10	9	7	7	10	15	13	10	6	7	10

3. Construct a circle graph showing the average annual outlays of the Federal government: **5-3**

Education, 5%; Health, 8%; Interest, 7%; National Defense, 33%; Income Security, 28%; Other, 19%

4. Make a frequency distribution table for the following list of scores: 72, 66, 75, 67, 73, 74, 69, 72, 68, 71, 70, 65, 66, 71, 68, 74, 65, 71, 66, 74, 70, 67, 75, 68, 74, 69, 68, 70, 72, 69, 74 **5-4**

5. Find the mean of the following scores: 37, 42, 39, 45, 41, 36, 47 41 **5-5**

6. Find the median of the following scores: 11, 15, 12, 18, 14, 13, 15, 17, 14, 16, 15, 11, 17, 12, 13 14 **5-5**

7. Find the mode of the following scores: 3, 5, 9, 6, 4, 8, 5, 2, 7, 9, 3, 5, 2, 6, 8, 4, 7, 3, 5, 6, 1, 7, 2, 5, 4 5 **5-5**

8. Find the range of the following scores: 42, 31, 57, 65, 48 34 **5-5**

9. What is Helen's percentile rank if she is ranked twelfth in her class of 40 students? 70th percentile **5-6**

10. If all the students above the upper quartile are to be admitted to a certain class without an entrance examination, who among the following would be admitted? **5-6**

A student with a percentile rank of: Students with scores 79, 85, and 76

a. 79 **b.** 85 **c.** 73 **d.** 76 **e.** 68

11. What is the probability of drawing at random on the first draw a yellow ball from a box containing 12 white balls and 18 yellow balls? $\frac{18}{30}$ or 60% **5-7**

12. What are the odds of selecting on the first draw a marked card from a hat holding 30 cards of which 12 are marked? **5-7**

2 to 3

GRAPHS

Bar Graph

OBJECTIVE: To use and construct bar graphs.

VOCABULARY: bar graph

The bar graph is used to compare the size of quantities, generally statistical information.

To construct a bar graph:

a. Draw a horizontal guide line on the bottom of the squared paper and a vertical guide line on the left.

b. Select a convenient scale for the numbers that are being compared, first rounding large numbers. For a vertical bar graph, write the number scale along the vertical guide line; for a horizontal bar graph, use the horizontal guide line. Label the scale.

c. Print names of the items opposite alternate squares along the other guide line. Label the items. Mark off for each item the height corresponding to the given number. Draw lines to complete the bars. All bars should have the same width. Select and print an appropriate title.

Basic: Prob. 1; *Average:* Prob. 1–2; *Enriched:* Prob. 1–3

PRACTICE PROBLEMS

1. Use the graph on the right to answer the following questions regarding the lengths of some foreign rivers.

a. How many miles does the side of a small square represent in the vertical scale of the bar graph? $\frac{1}{4}$ of the side of a square?

400 miles; 100 miles

b. Which river is the longest? The shortest?

Nile; Euphrates

c. Find the approximate length of each river.

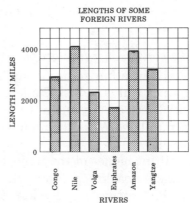

LENGTHS OF SOME FOREIGN RIVERS

LENGTH IN MILES

RIVERS

Congo 2,900 mi.; Nile 4,100; Volga 2,300; Euphrates 1,700 mi.; Amazon 3,900 mi., Tangtze 3,100 mi.

Related Resources: *Practical Applications in Mathematics:* Problems, p. 85

Construct bar graphs showing the following data:

2. Marilyn's final grades for the year:

English,	92;
Mathematics,	87;
Social Studies,	85;
Science,	90;
Art,	86;
Music,	82;
Physical and Health Education,	78.

3. The lengths of the channel spans of some famous suspension bridges in the United States:

Golden Gate,	4,200 ft.
Bear Mountain,	1,632 ft.
George Washington,	3,500 ft.
Delaware River,	1,750 ft.
Brooklyn,	1,595 ft.

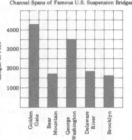

EXERCISE

5-2 | Line Graph

VOCABULARY: line graph

OBJECTIVE: To use and construct line graphs.

The line graph is used to show changes and the relationship between quantities.

To construct a line graph:

a. Draw a horizontal guide line on the bottom of the squared paper and a vertical guide on the left.

b. Select a convenient scale for the related numbers, first rounding large numbers. Write the number scale along one of the guide lines. Label the scale.

c. Print names of the items below the other guide line, using a separate line for each item. Label the items. On each of these lines mark with a dot the location of the value corresponding to the given number. Draw straight lines to connect successive dots. Select and print an appropriate title.

Related Resources: *Practical Applications in Mathematics:* Problems, p. 86

PRACTICE PROBLEMS

1. Use the line graph below to answer the following questions regarding lateness at the Wilson School.

  a. How many pupils late does the side of a small square indicate in the vertical scale of the line graph? $\frac{1}{2}$ of the side of a square? **2 pupils; 1 pupil**

  b. During what month was the punctuality poorest? best? **Feb.; June**

  c. How many pupils were late each month? **Sept., 6 pupils; Oct., 9 pupils; Nov., 13 pupils; Dec., 20 pupils; Jan., 21 pupils; Feb., 23 pupils; Mar., 15 pupils; Apr., 13 pupils; May, 7 pupils; June, 4 pupils**

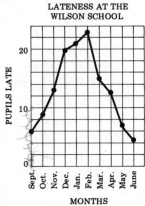

LATENESS AT THE
WILSON SCHOOL

PUPILS LATE

MONTHS

Construct line graphs showing the following data: **For ex. 2–3 see Extended Answers, p. T88.**

2. In twelve arithmetic progress tests, each containing 20 examples, George solved 8, 10, 13, 15, 12, 11, 14, 17, 16, 19, 18, and 20 examples correctly.

3. The monthly normal temperature (Fahrenheit) in some United States cities:

	Jan.	Feb.	Mar.	Apr.	May	June	July	Aug.	Sept.	Oct.	Nov.	Dec.
Atlanta	43	45	52	61	70	76	78	77	72	63	52	45
Chicago	24	26	35	47	58	67	72	72	65	54	40	29
Cincinnati	30	33	41	52	63	71	75	74	67	56	42	33
Los Angeles	55	56	58	59	62	66	70	71	69	65	61	57
New Orleans	54	57	63	69	75	81	82	82	79	71	62	56
New York	31	31	38	49	61	69	74	73	67	56	44	35
Philadelphia	33	34	41	52	63	71	76	75	68	58	46	36
St. Paul	13	16	29	46	58	67	72	69	61	49	32	19
Seattle	41	43	46	51	57	61	66	65	60	54	47	43

For a Class Activity see p. T33.

OBJECTIVE: To use and construct circle graphs.

The circle graph is used to show the relation of the parts to the whole and to each other.

To construct a circle graph:

a. Make a table showing: (1) given facts; (2) fractional part or percent each quantity is of the whole; (3) the number of degrees representing each fractional part or percent, obtained by multiplying 360° by the fraction or percent.

b. Draw a convenient circle. With a protractor construct successive central angles, using the number of degrees representing each part.

c. Label each sector (part of circle bounded by 2 radii). Select and print an appropriate title describing the graph.

Basic: Prob. 1–2; *Average:* Prob. 1–3; *Enriched:* Prob. 1–4

PRACTICE PROBLEMS

1. Use the graph on the right to answer the following questions regarding the family budget.

a. What fractional part of its entire income does the Harris family plan to spend for food? for shelter? $\frac{1}{4}; \frac{1}{5}$

b. Compare the percent to be spent for shelter to the percent of savings. Twice as much.

For ex. 1c–4 see Extended Answers, p. T89.

c. If the family income is $18,600 per year, what are the amounts to be spent for each item per year? per month? per week?

The Harris Family Budget

Construct circle graphs showing the following data:

2. A certain large city planned to spend its income as follows: schools, 30%; interest on debt, 25%; safety, 15%; health and welfare, 12%; public works, 10%; other services, 8%.

3. Charles spends his time as follows: 9 hours for sleep, 6 hours for school, 2 hours for study, 3 hours for recreation, and 4 hours for miscellaneous activities.

4. A mathematics test showed the following distribution of marks: A, 6 pupils; B, 8 pupils; C, 12 pupils; D, 10 pupils.

STATISTICS

Frequency Distribution; Histogram; Frequency Polygon

OBJECTIVE: To make frequency distributions, histograms, and frequency polygons.

VOC.: statistics   frequency distribution
frequency   frequency polygon
histogram

Statistics is the study of collecting, organizing, analyzing, and interpreting data. These data may be scores, measurements, or other numerical facts.

Data may be arranged in a tabular form. Thus test scores may first be tallied and then summarized in a table indicating the number of times each score occurs, called its frequency. Arrangement of data in this form is called a frequency distribution.

A frequency distribution can be pictured by a special bar graph called a histogram or by a special line graph called a frequency polygon.

A class of 34 pupils made the following scores on a test:
80, 95, 70, 80, 85, 65, 90, 75, 60, 90, 100, 65, 85, 75, 80, 95, 85, 65, 80, 70, 90, 70, 80, 85, 80, 60, 75, 80, 85, 95, 80, 70, 90, 85

FREQUENCY DISTRIBUTION TABLE

Score	Tally	Frequency
100	|	1
95	|||	3
90	||||	4
85	ⅢⅡ|	6
80	ⅢⅡ|||	8
75	|||	3
70	||||	4
65	|||	3
60	||	2
	Total	34

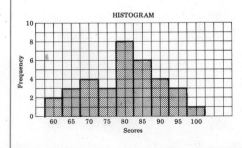

HISTOGRAM

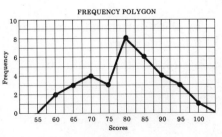

FREQUENCY POLYGON

531

PRACTICE PROBLEMS   For ex. 1–5 see Extended Answers, pp. T89–T91.

Make: **a.** A frequency distribution table
          b. A histogram
          c. A frequency polygon
for each of the following lists of scores:

1. 85, 60, 75, 40, 85, 70, 90, 100, 95, 85, 75, 80, 70, 70, 85, 90, 60, 65, 75, 80, 90, 75, 90, 80, 95

2. 9, 2, 7, 6, 4, 5, 8, 8, 6, 9, 7, 10, 5, 7, 2, 8, 3, 7, 4, 6, 5, 9, 1, 8, 7, 6, 9, 7, 5, 8

3. 45, 43, 44, 48, 45, 52, 47, 48, 46, 50, 45, 49, 49, 47, 51, 49, 52, 50, 43, 45, 48, 44, 47, 49, 45, 51, 50, 43, 45, 51, 46, 49, 50, 45, 44

4. 17, 14, 12, 15, 17, 19, 13, 12, 15, 14, 18, 11, 10, 14, 18, 16, 12, 15, 19, 14, 13, 17, 12, 11, 18, 14, 16, 19, 13, 18, 15, 15, 12, 17, 14, 13, 19, 14, 16, 14, 17

5. 6, 8, 4, 2, 3, 9, 6, 7, 5, 8, 4, 9, 1, 7, 4, 6, 9, 2, 5, 5, 8, 3, 10, 5, 2, 6, 7, 6, 8, 9, 5, 7, 4, 9, 2, 8, 8, 3, 5, 10, 6, 7, 4, 8, 9, 6, 8, 10, 8, 7

For a Class Activity see p. T35.

EXERCISE
5-5 | Averages— Measures of Central Tendency

OBJECTIVE: To find the mean, median, mode, and range of data.

VOCABULARY: average   arithmetic mean, mean
                    median   mode
                    range

An average is a measure of central tendency of data. Three commonly used averages are the arithmetic mean, the median, and the mode.

The arithmetic mean (or simply mean) of a group of numbers is determined by dividing the sum of the numbers by the number of items in the group.

Find the arithmetic mean of the following scores:	$\dfrac{78 + 83 + 91 + 82 + 86}{5}$
78, 83, 91, 82, and 86	$= \dfrac{420}{5}$
	$= 84$
	Answer: 84

To find the arithmetic mean by using the frequency distribution table, first multiply each score by its frequency and add these products. Then divide this sum by the number of scores.

The median of a group of numbers is the middle number when the numbers are arranged in order of size.
To determine the median count from either end. If the number of members is even, the median is determined by dividing the sum of the two middle numbers by 2.

The mode of a group of numbers is the number that occurs most frequently in the group. There may be more than one mode.
The range is the difference between the highest and the lowest numbers in the group of numbers.

Arrange the following group of scores in a frequency distribution table; then find the mean, median, mode, and range:
5, 7, 9, 8, 5, 8, 6, 5, 9, 8, 7, 8, 7, 5, 9, 5, 8, 5, 9

**Frequency Distribution
Table**

Score	Tally	Frequency
9	||||	4
8	⊬⊬⊤	5
7	|||	3
6	|	1
5	⊬⊬⊤|	6
	Total	19

Arithmetic Mean:

$9 \times 4 = 36$
$8 \times 5 = 40$
$7 \times 3 = 21$ $\frac{133}{19} = 7$
$6 \times 1 = 6$
$5 \times 6 = 30$
—————
Sum = 133

Arithmetic mean = 7

Median or Middle Score:

10th score from top or bottom = 7

Mode or most frequent score = 5

Range: $9 - 5 = 4$

PRACTICE PROBLEMS

1. Find the mean for each of the following lists of scores:

84 **a.** 83, 89, 80, 81, 84, 86, 85

7 **b.** 9, 5, 8, 4, 10, 3, 7, 8, 6, 7, 9, 9, 7, 3, 8, 9

46 **c.** 47, 43, 50, 48, 42, 47, 49, 43, 45, 47, 48, 46, 47, 44, 40, 44, 47, 49, 42, 44, 50, 47, 46, 49

27 **d.** 17, 24, 12, 28, 19, 16, 25, 28, 92, 17, 21, 18, 15, 24, 19, 22, 20, 26, 29, 16, 23, 25, 18, 14, 90, 23, 28, 86, 15, 19, 20, 18, 24

2. Find the median for each of the following lists of scores:

6 **a.** 8, 6, 7, 5, 9, 10, 3, 0, 5

4.6 **b.** 4.2, 5.6, 3.8, 4.7, 6.1, 4.5

80 **c.** 65, 85, 70, 95, 75, 80, 70, 85, 80, 85, 95, 60, 75, 80, 75, 95, 85, 60, 75, 90, 85

7 **d.** 9, 6, 8, 3, 4, 9, 2, 8, 3, 7, 7, 6, 8, 10, 4, 5, 9, 6, 7, 8, 8, 5, 6, 8, 10, 8, 5, 9, 6, 7, 6, 6, 9, 10, 10, 9, 5, 2, 6, 8, 7, 9, 4, 10

3. Find the mode for each of the following lists of scores:

15 **a.** 11, 15, 10, 19, 15, 12, 11, 17, 14, 15, 13, 12

53 and 58 **b.** 57, 49, 64, 53, 58, 55, 53, 60, 58, 52, 53, 64, 58, 49, 60

7 **c.** 5, 4, 9, 7, 6, 8, 6, 4, 5, 9, 7, 7, 5, 6, 9, 8, 3, 2, 5, 7, 4, 9, 3, 6, 2, 7, 8

45 **d.** 65, 25, 95, 80, 30, 85, 65, 30, 45, 60, 95, 70, 80, 95, 45, 90, 70, 45, 25, 85, 95, 75, 70, 45, 85, 70, 65, 80, 45, 60, 65, 80, 70, 45, 55

4. Find the range for each of the following lists of scores:

21 **a.** 31, 26, 29, 43, 34, 39, 47, 28, 45

55 **b.** 85, 90, 75, 65, 35, 80, 55, 85, 60, 75, 40, 80

15 **c.** 12, 7, 9, 15, 8, 4, 11, 6, 17, 9, 5, 10, 2, 16, 3

43 **d.** 23, 41, 37, 16, 27, 32, 19, 49, 36, 52, 28, 46, 59, 44, 27, 30

For ex. 5 see Extended Answers, p. T92.

5. Tally and arrange each of the following lists of scores in a frequency distribution table; then find the mean, median, and mode:

a. 6, 4, 8, 3, 7, 8, 6, 9, 8, 5, 7, 9, 7, 4, 9, 6, 7, 9, 8, 5, 8, 9, 8, 6, 9

b. 48, 47, 50, 43, 42, 47, 49, 43, 45, 47, 48, 46, 47, 44, 47, 44, 40, 49, 48

c. 18, 15, 12, 19, 14, 16, 14, 18, 13, 17, 16, 19, 16, 12, 14, 17, 15, 16, 20, 18, 11, 14, 19, 15, 17, 20, 19, 17, 18, 13, 14

d. 58, 62, 64, 59, 57, 61, 60, 59, 63, 58, 64, 60, 63, 59, 59, 56, 64, 61, 58, 62, 64, 60, 57, 59, 63, 55, 61, 56, 60, 59, 61, 58, 62, 59, 57, 62

e. 9, 6, 2, 5, 8, 3, 7, 6, 5, 9, 4, 4, 3, 1, 8, 10, 9, 6, 9, 4, 2, 5, 7, 3, 6, 8, 5, 9, 7, 2, 8, 4, 7, 5, 4, 6, 9, 8, 8, 3, 9, 10, 5, 7, 3, 9, 5

For a Class Activity see p. T35.

Percentiles and Quartiles

OBJECTIVE: To find
percentiles and quartiles.

VOCABULARY: percentile   quartile

Individual scores may be compared with all the other scores in a group by giving the score a positional standing or rank.

The percentile rank of a score indicates the percent of all the scores that are below this given score. If the rank of a particular score is the 60th percentile, it means that 60% of all the scores are lower than this score.

If Todd ranked fourth in a class of 16 students, there are 12 students of 16, or 75%, with a lower rank. He would have a percentile rank of 75 or a rank of the 75th percentile.

If Todd ranked fourth in a class of 40 students, there are 36 students of 40, or 90%, with a lower rank. He would have a percentile rank of 90 or a rank of the 90th percentile.

A quartile is the score at each point of division when an arranged group of scores (or other data) is divided into four equal parts. A percentile also refers to a score.

The upper quartile (or third quartile) is the score at the point below which 75% of all the scores fall. It corresponds to the 75th percentile.

The median (or second quartile) corresponds to the 50th percentile.

The lower quartile (or first quartile) is the score at the point below which 25% of all the scores fall. It corresponds to the 25th percentile.

Basic: Prob. 1–2; *Average:* Prob. 1–4; *Enriched:* Prob. 1–5

PRACTICE PROBLEMS

1. Find the percentile rank of each of the following persons:
   a. Elaine ranked sixth in a class of 20 students. 70th
   b. Angelina ranked fourteenth in a class of 35 students. 60th
   c. Mario ranked twenty-fourth in a class of 30 students. 20th
   d. Andrew ranked second in a class of 25 students. 92nd
   e. Kim ranked thirty-fifth in a class of 150 students. $76\frac{2}{3}$

2. Find the lower quartile, upper quartile, and 50th percentile for each of the following lists of scores:
   a. 1, 2, 4, 5, 6, 7, 9, 10 3; 8; 5.5
   b. 25, 26, 28, 29, 30, 31, 31, 34, 36, 37, 38, 40 28.5; 36.5; none
   c. 90, 92, 92, 94, 95, 95, 95, 96, 97, 97, 99, 99 93; none; none
   d. 65, 66, 66, 67, 67, 68, 68, 69, 69, 69, 70, 70, 70, 71, 71, 72 none; none; none
   e. 30, 30, 35, 35, 35, 40, 40, 40, 40, 45, 45, 45, 45, 50, 50, 55, 60, 60, 60, 70 36; 51; none

535

3. If all students above the upper quartile are exempt from writing a term paper, who among the following would not be exempt?

A student with a percentile rank of:

    a. 70   **b.** 76   **c.** 69   **d.** 74   **e.** 83

4. If the top 30 students of the graduating class of 250 students are to receive awards, who among the following would get an award?

A student whose percentile rank is:

    a. 87   **b.** 90   **c.** 85   **d.** 92   **e.** 89

5. In a class of 40 pupils Rosa has a percentile rank of 85. What is her rank in the class? Does she have a higher rank than Pablo who ranks tenth in this class? 6th; yes

---

EXERCISE

5-7 | Probability; Odds

OBJECTIVE: To find probability and odds.

VOCABULARY: probability   odds

In the study of probability, activities like tossing coins, drawing cards, and so on are used as experiments.

Tossing a coin so that it lands heads up is an example of an outcome, a particular way in which something happens. When a coin is tossed, there are two outcomes—heads and tails. If the desired result is heads, then there is only 1 favorable outcome—heads, and 1 unfavorable outcome—tails.

I. Probability

Probability is the numerical measure indicating the chance or likelihood for a particular event or outcome to occur. It is usually expressed as a ratio named by a common fraction or a decimal numeral or a percent.

> When a coin is tossed in the air, it is equally likely to land heads up as tails up.
>
> Thus the chance that this coin will land heads up is one out of two, and the probability ratio is $\frac{1}{2}$.
>
> This may be indicated by the notation: $P(H) = \frac{1}{2}$ which is read "the probability of heads is one half."

Probability is the ratio of favorable outcomes to the total number of possible outcomes.

> In a bag containing 12 marbles there are 12 possible outcomes when drawing a marble at random. If this bag contains 3 blue marbles, the probability of selecting a blue marble is three chances out of twelve or the ratio $\frac{3}{12}$ which is equivalent to $\frac{1}{4}$. This may be expressed as: P (blue marble) = $\frac{1}{4}$.

If there are no favorable outcomes, the probability is 0.
If all are favorable outcomes, the probability is 1.

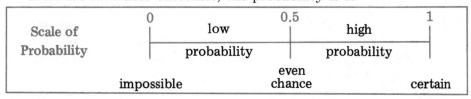

II. Odds

The odds for an event to happen against its failure to happen are the ratio of its favorable outcomes to unfavorable outcomes.

The odds against an event to happen are the ratio of its unfavorable outcomes to favorable outcomes.

> When there are 3 blue marbles and 9 marbles of other colors in a bag, the odds of selecting a blue marble are 3 to 9 or 1 to 3. The odds of *not* selecting a blue marble are 9 to 3 or 3 to 1.

Basic: Prob. 1–2; *Average:* Prob. 1–4; *Enriched:* Prob. 1–5

PRACTICE PROBLEMS

1. What is the probability of drawing at random on the first draw:
   a. A green ball from a box containing 9 balls of which 4 are green? $\frac{4}{9}$
   b. A marked card from a hat containing 15 cards of which 9 are marked? $\frac{9}{15}$ or $\frac{3}{5}$
   c. A red checker from a box containing 20 checkers of which 8 are red? $\frac{8}{20}$ or $\frac{2}{5}$
   d. A blue bead from a bag containing 48 beads of which 30 are blue? $\frac{30}{48}$ or $\frac{5}{8}$
   e. A yellow jelly bean from a bag containing 54 jelly beans of which 36 are yellow? $\frac{36}{54}$ or $\frac{2}{3}$

2. What is the probability of drawing at random on the first draw:

   a. A blue marble from a bag containing 7 blue marbles and 14 yellow marbles? $\frac{1}{3}$

   b. A black checker from a box containing 12 red checkers and 12 black checkers? $\frac{1}{2}$

   c. An orange gumdrop from a bag containing 18 green gumdrops and 24 orange gum drops? $\frac{4}{7}$

   d. A brown bead from a bag containing 42 brown beads and 28 purple beads? $\frac{3}{5}$

   e. A white ball from a box containing 56 yellow balls and 40 white balls? $\frac{5}{12}$

3. **a.** A die is a cube each of whose six faces is marked with a number of dots, from one through six. The plural of die is dice.

   When a die is rolled, what are the odds that the face with six dots will be up? $\frac{1}{5}$

   b. A hat contains 16 cards of which 6 are marked and the rest are blank. What are the odds of selecting at random on the first draw a blank card? $\frac{10}{6}$ or $\frac{5}{3}$

   c. A spinner has a circle with twelve sectors each of the same size numbered 1 through 12. What are the odds that the pointer on the first spin does not stop on a sector named by a number divisible by 4? $\frac{9}{3}$ or $\frac{3}{1}$

   d. What are the odds against selecting at random on the first try a yellow ball from a box containing 15 yellow balls and 30 white balls? $\frac{30}{15}$ or $\frac{2}{1}$

   e. What are the odds of selecting at random on the first try a nickel from your pocket when it contains 9 nickels and 12 quarters? $\frac{9}{12}$ or $\frac{3}{4}$

4. What is the probability that you will win the prize if you hold 50 raffle tickets of the 1,500 raffle tickets sold at a charity affair? What are the odds that you will not win the prize? $\frac{1}{30}$; 29 to 1

5. A standard deck of 52 playing cards consists of an ace, king, queen, jack, 10, 9, 8, 7, 6, 5, 4, 3, and 2 in four different suits: clubs, diamonds, hearts, and spades. Diamond and heart suits are in red; club and spade suits are in black. The king, queen, and jack cards are called picture cards.

   From a well-shuffled deck drawn at random,

   a. What is the probability that the first draw will be:

   (1) a red card? $\frac{26}{52}$ or $\frac{1}{2}$ (3) a king? $\frac{4}{52}$ or $\frac{1}{13}$

   (2) a picture card? $\frac{12}{52}$ or $\frac{3}{13}$ (4) the queen of spades? $\frac{1}{52}$

b. What is the probability that the first draw will not be:

(1) an ace? $\frac{48}{52}$ or $\frac{12}{13}$ (3) a picture card? $\frac{40}{52}$ or $\frac{10}{13}$

(2) a black card? $\frac{26}{52}$ or $\frac{1}{2}$ (4) a red 7? $\frac{50}{52}$ or $\frac{25}{26}$

c. What are the odds in favor of the first draw being:

(1) a black 10? $\frac{2}{50}$ or $\frac{1}{25}$ (3) the jack of diamonds? $\frac{1}{51}$

(2) a red card? $\frac{26}{26}$ or even (4) a picture card? $\frac{12}{40}$ or $\frac{3}{10}$

d. What are the odds against the first draw being:

(1) a black picture card? $\frac{46}{6}$ or $\frac{23}{3}$ (3) a queen? $\frac{48}{4}$ or 12 to 1

(2) the ace of clubs? $\frac{51}{1}$ or 51 to 1 (4) a spade card? $\frac{39}{13}$ or 3 to 1

For a Class
Activity see p. T35.
For a Test on Statistics and Probability see p. T58 or *Testing Program*, p. 36.

REVIEW OF PART 5 For ex. 1–3 see
Extended Answers, p. T93.

The numeral in color following each problem indicates the Exercise where explanatory material may be found.

1. Construct a circle graph showing how a family spends its monthly income of $1,800. It spends $540 for food, $360 for housing, $234 for clothing, $162 for transportation, $324 for miscellaneous expenses, and allows $180 for savings.   5-3

2. Make: **a.** A frequency distribution table
   b. A histogram
   c. A frequency polygon   5-4
   for the following list of scores: 6, 4, 5, 2, 1, 6, 8, 5, 2, 7, 3, 9, 5, 6, 7, 4, 2, 3, 8, 5, 1, 4, 7, 4, 6, 8, 5, 2, 8, 3, 5, 6

3. Tally and arrange the following list of scores in a frequency distribution table; then determine the mean, median, mode, and range: 19, 16, 18, 17, 14, 16, 17, 16, 19, 15, 20, 17, 16, 19, 18, 16, 14, 16, 19, 17, 19, 16, 18, 16, 17   5-5

4. **a.** Find the percentile rank of Carlos if he ranks seventh in a class of 28 students. 75th percentile   5-6
   b. Find the lower quartile, upper quartile, and 50th percentile for the following list of scores: 32, 34, 35, 37, 37, 38, 39, 40 35; 39; none

5. Suppose Mr. O'Leary has 8 one-dollar bills, 7 five-dollar bills, and 5 ten-dollar bills mixed in his wallet.   5-7
   a. What is the probability that he will select at random on the first try a one-dollar bill? $\frac{8}{20}$ or $\frac{2}{5}$
   b. What are the odds against his selecting a ten-dollar bill on the first try? $\frac{15}{5}$ or 3 to 1

COMPETENCY CHECK TEST

Solve each problem and select the letter corresponding to your answer. The numerals in red boxes indicate Exercises where help may be found.

1. The ratio of the number of *A* students to the number of *C* students is: 5-1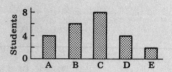
   a. $\frac{1}{4}$ b. $\frac{2}{1}$ <u>c.</u> $\frac{1}{2}$ d. $\frac{4}{1}$

2. The mean (average) of the scores: 18, 16, 24, 16, 16, 24 is: 5-5
   a. 16   b. 18   <u>c.</u> 19   d. 24

3. The median of the scores: 41, 43, 43, 44, 44, 47, 47, 47, 47 is: 5-5
   a. 41   b. 43   <u>c.</u> 44   d. 47

4. The mode of the scores: 9, 6, 5, 8, 4, 2, 6, 8, 5, 6, 7, 2 is: 5-5
   a. 9   b. 8   c. 5   <u>d.</u> 6

5. The range in the scores: 72, 59, 62, 85, 79, 67, 72, 62 is: 5-5
   a. 20   b. 13   <u>c.</u> 26   d. 29

6. The percentile rank of a student who ranks tenth in a class of 40 students is: 5-6
   a. 70   b. 80   <u>c.</u> 75   d. 85

7. The upper quartile for the following scores:
   55, 60, 65, 65, 70, 70, 70, 80, 80, 85, 85, 90 is: 5-6
   <u>a.</u> 85   b. 80   c. 70   d. 65

8. The probability of drawing at random on the first draw a green marble from a bag containing 24 green marbles and 32 yellow marbles is: 5-7
   a. $\frac{2}{3}$ b. $\frac{4}{7}$ c. $\frac{3}{4}$ <u>d.</u> $\frac{3}{7}$

9. The odds of drawing a blank card on the first draw from a hat containing 14 marked cards and 21 blank cards are: 5-7
   a. 2 to 5   b. 2 to 3   <u>c.</u> 3 to 2   d. 5 to 2

10. Two hundred fifty tickets for the door prize were sold at a charity affair. If you hold 5 of these tickets, the probability of your not winning the prize is: 5-7
    a. $\frac{99}{100}$ b. $\frac{9}{10}$ <u>c.</u> $\frac{49}{50}$ d. Answer not given

Everyday Applications– Problem Solving

6

INCOME, TAKE-HOME PAY, INCOME TAX

VOC.: salary
time-and-a-half
double time

OBJECTIVE: To compute income.

A person's income usually consists of earnings in the form of a salary or wages per hour, day, week, month, or year. Other types of income include commissions (see Exercise 6-25), profits (see Exercise 6-26), bonuses, fees, tips, interest (see Exercises 6-20 and 6-22), dividends, and pensions.

In some occupations wages are calculated at hourly rates of pay which vary depending on the job. In other occupations the employed persons have fixed salaries on a weekly, semi-monthly, or annual basis. A salary is a fixed sum of money paid for working a definite period of time.

Some people work where their earnings are based on the number of pieces or units of work that a person completes at a given pay rate per piece or unit. Salespersons work on commission or salary and commission. Sometimes bonuses are paid.

Repair persons and most professional people usually charge a fee for their work.

When a person works at an hourly rate more than 40 hours per week or more than 8 hours per day, time and a half is usually paid for the overtime hours and double time for all hours worked on holidays and Sundays.

To find the weekly wages when the hourly rate and the number of working hours are known, multiply the number of hours of work by the hourly rate, using the overtime rate for the time worked over 40 hours.

Basic: Prob. 1–3, 7–14, 19, 22–26, 31, 34–35, 38; *Average:* Prob. 1–5, 7–16, 19–20, 22–28, 31–32, 34–36, 38–39; *Enriched:* Prob. 1–40

6-1 COMPUTING INCOME   **543**

PRACTICE PROBLEMS

Hourly Wages

1. List several occupations in which wages are paid on an hourly basis. Answers will vary.

2. What is the weekly wage of a person who works 40 hours per week at each of the following rates?

   a. $4.10 $164   **b.** $7.95 $318   **c.** $12.80 $512   **d.** $5.65 $226   **e.** $10.75 $430

3. Find the weekly earnings of a person who works:

   a. 28 hours at $6.90 per hour $193.20
   b. 35 hours at $4.30 per hour $150.50
   c. 31 hours at $11.65 per hour $361.15
   d. 39 hours at $5.45 per hour $212.55
   e. $33\frac{1}{2}$ hours at $6.84 per hour $229.14
   f. $37\frac{1}{2}$ hours at $10.00 per hour $375.00
   g. $25\frac{1}{2}$ hours at $9.20 per hour $234.60
   h. $18\frac{1}{2}$ hours at $8.25 per hour $152.63

4. If a person is paid the overtime rate of $1\frac{1}{2}$ times the regular rate for all working time over 40 hours per week, find the weekly earnings of a person who works:

   a. 44 hours at $4.80 per hour $220.80
   b. 41 hours at $8.50 per hour $352.75
   c. 47 hours at $12.85 per hour $648.93
   d. $42\frac{1}{2}$ hours at $6.20 per hour $271.25
   e. 50 hours at $7.75 per hour $426.25
   f. $46\frac{1}{2}$ hours at $9.18 per hour $456.71
   g. $49\frac{1}{2}$ hours at $10.45 per hour $566.91
   h. 53 hours at $7.05 per hour $419.48

5. Find the weekly wages for each of the following persons:

| | Hours Worked | | | | | Total | Hourly | |
	M	T	W	T	F	Hours	Rate	Wages
Mr. A	8	7	8	8	8	39	$6.35	$247.65
Mr. B	7	7	8	7	7	36	$8.15	$293.40
Mrs. C	8	$6\frac{1}{2}$	8	4	7	33.5	$9.10	$304.85
Mr. D	5	5	7	$6\frac{1}{2}$	8	31.5	$5.75	$181.13
Ms. E	8	$4\frac{1}{2}$	$5\frac{1}{2}$	7	$7\frac{1}{2}$	32.5	$7.60	$247.00

6. For each of the following time cards, compute the number of hours worked each day and the total time for the week. Then find the amount of wages due each employee.

Time Card					
No. __25__ NAME: __LISA ROSS__					
WEEK ENDING _____					
DAY	IN	OUT	IN	OUT	HOURS
M.	8:00	12:00	1:00	5:00	
T.	8:30	12:00	1:00	5:00	
W.	8:00	11:00	12:30	5:00	
T.	9:00	12:30	1:00	5:30	
F.	8:00	1:00	2:00	5:00	
S.					
RATE __$8.35__ TOTAL HOURS __39__					
WAGES _____$325.65_____					

Time Card					
No. __57__ NAME: __PETER CHANG__					
WEEK ENDING _____					
DAY	IN	OUT	IN	OUT	HOURS
M.	9:00	12:00	12:30	5:00	
T.	8:30	11:30	12:00	4:00	
W.	8:30	12:00	1:30	6:00	
T.	8:00	12:30	1:00	4:00	
F.	8:30	12:00	1:00	5:30	
S.					
RATE __$5.40__ TOTAL HOURS __38__					
WAGES _____$205.20_____					

Weekly Salaries, Monthly Salaries, Annual Salaries

7. List several occupations in which salaries are paid on a: Answers will vary.
a. weekly basis   **b.** annual basis   **c.** monthly basis

8. Find the hourly rate of pay when a person's weekly earnings are:
a. $190 for 40 hours of work $4.75
b. $234 for 40 hours of work $5.85
c. $150.10 for 38 hours of work $3.95
d. $247.50 for 33 hours of work $7.50
e. $193.45 for $36\frac{1}{2}$ hours of work $5.30

9. Find the annual salary of a person earning each week:
a. $85   **b.** $425   **c.** $104.50   **d.** $257.50   **e.** $362.25

10. Find the annual salary of a person earning each month:
a. $700   **b.** $535   **c.** $1,000   **d.** $1,915   **e.** $2,450

11. Find the annual salary of a person earning:
a. $172.50 per week $8,970   **c.** $592.75 semi-monthly $14,226
b. $965 per month $11,580   **d.** $5.60 per hour, 40-hour week $11,648

9.a. $4,420   b. $22,100   c. $5,434   d. $13,390   e. $18,837
10.a. $8,400   b. $6,420   c. $12,000   d. $22,980   e. $29,400

12.a. $180 b. $350 c. $280 d. $133.50 e. $250 13.a. $675 b. $265 c. $1,110 d. $520.50
e. $875 14.a. $96 b. $216 c. $88.50 d. $173.08 e. $253.85 15.a. $300 b. $250
c. $525 d. $172.50 e. $425.75

6-1 COMPUTING INCOME   **545**

12. Find your weekly salary if your annual salary is:
   a. $9,360   **b.** $18,200   **c.** $14,560   **d.** $6,942   **e.** $13,000

13. Find your monthly salary if your annual salary is:
   a. $8,100   **b.** $3,180   **c.** $13,320   **d.** $6,246   **e.** $10,500

14. Find your weekly salary if your monthly salary is:
   a. $416   **b.** $936   **c.** $383.50   **d.** $750   **e.** $1,100

15. Find your semi-monthly salary if your annual salary is:
   a. $7,200   **b.** $6,000   **c.** $12,600   **d.** $4,140   **e.** $10,218

16. Find the weekly earnings of a person who works during the week:
   a. 5 days at $59 per day $295
   b. $2\frac{1}{2}$ days at $33 per day $82.50
   c. 4 days at $54.75 per day $219
   d. $3\frac{1}{2}$ days at $27.50 per day $96.25

17. Find the rate of pay per day when a person earns:
   a. $78 for 3 days' work $26
   b. $87.20 for 2 days' work $43.60
   c. $51 for $1\frac{1}{2}$ days' work $34
   d. $216 for $4\frac{1}{2}$ days' work $48

18. Which is a better wage:
   a. $180 per week or $750 per month?
   b. $15,400 per year or $1,250 per month?
   c. $7,700 per year or $145 per week?
   d. $510 semi-monthly or $240 per week?
   e. $225 per week or $11,500 per year?
   f. $1,250 per month or $300 per week?
   g. $175 per week or $9,000 per year?

Piece Work

19. A farm worker picked 49 bushels of apples per day at 68¢ a bushel. How much did he earn for the day? $33.32

20. A typist charges 95¢ per page and types on an average 40 pages per day. How much does the typist earn during a week of 5 workdays? $190.00

21. For each of the following persons, find the earnings for the week when the number of units and the rate per unit are as follows:

	Number of Units					Total Units	Unit Rate	Total Wages
	M	T	W	T	F			
CHARLOTTE	28	25	33	29	36	151	$.70	$105.70
MARILYN	32	41	37	45	39	194	$.65	$126.10
LEON	147	159	162	154	167	789	$.30	$236.70
RONALD	119	107	99	98	125	548	$.38	$208.24
ELAINE	206	195	189	214	179	983	$.13	$127.79

Commission and Bonuses
(See Exercise 6-25.)

22. At the rate of 5% commission on sales, how much will a salesperson receive for selling $3,827 worth of merchandise? $191.35

23. An auctioneer, charging 8% commission, sold goods amounting to $4,680. How much was the commission? $374.40

24. A real-estate agent sold a house for $37,400. If the agent charges 6% commission, how much money does she receive? $2,244

25. If an agent charges $4\frac{1}{2}$% commission for purchasing goods, how much will he receive for buying 890 boxes of honeydew melons at $4.40 a box? $176.22

26. If an insurance broker receives 18% of the premium as her commission, how much money is due her when the premium on a homeowners insurance policy is $183.50? $33.03

27. A salesman receives 3% commission on sales. How much did he earn during a week when he sold 4 refrigerators at $595.50 each, 3 television sets at $349.75 each, 7 electric irons at $15.95 each, 2 clothes dryers at $237.50 each, and 3 dishwashers at $297.50 each? $147.31

$350.66 **28.** Steve receives a weekly salary of $75 and 7% commission on sales. If his weekly sales are $3,938, what are his total earnings for the week?

29. Lisa receives a salary of $65 a week and a 3% bonus on all sales over the weekly quota of $1,250. If her sales for a certain week were $7,189, what were her total earnings for that week? $243.17

30. Scott receives a salary of $90 a week and a sliding scale bonus of 1% on all weekly sales over $3,000 up to and including $5,000, and 2% on all weekly sales over $5,000. What are his total weekly earnings when his sales for the week are $9,250? $195

Profit
(See Exercise 6-26.)

31. A storekeeper found at the end of the year that his sales for the year amounted to $74,256, the cost of the goods sold was $38,547, and his other business operating expenses were $17,695. What was his net profit for the year? $18,041

32. Sales for the entire year at the Miller Department Store amounted to $946,104.61. The cost of goods sold was $597,686.48 and the operating expenses were $158,574.75. Find the net profit. $189,843.38

33. Find the net profit in each of the following:

Sales	Cost of Goods Sold	Operating Expenses
a. $58,227	$36,912	$9,375 $11,940
b. $41,492	$23,748	$6,567 $11,177
c. $79,010.67	$52,826.75	$11,400.82 $14,783.10
d. $216,351.25	$161,695.40	$19,748.56 $34,907.29
e. $487,265.36	$293,869.73	$40,093.98 $153,301.65

Fees, Tips, Pensions

34. A TV repairman charges $4.50 to service a black-and-white TV set and $9.50 for a color set. What were his earnings for the day when he made 6 service calls for black-and-white sets and 5 service calls for color sets? $74.50

35. If a doctor sees 4 patients an hour and charges $15 an office visit, how much does she earn during a day when her office hours are from 2 P.M. to 5 P.M. and 7 P.M. to 9 P.M. and her appointment schedule is full? $300

36. If a bellhop averages $39 per day in tips, how much does he earn in tips for a week when he works 6 days? $234

37. If a retired worker receives a social security pension of $250 per month and a private employment pension of $310 per month, how much is her total monthly pension? Her annual pension? Her weekly pension? $560 $6,720 $129.23

Interest and Dividends
(See Exercises 6-20 and 6-22.)

38. How much interest should a person receive for the year if his savings account in the bank averages over the year:
   a. $2,000 and the bank pays $5\frac{1}{2}$% annual interest? $110
   b. $1,840 and the bank pays 6% annual interest? $110.40
   c. $900 and the bank pays $5\frac{3}{4}$% annual interest? $51.75
   d. $3,300 and the bank pays 6% annual interest? $198
   e. $10,000 and the bank pays 5.5% annual interest? $550

39. Sal's brother owns 100 shares of XYZ stock that pays a $37\frac{1}{2}$¢ dividend every three months. Find the total quarterly dividend and the total dividend for the year. $37.50; $150

40. Becky's mother has $25,000 invested at the rate of $10\frac{3}{4}$% per year. What is her annual income from this investment? $2,687.50

For Class Activities see p. T36.

6-2 Withholding Tax

OBJECTIVE: To compute withholding tax.   **VOC.: withholding tax**

Withholding tax is the amount of income tax that employers are required by the federal government to withhold from their employees' earnings. This money is forwarded to the Collector of Internal Revenue's offices, where it is entered on each employee's account.

The employer may use either the percentage method or the income tax withholding tables (page 549) to compute the amount of income tax to be deducted and withheld from a payment of wages to an employee.

Percentage Method

To compute the amount of weekly withholding tax for a married person by the percentage method, first find the difference between the total wage payment and the number of withholding allowances (exemptions claimed) times $14.40. Then use the following rates:

When this   Amount of
difference is: withholding tax:

Not over $61.......................... 0

Over—	But not over—		Of Excess over—
$61	—$105.......	15%	—$61
$105	—$223.......	$6.60, plus 18%	—$105
$223	—$278.......	$27.84, plus 22%	—$223
$278	—$355.......	$39.94, plus 25%	—$278
$355	—$432.......	$59.19, plus 28%	—$355
$432	—$509.......	$80.75, plus 32%	—$432
$509	....	$105.39, plus 36%	—$509

To compute *by the percentage method* the amount of income tax to be withheld each week when a married person's weekly wages are $245 and 4 allowances are claimed, do the following:

(1) Total wage payment		$245.00
(2) Amount of one allowance	$14.40	
(3) Number of allowances claimed	4	
(4) Line 2 multiplied by line 3		57.60
(5) Amount subject to withholding (line 1 minus line 4)		$187.40
(6) Tax to be withheld on $187.40, using above rates: Tax on first $105		$ 6.60
Tax on remainder $82.40 @ 18%		14.83
Total to be withheld		$ 21.43

By Table

To compute the amount of weekly withholding tax for a married person using the income tax withholding tables, locate the row indicating the weekly salary, then select the amount of tax from the column corresponding to the number of withholding allowances claimed.

> To compute *by using the table of withholding tax* the amount of income tax to be withheld each week when a married person's weekly wages are $178 and 2 allowances are claimed, first locate the row that reads "wages at least $170 but less than $180" and then use the column headed "2," meaning 2 allowances claimed, to obtain $14.10 as the weekly amount of withholding tax.

Table of withholding tax—For married persons—Weekly payroll periods

And the wages are—		And the number of withholding allowances claimed is—						
		0	1	2	3	4	5	6
At least	But less than	The amount of income tax to be withheld shall be—						
88	90	4.30	2.10	0	0	0	0	0
90	92	4.60	2.40	.20	0	0	0	0
92	94	4.90	2.70	.50	0	0	0	0
94	96	5.20	3.00	.80	0	0	0	0
96	98	5.50	3.30	1.10	0	0	0	0
98	100	5.80	3.60	1.40	0	0	0	0
100	105	6.30	4.10	2.00	0	0	0	0
105	110	7.10	4.90	2.70	.50	0	0	0
110	115	8.00	5.60	3.50	1.30	0	0	0
115	120	8.90	6.40	4.20	2.00	0	0	0
120	125	9.80	7.20	5.00	2.80	.60	0	0
125	130	10.70	8.10	5.70	3.50	1.40	0	0
130	135	11.60	9.00	6.50	4.30	2.10	0	0
135	140	12.50	9.90	7.30	5.00	2.90	.70	0
140	145	13.40	10.80	8.20	5.80	3.60	1.50	0
145	150	14.30	11.70	9.10	6.50	4.40	2.20	.10
150	160	15.70	13.10	10.50	7.90	5.50	3.30	1.20
160	170	17.50	14.90	12.30	9.70	7.10	4.80	2.70
170	180	19.30	16.70	14.10	11.50	8.90	6.30	4.20
180	190	21.10	18.50	15.90	13.30	10.70	8.10	5.70
190	200	22.90	20.30	17.70	15.10	12.50	9.90	7.30
200	210	24.70	22.10	19.50	16.90	14.30	11.70	9.10
210	220	26.50	23.90	21.30	18.70	16.10	13.50	10.90
220	230	28.40	25.70	23.10	20.50	17.90	15.30	12.70
230	240	30.60	27.50	24.90	22.30	19.70	17.10	14.50
240	250	32.80	29.60	26.70	24.10	21.50	18.90	16.30
250	260	35.00	31.80	28.60	25.90	23.30	20.70	18.10
260	270	37.20	34.00	30.80	27.70	25.10	22.50	19.90
270	280	39.40	36.20	33.00	29.80	26.90	24.30	21.70
280	290	41.80	38.40	35.20	32.00	28.90	26.10	23.50
290	300	44.30	40.70	37.40	34.20	31.10	27.90	25.30
300	310	46.80	43.20	39.60	36.40	33.30	30.10	27.10
310	320	49.30	45.70	42.10	38.60	35.50	32.30	29.10
320	330	51.80	48.20	44.60	41.00	37.70	34.50	31.30
330	340	54.30	50.70	47.10	43.50	39.90	36.70	33.50

PRACTICE PROBLEMS

1. Use the percentage method to compute the amount of income tax to be withheld each week if a married person's weekly wages are:
- **a.** $119 and 2 withholding allowances are claimed $4.38
- **b.** $97 and 1 withholding allowance is claimed $3.24
- **c.** $342 and 3 withholding allowances are claimed $45.14
- **d.** $260 and 4 withholding allowances are claimed $24.13
- **e.** $495 and 2 withholding allowances are claimed $91.69
- **f.** $82.50 and 1 withholding allowance is claimed $1.07
- **g.** $196.75 and 3 withholding allowances are claimed $15.34
- **h.** $439.50 and 5 withholding allowances are claimed $62.69
- **i.** $247.50 and 3 withholding allowances are claimed $24.47
- **j.** $385.75 and 2 withholding allowances are claimed $59.74

2. Use the table of withholding tax to compute the amount of income tax to be withheld each week if a married person's weekly wages are:
- **a.** $93 and 1 withholding allowance is claimed $2.70
- **b.** $110 and 3 withholding allowances are claimed $1.30
- **c.** $226 and 0 withholding allowance is claimed $28.40
- **d.** $314 and 2 withholding allowances are claimed $42.10
- **e.** $191 and 4 withholding allowances are claimed $12.50
- **f.** $112.50 and 5 withholding allowances are claimed 0
- **g.** $308.75 and 1 withholding allowance is claimed $43.20
- **h.** $211.25 and 6 withholding allowances are claimed $10.90
- **i.** $270.50 and 3 withholding allowances are claimed $29.80
- **j.** $162.25 and 2 withholding allowances are claimed $12.30

3. Use the table of withholding tax to compute the amount of income tax withheld for the year if a married person's weekly wages are:
- **a.** $98 and 1 withholding allowance is claimed $3.60
- **b.** $175 and 2 withholding allowances are claimed $14.10
- **c.** $312.50 and 5 withholding allowances are claimed $32.30
- **d.** $130.75 and 2 withholding allowances are claimed $6.50
- **e.** $274.25 and 3 withholding allowances are claimed $29.80

4. Use the percentage method to compute the amount of income tax withheld for the year if a married person's weekly wages are:
- **a.** $165 and 2 withholding allowances are claimed $12.22
- **b.** $220 and 3 withholding allowances are claimed $19.52
- **c.** $445 and 1 withholding allowance is claimed $80.36
- **d.** $187.75 and 4 withholding allowances are claimed $11.13
- **e.** $362.50 and 2 withholding allowances are claimed $53.87

OBJECTIVE: To compute social security tax.

Social Security

VOC.: social security tax

Social security tax is money deducted from employees' earnings to help pay old age pensions. Employees pay a tax at the rate of 6.7% of the first $32,400 of their wages. Employers pay the same amount, matching the amounts paid by the employees.

Basic: Prob. 1; *Average:* Prob. 1–2

PRACTICE PROBLEMS

1. Find the amount deducted from the employee's wages for social security tax if an employee earns each week:

a. $60 $4.02	**f.** $615 $41.21	**k.** $72.50 $4.86	**p.** $53.90 $3.61
b. $115 $7.71	**g.** $584 $39.13	**l.** $101.75 $6.82	**q.** $445.80 $29.87
c. $89 $5.96	**h.** $270 $6.82	**m.** $377.50 $25.29	**r.** $209.75 $14.05
d. $227 $15.21	**i.** $398 $29.87	**n.** $532.45 $35.68	**s.** $315.50 $21.14
e. $405 $27.14	**j.** $725 $134	**o.** $218.40 $14.63	**t.** $608.25 $40.75

2. Find the amount deducted from the employee's wages for social security tax for the given work period if an employee earns:

a. $395 semi-monthly $26.47

b. $2,000 monthly $134

c. $32,400 annually $2,170.80

d. $687.50 monthly $46.06

e. $14,925 annually $999.98

f. $581.75 semi-monthly $38.98

Related Resources: Practical Applications in Mathematics: Problems, p. 166

State or Local Income Tax— Wage Tax

OBJECTIVE: To compute state or local income tax.

VOC.: wage tax

There are a number of states and cities in the United States that have, in addition to the federal income tax, another tax on earnings. This income tax is sometimes called a wage tax. It is usually expressed as a fixed rate of percent of the earnings.

Basic: Prob. 1–3; *Average:* Prob. 1–4; *Enriched:* Prob. 1–5

PRACTICE PROBLEMS

1. If the wage tax rate is 2% of the wages, find the tax a person owes for the week if he earns weekly:

a. $85 $1.70 **b.** $116 $2.32 **c.** $220 $4.40 **d.** $174.25 $3.49 **e.** $369.50 $7.39

2. If the wage tax rate is 3% of the wages, find the tax a person owes for the week if she earns weekly:

a. $68 $2.04 **b.** $245 $7.35 **c.** $407 $12.21 **d.** $191.75 $5.75 **e.** $382.25 $11.47

Related Resources: Practical Applications in Mathematics: Problems, pp. 167–168

3. If the wage tax rate is $2\frac{1}{2}\%$ of the wages, find the tax a person owes for the semi-monthly pay period if he earns semi-monthly:

    a. \$182 $4.55 **b.** \$634 $15.85 **c.** \$570 $14.25 **d.** \$245.50 $6.14 **e.** \$406.75 $10.17

4. If the wage tax rate is $1\frac{3}{4}\%$ of the wages, find the tax a person owes for the month if she earns monthly:

    a. \$900 $15.75**b.** \$1,385$24.24 **c.** \$1,025$17.94**d.** \$652.75$11.42**e.** \$1,774.50 $31.05

5. If the wage tax rate is $2\frac{1}{4}\%$ of the wages, find the tax a person owes for the year if his annual salary is:

    a. \$6,200   **b.** \$8,300   **c.** \$11,000   **d.** \$14,850   **e.** \$12,695
      $139.50 $186.75 $247.50 $334.13 $285.64

EXERCISE

6-5 | Take-Home Pay

OBJECTIVE: To compute take-home pay.   **VOC.: take-home pay**

Take-home pay is the amount of a person's earnings remaining after federal withholding tax, social security taxes, and wage tax, if any, are deducted from the person's wages. Sometimes union dues and health insurance costs are also deducted.

To determine the take-home pay, first find the federal withholding tax, the social security tax, and, if any, the wage tax. Then subtract the sum of these taxes from the amount of the person's wages.

To determine the take-home pay of a married person whose weekly wages are \$215 and 3 withholding allowances are claimed, do as follows:

Withholding Tax		
by Table:	\$18.70	\$215.00 wages
Social Security Tax		33.11 deductions
(6.7% of \$215):	\$14.41	\$181.89 take-home pay
Sum of Taxes*	\$33.11	

* When there is also a wage tax, add it to this sum.

                            Answer: \$181.89 take-home pay

Related Resources: *Practical Applications in Mathematics:* Problems, p. 166

PRACTICE PROBLEMS

In the following problems, use the table of withholding tax (page 549) for all wages between $88 and $340; for all other wages use the rates given on page 548. The social security tax rate is 6.7%.

1. Mrs. Brown earns $182 each week and claims 3 withholding allowances. How much federal income tax is withheld each week? How much is deducted for social security? What is her take-home pay? $13.30; $12.19; $157.54

2. Find the take-home pay for each of the following weekly wages:
a. $94 with 4 withholding allowances claimed $88.24
b. $105 with 2 withholding allowances claimed $95.86
c. $320 with 1 withholding allowance claimed $252.18
d. $180 with 3 withholding allowances claimed $155.67
e. $214 with 2 withholding allowances claimed $179.58
f. $99.75 with 5 withholding allowances claimed $93.64
g. $308.50 with 4 withholding allowances claimed $256.29
h. $191.60 with 3 withholding allowances claimed $164.75
i. $205.25 with 1 withholding allowance claimed $170.57
j. $145.50 with 2 withholding allowances claimed $127.48

3. Which take-home pay is greater, a weekly wage of $140.25 with 1 withholding allowance claimed or $129.75 with 5 withholding allowances claimed? $129.75

4. Mr. Wilson earns $217.50 each week and claims 2 withholding allowances. He works in a city where a tax of $1\frac{1}{2}\%$ is levied on wages. How much federal income tax is withheld each week? How much is deducted for social security? What is his weekly city wage tax? What is his take-home pay? $21.30; $14.57; $3.26; $179.61

5. Find the take-home pay for each of the following weekly wages when a wage tax is also levied:
a. $96 with 1 withholding allowance claimed; wage tax, 2% $84.90
b. $207 with 2 withholding allowances claimed; wage tax, 1% $172.74
c. $193 with 4 withholding allowances claimed; wage tax, $1\frac{1}{2}\%$ $165.77
d. $100 with 2 withholding allowances claimed; wage tax, $2\frac{1}{4}\%$ $89.62
e. $318 with 3 withholding allowances claimed; wage tax, $1\frac{3}{4}\%$ $254.34
f. $93.50 with 1 withholding allowance claimed; wage tax, $1\frac{5}{8}\%$ $83.55
g. $209.25 with 2 withholding allowances claimed; wage tax, 2% $172.74
h. $191.60 with 4 withholding allowances claimed; wage tax, $1\frac{1}{2}\%$ $164.48
i. $327.50 with 3 withholding allowances claimed; wage tax, $2\frac{1}{2}\%$ $258.24

For a Class Activity see p. T33.

OBJECTIVE: To compute federal income tax.

An income tax is a tax on earnings. Page 556 contains a part of the tax table used by a married person who files a joint return. For the rest of this tax table and other tax tables, use the Internal Revenue Service tax publications. Income tax rates are subject to change.

To find the federal income tax owed, select the appropriate table, locate the row indicating the taxable income (actually the adjusted gross income less itemized deductions), then read across to the column headed by the total number of exemptions claimed.

To determine the federal income tax owed by a married person with a taxable income of $10,775 who files a joint return and claims 3 exemptions:

Locate in the table the row that reads, "At least $10,750 but less than $10,800."

Read across to the column headed "3" meaning 3 exemptions to obtain $765 as the amount of income tax owed.

Answer: $765

Basic: Prob. 1–3; *Average:* Prob. 1–4; *Enriched:* Prob. 1–5

PRACTICE PROBLEMS

In these problems use the earnings given as the taxable income.
Use Tax Table B on page 556.

1. How much federal income tax is owed by a married person, filing a joint return, who earns annually:
  a. $9,475 and claims 4 exemptions? $357
  b. $10,250 and claims 2 exemptions? $799
  c. $10,700 and claims 3 exemptions? $748
  d. $9,600 and claims 4 exemptions? $374
  e. $10,975 and claims 5 exemptions? $450
  f. $9,320 and claims 2 exemptions? $646
  g. $11,045 and claims 3 exemptions? $808
  h. $9,916 and claims 4 exemptions? $433
  i. $10,460 and claims 2 exemptions? $842
  j. $10,989 and claims 6 exemptions? $287

2. Find the federal income tax owed for the year by a married person, filing a joint return, who earns weekly:

a. $185 and claims 2 exemptions $697
b. $177 and claims 4 exemptions $314
c. $192 and claims 2 exemptions $757
d. $210 and claims 3 exemptions $791
e. $183 and claims 5 exemptions $207
f. $181.50 and claims 2 exemptions $633
g. $189.75 and claims 6 exemptions $108
h. $190.25 and claims 4 exemptions $425
i. $206.80 and claims 3 exemptions $765
j. $214.50 and claims 5 exemptions $488

3. The Williams family consists of a husband, a wife, and two young children. How many exemptions may Mrs. Williams claim? If she earns $193.50 each week, how much federal income tax is withheld from her pay each week? Use the table on page 549. How much federal income tax is withheld from her pay for the year? How much federal income tax does she owe for the entire year? Use the table on page 556. Does she still owe some taxes or did she overpay? How much?
4; $12.50; $650; $459; overpaid by $191

4. Mr. Bell earns $214.75 each week and claims his wife and himself as exemptions. Determine how much federal income tax is withheld from his pay for the year and how much tax he owes for the entire year. Does he still owe some taxes or did he overpay? How much?
$1,107.60 is withheld; overpaid $146.60

5. Using the income tax table on page 556 and the table of withholding tax on page 549, find the balance of the federal income tax due or the amount of overpayment for the year when a married person, filing a joint return, earns weekly:

a. $180 and claims 3 exemptions overpaid $189.60
b. $200 and claims 4 exemptions overpaid $230.00
c. $212 and claims 5 exemptions overpaid $243.00
d. $196 and claims 2 exemptions overpaid $129.40
e. $208 and claims 6 exemptions overpaid $212.20
f. $186.50 and claims 2 exemptions overpaid $120.80
g. $201.75 and claims 4 exemptions overpaid $211.60
h. $192.25 and claims 3 exemptions overpaid $170.20
i. $212.50 and claims 5 exemptions overpaid $243.00
j. $198.75 and claims 4 exemptions overpaid $146.00

For a Test on Income, Take-Home Pay, Income Tax see p. T59 or *Testing Program*, p. 38.

Tax Table B—Married Filing Joint Return

If taxable income is—		And the total number of exemptions claimed on line 7 is—							
Over	But not over	2	3	4	5	6	7	8	9
		Your tax is—							
9,200	9,250	629	477	314	159	9	0	0	0
9,250	9,300	638	485	323	167	16	0	0	0
9,300	9,350	646	494	331	175	24	0	0	0
9,350	9,400	655	502	340	183	31	0	0	0
9,400	9,450	663	511	348	191	39	0	0	0
9,450	9,500	672	520	357	199	46	0	0	0
9,500	9,550	680	529	365	207	54	0	0	0
9,550	9,600	689	539	374	215	61	0	0	0
9,600	9,650	697	548	382	223	69	0	0	0
9,650	9,700	706	558	391	231	76	0	0	0
9,700	9,750	714	567	399	239	84	0	0	0
9,750	9,800	723	577	408	247	92	0	0	0
9,800	9,850	731	586	416	255	100	0	0	0
9,850	9,900	740	596	425	263	108	0	0	0
9,900	9,950	748	605	433	271	116	0	0	0
9,950	10,000	757	615	442	279	124	0	0	0
10,000	10,050	765	624	450	288	132	0	0	0
10,050	10,100	774	634	459	296	140	0	0	0
10,100	10,150	782	643	467	305	148	0	0	0
10,150	10,200	791	653	476	313	156	4	0	0
10,200	10,250	799	662	485	322	164	11	0	0
10,250	10,300	808	672	494	330	172	19	0	0
10,300	10,350	816	681	504	339	180	26	0	0
10,350	10,400	825	691	513	347	188	34	0	0
10,400	10,450	833	700	523	356	196	41	0	0
10,450	10,500	842	710	532	364	204	49	0	0
10,500	10,550	850	719	542	373	212	57	0	0
10,550	10,600	859	729	551	381	220	65	0	0
10,600	10,650	867	738	561	390	228	73	0	0
10,650	10,700	876	748	570	398	236	81	0	0
10,700	10,750	884	757	580	407	244	89	0	0
10,750	10,800	893	765	589	415	253	97	0	0
10,800	10,850	901	774	599	424	261	105	0	0
10,850	10,900	910	782	608	432	270	113	0	0
10,900	10,950	918	791	618	441	278	121	0	0
10,950	11,000	927	799	627	450	287	129	0	0
11,000	11,050	935	808	637	459	295	137	0	0
11,050	11,100	944	816	646	469	304	145	0	0
11,100	11,150	952	825	656	478	312	153	0	0
11,150	11,200	961	833	665	488	321	161	6	0

SPENDING MONEY

Buying Food

OBJECTIVE: To find the cost of food.

When the purchase price of a food item is computed and the result contains any fractional part of a cent, this fraction is dropped and one cent is added to the price. For example, when 3 bars of soap sell for 67¢, the purchase price of 1 bar of soap is 23¢.

The ¢ symbol indicates cents. For easier computation you may replace the ¢ symbol with the dollar mark ($) and a decimal point. Thus, 95¢ would be written as $.95 and 36.8¢ as $.368.

Basic: Prob. 1–3; *Average:* Prob. 1–5; *Enriched:* Prob. 1–6

PRACTICE PROBLEMS

Grocery Price List	
Beverages, 2 bottles for 87¢	Peas, 6 cans for $1.98
Coffee, $2.29 a pound	Soap, 4 bars for 85¢
Corn, 5 cans for $1.49	Soups, 2 cans for 69¢
Flour, 10 lb. for $1.85	Sugar, 5 lb. for $1.89
Laundry detergent, $1.45 a pkg.	Tomato juice, 3 cans for $1.29
Peaches, 2 cans for $1.29	Tuna fish, 2 cans for $2.25

1. Find the cost of:

a. 1 can of corn $.30
b. 1 can of peaches $.65
c. 1 can of soup $.35
d. 2 pkgs. of detergent $2.90
e. 8 bars of soap $1.70
f. 6 cans of tuna fish $6.75
g. 10 cans of corn $2.98
h. 3 cans of peas $.99

i. 2 bars of soap $.43
j. 1 bottle of beverage $.44
k. 20 lb. of flour $3.70
l. 12 cans of soup $4.14
m. 2 cans of tomato juice $.86
n. 4 cans of corn $1.20
o. 1 can of tuna fish $1.13
p. 5 bottles beverage $2.18

Related Resources: *Practical Applications in Mathematics:* Problems, p. 170

557

Bakery Price List

Assorted cakes, $1.92 a dozen   Cookies, $2.40 a pound
Bread, 2 loaves for 99¢   Rolls, 6 for 78¢

---

2. Find the cost of:

a. 1 loaf of bread $.50

b. 1 cake $.16

c. 1 roll $.13

d. $\frac{1}{2}$ doz. cakes $.96

e. $\frac{3}{4}$ lb. cookies $1.80

f. $1\frac{1}{2}$ doz. rolls $2.34

g. 6 oz. cookies $.90

h. 3 loaves of bread $1.49

i. 16 cakes $2.55

j. 19 oz. cookies $2.85

Meats, Fish, and Dairy Price List

Bacon, $1.69 a pound   Bluefish, 1.79¢ a pound
Margarine, 96¢ a pound   Lamb, $2.24 a pound
Cheese, $2.56 a pound   Milk, 2 qt. for $1.25
Chicken, 95¢ a pound   Rib Roast, $2.80 a pound
Eggs, 89¢ a dozen   Sirloin, $2.56 a pound
Flounder, $1.92 a pound   Smoked ham, $1.76 a pound

---

3. Find the cost of:

a. 2 doz. eggs $1.78

b. 1 qt. milk $1.12

c. $1\frac{1}{2}$ lb. margarine $1.44

d. $2\frac{3}{4}$ lb. flounder $5.28

e. $3\frac{1}{4}$ lb. rib roast $9.10

f. $\frac{3}{4}$ lb. smoked ham $1.32

g. $3\frac{1}{2}$ lb. chicken $3.33

h. $2\frac{1}{4}$ lb. bluefish $4.03

i. 6 oz. cheese $.96

j. 11 oz. margarine $.66

k. 3 lb. 5 oz. sirloin $8.48

l. 4 lb. 14 oz. rib roast $13.65

m. 2 lb. 10 oz. lamb $5.88

n. $1\frac{1}{2}$ doz. eggs $1.34

Fruit and Produce Price List

Apples, 49¢ a pound   Carrots, 2 bunches for 51¢
Bananas, 32¢ a pound   Celery, 59¢ a stalk
Grapefruit, 3 for 69¢   Lettuce, 2 heads for 95¢
Grapes, 89¢ a pound   Onions, 3 lb. for 81¢
Lemons, 6 for 84¢   Peppers, 3 for 35¢
Oranges, 98¢ a dozen   Potatoes, 10 lb. for $1.49
Peaches, 39¢ a pound   String beans, 40¢ a pound
Pears, 48¢ a pound   Tomatoes, 59¢ a pound

4. Find the cost of:

a. 2 lb. string beans $.80

b. 1 lb. onions $.27

c. 1 head lettuce $.48

d. 1 bunch carrots $.26

e. $\frac{3}{4}$ lb. pears $.36

f. $1\frac{1}{2}$ lb. tomatoes $.89

g. 5 lb. potatoes $.75

h. 1 pepper $.12

i. $2\frac{1}{2}$ doz. oranges $2.45

j. $1\frac{1}{4}$ lb. grapes $1.11

k. 3 lemons $.42

l. 6 grapefruit $1.38

m. 2 lb. 5 oz. bananas $.74

n. 1 lb. 11 oz. pears $.81

o. 4 lb. 6 oz. apples $2.15

p. 3 oranges $.25

5. Using the price lists given in problems 1 through 4, find the cost of each of the following orders:

a. 1 loaf bread, 1 doz. eggs, $\frac{1}{2}$ lb. margarine, 10 lb. flour, 1 lb. coffee $6.01

b. $1\frac{1}{2}$ doz. oranges, 8 oz. margarine, 2 lb. 3 oz. flounder, $4\frac{1}{2}$ lb. chicken, 3 peppers, 1 lemon, 1 lb. string beans, 1 lb. onions $11.59

c. 4 cans tuna fish, $1\frac{1}{2}$ doz. rolls, 1 can tomato juice, 1 bar soap, 1 pkg. detergent, 1 can peaches, 2 grapefruit, 2 lb. 4 oz. smoked ham $14.00

d. 2 doz. eggs, 4 loaves bread, 3 qt. milk, 6 lb. onions, $1\frac{3}{4}$ lb. tomatoes, 9 oranges, $1\frac{3}{4}$ lb. bananas, 4 lb. 5 oz. sirloin, 10 oz. cheese $23.26

e. 5 lb. sugar, 1 can peaches, 3 cans corn, 8 rolls, 2 lemons, $1\frac{1}{2}$ doz. eggs, 12 cans soup, 4 bottles beverage, $4\frac{3}{4}$ lb. rib roast, 2 lb. 3 oz. flounder, 6 peppers, 1 lb. 10 oz. tomatoes $31.13

f. 2 bunches carrots, 3 heads of lettuce, $4\frac{1}{2}$ lb. chicken, $1\frac{1}{2}$ lb. string beans, 4 cans corn, 10 rolls, 1 loaf bread, $1\frac{1}{4}$ lb. peaches, 4 grapefruit, 12 oz. margarine, 1 lb. 8 oz. lamb, 2 lb. 15 oz. bananas $16.24

g. 1 qt. milk, $3\frac{1}{2}$ lb. bluefish, 4 oz. cheese, 1 lb. 4 oz. margarine, 3 lb. 9 oz. sirloin, 2 lb. 14 oz. apples, 1 can peaches, 2 pkgs. detergent, 12 bars soap, 3 bunches carrots, 12 peppers, 3 doz. eggs, 12 cans corn, 18 cans soup, 1 lb. 3 oz. cookies, 20 rolls $45.45

6. a. When 1 kg of beef costs $4.16, find the cost of:

(1) 1.6 kg of beef $6.66 (2) 900 g of beef $3.75 (3) 2.74 kg of beef $11.40

b. When 2 kg of apples cost $1.50, find the cost of:

(1) 3.4 kg of apples (2) 750 g of apples (3) 1.25 kg of apples

c. When 500 g of cake cost $2.48, find the cost of:

(1) 1.4 kg of cake $6.95 (2) 800 g of cake $3.97 (3) 450 g of cake $2.24

d. When 100 g of cheese cost 50¢, find the cost of:

(1) 600 g of cheese (2) 1.7 kg of cheese (3) 380 g of cheese

e. When 250 g of bologna cost $1.25, find the cost of:

(1) 540 g of bologna (2) 1.1 kg of bologna (3) 1.75 kg of bologna

f. When gasoline costs 39.9¢ per liter, find the cost of:

(1) 46 L of gasoline (2) 31.5 L of gasoline (3) 53.8 L of gasoline

6.b. (1) $2.55 (2) $.57 (3) $.94 6.d. (1) $3.00 (2) $8.50 (3) $1.90
6.e. (1) $2.70 (2) $5.50 (3) $8.75 6.f. (1) $18.35 (2) $12.57 (3) $21.47

For a Class Activity see p. T36.

Unit Pricing—
Determining the Better Buy

OBJECTIVE: To find the unit price and to
determine the better buy using unit pricing.

VOCABULARY: unit price

Generally when buying food or other commodities that are packaged
and sold in two different sizes, one must consider whether the larger size
will be used without waste. In the problems of this section we shall
disregard this consideration.

To determine the "better buy" when two different quantities of the
same item are priced at two rates, first find the unit price or the cost per
unit of the item at each rate and select the lower unit price as the better
buy. Although the larger size or larger quantity should be marked at the
lower unit price, because of faulty pricing not always is the larger size of
an item the better buy.

Which is the better buy: 10 oranges for 85¢ or 12 oranges for
99¢? Oranges at the rate of 10 for 85¢ cost 8.5¢ each, but at the
rate of 12 for 99¢ the cost is 8.25¢ each.

Therefore, 12 oranges for 99¢ is the better buy.

The unit price may be the cost per item, the cost per ounce, the cost
per pound, the cost per quart, the cost per liter, the cost per kilogram, etc.
The symbol @ means "at" and indicates the unit price. "3 shirts @ $11.95
each" or just "3 shirts @ $11.95" is read "three shirts at $11.95 each."

Sometimes in working with unit prices like cost per ounce, we use
parts of a cent represented by a numeral like $.467. Observe that the
numeral $.467 means 46.7 cents or 46.7¢.

Basic: Prob. 1–13; *Average:* Prob. 1–17; *Enriched:* Prob. 1–20

PRACTICE PROBLEMS

1. Rewrite each of the following, using a dollar mark $ and a decimal
point:

    a. 37¢ $.37   **c.** 4¢ $.04   **e.** 5.2¢ $.052   **g.** $6\frac{1}{2}$¢ $.065   **i.** 29.1¢ $.291

    b. 95¢ $.95   **d.** 8¢ $.08   **f.** 23.9¢ $.239   **h.** $30\frac{3}{4}$¢ $.3075   **j.** 43.57¢ $.4357

2. Rewrite each of the following, using the ¢ symbol:

    a. $.86 86¢   **c.** $.09 9¢   **e.** $.017 1.7¢   **g.** $.281 28.1¢   **i.** $.9402 94.02¢

    b. $.03 3¢   **d.** $.63 63¢   **f.** $.744 74.4¢   **h.** $.029 2.9¢   **j.** $.0075 .75¢

3. a. Read each of the following in terms of cents: For ex. 3 see Extended Answers,
p. T94.

    (1) 34.6¢   **(2)** 5.4¢   **(3)** $.713   **(4)** $.082   **(5)** $.6925

    b. Read each of the following in terms of dollars and cents:

    (1) $2.374   **(2)** $1.329   **(3)** $5.067   **(4)** $3.108   **(5).** 1.43\frac{9}{10}$

4. Find the unit price (cost per pound) to the nearest tenth of a cent for each of the following:

a. 2-lb. package bacon costing $2.98 $1.49

b. 8-oz. package sliced cheese costing $1.69 $3.38

c. 10-oz. jar jelly costing 72¢ $1.152

d. 1-lb. 6-oz. pie costing $1.49 $1.084

e. 3-lb. 2-oz. jar applesauce costing $1.25 $.40

5. Find the unit price (cost per quart) to the nearest tenth of a cent for each of the following:

a. 1 pt. vinegar costing 43¢ 86¢

b. 1 gal. bleach costing 96¢ 24¢

c. 48-oz. bottle cranberry juice cocktail costing $1.28 85.3¢

d. 1-pt. 6-oz. bottle liquid detergent costing $1.10 $1.60

e. 1-qt. 14-oz. can juice costing 83¢ 57.7¢

6. Find the unit price (cost per ounce) to the nearest tenth of a cent for each of the following:

a. 3-oz. jar olives costing 89¢ 29.7¢

b. 12-oz. box cereal costing $1.29 10.8¢

c. 10-oz. bottle beverage costing 25¢ 2.5¢

d. 6.5-oz. can tuna fish costing $1.09 16.8¢

e. 7¾-oz. can salmon costing $1.95 25.2¢

7. Find the unit price (cost per 50 pieces) to the nearest tenth of a cent for each of the following:

a. Package of 100 tea bags costing $1.80 90¢

b. Package of 10 soap pads costing 63¢ $3.15

c. Package of 75 storage bags costing $1.50 $1.00

d. Package of 20 trash bags costing $2.19 $5.475

e. Package of 160 paper napkins costing 81¢ 25.3¢

8. Find the unit price (cost per quart) for each of the following bottles of vegetable oil and determine the difference between the highest and lowest unit prices:

a. 1 gal. at $5.49 $1.3725   **b.** 1 pt. at 81¢ $1.62   **c.** 24 oz. at $1.13 $1.506 diff. 24.75¢

9. Find the unit price (cost per pound) for each of the following packages of cereal and determine the difference between the highest and lowest unit prices:

a. 12 oz. at $1.39 $1.85   **b.** 8 oz. at 97¢ $1.94   **c.** 15 oz. at $1.68 $1.79 diff. $.15

10. Find the unit price (cost per quart) for each of the following bottles of syrup and determine the difference between the highest and lowest prices:

a. 12 oz. at $1.12 $2.987   **b.** 1 pt. 8 oz. at $1.79 $2.387   **c.** 1 qt. 4 oz. at $2.25 $2.00 diff. 98.7¢

11. Find the unit price (cost per single item) in each of the following:

    a. 5 jars of baby food costing $1.45   29¢

    b. 10 oranges costing 95¢   9.5¢

    c. 2 shirts costing $37 $18.50

    d. 4 tires costing $165 $41.25

    e. 6 grapefruit costing 79¢ $13\frac{1}{6}$¢

12. In each of the following, the same item is priced at two rates. Which is the better buy?

    a. Pears: 2 for 25¢ or 5 for 49¢   5 for 49¢

    b. Apples: 4 for 39¢ or 10 for 89¢   10 for 89¢

    c. Doughnuts: 3 for 49¢ or 12 for $1.89   12 for $1.89

    d. Grass seed: 5 lb. for $6.50 or 25 lb. for $27.75   25 lb. for $27.75

    e. Soup: 7 cans for $2 or 3 cans for 95¢   7 cans for $2

    f. Aspirin: 100 tablets for $1.73 or 200 tablets for $2.69   200 tab. for $2.69

    g. Soap: 4 bars for 75¢ or 6 bars for 99¢   6 bars for 99¢

    h. Facial tissue: 2 boxes for 83¢ or 5 boxes for $2   5 boxes for $2

    i. Rolls: 6 for 59¢ or 99¢ per dozen   99¢ per dozen

    j. Fruit juice: 2 cans for 99¢ or 3 cans for $1.35   3 cans for $1.35

13. Find how much you save by buying the larger size of each of the following items instead of an equivalent quantity in the smaller size:

    a. Pretzels: 4-oz. bag at 33¢; 12-oz. bag at 89¢   10¢

    b. Bleach: 1-qt. bottle at 45¢; 1-pt. bottle at 27¢   9¢

    c. Paint: 1-qt. can at $7.49; 1-gal. can at $12.95 $17.01

    d. Spaghetti: 8-oz. box at 37¢; 1-lb. box at 69¢   5¢

    e. Beverage: 10-oz. bottle at 25¢; 1-pt. 14-oz. bottle at 59¢   16¢

14. How much do you save per ounce by buying the larger size instead of the smaller size?

    a. Peanut butter: 6-oz. jar at 85¢; 12-oz. jar at $1.59   .916¢

    b. Mayonnaise: 1-qt. jar at $1.69; 1-pt. jar at 95¢   .656¢

    c. Mouth-wash: 14-oz. bottle at $1.09; 20-oz. bottle at $1.39   .836¢

    d. Tomato juice: 1-pt. 2-oz. can at 40¢; 1 qt. 14-oz. can at 83¢   .418¢

    e. Tuna fish: 7-oz. can at $1.13; $3\frac{1}{2}$-oz. can at 67¢   3¢

15. Which is the better buy per unit price (cost per quart or cost per pound)?

    a. Milk: 1 qt. at 62¢; 1 half-gallon at $1.12   1 half-gal. at $1.12

    b. Ice cream: 1 half-gallon at $2.79; 1 pt. at 89¢   1 half-gal. at $2.79

    c. Vinegar: 1 qt. at 73¢; 1 pt. at 43¢   1 qt. at 73¢

    d. Cooking oil: 24-oz. bottle at $1.13; 16-oz. bottle at 81¢   24-oz. bot. at $1.13

    e. Mustard: 9-oz. jar at 35¢; 24-oz. jar at 69¢   24-oz. jar at 69¢

    f. Salad dressing: 8-oz. jar at 67¢; 1-pt. jar at $1.08   1-pt. jar at $1.08

g. 5-lb. 4-oz. pkg. at $3.34

g. Laundry detergent: 1-lb. 4-oz. pkg. at 99¢; 5-lb. 4-oz. pkg. at $3.34

h. Grape jelly: 10-oz. jar at 79¢; 20-oz. jar at $1.19 20-oz. jar at $1.19

i. Floor wax: 27-oz. can at $1.39; 46-oz. can at $2.29 46-oz. can at $2.29

j. Baked beans: 1-lb. can at 41¢; 1-lb. 10-oz. can at 59¢ 1-lb. 10-oz. can at 59¢

k. Beverage: 6 16-oz. bottles at $2.39; 8 10-oz. bottles at $2.49 8 10-oz. bottles

l. Cream cheese: 3-oz. pkg. at 43¢; 8-oz. pkg. at 87¢ 8-oz. pkg. at 87¢

m. Honey: 1-lb. jar at $1.49; 12-oz. jar at $1.19 1-lb. jar at $1.49

n. Face cream: 10.4-oz. jar at $3.73; 13.4-oz. jar at $4.79 13.4-oz. jar at $4.79

o. Toothpaste: 5-oz. tube at 99¢; 7-oz. tube at $1.25 7-oz. tube at $1.25

16. Find how much you save on each can when buying in quantity:

a. 1 dozen cans of peas for $4.20 or 41¢ each. 6¢

b. 7 cans of soup for $1.75 or 31¢ each. 6¢

c. 10 cans of tomato paste for $2.49 or 29¢ each. 4.1¢

d. 5 cans of beans for $1.75 or 37¢ each. 2¢

e. 8 cans of beets for $2.50 or 35¢ each. 3.75¢

f. 3 cans of peaches for $1.79 or 63¢ each. 3.33¢

g. 6 cans of tomato juice for $2.24 or 41¢ each. 3.66¢

h. 12 cans of beverage for $1.99 or 19¢ each. 2.42¢

17. Find the unit price (cost per kilogram) to the nearest tenth of a cent for each of the following:

a. 5 kg of potatoes costing $1.19 23.8¢

b. 2.27-kg bag of sugar costing $1.65 72.7¢

c. 283-g jar of jelly costing 69¢ $2.438

d. 964-g jar of applesauce costing 73¢ 75.7¢

18. Find the unit price (cost per liter) to the nearest tenth of a cent for each of the following:

a. 2-L bottle of beverage costing $1.25 62.5¢

b. 1.36-L can of fruit punch costing 64¢ 47.1¢

c. 532-mL can of tomato juice costing 45¢ 84.6¢

d. 946-mL carton of milk costing 68¢ 71.9¢

19. Which is the better buy per unit price (cost per kilogram or cost per liter)?

a. Flour: 2.26-kg bag at 95¢ or 907-g bag at 51¢ 2.26-kg bag at 95¢

b. Bleach: 950-mL bottle at 45¢ or 1.89-L bottle at 67¢ 1.89-L bot. at 67¢

c. Crackers: 454-g box at $1.05 or 340-g box at 93¢ 454-g box at $1.05

d. Pickles: 946-mL jar at $1.09 or 473-mL jar at 65¢ 946-mL jar at $1.09

20. Find how much you save buying the larger size of each of the following items instead of an equivalent quantity in the smaller size:

a. Fabric softener: .94-L bottle at 33¢; 3.76-L bottle at 87¢ $.45

b. Grapefruit sections: 227-g can at 31¢; 1.362-kg can at $1.29 $.57

Buying at a Sale

OBJECTIVE: To find sale price, rate of reduction, and regular price.

VOCABULARY: list price
net price
rate of reduction

A person buying an article at a sale expects to pay a reduced price for this article. The regular or full price of an article is generally called the marked price or list price. The reduced price is sometimes called the sale price or net price. See Exercise 6-24.

The percent taken off is the rate of reduction. *To find the rate of reduction,* determine what percent the reduction is of the regular price.

To find the regular price when the sale price and rate of reduction are given, subtract the given rate from 100%, then divide the answer into the sale price.

Larue's
DEPARTMENT STORE

FURNITURE—*Reduced 30%*

Tables *regularly* $79
Lamps *regularly* $54.50
Sofas *regularly* $425

Bedroom Sets *regularly* $899
Bookcases *regularly* $72.75
Mirrors *regularly* $37.98

CLOTHING—*Reduced 33⅓%*

Dresses *regularly* $39
Coats *regularly* $90
Shoes *regularly* $24.75
Suits *regularly* $97.50

Sweaters *regularly* $18.98
Skirts *regularly* $21.45
Ties *regularly* $3.50
Shirts *regularly* $14.99

APPLIANCES—*Reduced 20%*

Refrigerators *regularly* $595
Toaster-ovens *regularly* $39.95
Hair dryers *regularly* $36.75

Color TVs *regularly* $445
Blenders *regularly* $29.95
Can openers *regularly* $14.69

PRACTICE PROBLEMS

1.a. Tables, $55.30; lamps, $38.15; sofas, $297.50; b.r. sets, $629.30; book c., $50.93; mir., $26.59
   b. Dresses, $26; coats, $60; shoes, $16.50; suits, $65; sweat., $12.66; skirts, $14.30; ties, $2.34; shirts, $10
   c. Refrig., $476; t-ovens, $31.96; h-dryers, $29.40; TV, $356; blend., $23.96; can open., $11.75

1. Use the advertisement on the opposite page to solve the following problems:

a. Find the sale price of each of the listed furniture items.

b. Find the sale price of each of the listed items of clothing.

c. Find the reduced price of each of the listed appliances.

2. How much is a football reduced in price at a 25% reduction sale if its regular price is $15? What is its sale price? $3.75; $11.25

3. At a year-end clearance sale an automobile, regularly selling for $6,850, can now be purchased at a reduction of 17%. What is its sale price? $5,685.50

4. Find the sale price of each of the following articles:

a. Baseball; regular price, $3.50; reduced 40%. $2.10

b. Camera; regular price, $79.95; reduced 20%. $63.96

c. Stereo system; regular price; $225; $\frac{1}{3}$ off. $150

d. Sewing machine; regular price, $204.50; reduced 15%. $173.83

e. Automatic washer; regular price, $349; reduced 23%. $268.73

5. Find the rate of reduction allowed when a lawn mower that regularly sells for $185 was purchased at $148? 20%

6. Find the rate of reduction allowed on each of the following articles:

a. Typewriter; regular price, $240; sale price, $180 25%

b. Kitchen clock; regular price, $18; sale price, $15 $16\frac{2}{3}$%

c. Tire; regular price, $48; sale price, $42 $12\frac{1}{2}$%

d. Vacuum sweeper; regular price, $87.50; sale price, $75 14.3%

e. Blanket; regular price, $21.75; sale price $14.50 $33\frac{1}{3}$%

7. Find the regular price of a desk that sold for $149.85 at a 25% reduction sale. $199.80

8. Find the regular price of each of the following:

a. Battery, $28 sale price when reduced 30% $40

b. Tablecloth, $20.40 sale price when reduced 15% $24

c. Sprinkler, $9.60 sale price when reduced $37\frac{1}{2}$% $15.36

d. Freezer, $246.66 sale price when reduced $33\frac{1}{3}$% $370

e. Toaster, $17.55 sale price when reduced 10% $19.50

9. A calculator which regularly sells for $49.88 is now on sale at $\frac{1}{4}$ off. What is its sale price? $37.41

10. What is the regular price of a pair of slacks that sold for $23.60 at a 20% reduction sale? $29.50

For a Class Activity see p. T37.   For a Computer Activity see p. T40.

6-10 Making Change

OBJECTIVE: To find out how much change should be received.

Usually store clerks, in making change, will add the value of the coins and bills to the amount of the purchase until the sum of the change and the amount of the purchase equals the amount offered in payment.

Basic: Prob 1; *Average:* Prob. 1–2

PRACTICE PROBLEMS

1. Find how much change you should receive:

a. From $1 if your purchases cost:

   (1) $.79 $.21   (2) $.48 $.52   (3) $.16 $.84   (4) $.09 $.91   (5) $.83 $.17

b. From $5 if your purchases cost:

   (1) $1.34 $3.66 (2) $2.07 $2.93 (3) $4.58 $.42   (4) $.72 $4.28 (5) $3.96 $1.04

c. From $10 if your purchases cost:

   (1) $6.41 $3.59 (2) $3.15 $6.85 (3) $9.69 $.31   (4) $7.87 $2.13 (5) $.92 $9.08

d. From $20 if your purchases cost:

   (1) $8.08   (2) $17.94   (3) $5.16   (4) $11.29   (5) $15.47
   $11.92 $2.06 $14.84 $8.71 $4.53

2. For each of the following purchases, determine how much change should be given when the specified amount of money is offered in payment:

Articles Purchased	Amount Offered in Payment
a. 3 candy bars @ 25¢ each, 2 cones ice cream @ 45¢ each	$2.00 $.35
b. 1 shirt @ $17.95, 2 ties @ $5.50 each	$30.00 $1.05
c. 2 dresses @ $39.98 each, 3 sweaters @ $19.99 each	$150.00 $10.07
d. 48 L gasoline @ 33.9¢ per L, 1 L oil @ $2.25	$20.00 $1.48
e. 9 cans soup @ 33¢ each 3 cans beans @ 37¢ each	$10.00 $5.92
f. 8 grapefruit @ 19¢ each, 15 apples @ 9¢ each, 10 pears @ 11¢ each	$5.00 $1.03
g. 5 gal. paint @ $15.49 each, 2 brushes @ $2.69 each	$100.00 $17.17
h. 1 suit @ $109.50, 2 pr. slacks @ $19.95 each, 4 shirts @ $10.75 each, 1 tie @ $4.50	$200.00 $3.10

OBJECTIVE: To compute sales tax.

Many states and cities have a sales tax on the retail selling price of articles which consumers purchase and on some services that they require. If the resulting sales tax contains any fractional part of a cent, drop the fraction and use the next higher cent.

Basic: Prob 1; Average: Prob 1–2; Enriched: Prob. 1–3

PRACTICE PROBLEMS

1. Find the tax on each of the following sales when the tax rate is:

a. 2%:   (1) $5.00 $.10   (2) $10.40 $.21 (3) $84.25 $1.69 (4) $134.80 $2.70

b. 6%:   (1) $16.00 $.96   (2) $.48 $.03   (3) $8.97 $.54   (4) $107.55 $6.46

c. 5%:   (1) $3.00 $.15   (2) $9.50 $.48   (3) $57.30 $2.87 (4) $212.45 $10.63

d. 8%:   (1) $91.00 $7.28 (2) $6.89 $.56   (3) $731.75 $58.54 (4) $2,451.49 $196.12

e. 4%:   (1) $175.00 $7   (2) $5.12 $.21   (3) $27.64 $1.11 (4) $1,783.50 $71.34

f. $1\frac{1}{2}$%:   (1) $2.00 $.03   (2) $.79 $.02   (3) $45.80 $.69   (4) $362.75 $5.45

g. $3\frac{1}{2}$%:   (1) $12.00 $.42   (2) $9.60 $.34   (3) $78.35 $2.75 (4) $825.88 $28.91

2. Find the selling price, including sales tax, of each of the following articles:

a. Bag, $18.98, tax 7% $20.31   f. Basketball, $14.89, tax 2% $15.19

b. Calculator, $24.99, tax 3% $25.74 g. Wading pool, $39.95, tax 5% $41.95

c. Slacks, $29.50, tax 4% $30.68   h. Heating pad, $10.88, tax $3\frac{1}{2}$% $11.27

d. Bedspread, $31.75, tax 6% $33.66 i. Pajamas, $9.79, tax $2\frac{1}{2}$% $10.04

e. Crib, $87.49, tax $4\frac{1}{2}$% $91.43   j. Electric dryer, $269.95, tax $1\frac{1}{2}$% $274

Sales Tax Table—4%					
Purchase	Tax	Purchase	Tax	Purchase	Tax
$.01–$.10	$.00	$1.11–$1.25	$.05	$2.11–$2.25	$.09
.11– .25	.01	1.26– 1.50	.06	2.26– 2.50	.10
.26– .50	.02	1.51– 1.75	.07	2.51– 2.75	.11
.51– .75	.03	1.76– 2.10	.08	2.76– 3.10	.12
.76–1.10	.04				

3. Use the table above to find the tax on each of the following amounts of purchase:

a. $.28 $.02   d. $.75 $.03   g. $2.61 $.11   j. $.61 $.03

b. $.98 $.04   e. $1.47 $.06   h. $1.24 $.05   k. $2.51 $.11

c. $.55 $.03   f. $1.09 $.04   i. $2.98 $.12   l. $3.05 $.12

OBJECTIVE: To read meters and determine the cost of utilities and telephone services.

In order to check bills for the gas, electricity, and water services, one must be able to read a meter measuring the consumption. *To read a gas meter or an electric meter or a water meter,* select on each dial of the meter the numeral that was just passed by the dial pointer. *To find the consumption during a given period,* subtract the reading at the beginning of the period from the reading at the end of the period. *To find the cost,* apply the prevailing rate to the consumption.

The reading of the dials on the gas meter (see below) is expressed in *hundred cubic feet.* A reading such as 5486 represents 548,600 cubic feet.

The reading of the dials on the electric meter (see page 569) is expressed in *kilowatt hours.* A reading such as 8291 represents 8,291 kilowatt hours of electrical energy. A kilowatt is 1,000 watts. It is a unit measuring electric power. If an appliance requires the power of 1 kilowatt to be used for 1 hour, then 1 kilowatt hour of electrical energy is furnished. Observe also that if an appliance requires the power of 200 watts but is used for 5 hours, then 1 kilowatt hour of electricity is also used (since $200 \times 5 = 1,000$ watt hours $= 1$ kilowatt hour).

The reading of the dials on the water meter is expressed in *tens of gallons.* Although there is a small dial on the meter that measures from 0 to 10 gallons, it is not used as part of the reading. A reading such as 25736 represents 257,360 gallons of water.

Basic: Prob. 1–5; *Average:* Prob. 1–7; *Enriched:* Prob. 1–8

PRACTICE PROBLEMS

1. a. What is the reading of the gas meter on April 1? On May 1?

8352   8417

CUBIC FEET   CUBIC FEET

April 1   May 1

b. How many cubic feet of gas were consumed during the month of April? 6,500 cu. ft.

c. Find the cost of the gas consumed if the rates were:

First   200 cu. ft. or less $1.88
Next 2,200 cu. ft. $4.60 per 1,000 cu. ft.
Next 2,100 cu. ft. $4.26 per 1,000 cu. ft.
Over 4,500 cu. ft. $3.81 per 1,000 cu. ft. $28.57

2. Given the reading at the beginning and at the end of the period, use the above rates in each of the following to find the cost of the gas consumed during the period.

	READING				READING	
	At Beginning	**At End**			**At Beginning**	**At End**
a.	2418	2487 $23.22		**f.**	5523	5541 $7.24
b.	8594	8691 $31.22		**g.**	9905	0298 $115.88
c.	6956	7143 $56.96		**h.**	6182	6425 $72.98
d.	1009	1216 $62.68		**i.**	3527	3982 $133.61
e.	7231	7495 $78.99		**j.**	0064	0639 $167.93

3. a. What is the reading of the electric meter on November 1? on December 1? 6774   6529

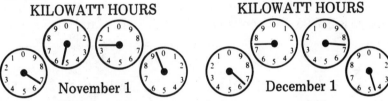

KILOWATT HOURS   KILOWATT HOURS

November 1   December 1

b. How many kilowatt hours of electricity were used during the month of November? 245 kWh

c. Find the cost of the electricity used if the rates were:

First   12 kWh or less $4.20
Next   42 kWh @ $.09
Next   46 kWh @ $.08
Over 100 kWh @ $.069 $21.67

4. Given the reading at the beginning and at the end of the period, use the above rates in each of the following to find the cost of the electricity used during the period.

	READING				READING	
	At Beginning	**At End**			**At Beginning**	**At End**
a.	4726	4773 $5.65		**f.**	8153	8163 $3.20
b.	6175	6266 $8.36		**g.**	2261	2345 $7.94
c.	1592	1704 $9.49		**h.**	0095	0506 $24.14
d.	3204	3381 $12.67		**i.**	7949	8482 $30.12
e.	9887	0069 $12.92		**j.**	5838	6097 $16.69

5. At the average cost of 8.6¢ per kilowatt hour, find the *hourly* cost of operating each of the following:

a. Toaster, 1,000 watts 8.6¢　　　**d.** Heater, 1,500 watts 12.9¢

b. Dryer, 4,000 watts 34.4¢　　　**e.** Iron, 1,200 watts 10.32¢

c. 100-watt light bulb .86¢　　　**f.** Radio, 40 watts .344¢

6. In many communities water bills are issued every 3 months. The reading on a certain water meter on February 1 was 43582, on May 1 it was 46106, on August 1 it was 51263, and on November 1 it was 54091.

a. How many gallons of water were used during the 3-month period ending: (1) May 1? (2) August 1? (3) November 1? 25,240 gal.; 51,570 gal.; 28,280 gal.

b. Find the cost of the water consumed during each of these periods using the rates: First 2,500 gallons or less, $7.58; Over 2,500 gallons, $1.77 for every 1,000 gallons. $47.83; $94.43; $53.21

7. Mr. Ross during the year bought 210 gallons of fuel oil at $1.249 per gallon, 185 gallons at $1.219 each, 235 gallons at $1.226 each, 220 gallons at $1.235 each, and 190 gallons at $1.253 each. How many gallons of fuel oil did he buy? What was the total cost of his fuel oil for the year? 1,040 gal.; $1,312.

8. What is the total amount due for telephone service if the monthly charge is $15.95, the itemized charges are $4.70, and the federal tax is $.62? $21.27

REFRESH YOUR SKILLS

1. Add:

976
3,452
58,879
6,986
——— 70,293

2. Subtract

106,237
78,189
———
28,048

3. Multiply:

708
905
———
640,740

4. Divide:

$87\overline{)52,983}$
609

5. Add:

$2\frac{7}{10} + 5\frac{3}{4}$ $8\frac{9}{20}$

6. Subtract:

$1\frac{2}{3} - \frac{1}{6}$ $1\frac{1}{2}$

7. Multiply:

$3\frac{1}{7} \times 2\frac{4}{5}$ $8\frac{4}{5}$

8. Divide:

$24 \div \frac{3}{4}$ 32

9. Add:

$2.16 + 4.9$ 7.06

10. Subtract:

$.759 - .7$ .059

11. Multiply:

800×62.5
50,000

12. Divide:

$56.48 \div 8$
7.06

13. Find 10.8% of $4,750. $513 **14.** What percent of $.60 is $.48? 80%

15. 9% of what amount is $162? $1,800　　**16.** Square 7.1 50.41

17. Find the square root of 6.4 to the nearest hundredth. 2.53

18. Write 28.05 *billion* as a complete numeral. 28,050,000,000

19. Express 6.93 kilograms as grams. 6,930 g

20. What part of an hour is 42 minutes? $\frac{7}{10}$

Owning an Automobile

OBJECTIVE: To solve several kinds of mathematical problems related to the ownership of an automobile.

Basic: Prob. 1–5; *Average:* Prob. 1–6; *Enriched:* Prob. 1–7

1. a. Mr. Delano bought a new car for $8,250. After using it for 4 years, he sold it for $1,725. What was the average yearly depreciation of the car? $1,631.25

b. At the end of the first year the odometer read 9425 miles, at the end of the second year 21132 miles, at the end of the third year 33051 miles, and at the end of the fourth year 47009 miles. How many miles did Mr. Delano drive during the second year? During the third year? During the fourth year? 11,707 mi.; 11,919 mi.; 13,958 mi.

c. During the first year his expenses in addition to the average yearly depreciation were: gasoline, 620 gallons at $1.359 per gal.; oil, 30 quarts at $2.10 per qt.; insurance, $385 per yr.; license fees, $36 per yr.; parking, $65 per yr.; miscellaneous expenses, $125. (1) How much did it cost Mr. Delano to run his car for the first year? (2) What was the average cost per month? (3) Per week? (4) Per mile? (1) $3,147.83 (2) $262.32 (3) $60.54 (4) 33¢

2. How far can a car go on a tankful of gasoline if it averages 15 miles on a gallon and the tank holds 22 gallons? 330 mi.

3. At a speed of 55 miles per hour, how many minutes will it take to go a distance of 22 miles? 24 min.

4. How long will it take a person, driving at an average speed of 50 m.p.h., to travel from Washington to Boston, a distance of 460 miles? If the car can average 14 miles on a gallon of gasoline, how many gallons are required to make the trip? 9.2 hr.; 32.86 gal.

5. Mrs. Potter insures her car costing $9,000 against fire and theft for 80% of its value at the annual rate of 85¢ per $100. If the charge per year for property damage insurance is $81.75, for liability insurance is $68.50, and for collision insurance is $76, what is her total annual premium? $287.45

6. If the driver's reaction time before applying the brakes is $\frac{3}{4}$ second, how many feet will the car go in $\frac{3}{4}$ second at the following speeds:

a. 60 m.p.h.? **b.** 45 m.p.h.? **c.** 30 m.p.h.? **d.** 65 m.p.h.? **e.** 70 m.p.h.?
66 ft.   49.5 ft.   33 ft.   71.5 ft.   77 ft.

7. The braking distance (distance an automobile travels before coming to a stop after the brakes are applied) on a dry, level road is 30 ft. at 20 m.p.h., 67.5 ft. at 30 m.p.h., 120 ft. at 40 m.p.h., and 187.5 ft. at 50 m.p.h. If the driver's reaction time is one second, find the total stopping distance (reaction time distance plus braking distance) at each of the above speeds. $59\frac{1}{3}$ ft.; $111\frac{1}{2}$ ft.; $178\frac{2}{3}$ ft.; $260\frac{5}{6}$ ft.

Buying Without Full Cash Payment

VOCABULARY: installment buying
finance charge
carrying charge
annual percentage rate
deferred payment

OBJECTIVE: To compute costs when buying without full cash payment.

Installment Buying

Installment buying or buying on credit is a purchase plan by which the customer usually pays a certain amount of cash at the time of purchase and then pays in equal installments at regular intervals, usually months, both the balance and the finance charge on this balance. This finance charge, also called carrying charge, is the interest (see page 586) which usually is applied on the entire balance as being owed for the full time of payment. The carrying charge or finance charge is the difference between the cash price and the total amount paid. However, the given percentage rate charge is not the true rate since in this type of purchase plan, part of the original balance owed is being repaid each month. The true rate charge is much higher since it is based on the lower average balance that is owed during the payment period. Actually, the true rate is almost twice the given rate.

Annual Percentage Rate

So that a person who buys on credit may know what this true rate is in order to compare finance charge rates, the federal government requires each credit purchase contract or loan contract (see page 573) to indicate the annual percentage rate which is the true finance charge for the year. Annual Percentage Rate Tables are generally used to determine the rates. (See page 575.)

Deferred Payment Purchase Plan

Many stores use deferred payment purchase plans. In these plans a specific monthly payment is made to reduce the balance owed. Sometimes these payments are fixed amounts. Other times these payments decrease as the bracket of the balance owed falls. However, a finance charge, usually $1\frac{1}{4}\%$ per month or $1\frac{1}{2}\%$ per month is applied to the previous monthly unpaid balance and is added (in addition to the new purchase charges) to this previous balance. These plans must also indicate the annual percentage rate.

To compute the annual percentage rate, multiply the rate per period by the number of periods per year. Thus, the finance charge of $1\frac{1}{2}\%$ per month is $12 \times 1\frac{1}{2}\%$ or an 18% annual percentage rate.

Credit card accounts are charged a late payment finance charge each month on past due amounts. Here also the annual percentage rates must be indicated.

PRACTICE PROBLEMS

1. Mrs. Smith bought an automatic dishwasher, paying $25 down and $46.50 a month for 8 months. How much more did the dishwasher cost on the installment plan if the cash price was $375? $22

2. Lisa's mother can purchase a refrigerator for the cash price of $489 or $39 down and 10 equal monthly payments of $48.38 each. How $33.80 much is the carrying charge? How much can be saved by paying cash? $33.80

3. Find the carrying charge when each of the following articles are purchased on the installment plan:

	Cash Price	Down Payment	Monthly Payment	Number of Monthly Payments
a. Piano	$950	$50	$63.00	15 $45.00
b. Calculator	$175	$25	$18.00	9 $12.00
c. Clothes dryer	$259	$19	$15.10	18 $31.80
d. Air conditioner	$245	$45	$27.50	8 $20.00
e. Typewriter	$179	$24	$16.43	10 $9.30

4. Find the monthly payment if the amount financed is $960 and the finance charge is $180, both to be paid in 30 equal installments. $38.00

5. A TV set can be purchased for the cash price of $395 or $35 down and the balance and the finance charge to be paid in 12 equal monthly payments. If 6% was charged on the full balance, how much is the monthly payment? $31.80

6. Find the monthly payment when each of the following articles are purchased on the installment plan and the finance charge interest rate is on the full balance:

	Cash Price	Down Payment	Finance Charge Interest Rate	Number of Monthly Payments
a. Camera	$118	$10	6%	12 $9.54
b. Oil furnace	$595	$55	6%	12 $47.70
c. Bedroom set	$869	$59	8%	18 $48.60
d. Sewing machine	$275	$35	$8\frac{1}{2}$%	6 $43.40
e. Microwave oven	$425	$25	$7\frac{1}{2}$%	10 $43.00
f. Fur coat	$585	$75	6%	12 $45.05
g. CB radio	$190	$15	8%	6 $31.50
h. Drum set	$250	$25	6%	8 $29.81
i. Desk	$220	$20	9%	12 $18.17

7. Compute the late finance charge on each of the following past due amounts at the specified rates:

    a. $58.50 past due; $1\frac{1}{2}$% monthly rate 88¢

    b. $84.96 past due; $\frac{3}{4}$ of 1% monthly rate 64¢

    c. $144 past due; $\frac{2}{3}$ of 1% monthly rate 96¢

8. Find the annual percentage rate if the monthly finance charge rate is:

    a. 1% 12%   **b.** $1\frac{3}{4}$% 21%   **c.** $2\frac{1}{2}$% 30%   **d.** $1\frac{2}{3}$% 20%   **e.** $2\frac{1}{4}$% 27%

9. Find the annual percentage rate if the monthly late payment finance charge rate on past due amounts is:

    a. 2% 24% **b.** $\frac{3}{4}$ of 1% 9%   **c.** $\frac{2}{3}$ of 1% 8%   **d.** $\frac{7}{8}$ of 1% 10.5% **e.** $\frac{5}{6}$ of 1% 10%

10. A department store uses the following deferred payment schedule:

Monthly Balance	Monthly Payments	Monthly Balance	Monthly Payments
$0 to $10	Full Balance	$300.01 to $350	$25
$10.01 to $200	$10	$350.01 to $400	$30
$200.01 to $250	$15	$400.01 to $500	$40
$250.01 to $300	$20	Over $500	$\frac{1}{10}$ of balance

If finance charges are computed at the monthly rate of $1\frac{1}{4}$% on the previous month's balance, find the monthly payment due on each of the following previous month's balances. First compute the finance charge, add it to the month's balance, and then find in the above schedule the monthly payment due.

    a. $79 $10   **b.** $198 $15   **c.** $600 $60.75   **d.** $265 $20   **e.** $349 $30

11. On page 575 is a condensed section of the Federal Reserve Board tables for finding annual percentage rates for monthly payment plans.

To find the annual percentage rate when the amount to be financed, the finance charge for this amount, and the number of monthly payments are known:

(1) Determine the finance charge per $100 of amount financed by multiplying the given finance charge by 100 and dividing this product by the amount financed.

(2) Follow down the left hand column of the table to the line indicating the required number of monthly payments; then move across this line until you reach the nearest number to the financed charge per $100.

(3) The rate at the top of this column is the annual percentage rate.

> For 15 monthly payments a finance charge of $10.11 per $100 financed is at the annual percentage rate of 14.75%.

NUMBER OF PAYMENTS	ANNUAL PERCENTAGE RATE										
	14.00%	14.25%	14.50%	14.75%	15.00%	15.25%	15.50%	15.75%	16.00%	17.00%	18.00%
	(FINANCE CHARGE PER $100 OF AMOUNT FINANCED)										
1	1.17	1.19	1.21	1.23	1.25	1.27	1.29	1.31	1.33	1.42	1.50
2	1.75	1.78	1.82	1.85	1.88	1.91	1.94	1.97	2.00	2.13	2.26
3	2.34	2.38	2.43	2.47	2.51	2.55	2.59	2.64	2.68	2.85	3.01
4	2.93	2.99	3.04	3.09	3.14	3.20	3.25	3.30	3.36	3.57	3.78
5	3.53	3.59	3.65	3.72	3.78	3.84	3.91	3.97	4.04	4.29	4.54
6	4.12	4.20	4.27	4.35	4.42	4.49	4.57	4.64	4.72	5.02	5.32
7	4.72	4.81	4.89	4.98	5.06	5.15	5.23	5.32	5.40	5.75	6.09
8	5.32	5.42	5.51	5.61	5.71	5.80	5.90	6.00	6.09	6.48	6.87
9	5.92	6.03	6.14	6.25	6.35	6.46	6.57	6.68	6.78	7.22	7.65
10	6.53	6.65	6.77	6.88	7.00	7.12	7.24	7.36	7.48	7.96	8.43
11	7.14	7.27	7.40	7.53	7.66	7.79	7.92	8.05	8.18	8.70	9.22
12	7.74	7.89	8.03	8.17	8.31	8.45	8.59	8.74	8.88	9.45	10.02
13	8.36	8.51	8.66	8.81	8.97	9.12	9.27	9.43	9.58	10.20	10.81
14	8.97	9.13	9.30	9.46	9.63	9.79	9.96	10.12	10.29	10.95	11.61
15	9.59	9.76	9.94	10.11	10.29	10.47	10.64	10.82	11.00	11.71	12.42
16	10.20	10.39	10.58	10.77	10.95	11.14	11.33	11.52	11.71	12.46	13.22
17	10.82	11.02	11.22	11.42	11.62	11.82	12.02	12.22	12.42	13.23	14.04
18	11.45	11.66	11.87	12.08	12.29	12.50	12.72	12.93	13.14	13.99	14.85
19	12.07	12.30	12.52	12.74	12.97	13.19	13.41	13.64	13.86	14.76	15.67
20	12.70	12.93	13.17	13.41	13.64	13.88	14.11	14.35	14.59	15.54	16.49
21	13.33	13.58	13.82	14.07	14.32	14.57	14.82	15.06	15.31	16.31	17.32
22	13.96	14.22	14.48	14.74	15.00	15.26	15.52	15.78	16.04	17.09	18.15
23	14.59	14.87	15.14	15.41	15.68	15.96	16.23	16.50	16.78	17.88	18.98
24	15.23	15.51	15.80	16.08	16.37	16.65	16.94	17.22	17.51	18.66	19.82
25	15.87	16.17	16.46	16.76	17.06	17.35	17.65	17.95	18.25	19.45	20.66
26	16.51	16.82	17.13	17.44	17.75	18.06	18.37	18.68	18.99	20.24	21.50
27	17.15	17.47	17.80	18.12	18.44	18.76	19.09	19.41	19.74	21.04	22.35
28	17.80	18.13	18.47	18.80	19.14	19.47	19.81	20.15	20.48	21.84	23.20
29	18.45	18.79	19.14	19.49	19.83	20.18	20.53	20.88	21.23	22.64	24.06
30	19.10	19.45	19.81	20.17	20.54	20.90	21.26	21.62	21.99	23.45	24.92
31	19.75	20.12	20.49	20.87	21.24	21.61	21.99	22.37	22.74	24.26	25.78
32	20.40	20.79	21.17	21.56	21.95	22.33	22.72	23.11	23.50	25.07	26.65
33	21.06	21.46	21.85	22.25	22.65	23.06	23.46	23.86	24.26	25.88	27.52
34	21.72	22.13	22.54	22.95	23.37	23.78	24.19	24.61	25.03	26.70	28.39
35	22.38	22.80	23.23	23.65	24.08	24.51	24.94	25.36	25.79	27.52	29.27
36	23.04	23.48	23.92	24.35	24.80	25.24	25.68	26.12	26.57	28.35	30.15
37	23.70	24.16	24.61	25.06	25.51	25.97	26.42	26.88	27.34	29.18	31.03
38	24.37	24.84	25.30	25.77	26.24	26.70	27.17	27.64	28.11	30.01	31.92
39	25.04	25.52	26.00	26.48	26.96	27.44	27.92	28.41	28.89	30.85	32.81
40	25.71	26.20	26.70	27.19	27.69	28.18	28.68	29.18	29.68	31.68	33.71

Using the table, find the annual percentage rate when:

	Total Amount Financed	Finance Charge on Total Amount	Number of Monthly Payments	
a.	$100	$6.65	10	14.25%
b.	$100	$14.82	21	15.5%
c.	$400	$120.60	36	18.0%
d.	$750	$93.45	16	17.0%
e.	$575	$33.35	8	15.25%

12. The annual percentage rate tables may be used to determine the finance charge and the monthly payment when the annual percentage rate is known.

To find the finance charge and the monthly payment:
(1) Use the column headed by the known annual percentage rate and read down this column until you reach the line corresponding to the given number of monthly payments. This is the finance charge per $100 of the amount to be financed.

(2) To determine the finance charge for the given amount financed, multiply the finance charge per $100 by the amount financed, then divide the product by 100.

(3) To find the monthly payment, add the finance charge to the amount financed, then divide this sum by the number of monthly payments.

Using the table, find the finance charge for each of the following amounts financed at the given annual percentage rates. Also find the monthly payment.

	Amount Financed	Annual Percentage Rate	Number of Monthly Payments	
a.	$100	15.00%	9	$6.35; $11.82
b.	$100	18.00%	5	$4.54; $20.91
c.	$900	14.25%	18	$104.94; $55.83
d.	$1,600	15.75%	24	$275.52; $78.15
e.	$4,500	14.50%	30	$891.45; $179. 72

Fire Insurance

OBJECTIVE: To compute the
cost of fire insurance.

VOCABULARY: fire insurance   face of the policy
term of the policy premium
policy

Fire insurance is a plan by which persons share risks so that each person is protected against financial loss. The insured person is charged a sum of money called a premium. The insurance company uses the fund created by these premiums to pay the insured persons who suffer losses.

The written contract between the insured person and the insurance company is called the policy. The amount of insurance specified in the policy is called the face of the policy. The length of time the insurance is in force is called the term of the policy.

To find the premium for a 1-year policy, multiply the number of $100 of fire insurance by the yearly rate per $100.

The prepaid premium rate on fire insurance for 3 years is 2.7 times the yearly rate. If this premium is paid by yearly installments, then 35% of the total premium is paid each year.

Basic: Prob. 1–3; *Average:* Prob. 1–5; *Enriched:* Prob. 1–6

PRACTICE PROBLEMS

1. What is the cost of $28,000 fire insurance for 1 year at the rate of $.18 per $100? $50.40

2. Find the premium for each of the following 1-year policies:

Face of policy	$25,000	$69,000	$14,000	$37,500
Yearly rate per $100	$.14	$.35	$.25	$.30

3. What is the premium for a 3-year policy if the yearly rate is $.20 per $100 and the face of the policy is $20,000? $108

4. Find for each of the following 3-year policies, **a.** the total premium and **b.** the yearly premium if paid by yearly installments:

Face of policy	$46,000	$11,000	$50,000	$35,000
Yearly rate per $100	$.16	$.32	$.18	$.26

5. How much does Mr. Collins save by buying a 3-year policy instead of three 1-year policies if he insures his house for $42,500 at the yearly rate of $.24 per $100? $30.60

6. Andrea's house cost $60,000 and the household contents cost $25,000. Her mother purchased 3-year policies, insuring the house for 80% of its value and the contents for 90% of their value. What was the total premium if the yearly rate on the building was $.22 per $100 and on the contents was $.18 per $100? $394.47

2. $35; $241.50; $35; $112.50
4.a. $198.72; $95.04; $243; $245.70
  b. $69.55; $33.26; $85.05; $86

Related Resources: *Practical Applications in Mathematics:* Problems, p. 183

EXERCISE

6-16 Life Insurance

VOCABULARY: life insurance
term insurance
ordinary life insurance

limited-payment life
insurance
endowment insurance

**OBJECTIVE: To compute the cost
of life insurance.**

Life insurance offers financial protection to the dependents of an insured person in the event of the insured person's death. The person who receives the money when the insured person dies is called the beneficiary.

1. Term insurance—Person is insured for a specified period of time. Premiums are paid only during that time. Beneficiary receives face of policy if insured person dies within the specified time.

2. Ordinary life insurance—Person is insured until death. Premiums are paid until death of the insured person. Beneficiary receives face of policy when insured person dies.

3. Limited-payment life insurance—Person is insured until death. Premiums are paid for a specified period of time. Beneficiary receives face of policy when insured person dies.

4. Endowment insurance—Person is insured for a specified period of time. Premiums are paid only during that time. Beneficiary receives face of policy if insured person dies within the specified time. The insured person or policy holder receives face of policy if alive at the end of the specified time.

To find the premium, select from the table on page 579 the premium for $1,000 insurance, then multiply it by the number of $1,000 of insurance.

Basic: Prob. 1–5; *Average:* Prob. 1–7; *Enriched:* Prob. 1–8

PRACTICE PROBLEMS

1. Average life expectation:

At the age of	Birth	1	5	10	15	20	25
Years male may expect to live	68.9	69.1	65.3	60.5	55.6	51.0	46.5
Years female may expect to live	76.6	76.6	72.7	67.8	62.9	58.1	53.3

At the age of	30	35	40	45	50	55	60
Years male may expect to live	41.8	37.2	32.6	28.1	24.0	20.1	16.5
Years female may expect to live	48.4	43.6	38.9	34.3	29.8	25.5	21.4

a. To what age may a 25-year-old female expect to live? 78.3 yr.

b. To what age may a 30-year-old male expect to live? 71.8 yr.

c. How many years longer may a 10-year-old female expect to live than a 10-year-old male? 7.3 yr.

Annual Life Insurance Premiums per $1,000

Age	5-Year Term		Ordinary Life		20-Payment Life		20-Year Endowment	
	Male	Female	Male	Female	Male	Female	Male	Female
18	3.77	3.74	13.48	12.95	23.97	23.22	45.17	45.17
20	4.01	4.01	14.20	13.62	24.91	24.11	45.30	45.30
22	4.20	4.20	15.00	14.35	25.89	25.05	45.46	45.46
24	4.40	4.40	15.86	15.15	26.94	26.03	45.64	45.64
25	4.50	4.50	16.34	15.59	27.49	26.55	45.75	45.75
26	4.60	4.60	16.83	16.04	28.06	27.09	45.86	45.85
28	4.83	4.83	17.90	17.02	29.26	28.21	46.14	46.10
30	5.09	5.05	19.09	18.11	30.54	29.42	46.48	46.40
35	6.08	5.86	22.71	21.43	34.24	32.88	47.72	47.49
40	8.03	7.44	27.20	25.56	38.49	36.85	49.45	49.00
45	11.10	10.13	32.79	30.73	43.34	41.40	51.81	51.06
50	16.00	14.45	40.07	37.50	49.28	46.94	55.40	54.18

2. Find the annual premium on each of the following policies:

a. 20-payment life for $10,000 issued to a woman at age 30. $294.20

b. 5-year term for $50,000 issued to a man at age 40. $401.50

c. 20-year endowment for $5,000 issued to a woman at age 25. $228.75

d. Ordinary life for $60,000 issued to a man at age 20. $852.00

e. 5-year term for $40,000 issued to a woman at age 22. $168.00

f. Ordinary life for $15,000 issued to a man at age 26. $252.45

g. 20-payment life for $80,000 issued to a woman at age 18. $1,857.60

h. 20-year endowment for $25,000 issued to a woman at age 35. $1,187.25

i. Ordinary life for $100,000 issued to a man at age 24. $1,586.00

j. 20-year endowment for $30,000 issued to a man at age 28. $1,384.20

3. How much more is the male's life insurance premium per $1,000 than the female's rate on each of the following policies:

a. 20-year endowment, issued at age 24? $0

b. Ordinary life, issued at age 30? $.98

c. 5-year term, issued at age 45? $.97

d. 20-payment life, issued at age 28? $1.05

e. Ordinary life, issued at age 20? $.58

4. Ms. Myers bought a 20-payment life insurance policy for $20,000 when she was 35 years old. What is the total amount of premium she must pay before the policy is fully paid up? $13,152

5. At the age of 25 Mr. Ryan bought an ordinary life policy for $10,000. What annual premium does he pay? How much will his beneficiary receive when he dies? $163.40; $10,000

6. Mrs. Miller purchased a 20-year endowment policy for $5,000 when she was 30 years old. She is now 50 years old. How much will she receive? What was the total premium she paid? $5,000; $4,640

7. When Mr. Walker was 40 years old he purchased a 5-year term policy for $25,000. He died after paying 3 annual premiums. How much more did the beneficiary receive than Mr. Walker paid? $24,397.75

8. Rosa's brother, now 25 years old, wishes to buy either a 20-payment life policy or an ordinary life policy. What will the difference be in total premiums paid per $1,000 insurance for these policies when he becomes 45 years of age? 55 years of age? 65 years of age? At age 45, 20-payment life is $223 more; at age 55, $59.60 more; at age 65, ordinary life is $103.80 more

EXERCISE

6-17 Property Tax

VOCABULARY: real property tax
real estate tax
assessed value

OBJECTIVE: To compute property tax.

A tax on buildings and land is called a real property or real estate tax. The value that tax officers place on property for tax purposes is the assessed value of the property.

The tax rate on property may be expressed as: cents per $1, mills per $1, dollars and cents per $100, dollars and cents per $1,000, and percent.

There are 10 mills in 1 cent. A mill is one-thousandth of a dollar ($.001) or one-tenth of a cent.

To find the amount of taxes on property, apply the tax rate to the assessed value of the property.

A tax rate of $.03 per $1 =
30 mills per $1 = $3 per $100 = $30 per $1,000 = 3%

Find the taxes on a house assessed for $9,800 when the tax rate is $2.70 per $100.

$$\begin{array}{r} 98 \\ \$100)\overline{\$9,800} \end{array}$$

$2.70 tax rate per $100
× 98 number of $100
21 60
243 0
$264.60 taxes

Answer: $264.60 taxes

Related Resources: *Practical Applications in Mathematics:* Problems, p. 180

PRACTICE PROBLEMS For ex. 1–5 see Extended Answers, p. T94.

1. Write as a decimal part of a dollar: 7 mills; 19 mills; 50 mills; 132 mills; 84 mills; 175 mills; 60 mills; 92 mills

2. Write as cents: 20 mills; 9 mills; 35 mills; 14 mills; 146 mills; 70 mills; 95 mills; 180 mills; 58 mills; 250 mills

3. Write as mills: $.04; $.08; $.005; $.026; $.061; $.105
Write each of the rates in problems 4 through 8 as:

4. Cents per $1: $4 per $100; 28 mills per $1; $43 per $1,000; 3.4%

5. Mills per $1: $5 per $100; $33 per $1,000; 4.5%; $.19 per $1

6. Dollars and cents per $100: $.08 per $1; 5.3%; 36 mills per $1; $19 per $1,000 $8 per $100; $5.30 per $100; $3.60 per $100; $1.90 per $100

7. Dollars and cents per $1,000: $4.50 per $100; 2.9%; 42 mills per $1; $.25 per $1 $45 per $1,000; $29 per $1,000; $42 per $1,000; $250 per $1,000

8. Percent: $4 per $100; $24.50 per $1,000; $.07 per $1; 61 mills per $1 4%; 2.45%; 7%; 6.1%

9. A community requires $1,000,000 in tax money for a year. The assessed valuation of the property in the community is $50,000,000. What percent of the assessed valuation is the tax needed? Express this tax rate in dollars per $100; in cents per $1; in mills per $1. 2%; $2 per $100; $.02 per $1; 20 mills per $1

10. How much must you pay for taxes on a house assessed for $10,200 if the tax rate is $5.60 per $100? $571.20

11. If the tax rate is 43 mills per $1, how much must you pay for taxes on a building assessed for $28,500? $1,225.50

12. Find the amount of taxes on property having the following assessed valuations and tax rates:

a.

Assessed valuation	$16,000	$9,500	$24,000	$41,700
Tax rate per $100	$2.90	$4.25	$3.80	$5.45
	$464	$403.75	$912	$2,272.65

b.

Assessed valuation	$14,800	$8,600	$45,000	$39,500
Tax rate per $1	$.04	$.10	$.058	$.07½
	$592	$860	$2,610	$2,962.50

c.

Assessed valuation	$20,000	$7,500	$43,200	$36,800
Tax rate per $1	26 mills	41 mills	75 mills	69 mills
	$520	$307.50	$3,240	$2,539.20

d.

Assessed valuation	$8,000	$32,000	$15,600	$40,500
Tax rate per $1,000	$25	$53	$91	$33
	$200	$1,696	$1,419.60	$1,336.50

e.

Assessed valuation	$5,000	$28,300	$16,500	$56,000
Tax rate	3%	5%	4.6%	2.8%
	$150	$1,415	$759	$1,568

13. Mr. Moore's house is assessed for 75% of its cost. The tax rate is $2.45 per $100. How much property tax does he pay if the house cost $42,400? $779.10

14. The assessed valuation of the Santini's house is $19,700. The tax rate is $3.40 per $100. What is the yearly property tax? A 3% discount is allowed if the tax is paid within the first quarter of the year. How much will the Santini family save and how much tax will they pay if they take advantage of the discount? $669.80; $20.09; $649.71

15. Which will produce a greater increase in taxes, an increase in the assessment of a property from $24,200 to $26,700 or an increase in the tax rate from $2.80 per $100 to $3.05 per $100? How much greater? increase in assessment; $9.50 greater

EXERCISE

6-18 | Buying and Renting a House

VOCABULARY: mortgage

OBJECTIVE: To compute the costs of buying and renting a house.

When borrowing money to buy a house, the borrower gives the lender a written claim (called a mortgage) to the property in case the borrower fails to repay the specified loan or pay the interest on the loan when due. *To find the amount of the mortgage,* subtract the down payment from the purchase price.

To determine the cost of owning a house, find the sum of the following expenses: interest (see page 586) on mortgage, cost of repairs, depreciation, cost of insurance on the house, property taxes, and interest lost on the principal paid. Although in the computation of the federal income tax we use the property tax and the interest paid on the mortgage as deductions and the interest received on the invested principal as income, we shall disregard these considerations in the following problems.

Related Resources: *Practical Applications in Mathematics:* Problems, p. 181

PRACTICE PROBLEMS

1. If no more than 25% of a person's income should be spent for the rent of a house or apartment, find the highest monthly rent someone earning each of the following incomes can afford to pay:
   a. $18,500 per year $385.42
   b. $750 per month $187.50
   c. $440 per week $476.67

2. If a family should not buy a house costing more than $2\frac{1}{2}$ times its annual income, what is the highest price a family can afford to pay if its income is:
   a. $27,250 per year? $68,125
   b. $1,250 per month? $37,500
   c. $250 per week? $32,500

3. A man bought a house for $25,495. He paid $6,495 down and gave a mortgage for the balance. What is the amount of his mortgage? $19,000

4. Todd's parents bought a house for $47,500, paying $7,500 down. They are charged $9\frac{1}{2}$% interest on the balance. The house is assessed for $20,600 and the tax rate is $5.20 per $100. The house is insured for 80% of its cost at $.28 per $100. The mortgage is to be paid off in 20 equal annual installments. The down payment of $7,500 was formerly invested at $5\frac{1}{4}$% interest.
   a. How much interest is due on the mortgage for the first year? $3,800
   b. What is the property tax? $1,071.20
   c. What does the insurance cost? $106.40
   d. How much interest do Todd's parents lose on the money used as the down payment? $393.75
   e. How much does home ownersip cost Todd's parents for the first year? What is the monthly cost? $5,371.35; $447.61
   f. How much is the annual installment due on the mortgage? $2,000

5. Compare the costs of renting and buying a house when the house can be rented for $385 per month or can be purchased for $39,900 with a down payment of $4,900. It is assessed for $22,500 with a tax rate of $3.85 per $100. Insurance costs $325 for a year. The principal used as a down payment could be invested at $5\frac{1}{2}$% interest. The interest on the mortgage is $8\frac{3}{4}$%. Repairs average $400 per year. The cost incurred during the first year when buying the house is $303.25 more than the cost of renting for the same year. Cost of buying: $4,923.50; renting $4,620.

For Class Activities see pp. T37 and T38.
For a Test on Spending Money see p. T60 or Testing Program, p. 40.

MANAGING MONEY

OBJECTIVE: To write out deposit
slips, checks, and check stubs.

VOCABULARY: savings account   check
checking account   stub
deposit slip

There are two main kinds of banks: savings banks and commercial banks. A third type of banking institution that has grown greatly in recent years is the federal and state savings and loan association, which is very much like a savings bank. Credit unions also render both savings and loan services, but they are usually offered only to particular groups of people, like city employees, factory workers, teachers, etc.

The life of a bank depends on the use of its depositors' money as loans to business people and others at rates of interest higher than the rates of interest that the bank pays to its depositors. The difference between these rates of interest produces money for the bank both to pay its expenses and to earn a profit.

The main service offered by the savings bank is the savings account, on which the depositor is paid interest. Savings banks also provide other services such as passbook loans, safe deposit boxes, and money orders. We use a deposit slip to deposit money and a withdrawal slip to withdraw money from our accounts.

The commercial bank provides almost similar services. However it also provides its most important service—the checking account, a service generally not furnished by the savings bank.

To deposit money in a checking account, we use a deposit slip, a form of which is shown on page 585.

To transfer funds from a checking account to persons or business establishments to whom we owe money, we use a form called a check (see page 585). This method facilitates the payment of bills without any cash being handled directly.

The check stub on the left of the check is filled out for our records.

PRACTICE PROBLEMS
All answers on Students Check Stubs and Deposit Slips.

Use two deposit slip forms and four check forms, or copy them as illustrated, to record the following:

1. On July 1, you have a balance of $1,593.07. Write this as the balance brought forward on first check stub.

2. On July 2, you deposit 4 $20 bills, 6 $5 bills, 12 $1 bills, and a check, numbered 2101–5, drawn on the Valley Bank for $284.75. Fill in deposit slip. Write the total deposit on check stub.

3. On July 5, you send a check to Peter Jaffe for $508.40 for purchases. Fill in the first check stub. Carry the balance to the second check stub. Write out the first check.

LOCAL TRUST COMPANY		
YOUR CITY _____ 19__		
ACCOUNT NO. _____		
NAME _____		
ADDRESS _____		
	DOLLARS	CENTS
CASH →		
CHECKS BY BANK NUMBER		
1		
2		
3		
4		
5		
6		
7		
TOTAL		

No.___ Date_____ 19__	Your City _____ 19__ No. _____
To _____	**LOCAL TRUST COMPANY**
For _____	
Dollars Cents	
Bal. Bro't For'd	Pay to the order of _____ $_____
Am't Deposited	
Total	_____ Dollars
Am't This Check	Memo_____ _____
Bal. Car'd For'd	95 646 718

4. On July 12, you send a check to the Stephen Furniture Co. in the amount of $241.95 for furniture purchased. Fill in the second check stub. Carry the balance to the third check stub. Write out the second check.

5. On July 16, you deposit 16 $10 bills, 9 $5 bills, 23 $1 bills, a check, numbered 2536–9, drawn on the First City Bank for $305.50, and a check, numbered 2158–6, drawn on the Second National Bank for $193.84. Fill in the second deposit slip. Write the total deposit on the third check stub.

6. On July 22, you send a check to the Scott Department Store for $485.69 for merchandise purchased. Fill in the third check stub. Carry the balance to the fourth check stub. Write out the third check.

7. On July 29, you send a check to Andrew Todd, Inc., for $94.95 for a desk purchased. Fill in the fourth check stub. Write the fourth check.

For a Class Activity see p. T38.

6-20 Simple Interest

OBJECTIVE: To find simple interest.

VOCABULARY: interest   principal
    simple interest amount
    rate of interest

Interest is money paid for the use of money. The money borrowed or invested, and on which interest is paid, is called principal. Interest paid on the principal only is called simple interest. The interest charged is generally expressed as a percent of the principal. This percent is called the rate of interest. The rate of interest is usually understood as the rate per year unless specified otherwise. The sum of the principal and the interest is called the amount.

a. *To find the interest,* multiply the principal by the rate of interest per year by the time expressed in years.

$$\text{Formula: } i = prt$$

When using the formula, first write the formula, then substitute the given quantities, and finally compute as required.

The rate may be expressed as a decimal and, when cancellation is possible, as a common fraction.

Find the interest on $900 for 3 yr. 6 mo. at 12%, using the formula.

$p = \$900$ $i = prt$

$r = 12\% = \frac{12}{100}$ $i = \$900 \times \frac{12}{100} \times 3\frac{1}{2}$

$t = 3\frac{1}{2}$ yr. $i = \$378$

$i = ?$

$$\frac{\$900}{1} \times \frac{\overset{6}{\cancel{12}}}{\cancel{100}} \times \frac{7}{\cancel{2}} = \$378$$

Answer: $378

---

b. *To find the amount,* add the interest to the principal.

Find the interest and amount due on $625 borrowed for 7 yr. at 5%.

$625 principal	$625.00 principal
× .05 rate	218.75 interest
———	———
$31.25 interest for 1 year	$843.75 amount
× 7	
———	
$218.75 interest for 7 years	

Answer: $218.75 interest; $843.75 amount

Related Resources: *Practical Applications in Mathematics:* Problems, pp. 176–177

c. Sixty-day or 6% method: *To find the interest for 60 days at 6%,* write the same figures as in the principal but move the decimal point two places to the left.

Find the interest on $582 for 60 days at 6%.   *Answer:* $5.82

---

d. *To find the annual rate of interest,* find what percent the interest for one year is of the principal.

What is the annual rate of interest if the annual interest on a principal of $250 is $20?

$$\frac{\$20}{\$250} = \$20 \div \$250 = \$250\overline{)\$20.00}^{\;.08\,=\,8\%}$$

Answer: 8%

---

e. *To find the principal* when the annual rate and annual interest are known, divide the interest by the rate.

Basic: Prob. 1–3; *Average:* Prob. 1–16; *Enriched:* Prob. 1–20

PRACTICE PROBLEMS

Find the interest on:

1. **a.** $200 for 1 yr. at 15% $30
    b. $750 for 1 yr. at 8% $60
    c. $1,000 for 1 yr. at $5\frac{1}{2}$% $55
    d. $2,800 for 1 yr. at 10.25% $287

2. **a.** $600 for 2 yr. at 7% $84
    b. $1,450 for 8 yr. at 14% $1,624
    c. $2,000 for 6 yr. at $9\frac{3}{4}$% $1,170
    d. $5,400 for 3 yr. at 11.5% $1,863

3. **a.** $900 for $3\frac{1}{2}$ yr. at 10% $315
    b. $500 for $\frac{3}{4}$ yr. at 6% $22.50
    c. $1,600 for $2\frac{1}{4}$ yr. at $13\frac{1}{2}$% $477
    d. $3,600 for $1\frac{2}{3}$ yr. at $8\frac{1}{4}$% $493.02

4. **a.** $700 for 3 mo. at 11% $19.25
    b. $4,800 for 11 mo. at 4% $176
    c. $10,000 for 8 mo. at 9.6% $640
    d. $6,000 for 1 mo. at $7\frac{3}{4}$% $38.75

5. **a.** $50 for 1 yr. 6 mo. at 15% $11.25
    b. $175 for 6 yr. 2 mo. at 6% $64.75
    c. $3,000 for 4 yr. 1 mo. at $9\frac{1}{2}$% $1,163.75
    d. $15,000 for 2 yr. 10 mo. at $8\frac{3}{4}$% $3,720.94

6. Find the semi-annual (6 mo.) interest on:

    6.c. $343.75
    6.d. $445

    a. $400 at 7% $14   **c.** $11,000 at $6\frac{1}{4}$%
    b. $1,250 at 2% $12.50 **d.** $10,000 at 8.9%
    e. $30,000 at 12.47% $1,870.50
    f. $40,000 at 9.383% $1,876.60

7. Find the quarterly (3 mo.) interest on:
    a. $900 at 20% $45   **b.** $8,000 at $10\frac{3}{4}$% $215   **c.** $1,600 at 8.64% $34.56

8. Find the monthly interest on:
    a. $1,200 at 8% $8   **b.** $900 at $12\frac{1}{2}$% $9.375   **c.** $12,000 at 9.791% $97.91

9. Using 1 year = 360 days, 1 month = 30 days, find the interest on:

$4 **a.** $800 for 30 da. at 6% $12 **g.** $500 for 72 da. at 12%

$6.75 **b.** $675 for 90 da. at 4% $21 **h.** $800 for 105 da. at 9%

$60 **c.** $2,400 for 60 da. at 15% $8.87 **i.** $600 for 1 mo. 8 da. at 14%

$106.67 **d.** $4,000 for 120 da. at 8% $10.80 **j.** $960 for 4 mo. 15 da. at 3%

$.50 **e.** $120 for 15 da. at 10% $33.11 **k.** $1,290 for 2 mo. 24 da. at 11%

$25.00 **f.** $1,000 for 45 da. at 20% $34.38 **l.** $1,500 for 3 mo. 10 da. at $8\frac{1}{2}$%

10. Find the exact interest (1 year = 365 days) on:

$5.95 **a.** $425 for 73 da. at 7% 39.5¢ **e.** $240 for 6 da. at 10%

$14.40 **b.** $2,920 for 15 da. at 12% $6.90 **f.** $1,500 for 21 da. at 8%

$24 **c.** $1,000 for 146 da. at 6% $2.79 **g.** $750 for 34 da. at 4%

$17.75 **d.** $2,400 for 30 da. at 9% $24.11 **h.** $5,000 for 11 da. at 16%

11. Use the 60-day, 6% method to find the interest on:

$45 **a.** $4,500 for 60 days at 6% $4.20 **c.** $840 for 30 days at 6%

$193.75 **b.** $19,375 for 60 days at 6% $10 **d.** $2,000 for 30 days at 6%

12. Find the interest and amount on the following loans:

$45; $345 **a.** $300 for 1 yr. at 15% **e.** $4,500 for 9 mo. at 10%

$570; $2,470 **b.** $1,900 for 4 yr. at $7\frac{1}{2}$% **f.** $540 for 3 yr. 5 mo. at 14%

$15; $33.75 **c.** $18.75 for 10 yr. at 8% **g.** $2,700 for 9 yr. 2 mo. at 8.85%

$5,040; $13,040 **d.** $8,000 for 6 yr. at $10\frac{1}{2}$% **h.** $12,000 for 1 yr. 11 mo. at 6%

13. Mr. Thompson bought a house for $43,200. He paid $33\frac{1}{3}$% down. If he gave a mortgage bearing $13\frac{1}{2}$% interest for the remainder, how much interest should he pay semi-annually? $1,903.50

14. Find the exact interest Mrs. Dupont owes if she borrowed $730 on her life insurance policy at 6% from Sept. 8 to Oct. 13. $4.32

15. Mr. Sanchez owns a $1,000 bond bearing 12.95% interest. How much interest does he receive every 6 months? $64.75

16. What is the amount due on $420 borrowed at 15% and repaid at the end of 2 yr. 8 mo.? $588

17. What is the annual rate of interest if the principal is $3,600 and the annual interest is $108? 3%

18. Find the annual rates of interest when the interest for

12% **a.** 1 yr. on $150 is $18 3% **c.** 6 yr. on $5,000 is $900

10% **b.** 3 yr. on $2,000 is $600 $4\frac{1}{2}$% **d.** 4 yr. on $1,600 is $288

19. Mr. Carroll receives $160 interest each year on an investment of $4,000. He also receives $180 semi-annual interest on an investment of $9,000. On which investment does he receive a higher rate of interest?

20. What sum, invested at 8%, will earn $20,000 per year? $250,000

12.e. $337.50; $4,837.50 f. $258.30; $798.30 g. $2,190.38; $4,890.38 h. $1,380; $13,380

19. Neither, the rates are the same.

Borrowing Money

VOCABULARY: collateral loans
demand loans   add-on-interest loan
discount loans   finance charge   amortizing

Money may be borrowed from other people, from banks, from loan companies, from savings and loan associations, from credit unions, and on your life insurance policy. When money is borrowed, there is a charge (called interest) for the use of this money. (See Exercise 6-20.)

Loans which are secured by stocks, bonds, property, etc. are called collateral loans. If the loan is not repaid at the specified time, then the lending institution can collect the amount due by selling the collateral. The rate of interest charged on collateral loans is less than the rate of interest charged on unsecured loans.

Some banks will lend money to customers, allowing them to keep the principal until a specified time but requiring payment of interest monthly. This type of loan is called a demand loan.

I. Discount Loan and Add-on-Interest Loan

The discount loan and the add-on-interest loan are the two types of loans most generally used. *When a discount loan is made,* the lending company immediately takes off the full interest from the principal (amount of the note) and the borrower receives only the difference (called net proceeds). The borrower, however, is required to repay the full principal in a specified number of equal monthly payments. See problem below.

When an add-on-interest loan is made, the borrower receives the full principal. The interest is added on to this principal and the borrower is required to repay the full amount (principal plus interest) in a specified number of equal monthly payments.

Mr. Elliot borrows $900 at 6% annual interest and is required to repay his loan in 12 equal monthly payments. How much money does he receive and what is his monthly payment if he takes a discount loan? If he takes an add-on-interest loan?

Discount Loan	Add-on-Interest Loan
$900 principal (amount of note)	$900 amount received
−54 interest (6% of $900)	$900 principal
$846 amount received	+54 interest
$75 monthly payment	$954 amount to be repaid
12)$900 amount to be repaid	$79.50 monthly payment
	12)$954.00 amount to be repaid

Answer: $846, amount received
$75, monthly payment

Answer: $900, amount received
$79.50, monthly payment

Observe in these loans that since part of the principal is being repaid monthly, the average amount that was borrowed is only about one-half of the original amount of the loan. Therefore the true rate of interest is almost double the stated rate of interest.

Under the terms of many life insurance policies the policyholder is permitted to borrow from the insurance company an amount of money not more than the loan value or the cash surrender value of the policy. The interest charged is simple interest for the period of time the money is borrowed.

Basic: Prob. 1–4; *Average:* 1–5; *Enriched:* Prob. 1–6

PRACTICE PROBLEMS

1.a. 12%; 3%; 18%; 7.2%; 9%   b. 52%; 26%; 91%; 10.4%; 13%
  c. 365%; 36.5%; $273\frac{3}{4}$%; $547\frac{1}{2}$%; $182\frac{1}{2}$%

1. What is the annual percentage rate if the rate of interest charged on a loan:

a. Per month is:	1%?	$\frac{1}{4}$%?	$1\frac{1}{2}$%?	0.6%?	$\frac{3}{4}$%?
b. Per week is:	1%?	$\frac{1}{2}$%?	$1\frac{3}{4}$%?	0.2%?	$\frac{1}{4}$%?
c. Per day is:	1%?	0.1%?	$\frac{3}{4}$%?	$1\frac{1}{2}$%?	$\frac{1}{2}$%?

2. What is the annual rate of interest you are paying when you:
a. Borrow $10 and pay back $12.50 at the end of a month? 300%
b. Borrow $1 and pay back $1.05 at the end of a week? 260%
c. Borrow $5 and pay back $5.10 the next day? 730%

3. What is the actual amount of money (net proceeds) the borrower receives and how much is the monthly payment in each of the following discount loans?

	Amount of Note	Interest Rate	Period of Loan
a.	$600 $564; $50	6%	12 months
b.	$1,500 $1,350; $62.20	10%	24 months
c.	$9,000 $7,650; $500	15%	18 months
d.	$2,700 $2,362.50; $75	$12\frac{1}{2}$%	36 months
e.	$1,800 $1,656; $200	8%	9 months
f.	$4,000 $3,450; $133.33	$5\frac{1}{2}$%	30 months
g.	$7,200 $6,300; $150	13%	48 months
h.	$10,000 $8,550; $833.33	$14\frac{1}{2}$%	12 months
i.	$14,700 $13,781.25; $350	$6\frac{1}{4}$%	42 months
j.	$3,200 $2,952; $53.33	$7\frac{3}{4}$%	60 months

4. What is the actual amount of money the borrower receives and how much is the monthly payment in each of the following add-on-interest loans?

	Principal	Interest Rate	Period of Loan
a.	$600 $636; $53	6%	12 months
b.	$7,200 $9,360; $260	10%	36 months
c.	$3,000 $4,200; $175	20%	24 months
d.	$900 $1,170; $39	12%	30 months
e.	$1,800 $1,800; $49.50	8%	48 months
f.	$6,000 $6,855; $380.83	$9\frac{1}{2}$%	18 months
g.	$2,400 $2,586; $431	$15\frac{1}{2}$%	6 months
h.	$7,500 $7,500; $171.70	$5\frac{1}{4}$%	54 months
i.	$5,000 $5,000; $111.46	$6\frac{3}{4}$%	60 months
j.	$3,800 $4,797.50; $114.23	$7\frac{1}{2}$%	42 months

5. How much interest is due on the following loans on insurance policies?

a. $1,500 borrowed for 6 months at 10%. $75

b. $10,000 borrowed for 4 months at $8\frac{1}{2}$%. $283.33

c. $900 borrowed for 90 days at 9.2%. $20.70

d. $4,000 borrowed for 1 week at 7.75%. $5.96

e. $2,400 borrowed for 1 year 8 months at 12%. $478.08

6. How much interest is due each month on each of the following collateral loans?

a. $4,000 borrowed for 1 year at 9%. $30

b. $8,600 borrowed for 3 months at $10\frac{1}{2}$%. $75.25

c. $10,000 borrowed for 9 months at $12\frac{3}{4}$%. $106.25

d. $50,000 borrowed for 6 months at 13.25%. $552.08

e. $12,500 borrowed for 15 months at $8\frac{1}{4}$%. $68.75

II. Financing a Car

Persons who finance the purchase of automobiles usually pay by monthly installments. (See Exercise 6-14.)

The finance charge is the difference between the cash price and the total amount paid.

The table on the following page indicates some of these monthly payments.

Amount to be Borrowed	Amount of Each Monthly Payment		
	24 months	30 months	36 months
$1,000	$45.83	$37.50	$31.94
1,200	55.00	45.00	38.33
1,500	68.75	56.25	47.91
1,800	82.50	67.50	57.50
2,000	91.66	75.00	63.88
2,500	114.58	93.75	79.86
3,000	137.50	112.50	95.83
3,500	160.41	131.25	111.80

Basic: Prob. 1–3; *Average:* Prob. 1–4; *Enriched:* Prob. 1–5

PRACTICE PROBLEMS

Use the above table in each of the following problems:

1. What is the monthly payment to finance a car if you borrow:

$45 **a.** $1,200 for 30 months? $68.75 **c.** $1,500 for 24 months?

$63.88 **b.** $2,000 for 36 months? $112.50 **d.** $3,000 for 30 months?

2. Find the total payment if you borrow:

$1,980 **a.** $1,800 for 24 months. $2,812.50 **b.** $2,500 for 30 months.

3. Find the total finance charge if you borrow:

$349.84 **a.** $3,500 for 24 months. $250 **c.** $2,000 for 30 months.

$224.76 **b.** $1,500 for 36 months. $449.88 **d.** $3,000 for 36 months.

4. How much more is the total finance charge on the 36-month loan of $2,500 than on the 24-month loan? $125.04

5. How much less is the total finance charge when you borrow $3,500 for 30 months instead of 36 months? $87.30

III. Amortizing a Loan

When a loan is made to purchase a house, each equal monthly payment to pay off the loan consists of (1) a reduction in principal and (2) the interest on the new reduced balance of principal owed each month instead of the entire amount of the loan, since part of each equal monthly payment goes to pay off the principal.

The repayment of a mortgage loan in this way is sometimes described as amortizing the loan. This is different from what we found with discount loans and add-on-interest loans where the borrower pays interest on the full amount of the loan for the entire period of the loan instead of on the balance owed on the loan each month.

The following partial schedule indicates the monthly payments necessary to amortize a loan that was made at $8\frac{1}{2}$% annual interest:

Monthly Payment
Necessary to Amortize a Loan Made at $8\frac{1}{2}$% Annual Interest

TERM AMOUNT	10 Years	15 Years	20 Years	25 Years	30 Years	35 Years	40 Years
$ 5,000	62.00	49.24	43.40	40.27	38.45	37.35	36.66
10,000	123.99	98.48	86.79	80.53	76.90	74.69	73.31
15,000	185.98	147.72	130.18	120.79	115.34	112.03	109.97
20,000	247.98	196.95	173.57	161.05	153.79	149.38	146.62
25,000	309.97	246.19	216.96	201.31	192.23	186.72	183.28
30,000	371.96	295.43	260.35	241.57	230.68	224.06	219.93
35,000	433.95	344.66	303.74	281.83	269.12	261.41	256.59
40,000	495.95	393.90	347.13	322.10	307.57	298.75	293.24
45,000	557.94	443.14	390.53	362.36	346.02	336.09	329.90
50,000	619.93	492.37	433.92	402.62	384.46	373.44	366.55

Basic: Prob. 1–4; *Average:* Prob. 1–5; *Enriched:* Prob. 1–6

PRACTICE PROBLEMS

Use the above table of monthly payments in the following problems:

1. If a person borrowed $40,000 at $8\frac{1}{2}$% annual interest, what is the monthly payment when the loan is to be amortized in:

a. 25 years? $322.10   **c.** 15 years? $393.90   **e.** 20 years? $347.13
b. 40 years? $293.24   **d.** 30 years? $307.57   **f.** 35 years? $298.75

2. If a loan at $8\frac{1}{2}$% annual interest is to be amortized in 25 years, what is the monthly payment when the amount of this loan is:

a. $50,000? $402.62   **c.** $25,000? $201.31   **e.** $15,000? $120.79
b. $35,000? $281.83   **d.** $40,000? $322.10   **f.** $45,000? $362.36

3. What is the monthly payment necessary to amortize a loan of $25,000 at $8\frac{1}{2}$% annual interest in 30 years? What is the payment per year? What is the total payment for 30 years? How much of this total payment is interest on the loan? $192.23; $2,306.76; $69,202.80; $44,202.80

4. Find the total interest that was paid on each of the following loans at $8\frac{1}{2}\%$ annual interest when the loans were amortized in the specified terms:

Amount	Term
a. $10,000	15 years $7,726.40
b. $30,000	40 years $75,566.40
c. $50,000	35 years $106,844.80
d. $45,000	20 years $48,727.20
e. $20,000	10 years $9,757.60
f. $35,000	30 years $61,883.20

5. Find the total payment over the full term required on a loan of $45,000 at $8\frac{1}{2}\%$ annual interest when it is amortized in 30 years. When it is amortized in 40 years. For which term is the interest less? How much less? **$124,567.20; $158,352; 30 yr.; $33,784.80**

6. Mrs. Rizzo purchased a house for $42,500. She paid $7,500 in cash and obtained a mortgage loan at $8\frac{1}{2}\%$ annual interest for the balance. What is the monthly payment due if it is to be amortized in 35 years? **$261.41**

EXERCISE

6-22 Compound Interest

VOCABULARY: compound interest

OBJECTIVE: To find compound interest.

Savings and loan associations and some savings banks advertise that savings left with them earn interest compounded semiannually or quarterly (every 3 months). Some banks even advertise interest compounded daily on passbook savings or on savings certificates.

Compound interest is interest paid on both the principal and the interest earned previously.

To find how much a given principal will amount to compounded at a given rate for a certain period of time by the use of the table, select from the table on page 595 how much $1 will amount to based on the given rate and the given number of years (periods) when the interest is compounded annually; on half the given rate and twice as many periods as the given number of years when the interest is compounded semiannually; on one-fourth the given rate and four times as many periods as the given number of years when the interest is compounded quarterly; and on one-twelfth the given rate and twelve times as many periods as the given number of years when the interest is compounded monthly. Then multiply this amount by the given principal.

Related Resources: *Practical Applications in Mathematics:* Problems, p. 178

COMPOUND INTEREST TABLE
Showing How Much $1 Will Amount to at Various Rates

Periods	1%	2%	3%	4%	5%	6%
1	1.01000	1.02000	1.03000	1.04000	1.05000	1.06000
2	1.02010	1.04040	1.06090	1.08160	1.10250	1.12360
3	1.03030	1.06120	1.09272	1.12486	1.15762	1.19101
4	1.04060	1.08243	1.12550	1.16985	1.21550	1.26247
5	1.05101	1.10408	1.15927	1.21665	1.27628	1.33822
6	1.06152	1.12616	1.19405	1.26531	1.34009	1.41851
7	1.07213	1.14868	1.22987	1.31593	1.40709	1.50362
8	1.08285	1.17165	1.26676	1.36856	1.47745	1.59384
9	1.09368	1.19509	1.30477	1.42330	1.55132	1.68947
10	1.10462	1.21899	1.34391	1.48024	1.62889	1.79084
11	1.11566	1.24337	1.38423	1.53945	1.71033	1.89829
12	1.12682	1.26823	1.42575	1.60102	1.79585	2.01219
13	1.13809	1.29360	1.46852	1.66506	1.88564	2.13292
14	1.14947	1.31947	1.51258	1.73167	1.97992	2.26089
15	1.16096	1.34586	1.55796	1.80093	2.07892	2.39654
16	1.17257	1.37277	1.60469	1.87297	2.18286	2.54034
17	1.18429	1.40023	1.65283	1.94789	2.29201	2.69276
18	1.19614	1.42823	1.70242	2.02580	2.40661	2.85432
19	1.20810	1.45680	1.75349	2.10683	2.52694	3.02558
20	1.22018	1.48593	1.80610	2.19111	2.65328	3.20712

Periods	8%	10%	12%	14%	16%	20%
1	1.08000	1.10000	1.12000	1.14000	1.16000	1.20000
2	1.16640	1.21000	1.25440	1.29960	1.34560	1.44000
3	1.25971	1.33100	1.40492	1.48154	1.56089	1.72800
4	1.36048	1.46410	1.57351	1.68896	1.81063	2.07360
5	1.46932	1.61051	1.76234	1.92541	2.10034	2.48832
6	1.58687	1.77156	1.97382	2.19497	2.43638	2.98598
7	1.71382	1.94871	2.21067	2.50226	2.82621	3.58318
8	1.85092	2.14358	2.47596	2.85258	3.27841	4.29981
9	1.99900	2.35794	2.77307	3.25194	3.80295	5.15977
10	2.15892	2.59374	3.10584	3.70721	4.41143	6.19173
11	2.33163	2.85311	3.47854	4.22622	5.11725	7.43008
12	2.51816	3.13842	3.89597	4.81789	5.93601	8.91609
13	2.71961	3.45226	4.36348	5.49240	6.88578	10.6993
14	2.93718	3.79749	4.88710	6.26133	7.98750	12.8391
15	3.17216	4.17724	5.47355	7.13792	9.26550	15.4070
16	3.42593	4.59496	6.13038	8.13723	10.7479	18.4884
17	3.70000	5.05446	6.86602	9.27644	12.4676	22.1860
18	3.99600	5.55990	7.68995	10.5751	14.4624	26.6232
19	4.31568	6.11589	8.61274	12.0556	16.7764	31.9479
20	4.66094	6.72748	9.64627	13.7434	19.4606	38.3375

To find the compound interest, subtract the principal from the amount.

Find the amount and the interest earned on $800 deposited for 3 years at 12% when the interest is:

Compounded Annually; Rate, 12%; Periods, 3.	Compounded Quarterly; Rate, 3%; Periods, 12.
1.40492 × $800	1.42575 × $800
$1,123.93600	$1,140.60000
= $1,123.94 amount	= $1,140.60 amount
$1,123.94 amount − 800.00 principal	$1,140.60 amount − 800.00 principal
$323.94 interest	$340.60 interest
Answer: $1,123.94, amount $323.94, interest	*Answer:* $1,140.60, amount $340.60, interest

---

Basic: Prob. 1–4; *Average:* Prob. 1–5; *Enriched:* Prob. 1–6

PRACTICE PROBLEMS

1. Find the amount and the interest earned when the interest is compounded annually on:

   a. $100 for 6 yr. at 14% $219.50; $119.50
   b. $900 for 11 yr. at 6% $1,708.48; $808.47
   c. $1,400 for 3 yr. at 16% $2,185.25; $785.25
   d. $10,000 for 15 yr. at 5% $20,789.28; $10,789.28
   e. $8,500 for 7 yr. at 8% $14,567.47; $6,067.47
   f. $2,750 for 20 yr. at 10% $18,500.57; $15,750.57

2. Find the amount and the interest earned when the interest is compounded semiannually on:

   a. $600 for 1 yr. at 12% $674.16; $74.16
   b. $2,300 for 7 yr. at 16% $6,755.51; $4,455.51
   c. $9,000 for 4 yr. at 6% $11,400.93; $2,400.93
   d. $15,000 for 10 yr. at 4% $22,288.95; $7,288.95
   e. $7,450 for 8 yr. at 8% $13,953.63; $6,503.63
   f. $20,000 for 5 yr. at 10% $32,577.80; $12,577.80

3. Find the amount and the interest earned when the interest is compounded quarterly on:

a. $350 for 2 yr. at 12% $443.37; $93.37

b. $2,000 for 4 yr. at 4% $2,345.16; $345.16

c. $1,500 for 1 yr. at 8% $1,623.65; $123.65

d. $10,000 for 5 yr. at 20% $26,532.80; $16,532.80

e. $6,300 for 3 yr. at 8% $7,989.85; $1,689.85

f. $50,000 for 4 yr. at 16% $93,648.50; $43,648.50

4a. $510; $520; $530; $540; $560; $580; $600; $620; $650; $660; $670; $800

4.b. $20; $40; $60; $80; $120; $140; $160; $200; $240; $300; $320; $360; $400

4. A bank advertises an 8% interest rate compounded quarterly on all deposits.

a. What would a deposit of $500 amount to at the end of 3 months? 6 months? 9 months? 1 year? 15 months? 18 months? 2 years? 30 months? 3 years? 42 months? 4 years? 51 months? 5 years?

b. What is the interest earned on a $1,000 deposit at the end of 3 months? 6 months? 9 months? 1 year? 18 months? 21 months? 2 years? 30 months? 3 years? 45 months? 4 years? 54 months? 5 years?

5. At the interest rate of 12% compounded monthly (12 times a year):

a. What will each of the following deposits amount to at the end of one year? (1) $700 (2) $1,250 (3) $4,800 (4) $10,000 $788.77; $1,408.53; $5,408.74; $11,268.20

b. How much interest is earned at the end of 1 year on each of the following deposits? (1) $250 (2) $980 (3) $3,600 (4) $5,000 $31.71; $124.28; $456.55; $634.10

c. What will each of the following deposits amount to?

(1) A deposit of $75 at the end of 8 months $81.28

(2) A deposit of $400 at the end of 3 months $412.68

(3) A deposit of $1,800 at the end of 1 month $1,819.03

d. How much interest is earned on:

(1) A deposit of $60 at the end of 9 months? $5.71

(2) A deposit of $840 at the end of 6 months? $53.26

(3) A deposit of $3,000 at the end of 2 months? $63.41

6. Find the interest earned on a $5,000 deposit at the end of 1 year at each of the following rates:

a. 12% compounded quarterly $627.50

b. 12% compounded annually $600.00

c. 12% compounded semiannually $618.00

d. 12% compounded monthly $634.10

Which rate brings the greatest amount of interest? compounded monthly

Which rate brings the smallest amount of interest? compounded annually

For a Class Activity see p. T37.

Stocks; Bonds; Mutual Funds;
Money Rates

VOCABULARY: price-to-earnings ratio   certificate of deposit
bond   rate of income
date of maturity   yield
face value   mutual fund
par value   net asset fund
market value   no load

I. Stocks

A person buying a share of stock becomes a part-owner of the business and receives dividends from the profits earned by the business.

Stock quotations are in terms of dollars.

> $36\frac{1}{2}$ means $36.50

The P.E. (price to earnings) ratio column heading in the stock quotation refers to an approximate ratio comparing the closing price of a share of each specific stock to its annual earnings. For example, if the P.E. ratio of a certain stock is 9, then the price of the stock is approximately 9 times its annual earnings. To find the annual earnings, divide the closing price by the P.E. ratio number.

Fees are paid to brokers for the service of buying or selling stocks. These fees are subject to change and at present there is no one set of fixed fees for all brokers.

A Typical Fee Schedule

Showing the commissions charged for buying or selling shares of stock when the cost of the stock is:

$100 but under $2,500	1.3% of cost of stock plus $16
$2,500 but under $20,000	0.9% of cost of stock plus $28
$20,000 but under $30,000	0.6% of cost of stock plus $90
Over $30,000	0.4% of cost of stock plus $152

II. Bonds

A bond is a written promise of a private corporation or of a local, state, or the national government to pay a given rate of interest at stated times and to repay the face value of the bond at a specified time (date of maturity).

A person buying a bond is lending money to the business or government and receives interest on the face value of the bond.

The original value of a bond is called par value or face value. The value at which it sells at any given time is called market value. The value is *above par* when the market value is more than the par value and *below par* when the market value is less than the par value.

Bond quotations are in terms of percent. $101\frac{1}{4}$ means $101\frac{1}{4}\%$

In the following bond quotation: MGM $9\frac{3}{4}$s 86 $98\frac{1}{2}$

$9\frac{3}{4}$s   represents a $9\frac{3}{4}$% annual rate of interest.

86   is the date of maturity, 1986.

$98\frac{1}{2}$ is the market price, indicating $98\frac{1}{2}$% of the face value of bond.

To find the rate of income or yield of a bond (see column marked Cur. Yld. in the bond quotations on page 602), find what percent the annual interest is of the market price of the bond.

Varying fees are paid to the brokers for buying and selling bonds.

III. Mutual Funds

A mutual fund is an investment company. It obtains money for investment by selling shares of stock of the company. Mutual fund quotations are expressed as a Net Asset Value (NAV) and an offer price. When we buy shares, we pay the offer price; when we sell shares, we receive the Net Asset Value (NAV) of each share.

N.L. means no load or no commission charged for buying shares. When N.L. is indicated, the offer price is the same as the NAV.

IV. Money Rates

Prime Rate: The prime rate is the base rate commercial banks charge their most valued customers. In recent years this rate ranged from 8% to $21\frac{1}{2}$%.

Certificates of Deposit: Banks and savings and loan associations issue savings certificates called certificates of deposit (CD) paying varying rates of interest depending upon the term of the certificate. These rates can change.

(1) The rates below are typical for the minimum deposit of $500:

Interest Rate		Annual Yield
15.40%	30-month certificate	16.89%
14.00%	4-year certificate	15.25%
13.75%	6-year certificate	14.96%
13.50%	8-year certificate	14.63%

Here the annual yield shows the actual interest rate being earned due to the daily compounding of the savings.

(2) The rates below are typical for the minimum investment of $100,000. These rates change daily.

| 14–29 days, 14.55% | 60–89 days, 14.80% | 120–179 days, 15.15% |
| 30–59 days, 14.65% | 90–119 days, 15.00% | 180–360 days, 14.75% |

(3) Banks also issue a special 6-month certificate of deposit that pays 0.25% interest higher than the yield of the last auction on federal government 6-month (26 weeks) bills or the average yield at the four most recent auctions.

Money Market Funds: Some investment firms manage money market funds that pay comparable interest rates to the CDs. However, the investment is not government issued and the interest rates change daily.

Passbook Interest: Passbook interest on savings is paid by savings institutions at a fixed annual rate of 5.5%. Commercial banks pay 5.25%.

V. Government Bonds; Treasury Bills, Notes, and Bonds

The United States government borrows money by selling savings bonds and Treasury bills (3 mo., 6 mo., or 1 yr.), notes (2 yr. to 10 yr.), or bonds (over 10 yr.).

The EE savings bonds mature in 8 years and sell at 50% of face amount, starting with a $50 bond minimum. They pay 9% if held to maturity. The HH bonds mature in 10 years and pay $8\frac{1}{2}$% if held to maturity with denominations starting at $500.

The Treasury bills, notes, and bonds can be sold at any time. The rates of interest vary. Treasury bills are discounted. Investors pay the auction price (face amount less interest) but receive the full face amount at maturity.

Basic: Prob. 1–10; *Average:* Prob. 1–2; *Enriched:* Prob. 1–15

PRACTICE PROBLEMS

Stock Quotations										
52 Weeks				Yld.	P-E	Sales				Net
High	Low	Stock	Div.	%	Ratio	100s	High	Low	Close	Chg.
$47\frac{7}{8}$	34	INACp	2.60	6.5	5	106	$40\frac{1}{2}$	40	$40\frac{1}{4}$	$-\frac{1}{8}$
$27\frac{1}{2}$	24	IdahoP	2.28	9.1	8	31	$25\frac{1}{8}$	$24\frac{3}{4}$	25	$+\frac{1}{4}$
$25\frac{7}{8}$	$21\frac{7}{8}$	IllPowr	2.28	10	8	314	$22\frac{5}{8}$	22	$22\frac{1}{4}$	. . .
$64\frac{1}{4}$	$44\frac{1}{4}$	IngerR	3.16	6.4	7	197	$50\frac{1}{4}$	$49\frac{3}{4}$	$49\frac{3}{4}$	$-\frac{1}{2}$
$41\frac{7}{8}$	$33\frac{1}{2}$	InIndStl	2.80	8.0	5	200	$35\frac{1}{2}$	$34\frac{5}{8}$	35	$-\frac{3}{4}$
$72\frac{3}{4}$	$50\frac{5}{8}$	IBM	3.44	6.7	9	4248	$51\frac{1}{8}$	$50\frac{3}{4}$	51	$-\frac{1}{8}$
$44\frac{7}{8}$	$26\frac{1}{4}$	IntHarv	2.30	6.0	6	275	$38\frac{3}{8}$	$37\frac{3}{4}$	$38\frac{3}{8}$	$+\frac{3}{8}$
$43\frac{1}{2}$	34	IntMin	2.60	6.2	7	144	$41\frac{5}{8}$	$41\frac{1}{4}$	$41\frac{5}{8}$	$+\frac{1}{8}$
$49\frac{1}{4}$	$35\frac{1}{8}$	IntPaper	2.60	5.2	9	633	$42\frac{5}{8}$	$41\frac{3}{4}$	$42\frac{1}{4}$	$+\frac{5}{8}$
$34\frac{3}{8}$	$26\frac{3}{8}$	IntTT	2.20	7.9	6	961	$28\frac{3}{8}$	$27\frac{3}{4}$	$27\frac{7}{8}$	$-\frac{1}{8}$
28	$23\frac{7}{8}$	IowaPS	2.40	9.8	7	133	$24\frac{3}{4}$	$24\frac{1}{2}$	$24\frac{1}{2}$	$-\frac{1}{4}$
$32\frac{1}{2}$	14	ItekCp	. . .		13	103	$20\frac{3}{8}$	$19\frac{1}{2}$	$20\frac{3}{8}$	$+1$

1. Using the lowest prices listed in the stock quotations, find the cost of each of the following purchases, excluding the commission.

$4,000 **a.** 100 shares of INA Cp $1,268.75 **f.** 25 shares of IBM

$1,950 **b.** 100 shares of Itek Cp $770 **g.** 35 shares of Ill Powr

$4,175 **c.** 100 shares of Int Paper $8,225 **h.** 200 shares of Int Min

$980 **d.** 40 shares of Iowa PS $1,887.50 **i.** 50 shares of Int Harv

$2,081.25 **e.** 75 shares of Int TT $14,925 **j.** 300 shares of Inger R

2. Using the highest prices listed in the stock quotations, find the total cost of each of the following purchases, including the commission.

a. 100 shares of Int Paper $4,328.86   **f.** 50 shares of Iowa PS $1,269.59

b. 100 shares of Int TT $2,891.04   **g.** 75 shares of Idaho P $1,924.88

c. 100 shares of Ill Powr $2,307.91   **h.** 200 shares of INA Cp $8,200.90

d. 100 shares of Int Harv $3,900.04   **i.** 60 shares of Inlnd Stl $2,173.69

e. 100 shares of Int Min $4,227.96   **j.** 400 shares of Itek Cp $8,251.35

3. Using the closing prices listed in the stock quotations, find the net amount due (selling price of stock less commission) on each sale of stock.

a. 100 shares of Int Min $4,097.04   **f.** 200 shares of Int TT $5,496.83

b. 100 shares of Iowa PS $2,402.15   **g.** 30 shares of IBM $1,494.11

c. 100 shares of Idaho P $2,449.50   **h.** 50 shares of Itek Cp $989.51

d. 100 shares of Ill Powr $2,180.08   **i.** 70 shares of INA Cp $2,764.14

e. 100 shares of Int Paper $4,158.98   **j.** 300 shares of Int Harv $11,380.89

For ex. 4–6 see Extended Answers, p. T94.

4. How many points above the year low and below the year high is each of the following stock's closing price?

a. Idaho P   **c.** Int Harv   **e.** IBM

b. Int TT   **d.** Int Min   **f.** Iowa PS

5. The dividend listed for each stock in the stock quotations is for a year. Determine for each stock the amount of the quarterly dividend.

6. Find the annual earnings (to nearest cent) for each of the stocks listed in the stock quotations on page 600.

7. Include the commission in each of the following and find the amount of profit:

a. 100 shares of XYZ, purchased at 24, sold for 32 $696

b. 100 shares of EIS, purchased at $18\frac{1}{2}$, sold for $27\frac{1}{4}$ $782.43

c. 50 shares of LJL, purchased at $33\frac{3}{4}$, sold for 40 $232.56

d. 85 shares of MSL, purchased at $45\frac{7}{8}$, sold for $51\frac{3}{4}$ $368.69

e. 30 shares of ASP, purchased at $37\frac{5}{8}$, sold for $46\frac{1}{8}$ $190.34

8. Include the commission in each of the following and find the amount of loss:

a. 100 shares of ZZZ, purchased at 47, sold for 38 **$1,032.50**
b. 100 shares of EBS, purchased at $25\frac{1}{4}$, sold for 20 **$617.73**
c. 60 shares of SHJ, purchased at $18\frac{5}{8}$, sold for $12\frac{1}{2}$ **$423.78**
d. 25 shares of NMN, purchased at $41\frac{3}{8}$, sold for $29\frac{3}{4}$ **$345.75**
e. 78 shares of PQR, purchased at $62\frac{1}{8}$, sold for $47\frac{7}{8}$ **$1,244.72**

Mutual Funds					
	NAV	Offer Price		NAV	Offer Price
Bullock Fd	15.08	16.70	Mass Fd	11.87	12.97
Delaware	15.17	16.58	Phila Fd	9.59	10.48
Fidelity Fd	17.84	N.L.	Pilgrim Fd	13.97	15.06
Ivest	12.26	N.L.	Value Line	15.27	15.66

9. Use the mutual fund quotations above to do the following:

a. Find the cost of buying:
   (1) 100 shares of Pilgrim Fd **$1,506** (3) 50 shares of Phila Fd **$524**
   (2) 100 shares of Ivest **$1,226**   (4) 75 shares of Mass Fd **$972.75**

b. How much money should you receive when you sell each of the following?
   (1) 100 shares of Bullock Fd **$1,508** (3) 100 shares of Fidelity Fd **$1,784**
   (2) 40 shares of Value Line **$610.80** (4) 85 shares of Delaware **$1,289.45**

Bonds	Cur. Yld.	Vol.	High	Low	Close	Net Chg.
Alcoa $5\frac{1}{4}$s 91	cv*	100	$100\frac{1}{2}$	$100\frac{1}{4}$	$100\frac{1}{2}$	$-\frac{1}{2}$
ATT $8\frac{5}{8}$s 07	9.2	12	$94\frac{1}{4}$	94	94	$+\frac{1}{4}$
Digit $4\frac{1}{2}$s 02	cv*	87	$108\frac{1}{2}$	$108\frac{1}{4}$	$108\frac{1}{4}$	$-\frac{3}{4}$
Duke P $9\frac{1}{2}$s 05	9.7	10	$97\frac{1}{2}$	$97\frac{1}{2}$	$97\frac{1}{2}$	$-\frac{1}{2}$
Exxon $6\frac{1}{2}$s 98	8.2	19	79	$78\frac{5}{8}$	79	$+\frac{3}{8}$
Ford $9\frac{1}{4}$s 94	9.3	5	100	100	100	$+\frac{1}{2}$
GMA 8s 93	9.1	28	89	$88\frac{1}{4}$	$88\frac{3}{8}$	$-\frac{7}{8}$
PGE $9\frac{1}{2}$s 85	9.5	35	100	$99\frac{7}{8}$	$99\frac{7}{8}$	$-\frac{3}{8}$
Phila El $11\frac{5}{8}$s 00	11	3	$110\frac{1}{4}$	$110\frac{1}{4}$	$110\frac{1}{4}$	$+1\frac{3}{4}$
St O Ind 9.2s 04	9.2	4	$99\frac{3}{4}$	$99\frac{1}{2}$	$99\frac{3}{4}$	$+\frac{1}{4}$

*The abbreviation "cv" indicates convertible bonds.

10. Find the date of maturity and the interest rate for each of the following bonds:

   a. PGE 5s 89 1989; 5%

   b. Beth Steel 9s 00 2000; 9%

   c. Gulf Oil $8\frac{1}{2}$s 95 1995; $8\frac{1}{2}$%

   d. Rydr $11\frac{1}{2}$s 90 1990; $11\frac{1}{2}$%

   e. Dow 6.7s 98 1998; 6.7%

   f. Duke P $7\frac{3}{4}$s 02 2002; $7\frac{3}{4}$%

11. a. Find the cost of each of the following purchases, excluding commissions, using the highest prices in the bond quotations.

   b. Determine how much interest the purchaser will receive semi-annually.

   (1) One GMA 8s 93 bond, face value $1,000 $890; $40

   (2) Ten Ford $9\frac{1}{4}$s 94 bonds, face value $1,000 each $10,000; $462.50

   (3) Forty Phila El $11\frac{5}{8}$s 00 bonds, face value $1,000 each $44,100; $2,325

   (4) Five St 0 Ind 9.2s 04 bonds, face value $1,000 each $4,987.50; $230

   (5) Twenty ATT $8\frac{5}{8}$s 07 bonds, face value $1,000 each $18,850; $862.50

   (6) Fifty Alcoa $5\frac{1}{4}$s 91 bonds, face value $1,000 each $50,250; $1,312.50

12. For each bond listed above in the bond quotations, find the current yield (percent the annual interest is of the closing market price) to the nearest tenth of a percent, and check with the current yield column (cur. yld.). The face value of each bond is $1,000.

See Bond Quotations

13. Using the annual yield rates found on page 599, find the total interest earned by each of the following:

   a. A 4-year certificate of deposit for $10,000 $1,525

   b. A $2\frac{1}{2}$-year certificate of deposit for $25,000 $4,222.50

14. What rate of interest (see page 600) is effective on the special 6-month certificate of deposit issued by savings banks when the Treasury announces on the 26 week treasury bill an interest rate of:

   a. 13.906%? 14.156% **b.** 15.072%? 15.322% **c.** 9.452%? 9.702% **d.** 10.387%? 10.637%

15. Mrs. Berger owns three $10,000 Treasury Notes that pay an annual rate of $8\frac{7}{8}$%. How much interest does she receive semiannually?
$1,331.25

16. Using the rates found on page 600, find the interest earned on a $100,000 CD purchased for 20 days. For 75 days. For 90 days. $797.26; $3,041.10; $3,698.63

For a Class Activity see p. T33.

For a Test on Managing Money see p. T61 or *Testing Program*, p. 42.

APPLICATIONS IN BUSINESS AND TRAVEL

VOCABULARY: discount   cash discount
                   list price   net amount
                   marked price
                   net price
                   sale price
                   rate of discount
                   successive discount
                   trade discount

EXERCISE 6-24 | Discount

OBJECTIVE: To solve problems involving discount.

Discount is the amount that an article is reduced in price. The regular or full price of the article is called the list price or marked price. The price of the article after the discount is deducted is called the net price or sale price. The percent taken off is called the rate of discount. This rate is sometimes expressed as a common fraction. Manufacturers and wholesale houses sometimes allow trade discounts.

Often when a bill is paid at the time of purchase or within a specified time thereafter, a reduction called cash discount is allowed. The terms of sale are usually stated on the bill. The terms 2/10, $n/30$ mean that a 2% discount is allowed if the bill is paid within 10 days and the full amount is due by the 30th day. The net amount is the amount of the bill after the cash discount is deducted.

When two or more discounts are given, they are called successive discounts.

(1) *To find the discount when the list price and the rate of discount are given,* multiply the list price by the rate of discount.
(2) *To find the net price,* subtract the discount from the list price.

Find the discount and the net price when the list price is $290 and the rate of discount is 15%.

$290 list price	$290.00 list price
×.15 rate	−43.50 discount
1450	$246.50 net price
290	
$43.50 discount	

Answer: $43.50 discount; $246.50 net price

(3) *To find the rate of discount,* find what percent the discount is of the list price.

What is the rate of discount when the list price is $48 and the discount is $12?

$$\frac{\$12}{\$48} = \frac{1}{4} = 25\%$$

Answer: 25%

(4) *To find the list price when the net price and rate of discount are given,* subtract the given rate from 100%, then divide the answer into the net price.

(5) *To find the net amount,* subtract the cash discount from the original amount of the bill.

When two or more successive discount rates are given, do not add the rates but use them one at a time.

Basic: Prob. 1–10; *Average:* Prob. 1–12; *Enriched:* Prob. 1–15

PRACTICE PROBLEMS

1. Mr. Warner bought a camera at a 15% discount. If the regular price was $56, how much discount was he allowed? What net price did he pay? $8.40; $47.60

2. Find the discount and the net price:

List price	$43	$.96	$31.68	$170.94	$95.20
Rate of discount	8%	6%	25%	42%	37½%

3. How much trade discount is allowed if the catalog lists a metric tool set at $129.95 and the discount sheet shows a 16% discount? What is the net price? $20.79; $109.16

2. $3.44; $39.56; $.06; $.90; $7.92; $23.76; $71.79; $99.15; $35.70; $59.50

4. Find the trade discount and the net price:

Catalog list price	$183	$13.49	$93.75	$514	$262.50
Rate of trade discount	27%	12%	20%	$16\frac{2}{3}$%	9%

$49.41, $133.59; $1.62, $11.87; $18.75, $75; $85.67, $428.33; $23.63, $238.87

$2.92; $94.48 **5.** Find the cash discount received and the net amount paid if a bill for $97.40 is paid upon receipt. The terms of payment were 3/10, $n/30$.

6. Find the cash discount and net amount paid:

Amount of bill	$305	$179.60	$1,500	694.75	$843.50
Terms of payment	2/10, n/30	9/15, n/60	3/10, n/30	5/15, n/60	1/10, n/30
Date of bill	May 11	July 8	April 17	June 9	March 14
Date of payment	May 16	July 20	April 19	June 22	March 28

$6.10, $298.90; $16.16, $163.44; $45, $1,455; $34.74, $660.01; 0, $843.50

7. What is the net price of a drill press listed at $342.95 with a trade discount of 12% and an additional cash discount of 5%? $286.71

8. Find the net price:

List price	$160	$539.80	$316.50	$875	$920.25
Rates of discount	25%, 4%	30%, 2%	15%, 10%	20%, 8%	10%, 3%

$115.20 $370.30 $242.12 $644 $803.38

24% **9.** Find rate of discount when a radio, listed for $125, is sold for $95.

10. Find the rate of discount:

List price	$80	$175	$12.50	$248	$58.29
Net price	$68	$140	$7.50	$210.80	$38.86

15%   20%   40%   15% $33\frac{1}{3}$%

11. What is the rate of trade discount allowed when the catalog list price of a water heater is $190 and the net price is $161.50? 15%

$900 **12.** On what list price is the discount $63 when rate of discount is 7%?

13. Find the list price when the net price is $280 and the rate of discount is $12\frac{1}{2}$%. $320

14. Find the list price:

Net price	$42	$368	$73.80	$31.32	$208.95
Rate of discount	25%	8%	10%	$16\frac{2}{3}$%	30%

$56 $400 $82 $37.58 $298.50

15. The net price of a bicycle is $102 when a 15% discount is allowed. What is its list price? $120

For a Computer Activity see p. T40.

Commission

OBJECTIVE: To solve problems involving commission.

VOCABULARY: commission   **net proceeds**
           rate of commission   **agents**

A salesperson who sells goods for another person is usually paid a sum of money based on the value of the goods sold. This money which is received for the services is called commission. When it is expressed as a percent, it is called the rate of commission. The amount that remains after the commission is deducted from the total selling price is called net proceeds. When a buyer purchases goods for another person, the buyer also is usually paid a commission for the services based on the cost of the goods. Salespeople and buyers are sometimes called agents.

(1) Selling: *To find the commission when the amount of sales and rate of commission are given,* multiply the amount of sales by the rate of commission.

To find the net proceeds, subtract the commission from the amount of sales.

Find the commission and net proceeds when the sales are $364.89 and the rate of commission is 4%:

$364.89 sales
×.04 rate
‾‾‾‾‾‾‾‾‾
$14.5956
= $14.60 commission

$364.89 sales
−14.60 commission
‾‾‾‾‾‾‾‾‾‾‾‾‾‾
$350.29 net proceeds

Answer: $14.60 commission; $350.29 net proceeds

---

(2) Buying: *To find the commission when the cost of goods purchased and rate of commission are given,* multiply the cost of goods purchased by the rate of commission.

To find the total cost, add the commission to the cost of goods purchased.

(3) *To find the rate of commission,* find what percent the commission is of the amount of sales (or cost of goods purchased).

Find the rate of commission if the commission is $18.40 on sales of $230:

$$\frac{\$18.40}{\$230} = \$230\overline{)\$18.40} \quad .08 = 8\%$$
$$\underline{1840}$$

Answer: 8% commission

---

(4) *To find the sales when the commission and rate of commission are given,* divide the commission by the rate.

Basic: Prob. 1–10; *Average:* Prob. 1–12; *Enriched:* Prob. 1–15

PRACTICE PROBLEMS

1. A real estate agent sold a house for $43,500 and was paid a commission of 7%. What was the amount of the agent's commission? What were the net proceeds the owner received? **$3,045; $40,455**

2. Find the commission and net proceeds:

Sales	$426	$93.40	$6,570	$39.15	$229.48
Rate of commission	14%	5%	6%	20%	9%

3. Find the rate of commission if the commission is $78 on sales amounting to $650. **12%**

4. Find the rate of commission: **15%; 18%; 20%; 4%; 9.055%**

Sales	$860	$965	$824.60	$748.50	$3,267
Commission	$129	$173.70	$164.92	$29.94	$295.83

5. What is the rate of commission when sales amount to $6,300 and net proceeds are $6,111? **3%**

6. What must the sales be for a 6% rate of commission to bring a commission of $150? **$2,500**

7. Find the amount of sales: **$2,300; $520; $136.60; $390; $65.80**

Commission	$253	$65	$6.83	$54.60	$9.87
Rate of commission	11%	$12\frac{1}{2}$%	5%	14%	15%

8. Find the commission and total cost of goods purchased:

Cost of goods	$628	$879	$329.75	$956.40	$1,297.50
Rate of commission	9%	$5\frac{1}{2}$%	7%	$8\frac{1}{4}$%	10%

9. Alice's mother works in a department store. She receives $105 per week and 1% commission on sales. How much did she earn when she sold $12,028 worth of merchandise? **$225.28**

10. A salesman sold 9 window fans at $77.99 each and 7 air conditioners at $224.95 each. At 4% commission, how much did he earn? **$91.06**

11. A commission merchant, charging $7\frac{1}{2}$% commission, sold for a grower 290 boxes of melons at $5.65 a box and 475 boxes of plums at $6.40 a box. What net proceeds should the grower receive? **$4,327.61**

2. **$59.64, $366.36; $4.67, $88.73; $394.20, $6,175.80; $7.83, $31.32; $20.65, $208.83**
8. **$56.52, $684.52; $48.35, $927.35; $23.08, $352.83; $78.90, $1,035.30; $129.75, $1,427.25**

12. A lawyer collected a debt of $824 for a client, charging $123.60 for her services. What rate of commission did she charge? 15%

13. Abe's brother receives $90 per week as salary and an additional 2% commission on the amount of his sales. If he earned $261.34 as his total income in a week, find the amount of his sales. $8,567

14. If an agent charges $4\frac{1}{2}$% commission for purchasing goods, how much will he receive for buying 625 boxes of grapes at $8.20 a box? $230.63

15. Chad's sister is paid a weekly salary of $150 and a 3% commission on all sales over the first $1,000 sales each week. During the week when her sales were $4,297, what were her total earnings? $248.91

EXERCISE

Profit and Loss **6-26**

OBJECTIVE: To solve problems involving profit and loss.

Merchants are engaged in business to earn a profit. To determine the profit, each merchant must consider the cost of goods, the selling price of the same goods, and the operating expenses. The cost is the amount the merchant pays for the goods. The selling price is the amount the merchant receives for selling the goods. The operating expenses or overhead are expenses of running the business. They include wages, rent, heat, light, telephone, taxes, insurance, advertising, repairs, supplies, and delivery costs.

The difference between the selling price and the cost is called the margin. The terms gross profit, spread, and markup sometimes are also used. The margin is often erroneously thought of as the profit earned. Actually, the profit earned, called net profit, is the amount that remains after both the cost and the operating expenses are deducted from the selling price.

To earn a profit the merchant must sell goods at a price which is greater than the sum of the cost of the goods and the operating expenses. If the selling price is less than the sum of the cost and operating expenses, the goods are sold at a loss.

The profit may be expressed as a percent of either the cost or the selling price. This percent is called the rate of profit. (The loss may also be expressed in a similar way.)

Related Resources: *Practical Applications in Mathematics:* Problems, pp. 193–195

To determine the selling price of the articles they sell, some merchants use the percent markup (or gross profit) based on the cost while other merchants use the rate based on the selling price. This percent markup is an estimated rate large enough to cover the proportionate amount of overhead expenses to be borne by each article and the amount of profit desired. The percent markup (or gross profit) is the percent the margin is of the cost or of the selling price.

To find the selling price: (1) Add the cost, operating expenses, and net profit; or (2) add the cost and margin; or (3) multiply the cost by the percent markup on the cost, then add the product to the cost.

If the rate of markup on cost is 38%, what is the selling price of an article that costs $65?

$$\begin{array}{r} \$65 \text{ cost} \\ \times .38 \text{ rate} \\ \hline 520 \\ 195 \\ \hline \$24.70 \text{ markup} \end{array}$$

$$\begin{array}{r} \$65.00 \text{ cost} \\ +24.70 \text{ markup} \\ \hline \$89.70 \text{ selling price} \end{array}$$

Answer: $89.70 selling price

To find the selling price when the cost and percent markup (or gross profit) on the selling price are given, subtract the percent markup from 100%, then divide the answer into the cost.

Find the selling price of an article that costs $54 and the percent markup on the selling price is 28%.

$$\begin{array}{r} 100\% = \text{selling price} \\ -28\% = \text{markup} \\ \hline 72\% = \text{cost} \end{array}$$

$$\begin{array}{r} \$75 \\ .72_\wedge \overline{)\$54.00_\wedge} \\ \underline{50\ 4} \\ 3\ 60 \\ 3\ 60 \end{array}$$

Answer: $75 selling price

To find the margin or gross profit or markup: (1) Subtract the cost from the selling price; or (2) multiply the cost by the percent markup on cost; or (3) add the operating expenses and net profit.

To find the net profit: (1) Subtract the operating expenses from the margin; or (2) add the cost and the operating expenses, and subtract this sum from the selling price.

To find the loss: Add the cost and the operating expenses, and subtract the selling price from this sum.

To find the percent markup (or gross profit) on cost, find what percent the margin is of the cost. When based on selling price, find what percent the margin is of the selling price.

Find the rate of markup on cost if an article costs $60 and sells for $80.

$80 selling price
-60 cost
$20 markup

$$\frac{\$20}{\$60} = \frac{1}{3} = 33\frac{1}{3}\%$$

Answer: $33\frac{1}{3}\%$ markup on cost

To find the rate of net profit (or loss) on the selling price, find what percent the net profit (or loss) is of the selling price. When based on the cost, find what percent the profit (or loss) is of the cost.

Find the net profit and rate of net profit on the selling price if an article costs $25.50, with operating expenses of $2.50 and selling price of $32.

$25.50 cost
$+2.50$ operating expenses
$28.00 total cost

$32 selling price
-28
$4 net profit

$$\frac{\$4}{\$32} = \frac{1}{8} = 12\frac{1}{2}\%$$

Answer: $4 net profit; $12\frac{1}{2}\%$ net profit on selling price

PRACTICE PROBLEMS

1. A video cassette recorder costs $590. If the rate of markup on cost is 35%, what is the selling price? $796.50

2. Find the selling price: $89.78; $1,051.70; $186.90; $615.94; $72.60

Cost	$67	$809	$133.50	$492.75	$52.80
Markup on cost	34%	30%	40%	25%	$37\frac{1}{2}$%

3. Find the net profit: $28; $29.85; $1.82; $55.40; $128.35

Cost	$72	$61.90	$3.98	$203	$429.75
Operating expenses	$9	$6.75	$.45	$8.59	$31.40
Selling price	$109	$98.50	$6.25	$266.99	$589.50

4. Find the loss if the cost is $96.50, operating expenses are $3.60, and the selling price is $89.95. $10.15

5. What is the selling price of a stereo digital clock radio if it costs $63 and the rate of markup on the selling price is 30%? $90

6. Find the selling price: $48; $12.50; $.56; $260.64; $131.25

Cost	$36	$7.50	$.28	$162.90	$87.50
Markup on selling price	25%	40%	50%	$37\frac{1}{2}$%	$33\frac{1}{3}$%

7. Find the rate of markup (margin or gross profit) on cost:

Cost	$150	$316	$24.90	$496.50	$59.80
Selling price	$240	$417.12	$37.35	$695.10	$74.75

8. Find the rate of gross profit on selling price: 20%; 36%; 30%; 35%; $33\frac{1}{3}$%

Cost	$280	$.48	$6.65	$508.69	$86.50
Selling price	$350	$.75	$9.50	$782.60	$129.75

9. What is the rate of net profit on the selling price if the net profit is $3.50 and the selling price is $28? $12\frac{1}{2}$%

10. Find the rate of net profit on the selling price:

Cost	$53	$93.40	$2.98	$69.89	$415.36
Operating expenses	$7	$14.60	$.24	$4.61	$52.44
Selling price	$75	$120	$3.50	$89.40	$584.75

20%   10%   8% $16\frac{2}{3}$%   20%

7. 60%; 32%; 50%; 40%; 25%

11. What is the rate of loss on the selling price if the loss is $10.50 and the selling price is $50? 21%

12. What is the rate of net profit on the cost:

Cost	$60	$425	$19.40	$.80	$1,000
Operating expenses	$10	$37.50	$2.50	$.15	$60.40
Selling price	$85	$550	$27.75	$1.35	$1,149.50
	25%	20.59%	30.15%	50%	8.91%

13. Find the selling price of each of the following articles:
a. Watch, costing $86.50 and 40% markup on cost. $121.10
b. Refrigerator, costing $270 and 25% markup on cost. $337.50
c. Set of golf clubs, costing $140 and 30% markup on selling price. $200
d. Shoes, costing $24 and $33\frac{1}{3}$% markup on selling price. $36

14. Find the rate of gross profit on the selling price of each of the following:
a. Desk which costs $160 and sells for $240. $33\frac{1}{3}$%
b. Tire which costs $37.50 and sells for $45. $16\frac{2}{3}$%
c. Mirror which costs $38.10 and sells for $57.15. $33\frac{1}{3}$%
d. Dress which costs $18.60 and sells for $31.89. 41.67%

15. At what price should a dealer sell a hand calculator costing $63 to realize a profit of 40% of the selling price? $105

16. Shirts cost a merchant $54 a dozen. At what price must each shirt be sold to make a profit of $33\frac{1}{3}$% on the total cost? $6

17. A dealer bought a table for $76.80 and marked it to sell for a profit of 35% on the cost. If it was finally sold at a reduction of 10% on the marked price, what was the selling price and the amount of profit? $93.31; $16.51

18. Barry's aunt bought a car for $4,160. She sold it for $3,640. What was her rate of loss on the cost of the car? $12\frac{1}{2}$%

19. A storekeeper bought baseballs at $19.20 per dozen. If his overhead expenses are $12\frac{1}{2}$% of the cost and he wishes to make a net profit of 25% on the cost, how much should he charge for each ball? $2.20

20. A dealer wishes to make a 35% profit on the selling price of a rug that cost her $130. What should the selling price be? $200

For a Class Activity see p. T38.

OBJECTIVE: To read a road map.

Basic: Prob. 1
Average: Prob. 1–2

LEGEND

The route symbol ⬡ indicates an interstate highway.

The route symbol ⬡ indicates a federal highway.

The route symbol ◯ indicates a state highway.

The symbol ▮ is paired with a like symbol to indicate mileage between two points.

1.a. ⬡52 to ⬡74 ;
Shelbyville, Rushville

b. ⬡65 all the way;
Seymour, Franklin

c. ⬡71 all the way;
Carrollton

d. Indianapolis to Cincinnati:
⬡74 all the way;

Louisville to Indianapolis:
⬡31 all the way;

Cincinnati to Louisville:
⬡42 all the way

PRACTICE PROBLEMS

1. Listing highway markers and towns, what route would you take to drive by car from:

a. Indianapolis to Cincinnati?
b. Louisville to Indianapolis?
c. Cincinnati to Louisville?
d. What alternate routes could you take for each?

2. How far is it by highway from:

a. Shelbyville to Connersville?
b. Columbus to Madison?
c. Seymour to Aurora?

2.a. 36 mi.
b. 47 mi.
c. 50 mi.

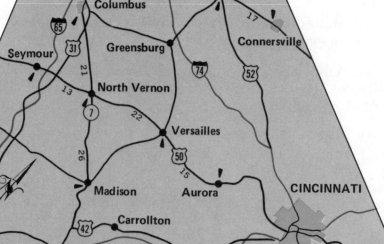

Reading a Timetable

OBJECTIVE: To read a timetable.

As you will see, timetables vary in appearance.

The *train timetable* (below) lists in columns the time of arrival or departure of the train at each stop.

The *bus timetable* (page 616) lists the times for a round trip. It is read down on the left side and read up on the right side.

The *airlines schedule* (page 617) gives separate listings for each city.

Basic: Prob. 1–3, 6–8, 11–13; *Average:* Prob. 1–4, 6–9, 11–14; *Enriched:* Prob. 1–15

PRACTICE PROBLEMS

I. Train Timetable

Train Number		123	177	179	193	195
Train Name		Metro-liner Service	The Senator	Mer-chants Limited	Evening Liberty Express	Ben-jamin Franklin
Frequency of Operation		Daily	Daily	Daily	Daily	Daily

km	mi.			123	177	179	193	195
0	0	Boston, MA	Dp		1 45 P	4 00 P	5 30 P	7 00 P
19	12	Route 128, MA			2 03 P	4 18 P	5 48 P	7 18 P
71	44	Providence, RI			2 39 P	4 54 P	6 25 P	7 54 P
114	71	Kingston, RI				5 19 P		8 20 P
171	106	New London, CT			3 44 P	6 02 P	7 25 P	9 01 P
253	157	New Haven, CT	Ar		4 48 P	7 07 P	8 17 P	10 00 P
			Dp		4 58 P	7 17 P	8 27 P	10 10 P
280	174	Bridgeport, CT			5 18 P	7 37 P		10 30 P
315	196	Stamford, CT			5 41 P	8 00 P	9 10 P	10 55 P
373	232	New York, NY	Ar		6 40 P	8 50 P	10 00 P	11 47 P
			Dp	5 30 P	7 00 P	9 00 P	10 10 P	12 01 A
389	242	Newark, NJ		5 45 P	7 15 P	9 17 P	10 25 P	12 17 A
452	281	Princeton Jct., NJ					11 02 P	1 01 A
467	290	Trenton, NJ			8 02 P	9 59 P	11 13 P	1 11 A
518	322	Philadelphia, PA		6 51 P	8 44 P	10 33 P	11 50 P	1 55 A
560	348	Wilmington, DE		7 17 P	9 14 P	11 00 P		
669	416	Baltimore, MD		8 11 P	10 17 P	12 04 A		
734	456	Washington, DC	Ar	8 53 P	11 02 P	12 45 A		

The letter A indicates A.M. The letter P indicates P.M.

1. 7:00 P.M.; 10:00 P.M.; 12:17 A.M.; does not arrive in Wilmington

1. At what time does train 195 leave Boston? At what time does it arrive in New Haven? In Newark? In Wilmington?

2. At what time does train 193 leave Providence? At what time does it arrive in New York? In Trenton? In Baltimore?

3. How far (in kilometers) is it by rail from Boston to Philadelphia? How long does it take train 177 to go from Boston to Philadelphia? What is its average rate of speed during this trip? 518 km; 6 hr. 59 min.; 74.2 km/h

2. 6:25 P.M.; 10:00 P.M.; 11:13 P.M.; does not arrive in Baltimore

4. Train 179 takes 22 min. longer.

4. Compare the time it takes the Metroliner to travel from New York to Washington to that of train 179. What is the difference in time?

5. If you wish to travel from Boston to Washington, arriving in Washington just before midnight, what train should you take? At what time does it leave Boston? At what time does it arrive in Washington? How far (in miles) is it by rail from Boston to Washington?

5. The Senator; 1:45 P.M.; 11:02 P.M.; 456 miles

II. Bus Timetable

Read Down							Read Up		
824	822	820 ←		— Bus Number —		→ 823	825	827
3 30	9 30	**3 15**	Lv	Atlanta, Ga.	EST	Ar	7 00	10 55	6 35
4 25	10 25	↓	Lv	Anniston	CST	Ar	**3 30**	8 05	↑
6 00	11 59	5 05	Ar	Birmingham, Ala.		Lv	**1 45**	6 40	**2 45**
6 30	**12 45**	6 00	Lv	Birmingham		Ar	**1 05**	6 00	**2 20**
10 00	**4 35**	9 20		Tupelo, Miss.			9 35	**2 15**	**10 50**
12 30	**7 20**	**12 15**	Ar	Memphis, Tenn.		Lv	7 00	11 30	**8 20**
1 10	**8 30**	**1 30**	Lv	Memphis		Ar	6 00	10 50	**7 00**
4 55	**12 15**	**5 15**	Ar	Little Rock, Ark.		Lv	**2 15**	7 05	**3 15**
5 15	**12 30**	**6 00**	Lv	Little Rock		Ar	**1 55**	6 40	**2 40**
9 00	**3 40**	**9 10**	Ar	Fort Smith, Ark.		Lv	10 45	**3 20**	10 55
9 10	**3 55**	**10 10**	Lv	Fort Smith		Ar	10 30	**2 55**	10 30
12 10	6 30	12 55		McAlester, Okla.			8 00	↑	↑
3 30	9 30	**3 45**	Ar	Oklahoma City		Lv	**4 15**	11 10	**5 45**
5 00	11 40	**4 45**	Lv	Oklahoma City		Ar	**2 40**	10 45	**4 40**
11 25	**6 10**	10 35	Ar	Amarillo, Tex.		Lv	8 15	**4 30**	11 25
11 55	**6 50**	11 00	Lv	Amarillo	CST	Ar	7 30	**4 15**	11 10
4 50	**11 45**	**4 10**	Ar	Albuquerque, N.M.	MST	Lv	12 35	9 00	**3 45**
6 30	**1 00**	**5 15**	Lv	Albuquerque		Ar	11 45	8 00	**2 30**
9 15	**3 40**	**8 05**		Gallup, N.M.			9 00	5 15	11 45
1 20	**7 40**	**12 10**	Ar	Flagstaff, Ariz.		Lv	**4 40**	1 15	**7 45**
5 55	11 40	**3 40**	Ar	Kingman, Ariz.	MST	Lv	**1 05**		**3 50**
9 40	**3 20**	**6 50**	Ar	Barstow, Cal.	PST	Lv	7 00		**10 30**
10 15	**4 50**	**7 25**	Lv	Barstow		Ar	6 20		9 50
1 10	**7 50**	**10 25**	Ar	Bakersfield		Lv	**3 10**		**6 50**
3 30	**10 55**	**1 00**	Ar	Fresno		Lv	12 45		**4 00**
9 35	**4 55**	**6 05**	Ar	Oakland		Lv	8 55		↑
10 00	**5 20**	**7 00**	Ar	San Francisco, Cal.		Lv	8 30		10 45

Times are given in local time.

Lightface type indicates A.M. time. **Boldface** type indicates P.M. time.

EST—Eastern Standard Time   MST—Mountain Standard Time

CST—Central Standard Time   PST—Pacific Standard Time

6. 9:30 A.M., EST; 11:59 A.M., CST; 7:20 P.M., CST; 12:15 A.M., CST
7. 10:45 A.M., PST; 4:00 P.M., PST; 9:50 P.M., PST; 7:45 A.M., MST

6-28 READING A TIMETABLE   **617**

6. At what time does Bus 822 leave Atlanta? At what time does it arrive in Birmingham? In Memphis? In Little Rock?

7. At what time does Bus 827 leave San Francisco? At what time does it arrive in Fresno? In Barstow? In Flagstaff?

8. At what time does Bus 824 leave Oklahoma City? At what time does it arrive in Albuquerque? How long does it take to make this trip?

9. How long does it take Bus 825 to go from Amarillo to Memphis? 18 hr. 20 min.

10. Compare the time it takes Bus 820 to travel from Atlanta to San Francisco with the time it takes Bus 824 to make the same trip.

III. Airlines Schedule

Bus 820 takes 66 hr. 45 min.; Bus 824 takes 69 hr. 30 min.

Leave	Arrive	Flight	Stops	Meals	Leave	Arrive	Flight	Stops	Meals
From: DALLAS–FORT WORTH (CST)					From: DALLAS–FORT WORTH (CST)				
TO: CHICAGO (CST)					TO: MIAMI (EST)				
6 30a	9 15a	242	One-Stop	M	8 35a	12 00n	79	Non-Stop	M
8 20a	10 15a	54	Non-Stop	M	12 05p	3 25p	63	Non-Stop	M
10 20a	2 10p	248	Two-Stops	S	5 00p	9 10p	169	One-Stop	M
2 40p	5 30p	140	One-Stop	S	8 10p	12 15a	405	One-Stop	S
3 20p	5 20p	44	Non-Stop	S	TO: NEW ORLEANS (CST)				
5 20p	7 20p	36	Non-Stop	M	8 35a	9 45a	157	Non-Stop	M
7 15p	10 45p	120/156	Kansas City	M	11 30a	12 40p	235	Non-Stop	M
					4 40p	5 50p	133	Non-Stop	S
TO: DENVER (MST)					TO: SEATTLE-TACOMA (PST)				
9 25a	10 05a	62	Non-Stop	M	11 30a	1 10p	95	Non-Stop	M
11 25a	10 05p	68	Non-Stop	M	6 40p	8 20p	182	Non-Stop	M
4 25p	5 30p	66	One-Stop	S	7 45p	10 20p	184	One-Stop	M
6 25p	7 05p	78	Non-Stop	M					

Times are given in local time. The letter a indicates A.M.; p indicates P.M. M—Meal; S—Snack.

11. At what time does Flight 54 leave Dallas–Fort Worth for Chicago? At what time does it arrive in Chicago? How long does the flight take? Are there any stops on the way? 8:20 A.M., CST; 10:15 A.M., CST; 1 hr. 55 min.; no

12. What is the flight number of the airplane leaving Dallas–Fort Worth for New Orleans at 11:30 A.M.? When does it arrive in New Orleans? Is a meal or a snack served? 235; 12:40 P.M., CST; meal

13. How long does Flight 63 take from Dallas–Fort Worth to Miami? 2 hr. 20 min.

14. How much longer does it take Flight 66 to fly from Dallas–Fort Worth to Denver than Flight 78? 25 min.

15. How long does it take Flight 95 to fly from Dallas–Fort Worth to Seattle-Tacoma? How much faster is it than Flight 184? 3 hr. 40 min.; 55 min.

8. 5:00 P.M., CST; 4:50 A.M., MST; 12 hr. 50 min.

For a Class Activity see p. T38.
For a Test on Applications in Business and Travel see p. T62 or *Testing Program*, p. 44.

PROBLEM SOLVING COMPETENCY CHECK TEST

Solve each problem and select the letter corresponding to your answer. The numerals in red boxes indicate Exercises where help may be found.

1. The weekly earnings of a person who works 40 hours at $5.65 per hour are:

   a. $216   b. $224.60   c. $242.30   d. $226 [6-1]

2. Esther receives $185 per week and 2% commission on sales. One week her sales totaled $8,509. Her total earnings that week were:

   a. $202.02   b. $287.48   c. $355.18   d. $306.52 [6-1] [6-25]

3. If the withholding tax is $17.70 and social security tax is $13.07 on your weekly salary of $195, your take-home pay is:

   a. $164.23   b. $183.05   c. $224.65   d. $177.30 [6-5]

4. Which is the best buy of the same kind of dinner rolls?

   a. 6 for 59¢   b. 8 for 75¢   c. 12 for $1   d. 20 for $1.69 [6-8]

5. What is the unit price (cost per kilogram) of a jar of applesauce which weighs 1.25 kg and costs $1?

   a. $1.25   b. $.80   c. $.25   d. $2.25 [6-8]

6. A camera which regularly sells for $62.50 is reduced 18%. Its sale price is:

   a. $44.50   b. $51.25   c. $62.32   d. Answer not given [6-9]

7. An automobile that regularly sells for $6,000 was purchased for $5,100. The rate of reduction allowed is:

   a. 9%   b. 12%   c. 15%   d. 19% [6-9]

8. If a purchase is $9.27, the change from a $20 bill should be:

   a. $11.83   b. $10.83   c. $10.73   d. $11.73 [6-10]

9. If the rate is 6%, the sales tax on a radio selling for $49.50 is:

   a. $6.00   b. $2.97   c. $4.95   d. $29.70 [6-11]

10. Andrew's mother can purchase a clothes washer for the cash price $389.95 or $25 down and 12 equal monthly payments of $34.06. By paying cash she can save:

    a. $34.06   b. $37.00   c. $55.89   d. $43.77 [6-14]

11. The cost of $40,000 fire insurance for 1 year at $.28 per $100 is:

    a. $28   b. $112   c. $280   d. $400 [6-15]

Continued on page 619.

12. If the tax rate is $7.50 per $100, the property tax on a house costing $52,000 and assessed for $35,000 is:

    a. $520   b. $3,900   c. $2,625   d. $1,325 6-17

13. Joe's father bought a house for $45,000. He paid 20% down and gave a mortgage for the remainder. What is the amount of the mortgage?

    a. $36,000   b. $20,000   c. $40,000   d. $9,000 6-18

14. Elaine Bauman paid the Vining Furniture Co. by check in the amount of $59.75 on July 18, 1980.

FIRST STATE BANK   **650**

_____ 19_____

PAY TO THE
ORDER OF _____ $_____
(1)
_____ DOLLARS

MEMO _____   _____

95 646 718

On line (1) should be written: 6-19

    a. Elaine Bauman   c. July 18, 1980
    b. Vining Furniture Co.   d. Fifty-nine and $\frac{75}{100}$

15. If you have $927.56 in your checking account and make a deposit of $83.49, your new balance is:

    a. $1,101.05   b. $844.07   c. $834.07   d. $1,011.05 6-19

16. The interest on $2,600 for 3 years at $7\frac{1}{2}$% annual rate is:

    a. $585   b. $260   c. $195   d. $695 6-20

17. If you borrow $3,000 for 24 months to buy a car and pay $137.50 per month, the finance charge is:

    a. $137.50   b. $350   c. $300   d. $287.50 6-21

18. What is the monthly payment on an add-on-interest loan, if $6,000 is borrowed at 10% for 12 months?

    a. $600   b. $550   c. $500   d. $720 6-21

19. If you purchase 100 shares of HMV at 24 and sell at $30\frac{1}{2}$, your profit, excluding commissions, is:

    a. $750   b. $700   c. $650   d. $550 6-23

Continued on page 620.

20. Ms. Costello owns a $10,000 Treasury Note that pays an annual rate of $9\frac{1}{4}\%$ interest. She receives semiannually:

   a. $1,000   **b.** $925   **c.** $462.50   **d.** $437.50 6-23

21. When the catalog lists a table saw at $275 and the discount sheet shows a 17% discount, the trade discount allowed is:

   a. $46.75   **b.** $17.00   **c.** $27.50   **d.** $228.25 6-24

22. At what price should a dealer sell a watch costing $96 to make a profit of 40% on the selling price?

   a. $134.40   **b.** $384   **c.** $136   **d.** $160 6-26

23. A train leaves New York at 7 P.M. and arrives in Baltimore, 297 km away, at 10 P.M. The average speed of this train is:

   a. 85 km/h   **b.** 94 km/h   **c.** 99 km/h   **d.** 105 km/h 6-28

24. When the monthly finance charge rate is $1\frac{1}{4}\%$, the annual rate is:

   a. 10%   **b.** 12%   **c.** 15%   **d.** 18% 6-14 6-21

25. At the average cost of 6.5¢ per kilowatt hour, the weekly cost of operating 4 hours each day a TV set that uses 200 watts of power is:

   a. 36.4¢   **b.** $3.64   **c.** 52¢   **d.** $5.20 6-12

For an alternate Competency Check Test (Problem Solving) see p. T63 or *Testing Program*, p. 46. For two Comprehensive Final Tests see pp. 64–67 or *Testing Program*, pp. 48–54.

TABLE OF SQUARES AND SQUARE ROOTS

No.	Square	Square Root	No.	Square	Square Root	No.	Square	Square Root
1	1	1.000	34	1,156	5.831	67	4,489	8.185
2	4	1.414	35	1,225	5.916	68	4,624	8.246
3	9	1.732	36	1,296	6.000	69	4,761	8.307
4	16	2.000	37	1,369	6.083	70	4,900	8.367
5	25	2.236	38	1,444	6.164	71	5,041	8.426
6	36	2.449	39	1,521	6.245	72	5,184	8.485
7	49	2.646	40	1,600	6.325	73	5,329	8.544
8	64	2.828	41	1,681	6.403	74	5,476	8.602
9	81	3.000	42	1,764	6.481	75	5,625	8.660
10	100	3.162	43	1,849	6.557	76	5,776	8.718
11	121	3.317	44	1,936	6.633	77	5,929	8.775
12	144	3.464	45	2,025	6.708	78	6,084	8.832
13	169	3.606	46	2,116	6.782	79	6,241	8.888
14	196	3.742	47	2,209	6.856	80	6,400	8.944
15	225	3.873	48	2,304	6.928	81	6,561	9.000
16	256	4.000	49	2,401	7.000	82	6,724	9.055
17	289	4.123	50	2,500	7.071	83	6,889	9.110
18	324	4.243	51	2,601	7.141	84	7,056	9.165
19	361	4.359	52	2,704	7.211	85	7,225	9.220
20	400	4.472	53	2,809	7.280	86	7,396	9.274
21	441	4.583	54	2,916	7.348	87	7,569	9.327
22	484	4.690	55	3,025	7.416	88	7,744	9.381
23	529	4.796	56	3,136	7.483	89	7,921	9.434
24	576	4.899	57	3,249	7.550	90	8,100	9.487
25	625	5.000	58	3,364	7.616	91	8,281	9.539
26	676	5.099	59	3,481	7.681	92	8,464	9.592
27	729	5.196	60	3,600	7.746	93	8,649	9.644
28	784	5.292	61	3,721	7.810	94	8,836	9.695
29	841	5.385	62	3,844	7.874	95	9,025	9.747
30	900	5.477	63	3,969	7.937	96	9,216	9.798
31	961	5.568	64	4,096	8.000	97	9,409	9.849
32	1,024	5.657	65	4,225	8.062	98	9,604	9.899
33	1,089	5.745	66	4,356	8.124	99	9,801	9.950

TABLES OF MEASURE
Measure of Length—Metric

10 millimeters (mm) = 1 centimeter (cm)
10 centimeters (cm) = 1 decimeter (dm)
10 decimeters (dm) = 1 meter (m)
10 meters (m) = 1 dekameter (dam)
10 dekameters (dam) = 1 hectometer (hm)
10 hectometers (hm) = 1 kilometer (km)

1 centimeter (cm) = 10 millimeters (mm)
1 meter (m) = 100 centimeters (cm) = 1,000 millimeters (mm)
1 kilometer (km) = 1,000 meters (m)

Measure of Capacity—Metric

10 milliliters (mL) = 1 centiliter (cL)
10 centiliters (cL) = 1 deciliter (dL)
10 deciliters (dL) = 1 liter (L)
10 liters (L) = 1 dekaliter (daL)
10 dekaliters (daL) = 1 hectoliter (hL)
10 hectoliters (hL) = 1 kiloliter (kL)

Measure of Weight—Metric

10 milligrams (mg) = 1 centigram (cg)
10 centigrams (cg) = 1 decigram (dg)
10 decigrams (dg) = 1 gram (g)
10 grams (g) = 1 dekagram (dag)
10 dekagrams (dag) = 1 hectogram (hg)
10 hectograms (hg) = 1 kilogram (kg)
1,000 kilograms (kg) = 1 metric ton (t)

Metric Measures of Area, Volume, and Miscellaneous Equivalents

See pages 298–300.

Measure of Length—Customary

1 foot (ft.) = 12 inches (in.)
1 yard (yd.) = 3 feet (ft.) = 36 inches (in.)
1 rod (rd.) = $16\frac{1}{2}$ feet (ft.) = $5\frac{1}{2}$ yards (yd.)
1 statute mile (stat. mi.) = 5,280 feet (ft.) = 1,760 yards (yd.) = 320 rods (rd.)
1 statute mile (stat. mi.) $\approx$ 0.87 nautical mile (0.8684 naut. mi.)
1 nautical mile (naut. mi.) $\approx$ 6,080 feet (6,080.2 ft.)
$\approx$ 1.15 statute miles (1.1515 stat. mi.)
1 fathom (fath.) $\approx$ 6 feet (ft.)

Measure of Area—Customary

1 square foot (sq. ft.) = 144 square inches (sq. in.)
1 square yard (sq. yd.) = 9 square feet (sq. ft.)
1 square rod (sq. rd.) = 30.25 square yards (sq. yd.)
1 acre = 160 square rods (sq. rd.)
= 4,840 square yards (sq. yd.)
= 43,560 square feet (sq. ft.)
1 square mile (sq. mi.) = 640 acres

Measure of Volume—Customary

1 cubic foot (cu. ft.) = 1,728 cubic inches (cu. in.)
1 cubic yard (cu. yd.) = 27 cubic feet (cu. ft.)

Liquid Measure—Customary

1 pint = 16 ounces (oz.)
= 4 gills (gi.)
1 quart (qt.) = 2 pints (pt.)
1 gallon (gal.) = 4 quarts (qt.)

Dry Measure—Customary

1 quart (qt.) = 2 pints (pt.)
1 peck (pk.) = 8 quarts (qt.)
1 bushel (bu.) = 4 pecks (pk.)

Measure of Weight—Avoirdupois—Customary

1 pound (lb.) = 16 ounces (oz.)
1 short ton (sh. tn. or T.) = 2,000 pounds (lb.)
1 long ton (l. ton) = 2,240 pounds (lb.)

Volume, Capacity, and Weight Equivalents—Customary

1 gallon (gal.) $\approx$ 231 cubic inches (cu. in.)
1 cubic foot (cu. ft.) $\approx 7\frac{1}{2}$ gallons (gal.)
1 bushel (bu.) $\approx 1\frac{1}{4}$ cubic feet (cu. ft.)
$\approx$ 2,150.42 cubic inches (cu. in.)
1 cu. ft. of fresh water weighs $62\frac{1}{2}$ pounds (lb.)
1 cu. ft. of sea water weighs 64 pounds (lb.)

Measure of Time

1 minute (min.) = 60 seconds (sec.)
1 hour (hr.) = 60 minutes (min.)
1 day (da.) = 24 hours (hr.)
1 week (wk.) = 7 days (da.)
1 year (yr.) = 12 months (mo.)
= 52 weeks (wk.)
= 365 days (da.)

Angles and Arcs

1 circle = 360 degrees (°)
1 degree (°) = 60 minutes (′)
1 minute (′) = 60 seconds (″)

Measure of Speed

1 knot = 1 nautical m.p.h.

Metric—Customary Equivalents

1 meter = 39.37 inches
$\approx$ 3.28 feet (3.2808)
$\approx$ 1.09 yards (1.0936)
1 centimeter $\approx$ .39 or .4 inch (.3937)
1 millimeter $\approx$ .04 inch (.03937)
1 kilometer $\approx$ .62 mile (.6214)
1 liter $\approx$ 1.06 liquid quarts (1.0567)
1 liter $\approx$ .91 dry quart (.9081)
1 gram $\approx$ .04 ounce (.0353)
1 kilogram $\approx$ 2.2 pounds (2.2046)
1 metric ton $\approx$ 2,204.62 pounds
1 inch = 25.4 millimeters
1 foot $\approx$ .3 meter (.3048)
1 yard $\approx$ .91 meter (.9144)
1 mile $\approx$ 1.61 kilometers (1.6093)
1 liquid quart $\approx$ .95 liter (.9463)
1 dry quart $\approx$ 1.1 liters (1.1012)
1 ounce $\approx$ 28.35 grams (28.3495)
1 pound $\approx$ .45 kilogram (.4536)
1 short ton $\approx$ .91 metric ton (.9072)
1 square inch $\approx$ 6.45 square centimeters (6.4516)
1 cubic inch $\approx$ 16.39 cubic centimeters (16.3872)

Index

ANSWERS TO ODD-NUMBERED PROBLEMS

PART 1 BASIC SKILLS IN ARITHMETIC AND PROBLEM SOLVING

Basic Skills—Inventory Test I, Page 10

1-1. Ninety-five thousand, two hundred thirty-seven **1-3.** 6,800 **1-5.** 6,174 **1-7.** 27 **1-9.** $1\frac{3}{8}$
1-11. 6 **1-13.** $3\frac{5}{12}$ **1-15.** $\frac{9}{2}$ **1-17.** $4\frac{1}{2}$ **1-19.** 144 **1-21.** .96 **1-23.** 2.89 **1-25.** .65 **1-27.** 34
1-29. .49 **1-31.** $\frac{4}{5}$ **1-33.** 40 **1-35.** 16.3 million **1-37.** .09 **1-39.** $\frac{4}{10}$ **1-41.** 23.25 **1-43.** 90
1-45. 62

Basic Skills—Inventory Test II, Page 11

1-1. Five million, three hundred seventy-four thousand, eighty-nine **1-3.** 4,827,000 **1-5.** 54,639
1-7. 97 **1-9.** $2\frac{1}{12}$ **1-11.** 12 **1-13.** $\frac{3}{8}$ **1-15.** $\frac{10}{3}$ **1-17.** $5\frac{1}{4}$ **1-19.** 27 **1-21.** .504 **1-23.** 10.58
1-25. .03 **1-27.** 50 **1-29.** .0689 **1-31.** $\frac{1}{4}$ **1-33.** 30 **1-35.** 6.4 trillion **1-37.** .64 **1-39.** $\frac{5}{6}$
1-41. 19.43 **1-43.** 20 **1-45.** 156

Basic Skills—Inventory Test III, Page 12

1-1. One hundred sixty-three million, nine hundred eighty-seven **1-3.** 18,300,000 **1-5.** 2,861
1-7. 43.56 **1-9.** $2\frac{2}{3}$ **1-11.** 24 **1-13.** $4\frac{3}{5}$ **1-15.** $\frac{11}{6}$ **1-17.** $1\frac{9}{32}$ **1-19.** 72 **1-21.** 200.072
1-23. 23.04 **1-25.** .247 **1-27.** 14.2 **1-29.** .5765 **1-31.** $4\frac{3}{4}$ **1-33.** 18 **1-35.** 75.28 billion
1-37. .3 or .30 **1-39.** $\frac{1}{8}$ **1-41.** \$.66 **1-43.** 600 **1-45.** .86

Basic Skills—Inventory Test IV, Page 13

1-1. Five trillion, four hundred ninety-three billion, two hundred sixty-eight million, three thousand, seven hundred fifty-one **1-3.** 840,000,000 **1-5.** 1,370,954 **1-7.** 846 **1-9.** $3\frac{2}{9}$ **1-11.** 60
1-13. $4\frac{31}{48}$ **1-15.** $\frac{47}{7}$ **1-17.** $\frac{5}{11}$ **1-19.** 320 **1-21.** 60.0009 **1-23.** 57.878 **1-25.** Yes **1-27.** 120
1-29. .63407 **1-31.** $\frac{1}{6}$ **1-33.** 192 **1-35.** 1.839 trillion **1-37.** 1.03 **1-39.** $2\frac{1}{3}$ **1-41.** \$9,045
1-43. 36 **1-45.** .015

Problem Solving—Inventory Test, Page 15

1. 3 hours **3.** 646 miles **5.** 5 inches **7.** $2\frac{7}{8}$ **9.** $\frac{2}{100}$ or $\frac{1}{50}$ **11.** 897.93 **13.** \$3.41 **15.** 22%
17. \$50,000

UNIT One WHOLE NUMBERS

EXERCISE 1-1 Reading Numerals Naming Whole Numbers

Diagnostic Test, Page 20

1. 29,740,059,265,040 **3.** Sixty-eight **5.** Four thousand, nine hundred seventy-five **7.** Ten thousand, five hundred **9.** Four million, nine hundred thousand **11.** Three million, five hundred eighty-two thousand, nine hundred forty-two **13.** Nine hundred ninety-five million, twenty-eight thousand **15.** Nine hundred four trillion, sixty billion, three hundred eighty-one million, two hundred sixty-nine thousand, eight **17.** Seven hundred forty thousand, five hundred nine **19.** Two hundred eighteen billion, nine hundred four million, two hundred twenty-six thousand, seven **21.** Ninety-four million **23.** Sixteen trillion **25. a.** Five thousand, three hundred **b.** Fifty-three hundred

Related Practice Examples, Pages 21–22

Set 1: **1.** 793,058 **2.** 96,400,127 **3.** 592,473,085,604 **4.** 23,100,000,000,000 **5.** 8,335,000,000,000,000

Set 3: **1.** Ninety-six **2.** Forty-seven **3.** Fifty-three **4.** Eighty-four **5.** Seventy-nine

Set 5: **1.** Eight thousand, two hundred seventy-eight **2.** Three thousand, nine hundred twenty-six **3.** Two thousand, four hundred seventy-four **4.** Four thousand, two hundred six **5.** Seven thousand, nine hundred fifty-nine

Set 7: **1.** Twenty-seven thousand, four hundred thirty-two **2.** Fifty-nine thousand, one hundred seventeen **3.** Twenty-thousand, seven hundred thirty **4.** Forty-five thousand sixty-three **5.** Eighty-one thousand, eight hundred ninety-six

Set 9: **1.** Three million **2.** Eight million **3.** Two million, five hundred thousand **4.** Seven million, six hundred thousand **5.** Four million, two hundred thousand

Set 11: **1.** One million, five hundred ninety-six thousand, two hundred seventy-one **2.** Seven million, one hundred twenty-two thousand, eight hundred forty-three **3.** Eight million, nine hundred thousand, five hundred twenty-seven **4.** Five million, four hundred sixteen thousand, eighty-four **5.** Six million, four thousand, nine

Set 13: **1.** Seven hundred million **2.** Two hundred forty-one million, eight hundred forty-nine thousand **3.** Nine hundred twenty-eight million, three hundred seventy-six thousand, five hundred **4.** Four hundred six million, nine hundred twenty-three thousand, six hundred eighty-five **5.** Eight hundred forty-three million, two hundred ninety-seven thousand, eight hundred sixty-one

Set 15: **1.** Nine trillion **2.** Twenty-three trillion, five hundred eighty-seven billion, nine hundred fourteen million, sixty thousand **3.** Sixty-seven trillion, three hundred forty-six billion, five hundred twenty-eight million, nine hundred fifteen thousand, eight hundred ninety-eight **4.** One hundred thirty-nine trillion, five billion, four hundred seven million, eight hundred sixty-two thousand, one hundred **5.** Seven hundred sixty-eight trillion, eight hundred forty-one billion, two hundred fifty-three million, one hundred ninety-six thousand, three hundred sixty-nine

Set 17: **1.** Twenty-eight thousand, nine hundred **2.** Nine hundred eighty-three thousand, two hundred twenty-five **3.** Seven thousand, three **4.** Four hundred fifty-five thousand, six hundred seven **5.** Sixty-one thousand, four hundred fifty-nine

Set 19: **1.** Seventy-five billion **2.** Eight billion, four hundred ninety-seven million, five hundred thousand **3.** Four hundred sixty billion, one hundred twenty-two million, eight hundred forty-nine thousand **4.** Ninety-three billion, six hundred seventy-four million, three hundred seventeen thousand, five hundred five **5.** Two hundred forty-five billion, three hundred ninety million, six thousand, seven hundred

Set 21: **1.** Forty-seven million **2.** Nine million **3.** Five hundred sixty million **4.** Eighty-eight million **5.** One billion, seven hundred twenty-five million

Set 23: **1.** Ten trillion **2.** Eight trillion **3.** Two hundred trillion **4.** Eighty-seven trillion **5.** Three hundred seventy-five trillion

Set 25: **1. a.** Six thousand, nine hundred **b.** Sixty-nine hundred **2. a.** Four thousand, one hundred **b.** Forty-one hundred **3. a.** Two thousand, eight hundred **b.** Twenty-eight hundred **4. a.** Eight thousand, seven hundred **b.** Eighty-seven hundred **5. a.** Seven thousand, four hundred **b.** Seventy-four hundred

Practical Applications, Page 22

1. a. One hundred sixty-five million, two hundred forty-six thousand, three hundred twenty square kilometers **b.** Eighty-two million, four hundred forty-one thousand, five hundred sixty square kilometers **c.** Seventy-three million, four hundred forty-two thousand, seven hundred fifty-five square kilometers **d.** Fourteen million, ninety thousand, one hundred ten square kilometers **3. a.** Ten million, eight hundred twenty-thousand **b.** Nine million, five thousand **c.** Seven million, eight hundred ninety-five thousand, five hundred forty-three **d.** Seven million,

fifty thousand **e.** Five million, nine hundred forty-eight thousand, five hundred forty-six
5. a. Four hundred ninety-nine billion dollars **b.** Nineteen billion dollars

EXERCISE 1-2 Writing Numerals Naming Whole Numbers

Diagnostic Test, Page 23

1. 652 **3.** 30,405,018 **5.** 10,021,300,000,350 **7.** 503,000,000,000 **9.** 608,000 **11.** 460 000 000

Related Practice Examples, Pages 24-25

Set 1: **1.** 400 **2.** 833 **3.** 709 **4.** 917 **5.** 248
Set 3: **1.** 3,900,000 **2.** 85,006,240 **3.** 608,150,092 **4.** 35,077,968 **5.** 914,700,050
Set 5: **1.** 4,950,000,000,000 **2.** 11,800,236,000,000 **3.** 300,040,005,089,201 **4.** 1,682,407,016,000
       5. 500,093,004,000,988
Set 7: **1.** 3,000,000,000 **2.** $500,000,000,000 **3.** 18,000,000,000 **4.** $1,490,000,000,000
       5. $67,000,000,000
Set 9: **1.** 675,000 **2.** 54,000 **3.** $5,000 **4.** 70,000 **5.** $830,000
Set 11: **1.** 14,600 **2.** 3,080,923 **3.** 19,075,640,000 **4.** 13,000,000,000,000 **5.** 8,100,050,000,000

Miscellaneous Examples, Page 25

1. a. ten thousands **b.** millions **c.** hundred millions **d.** ten billions **3. a.** 952; 259 **b.** 7,630;
367 **c.** 854,200; 2,458 **d.** 9,863,110; 113,689 **5.** 3,578,524; 3,578,542; 3,582,745; 3,587,425

Practical Applications, Pages 26-27

1. a. $2,000,000,000 **b.** 106,000,000 **c.** $500,000,000 **d.** $53,000,000,000 **e.** $85,000,000
3. a. 3,788,426,000; 2,131,314,000; 1,301,421,000; 921,068,000; 311,464,000; 4,615,621,000;
8,049,566,000 **b.** $41,082,000,000; $43,356,000,000; $35,903,000,000; $23,604,000,000; $30,541,000,000;
$39,459,000,000; $2,690,000,000; $1,671,000,000; $4,396,000,000; $12,547,000,000

EXERCISE 1-3 Rounding Whole Numbers

Diagnostic Test, Page 28

1. a. 60 **b.** 390 **3. a.** 5,000 **b.** 63,000 **5. a.** 200,000 **b.** 1,700,000 **7. a.** 24,000,000,000
b. 171,000,000,000

Related Practice Examples, Page 29

Set 1a: **1.** 30 **2.** 50 **3.** 100 **4.** 420 **5.** 2,540
Set 1b: **1.** 30 **2.** 80 **3.** 150 **4.** 680 **5.** 3,840
Set 3a: **1.** 9,000 **2.** 4,000 **3.** 25,000 **4.** 58,000 **5.** 11,000
Set 3b: **1.** 9,000 **2.** 5,000 **3.** 36,000 **4.** 40,000 **5.** 878,000
Set 5a: **1.** 100,000 **2.** 300,000 **3.** 1,500,000 **4.** 3,800,000 **5.** 22,400,000
Set 5b: **1.** 600,000 **2.** 300,000 **3.** 2,400,000 **4.** 5,500,000 **5.** 19,800,000
Set 7a: **1.** 6,000,000,000 **2.** 17,000,000,000 **3.** 39,000,000,000 **4.** 253,000,000,000
       5. 598,000,000,000
Set 7b: **1.** 9,000,000,000 **2.** 22,000,000,000 **3.** 68,000,000,000 **4.** 821,000,000,000
       5. 909,000,000,000

EXERCISE 1-4 Addition of Whole Numbers

Preliminary Examples, Page 31

1. 3; 8; 13; 9; 3; 7; 8; 15; 8; 16 **3.** 8; 2; 11; 4; 11; 14; 12; 2; 12; 14 **5.** 11; 9; 10; 9; 9; 10; 11; 10;
10; 1 **7.** 6; 12; 4; 15; 4; 15; 9; 7; 4; 14 **9.** 6; 7; 5; 16; 16; 6; 13; 5; 6; 10

Diagnostic Test, Page 32

1. 77 **3.** 126 **5.** 128 **7.** 242 **9.** 269 **11.** 6,330 **13.** 1,830 **15.** 75,292 **17.** 19,800 **19.** 2,290
21. 363,136 **23.** 11,690,953 **25.** 70,771

Related Practice Examples, Pages 32-37

Set 1: 59; 99; 78; 86; 89
Set 3: 142; 100; 143; 114; 194
Set 5: 88; 158; 236; 166; 201
Set 7: 247; 229; 257; 57; 337
Set 9: 94; 267; 201; 322; 403
Set 11: 9,898; 6,493; 15,713; 6,691; 17,235
Set 13: 939; 2,183; 2,056; 1,599; 2,034
Set 15: 89,986; 109,116; 104,455; 78,435; 172,077
Set 17: 8,783; 12,291; 22,824; 8,085; 31,714
Set 19: 1,525; 3,353; 1,324; 1,186; 3,723
Set 21: 105,005; 334,630; 51,251; 125,742; 291,329
Set 23a: 39; 421; 1,847; 34,757; 181,062
Set 23b: 828,386; 2,560,934; 4,999,246; 13,116,101; 37,461,741
Set 25a: **1.** 15 **2.** 32 **3.** 193 **4.** 306 **5.** 351 **6.** 52 **7.** 139 **8.** 174 **9.** 229 **10.** 176
Set 25b: **1.** 1,315 **2.** 893 **3.** 1,175 **4.** 2,529 **5.** 1,194 **6.** 1,149 **7.** 1,833 **8.** 4,214 **9.** 2,211
  10. 4,251
Set 25c: **1.** 13,727 **2.** 113,992 **3.** 177,677 **4.** 21,159 **5.** 56,580 **6.** 69,723 **7.** 48,687 **8.** 73,602
  9. 115,567 **10.** 149,380

Miscellaneous Examples, Pages 37-38

1. Horizontal: 19; 17; 16; 23 Vertical: 26; 23; 26 Total: 75 **3.** Horizontal: 224; 93; 229; 239
Vertical: 321; 320; 144 Total: 785 **5.** Horizontal: 19; 24; 21; 18; 19 Vertical: 28; 37; 36
Total: 101 **7.** Horizontal: 170; 71; 168; 164; 164 Vertical: 186; 227; 324 Total: 737
9. Horizontal: 421; 1,062; 946; 432; 364 Vertical: 1,754; 945; 526 Total: 3,225

Problem Solving, Pages 38-40

1. b **3.** c **5.** b **7.** Jack **9.** 11,554 spectators
11. Football: Washington 34, Fernwood 44; Carson 46, Greenville 31; Baseball: Carson 12,
Springdale 10; Basketball: Greenville 66, Jefferson 67; Washington 63, Hilton 58 **13.** Jones 40,
Williams 41; Jones wins **15.** 32,863 seats **17.** 826,508 square kilometers **19.** 513,295 farms

Refresh Your Skills, Page 40

1. Six million, eight hundred sixty-two thousand, nine hundred twenty-five **3.** 236,819
5. 331,159

EXERCISE 1-5 Subtraction of Whole Numbers

Preliminary Examples, Page 42

1. 3; 0; 2; 2; 9; 7; 5; 4; 7; 4 **3.** 1; 2; 3; 9; 1; 7; 0; 5; 8; 4 **5.** 7; 3; 8; 7; 2; 3; 8; 3; 6; 6 **7.** 4; 8;
0; 9; 8; 3; 0; 3; 3; 1 **9.** 6; 9; 4; 9; 9; 9; 7; 7; 7; 0

Diagnostic Test, Page 43

1. 23 **3.** 62 **5.** 20 **7.** 21 **9.** 334 **11.** 106 **13.** 15 **15.** 3,123 **17.** 2,928 **19.** 1,208 **21.** 62,211
23. 23,852 **25.** 30,789 **27.** 45,138 **29.** 3,179,537 **31.** 13,673 **33.** 5,766

Related Practice Examples, Pages 43-46

Set 1: 55; 82; 93; 71; 64
Set 3: 24; 31; 72; 34; 34
Set 5: 30; 30; 60; 20; 70
Set 7: 23; 24; 7; 32; 25
Set 9: 512; 416; 155; 291; 154
Set 11: 210; 405; 300; 607; 407
Set 13: 85; 7; 87; 63; 49
Set 15: 6,223; 1,293; 2,410; 3,513; 4,037

Set 17: 3,737; 2,446; 3,808; 2,806; 718
Set 19: 2,349; 3,257; 3,466; 3,495; 573
Set 21: 12,241; 32,003; 3,402; 13,221; 43,512
Set 23: 22,166; 13,891; 40,901; 34,850; 5,755
Set 25: 68,668; 16,766; 1,789; 19,884; 7,425
Set 27: 65,539; 45,103; 16,765; 86,405; 87,076
Set 29: 5,524,025; 1,239,342; 7,608,045; 741,375; 1,049,034
Set 31a: **1.** 7 **2.** 263 **3.** 89 **4.** 202 **5.** 1,288 **6.** 707 **7.** 23,908 **8.** 3,189 **9.** 357,289
    10. 270,855
Set 31b: **1.** 55 **2.** 351 **3.** 216 **4.** 4,489 **5.** 2,879 **6.** 1,494 **7.** 48,850 **8.** 31,929 **9.** 505,843
    10. 742,950
Set 33: **1.** 261 **2.** 1,008 **3.** 23 **4.** 962 **5.** 2,162 **6.** 1,348 **7.** 1,382 **8.** 4,759 **9.** 59,437
    10. 35,626

Problem Solving, Pages 46-48

1. c **3.** d **5.** d **7.** 228 votes **9.** $975; $36,850; $37,825
11. $27,254 **13.** 1,644 meters **15.** 553 meters **17.** 4,812
19. 623,651 square kilometers

Refresh Your Skills, Page 48

1. Two hundred eight million, seventy thousand, nine **3.** 226,142 **5.** 135,908

EXERCISE 1-6 Multiplication of Whole Numbers

Preliminary Examples, Page 50

1. 15; 8; 10; 54; 63; 2; 42; 0; 16; 0 **3.** 0; 18; 49; 27; 0; 15; 48; 30; 1; 32 **5.** 28; 9; 2; 72; 0; 72;
6; 20; 0; 10 **7.** 20; 14; 30; 9; 0; 24; 16; 21; 0; 24 **9.** 0; 40; 8; 36; 6; 28; 7; 18; 12; 45

Diagnostic Test A, Page 51

1. 69 **3.** 72 **5.** 462 **7.** 5,244 **9.** 205,602 **11.** 180 **13.** 604 **15.** 15,000 **17.** 203,200
19. 12,805 billion **21.** 2,512

Related Practice Examples, Pages 51-53

Set 1: 64; 88; 86; 88; 63
Set 3: 92; 84; 95; 90; 92
Set 5: 963; 288; 1,084; 2,469; 3,648
Set 7: 752; 4,151; 4,210; 8,442; 5,576
Set 9: 96,956; 624,438; 347,641; 234,870; 676,592
Set 11: 160; 300; 280; 400; 450
Set 13: 208; 609; 608; 9,036; 68,064
Set 15: 42,240; 7,040; 66,500; 410,000; 180,000
Set 17: 21,280; 24,420; 30,400; 456,300; 320,240
Set 19: **1.** 28 trillion **2.** 232 billion **3.** 912 million **4.** 756 billion **5.** 714 billion **6.** 22 billion,
    nine hundred eighty million
Set 21: **1.** 1,296 **2.** 1,386 **3.** 5,120 **4.** 1,600 **5.** 4,000 **6.** 15,525 **7.** 22,014 **8.** 42,560
    9. 150,300 **10.** 162,082

Diagnostic Test B, Page 53

1. 888 **3.** 44,620 **5.** 2,980,055 **7.** 136,752 **9.** 13,329,774 **11.** 727,665,698 **13.** 459,000
15. 29,952 **17.** 345,621 **19.** 2,448,306 **21.** 229,248 **23.** 3,675 **25.** 1,800 million

Related Practice Examples, Pages 53-55

Set 1: 276; 700; 882; 703; 742
Set 3: 3,312; 22,092; 34,812; 10,857; 78,936
Set 5: 2,235,672; 3,284,570; 3,217,573; 5,335,848; 3,930,264

Set 7: 46,656; 622,050; 80,157; 561,429; 617,542
Set 9: 7,382,928; 38,551,152; 57,456,495; 26,500,868; 15,474,883
Set 11: 21,332,160; 286,707,552; 476,949,429; 333,825,108; 850,907,455
Set 13: 3,100; 8,100; 144,000; 85,120; 1,350,000
Set 15: 13,026; 13,088; 35,577; 115,632; 1,522,375
Set 17: 102,204; 348,783; 2,192,920; 19,657,915; 14,081,056
Set 19: 820,615; 18,300,738; 49,084,036; 20,238,522; 5,920,468,008
Set 21: **1.** 840 **2.** 72,384 **3.** 278,488 **4.** 935,952 **5.** 132,444 **6.** 3,630,666 **7.** 37,380,607
8. 836,400 **9.** 100,600 **10.** 13,587,120
Set 23: **1.** 512 **2.** 1,450 **3.** 810 **4.** 24,568 **5.** 13,968 **6.** 16,328 **7.** 50,000 **8.** 355,368
9. 577,962 **10.** 253,000
Set 25: **1.** 980 trillion **2.** 2,450 billion **3.** $28,288 million **4.** $3,900 billion **5.** 45,167 million
6. 294,500 billion

Problem Solving, Pages 56-57

1. c **3.** d **5.** c **7.** 10,150 kilometers **9.** 9,500 sheets **11.** Harrison with 90 points
13. $816 **15.** 29,406 **17.** 1,195,200 meters per hour **19.** Yes

Refresh Your Skills, Page 57

1. Nineteen trillion, two hundred seven billion, four hundred eighty-five million
3. 75,000,000,000 **5.** 87,147 **7.** 491,463

EXERCISE 1-7 Division of Whole Numbers

Preliminary Examples, Page 60

1. 2; 7; 1; 9; 0; 4; 1; 4; 6 **3.** 7; 1; 5; 6; 3; 0; 7; 3; 3 **5.** 0; 3; 1; 9; 3; 9; 9; 5; 2 **7.** 4; 8; 2; 0; 1;
5; 5; 4; 9 **9.** 5; 9; 5; 4; 5; 8; 0; 3; 4

Diagnostic Test A, Page 60

1. 34 **3.** 423 **5.** 139 **7.** 92 **9.** 2,842 **11.** 491 **13.** 5,219 **15.** 201 **17.** 3,004 **19.** 22 R2
21. $155\frac{5}{6}$ **23.** $120\frac{3}{7}$ **25.** 430

Related Practice Examples, Pages 61-63

Set 1: 31; 23; 22; 21; 14
Set 3: 231; 341; 122; 323; 122
Set 5: 236; 149; 153; 287; 126
Set 7: 72; 87; 78; 63; 76
Set 9: 1,651; 1,241; 1,149; 3,246; 2,392
Set 11: 749; 359; 849; 848; 764
Set 13: 9,122; 7,492; 5,856; 4,865; 7,496
Set 15: 304; 301; 2,041; 2,102; 3,020
Set 17: 3,001; 4,002; 30,012; 10,002; 10,001
Set 19: 2 R4; 184 R1; 763 R5; 3,949 R2; 24,655 R2
Set 21: $3\frac{8}{9}$; $14\frac{4}{5}$; $171\frac{1}{4}$; $776\frac{1}{2}$; $11,218\frac{5}{8}$
Set 23: $60\frac{5}{6}$; $80\frac{1}{2}$; $270\frac{2}{3}$; $1,090\frac{5}{9}$; $12,180\frac{4}{7}$
Set 25: **1.** 233 **2.** $724\frac{3}{8}$ **3.** $386\frac{5}{6}$ **4.** 27,669 **5.** $47,037\frac{1}{2}$ **6.** 8,568 **7.** $8,563\frac{2}{3}$ **8.** 8,178 **9.** $4,330\frac{1}{9}$
10. 10,008

Diagnostic Test B, Page 63

1. 4 **3.** 21 **5.** 314 **7.** 1,283 **9.** 483 **11.** 6,400 **13.** 429 R17 **15.** 6 **17.** 37 **19.** 857 **21.** 801
23. 845 **25.** $527\frac{2}{3}$ **27.** 59 **29.** 4,522 **31.** 94 **33.** 21 million

Related Practice Examples, Pages 64-66

Set 1: 5; 4; 3; 3; 4

Set 3: 28; 14; 23; 27; 41
Set 5: 123; 343; 216; 198; 186
Set 7: 2,415; 1,562; 2,876; 3,253; 1,562
Set 9: 189; 749; 872; 523; 926
Set 11: 220; 900; 4,800; 7,800; 8,700
Set 13: 862 R25; 258 R16; 463 R15; 769 R11; 388 R37
Set 15: $3\frac{18}{71}$; 7; 8; 3; 6
Set 17: 48; 58; 56; 69; 75
Set 19: 754; 584; 399; 565; 746
Set 21: 301; 608; 407; 708; 409
Set 23: 426; 967; 584; 609; 820
Set 25: $467\frac{1}{4}$; $816\frac{2}{21}$; $285\frac{1}{3}$; $471\frac{2}{5}$; $740\frac{1}{2}$
Set 27: 42; 67; 57; 85; 29
Set 29: 5,485; 3,756; 9,842; 4,582; 7,510
Set 31: **1.** 17 **2.** 2,684 **3.** 7,006 **4.** 752 **5.** 76
Set 33: **1.** 12 billion **2.** 23 million **3.** 46 million **4.** 79 billion **5.** 61 **6.** 520

Practice Problems, Pages 67–68

1. a. Yes; yes; no; yes **b.** No; yes; yes; yes **c.** No; yes; yes; no **d.** Yes; yes; yes; yes **e.** Yes; no; no; yes

Problem Solving, Page 68

1. c **3.** b **5.** 59 hours

Averages, Pages 68–70

1. 6 yards **3.** $318\frac{2}{3}$ kilometers per hour **5.** 6 kilometers per liter **7.** 83 **9.** 16° **11.** Johnson, 12 yd.; Bell, 8 yd.; Leonard, 5 yd.; Tompkins, 12 yd.; team's average $9\frac{7}{29}$ yd.

Refresh Your Skills, Page 70

1. 409,300 **3.** 3,623 **5.** 708

Estimating Answers, Page 70

1. 1,800 **3.** 3,500 **5.** 60,000 **7.** 2,000 **9.** 8,000 **11.** 630 **13.** 1,000,000 **15.** 200 **17.** 13,000 **19.** 60,000,000

Check by Calculator, Page 71

1. b and c are incorrect **3.** a and d are incorrect

Competency Check Test, Page 71

1. a **3.** c **5.** c **7.** c

Review of Unit One, Pages 72–73

1. a. Four million, nine hundred ninety-eight thousand, two hundred eighty-one **b.** Sixty-seven million, fifty thousand, three **c.** Seven hundred three million, four hundred twenty-eight thousand, nine hundred eighty-six **d.** Two billion, eighty-three thousand **e.** Four hundred one billion, five hundred seventy-five million, seven hundred thousand **3. a.** 928,000,000,000 **b.** 17,000,000,000,000 **c.** 2,509,000,000 **5. a.** 38,594 **b.** 219,985 **c.** 269,499 **d.** 3,494,996 **7. a.** 4,464 **b.** 186,300 **c.** 680,064 **d.** 4,653,756 **e.** 30,825 **9.** d **11.** $17,630 **13.** 4,332,180

Keyed Achievement Test, Page 73

1. Four million, ten thousand, one **3.** 535,000,000 **5.** 394,655 **7.** 4,874,436

UNIT Two   COMMON FRACTIONS

EXERCISE 1-8   Expressing Fractions in Lowest Terms

Diagnostic Test, Page 76

1. $\frac{1}{9}$ 3. $\frac{2}{3}$ 5. $\frac{2}{3}$ 7. $\frac{1}{4}$

Related Practice Examples, Pages 76–77

Set 1: 1. $\frac{1}{2}$ 2. $\frac{1}{3}$ 3. $\frac{1}{3}$ 4. $\frac{1}{4}$ 5. $\frac{1}{5}$ 6. $\frac{1}{6}$ 7. $\frac{1}{6}$ 8. $\frac{1}{8}$ 9. $\frac{1}{2}$ 10. $\frac{1}{3}$
Set 3: 1. $\frac{4}{5}$ 2. $\frac{2}{3}$ 3. $\frac{3}{10}$ 4. $\frac{2}{5}$ 5. $\frac{3}{4}$ 6. $\frac{2}{5}$ 7. $\frac{5}{8}$ 8. $\frac{7}{9}$ 9. $\frac{3}{4}$ 10. $\frac{9}{16}$
Set 5: 1. $\frac{2}{3}$ 2. $\frac{4}{5}$ 3. $\frac{2}{3}$ 4. $\frac{3}{4}$ 5. $\frac{3}{4}$ 6. $\frac{4}{7}$ 7. $\frac{2}{3}$ 8. $\frac{2}{7}$ 9. $\frac{7}{8}$ 10. $\frac{5}{6}$
Set 7: 1. $\frac{1}{4}$ 2. $\frac{1}{3}$ 3. $\frac{4}{5}$ 4. $\frac{1}{5}$ 5. $\frac{1}{4}$ 6. $\frac{3}{10}$ 7. $\frac{3}{5}$ 8. $\frac{3}{10}$ 9. $\frac{7}{10}$ 10. $\frac{5}{9}$

Miscellaneous Examples, Page 77

1. Set 1: **a.** $\frac{1}{2}$ in. **b.** $\frac{7}{8}$ in. **c.** $\frac{3}{4}$ in. **d.** $\frac{5}{8}$ in. **e.** $\frac{9}{16}$ in.
   Set 2: **a.** $\frac{2}{3}$ ft. **b.** $\frac{1}{4}$ ft. **c.** $\frac{5}{6}$ yd. **d.** $\frac{1}{8}$ mi. **e.** $\frac{3}{4}$ mi.
   Set 3: **a.** $\frac{5}{8}$ lb. **b.** $\frac{3}{10}$ ton **c.** $\frac{3}{8}$ ton **d.** $\frac{3}{4}$ qt. **e.** $\frac{1}{2}$ pk.
   Set 4: **a.** $\frac{5}{6}$ hr. **b.** $\frac{3}{4}$ min. **c.** $\frac{3}{8}$ day **d.** $\frac{5}{6}$ yr. **e.** $\frac{1}{5}$ yr.
3. **a.** $\frac{26}{32}$ inch **b.** $\frac{3}{8}$ inch 5. **a.** $\frac{18}{81}$ **b.** $\frac{63}{72}$

EXERCISE 1-9   Changing Improper Fractions and Simplifying Mixed Numbers

Preliminary Examples, Page 79

1. $8\frac{1}{4}$; $3\frac{11}{12}$; $1\frac{4}{5}$ 3. $\frac{7}{6}$; $\frac{4}{2}$; $\frac{5}{8}$; $\frac{45}{24}$

Diagnostic Test, Page 79

1. 3 3. $1\frac{1}{2}$ 5. $3\frac{1}{2}$ 7. 8 9. $10\frac{3}{10}$

Related Practice Examples, Pages 79–80

Set 1: 1. 1 2. 1 3. 3 4. 2 5. 3 6. 4 7. 3 8. 6 9. 8 10. 4
Set 3: 1. $1\frac{1}{2}$ 2. $1\frac{2}{3}$ 3. $1\frac{1}{4}$ 4. $1\frac{3}{4}$ 5. $1\frac{3}{4}$ 6. $1\frac{4}{5}$ 7. $1\frac{1}{2}$ 8. $1\frac{7}{9}$ 9. $1\frac{5}{9}$ 10. $1\frac{7}{8}$
Set 5: 1. $2\frac{1}{2}$ 2. $3\frac{1}{2}$ 3. $5\frac{2}{3}$ 4. $3\frac{3}{4}$ 5. $2\frac{1}{4}$ 6. $2\frac{5}{8}$ 7. $3\frac{2}{5}$ 8. $3\frac{1}{8}$ 9. $4\frac{1}{6}$ 10. $3\frac{1}{3}$
Set 7: 1. 7 2. 12 3. 10 4. 20 5. 21
Set 9: 1. $7\frac{3}{5}$ 2. $6\frac{3}{4}$ 3. $11\frac{1}{2}$ 4. $10\frac{2}{3}$ 5. $3\frac{5}{8}$ 6. $12\frac{1}{6}$ 7. $24\frac{9}{10}$ 8. $41\frac{7}{9}$ 9. $19\frac{3}{16}$ 10. $37\frac{5}{12}$

Miscellaneous Examples, Page 80

1. Set 1: **a.** $1\frac{1}{4}$ in. **b.** $2\frac{1}{4}$ in. **c.** $1\frac{1}{4}$ in. **d.** $3\frac{1}{2}$ in. **e.** $1\frac{9}{32}$ in.
   Set 2: **a.** $2\frac{1}{4}$ ft. **b.** $2\frac{11}{12}$ ft. **c.** $1\frac{7}{36}$ yd. **d.** $1\frac{3}{4}$ mi. **e.** $1\frac{1}{2}$ mi.
   Set 3: **a.** $1\frac{7}{16}$ lb. **b.** $1\frac{3}{4}$ tons **c.** $2\frac{1}{4}$ tons **d.** $2\frac{1}{8}$ gal. **e.** $3\frac{1}{2}$ bu.
   Set 4: **a.** $1\frac{1}{4}$ hr. **b.** $1\frac{4}{5}$ min. **c.** $1\frac{3}{4}$ da. **d.** $1\frac{1}{4}$ yr. **e.** $1\frac{3}{4}$ yr.
3. $\frac{72}{32}$ in.
5. Set 1: **a.** $5\frac{1}{4}$ in. **b.** $3\frac{3}{8}$ in. **c.** $1\frac{7}{16}$ in. **d.** $7\frac{1}{4}$ in. **e.** 10 in.
   Set 2: **a.** $6\frac{1}{3}$ ft. **b.** $2\frac{5}{12}$ ft. **c.** $4\frac{5}{9}$ yd. **d.** $2\frac{1}{4}$ mi. **e.** $3\frac{3}{8}$ mi.
   Set 3: **a.** $4\frac{1}{2}$ lb. **b.** $5\frac{7}{10}$ tons **c.** $2\frac{7}{8}$ tons **d.** $10\frac{3}{4}$ gal. **e.** $7\frac{3}{16}$ bu.
   Set 4: **a.** $2\frac{3}{10}$ hr. **b.** $10\frac{1}{5}$ min. **c.** $5\frac{2}{3}$ da. **d.** $3\frac{1}{3}$ yr. **e.** $1\frac{1}{4}$ yr.

EXERCISE 1-10   Changing a Fraction to Higher Terms

Diagnostic Test, Page 81

1. $\frac{3}{12}$ 3. $\frac{90}{100}$ 5. $\frac{60}{10}$

Related Practice Examples, Page 82

Set 1: **1.** $\frac{4}{8}$ **2.** $\frac{5}{15}$ **3.** $\frac{4}{20}$ **4.** $\frac{7}{28}$ **5.** $\frac{9}{18}$ **6.** $\frac{6}{48}$ **7.** $\frac{6}{72}$ **8.** $\frac{7}{42}$ **9.** $\frac{4}{64}$ **10.** $\frac{7}{70}$

Set 3: **1.** $\frac{70}{100}$ **2.** $\frac{75}{100}$ **3.** $\frac{40}{100}$ **4.** $\frac{45}{100}$ **5.** $\frac{8}{100}$ **6.** $\frac{86}{100}$ **7.** $\frac{84}{100}$ **8.** $\frac{37\frac{1}{2}}{100}$ **9.** $\frac{66\frac{2}{3}}{100}$ **10.** $\frac{34}{100}$

Set 5: **1.** $\frac{20}{10}$ **2.** $\frac{48}{12}$ **3.** $\frac{120}{24}$ **4.** $\frac{42}{6}$ **5.** $\frac{180}{60}$ **6.** $\frac{40}{8}$ **7.** $\frac{15}{5}$ **8.** $\frac{700}{100}$ **9.** $\frac{120}{10}$ **10.** $\frac{648}{72}$

Miscellaneous Examples, Page 82

1. $\frac{2}{16}''$ **3.** $\frac{2}{8}'' = \frac{4}{16}''$ **5.** $\frac{2}{4}'' = \frac{4}{8}'' = \frac{8}{16}''$

EXERCISE 1-11 Finding the Lowest Common Denominator and Changing Fractions to Equivalent Fractions

Diagnostic Test, Page 84

1. $8, \frac{4}{8}, \frac{3}{8}$ **3.** $80, \frac{56}{80}, \frac{45}{80}$

Related Practice Examples, Page 85

Set 1: **1.** $4, \frac{2}{4}, \frac{1}{4}$ **2.** $10, \frac{6}{10}, \frac{9}{10}$ **3.** $16, \frac{4}{16}, \frac{9}{16}$ **4.** $20, \frac{17}{20}, \frac{16}{20}$ **5.** $6, \frac{4}{6}, \frac{1}{6}$ **6.** $16, \frac{8}{16}, \frac{13}{16}$ **7.** $8, \frac{7}{8}, \frac{6}{8}$
8. $24, \frac{11}{24}, \frac{15}{24}$ **9.** $32, \frac{8}{32}, \frac{21}{32}$ **10.** $18, \frac{15}{18}, \frac{7}{18}$

Set 3: **1.** $12, \frac{3}{12}, \frac{2}{12}$ **2.** $40, \frac{35}{40}, \frac{36}{40}$ **3.** $24, \frac{9}{24}, \frac{9}{24}$ **4.** $20, \frac{15}{20}, \frac{14}{20}$ **5.** $48, \frac{8}{48}, \frac{45}{48}$ **6.** $160, \frac{16}{160}, \frac{45}{160}$
7. $28, \frac{10}{28}, \frac{7}{28}$ **8.** $60, \frac{39}{60}, \frac{35}{60}$ **9.** $72, \frac{57}{72}, \frac{52}{72}$ **10.** $96, \frac{88}{96}, \frac{81}{96}$

Miscellaneous Examples, Page 85

1. Group (b) **3.** Group (c) **5.** Group (c)

EXERCISE 1-12 Addition of Fractions and Mixed Numbers

Diagnostic Test, Page 87

1. $\frac{2}{3}$ **3.** 1 **5.** $1\frac{1}{2}$ **7.** $1\frac{3}{20}$ **9.** $1\frac{3}{5}$ **11.** $7\frac{3}{4}$ **13.** $5\frac{2}{5}$ **15.** $17\frac{1}{5}$ **17.** 9 **19.** $14\frac{1}{4}$ **21.** $25\frac{1}{2}$ **23.** $5\frac{1}{8}$ **25.** $1\frac{5}{8}$
27. $17\frac{7}{16}$

Related Practice Examples, Pages 87–90

Set 1: $\frac{4}{5}; \frac{5}{7}; \frac{14}{25}; \frac{5}{6}; \frac{9}{16}$
Set 3: 1; 1; 1; 1; 1
Set 5: $1\frac{2}{3}; 1\frac{1}{4}; 1\frac{1}{4}; 1\frac{1}{2}; 1\frac{3}{8}$
Set 7: $1\frac{5}{12}; \frac{5}{6}; \frac{17}{30}; 1\frac{39}{80}; 1\frac{19}{40}$
Set 9: 1; $2\frac{1}{4}; \frac{9}{16}; \frac{43}{48}; 1\frac{137}{240}$
Set 11: $13\frac{3}{5}; 13\frac{5}{8}; 15\frac{1}{2}; 25\frac{3}{8}; 31\frac{9}{32}$
Set 13: $13\frac{3}{4}; 9\frac{1}{2}; 26\frac{7}{8}; 28\frac{2}{3}; 61\frac{1}{2}$
Set 15: $4\frac{1}{3}; 12\frac{2}{5}; 46\frac{4}{9}; 36\frac{1}{8}; 10\frac{3}{16}$
Set 17: $6\frac{2}{5}; 9\frac{1}{2}; 8; 2\frac{1}{2}; 18\frac{7}{8}$
Set 19: $6\frac{1}{2}; 12\frac{2}{3}; 11\frac{2}{3}; 14\frac{1}{2}; 39\frac{3}{4}$
Set 21: $11\frac{1}{2}; 11\frac{1}{4}; 8\frac{1}{3}; 16\frac{1}{8}; 59\frac{1}{4}$
Set 23: $4; 14\frac{7}{16}; 12\frac{3}{16}; 8\frac{59}{60}; 12\frac{1}{8}$
Set 25: **1.** $\frac{3}{8}$ **2.** $\frac{5}{6}$ **3.** $1\frac{3}{16}$ **4.** 1 **5.** $\frac{7}{12}$ **6.** $2\frac{1}{8}$ **7.** $1\frac{11}{16}$ **8.** 2 **9.** $2\frac{1}{8}$ **10.** $1\frac{1}{2}$
Set 27: **1.** $1\frac{1}{4}$ **2.** 13 **3.** $10\frac{3}{10}$ **4.** $11\frac{5}{8}$ **5.** 20 **6.** $23\frac{15}{16}$ **7.** $11\frac{3}{16}$ **8.** $22\frac{11}{16}$ **9.** 33 **10.** $75\frac{1}{16}$

Problem Solving, Pages 90–91

1. b **3.** b **5.** c **7.** Sophomores with 29 points **9.** 7 games **11.** $17\frac{7}{16}$ inches
13. $5\frac{3}{16}$ in. by $5\frac{7}{8}$ in. by $7\frac{1}{4}$ in. **15.** $1\frac{7}{16}$ in.

Refresh Your Skills, Page 91

1. 6,281,000 **3.** 22,780 **5.** 996 **7.** $\frac{20}{24}$ **9.** $11\frac{4}{15}$

EXERCISE 1-13 Subtraction of Fractions and Mixed Numbers

Preliminary Examples—Regrouping, Page 93

Set 1: **1.** $3\frac{2}{2}$ **2.** $2\frac{6}{6}$ **3.** $6\frac{4}{4}$ **4.** $8\frac{3}{3}$ **5.** $11\frac{8}{8}$ **6.** $7\frac{10}{10}$ **7.** $5\frac{24}{24}$ **8.** $4\frac{16}{16}$ **9.** $13\frac{32}{32}$ **10.** $1\frac{20}{20}$

Set 3: **1.** $2\frac{5}{3}$ **2.** $7\frac{15}{8}$ **3.** $1\frac{3}{2}$ **4.** $3\frac{9}{5}$ **5.** $8\frac{7}{4}$ **6.** $14\frac{11}{6}$ **7.** $11\frac{17}{9}$ **8.** $5\frac{13}{10}$ **9.** $6\frac{25}{16}$ **10.** $9\frac{23}{12}$

Diagnostic Test, Page 93

1. $\frac{1}{3}$ **3.** $\frac{1}{3}$ **5.** $\frac{1}{20}$ **7.** $3\frac{5}{8}$ **9.** $4\frac{1}{2}$ **11.** $2\frac{15}{32}$ **13.** $8\frac{1}{5}$ **15.** $6\frac{5}{12}$ **17.** $8\frac{7}{8}$ **19.** $2\frac{59}{80}$ **21.** $\frac{5}{12}$ **23.** $8\frac{7}{12}$

Related Practice Examples, Pages 94–96

Set 1: $\frac{1}{5}$; $\frac{2}{7}$; $\frac{3}{8}$; $\frac{1}{4}$; $\frac{9}{32}$

Set 3: $\frac{11}{16}$; $\frac{1}{2}$; $\frac{1}{3}$; $\frac{1}{4}$; $\frac{9}{32}$

Set 5: $\frac{1}{12}$; $\frac{11}{24}$; $\frac{19}{48}$; $\frac{13}{24}$; $\frac{19}{80}$

Set 7: $2\frac{1}{4}$; $3\frac{3}{8}$; $2\frac{1}{6}$; $5\frac{3}{8}$; $12\frac{1}{5}$

Set 9: $5\frac{1}{4}$; $8\frac{3}{4}$; $2\frac{2}{8}$; $1\frac{7}{4}$; $3\frac{1}{4}$

Set 11: $2\frac{1}{2}$; $3\frac{3}{8}$; $6\frac{3}{8}$; $16\frac{9}{10}$; $25\frac{25}{32}$

Set 13: $4\frac{5}{8}$; $8\frac{5}{12}$; $3\frac{1}{2}$; $11\frac{1}{10}$; $9\frac{1}{3}$

Set 15: $3\frac{2}{15}$; $8\frac{13}{80}$; $6\frac{1}{6}$; $6\frac{7}{48}$; $32\frac{1}{20}$

Set 17: $2\frac{3}{4}$; $3\frac{13}{16}$; $2\frac{9}{16}$; $7\frac{7}{5}$; $4\frac{19}{32}$

Set 19: $7\frac{17}{20}$; $1\frac{19}{24}$; $5\frac{43}{48}$; $11\frac{23}{24}$; $9\frac{53}{60}$

Set 21: $4\frac{3}{4}$; $2\frac{11}{16}$; $\frac{13}{15}$; $\frac{5}{6}$; $\frac{13}{16}$

Set 23: **1.** $4\frac{1}{4}$; **2.** $\frac{3}{4}$ **3.** $3\frac{17}{20}$ **4.** $9\frac{1}{6}$ **5.** $5\frac{11}{40}$ **6.** $15\frac{13}{16}$ **7.** $8\frac{1}{20}$ **8.** $\frac{1}{4}$ **9.** $\frac{15}{16}$ **10.** $\frac{1}{2}$

Problem Solving, Pages 96–97

1. c **3.** c **5.** c **7.** $1\frac{1}{8}$ pounds **9.** 55 min. or $\frac{11}{12}$ hr. **11.** $1\frac{1}{3}$ hr.
13. 480 kilometers per hr. **15.** $2\frac{1}{4}$ in.

Refresh Your Skills, Page 97

1. 366,909 **3.** 693,462 **5.** 270,000,000 **7.** $\frac{48}{64}$ **9.** $2\frac{1}{2}$

EXERCISE 1-14 Comparing Fractions

Diagnostic Test, Page 98

1. $\frac{1}{2}$ **3.** $\frac{1}{10}$ **5.** $\frac{13}{16}$, $\frac{3}{8}$, $\frac{7}{32}$ **7.** Yes

Related Practice Examples, Page 99

Set 1: **1.** $\frac{1}{3}$ **2.** $\frac{1}{6}$ **3.** $\frac{1}{2}$ **4.** $\frac{1}{5}$ **5.** $\frac{1}{10}$

Set 3: **1.** $\frac{1}{6}$ **2.** $\frac{1}{10}$ **3.** $\frac{1}{16}$ **4.** $\frac{1}{5}$ **5.** $\frac{1}{8}$

Set 5: **1.** $\frac{1}{2}$, $\frac{1}{3}$, $\frac{1}{5}$ **2.** $\frac{2}{5}$, $\frac{3}{8}$, $\frac{3}{5}$ **3.** $\frac{3}{4}$, $\frac{7}{12}$, $\frac{9}{16}$ **4.** $\frac{7}{8}$, $\frac{5}{6}$, $\frac{4}{5}$ **5.** $\frac{3}{8}$, $\frac{5}{16}$, $\frac{1}{4}$

Set 7: **1.** True **2.** True **3.** False **4.** True **5.** False

Miscellaneous Examples, Page 99

1. Set 1: **a.** 4 **b.** 9 **c.** 7 **d.** 12 **e.** 28
  Set 2: **a.** 4 **b.** 6 **c.** 9 **d.** 17 **e.** 35
  Set 3: **a.** 2 **b.** 6 **c.** 3 **d.** 28 **e.** 53

EXERCISE 1-15 Changing Mixed Numbers to Improper Fractions

Diagnostic Test, Page 100

1. $\frac{4}{3}$ **3.** $\frac{15}{2}$

Related Practice Examples, Page 100

Set 1: **1.** $\frac{6}{5}$ **2.** $\frac{5}{4}$ **3.** $\frac{3}{2}$ **4.** $\frac{9}{8}$ **5.** $\frac{17}{16}$
Set 3: **1.** $\frac{41}{5}$ **2.** $\frac{22}{7}$ **3.** $\frac{55}{6}$ **4.** $\frac{49}{3}$ **5.** $\frac{57}{4}$

Miscellaneous Examples, Page 101

1. 11; 19 **3.** 45; 25 **5.** 49; 101

Practical Applications, Page 101

1. 75 bags **3.** 123 neckties **5.** 93 tea bags

Refresh Your Skills, Page 101

1. 335,667 **3.** 4,778,400 **5.** 7,900,000 **7.** $\frac{36}{60}$ **9.** $16\frac{1}{16}$

EXERCISE 1-16 Multiplication of Fractions and Mixed Numbers

Diagnostic Test, Page 103

1. $\frac{1}{15}$ **3.** $\frac{1}{3}$ **5.** $\frac{2}{5}$ **7.** $8\frac{1}{3}$ **9.** 4 **11.** $8\frac{1}{3}$ **13.** 9 **15.** $3\frac{1}{2}$ **17.** $\frac{15}{16}$ **19.** $12\frac{2}{3}$ **21.** 22 **23.** $22\frac{3}{4}$ **25.** $2\frac{11}{32}$
27. $1\frac{29}{96}$ **29.** $6\frac{3}{10}$ **31.** $\frac{2}{9}$ **33.** 132 **35.** $5\frac{1}{32}$ **37.** 90

Related Practice Examples, Pages 104–107

Set 1: $\frac{1}{8}$; $\frac{1}{24}$; $\frac{1}{20}$; $\frac{1}{20}$; $\frac{1}{64}$
Set 3: $\frac{1}{5}$; $\frac{3}{5}$; $\frac{1}{10}$; $\frac{1}{3}$; $\frac{5}{7}$
Set 5: $\frac{1}{6}$; $\frac{3}{5}$; $\frac{2}{3}$; $\frac{2}{3}$; $\frac{1}{6}$
Set 7: $\frac{5}{6}$; $2\frac{4}{5}$; 1; $2\frac{1}{4}$; $2\frac{1}{12}$
Set 9: 6; 10; 84; 15; 36
Set 11: $2\frac{1}{4}$; $11\frac{1}{4}$; $12\frac{2}{3}$; $18\frac{1}{5}$; $3\frac{1}{3}$
Set 13: 3; 5; 13; 29; 15
Set 15: $\frac{3}{4}$; $1\frac{1}{4}$; $3\frac{1}{2}$; $1\frac{1}{2}$; $3\frac{1}{4}$
Set 17: $2\frac{2}{3}$; $7\frac{1}{5}$; $3\frac{3}{8}$; $2\frac{11}{12}$; $11\frac{7}{10}$
Set 19: $8\frac{1}{2}$; $35\frac{1}{4}$; $31\frac{1}{3}$; $22\frac{4}{5}$; $136\frac{1}{4}$
Set 21: 42; 42; 46; 87; 142
Set 23: $12\frac{2}{3}$; $21\frac{7}{8}$; $28\frac{1}{2}$; $48\frac{15}{16}$; $12\frac{1}{4}$
Set 25: $1\frac{7}{8}$; $3\frac{13}{24}$; $\frac{11}{16}$; $2\frac{37}{64}$; $3\frac{31}{96}$
Set 27: $1\frac{3}{32}$; $1\frac{11}{16}$; $\frac{13}{15}$; $1\frac{7}{20}$; $1\frac{41}{64}$
Set 29: $2\frac{1}{4}$; $5\frac{7}{9}$; $2\frac{17}{20}$; $2\frac{17}{72}$; $5\frac{5}{6}$
Set 31: $\frac{1}{12}$; $\frac{1}{8}$; $\frac{1}{16}$; $\frac{9}{32}$; $\frac{9}{40}$
Set 33: 126; 92; $266\frac{7}{8}$; $52\frac{4}{5}$; $77\frac{1}{3}$
Set 35: **1.** $29\frac{1}{4}$ **2.** $12\frac{3}{8}$ **3.** $5\frac{1}{2}$ **4.** 6 **5.** $5\frac{1}{4}$
Set 37: **1.** 880 **2.** 4 **3.** $346\frac{1}{2}$ **4.** 12 **5.** 24 **6.** 176 **7.** 45 **8.** $3\frac{9}{19}$ **9.** $146\frac{2}{3}$ **10.** $13\frac{9}{11}$

Problem Solving, Pages 107–109

1. c **3.** a **5.** c **7.** 15 cups cake flour; $13\frac{1}{2}$ teaspoons baking powder; 3 cups of shortening; $7\frac{1}{2}$ cups sugar **9.** $1.41 **11. a.** 195 miles **b.** 112.50 miles **13. a.** 20°C **b.** 40°C **c.** 5°C **d.** 0°C **e.** 100°C **15. a.** 46°F **b.** $69\frac{1}{2}$°F **17.** $183\frac{3}{4}$ gal. **19.** $115\frac{5}{8}$ naut. mi.

Refresh Your Skills, Page 109

1. 687,500 **3.** 93,906 **5.** 409 **7.** $10\frac{7}{60}$ **9.** $\frac{21}{32}$

EXERCISE 1-17 Division of Fractions and Mixed Numbers

Diagnostic Test, Page 111

1. $\frac{4}{9}$ **3.** $\frac{2}{3}$ **5.** $1\frac{3}{7}$ **7.** $\frac{7}{16}$ **9.** $5\frac{1}{3}$ **11.** $\frac{1}{4}$ **13.** $\frac{5}{12}$ **15.** $1\frac{1}{3}$ **17.** $2\frac{6}{11}$ **19.** $4\frac{3}{8}$ **21.** $\frac{1}{2}$ **23.** 4 **25.** $\frac{1}{8}$ **27.** 1
29. $\frac{3}{28}$ **31.** $1\frac{1}{15}$

Related Practice Examples, Pages 112-114

Set 1: $\frac{3}{4}$; $\frac{15}{16}$; $\frac{3}{4}$; $\frac{32}{45}$; $\frac{36}{55}$
Set 3: $\frac{1}{3}$; $\frac{3}{4}$; $\frac{9}{20}$; $\frac{5}{6}$; $\frac{20}{21}$
Set 5: $1\frac{1}{7}$; $1\frac{1}{3}$; $2\frac{1}{10}$; $1\frac{1}{7}$; $2\frac{2}{9}$
Set 7: $\frac{3}{40}$; $\frac{1}{20}$; $\frac{7}{48}$; $\frac{11}{60}$; $\frac{5}{32}$
Set 9: 20; $7\frac{1}{2}$; 15; $12\frac{4}{5}$; $22\frac{6}{7}$
Set 11: $\frac{1}{2}$; $\frac{1}{3}$; $\frac{4}{5}$; $\frac{3}{16}$; $\frac{3}{8}$
Set 13: $\frac{13}{16}$; $\frac{33}{64}$; $2\frac{5}{6}$; $1\frac{3}{20}$; $\frac{19}{40}$
Set 15: $3\frac{1}{3}$; $5\frac{1}{3}$; $10\frac{2}{3}$; $7\frac{1}{2}$; $2\frac{2}{3}$
Set 17: $3\frac{3}{5}$; $\frac{20}{33}$; $4\frac{4}{9}$; $\frac{3}{7}$; $1\frac{11}{13}$
Set 19: $2\frac{2}{5}$; $2\frac{2}{3}$; $4\frac{2}{5}$; $3\frac{9}{10}$; $2\frac{13}{36}$
Set 21: $\frac{1}{20}$; $\frac{3}{4}$; $\frac{1}{4}$; $\frac{7}{30}$; $\frac{2}{3}$
Set 23: 5; 6; 4; 10; 12
Set 25: $\frac{2}{3}$; $\frac{1}{2}$; $\frac{29}{36}$; $\frac{5}{9}$; $\frac{7}{12}$
Set 27: 1; 1; 1; 1; 1
Set 29a: **1.** $7\frac{1}{9}$ **2.** $2\frac{1}{2}$ **3.** 2 **4.** $\frac{8}{9}$ **5.** $\frac{5}{9}$
Set 29b: **1.** $1\frac{1}{8}$ **2.** $\frac{21}{80}$ **3.** 16 **4.** $1\frac{1}{5}$ **5.** $2\frac{1}{3}$
Set 31a: $1\frac{3}{5}$; 3; $\frac{21}{40}$; $\frac{32}{35}$; $1\frac{13}{32}$
Set 31b: $\frac{1}{2}$; $\frac{3}{4}$; $1\frac{1}{2}$; $2\frac{1}{2}$; $1\frac{3}{5}$
Set 31c: $\frac{5}{16}$; $\frac{41}{96}$; $\frac{11}{18}$; $2\frac{3}{8}$; $\frac{17}{80}$
Set 31d: $\frac{7}{8}$; $\frac{1}{3}$; $\frac{1}{8}$; $\frac{5}{8}$; $\frac{1}{6}$
Set 31e: 10; 4; $1\frac{1}{3}$; $\frac{3}{8}$; $1\frac{21}{43}$
Set 31f: 2; $\frac{1}{2}$; $8\frac{2}{3}$; $\frac{48}{79}$; $2\frac{7}{24}$

Problem Solving, Pages 114-116

1. b **3.** d **5.** c **7.** $\frac{1}{8}$ cup flour, $1\frac{1}{4}$ tablespoons lemon juice, $\frac{1}{2}$ cup milk, and $\frac{3}{8}$ cup sugar
9. $10\frac{4}{5}$ knots **11. a.** 1 in. = 48 mi. **b.** 1 in. = 56 mi.; 1 in. = 40 mi.; 1 in. = 64 mi.;
1 in. = 1,250 yd.; 1 in. = 50 mi. **13. a.** 3,000 ft. **b.** 6,000 ft. **c.** 4,000 ft.
15. a. 8 lengths **b.** 128 boards **c.** 800 pieces

Refresh Your Skills, Page 116

1. 360,018 **3.** 3,258,000 **5.** 18,640,000 **7.** $5\frac{5}{6}$ **9.** $2\frac{11}{12}$

EXERCISE 1-18 Finding What Fractional Part One Number Is of Another

Diagnostic Test, Page 117

1. $\frac{1}{6}$ **3.** $\frac{5}{8}$ **5.** $\frac{3}{5}$ **7.** $\frac{11}{8}$ or $1\frac{3}{8}$ **9.** 8

Related Practice Examples, Page 118

Set 1: **1.** $\frac{1}{4}$ **2.** $\frac{1}{5}$ **3.** $\frac{7}{8}$ **4.** $\frac{13}{24}$ **5.** $\frac{5}{9}$
Set 3a: **1.** $\frac{1}{2}$ **2.** $\frac{1}{6}$ **3.** $\frac{1}{4}$ **4.** $\frac{1}{16}$ **5.** $\frac{1}{3}$
Set 3b: **1.** $\frac{2}{3}$ **2.** $\frac{4}{5}$ **3.** $\frac{7}{8}$ **4.** $\frac{3}{4}$ **5.** $\frac{5}{6}$
Set 3c: **1.** $\frac{5}{12}$ **2.** $\frac{13}{15}$ **3.** $\frac{3}{16}$ **4.** $\frac{11}{18}$ **5.** $\frac{15}{16}$
Set 5: **1.** $\frac{1}{8}$ **2.** $\frac{3}{7}$ **3.** $\frac{5}{6}$ **4.** $\frac{1}{2}$ **5.** $\frac{12}{25}$
Set 7: **1.** $1\frac{1}{6}$ **2.** $1\frac{5}{8}$ **3.** $2\frac{1}{4}$ **4.** $1\frac{2}{3}$ **5.** $3\frac{2}{5}$
Set 9: **1.** 2 **2.** 5 **3.** 10 **4.** 50 **5.** 28

Problem Solving, Page 119

1. c **3.** d **5.** c **7.** $\frac{7}{8}$ **9.** $\frac{23}{25}$ **11.** $\frac{1}{3}$ **13.** $\frac{1}{3}$; 6 quarts

Ratio, Page 120

1. Set 1: **a.** $\frac{2}{3}$ **b.** $\frac{3}{4}$ **c.** $\frac{5}{7}$ **d.** $\frac{1}{4}$ **e.** $\frac{5}{8}$
  Set 2: **a.** $\frac{3}{2}$ or $1\frac{1}{2}$ **b.** $\frac{8}{5}$ or $1\frac{3}{5}$ **c.** $\frac{8}{1}$ or 8 **d.** $\frac{5}{3}$ or $1\frac{2}{3}$ **e.** $\frac{9}{8}$ or $1\frac{1}{8}$
  Set 3: **a.** $\frac{1}{4}$ **b.** $\frac{2}{6}$ **c.** $\frac{4}{9}$ **d.** $\frac{5}{3}$ or $1\frac{2}{3}$ **e.** $\frac{3}{10}$

3. $\frac{4}{3}$ or $1\frac{1}{3}$ **5.** 6 eggs, 1 qt. milk, $\frac{1}{2}$ teaspoon salt, 1 teaspoon vanilla, $\frac{1}{2}$ cup sugar; $1\frac{1}{2}$ eggs, $\frac{1}{2}$ pint milk, $\frac{1}{8}$ teaspoon salt, $\frac{1}{4}$ teaspoon vanilla, $\frac{1}{8}$ cup sugar; 1 egg, $\frac{1}{3}$ pt. milk, $\frac{1}{12}$ teaspoon salt, $\frac{1}{6}$ teaspoon vanilla, $\frac{1}{12}$ cup sugar; $4\frac{1}{2}$ eggs, $1\frac{1}{2}$ pt. milk, $\frac{3}{8}$ teaspoon salt, $\frac{3}{4}$ teaspoon vanilla, $\frac{3}{8}$ cup sugar; 2 eggs, $\frac{2}{3}$ pt. milk, $\frac{1}{6}$ teaspoon salt, $\frac{1}{3}$ teaspoon vanilla, $\frac{1}{6}$ cup sugar **7.** $\frac{2}{5}$

EXERCISE 1-19 Finding a Number When a Fractional Part of It Is Known

Diagnostic Test, Page 121

1. 48

Related Practice Examples, Page 122

Set 1a: **1.** 10 **2.** 24 **3.** 76 **4.** 550 **5.** 1,000
Set 1b: **1.** 12 **2.** 54 **3.** 32 **4.** 108 **5.** 1,000
Set 1c: **1.** 16 **2.** 64 **3.** 152 **4.** 160 **5.** 336
Set 1d: **1.** 42 **2.** 144 **3.** 312 **4.** 432 **5.** 2,400
Set 1e: **1.** 30 **2.** 625 **3.** 400 **4.** 900 **5.** 3,000

Miscellaneous Examples, Page 123

Set 1: **1.** 32 **2.** 105 **3.** 92 **4.** 72 **5.** 480

Problem Solving, Page 123

1. b **3.** c **5.** 1,480 students **7. a.** 98¢ **b.** 92¢ **c.** 96¢ **d.** 48¢

Review of Unit Two, Pages 124–125

1. a. $\frac{3}{4}$ **b.** $\frac{4}{9}$ **c.** $\frac{3}{5}$ **d.** $\frac{7}{9}$ **e.** $\frac{3}{4}$ **3. a.** 9 **b.** 12 **c.** $2\frac{3}{5}$ **d.** $6\frac{1}{5}$ **e.** $13\frac{3}{4}$ **5. a.** 10ths **b.** 6ths
c. 12ths **d.** 16ths **e.** 120ths **7. a.** $5\frac{8}{9}$ **b.** $\frac{17}{16}$ **c.** $9\frac{7}{6}$ **d.** $7\frac{12}{16} = 6\frac{28}{16}$ **e.** $5\frac{16}{24} = 4\frac{40}{24}$ **9. a.** $\frac{1}{5}$; $\frac{2}{3}$
b. $\frac{1}{8}$; $\frac{1}{2}$ **c.** $\frac{7}{12}$, $\frac{11}{16}$, $\frac{3}{4}$ **11. a.** $\frac{39}{64}$ **b.** 87 **c.** $5\frac{1}{10}$ **d.** 14 **e.** $188\frac{1}{2}$ **13. a.** $\frac{9}{16}$ **b.** $\frac{5}{8}$ **15.** c
17. b **19.** $\frac{3}{1}$ or 3; $\frac{1}{3}$

Competency Check Test, Pages 125–126

1. b **3.** b **5.** a **7.** b **9.** d **11.** a

Keyed Achievement Test, Page 126

1. One billion, nine hundred seventy-eight million, two hundred forty-six thousand, five hundred twenty-three **3.** 278,274 **5.** 6,412,032 **7.** $\frac{9}{10}$ **9.** $\frac{2}{3}$ **11.** $13\frac{5}{8}$ **13.** $\frac{1}{12}$ **15.** 104

UNIT Three DECIMAL FRACTIONS

EXERCISE 1-20 Reading Decimals

Diagnostic Test, Page 128

1. Two tenths **3.** Fifty-eight hundredths **5.** Three and seventy-three hundredths **7.** Seventy-six thousandths **9.** Fourteen and seven hundred eight thousandths **11.** Seventeen thousand nine hundred twenty-five hundred-thousandths **13.** One hundred twenty-nine and four thousand two hundred sixty-one ten-thousandths **15.** Twenty-four and seventy-eight hundredths; Twenty-four point seventy-eight **17.** Sixteen trillion, nine hundred twenty-seven billion dollars

Related Practice Examples, Page 129

Set 1: **1.** Eight tenths **2.** One tenth **3.** Five tenths **4.** Four tenths **5.** Nine tenths

Set 3: **1.** Twenty-four hundredths **2.** Eighty-five hundredths **3.** Ninety-one hundredths **4.** Sixty hundredths **5.** Thirty-seven hundredths

Set 5: **1.** Two and fifty-one hundredths **2.** Seven and thirty-seven hundredths **3.** Thirty and forty-six hundredths **4.** Eighty-nine and three hundredths **5.** Two hundred forty-eight and nineteen hundredths

Set 7: **1.** Twenty-four thousandths **2.** Sixty-three thousandths **3.** Eighty thousandths **4.** Ninety-two thousandths **5.** Thirty-five thousandths

Set 9: **1.** Six and five thousandths **2.** Twenty-one and seven hundred sixty-nine thousandths **3.** Thirty-four and ninety-four thousandths **4.** One hundred eighty-six and five hundred twenty-eight thousandths **5.** Two hundred and forty-two thousandths

Set 11: **1.** Six hundred-thousandths **2.** Three hundred ninety-two hundred-thousandths **3.** Nine thousand four hundred thirteen hundred-thousandths **4.** Twenty-five thousand four hundred-thousandths **5.** Seventy-two thousand eight hundred fifteen hundred-thousandths

Set 13: **1.** Eight and twenty-five ten-thousandths **2.** Twenty-three and nine thousand three hundred seventeen ten-thousandths **3.** Forty-nine and eight thousand three hundred twenty-nine hundred-thousandths **4.** Three hundred fifty-four and three thousand six hundred twenty-five ten-thousandths **5.** Five hundred seventy-one and five thousand eight hundred seventy-five hundred-thousandths

Set 15: **1.** Thirty-six point eighty-nine; Thirty-six and eighty-nine hundredths **2.** Seventeen point six; Seventeen and six tenths **3.** Five point nine two five; Five and nine hundred twenty-five thousandths **4.** Two hundred ninety-four point zero one three; Two hundred ninety-four and thirteen thousandths **5.** Eighty-seven point four six six two; Eighty-seven and four thousand six hundred sixty-two ten-thousandths

Set 17: **1.** Thirty-three point four billion dollars **2.** Sixty-one point five six million dollars **3.** Two point eight nine eight trillion dollars **4.** One thousand, seven hundred fifty point zero eight billion dollars **5.** Nine hundred twenty-six point eight seven two million dollars

EXERCISE 1-21 Writing Decimals

Diagnostic Test, Page 131

1. .3 **3.** .07 **5.** 109.84 **7.** .094 **9.** 700.093 **11.** .00842 **13.** 260.0347 **15.** 519.9 **17.** $90.329 billion

Related Practice Examples, Pages 131–134

Set 1: **1.** .4 **2.** .8 **3.** .2 **4.** .9 **5.** .1
Set 3: **1.** .08 **2.** .02 **3.** .06 **4.** .04 **5.** .05
Set 5: **1.** 6.04 **2.** 5.62 **3.** 73.18 **4.** 200.05 **5.** 407.25
Set 7: **1.** .069 **2.** .047 **3.** .024 **4.** .016 **5.** .083
Set 9: **1.** 2.017 **2.** 800.035 **3.** 36.253 **4.** 57.784 **5.** 19.122
Set 11: **1.** .00003 **2.** .00042 **3.** .00456 **4.** .06822 **5.** .15786
Set 13: **1.** 500.0058 **2.** 2.0325 **3.** 70.2512 **4.** 85.00732 **5.** 643.000067
Set 15: **1.** 61.2 **2.** .641 **3.** 258.35 **4.** 880.1613 **5.** 3,010.79
Set 17: **1.** $12.4 trillion **2.** $5.864 million **3.** $70.38 billion **4.** $389.07 million **5.** $4,930.394 billion

Refresh Your Skills, Page 134

1. 2,236,021 **3.** 52,506,561 **5.** $11\frac{1}{4}$ **7.** $45\frac{1}{2}$ **9.** 806,000,000

EXERCISE 1-22 Rounding Decimals

Diagnostic Test, Page 136

1. a. .7 **b.** 5.4 **3. a.** 1.969 **b.** 3.247 **5. a.** .00592 **b.** 54.63286 **7. a.** $2.39 **b.** $4.84 **c.** $7.24 **d.** $1.68

Related Practice Examples, Pages 136–137

Set 1a: **1.** .3 **2.** .9 **3.** 1.0 **4.** 4.2 **5.** 2.4
Set 1b: **1.** .1 **2.** .3 **3.** 1.8 **4.** 3.7 **5.** 5.2
Set 3a: **1.** .215 **2.** .400 **3.** 3.082 **4.** 1.593 **5.** 16.769
Set 3b: **1.** .545 **2.** .929 **3.** 5.034 **4.** 2.008 **5.** 24.825
Set 5a: **1.** .00008 **2.** .00514 **3.** 1.07932 **4.** 25.41529 **5.** 83.59124
Set 5b: **1.** .00008 **2.** .03059 **3.** 2.00047 **4.** 23.07531 **5.** 71.11425
Set 7a: **1.** $.27 **2.** $.59 **3.** $1.32 **4.** $5.87 **5.** $24.68
Set 7b: **1.** $.64 **2.** $.83 **3.** $6.57 **4.** $14.96 **5.** $30.21
Set 7c: **1.** $.63 **2.** $.29 **3.** $3.85 **4.** $9.21 **5.** $15.15
Set 7d: **1.** $.46 **2.** $.73 **3.** $1.57 **4.** $4.89 **5.** $12.30

Practical Applications, Page 137

1. a. 1.15 **b.** 1.09 **c.** 2.20 **3. a.** 0.868 **b.** 6.452 **c.** 0.946

EXERCISE 1-23 Addition of Decimals

Diagnostic Test, Pages 139–140

1. .9 **3.** .06 **5.** .83 **7.** 1.00 **9.** 23.2 **11.** 12.37 **13.** 3.68 **15.** 10.484 **17.** 1,226.44 **19.** $3.08
21. 80.93 billion **23.** 11.282 **25.** 10.246

Related Practice Examples, Pages 140–143

Set 1: .8; .7; .8; .9; .9
Set 3: .07; .08; .09; .09; .09
Set 5: .94; .79; .96; .81; .98
Set 7: 1.00 or 1; .90 or .9; 2.50 or 2.5; 2.20 or 2.2; 2.00 or 2
Set 9: 14.3; 13.0 or 13; 19.9; 14.5; 32.7
Set 11: 5.41; 14.86; 12.66; 15.90 or 15.9; 17.23
Set 13: 6.18; 10.02; 4.84; 10.23; 16.90 or 16.9
Set 15: 8.009; 15.527; 18.337; 12.280 or 12.28; 23.489
Set 17: 391.77; 845.59; 16.9594; .95000 or .95; 1,807.58
Set 19: $2.78; $2.27; $2.31; $3.89; $3.93
Set 21: 84.693 billion; 10.9 trillion; $60 million; $193.398 billion; 1,141.28 million; $763 billion
Set 23: **1.** 1.08 **2.** 10.38 **3.** 12.329 **4.** 51.16 **5.** 8.806 **6.** 9.151 **7.** 119.073 **8.** 42.77
      9. 12.174 **10.** 221.376
Set 25: **1.** 15.462 **2.** 14.594 **3.** 27.23 **4.** 8.84 **5.** 6.07 **6.** 6.976 **7.** $11.82 **8.** $27.35
      9. $21.65 **10.** $10.22

Problem Solving, Page 144

1. b **3.** a **5.** c **7.** $687.88 **9. a.** $354.18 **b.** $334.03

Micrometer Caliper Readings, Page 145

1. 5.64 mm **3.** 14.98 mm **5.** 11.87 mm

Refresh Your Skills, Page 145

1. $10\frac{3}{10}$ **3.** $37\frac{1}{2}$ **5.** Forty-three point six; forty-three and six-tenths **7.** 8.28

EXERCISE 1-24 Subtraction of Decimals

Diagnostic Test, Page 147

1. .6 **3.** .19 **5.** .07 **7.** 1.5 **9.** 4.0 **11.** 1.77 **13.** .1615 **15.** .20021 **17.** .00535 **19.** 6.258
21. .165 **23.** 3.828 **25.** 6.86 billion **27.** .175 **29.** 3.37 **31.** $14.50

Related Practice Examples, Pages 148–150

Set 1: .4; .1; .5; .5; .1
Set 3: .18; .17; .16; .47; .29
Set 5: .09; .08; .09; .08; .05
Set 7: 4.2; 2.2; 5.1; 2.4; 5.1
Set 9: 2.0 or 2; 5.0 or 5; 3.0 or 3; 1.0 or 1; 4.0 or 4
Set 11: 4.11; 2.19; 3.73; 2.53; 1.17
Set 13: .0863; .0108; 17.69; 184.9; 1.868
Set 15: .3007; .3005; 2.0094; .40008; 8.007
Set 17: .0021; .00071; .00070 or .0007; .00712; .00082
Set 19: 2.247; 4.932; 0.0675; 1.531; 10.693
Set 21: .326; .6239; .0072; .11; .397
Set 23: 4.4; 8.91; 2.247; 1.06; 4.093
Set 25: 15.2 million; $288 billion; $1.26 million; $40.08 billion; 18.274 million; 14.506 billion
Set 27: **1**. .05 **2**. .54 **3**. .37 **4**. .184 **5**. .058 **6**. .029 **7**. .3482 **8**. .0656 **9**. .4892 **10**. .0004
Set 29: **1**. 3.36 **2**. 3.87 **3**. 21.65 **4**. 5.816 **5**. 5.723 **6**. 11.486 **7**. 77.02 **8**. 3.3944 **9**. 2.5
         10. 18.0098
Set 31: **1**. $.25 **2**. $.69 **3**. $1.88 **4**. $1.10 **5**. $4.40 **6**. $6.42 **7**. $15.47 **8**. $66.38 **9**. $167.09
         10. $561.25

Problem Solving, Pages 151–152

1. b **3**. d **5**. c **7**. $6.32 **9**. $25.74 **11**. 2.209 mm

Refresh Your Skills, Page 152

1. 2,471,780 **3**. 880,000 **5**. 30,000,000 **7**. $\frac{60}{100}$ **9**. 12.5 **11**. 176 **13**. $\frac{1}{3}$ **15**. Eight hundred and
eight hundredths **17**. $5.83 **19**. 3.77

EXERCISE 1-25 Comparing Decimals

Diagnostic Test, Page 153

1. .5 **3**. .699 **5**. 1.4, .388, .19 and .06 **7**. Yes

Related Practice Examples, Page 154

Set 1: **1**. .3 **2**. .04 **3**. .91 **4**. .7 **5**. .1561
Set 3: **1**. .89 **2**. .2 **3**. .05 **4**. .36 **5**. .0051
Set 5: **1**. .1, .01, .001 and .0001 **2**. 2.25, 2.249, .253 and .2485 **3**. 1.5, .475, .38 and .0506
        4. 5.02, .503, .1483 and .006 **5**. .98, .934, .9 and .89
Set 7: **1**. True **2**. False **3**. True **4**. False **5**. False

Practical Applications, Page 154

1. a. No b. Yes c. No **3**. a. No b. No c. Yes

EXERCISE 1-26 Multiplication of Decimals

Diagnostic Test, Page 157

1. 2.4 **3**. 30.186 **5**. 36 **7**. .06 **9**. .008 **11**. .12 **13**. .348 **15**. .2072 **17**. .729 **19**. .00441
21. .003776 **23**. 188.86 million **25**. $91.20 **27**. 54¢ **29**. $2.16 **31**. $9.61

Related Practice Examples, Pages 157–160

Set 1: .4; 4.2; 9.6; 1.5; 4.5
Set 3: 1.235; 15.504; 1.227; 90.948; 20.076
Set 5: 10.0 or 10; 19.00 or 19; 2.10 or 2.1; 1.000 or 1; 153.750 or 153.75
Set 7: .08; .09; .08; .06; .05
Set 9: .009; .003; .0035; .0078; .0056
Set 11: .72; .3; 2.52; 24.6; 14.25

Set 13: .102; .38; .462; 4.176; 5.98
Set 15: .2072; 5.1591; 2.2095; .7512; 7.7037
Set 17: .0012; .0014; .525; 1.8048; .3294
Set 19: .00004; .00716; .02166; .00628; .049
Set 21: .000048; .000027; .00026; .000007; .039006
Set 23: **1.** 24.8 trillion **2.** 8.12 billion **3.** $13.137 million **4.** $28.875 billion **5.** 57.08 million **6.** 216.07 billion
Set 25: $14.91; $138.24; $630; $1,565.76; $612
Set 27: $525; $3.40; $.70; $1,093.75; $1.35
Set 29: **1.** $.20 **2.** .435 **3.** $.20 **4.** $.97 **5.** $5.83 **6.** $1.13 **7.** $2.35 **8.** 16.016 **9.** $46.15 **10.** $46.13
Set 31: $10; $135; $.20; $.09; $1.06

Problem Solving, Pages 160–161

1. b **3.** a **5.** b **7. a.** $355.20 **b.** $431.28 **c.** $178.32
9. $108.90, $8.90; $117, $17; $125.46, $25.46

Refresh Your Skills, Page 161

1. 114,300 **3.** 821,688 **5.** $1\frac{11}{24}$ **7.** $46\frac{1}{2}$ **9.** 2.568 **11.** 150.7968

EXERCISE 1-27 Division of Decimals

Diagnostic Test A, Page 164

1. 2.3 **3.** 2.664 **5.** .854 **7.** 2.54 **9.** .02 **11.** 16.78 billion **13.** $.21

Related Practice Examples, Pages 165–166

Set 1: 2.3; 12.3; 15.6; 149.3; 1,127.6
Set 3: .121; .173; 3.971; 1.054; 24.638
Set 5: .84; .652; .403; .7653; .8391
Set 7: .28; 3.3; .487; 4.21; .78
Set 9: .05; .25; .03; 1.125; .012
Set 11: **1.** 15.63 million **2.** 5.5 trillion **3.** $1.389 billion **4.** $5.949 million **5.** 6.7 **6.** 1.258
Set 13: **a.** $.24; $.90; $.80; $.34; $.08; **b.** $.16; $.21; $.23; $3.25; $7.33

Diagnostic Test B, Page 166

1. 826 **3.** 1.12 **5.** .017 **7.** 20 **9.** 5 **11.** 6.7

Related Practice Examples, Pages 166–167

Set 1: 19; 179; 589; 3,007; 34,673
Set 3: .42; 7.56; 12.24; 209.76; 37.45
Set 5: .002; .032; .007; .003; .092
Set 7: 35; 25; 17.5; 16.25; 9.375
Set 9: 5; 2.5; 2.5; 1.25; 3.125
Set 11: **a.** 1.1; 2.4; 60.8; 3.2; 2,272.7 **b.** 6.67; 4.29; .09; 2.73; .58 **c.** .563; 284.615; 21.872; 35.088; 435.733

Diagnostic Test C, Page 168

1. 26,078 **3.** 4.37 **5.** .03 **7.** 300 **9.** 200 **11.** 46

Related Practice Examples, Pages 168–169

Set 1: 17; 251; 27; 53; 648
Set 3: 234.77; 9.02; 3.38; 4.58; .63
Set 5: .03; .05; .004; .062; .009
Set 7: 1,900; 50; 75; 87.5; 62.5
Set 9: 300; 500; 25; 37.5; 87.5

Set 11a: 173; 80; 72; 144; 24
Set 11b: (To nearest hundredth): 15.5; 15.6; 21.29; 24.56; 36.67

Diagnostic Test D, Page 169

1. 12,483   **3.** 934.16   **5.** 2.7   **7.** 430   **9.** 8,000

Related Practice Examples, Pages 169–170

Set 1:   109; 5,954; 38; 283; 147
Set 3:   14.91; 4.36; .69; 2.41; 3.36
Set 5:   3.2; .23; 5.7; 1.48; 24.2
Set 7:   270; 11,270; 640; 2,340; 860
Set 9:   3,000; 9,000; 30,000; 4,000; 6,000

Problem Solving, Pages 170–171

1. a   **3.** c   **5.** c   **7. a.** 5 kilograms   **b.** 6 dozen   **c.** 20 kilograms
9. 1.9998 in.; .0002 in.   **11.** 6 min.

Refresh Your Skills, Page 172

1. 2,310,548   **3.** 12,960,000   **5.** $2\frac{1}{2}$ **7.** $1\frac{31}{32}$ **9.** 2.26   **11.** 1.296   **13.** 48,300,000   **15.** $\frac{55}{60}$ **17.** $23.60
19. 96.02

United States Money—Review, Page 172

1. $.68; $2.06; $9.53; $22.06   **3.** $4.00; $506.25; $1.41; $5.71; $4.12

EXERCISE 1-28   Multiplying Whole Numbers and Decimals by 10, 100, 1000, etc.

Diagnostic Test, Page 173

1. 80   **3.** 4   **5.** .6   **7.** 2,600   **9.** 83   **11.** 9.87   **13.** 9,000   **15.** 365   **17.** 857.4

Related Practice Examples, Page 174

Set 1:   **1.** 50   **2.** 90   **3.** 270   **4.** 850   **5.** 7,630
Set 3:   **1.** 3   **2.** 1   **3.** 5   **4.** 9   **5.** 8
Set 5:   **1.** .3   **2.** .9   **3.** .85   **4.** .07   **5.** .625
Set 7:   **1.** 700   **2.** 5,100   **3.** 3,800   **4.** 42,300   **5.** 956,400
Set 9:   **1.** 42   **2.** 33   **3.** 19   **4.** 67   **5.** 95
Set 11:   **1.** 72.1   **2.** 3.9   **3.** 52.57   **4.** 4.16   **5.** 85.47
Set 13:   **1.** 3,000   **2.** 62,000   **3.** 597,000   **4.** 2,055,000   **5.** 4,682,000
Set 15:   **1.** 657   **2.** 942   **3.** 485   **4.** 76   **5.** 189
Set 17:   **1.** 265.3   **2.** 592.9   **3.** 35.7   **4.** 174.25   **5.** .72

Practical Applications, Page 174

1. a. 8,000 watts   **b.** 40,000 watts   **c.** 6,500 watts   **d.** 139,000 watts   **e.** 27,300 watts
3. a. 500 cm   **b.** 2,000 cm   **c.** 370 cm   **d.** 1,508 cm   **e.** 938.5 cm   **5.** $32.80

EXERCISE 1-29   Dividing Whole Numbers and Decimals by 10, 100, 1000, etc.

Diagnostic Test, Page 175

1. 8   **3.** .6   **5.** 1.5683   **7.** 8.75   **9.** .08   **11.** 1.972   **13.** 2.973   **15.** .072   **17.** .5273

Related Practice Examples, Page 176

Set 1:   **1.** 2   **2.** 5   **3.** 60   **4.** 30   **5.** 100
Set 3:   **1.** .5   **2.** .3   **3.** .9   **4.** .7   **5.** .4
Set 5:   **1.** .35   **2.** .982   **3.** 2.746   **4.** 3.9239   **5.** 14.5948

Set 7: **1.** 3.82 **2.** 8.29 **3.** 45.2 **4.** 39.48 **5.** 657.26
Set 9: **1.** .04 **2.** .09 **3.** .02 **4.** .03 **5.** .07
Set 11: **1.** .295 **2.** 1.2893 **3.** 5.0286 **4.** .6824 **5.** 15.0075
Set 13: **1.** 3.725 **2.** 2.89 **3.** 8.575 **4.** 15.925 **5.** 18.464
Set 15: **1.** .085 **2.** .093 **3.** .006 **4.** .008 **5.** .038
Set 17: **1.** .2849 **2.** .50074 **3.** .79582 **4.** 1.5261 **5.** 2.96345

Practical Applications, Pages 176–177

1. a. 20 kWh **b.** 135 kWh **c.** 27.5 kWh **3. a.** $427.20 **b.** $193.80; $1,128.75; $1,864.80; $1,912.50; $1,763.68 **c.** Reducing the rate; $1.35 **5. a.** $81.60 **b.** $54; $67.50; $115.00; $137.75; $344.40

Refresh Your Skills, Page 177

1. 3,507,822 **3.** 77,737,130 **5.** $8\frac{43}{60}$ **7.** $7\frac{1}{32}$ **9.** 95.608 **11.** 1

EXERCISE 1-30 Changing Common Fractions to Decimals

Diagnostic Test, Page 179

1. .9 **3.** .27 **5.** $.87\frac{1}{2}$ **7.** 1.25 **9.** $.88\frac{8}{9}$ **11.** $.87\frac{1}{2}$ **13.** 1.75 **15.** $1.28\frac{4}{7}$ **17.** .893

Related Practice Examples, Page 179

Set 1: **1.** .1 **2.** .7 **3.** .2 **4.** .8 **5.** .6
Set 3: **1.** .39 **2.** .54 **3.** .03 **4.** .06 **5.** .91
Set 5: **1.** $.37\frac{1}{2}$ **2.** $.62\frac{1}{2}$ **3.** $.12\frac{1}{2}$ **4.** $.56\frac{1}{4}$ **5.** $.06\frac{1}{4}$
Set 7: **1.** 1.15 **2.** 1.75 **3.** 1.83 **4.** 1.5 or 1.50 **5.** 2.34
Set 9: **1.** $.57\frac{1}{7}$ **2.** $.27\frac{3}{11}$ **3.** $.77\frac{7}{9}$ **4.** $.86\frac{2}{3}$ **5.** $.69\frac{3}{13}$
Set 11: **1.** $.37\frac{1}{2}$ **2.** $.83\frac{1}{3}$ **3.** $.66\frac{2}{3}$ **4.** $.87\frac{1}{2}$ **5.** $.62\frac{1}{2}$
Set 13: **1.** 1.5 or 1.50 **2.** $1.62\frac{1}{2}$ **3.** $1.42\frac{6}{7}$ **4.** 1.4 or 1.40 **5.** $1.43\frac{3}{4}$
Set 15: **1.** $1.55\frac{5}{9}$ **2.** $1.42\frac{6}{7}$ **3.** $1.26\frac{2}{3}$ **4.** $1.85\frac{5}{7}$ **5.** $1.11\frac{1}{9}$
Set 17: **1.** .571 **2.** .386 **3.** .049 **4.** .008 **5.** .647

Miscellaneous Examples, Page 180

1. a. .375 in. **b.** .125 in. **c.** .75 in. **d.** .625 in. **e.** .875 in. **3. a.** .21875 in. **b.** .78125 in. **c.** .28125 in. **d.** .59375 in. **e.** .40625 in. **5. a.** 1.53125 in. **b.** 5.5625 in. **c.** 2.359375 in. **d.** 8.0625 in. **e.** 3.84375 in.

Practical Applications, Page 180

1. a. $\frac{5}{8}$ inch **b.** .5347 inch **3.** Todd .375; Maria .333; José .317 **5.** Dave .920; Karen .917; Felipe .900; Frank .857; Rosa .848

EXERCISE 1-31 Changing Decimals to Common Fractions

Diagnostic Test, Page 182

1. $\frac{3}{10}$ **3.** $\frac{1}{25}$ **5.** $\frac{2}{3}$ **7.** $2\frac{17}{20}$ **9.** $\frac{84}{125}$ **11.** $3\frac{1}{8}$ **13.** $\frac{3}{400}$

Related Practice Examples, Page 182

Set 1: **1.** $\frac{3}{5}$ **2.** $\frac{1}{5}$ **3.** $\frac{1}{2}$ **4.** $\frac{1}{10}$ **5.** $\frac{9}{10}$
Set 3: **1.** $\frac{1}{50}$ **2.** $\frac{7}{100}$ **3.** $\frac{1}{100}$ **4.** $\frac{2}{25}$ **5.** $\frac{3}{50}$
Set 5: **1.** $\frac{1}{6}$ **2.** $\frac{7}{8}$ **3.** $\frac{1}{16}$ **4.** $\frac{5}{8}$ **5.** $\frac{5}{8}$
Set 7: **1.** $1\frac{1}{4}$ **2.** $2\frac{21}{50}$ **3.** $2\frac{67}{100}$ **4.** $1\frac{8}{25}$ **5.** $3\frac{21}{25}$
Set 9: **1.** $\frac{1}{8}$ **2.** $\frac{7}{8}$ **3.** $\frac{473}{500}$ **4.** $\frac{48}{125}$ **5.** $\frac{24}{125}$
Set 11: **1.** $1\frac{3}{8}$ **2.** $1\frac{31}{125}$ **3.** $2\frac{241}{250}$ **4.** $3\frac{151}{200}$ **5.** $1\frac{9}{200}$
Set 13: **1.** $\frac{1}{400}$ **2.** $\frac{27}{5,000}$ **3.** $\frac{17}{2,500}$ **4.** $\frac{3}{2,000}$ **5.** $\frac{3}{4,000}$

Miscellaneous Examples, Page 182

1. True   **3.** True   **5.** True

EXERCISE 1-32   What Decimal Part One Number Is of Another

Diagnostic Test, Page 185

1. .3   **3.** .21   **5.** .875   **7.** .643   **9.** .33   **11.** .05   **13.** .3

Related Practice Examples, Pages 185–186

Set 1:   **1.** .7   **2.** .5   **3.** .8   **4.** .1   **5.** .6
Set 3:   **1.** .07   **2.** .59   **3.** .83   **4.** .31   **5.** .97
Set 5:   **1.** .375   **2.** .6875   **3.** .625   **4.** .5625   **5.** .65625
Set 7:   **1.** .444   **2.** .714   **3.** .727   **4.** .133   **5.** .722
Set 9:   **1.** .83   **2.** .33   **3.** .42   **4.** .71   **5.** .83
Set 11:   **1.** 2.5   **2.** .25   **3.** .6   **4.** .667   **5.** .9
Set 13:   **1.** .05   **2.** .29   **3.** .8   **4.** .62   **5.** .05

EXERCISE 1-33   Finding the Number When a Decimal Part of It Is Known

Diagnostic Test, Page 187

1. 70   **3.** $2\frac{1}{2}$ **5.** 800   **7.** 800   **9.** 13.75   **11.** 640   **13.** 8   **15.** 300

Related Practice Examples, Pages 188–189

Set 1:   **1.** 40   **2.** 60   **3.** 600   **4.** 500   **5.** 90
Set 3:   **1.** 5   **2.** 5   **3.** 5   **4.** 3.33   **5.** 10
Set 5:   **1.** 700   **2.** 475   **3.** 800   **4.** 1,440   **5.** 7,000
Set 7:   **1.** 56   **2.** 1,216   **3.** 712   **4.** 4,000   **5.** 6,000
Set 9:   **1.** 37.5   **2.** 16.667   **3.** 56   **4.** 266.667   **5.** 342.857
Set 11:   **1.** 72   **2.** 1,800   **3.** 8,000   **4.** 16,000   **5.** 1,440
Set 13:   **1.** 23   **2.** 6   **3.** 39   **4.** 14.4   **5.** 8
Set 15:   **1.** 400   **2.** 300   **3.** $12.30   **4.** 5,400   **5.** 28

Estimating Answers, Page 189

1. 2.5   **3.** 67   **5.** 8.2   **7.** 91   **9.** $46   **11.** $12   **13.** 1.5   **15.** .05   **17.** $45   **19.** .03

Check by Calculator, Page 190

1. b and c are incorrect   **3.** b and d are incorrect   **5.** b and d are incorrect   **7.** b and c are incorrect

Review of Unit Three, Pages 191–192

1. a. .09   **b.** 500.043   **c.** 8.5   **d.** .000011   **e.** .7822   **3. a.** 1.0   **b.** 8.29   **c.** 25.143   **d.** $4.28   **e.** $70.16   **5. a.** 7.677   **b.** .189   **c.** .425   **d.** 9.96   **e.** $24.75   **7. a.** .008   **b.** 11.34   **c.** .0009   **d.** $3.95   **e.** $24.11   **9. a.** 9,000   **b.** 35.6   **c.** 400   **d.** 8   **e.** 28,950   **11. a.** .6   **b.** .83   **c.** .68   **d.** .875   **e.** .8125   **13. a.** 1.55 or $1\frac{11}{20}$ **b.** $.33\frac{1}{3}$ or $\frac{1}{3}$ **c.** .301   **d.** 72   **e.** 4   **15.** d   **17.** c   **19.** $879.10

Cumulative Practice, Page 193

1. 84,874   **3.** 83,947,336   **5.** $16\frac{15}{16}$ **7.** $759\frac{1}{2}$ **9.** 23.05   **11.** .1

Keyed Achievement Test, Page 193

1. 273,524   **3.** 470,028   **5.** $11\frac{1}{2}$ **7.** $28\frac{1}{8}$ **9.** $\frac{3}{4}$ **11.** $\frac{2}{3}$ **13.** 7.81   **15.** .15   **17.** 29.43   **19.** $\frac{7}{8}$

Competency Check Test, Page 194

1. d **3.** d **5.** b **7.** b **9.** d **11.** b **13.** b **15.** d

UNIT Four   SHORTENED NAMES FOR LARGE WHOLE NUMBERS

Review Exercises, Page 195

1. a. Eighty-five million **b.** Six hundred ninety-one billion dollars **c.** Three million four hundred eighty-seven thousand **d.** Fourteen trillion **3. a.** Four point ninety-five billion **b.** Twenty-three point zero nine million dollars **c.** One point forty-two trillion dollars **d.** Seventy-four point thirty-eight thousand **5. a.** Eight point six eight six billion **b.** Three hundred eighty-one point two nine nine million **c.** Eight point zero five six trillion dollars **d.** Seven hundred thirty-six point eight four seven billion

Practical Applications, Page 196

1. a. Twenty-five million people **b.** Thirty-seven billion dollars **c.** Two hundred eighteen million consumers; three billion dollars **d.** Three hundred thirty-five billion dollars **e.** Four billion dollars **3. a.** Two point five million sets **b.** One point one two million bales **c.** Two point three eight million metric tons **d.** One point four three billion dollars **e.** One point three seven five trillion dollars

EXERCISE 1-34   Writing Complete Numerals for the Shortened Names

Diagnostic Test, Page 197

1. 725,000,000 **3.** 483,280,000 **5.** $34,560,000 **7.** 98,400,000,000 **9.** 350,742,000,000 **11.** 31,000,000,000,000 **13.** 408,570,000,000,000 **15.** $70,040,000,000,000 **17.** 18,947 **19.** 182,600

Related Practice Examples, Pages 198–199

Set 1: **1.** 83,000,000 **2.** 9,000,000 **3.** 460,000,000 **4.** 2,805,000,000 **5.** 589,000,000
Set 3: **1.** 17,450,000 **2.** 5,070,000 **3.** 227,930,000 **4.** 1,540,160,000 **5.** 24,820,000
Set 5: **1.** $4,500,000 **2.** $27,400,000 **3.** 75,090,000 **4.** 408,263,000 **5.** 667,850,000
Set 7: **1.** 9,400,000,000 **2.** 72,800,000,000 **3.** 403,700,000,000 **4.** 2,576,300,000,000
     5. 80,200,000,000
Set 9: **1.** 7,568,000,000 **2.** 31,047,000,000 **3.** 564,382,000,000 **4.** 1,067,509,000,000
     5. 81,635,000,000
Set 11: **1.** 4,000,000,000,000 **2.** 17,000,000,000,000 **3.** 106,000,000,000,000 **4.** 58,000,000,000,000
     5. 292,000,000,000,000
Set 13: **1.** 5,920,000,000,000 **2.** 60,070,000,000,000 **3.** 139,780,000,000,000 **4.** 17,050,000,000,000
     5. 308,590,000,000,000
Set 15: **1.** $9,600,000,000,000 **2.** $3,280,000,000,000 **3.** $87,920,000,000,000
     4. $40,071,000,000,000 **5.** $156,435,000,000,000
Set 17: **1.** 8,700 **2.** 28,500 **3.** 63,090 **4.** 592,160 **5.** 860,442
Set 19: **1.** 3,800 **2.** 95,200 **3.** 243,300 **4.** $7,700 **5.** $60,500

Miscellaneous Examples, Page 199

Set 1: **1.** 6; 547 **2.** 29; 380 **3.** 820; 600 **4.** 52; 400 **5.** 17; 75
Set 3: **1.** 4; 200 **2.** 1; 480 **3.** 16; 952 **4.** 8; 500 **5.** 21; 760

EXERCISE 1-35   Writing Shortened Names for Large Whole Numbers

Diagnostic Test, Page 200

1. 214 hundred **3.** 87.45 hundred **5.** 750.6 thousand **7.** 48.409 thousand **9.** 4.7 million **11.** 80.557 million **13.** 4.2 billion **15.** 50.088 billion **17.** 8.1 trillion **19.** 724.525 trillion **21.** $838.25 million **23.** $4.202 trillion

Related Practice Examples, Pages 201–202

Set 1: **1.** 9 hundred **2.** 32 hundred **3.** 71 hundred **4.** 125 hundred **5.** 493 hundred
Set 3: **1.** 25.68 hundred **2.** 6.93 hundred **3.** 57.56 hundred **4.** 334.97 hundred
5. 99.28 hundred
Set 5: **1.** 8.5 thousand **2.** 14.4 thousand **3.** 260.8 thousand **4.** 53.7 thousand
5. 4,839.1 thousand
Set 7: **1.** 2.654 thousand **2.** 9.068 thousand **3.** 13.852 thousand **4.** 824.905 thousand
5. 1,303.056 thousand
Set 9: **1.** 7.3 million **2.** 14.6 million **3.** 256.8 million **4.** 69.2 million **5.** 3,108.5 million
Set 11: **1.** 28.063 million **2.** 4.709 million **3.** 85.458 million **4.** 396.677 million
5. 54.096 million
Set 13: **1.** 7.5 billion **2.** 15.4 billion **3.** 32.7 billion **4.** 184.3 billion **5.** 1,421.9 billion
Set 15: **1.** 51.366 billion **2.** 4.592 billion **3.** 608.028 billion **4.** 79.009 billion
5. 4,312.454 billion
Set 17: **1.** 4.7 trillion **2.** 26.4 trillion **3.** 870.5 trillion **4.** 329.9 trillion **5.** 53.6 trillion
Set 19: **1.** 17.526 trillion **2.** 2.905 trillion **3.** 56.847 trillion **4.** 290.061 trillion
5. 388.519 trillion
Set 21: **1.** \$84 million **2.** \$9.36 million **3.** \$450.844 million **4.** \$17.525 million
5. \$2,655.007 million
Set 23: **1.** \$6.5 trillion **2.** \$48.32 trillion **3.** \$210.68 trillion **4.** \$36.493 trillion
5. \$9.057 trillion

Operations with Shortened Names, Pages 202–203

1. 223 million; 579 billion; 13.6 trillion; 401.33 million **3.** 409 billion; 48.4 thousand; 1.59 million;
7.489 trillion **5. a.** \$240 billion **b.** \$37.08 million **c.** 13.6 trillion **d.** \$766 billion
7. 156 billion dollars **9.** 211 **11.** 2.62 million shares **13.** 10.1 million dollars **15.** \$8.3 million;
\$1 billion

Review of Unit Four, Page 204

1. a. Six hundred fifty-seven point nine billion **b.** Seventy-one point zero four eight million
c. Six point seven three trillion dollars **d.** Nine hundred three point five thousand
3. a. \$18,380,000 **b.** 257,400 **c.** 220,860,000,000 **d.** \$8,529,000,000,000 **5. a.** 1,372.6 million
b. \$219.9 billion **c.** \$17.28 trillion **d.** 2.3 million **7.** \$51 billion **9.** 511.4 million tons

Competency Check Test, Page 205

1. d **3.** c **5.** c **7.** c **9.** a **11.** b **13.** c **15.** b

UNIT Five PERCENT

EXERCISE 1-36 The Meaning of Percent

Diagnostic Test, Page 207

1. 18% **3.** 35 hundredths

Related Practice Examples, Page 208

Set 1: **1.** 3% **2.** 9% **3.** 16% **4.** 30% **5.** 74% **6.** 200% **7.** $5\frac{1}{2}$% **8.** $37\frac{1}{2}$% **9.** 62.5% **10.** $\frac{1}{4}$%
Set 3: **1.** 4 hundredths **2.** 7 hundredths **3.** 19 hundredths **4.** 45 hundredths
5. 60 hundredths **6.** 140 hundredths **7.** $87\frac{1}{2}$ hundredths **8.** $4\frac{3}{4}$ hundredths
9. 3.5 hundredths **10.** $\frac{1}{2}$ hundredth

Miscellaneous Examples, Page 208

1. Cloth content is 55 hundredths Dacron and 45 hundredths wool

EXERCISE 1-37 Changing Percents to Decimals

Diagnostic Test, Page 209

1. .08 **3.** .90 **5.** 1.60 **7.** .125 or .12$\frac{1}{2}$ **9.** .40625 or .40$\frac{5}{8}$ **11.** .275 **13.** .086 **15.** 3.52875
17. .0075 or .00$\frac{3}{4}$

Related Practice Examples, Page 210

Set 1: **1.** .06 **2.** .09 **3.** .01 **4.** .03 **5.** .04
Set 3: **1.** .40 or .4 **2.** .70 or .7 **3.** .20 or .2 **4.** .60 or .6 **5.** .10 or .1
Set 5: **1.** 1.30 or 1.3 **2.** 1.20 or 1.2 **3.** 1.80 or 1.8 **4.** 1.50 or 1.5 **5.** 1.40 or 1.4
Set 7: **1.** .37$\frac{1}{2}$ or .375 **2.** .62$\frac{1}{2}$ or .625 **3.** .83$\frac{1}{3}$ **4.** .16$\frac{2}{3}$ **5.** .18$\frac{3}{4}$ or .1875
Set 9: **1.** .60$\frac{1}{2}$ or .605 **2.** .20$\frac{1}{2}$ or .205 **3.** .10$\frac{3}{8}$ or .10375 **4.** .50$\frac{2}{3}$ **5.** .70$\frac{1}{4}$ or .7025
Set 11: **1.** .875 **2.** .125 **3.** .464 **4.** .573 **5.** .951
Set 13: **1.** .035 **2.** .028 **3.** .047 **4.** .093 **5.** .064
Set 15: **1.** .26375 **2.** .31625 **3.** 1.28333 **4.** 4.32125 **5.** 8.95667
Set 17: **1.** .00$\frac{1}{2}$ or .005 **2.** .00$\frac{1}{4}$ or .0025 **3.** .00$\frac{5}{8}$ or .00625 **4.** .00$\frac{2}{3}$ **5.** .00$\frac{7}{16}$

Miscellaneous Examples, Page 210

1. .08 **3.** 1.00$\frac{5}{8}$ or 1.00625

EXERCISE 1-38 Changing Decimals to Percents

Diagnostic Test, Page 211

1. 6% **3.** 70% **5.** 140% **7.** 4$\frac{1}{2}$% **9.** 166$\frac{2}{3}$% **11.** 4$\frac{2}{3}$% **13.** 287.5% **15.** $\frac{3}{8}$%

Related Practice Examples, Page 212

Set 1: **1.** 1% **2.** 8% **3.** 4% **4.** 3% **5.** 5%
Set 3: **1.** 60% **2.** 30% **3.** 80% **4.** 10% **5.** 90%
Set 5: **1.** 120% **2.** 180% **3.** 170% **4.** 130% **5.** 160%
Set 7: **1.** 1$\frac{1}{4}$% **2.** 5$\frac{1}{2}$% **3.** 3$\frac{3}{4}$% **4.** 4$\frac{2}{3}$% **5.** 6$\frac{5}{6}$%
Set 9: **1.** 137$\frac{1}{2}$% **2.** 183$\frac{1}{4}$% **3.** 100$\frac{1}{3}$% **4.** 116$\frac{2}{3}$% **5.** 105$\frac{3}{4}$%
Set 11: **1.** 26.25% **2.** 5.25% **3.** 8.75% **4.** 2.33% **5.** 65.125%
Set 13: **1.** 124.5% **2.** 137.5% **3.** 266.7% **4.** 372.75% **5.** 296.25%
Set 15: **1.** $\frac{1}{4}$% **2.** $\frac{1}{2}$% **3.** $\frac{3}{5}$% **4.** $\frac{5}{8}$% **5.** $\frac{2}{3}$%

Practical Applications, Page 212

1. 62$\frac{1}{2}$% **3.** 20% **5.** 12$\frac{1}{2}$%

EXERCISE 1-39 Changing Percents to Common Fractions

Diagnostic Test, Page 213

1. $\frac{3}{4}$ **3.** $\frac{1}{50}$

Related Practice Examples, Pages 213–214

Set 1: **1.** $\frac{1}{2}$ **2.** $\frac{1}{4}$ **3.** $\frac{1}{10}$ **4.** $\frac{4}{5}$ **5.** $\frac{9}{10}$ **6.** $\frac{3}{5}$ **7.** $\frac{3}{10}$ **8.** $\frac{2}{5}$ **9.** $\frac{7}{10}$ **10.** $\frac{1}{5}$
Set 3: **1.** $\frac{3}{50}$ **2.** $\frac{1}{25}$ **3.** $\frac{1}{20}$ **4.** $\frac{9}{100}$ **5.** $\frac{23}{50}$ **6.** $\frac{6}{25}$ **7.** $\frac{13}{25}$ **8.** $\frac{9}{50}$ **9.** $\frac{17}{20}$ **10.** $\frac{9}{25}$

Practical Applications, Page 214

1. $\frac{3}{10}$ **3.** $\frac{2}{5}$ **5.** $\frac{5}{6}$

Refresh Your Skills, Page 214

1. 157,156 **3.** 546,448 **5.** 13$\frac{1}{12}$ **7.** 8.8125 or 8$\frac{13}{16}$ **9.** 30.35 **11.** .0203 **13.** $18.70 **15.** $700

EXERCISE 1-40   Changing Common Fractions to Percents

Diagnostic Test, Page 216

1. 75%   **3.** 3%   **5.** 2%   **7.** $77\frac{7}{9}$%   **9.** 100%   **11.** 175%   **13.** $266\frac{2}{3}$%

Related Practice Examples, Pages 216–217

Set 1:   **1.** 25%   **2.** 40%   **3.** 50%   **4.** 70%   **5.** 80%
Set 3:   **1.** 7%   **2.** 39%   **3.** 145%   **4.** $87\frac{1}{2}$%   **5.** 81%
Set 5:   **1.** 9%   **2.** 5%   **3.** 4%   **4.** 12%   **5.** 15%
Set 7:   **1.** $71\frac{3}{7}$%   **2.** $72\frac{8}{11}$%   **3.** $44\frac{4}{9}$%   **4.** $46\frac{2}{3}$%   **5.** $27\frac{7}{9}$%
Set 9:   **1.** 100%   **2.** 100%   **3.** 100%   **4.** 100%   **5.** 100%
Set 11:   **1.** $166\frac{2}{3}$%   **2.** $162\frac{1}{2}$%   **3.** $114\frac{2}{7}$%   **4.** $122\frac{2}{9}$%   **5.** 180%
Set 13:   **1.** $262\frac{1}{2}$%   **2.** 380%   **3.** $466\frac{2}{3}$%   **4.** $314\frac{2}{7}$%   **5.** 275%

Practical Applications, Page 217

1. 60%   **3.** 30%   **5.** 25%

Refresh Your Skills, Page 217

1. 4,237,862   **3.** 53,143,872   **5.** $12\frac{9}{32}$ **7.** $14\frac{2}{3}$ **9.** $186.68   **11.** $312.50   **13.** $\frac{4}{5}$ **15.** 6.13%

EXERCISE 1-41   Finding a Percent of a Number

Diagnostic Test, Page 219

1. 8.28   **3.** 2.6325   **5.** 9   **7.** 943.61   **9.** 150   **11.** 426.3   **13.** 3.51   **15.** $1.45

Related Practice Examples, Pages 219–220

Set 1:   **1.** 12.48   **2.** 35.67   **3.** 47.25   **4.** 11.2   **5.** 44.16   **6.** 38.4   **7.** 552.24   **8.** 368.55   **9.** 930.24
      10. 920.64
Set 3:   **1.** .072   **2.** .412   **3.** .2886   **4.** .846   **5.** 2.3571   **6.** .154   **7.** 1.665   **8.** 14.754   **9.** 18.81
      10. 95.6422
Set 5:   **1.** 3   **2.** 31   **3.** 26   **4.** 69   **5.** 15   **6.** 120   **7.** 666   **8.** 82   **9.** 147   **10.** 980
Set 7:   **1.** 32.48   **2.** 59.85   **3.** 165.6   **4.** 1,212.75   **5.** 3,525
Set 9:   **1.** 72   **2.** 2,000   **3.** 650   **4.** 8,000   **5.** 11,562
Set 11:   **1.** 4.9   **2.** 91.36   **3.** 3.78   **4.** 319.725   **5.** 4,943.75
Set 13:   **1.** .9   **2.** 39.48   **3.** 48.9375   **4.** 10.8   **5.** 369
Set 15a: **1.** $.20   **2.** $1.88   **3.** $16.38   **4.** $30   **5.** $5   **6.** $96   **7.** $4.37   **8.** $570   **9.** $2,300
      10. $2,313.63
Set 15b: **1.** $47.46   **2.** $396   **3.** $7.50   **4.** $33.30   **5.** $800   **6.** $1,479   **7.** $.14   **8.** $245.20   **9.** $15
      10. $27

Problem Solving, Pages 221–222

1. b   **3.** c   **5.** c   **7.** Food, $7,200; Shelter, $5,760; Transportation, $5,184; Clothing, $4,320; Savings, $2,880; Miscellaneous, $3,456   **9.** $15   **11.** 177 lb.   **13.** 4,655 ft.   **15. a.** $29.70   **b.** $58.33
c. $105.00   **d.** $43.98

Refresh Your Skills, Page 222

1. 181,818   **3.** 219,114   **5.** $9\frac{15}{16}$ **7.** $14\frac{2}{3}$ **9.** 19   **11.** $1.05   **13.** 4,000,000   **15.** 18%   **17.** .56
19. $1.69

EXERCISE 1-42   Finding What Percent One Number Is of Another

Diagnostic Test, Page 224

1. 80%   **3.** $87\frac{1}{2}$%   **5.** $28\frac{4}{7}$%   **7.** 100%   **9.** 125%   **11.** 60%   **13.** $33\frac{1}{3}$%   **15.** 52%

Related Practice Examples, Pages 224-226

Set 1a: **1.** 60% **2.** 25% **3.** 75% **4.** 20% **5.** 90%
Set 1b: **1.** 25% **2.** 40% **3.** 50% **4.** 70% **5.** 30%
Set 3: **1.** $12\frac{1}{2}$% **2.** $83\frac{1}{3}$% **3.** $66\frac{2}{3}$% **4.** $37\frac{1}{2}$% **5.** $62\frac{1}{2}$%
Set 5: **1.** $42\frac{6}{7}$% **2.** $41\frac{2}{3}$% **3.** $77\frac{7}{9}$% **4.** $61\frac{7}{13}$% **5.** $94\frac{4}{9}$%
Set 7: **1.** 100% **2.** 100% **3.** 100% **4.** 100% **5.** 100%
Set 9: **1.** 120% **2.** 175% **3.** 225% **4.** 130% **5.** 180%
Set 11: **1.** 75% **2.** $66\frac{2}{3}$% **3.** $62\frac{1}{2}$% **4.** 8% **5.** 40%
Set 13: **1.** 25% **2.** 20% **3.** 75% **4.** $66\frac{2}{3}$% **5.** 60%
Set 15: **1.** 54% **2.** 35% **3.** 68% **4.** 84% **5.** 8%

Miscellaneous Examples, Page 226

Set 1: **1.** 55.6% **2.** 61.9% **3.** 73.3% **4.** 6.3% **5.** 58.3%

Problem Solving, Pages 226-227

1. c **3.** d **5.** b **7.** $12\frac{1}{2}$% **9. a.** 20% increase **b.** $16\frac{2}{3}$% decrease **c.** $33\frac{1}{3}$% increase
d. 25% decrease **e.** 25% increase

Refresh Your Skills, Page 227

1. 246,510 **3.** 325,221 **5.** $12\frac{1}{2}$ **7.** $10\frac{1}{16}$ **9.** 29.27 **11.** 251.20 **13.** 4.872 **15.** $44\frac{4}{9}$%

EXERCISE 1-43 Finding a Number When a Percent of It Is Known

Diagnostic Test, Page 229

1. 200 **3.** 24 **5.** 200 **7.** 70 **9.** 31.5 **11.** 2,000 **13.** 800 **15.** $2,250

Related Practice Examples, Pages 229-230

Set 1: **1.** 200 **2.** 400 **3.** 500 **4.** 120 **5.** 900
Set 3: **1.** 15 **2.** 120 **3.** 30 **4.** 130 **5.** 144
Set 5: **1.** 500 **2.** 650 **3.** $1,583\frac{1}{3}$ **4.** 1,500 **5.** 900
Set 7: **1.** 59 **2.** 80 **3.** 5 **4.** 18 **5.** 180
Set 9: **1.** 275 **2.** 78 **3.** 158 **4.** 640 **5.** 954
Set 11: **1.** 400 **2.** 200 **3.** 720 **4.** 250 **5.** 5,000
Set 13: **1.** 500 **2.** 8,000 **3.** 2,000 **4.** 3,600 **5.** 1,000
Set 15: **1.** $200 **2.** $1,300 **3.** $1.20 **4.** $5,000 **5.** $21.68

Problem Solving, Page 231

1. c **3.** $50,000 **5. a.** $55.00 **b.** $240 **c.** $15.90 **d.** $239

Refresh Your Skills, Page 231

1. 309,347 **3.** 72,436,200 **5.** $14\frac{1}{16}$ **7.** $52\frac{1}{2}$ **9.** $55.62 **11.** .0084 **13.** $150 **15.** 250

Practice Problems, Pages 232-233

1. a. $\frac{6}{100}$ or $\frac{3}{50}$ **b.** $\frac{18}{100}$ or $\frac{9}{50}$ **c.** $\frac{93}{100}$ **d.** $\frac{70}{100}$ or $\frac{7}{10}$ **e.** $\frac{4}{100}$ or $\frac{1}{25}$ **f.** $\frac{120}{100}$ or $\frac{6}{5}$ **g.** $\frac{300}{100}$ or $\frac{3}{1}$ **h.** $\frac{62\frac{1}{2}}{100}$ or $\frac{5}{8}$
i. $\frac{51}{100}$ or $\frac{21}{400}$ **j.** $\frac{24}{100}$ or $\frac{3}{125}$ **k.** $\frac{\frac{1}{2}}{100}$ or $\frac{1}{200}$ **l.** $\frac{25}{100}$ or $\frac{1}{4}$ **m.** $\frac{57}{100}$ **n.** $\frac{33\frac{1}{3}}{100}$ or $\frac{1}{3}$ **o.** $\frac{0.7}{100}$ or $\frac{7}{1,000}$
3. a. 40% **b.** $33\frac{1}{3}$% **c.** $66\frac{2}{3}$% **d.** 75% **e.** $87\frac{1}{2}$% **f.** $83\frac{1}{3}$% **g.** 200% **h.** 125% **i.** 80% **j.** $63\frac{7}{11}$%
k. 60% **l.** $16\frac{2}{3}$%

Estimating Answers, Page 233

1. 12 **3.** $3.10 **5.** $28 **7.** 9% **9.** 65%

Check by Calculator, Page 234

1. b and c are incorrect **3.** c and d are incorrect **5.** b and d are incorrect **7.** a, b and d are incorrect **9.** b and c are incorrect

Review of Unit Five, Page 235

1. a. 7%, .07, $\frac{7}{100}$ **b.** 56%, .56, $\frac{56}{100}$ **c.** 53%, .53, $\frac{53}{100}$ **d.** 91%, .91, $\frac{91}{100}$ **e.** $39\frac{1}{3}$%, $.39\frac{1}{3}$, $\frac{39\frac{1}{3}}{100}$ **3. a.** 3%
b. 30% **c.** 118% **d.** 94.2% **e.** $66\frac{2}{3}$% **5. a.** 60% **b.** 44% **c.** 125% **d.** 100% **e.** $366\frac{2}{3}$%
7. a. $66\frac{2}{3}$% **b.** 58% **c.** 15% **d.** 250% **e.** 44.4% **9.** $282

Cumulative Practice, Page 236

1. 2,444,109 **3.** 3,536,463 **5.** $10\frac{13}{24}$ **7.** $1\frac{91}{128}$ **9.** $9.06 **11.** $.13 **13.** $34.16 **15.** 40

Keyed Achievement Test, Pages 236–237

1. 93,483,000 **3.** 143,385 **5.** 42,349,344 **7.** $14\frac{2}{3}$ **9.** $22\frac{1}{2}$ **11.** 102.177 **13.** .00048 **15.** .375
17. 6.9% **19.** $62\frac{1}{2}$%

Meanings and Understandings, Page 237

1. a. remainder or difference **b.** product **c.** quotient **d.** sum **3. a.** equal **b.** greater
c. smaller **5. a.** No **b.** No

Competency Check Test, Page 238

1. b **3.** a **5.** d **7.** a **9.** a **11.** d **13.** b **15.** d

UNIT Six　SQUARES AND SQUARE ROOTS

EXERCISE 1-44　Squaring a Number

Diagnostic Test, Page 240

1. 64 **3.** .0625 **5.** $\frac{81}{16}$ or $5\frac{1}{16}$

Related Practice Examples, Page 241

Set 1: **1.** 16 **2.** 1 **3.** 81 **4.** 49 **5.** 100 **6.** 9 **7.** 36 **8.** 4 **9.** 144 **10.** 121
Set 3: **1.** .04 **2.** .49 **3.** .0001 **4.** .0081 **5.** .5625 **6.** .0324 **7.** 2.1609 **8.** 4.2025 **9.** 9.8596
　　　 10. 350.8129
Set 5: **1.** $\frac{9}{4}$ or $2\frac{1}{4}$ **2.** $\frac{25}{9}$ or $2\frac{7}{9}$ **3.** $\frac{289}{64}$ or $4\frac{33}{64}$ **4.** $\frac{121}{16}$ or $7\frac{9}{16}$ **5.** $\frac{100}{9}$ or $11\frac{1}{9}$ **6.** $\frac{25}{16}$ or $1\frac{9}{16}$
　　　 7. $\frac{81}{4}$ or $20\frac{1}{4}$ **8.** $\frac{169}{64}$ or $2\frac{41}{64}$ **9.** $\frac{144}{25}$ or $5\frac{19}{25}$ **10.** $\frac{49}{36}$ or $1\frac{13}{36}$

Practical Applications, Page 241

1. a. 49 sq. in. **b.** 169 mm^2 **c.** $6\frac{1}{4}$ sq. ft. **d.** 625 m^2 **e.** 361 cm^2 **3. a.** 25 **b.** 144 **c.** 900
d. 729 **e.** 2,025

EXERCISE 1-45　Square Root

Diagnostic Test, Page 244

1. 8 **3.** .3 **5.** 78 **7.** 586 **9.** 1,635 **11.** 6,004 **13.** .94 **15.** 27.72

Related Practice Examples, Pages 244–245

Set 1: **1.** 2 **2.** 7 **3.** 5 **4.** 3 **5.** 9 **6.** 4 **7.** 6 **8.** 1 **9.** 8 **10.** 10
Set 3: **1.** .2 **2.** .8 **3.** .9 **4.** .5 **5.** .1 **6.** .4 **7.** .7 **8.** .03 **9.** .06 **10.** .01

Set 5: **1.** 51 **2.** 66 **3.** 77 **4.** 94 **5.** 59 **6.** 37 **7.** 86 **8.** 99 **9.** 75 **10.** 38
Set 7: **1.** 375 **2.** 826 **3.** 631 **4.** 419 **5.** 937 **6.** 584 **7.** 457 **8.** 698 **9.** 752 **10.** 894
Set 9: **1.** 1,276 **2.** 1,978 **3.** 2,345 **4.** 1,053 **5.** 2,573 **6.** 1,257 **7.** 2,386 **8.** 3,142 **9.** 2,971
10. 1,849
Set 11: **1.** 1,007 **2.** 2,004 **3.** 3,005 **4.** 8,006 **5.** 4,003 **6.** 5,009 **7.** 6,002 **8.** 9,001 **9.** 7,008
10. 6,006
Set 13: **1.** .36 **2.** .18 **3.** .29 **4.** .74 **5.** .97 **6.** 13.8 **7.** 1.74 **8.** 9.41 **9.** 84.6 **10.** 32.72
Set 15: **1.** .32 **2.** .77 **3.** 12.15 **4.** 1.87 **5.** 14.41 **6.** 2.51 **7.** 5.09 **8.** .96 **9.** 9.73 **10.** 22.44

Practice Problems, Page 246

a. 2.449 **c.** 4.583 **e.** 6.164 **g.** 10.247 **i.** 6.403 **k.** 11.180 **m.** 23.452 **o.** 26.401

Practice Problems, Page 247

1. a. 4.359 **b.** 7.550 **c.** 9.747 **d.** 8.544 **e.** 6.928 **f.** 5.385 **g.** 8.124 **h.** 9.165 **i.** 5.745 **j.** 9.539

Practical Applications, Page 247

1. a. 5 cm **b.** 12 in. **c.** 83 m **d.** 69.6 yd. or 208.7 ft. **3. a.** 5.292 mm **b.** 6.481 ft. **c.** 7.937 cm
d. 11.832 m **5. a.** 10 in. **b.** 50 m **c.** 20 cm **d.** 30 in.

Review of Unit Six, Page 248

1. a. 169 **b.** 62,500 **c.** .000001 **d.** 24.01 **e.** .098 **f.** 3.063 **3. a.** 97 **b.** 293 **c.** 7,009 **d.** 5,934
e. 7.87 **f.** 20.14 **5. a.** 6.63 **b.** 4.80 **c.** 8.31 **d.** 94 **e.** 29 **f.** 78

Cumulative Practice, Page 248

1. 333,533 **3.** 44,592,342 **5.** 26 **7.** 12.375 **9.** 61.416 **11.** .00064 **13.** $533.80 **15.** 650

Competency Check Test, Page 249

1. c **3.** a **5.** b **7.** a **9.** c **11.** d **13.** b **15.** d **17.** b

UNIT Seven SUPPLEMENTARY TOPICS

EXERCISE 1-46 Roman System of Numeration

Practice Problems, Page 251

1. a. 2; 8; 9; 17; 14 **b.** 26; 29; 67; 48; 94 **c.** 73; 99; 166; 114; 390 **d.** 411; 227; 741; 1,669;
1,254 **e.** 1,450; 1,965; 1,062; 1,993; 1,444 **3.** 1919 **5. a.** IV **b.** XX **c.** XXV **d.** C **e.** XL
f. L

EXERCISE 1-47 Reading and Writing Shortened Lists of Numerals

Practice Problems, Page 252

1. a. zero, one, two, three and so on without end **b.** zero, two, four, six and so on without end
c. one, three, five, seven and so on without end **d.** two, three, five, seven and so on without
end **e.** zero, five, ten, fifteen and so on without end **f.** one, four, nine, sixteen and so on without end **3. a.** 0, 3, 6, 9, . . . **b.** 0, 6, 12, 18, . . . , 72 **c.** 2, 3, 5, 7, . . . , 89 **d.** 0, 20, 40, 60, . . .
e. 1, 2, 3, 4, . . . , 100 **f.** 3, 9, 27, 81, . . .

EXERCISE 1-48 Exponents

Practice Problems, Page 253

1. a. Three to the fourth power **b.** Seven to the eighth power **c.** Two to the eleventh power

d. Ten to the fifth power **e.** Six to the first power **f.** Eleven to the ninth power **g.** Five to the third power or five cubed **h.** Twenty to the second power or twenty squared **3. a.** 7 **b.** 9 **c.** 1 **d.** 12 **e.** 4 **5. a.** 8 **b.** 12 **c.** 7 **d.** 15 **e.** 30 **7. a.** $7^3 = 7 \times 7 \times 7$ **b.** $6^4 = 6 \times 6 \times 6 \times 6$ **c.** $3^{10} = 3 \times 3 \times 3 \times 3 \times 3 \times 3 \times 3 \times 3 \times 3 \times 3$ **d.** $8^7 = 8 \times 8 \times 8 \times 8 \times 8 \times 8 \times 8$ **e.** $12^9 = 12 \times 12 \times 12 \times 12 \times 12 \times 12 \times 12 \times 12 \times 12$ **f.** $9^{12} = 9 \times 9 \times 9 \times 9 \times 9 \times 9 \times 9 \times 9 \times 9 \times 9 \times 9 \times 9$ **9. a.** $36 = 6 \times 6 = 6^2$ **b.** $25 = 5 \times 5 = 5^2$ **c.** $8 = 2 \times 2 \times 2 = 2^3$ **d.** $81 = 9 \times 9 = 9^2$ **e.** $121 = 11 \times 11 = 11^2$ **f.** $625 = 25 \times 25 = 25^2$

EXERCISE 1-49 Powers of Ten

Practice Problems, Page 254

1. a. 10^3 **b.** 10^5 **c.** 10^7 **d.** 10^9 **e.** 10^{14} **f.** 10^{16} **3. a.** 7×10^6 **b.** 8×10^{12} **c.** 5×10^9 **d.** 4×10^{15} **e.** 9×10^{21} **f.** 3×10^{18}

EXERCISE 1-50 Writing Decimal Numerals as Polynomials—Expanded Notation

Practice Problems, Page 256

1. a. 67,294 **b.** 356,827 **3. a.** .734 **b.** .6175 **c.** 93.68 **d.** 207.49 **e.** 48,368.027 **f.** 6,492.50381

EXERCISE 1-51 Number Sentences

Practice Problems, Page 258

1. a. Thirty-four equals thirty-four **c.** Fifty-three is greater than forty-nine **e.** Sixteen is less than twenty **g.** Seventeen is not equal to ten plus eight **i.** Fifteen is not greater than twenty-three **k.** Two plus three is not less than three plus two **m.** Fifteen times ten is equal to ten times fifteen **o.** Eleven minus four is greater than two times three **q.** Thirty-seven minus eighteen is not less than fifteen minus fifteen **s.** Sixty-three divided by seven is less than sixty-three minus seven **u.** One times one is equal to one divided by one **w.** Point six minus point six is less than point six divided by point six **3. a.** True **b.** True **c.** True **d.** False **e.** False **f.** True **g.** True **h.** False **i.** True **j.** False **k.** False **l.** True

EXERCISE 1-52 Even and Odd Numbers

Practice Problems, Pages 259-260

1. a. even: 10; 284 odd: 49; 97; 565; 2,183 **b.** even: 36; 78; 300; 902 odd: 81; 5,749 **3. a.** forty-eight, fifty, fifty-two, fifty-four, fifty-six, fifty-eight, sixty, and sixty-two **b.** ninety-five, ninety-seven, ninety-nine, one hundred one, one hundred three, one hundred five **5. a.** even; $2 \times 2 = 4$ **b.** odd; $5 \times 3 = 15$ **c.** even; $3 \times 2 = 6$ **7. a.** No **b.** No **c.** An even number **9. a.** An even number. Because the product of the multiplication of two even numbers is an even number. **b.** An odd number. Because the product of the multiplication of two odd numbers is an odd number.

Refresh Your Skills, Page 260

1. 49,625 **3.** 435,984 **5.** $8\frac{1}{3}$ **7.** 8 **9.** 1.384 **11.** .028 **13.** $840 **15.** 620 **17.** .95

EXERCISE 1-53 Factors

Practice Problems, Pages 261-262

1. a. Yes **b.** Yes **c.** No **d.** Yes **3. a.** 1, 2, 13, 26 **b.** 1, 2, 3, 6, 9, 18 **c.** 1, 17 **d.** 1, 3, 5, 9, 15, 45 **e.** 1, 2, 3, 6, 9, 18, 27, 54 **f.** 1, 2, 5, 7, 10, 14, 35, 70

g. 1, 2, 3, 4, 6, 8, 12, 16, 24, 48 **h.** 1, 2, 3, 4, 6, 7, 12, 14, 21, 28, 42, 84 **i.** 1, 2, 5, 10, 25, 50, 125, 250
j. 1, 2, 3, 4, 6, 8, 9, 12, 18, 24, 36, 72 **k.** 1, 2, 4, 5, 8, 10, 16, 20, 25, 40, 50, 80, 100, 200, 400
l. 1, 2, 3, 4, 6, 11, 12, 22, 33, 44, 66, 132 **5. a.** 1, 2, 3, 4, 6, 9, 12, 18, 36
b. 1, 2, 3, 4, 6, 9, 12, 18, 27, 36, 54, 108 **c.** 1, 3, 9, 27, 81 **d.** 1, 3, 9 **7. a.** 1, 2 **b.** 1, 3
c. 1, 2, 4, 8 **d.** 1, 3, 5, 9, 15, 45 **e.** 1, 2 **f.** 1, 2, 3, 4, 6, 12 **g.** 1, 13
h. 1, 2, 3, 4, 6, 7, 12, 14, 21, 28, 42, 84 **i.** 1, 3, 5, 15, 25, 75 **9. a.** 1, 2, 3, 4, 6, 8, 12, 16, 24, 48
b. 1, 2, 4, 5, 8, 10, 16, 20, 40, 80 **c.** 1, 2, 4, 7, 8, 14, 16, 28, 56, 112 **d.** 1, 2, 4, 8, 16 **e.** 16

EXERCISE 1-54 Prime and Composite Numbers

Practice Problems, Page 264

1. 23; 79; 97 **3. a.** 2, 3, 5, 7 **b.** 2 **c.** 3, 5, 7 **5. a.** No; 2 **b.** No; 9, 15, 21, 25, 27, 33, 35, etc.
7. a. Relatively prime **b.** Relatively prime **c.** Not relatively prime **d.** Relatively prime
e. Not relatively prime **f.** Relatively prime **9. a.** 3 and 5 **b.** 3 and 11; 7 and 7 **c.** 3 and 29;
13 and 19 **d.** 3 and 73; 5 and 71; 17 and 59; 23 and 53; 29 and 47 **e.** 19 and 79; 31 and 67;
37 and 61 **f.** 3 and 107; 7 and 103; 13 and 97; 31 and 79; 37 and 73; 43 and 67 **g.** 3 and 59;
19 and 43; 31 and 31

EXERCISE 1-55 Factoring a Natural Number—Complete Factorization

Practice Problems, Page 266

1. a. 1×24, 24×1, 2×12, 12×2, 3×8, 8×3, 4×6, 6×4 **b.** 1×19, 19×1 **c.** 1×42,
42×1, 2×21, 21×2, 3×14, 14×3, 6×7, 7×6 **d.** 1×144, 144×1, 2×72, 72×2, 3×48,
48×3, 4×36, 36×4, 6×24, 24×6, 8×18, 18×8, 9×16, 16×9, 12×12 **e.** 1×108,
108×1, 2×54, 54×2, 3×36, 36×3, 4×27, 27×4, 6×18, 18×6, 9×12, 12×9
f. 1×400, 400×1, 2×200, 200×2, 4×100, 100×4, 5×80, 80×5, 8×50, 50×8, 10×40,
40×10, 16×25, 25×16, 20×20 **g.** 1×360, 360×1, 2×180, 180×2, 3×120, 120×3,
4×90, 90×4, 5×72, 72×5, 6×60, 60×6, 9×40, 40×9, 10×36, 36×10, 12×30,
30×12, 15×24, 24×15, 18×20, 20×18 **3. a.** $2 \times 2 \times 2 \times 5$ arranged in different orders
b. $2 \times 2 \times 2 \times 9$ or $2 \times 2 \times 3 \times 6$ or $2 \times 3 \times 3 \times 4$ arranged in different orders
c. $2 \times 3 \times 3 \times 3$ arranged in different orders **d.** $3 \times 3 \times 5 \times 5$ arranged in different orders
e. $2 \times 2 \times 2 \times 20$ or $2 \times 2 \times 4 \times 10$ or $2 \times 2 \times 5 \times 8$ or $2 \times 4 \times 4 \times 5$ arranged in different
orders **f.** $2 \times 2 \times 3 \times 35$ or $2 \times 2 \times 5 \times 21$ or $2 \times 2 \times 7 \times 15$ or $2 \times 3 \times 5 \times 14$ or
$2 \times 3 \times 7 \times 10$ or $2 \times 5 \times 6 \times 7$ or $3 \times 4 \times 5 \times 7$ arranged in different orders **g.** Any of the
following arranged in different orders: $2 \times 2 \times 2 \times 150$, $2 \times 2 \times 3 \times 100$, $2 \times 2 \times 4 \times 75$,
$2 \times 2 \times 5 \times 60$, $2 \times 2 \times 6 \times 50$, $2 \times 2 \times 10 \times 30$, $2 \times 2 \times 12 \times 25$, $2 \times 2 \times 15 \times 20$,
$2 \times 3 \times 4 \times 50$, $2 \times 3 \times 5 \times 40$, $2 \times 3 \times 8 \times 25$, $2 \times 3 \times 10 \times 20$, $2 \times 4 \times 5 \times 30$,
$2 \times 4 \times 6 \times 25$, $2 \times 4 \times 10 \times 15$, $2 \times 5 \times 5 \times 24$, $2 \times 5 \times 6 \times 20$, $2 \times 5 \times 8 \times 15$,
$2 \times 5 \times 10 \times 12$, $2 \times 6 \times 10 \times 10$, $3 \times 4 \times 4 \times 25$, $3 \times 4 \times 5 \times 20$, $3 \times 4 \times 10 \times 10$,
$3 \times 5 \times 5 \times 16$, $3 \times 5 \times 8 \times 10$, $4 \times 4 \times 5 \times 15$, $4 \times 5 \times 5 \times 12$, $4 \times 5 \times 6 \times 10$,
$5 \times 5 \times 6 \times 8$ **5. a.** $2 \times 2 \times 5$ **b.** $2 \times 3 \times 3 \times 3$ **c.** 3×7 **d.** $2 \times 2 \times 2 \times 2 \times 3$
e. $2 \times 2 \times 5 \times 5$ **f.** $2 \times 2 \times 2 \times 3 \times 3$ **g.** $2 \times 3 \times 5 \times 5$

EXERCISE 1-56 Number Multiples

Practice Problems, Page 268

1. a. 28; 52; 148 **b.** 84; 91; 119 **c.** 42; 51; 102 **3.** Yes **5. a.** 0, 8, 16, 24, 32, . . . **b.** 0, 10, 20,
30, 40, . . . **c.** 0, 11, 22, 33, 44, . . . **d.** 0, 32, 64, 96, 128, . . . **e.** 0, 40, 80, 120, 160, . . . **f.** 0, 17,
34, 51, 68, . . . **g.** 0, 75, 150, 225, 300, . . . **h.** 0, 96, 192, 288, 384, . . . **i.** 0, 120, 240, 360, 480,
. . . **j.** 0, 275, 550, 825, 1,100, . . . **7.** Yes, Yes, Yes **9. a.** 0, 15, 30, 45, . . . **b.** 0, 8, 16, 24, . . .
c. 0, 60, 120, 180, . . . **d.** 0, 144, 288, 432, . . . **e.** 0, 100, 200, 300, . . . **f.** 0, 40, 80, 120, . . .
g. 0, 16, 32, 48, . . . **h.** 0, 180, 360, 540, . . . **i.** 0, 210, 420, 630, . . . **11. a.** 24 **b.** 12 **c.** 60
d. 48 **e.** 100 **f.** 168 **g.** 672 **h.** 300 **i.** 1,050 **j.** 12 **k.** 120 **l.** 600

EXERCISE 1-57   Equivalent Fractions

Practice Problems, Page 270

1. **a.** $\frac{1}{2}, \frac{2}{4}, \frac{3}{6}, \frac{4}{8}, \cdots$ **b.** $\frac{1}{9}, \frac{2}{18}, \frac{3}{27}, \frac{4}{36}, \cdots$ **c.** $\frac{3}{5}, \frac{6}{10}, \frac{9}{15}, \frac{12}{20}, \cdots$ **d.** $\frac{2}{3}, \frac{4}{6}, \frac{6}{9}, \frac{8}{12}, \cdots$ **e.** $\frac{5}{8}, \frac{10}{16}, \frac{15}{24}, \frac{20}{32}, \cdots$
f. $\frac{13}{16}, \frac{26}{32}, \frac{39}{48}, \frac{52}{54}, \cdots$ **g.** $\frac{11}{12}, \frac{22}{24}, \frac{33}{36}, \frac{44}{48}, \cdots$ **h.** $\frac{5}{6}, \frac{10}{12}, \frac{15}{18}, \frac{20}{24}, \cdots$ **i.** $\frac{3}{4}, \frac{6}{8}, \frac{9}{12}, \frac{12}{16}, \cdots$ **j.** $\frac{7}{10}, \frac{14}{20}, \frac{21}{30}, \frac{28}{40}, \cdots$
k. $\frac{9}{13}, \frac{18}{26}, \frac{27}{39}, \frac{36}{52}, \cdots$ **l.** $\frac{8}{15}, \frac{16}{30}, \frac{24}{45}, \frac{32}{60}, \cdots$ **m.** $\frac{4}{11}, \frac{8}{22}, \frac{12}{33}, \frac{16}{44}, \cdots$ **n.** $\frac{19}{25}, \frac{38}{50}, \frac{57}{75}, \frac{76}{100}, \cdots$
o. $\frac{17}{20}, \frac{34}{40}, \frac{51}{60}, \frac{68}{80}, \cdots$ **p.** $\frac{31}{50}, \frac{62}{100}, \frac{93}{150}, \frac{124}{200}, \cdots$ 3. **a.** Yes **b.** No **c.** Yes **d.** No **e.** No **f.** Yes
g. No **h.** No **i.** Yes   5. **a.** Yes **b.** No **c.** No **d.** Yes **e.** No **f.** Yes **g.** Yes **h.** Yes **i.** No

EXERCISE 1-58   Comparing Fractions

Practice Problems, Page 271

1. **a.** True **b.** False **c.** True **d.** True **e.** True   3. **a.** True **b.** False **c.** True **d.** False
e. False

EXERCISE 1-59   Multiplicative Inverse

Practice Problems, Page 272

1. **a.** $\frac{1}{6}$ **b.** $\frac{1}{3}$ **c.** 1 **d.** $\frac{1}{9}$ **e.** $\frac{1}{25}$ **f.** none **g.** $\frac{1}{12}$ 3. **a.** $\frac{6}{5}$ **b.** $\frac{12}{7}$ **c.** $\frac{8}{3}$ **d.** $\frac{5}{2}$ **e.** $\frac{10}{9}$ **f.** $\frac{16}{11}$ **g.** $\frac{24}{23}$
5. **a.** $\frac{4}{5}$ **b.** $\frac{3}{20}$ **c.** $\frac{4}{7}$ **d.** $\frac{8}{35}$ **e.** $\frac{12}{17}$ **f.** $\frac{5}{32}$ **g.** $\frac{16}{57}$

EXERCISE 1-60   Scientific Notation

Practice Problems, Page 274

1. **a.** 5×10; 8.7×10; 4.00×10^2; 9.14×10^2; 8.0×10^3; 3.06×10^3 **b.** 6×10^4; 7.9×10^4;
5.2×10^5; 3.6×10^7; 4.875×10^6 **c.** 2×10^9; 3.08×10^{11}; 8.45×10^{13} **d.** 9.6×10^8; 5.1×10^{11};
4.23×10^{17} **e.** 7×10^{18}; 6.7×10^{21} 3. **a.** 5×10^{23} horsepower **b.** 3.86×10^9 kilometers
c. 6.6×10^{21} tons **d.** 7.79×10^8 kilometers **e.** (1) 4.9×10^{16} (2) 5.28×10^{20}

EXERCISE 1-61   Repeating Decimals

Practice Problems, Page 275

1. $.\overline{6}$ 3. $.1\overline{6}$ 5. $.\overline{5}$ 7. $.41\overline{6}$ 9. $.\overline{63}$ 11. $.8\overline{0}$ 13. $.4\overline{6}$ 15. $.\overline{296}$ 17. $.\overline{81}$ 19. $.9\overline{3}$ 21. $.91\overline{6}$
23. $.5\overline{6}$ 25. $.\overline{3571428}$

EXERCISE 1-62   Expressing a Repeating Decimal as a Common Fraction

Practice Problems, Page 276

1. $\frac{4}{9}$ 3. $\frac{6}{11}$ 5. $\frac{4}{27}$ 7. $\frac{17}{18}$ 9. $\frac{19}{24}$ 11. $\frac{3}{8}$ 13. $\frac{3}{20}$ 15. $\frac{3}{7}$

Review of Unit Seven, Page 277

1. 1944   3. **a.** 243 **b.** 8^2 5. $(7 \times 10^2) + (2 \times 10^1) + (5 \times 10^0) + (9 \times 10^{-1}) + (6 \times 10^{-2})$
7. **a.** False **b.** True **c.** True **d.** True   9. 216   11. $2 \times 2 \times 7 \times 7$ 13. Yes   15. **a.** False
b. True   17. 2.9×10^8 19. $.\overline{4}$

Basic Skills—Achievement Test I, Page 279

1. Eighty-five million, two hundred sixty-seven thousand, four hundred twenty-eight   3. 582,000
5. 3,362   7. 703   9. $4\frac{1}{2}$ 11. 16   13. $1\frac{11}{12}$ 15. $\frac{20}{3}$ 17. .125   19. 80   21. 15.7   23. \$20.15   25. .462
27. 124   29. .0195   31. $\frac{13}{20}$ 33. 300   35. 63.7 billion   37. .07   39. $\frac{1}{5}$ 41. 196   43. 80   45. 539

Basic Skills—Achievement Test II, Page 280

1. Seven million, fifteen thousand, six hundred seventy-three **3.** 496,000,000 **5.** 100,864
7. 8,006 **9.** $1\frac{3}{4}$ **11.** 12 **13.** $\frac{5}{8}$ **15.** $\frac{69}{16}$ **17.** .625 or $\frac{5}{8}$ **19.** 102 **21.** 500.006 **23.** 46.271 **25.** .92
27. .003 **29.** .01853 **31.** $\frac{3}{8}$ **33.** 2,400 **35.** 594.2 million **37.** .045 **39.** $1\frac{3}{4}$ **41.** $42.00 **43.** 400
45. 2.45

Basic Skills—Achievement Test III, Page 281

1. Five billion, nine hundred sixty-five million, four hundred eight thousand **3.** 147,500,000
5. 10,891 **7.** 42 **9.** $3\frac{3}{16}$ **11.** 60 **13.** $7\frac{5}{16}$ **15.** $\frac{43}{12}$ **17.** 2 **19.** 80 **21.** .471 **23.** $24.70 **25.** No
27. 18 **29.** 3.72 **31.** $\frac{2}{125}$ **33.** 650 **35.** 4.129 trillion **37.** 1.5 or 1.50 **39.** $\frac{2}{25}$ **41.** $1.25 **43.** 56
45. .95

Basic Skills—Achievement Test IV, Page 282

1. Four trillion, three hundred six billion, nine million, eighty-five thousand **3.** 6,400,000,000
5. 4,962,011 **7.** 8,075 **9.** $4\frac{1}{6}$ **11.** 120 **13.** $2\frac{13}{20}$ **15.** $\frac{79}{10}$ **17.** $33\frac{1}{3}$ **19.** 126 **21.** 50.09 **23.** 6.66
25. .885 **27.** 50 **29.** .06349 **31.** $2\frac{3}{4}$ **33.** $4,000.00 **35.** 199.56 billion **37.** .074 **39.** $2\frac{1}{2}$
41. $1.65 **43.** $150 **45.** 2.07

PART **2**   BASIC UNITS OF MEASURE

Inventory Test, Page 284

1. a. 420 mm **b.** .690 km **3. a.** 700 cL **b.** .665 L **5. a.** 8,300,000 cm^3 **b.** 9.250 dm^3
c. 6,750 cm^3; 6.75 kg **7.** $\frac{3}{4}$ **9.** 24 **11.** 189 cu. ft. **13.** 76° 18′ **15.** No **17.** 8 A.M. **19. a.** 4
significant digits 9, 0, 0, 8 **b.** 2 significant digits 4, 0 **c.** 8 significant digits 8, 9, 0, 2, 6, 5, 7, 1
d. 5 significant digits 6, 7, 1, 3, 9 **e.** 2 significant digits 3, 8 **f.** 3 significant digits 5, 6, 3

METRIC SYSTEM OF MEASURE

EXERCISE 2-1   Measure of Length—Metric

Practice Problems, Pages 290–293

1. a. kilometer **b.** centimeter **c.** meter **d.** millimeter **e.** decimeter **f.** dekameter
g. hectometer **3. a.** 10 **b.** 10 **c.** 10 **d.** 10 **e.** 10 **f.** 10 **g.** $\frac{1}{10}$ **h.** $\frac{1}{10}$ **i.** $\frac{1}{10}$ **j.** $\frac{1}{10}$ **k.** $\frac{1}{10}$ **l.** $\frac{1}{10}$
5. a. 7,583 m **b.** 8.265 m **c.** 6.049 m **d.** 3,005.016 m **e.** 9,427.375 m **7.** 47 cm; 640 cm; 85.1 cm;
800,000 cm; 930 cm **9. a.** 780 m; 42.06 m; 390 m; 6.04 m; .569 m **b.** 570 m; 214 m; .05 m; 12.6 m;
9,625 m **11.** 600 hm; .164 hm; 40 hm; 2.2538 hm; 26 hm **13. a.** .7 km, 7 hm, 70 dam, 700 m,
7,000 dm, 70,000 cm, 700,000 mm **b.** 3 km, 30 hm, 300 dam, 3,000 m, 30,000 dm, 300,000 cm,
3,000,000 mm **c.** .05 km, .5 hm, 5 dam, 50 m, 500 dm, 5,000 cm, 50,000 mm **d.** 6.8 km, 68 hm,
680 m, 6,800 dm, 68,000 cm, 6,800,000 mm **15. a.** 482 cm **b.** 2,065 mm
c. 100,903 cm **d.** 7,054 mm **17. a.** 6.094 m **b.** 4.01752 km **c.** 8.00609 km **d.** 5.026 m
19. 10 meters **21.** .306 meter or 306 millimeters **23. a.** 7 m **b.** 40 cm **c.** 260,000 mm
d. 6.9 cm **e.** 2.7 km **f.** 849 mm **25.** 84.5 mm, 845 m, 8.45 km, 8,450,000 cm

EXERCISE 2-2   Measure of Mass or Weight—Metric

Practice Problems, Pages 294–296

1. a. milligram **b.** kilogram **c.** decigram **d.** gram **e.** centigram **f.** hectogram **g.** dekagram
3. a. 1,000 **b.** 100 **c.** $\frac{1}{100}$ **d.** $\frac{1}{1,000}$ **e.** 1,000 **f.** $\frac{1}{1,000}$ **g.** 1,000; 1,000,000 **h.** $\frac{1}{1,000}$; $\frac{1}{1,000,000}$ **i.** 10
j. 100 **k.** 10,000 **l.** 100,000 **m.** 1,000,000 **n.** $\frac{1}{10}$ **o.** $\frac{1}{100}$ **p.** $\frac{1}{10,000}$ **q.** $\frac{1}{100,000}$ **r.** $\frac{1}{1,000,000}$
5. 3.9 cg; 57.4 cg; 6,200 cg; 140,000 cg; 2,890 cg **7.** 600 g; 53 g; 9,000 g; .258 g; .750 g **9.** 53.3 hg;

70 hg; 400 hg; 4.519 hg; 65 hg **11.** 3 metric tons; .57 metric tons; 6.3 metric tons; 1.9 metric tons; 82.6 metric tons **13. a.** 56 mg **b.** 8,130 g **c.** 725 cg **d.** 2,084 mg **e.** 39 mg **f.** 6,370 g **15. a.** 5.2 cg **b.** 15.260 kg **c.** 9.04 g **d.** 7.001 g **e.** 8.007 kg **f.** 19.058 g **17. a.** 901.6 cg **b.** 2,008.04 g **c.** .053 g **d.** .65047 kg **19. a.** the two are equal **b.** 6.1 cg **c.** 19 g **d.** 8 kg **e.** 5.2 g **f.** 4.9 kg **21.** 625,000 mg, 67,000 cg, 5.9 kg, 6,500 g

EXERCISE 2-3   Measure of Capacity—Metric

Practice Problems, Pages 296-298

1. a. kiloliter **b.** liter **c.** milliliter **d.** centiliter **e.** hectoliter **f.** deciliter **g.** dekaliter **3. a.** 1,000 **b.** 100 **c.** 1,000 **d.** $\frac{1}{1,000}$ **e.** $\frac{1}{100}$ **f.** $\frac{1}{1,000}$ **g.** 1,000; 1,000,000 **h.** $\frac{1}{1,000}$; $\frac{1}{1,000,000}$ **5.** 8,500 cL; 90 cL; 1.8 cL; 670 cL; 4.42 cL **7.** .9 L; 6,000 L; 21,600 L; 4 L; 8.405 L **9.** 79 hL; .52 hL; 60.08 hL; 215.7 hL; 8.251 hL
11.

	L	dL	cL	mL
a.	7	70	700	7,000
b.	64.5	645	6,450	64,500
c.	.9	9	90	900
d.	3.8	38	380	3,800
e.	5.452	54.52	545.2	5,452

13. a. 6,019 mL **b.** 1,578 mL **15. a.** 8.037 L **b.** 3.068 L **17.** 5 liters **19.** .639 L and 6.39 dL are equal, 63.8 cL, 637 mL

EXERCISE 2-4   Measure of Area—Metric

Practice Problems, Page 299

1. a. square centimeter **b.** square kilometer **c.** square meter **d.** square millimeter **e.** hectare **f.** square decimeter **g.** square dekameter **3.** 600 mm^2; 2,150,000 mm^2; 17,000 mm^2; 4,190 mm^2; 850,000 mm^2 **5.** 8.19 dm^2; 3,000 dm^2; 60 dm^2; 750 dm^2; 2.407 dm^2 **7.** 80 dam^2; 3,100 dam^2; .0277 dam^2; 680 dam^2; 123 dam^2 **9.** 5 km^2; 3.825 km^2; .726 km^2; .60922 km^2

EXERCISE 2-5   Measure of Volume—Metric

Practice Problems, Pages 301-302

1. a. cubic meter **b.** cubic millimeter **c.** cubic kilometer **d.** cubic centimeter **e.** cubic decimeter **f.** cubic hectometer **g.** cubic dekameter **3.** 8,000 mm^3; 15,000,000,000 mm^3; 6,100,000 mm^3; 25,800 mm^3; 53,600,000,000 mm^3 **5.** 63,000 dm^3; 2.930 dm^3; 17,900 dm^3; .073250 dm^3; .0864 dm^3 **7. a.** 21 cm^3 **b.** 9,000 cm^3 **c.** 430 mL **d.** 18.6 L **e.** 7.9 L **f.** 5 kg **g.** 67 g **h.** 820 g **i.** 9.1 kg **j.** 615 g **k.** 4 L **l.** 75 mL **m.** .545 L **n.** 26 dm^3 **o.** 19 cm^3 **p.** 8,800 cm^3 **9. a.** 13 mL; 2.5 mL; .8 mL; 7,000 mL; 90,000 mL **b.** .8 cL; 318,000 cL; 260 cL; .507 cL; .91 cL **c.** 7 L; 28.3 L; 5,900 L; 6 L; .043 L **11. a.** 6,000 cm^3; 30 cm^3; 600 cm^3; .950 cm^3; 8.74 cm^3 **b.** 15 L; 4.1 L; .520 L; .004 L; .00709 L **c.** 7 mL; 138.51 mL; 1,800 mL; 6.270 mL; 95 mL **d.** 2.6 cL; 570 cL; 61.5 cL; 17.4 cL; .9 cL **13.** 171,000 kilograms **15.** 7,540 L; 7,540 kg

CUSTOMARY SYSTEM OF MEASURE

EXERCISE 2-6   Measure of Length—Customary

Practice Problems, Pages 305-306

1. a. 96 in.; 75 in.; 42 in.; 600 in. **b.** 324 in.; 63 in.; 168 in.; 576 in. **c.** 67 in.; 111 in.; 47 in.; 201 in. **3. a.** 12,320 yd.; 23,056 yd.; 7,920 yd.; 66,880 yd. **b.** 49.5 yd.; 25.3 yd.; 34.375 yd.; 137.50 yd. **c.** 7,240 yd.; 27,200 yd.; 21.5 yd.; 74.5 yd. **d.** 5 yd.; 21 yd.; 1.4 yd.; 16 yd. **e.** 7 yd.;

19 yd.; $29\frac{2}{3}$ yd.; 2.5 yd. **5. a.** 3 mi.; 7 mi.; $\frac{5}{8}$ mi.; 25 mi. **b.** 5 mi.; 40 mi.; $2\frac{3}{4}$ mi.; 3.41 mi. **7.** 13 ft. 11 in.; 9 yd. 10 in.; 9 mi. 110 yd.; 25 yd. **9.** 20 ft. 8 in.; 37 mi. 940 yd.; 18 yd.; 27 yd. 2 ft. 6 in. **11. a.** 36,480 ft.; 17,024 ft.; 25,840 ft.; 243,200 ft. **b.** $22,293\frac{1}{3}$ yd.; $17,733\frac{1}{3}$ yd.; $1,773\frac{1}{3}$ yd.; $293,866\frac{2}{3}$ yd. **c.** 8.05 mi.; 10.695 mi.; 41.975 mi.; 166.75 mi. **13.** b **15.** a

EXERCISE 2-7   Measure of Weight—Customary

Practice Problems, Pages 307–308

1. a. 96 oz.; $35\frac{1}{5}$ oz.; 92 oz.; 592 oz. **b.** 18 oz.; 72 oz.; 151 oz.; 301 oz. **3.** 3 short tons; 12 short tons; $3\frac{1}{2}$ short tons **5. a.** $\frac{3}{4}$; $\frac{1}{2}$; $\frac{7}{8}$; $\frac{5}{16}$ **b.** $\frac{3}{5}$; $\frac{1}{4}$; $\frac{3}{8}$; $\frac{9}{10}$ **c.** $\frac{1}{4}$; $\frac{1}{2}$; $\frac{3}{8}$; $\frac{7}{8}$ **7.** 3 lb. 9 oz.; 7 oz.; 1 s.t. 800 lb.; 1 lb. 10 oz. **9.** 2 lb. 3 oz.; 1 l.t. 600 lb.; 2 lb. 6 oz.; 1 s.t. 525 lb. **11.** b **13.** d **15.** c

EXERCISE 2-8   Liquid Measure—Customary

Practice Problems, Pages 308–310

1. a. 144 oz.; 76 oz.; 88 oz.; 272 oz. **b.** 160 oz.; 35.2 oz.; 25.6 oz.; 768 oz. **c.** 896 oz.; 32 oz.; 864 oz.; 4,992 oz. **d.** 40 oz.; 172 oz.; 176 oz. **3. a.** 32 qt.; 41 qt.; 70 qt.; 400 qt. **b.** 27 qt.; 29 qt.; 6 qt. **c.** 7 qt.; $3\frac{1}{4}$ qt.; 4.5 qt.; 13.5 qt. **d.** 3 qt.; 10 qt.; 6 qt.; $21\frac{7}{8}$ qt. **5. a.** $\frac{1}{2}$; $\frac{3}{4}$; $\frac{7}{16}$; $\frac{1}{4}$; $\frac{3}{8}$; $\frac{5}{8}$ **b.** $\frac{1}{2}$; $\frac{3}{4}$; $\frac{1}{2}$; $\frac{1}{4}$; $\frac{3}{4}$; $\frac{7}{8}$ **c.** $\frac{1}{2}$; $\frac{3}{4}$; $\frac{1}{4}$; $\frac{1}{8}$; $\frac{3}{4}$; $\frac{7}{8}$ **7.** 3 qt.; 10 oz.; 1 gal. 2 qt.; 1 qt. 1 pt. 10 oz. **9.** 4 gal. 1 qt.; 3 pt. 4 oz.; 1 qt. 11 oz.; 2 gal. 1 qt. **11.** c

EXERCISE 2-9   Dry Measure—Customary

Practice Problems, Page 310

1. 8 pt.; 60 pt.; 7.5 pt.; 19 pt. **3. a.** 36 pk.; 112 pk.; 3 pk.; 30 pk. **b.** 9 pk.; 71 pk.; 5 pk.; $2\frac{3}{4}$ pk. **5. a.** $\frac{1}{2}$; $\frac{1}{4}$; $\frac{3}{4}$; $\frac{1}{8}$ **b.** $\frac{3}{4}$; $\frac{1}{2}$; $\frac{5}{8}$; $\frac{3}{8}$ **7.** 4 pk. 5 qt.; 2 bu. 1 pk.; 6 pk. 3 qt.; 3 bu. 3 pk. **9.** 3 pk. 1 qt.; 2 bu. 3 pk.

EXERCISE 2-10   Measure of Area—Customary

Practice Problems, Page 311

1. 2,304 sq. in.; 6,480 sq. in.; 2,124 sq. in.; 864 sq. in. **3.** 5 sq. yd.; 7 sq. yd.; 33,880 sq. yd.; 10,067,200 sq. yd. **5.** 5 acres; 2,304 acres; 5 acres; $\frac{3}{10}$ acre **7.** $\frac{1}{2}$ sq. ft.; $\frac{7}{8}$ sq. ft.; $\frac{9}{16}$ sq. ft. **9.** $\frac{3}{8}$ sq. mi.; $\frac{1}{50}$ sq. mi.; $\frac{1}{100}$ sq. mi. **11.** c

EXERCISE 2-11   Measure of Volume—Customary

Practice Problems, Page 312

1. 65,664 cu. in.; 34,992 cu. in.; 12,528 cu. in. **3.** 1,080 cu. ft.; $175\frac{1}{2}$ cu. ft.; 3 cu. ft. **5.** 5 cu. yd.; 2 cu. yd.; $2\frac{1}{3}$ cu. yd. **7.** $\frac{1}{4}$ cu. ft.; $\frac{7}{16}$ cu. ft.; $\frac{5}{6}$ cu. ft. **9.** $\frac{1}{3}$ cu. ft.; $\frac{1}{16}$ cu. ft.; $\frac{2}{3}$ cu. ft. **11.** d

EXERCISE 2-12   Volume, Capacity, and Weight Relationships—Customary

Practice Problems, Page 313

1. 14 gal. **3.** 390 gal. **5.** 3,072 lb. **7.** 15,625 lb. **9.** 80 bu. **11. a.** 8 gal.; $157\frac{1}{2}$ gal; $36\frac{3}{4}$ gal. **b.** 64 bu.; 10 bu.; 756 bu. **c.** 6 cu. ft.; 120 cu. ft.; $257\frac{1}{2}$ cu. ft. **d.** 1,617 cu. in.; $3,649\frac{4}{5}$ cu. in.; $1,039\frac{1}{2}$ cu. in. **e.** 75 cu. ft.; 96 cu. ft. **f.** $4,687\frac{1}{2}$ lb.; 7,680 lb. **13.** 120,000,000 gal. per min.

OTHER BASIC MEASURES

EXERCISE 2-13   Measure of Time

Practice Problems, Pages 314–316

1. a. 1,440 sec.; 498 sec.; 975 sec.; 3,120 sec.   **b.** 36,000 sec.; 22,500 sec.; 5,400 sec.; 61,200 sec.
c. 320 sec.; 872 sec.; 3,375 sec.; 8,106 sec.   **3. a.** 96 hr.; 196 hr.; 87 hr.; 720 hr.   **b.** 30 hr.; 133 hr.;
273 hr.; 213 hr.   **c.** 8 hr.; 53 hr.; $1\frac{2}{3}$ hr.; 3 hr.   **5. a.** 84 mo.; 69 mo.; 30 mo.; 120 mo.   **b.** 52 mo.;
18 mo.; 46 mo.; 149 mo.   **7. a.** 7 yr.; $2\frac{11}{12}$ yr.; 13 yr.; 25 yr.   **b.** 3 yr.; 5 yr.; $\frac{3}{4}$ yr.; 4 yr.   **c.** 3 yr.;
$\frac{3}{5}$ yr.; $1\frac{2}{5}$ yr.; 7 yr.   **9.** 23 hr.; 9 wk. 2 da.; 19 yr. 11 mo.; 20 hr.   **11.** 15 yr. 9 mo.; 11 da. 11 hr.;
20 hr.; 25 hr. 32 min. 30 sec.   **13. a.** half hour   **b.** three-quarters of a year   **15. a.** 9 P.M.
b. 11 A.M.   **c.** 2:30 P.M.   **d.** 6:15 A.M.   **e.** 9:35 P.M.   **17. a.** 16 hr.   **b.** $8\frac{1}{2}$ hr.   **c.** 14 hr. 44 min.
19. a. 31 days   **b.** 31 days   **c.** 30 days   **d.** 31 days   **e.** 30 days   **f.** 30 days   **g.** 28 days
h. 31 days   **21.** Answers will vary   **23.** c   **25.** d

EXERCISE 2-14   Measure of Angles and Arcs

Practice Problems, Page 317

1. 2,700′; 1,926′; 49′; 780′   **3.** 3°; 72°; 3°; 10°   **5.** 53° 29′; 51° 15′; 21° 51′ 17″; 84° 35′ 29″
7. 7° 3′; 35′; 7° 23′ 27″; 8° 34′ 46″   **9.** 172°   **11.** 12° 6′

EXERCISE 2-15   Rates of Speed

Practice Problems, Page 318

1. 4 m/s; 41.67 m/s; 0.01 m/s   **3.** 2,500 cm/s; 141.67 cm/s; 54 cm/s   **5.** 37.84 knots;
1,167.57 knots; 135.14 knots   **7.** 240 m.p.h.; 103.5 m.p.h.; 20.45 m.p.h.   **9.** 69.2 knots; 74.68 knots;
78.16 knots

EXERCISE 2-16   Temperature

Practice Problems, Page 320

1. a. 323 K   **b.** 369 K   **c.** 281 K   **d.** 273 K   **e.** 433 K   **3. a.** 30°C   **b.** 0°C   **c.** 45°C   **d.** 100°C
e. 15.56°C   **5.** 20°C

EXERCISE 2-17   Twenty-Four Hour Clock

Practice Problems, Page 321

1. a. 1126   **b.** 1009   **c.** 0945   **d.** 0338   **e.** 0600   **f.** 0200   **g.** 1900   **h.** 2200   **i.** 1629   **j.** 2114
k. 2237   **l.** 2302   **m.** 0018   **n.** 0054   **o.** 1223   **p.** 1237   **q.** 0126   **r.** 1735   **s.** 1140   **t.** 2025
3. a. 4 hr. 27 min.   **b.** 10 hr. 27 min.   **c.** 11 hr. 25 min.   **d.** 9 hr. 12 min.   **e.** 11 hr. 28 min.
f. 1 hr. 7 min.   **g.** 4 hr. 54 min.   **h.** 8 hr. 34 min.   **i.** 11 hr. 56 min.   **j.** 22 hr. 49 min.
5. 31 hr. 35 min.

EXERCISE 2-18   Time Zones

Practice Problems, Page 322

1. Answers will vary   **3. a.** 7 P.M.   **b.** 9 P.M.   **c.** 8 P.M.   **d.** 6 P.M.   **e.** 9 P.M.   **f.** 8 P.M.
g. 7 P.M.   **h.** 6 P.M.   **i.** 9 P.M.   **j.** 9 P.M.   **k.** 8 P.M.   **l.** Answers will vary   **5.** 3 hr. 38 min.

MEASUREMENT

EXERCISE 2-19   Precision and Accuracy in Measurement

Practice Problems, Page 325

1. a. 1 mm **b.** hectometer **c.** .1 mg **d.** .001 mm **e.** dekagram **f.** liter **g.** $\frac{1}{16}$ in. **h.** $\frac{1}{5}$ hr.
i. $\frac{1}{2}$ pt. **j.** $\frac{1}{4}$ lb. **k.** $\frac{1}{10}$ sec. **l.** centimeter **m.** .1 mg **n.** gram **o.** min. **p.** $\frac{1}{4}$ in. **3.a.** equally
precise **b.** $3\frac{17}{32}$ in. **c.** 3.9 g **d.** $5\frac{2}{3}$ yd. **e.** 10 hr. 8 min. **f.** 3 ft. 7 in. **5.a.** $\frac{1}{130}$ **b.** $\frac{1}{92}$ **c.** $\frac{1}{4,114}$
d. $\frac{1}{16}$ **e.** $\frac{1}{946}$ **f.** $\frac{1}{74}$ **g.** $\frac{1}{94}$ **h.** $\frac{1}{1,734}$ **i.** $\frac{1}{190,536}$ **j.** $\frac{1}{114}$ **k.** $\frac{1}{1,520}$ **l.** $\frac{1}{800}$ **m.** $\frac{1}{54,000}$ **n.** $\frac{1}{1,160}$ **o.** $\frac{1}{78}$
p. $\frac{1}{1,218}$ **q.** $\frac{1}{70}$ **r.** $\frac{1}{40}$ **s.** $\frac{1}{278}$ **t.** $\frac{1}{4,120}$ **7. a.** 6.52 cm **b.** same **c.** 250 km **d.** 75,00̲0 km **e.** same
f. $10\frac{7}{8}$ mi. **9. a.** precision: .009 mm, .09 mm, .9 mm, 9 mm, 90 mm; all have the same accuracy
b. precision: 7.26 in., 9.375 yd., 63.4 ft., 83.0 mi., 500 mi., accuracy: 9.375 yd., 83.0 mi., 7.26 in.,
63.4 ft., 500 mi.

EXERCISE 2-20   Significant Digits

Practice Problems, Page 327

1. two; 2, 8 **3.** four; 4, 2, 7, 1 **5.** one; 9 **7.** two; 9, 5 **9.** three; 1, 3, 0 **11.** four; 2, 3, 0, 5
13. four; 8, 7, 0, 0 **15.** two; 6, 3 **17.** two; 2, 5 **19.** three; 6, 8, 1 **21.** five; 3, 1, 4, 1, 6
23. four; 9, 5, 2, 8 **25.** three; 4, 0, 0 **27.** nine; 9, 0, 0, 0, 0, 0, 0, 2, 7 **29.** five; 4, 0, 0, 6, 0

EXERCISE 2-21   Approximate Numbers

Practice Problems, Page 328

1. a. 10.2 **b.** 19.0 **c.** 29 **3. a.** 40 **b.** 43 **c.** 85.13

Review of Part 2, Pages 329–330

1. a. 6,380 cm **b.** 41.9 cm **c.** 52,000 m **d.** 29.65 m **3. a.** 85 mL **b.** .759 L **c.** 1,200 cL
d. 4.6 L **5. a.** (1) .01495 dm^3 (2) 6,300 mm^3 (3) 83,900 dm^3 (4) 2,800,000 cm^3 **b.** 33,400 cm^3;
33.4 kg **c.** 600,000 L **7.** 42 oz. **9.** 42 pecks **11.** 11 cu. yd. **13.** 935° **15.** No **17.** 1 A.M.;
11 P.M.; 10 P.M. **19. a.** three **b.** eight **c.** two **d.** two **e.** two **f.** four

Competency Check Test, Page 330

1. b **3.** a **5.** d **7.** b **9.** d

PART 3   FACTS AND SKILLS IN GEOMETRY

Inventory Test, Pages 332–335

1. Line segment BC, line NR, ray AT, line CF, line segment SY, ray PQ **3. a.** E **b.** No
c. Yes **d.** Yes, because a straight line can be drawn through any two points and points that lie
on the same straight line are collinear. **e.** Yes **5.** Construction **7.** 125 miles **9.** Construction
11. Construction **13.** Construction **15.** Construction **17.** Construction **19.** 21°
21. $\angle 1 = 56°$; $\angle 4 = 105°$ **23.** $\angle 1 = 85°$; $\angle 2 = 95°$; $\angle 3 = 95°$; $\angle 4 = 85°$; $\angle 5 = 85°$;
$\angle 7 = 95°$; $\angle 8 = 85°$ **25.** 424 mm **27.** 308 ft. **29.** 54,205 m^2 **31.** 10,486 sq. ft. **33.** 3,968 m^2
35. 2,464 sq. in. **37.** $18\frac{26}{27}$ cu. ft. or 32,768 cu. in. **39.** 47.69 m^3 **41.** $1,109\frac{1}{3}$ cu. yd.
43. 518.94 m^2 **45.** 22,176 mm^2 **47.** 400 ft. **49.** 75 ft.

APPLICATIONS USING UNITS OF LENGTH

EXERCISE 3-1 Points and Lines

Practice Problems, Pages 339–342

1. B, C **3.** Line segment EF; line segment r; line segment k; line segment YZ **5.** Ray AB, $\overrightarrow{AB}$; ray ML, $\overleftarrow{ML}$; ray FD, $\overrightarrow{FD}$; ray TN, $\overrightarrow{TN}$ **7.** Line CN; ray SV; line segment KM; ray BR; line GT; line segment DS; line NJ; ray FL; line segment ME; line AD **9.** Line segment RS, line segment ST **11.** Line segment BD, line segment DC **13.** Line segment DF, line segment FG and line segment GE **15. a.** curved **b.** slanting **c.** curved **d.** perpendicular **e.** broken **f.** straight **g.** curved **h.** horizontal **i.** broken **j.** curved **17. a.** intersecting line and ray **b.** intersecting lines **c.** parallel lines **d.** perpendicular line and ray **e.** perpendicular rays **f.** parallel lines **g.** parallel lines **h.** intersecting rays **i.** intersecting lines **j.** perpendicular line and ray **19. a.** $\overrightarrow{CE}$ **b.** $\overrightarrow{CF}$ **21. a.** D **b.** C **c.** B **23. a.** Yes **b.** No; yes; yes; yes **c.** Yes; no; no; yes **25.** One; yes; yes **27.** The straight line; yes

EXERCISE 3-2 Drawing and Measuring Line Segments

Practice Problems, Pages 343–344

1. Construction **3.** Construction **5. a.** $m\overline{BR} = 44$ mm; $m\overline{AD} = 3.5$ cm; $m\overline{ST} = 42$ mm; $m\overline{NH} = 35$ mm **b.** Yes **c.** No **7. a.** Construction **b.** Construction

Refresh Your Skills, Page 344

1. 22,168 **3.** 425,072 **5.** $5\frac{1}{2}$ **7.** 30 **9.** 1.696 **11.** .0003 **13.** \$1,020 **15.** \$6.00 **17.** 24.5

EXERCISE 3-3 Scale

Practice Problems, Pages 346–348

1. a. 105 m **b.** 195 m **c.** 975 m **d.** 615 m **e.** 1,590 m **3. a.** 240 mi. **b.** 312 mi. **c.** 132 mi. **d.** 174 mi. **e.** 93 mi. **5. a.** 4 ft. **b.** 13 ft. **c.** $18\frac{1}{2}$ ft. **d.** $7\frac{3}{4}$ ft. **e.** $15\frac{3}{8}$ ft. **7.** 5,000,000 actual mm; 5,000 m; 5 km **9.** 75,000 actual mm; 1 : 75,000 **11. a.** 1 : 4,000,000 **b.** 1 : 4,500,000 **c.** 1 : 2,500 **d.** 1 : 90,000 **13. a.** 2,400 m or 2.4 km **b.** 16,200 cm or 162 m **c.** 5,580 cm or 55.8 m **d.** 37,800 mm or 37.8 m **e.** 92,400 mm or 92.4 m **15. a.** 3 in. **b.** $2\frac{1}{2}$ in. **c.** $4\frac{3}{4}$ in. **d.** $1\frac{5}{8}$ in. **e.** $5\frac{13}{16}$ in. **17. a.** 5 in. **b.** $13\frac{1}{4}$ in. **c.** $\frac{3}{4}$ in. **d.** $29\frac{1}{4}$ in. **e.** $7\frac{3}{8}$ in. **19. a.** 4 mm **b.** 15 mm **c.** 58 mm **d.** 49 mm **e.** $23\frac{1}{2}$ mm **21.** Construction **23.** Construction **25.** Construction **27. a.** 60 km **b.** 25 km **c.** 37 km **d.** 74 km **e.** 93 km **29.** 112 mi.

EXERCISE 3-4 Geometric Figures

Practice Problems, Pages 353–355

1. Plane ABC; plane PRSQ; plane LM; plane T **3. a.** equilateral **b.** isosceles **c.** scalene **d.** equiangular **e.** right **f.** obtuse **5. a.** three **b.** one **c.** two **d.** five **e.** one **7.** Yes; no; yes **9.** Straight line; yes **11.** Yes

EXERCISE 3-5 Perimeter

I. Rectangle

Practice Problems, Pages 357–359

1. 136 mm **3.** 196 m; \$1,175.02; **5.** 7 ft. **7.** \$17.93 **9.** 5 times

II. Square

Practice Problems, Pages 358–359

1. 32 kilometers   **3.** 120 yards

III. Triangle

Practice Problems, Page 359

1. a. 42 m   **b.** 67 cm   **c.** 14.52 km   **d.** $18\frac{3}{8}$ mi.   **e.** 7 ft. 5 in. or 2 yd. 1 ft. 5 in.   **3.** 38 cm
5. 42 bricks (with one broken); $8.40

EXERCISE 3-6   Circumference of a Circle

Practice Problems, Page 361

1. a. 14 cm   **b.** 46 m   **c.** 13 km   **d.** $21\frac{3}{8}$ in., or 1 ft. $9\frac{3}{8}$ in.   **e.** $4\frac{1}{2}$ ft.   **3.** 188.4 m   **5.** 163.28 m
7. 91 cm*   **9.** 132 in.*   **11.** 150.72 cm   **13.** 88 in.*; 720 times each   **15.** 56 cm

Review, Page 362

1. Line RT; line segment OM; ray DG   **3.** None; N   **5.** at N   **7.** Yes   **9.** Construction
11. 454 m   **13.** $30\frac{1}{2}$ in.   **15.** 10.4 m   **17.** bridge table   **19.** An infinite number; one

APPLICATIONS USING ANGLES AND ARCS

EXERCISE 3-7   Facts About Angles

Practice Problems, Page 364

1. Side ED and side DF; vertex D   **3.** Angle O, angle TOP, angle POT   **5.** Angle Y, angle 2
7. Angle EFH; angle EFG; angle GFH   **9.** Angle ATN, angle NTB, angle BTM and angle MTA

Practice Problems, Page 366

1. degrees; °   **3.** 90°; $\frac{1}{4}$ **5.** 90°; 180°   **7.** 67°; 89°   **9. a.** C, E, G   **b.** A, D   **c.** B   **d.** F

EXERCISE 3-8   Measuring and Drawing Angles

I. Measuring Angles

Practice Problems, Pages 367–368

1. a. 80°   **b.** 130°   **c.** 90°   **d.** 45°   **e.** 75°   **f.** 150°   **3. a.** angle s   **b.** sides   **5. a.** 180°   **b.** 60°
c. 90°   **d.** 150°   **e.** 240°   **f.** 360°

II. Drawing Angles

Practice Problems, Page 369

1. Construction   **3.** Construction

*$\pi = 3\frac{1}{7}$ was used to determine answers marked with asterisk. Otherwise, 3.14 was used.

EXERCISE 3-9   Angles in Navigation

Practice Problems, Page 371

1. 270° **3.** 135° **5.** Construction **7.** Construction **9.** Construction

EXERCISE 3-10   Sum of Angles of Polygons

I. Triangle

Practice Problems, Page 372

1. a. 64° **b.** 45° **c.** 23° **d.** 45° **e.** 41° **f.** 41° **3.** 62° **5.** Since the sum of the three angles of any triangle is 180°, the measure of any two angles cannot equal or exceed 180°; Acute **7.** The third angle in each triangle equals 180° minus the sum of the other two angles. Since the sum of the two given angles in each triangle is the same, the third angles are equal.

II. Quadrilateral

Practice Problems, Page 373

1. 360°; 360° **3.** 65°; 115°; 115°

Refresh Your Skills, Page 373

1. 32,926 **3.** 395,430 **5.** $8\frac{4}{15}$ **7.** $23\frac{1}{2}$ **9.** 7.78 **11.** $\frac{3}{5}$ **13.** $517.99 **15.** 56 **17.** .96

EXERCISE 3-11   Pairs of Angles

I. Complementary Angles

Practice Problems, Page 374

1. a, b, c and f **3. a.** $m\angle 1 = 64°$ **b.** $m\angle 2 = 37°$ **c.** $m\angle 1 = 76°$ **d.** $m\angle 2 = 23°$ **e.** $m\angle 1 = 47° \ 30'$ **5.** An acute angle is an angle greater than 0° but less than 90° An acute angle subtracted from a right angle must result in an angle that is greater than 0° but less than 90°, thus another acute angle.

II. Supplementary Angles

Practice Problems, Page 375

1. a, d and e **3.** Yes; yes **a.** 148° **b.** 64° **c.** 90° **d.** 35° **e.** 103°30′

III. Opposite or Vertical Angles

Practice Problems, Page 376

1. $\angle 3$; yes; $\angle 2$; yes **3.** $m\angle 2 = 110°$; $m\angle 1 = 70°$; $m\angle 3 = 70°$ **5.** $m\angle 3 = 35°$; $m\angle 4 = 145°$; $m\angle 1 = 35°$

IV. Exterior Angle of a Triangle

Practice Problems, Page 376

1. $m\angle 4 = 111°$; $m\angle 3 = 69°$ **3.** $m\angle 1 = 61°$; $m\angle 4 = 99°$ **5.** $m\angle 3 = 49°$; $m\angle 1 = 86°$

EXERCISE 3-12   Parallel Lines and Angle Relationships

Practice Problems, Pages 377–378

1. a. transversal **b.** alternate-interior angles **c.** alternate-interior angles **d.** corresponding angles **e.** corresponding angles **3.** equal; $\angle 2 = \angle 6$; $\angle 7 = \angle 3$; $\angle 1 = \angle 5$; $\angle 8 = \angle 4$ **5.** 180°; supplementary angles **7. a.** $\angle 6$ and $\angle 7$, and $\angle 3$ and $\angle 2$ are opposite angles and hence are equal. ∠3 and ∠6 are alternate-interior angles and hence are equal. Therefore, if ∠3 and ∠6 are equal, their opposite angles, ∠2 and ∠7 are equal. **b.** ∠2 and ∠6 are corresponding angles and hence are equal. ∠1 and ∠2 are supplementary angles. Therefore, since ∠2 = ∠6 and ∠1 and ∠2 are supplementary, ∠1 and ∠6 are supplementary. **c.** ∠5 and ∠4 are alternate-interior angles and hence are equal. ∠1 and ∠4, and ∠5 and ∠8 are opposite angles and hence are equal. If ∠5 = ∠4, ∠1 = ∠4 and ∠5 = ∠8, then ∠1 = ∠8. **d.** ∠3 and ∠6 are opposite-interior angles and therefore are equal. ∠5 and ∠6 are supplementary angles. Since ∠3 = ∠6 and ∠5 and ∠6 are supplementary, ∠3 and ∠5 must be supplementary. **e.** ∠6 and ∠8 are supplementary angles. ∠2 and ∠6 are equal because they are corresponding angles. Since ∠2 = ∠6 and ∠6 and ∠8 are supplementary, ∠2 and ∠8 are supplementary. **9.** parallel

EXERCISE 3-13   Congruent Triangles

Practice Problems, Pages 379-380

1. Triangles ABC and JKL are congruent because three sides of triangle ABC are equal to three sides of triangle JKL. **3.** Triangles DEF and JKL are congruent because two sides and an included angle of triangle DEF are equal respectively to two sides and an included angle of triangle JKL. **5.** m∠a = 25°; side b = 55'; side c = 70' **7.** It is given that side EF = side HG and side EH = side FG. Since both triangles share a common side EG, the two triangles EHG and EFG have three equal sides and are therefore congruent.

EXERCISE 3-14   Similar Triangles

Practice Problems, Pages 381-382

1. Triangles DEF and JKL are similar because two angles of triangle DEF are equal to two angles of triangle JKL. **3.** Triangles ABC and JKL are similar because two sides of triangle ABC are proportional to two corresponding sides of triangle JKL and the included angles are equal. **5.** Side a = 144'; side b = 33' **7.** side a = 144'

EXERCISE 3-15   Longitude, Latitude, and Nautical Mile

Practice Problems, Pages 384-385

Set 1: **1.** 45° **2.** 34° **3.** 51° **4.** 89° **5.** 21°
Set 3: **1.** 57° **2.** 44° **3.** 144° **4.** 45° **5.** 69°
Set 5: **1.** 1,680 naut. mi. **2.** 780 naut. mi. **3.** 2,160 naut. mi. **4.** 942 naut. mi.
       5. 2,709 naut. mi.
Set 7: **1.** 110 naut. mi.; $5\frac{1}{2}$ hr. **2.** 600 naut. mi. **3.** 360 naut. mi.; 414 stat. mi.; 200 knots; 230 stat. m.p.h. **4.** 775 naut. mi. **5.** 2 hr. 12 min.

EXERCISE 3-16   Longitude and Time

Practice Problems, Pages 387-388

Set 1: **1.** 3 hr. **2.** 5 hr. **3.** 12 hr. **4.** 7 hr. **5.** 20 hr.
Set 3: **1.** 4 hr. 3 min. **2.** 9 hr. 2 min. **3.** 3 min. 1 sec. **4.** 1 hr. 2 min. 3 sec. **5.** 8 hr. 1 min. 2 sec.
Set 5: **1.** 32 min. **2.** 56 min. **3.** 20 sec. **4.** 44 sec. **5.** 8 min.
Set 7: **1.** 3 hr. 17 min. 44 sec. **2.** 1 hr. 53 min. 20 sec. **3.** 27 min. 24 sec. **4.** 6 min. 52 sec. **5.** 10 min. 36 sec.
Set 9: **1.** 60° **2.** 105° **3.** 30° **4.** 345° **5.** 150°
Set 11: **1.** 3'45'' **2.** 10' **3.** 3' **4.** 7'30'' **5.** 12'15''
Set 13: **1.** 56° 22' 30'' **2.** 33° 12'' **3.** 22° 17' 30'' **4.** 120° 58' 30'' **5.** 84° 21' 15''

Set 15: **1.** 15° W.; 45° E.; 90° W.; 80° E.; 123° W. **2.** 10° E.; 26° W.; 130° W.; 113° E.; 46° W. **3.** 18° 45′ E. **4.** 175° 52′ W. **5.** 90° 31′ W.

EXERCISE 3-17   The Magnetic Compass

Practice Problems, Pages 390–391

Set 1: **1.** 252° **2.** 36° **3.** 89° **4.** 131° **5.** 73°
Set 3: **1.** 240° **2.** 333° **3.** 83° **4.** 43° **5.** 24°
Set 5: **1.** 105° **2.** 277° **3.** 59° **4.** 166° **5.** 350° **6.** 245° **7.** 190° **8.** Variation = 7° W.; deviation = 3° E. **9.** 118° **10.** 63°

Refresh Your Skills, Page 391

1. 58,594 **3.** 4,286,400 **5.** $17\frac{7}{16}$ **7.** $6\frac{3}{16}$ **9.** \$31.36 **11.** \$1,149.12 **13.** \$51.91 **15.** 45

Review, Page 392

1. Angle LMN, angle NML, angle M **3.** 70° **5.** Construction **7.** 49° **9.** $m\angle 3 = 55°$; $m\angle 2 = 93°$ **11.** $m\angle 4 = 79°$; $m\angle 6 = 101°$; $m\angle 3 = 101°$; $m\angle 2 = 101°$; $m\angle 8 = 79°$; $m\angle 1 = 79°$; $m\angle 7 = 101°$ **13.** 0800 **15.** Construction

CONSTRUCTIONS

EXERCISE 3-18   Copying a Line Segment

Practice Problems, Page 393

1. Construction **3.** Construction

EXERCISE 3-19   Constructing Triangles

I. Three Sides

Practice Problems, Pages 394–395

1.–7. Constructions

II. Two Sides and an Included Angle

Practice Problems, Pages 395–396

1.–5. Constructions

III. Two Angles and an Included Side

Practice Problems, Pages 396–397

1. Construction **3.** Construction; yes; yes **5.** Construction **7.** Construction; no; yes

EXERCISE 3-20   Constructing Regular Polygons and Other Polygons

Practice Problems, Page 398

1.–5. Constructions

EXERCISE 3-21 Constructing a Perpendicular to a Line At or Through a Point on the Line

Practice Problems, Page 399

1. Construction **3.** Construction **5.** Construction; yes

EXERCISE 3-22 Constructing a Perpendicular to a Line From or Through a Point Not on the Line

Practice Problems, Page 400

1. Construction **3.** Construction; altitude; yes **5.** Construction

EXERCISE 3-23 Bisecting a Line Segment

Practice Problems, Page 401

1. Construction **3.** Construction; perpendicular bisector **5.** Construction; yes; yes

EXERCISE 3-24 Copying a Given Angle

Practice Problems, Page 402

1. Construction **3.** Construction **5.** Construction

EXERCISE 3-25 Bisecting an Angle

Practice Problems, Pages 403–404

1. Construction **3.** Construction **5.** Construction; yes; yes **7.** Construction; yes; yes; yes
9. Construction; no

EXERCISE 3-26 Constructing a Line Parallel to a Line Through a Point Not on the Line

Practice Problems, Page 405

1. Construction **3.** Construction; yes

Review, Page 406

1.–9. Constructions

Refresh Your Skills, Page 406

1. 300,131 **3.** 1,641,600 **5.** $6\frac{71}{80}$ **7.** $26\frac{2}{5}$ **9.** 1.47 **11.** 4.01 **13.** \$206.25 **15.** 25

INDIRECT MEASUREMENT

EXERCISE 3-27 Rule of Pythagoras

Practice Problems, Page 409

1. 15 m; 17 cm; 65 km; 65 ft.; 212 m; 289 yd. **3.** 7 cm; 60 m; 55 mm; 35 yd.; 135 ft.; 405 km
5. 127.3 ft. **7.** 49.5 m **9.** 76.3 ft.

EXERCISE 3-28 Similar Triangles

Practice Problems, Page 411

1. EB; EA; CD; 85 m **3.** 10 m **5.** 160 m

EXERCISE 3-29 Numerical Trigonometry

Practice Problems, Pages 414–416

1. a. .9325; 6.3138; 2.0065; .7744 **b.** .9816; .1219; .6626; .4120 **c.** .9511; .5592; .9940; .3145
3. a. 87°; 42°; 51° 40′; 31° 30′ **b.** 41°; 87°; 30° 30′; 71° 15′ **c.** 84°; 1°; 63° 45′; 3° 12′
5. a. 392.5 m **b.** 303.1 km **c.** 20° **7. a.** 21.1 ft. **b.** 116.7 km **c.** 31° **9. a.** 223.7 yd.
b. 250 cm **c.** 30° **11.** 6,004.6 ft. **13.** 122.56 ft. **15.** 20°

Review, Page 418

1. 95 m **3.** 112 cm **5.** If the umbrella is fitted into the suitcase diagonally it will form the hypotenuse of a right triangle with the sides of the suitcase. It should fit by .017 of an inch.
7. 606.2 yd. **9.** 128.67 km **11.** 47°

APPLICATIONS MEASURING AREA AND VOLUME

EXERCISE 3-30 Measuring Area

I. Rectangle

Practice Problems, Page 420

1. 126 m^2 **3. a.** $247.50 **b.** $229.50 **c.** $97.13 **d.** $217.00 **e.** $95.82 **5.** $81,250 **7.** $5.40
9. 1.3 acres

II. Square

Practice Problems, Page 421

1. 529 sq. ft. **3.** 210.25 cm^2 **5.** 12.78 acres

III. Parallelogram

Practice Problems, Page 422

1. 48 m^2 **3.** $7.49

IV. Triangle

Practice Problems, Pages 422–423

1. 40 sq. in. **3.** 62$\frac{1}{2}$ sq. ft.

V. Trapezoid

Practice Problems, Page 423

1. 77 sq. in. **3.** 81 sq. ft.

VI. Circle

Practice Problems, Pages 424–425

1. 113.04 cm^2 **3.** 452.16 sq. ft. **5.** 5,538.96 km^2 **7.** 153.86 m^2 **9.** area of square; 7.74 cm^2
11. 9$\frac{5}{8}$ sq. in.* **13.** $268.47 **15.** 3 pipes

*$\pi = 3\frac{1}{7}$ was used to determine answers marked with asterisk. Otherwise, 3.14 was used.

VII. Total Area of a Rectangular Solid

Practice Problems, Page 426

1. 2,050 cm^2; 9,414 mm^2; 171.18 m^2; 515$\frac{1}{8}$ sq. in.; 27,228 sq. in. or 189$\frac{1}{12}$ sq. ft. **3.** 800 sq. ft.;
4 gal.; $58.76

VIII. Total Area of a Cube

Practice Problems, Page 426

1. **a.** 8,214 mm^2 **b.** 2,400 cm^2 **c.** 507.84 m^2 **d.** 90$\frac{3}{32}$ sq. in. **e.** 3,174 sq. in. or 22$\frac{1}{24}$ sq. ft.

IX. Lateral Area and Total Area of a Right Circular Cylinder

Practice Problems, Page 428

1. **a.** 125.6 m^2; 8,478 cm^2; 2,816 mm^2*; 57$\frac{3}{4}$ sq. in.*; 34$\frac{4}{7}$ sq. ft.* **b.** 94.2 mm^2; 1,760 m^2*;
4,752 cm^2*; 138$\frac{2}{7}$ sq. in.*; 284.95 sq. ft.* or 40,995.84 sq. in. **3. a.** 75.36 cm^2; 836 m^2*;
10,362 mm^2; 670.31 sq. in.*; 42.39 sq. ft. **b.** 164.85 m^2; 50.24 mm^2; 6,325 cm^2*; 64.82 sq. in.*;
3,141.57 sq. in.

X. Sphere

Practice Problems, Page 428

1. **a.** 1,808.64 m^2 **b.** 9,856 mm^2* **c.** 498.96 cm^2* **d.** 1,164$\frac{5}{8}$ sq. in.* **e.** 46,735.76 sq. in.
3. 195,967,400 sq. mi.

EXERCISE 3-31 Measuring Volume

I. Rectangular Solid

Practice Problems, Pages 429–430

1. 252 cu. in. **3.** 378 m^3; 63 pupils **5.** 40$\frac{5}{6}$ lb. **7.** 115.32 m^3

II. Cube

Practice Problems, Page 430

1. 15,625 cm^3 **3.** 216 cu. ft.

III. Right Circular Cylinder

Practice Problems, Pages 431–432

1. 169.56 m^3 **3. a.** 628 mm^3; 6,160 m^3* (or 6,154.4 m^3); 28,338.5 cu. ft.; 162.27 cu. in.;
64.76 cu. ft. **b.** 75.36 cm^3; 49,280 mm^3* (or 49,235.2 mm^3); 3,168 m^3* (or 3,165.12 m^3);
265$\frac{37}{128}$ cu. in.*; 4,158 cu. in. or 2$\frac{13}{32}$ cu. ft.* **5.** 24$\frac{3}{4}$ cu. yd.* **7.** 62.8 cu. in. **9.** .317925 m^3;
317.925 L; 317.925 kg

IV. Sphere

Practice Problems, Page 432

1. **a.** 113,142$\frac{6}{7}$ m^3* **b.** 735,914$\frac{2}{3}$ cu. ft.* **c.** 4,709.43 cm^3 **d.** 1,054.06 cu. in.* **e.** 137,188.69 cu. in.
or 79.39 cu. ft. **3.** .3 cu. ft.

V. Right Circular Cone

Practice Problems, Page 433

1. **a.** 1,540.69 cm^3; 2,464 mm^3*; 398.36 m^3*; 113.43 cu. ft. **b.** 2,453.13 m^3; 26,796 cm^3*;
57$\frac{3}{4}$ cu. ft.*; 1,814.92 cu. in. **3.** 267.95 bu.

*$\pi = 3\frac{1}{7}$ was used to determine answers marked with asterisk. Otherwise, 3.14 was used.

VI. Pyramid

Practice Problems, Page 433

1. 14,400 m^3 **3.** 1,200 cu. ft.

Review, Pages 433–434

1. 2,795 m^2 **3.** 441 sq. ft. **5.** 60.1 sq. in. **7.** 4,928 sq. ft. **9.** 3,530 sq. yd. **11.** 452.16 m^2
13. 2$\frac{1}{4}$ times larger **15.** 5,024 cu. ft. **17.** 381.51 m^3 **19.** 781.25 m^3

Refresh Your Skills, Page 434

1. 681,120 **3.** 3,770,712 **5.** 10$\frac{7}{16}$ **7.** 1$\frac{31}{32}$ **9.** 73.456 **11.** .7854

Review of Part 3 Page 435

1. a. Line segment BR; line MY, ray LT **b.** (1) line CN (2) ray ZG (3) line segment RM
(4) angle E (5) triangle RPT (6) plane DF **3. a.** 3 **b.** 1 **c.** 2 **5. a.** 8 **b.** 7 **7.** Construction
9. Construction **11.** Construction **13.** Construction **15.** Construction **17.** Construction
19. 83° **21.** m∠1 = 74°; m∠3 = 59° **23.** m∠5 = 56°; m∠6 = 124°; m∠7 = 124°;
m∠3 = 124°; m∠4 = 56°; m∠1 = 56°; m∠2 = 124° **25.** 14.0625 in. or 14$\frac{1}{16}$ in. **27.** 87.92 mm
29. 2,209 sq. ft. **31.** 254.34 sq. in. **33.** 1,176 m^2 **35.** 1,000 mm^2 **37.** 17,295.12 cm^2
39. 58.374 m^3 **41.** 310,181.76 cm^3 **43.** 65 cm **45.** a = 49 mm; b = 64 mm **47.** 36 m
49. 246.28 ft.

Competency Check Test, Pages 439–440

1. c **3.** b **5.** a **7.** b **9.** d **11.** c **13.** c **15.** b **17.** c **19.** d **21.** b **23.** d **25.** d

PART 4 ELEMENTS OF ALGEBRA

Inventory Test, Pages 442–445

1. Positive: 86, 1.7; Negative: $-\frac{2}{3}$, $-.625$, $-5\frac{1}{2}$, -300 **3. a.** 63 **b.** $\frac{4}{5}$ **5.** -5, -2, 0, 3, 6
7. a. False **b.** True **c.** False **d.** True **9. a.** $+18$ **b.** -23 **c.** $-(-10) = +10$ **d.** $+2$
11. a. 14 **b.** -2 **c.** -1 **d.** -19 **e.** 4 **f.** 10 **g.** -4 **13. a.** 5 **b.** -8 **c.** -24 **d.** 4 **e.** -1
f. 8 **g.** -3 **15. a.** $y - 5 = 12$ **b.** $x + 2 > 8$ **c.** $10 \times n \le 14$ **d.** $-3 < s < 9$ **17. a.** 35
b. 14 **c.** 59 **d.** 6 **19. a.** 2, 3, 5 **b.** 2, 3, 5 **c.** 3 **21. a.** $x = 7$ **b.** $y = -9$ **c.** $r = 3$
d. $b = 11$ **e.** $a = 162$ **f.** $n = 3$ **g.** $x = 2.5$ **h.** $m = 19$ **23. a.** $n = +10$ or -10
b. $y = +9$ or -9
25. a.

b.

c.

27. a. (1) 6 to 1 (2) 2 to 5 (3) 1 to 10 **b.** $n = 39$ **29. a.** all integers greater than 5 **b.** all
multiples of 3 that are 12 or greater **c.** no solution

POSITIVE AND NEGATIVE NUMBERS

EXERCISE 4-1   Integers

Practice Problems, Page 449

1. a. negative six **b.** positive ten **c.** negative seven **d.** positive fifteen **e.** negative thirty-three **f.** positive fifty **g.** positive sixty-three **h.** negative one hundred forty-seven **i.** negative seventy-four **3. a** -7 **b.** $+3$ **c.** no opposite **d.** $+19$ **e.** -46 **f.** $+22$ **g.** $+100$ **h.** -79 **i.** $+61$ **5. a.** $-11, -10, -9, -8, -7, -6, -5, -4, -3, -2, -1, 0, 1, 2, 3$ **b.** $-9, -7, -5, -3, -1, 1, 3, 5, 7, 9$ **c.** $-12, -10, -8, -6, -4, -2, 0, 2, 4, 6, 8, 10$

EXERCISE 4-2   Rational Numbers; Real Numbers

Practice Problems, Page 451

1. a. positive five sixths **b.** negative thirteen fourths **c.** negative six and one third **d.** positive eighty-nine hundredths **e.** negative seventeen point three **f.** positive two and two fifths **g.** negative eleven sixteenths **h.** negative four thousandths **3. a.** $-\frac{5}{8}$ **b.** $+\frac{9}{4}$ **c.** $-.07$ **d.** $+2.75$ **e.** $+1\frac{3}{5}$ **f.** $-.102$ **g.** $-6\frac{4}{7}$ **h.** $+58.3$ **5. a.** Integers: $-17, \frac{16}{2}$ Rational: $-17, +\frac{3}{4}, -.05, \frac{16}{2}, -7\frac{1}{3}$ Irrational: $-\sqrt{29}$ Real: $-17, +\frac{3}{4}, -.05, -\sqrt{29}, \frac{16}{2}, -7\frac{1}{3}$ **b.** Integers: $+100, -\frac{18}{6}$ Rational: $-\frac{4}{7}, +100, -5.08, -2\frac{3}{4}, -\frac{18}{6}$ Irrational: $+\sqrt{13}$ Real: $-\frac{4}{7}, +100, -5.08, -2\frac{3}{4}, +\sqrt{13}, -\frac{18}{6}$ **c.** Integers: $-\sqrt{16}, -69, \frac{28}{4}$ Rational: $-.375, -\frac{5}{8}, -\sqrt{16}, -69, +\frac{28}{4}, +3\frac{5}{6}$ Irrational: none Real: $-.375, -\frac{5}{8}, -\sqrt{16}, -69, +\frac{28}{4}, +3\frac{5}{6}$ **d.** Integers: $+97$ Rational: $-\frac{15}{16}, +97, +.18, -4\frac{7}{12}, +\frac{21}{4}$ Irrational: $-\sqrt{5}$ Real: $-\sqrt{5}, -\frac{15}{16}, +97, +.18, -4\frac{7}{12}, +\frac{21}{4}$

EXERCISE 4-3   Absolute Value

Practice Problems, Page 452

1. a. The absolute value of positive eleven **b.** The absolute value of negative three **c.** The absolute value of positive thirty-five **d.** The absolute value of negative seven tenths **e.** The absolute value of negative nine hundredths **f.** The absolute value of positive five and two-thirds **3. a.** 7 **b.** 3 **c.** 6 **d.** 0 **e.** .8 **f.** $\frac{2}{3}$ **g.** $2\frac{1}{2}$ **h.** 27 **i.** 19 **j.** .82 **k.** $\frac{9}{4}$ **l.** 1.75 **5. a.** 6 **b.** $\frac{1}{10}$ **c.** $5\frac{1}{6}$ **d.** 2.35 **7. a.** 6 **b.** 1.5625 or $1\frac{9}{16}$ **c.** $2\frac{8}{19}$ **d.** 70

EXERCISE 4-4   The Real Number Line; Graph of a Group of Numbers

Practice Problems, Page 455

1. a. $-2; +2; -7; +5; 0; -9; +8; +3; -6; +10$ **b.** O; U; B; D; F; I; N; Q; A; T
3. a.

   b.

   c.

EXERCISE 4-5   Opposite Meanings

Practice Problems, Page 456

1. 60 meters below sea level **3.** -4% **5.** a head wind of 23 m.p.h. **7.** 72 degrees above zero **9.** $+58$ items **11.** 9 pounds underweight **13.** $+5$ yards **15.** $+18$ meters

EXERCISE 4-6   Comparing Integers

Practice Problems, Pages 457-458

1. a. T b. V c. Q d. U 3. a. $+9$ b. -5 c. $+3$ d. 0 e. $+8$ f. $+4$ g. $+2$ h. $+5$
5. a. True b. False c. True d. False e. True f. False g. True h. False i. False
j. False k. True l. True m. True n. True o. True p. False 7. a. $-9, -6, -4, -3, -2,$
$0, +4, +5, +7, +10$ b. $-10, -7, -5, -3, -1, 0, +2, +8, +10, +11$ 9. a. absolute value of
-8 b. both are equal to nine

EXERCISE 4-7   Opposite Directions; Vectors

Practice Problems, Pages 460-461

1. a. 7 units to the left b. 9 units to the right c. 2 units to the left d. 12 units to the right
e. 1 unit to the right f. $4\frac{1}{4}$ units to the left g. $2\frac{1}{4}$ units to the left h. $\frac{2}{3}$ of a unit to the right
3. a. -5 b. $+4$ c. $+9$ d. -8 5. a. $+8$ b. -9 c. -10 d. $+9$

EXERCISE 4-8   Additive Inverse of a Number; More About Opposites

Practice Problems, Pages 462-463

1. a. $+7$ b. -26 c. 0 d. $+\frac{5}{8}$ e. -3.08 f. $+4\frac{1}{2}$ g. $-\frac{5}{6}$ h. $+\frac{11}{3}$ 3. a. The opposite of nine
or negative nine b. The opposite of positive two or the negative of positive two c. The
opposite of negative five sixths or the negative of negative five sixths d. The opposite of
negative eleven or the negative of negative eleven e. The opposite of negative eight or the
negative of negative eight is equal to positive eight f. The opposite of positive ten or the
negative of positive ten is equal to negative ten g. The opposite of fifteen or the negative of
positive fifteen is equal to negative fifteen h. The opposite of negative six point four or the
negative of negative six point four is equal to positive six point four 5. a. -7; $+6$; $+\frac{3}{8}$; $-.9$;
$-3\frac{1}{2}$ b. The opposite of the opposite of a number is equal to the original number.

Refresh Your Skills, Page 463

1. 19,044 3. 1,251,540 5. $11\frac{1}{5}$ 7. $1\frac{3}{4}$ 9. 2.889 11. 1 13. \$3 15. $122.\overline{2}$

COMPUTATION WITH POSITIVE AND NEGATIVE NUMBERS

EXERCISE 4-9   Addition of Integers and Rational Numbers

Practice Problems, Pages 466-467

1. $+8, -5, +4, -4, -6, +4, 0, -9$ 3. a. -7 b. -7 c. $+10$ d. -2 e. -17 f. -7
g. -20 h. $+1$ i. $+2$ j. 0 5. a. 11 b. 2 c. 9 d. 6 e. 17 f. 8 g. -9 h. -8 i. -1
j. -2

EXERCISE 4-10   Subtraction of Integers and Rational Numbers

Practice Problems, Pages 469-470

1. $+3, -7, -4, +4, +10, -7, +8, 0$ 3. a. -11 b. $+17$ c. 4 d. $+5$ e. -3 f. $+22$ g. 0
h. $+15$ i. -1 j. $+1$ 5. a. -5 b. -1 c. -1 d. 5 e. -6 f. 4

EXERCISE 4-11   Multiplication of Integers and Rational Numbers

Practice Problems, Page 472

1. a. $+45, +48, +49, +24, +6, +36, +42, +40$ b. $-56, -18, -48, -65, -36, -96, -20,$
-54 c. $0, 0, 0, 0, +.08, -3.5, -7\frac{7}{8}, +20$ 3. a. 9 b. 36 c. -125 d. 16 e. $-1,024$ f. $+1$

g. $+81$ **h.** $+27$ **5. a.** $-\frac{1}{6}$ **b.** $\frac{1}{14}$ **c.** $-\frac{5}{2}$ **d.** $-\frac{4}{17}$ **e.** $\frac{6}{5}$ **f.** $-\frac{2}{7}$

EXERCISE 4-12 Division of Integers and Rational Numbers

Practice Problems, Page 473

1. a. $+6, +4, +6, +6, +4, +3, +5, +6$ **b.** $-8, -6, -9, -6, -6, -9, -8, -9$ **c.** $+1, -1,$ $+15, -40, 0, 0, 0, 0$ **3. a.** -7 **b.** $+8$ **c.** $+9$ **d.** -10 **e.** -8 **f.** -3 **5. a.** 2 **b.** 12 **c.** 6 **d.** 12 **e.** -28 **f.** 2

LANGUAGE OF ALGEBRA

EXERCISE 4-13 Symbols—Review

Practice Problems, Page 475

1. a. Negative five plus positive seven **b.** Thirty-six times twenty-eight **c.** Fifty-three squared **d.** Forty-two divided by six **e.** The square root of sixty-five **f.** Positive twelve minus positive thirty **g.** Twenty-four times nineteen **h.** Seventeen cubed or seventeen to the third power **i.** Negative fifteen times negative four **j.** Six to the power of eight **k.** The absolute value of negative thirteen **l.** Seven times the sum of twenty-one and nine **m.** Eighteen minus the difference of positive twenty minus positive seven **n.** Nine times the product of twelve times five **o.** The negative of negative twenty **p.** The absolute value of negative six plus negative three

EXERCISE 4-14 Mathematical Expressions or Mathematical Phrases

Practice Problems, Pages 476–478

1. a. $4 + 6$ or $6 + 4$ **b.** 12×19 or $12 \cdot 19$ **c.** $10 - 2$ **d.** $\frac{15}{3}$ or $15 \div 3$ **e.** 9^2 **f.** $\sqrt{20}$ **g.** $8 + 3$ **h.** $6 - 2$ **i.** 9×5 or $9 \cdot 5$ **j.** $\frac{12}{4}$ or $12 \div 4$ **k.** $b + x$ **l.** ay **m.** $8g$ **n.** $c - g$ **o.** d/r or $d \div r$ **p.** $2(l + w)$ **q.** r^3 **r.** $\sqrt{s}$ **s.** bh **t.** $3(t - 9)$ **3. a.** $2r$ **b.** $3s$ **c.** $p + i$ **d.** $90° - B$ **e.** $A + B + C$ **f.** $c \div \pi$ or c/π **g.** lwh **h.** $c + p$ **i.** $\frac{1}{2}ab$ **j.** $180° - B$ **k.** s^2 **l.** $bh \div 2$ or $bh/2$ **m.** $2l + 2w$ **n.** e^3 **o.** $i \div p$ or i/p **p.** $\dfrac{b_1 + b_2}{2}$ **q.** $\frac{4}{3}\pi r^3$ **r.** $a^2 + b^2$ **s.** $2\pi r(h + r)$ **t.** $\sqrt{h^2 - b^2}$

EXERCISE 4-15 Mathematical Sentences

Practice Problems, Page 479

1. a. Each number n is less than sixteen **b.** Each number x is greater than twenty-nine **c.** Three times each number y is not equal to seventeen **d.** Each number b plus six is greater than twenty-five **e.** Each number a minus five is less than fifty-three **f.** Seven times each number t is not less than negative fourteen **g.** Nine times some number m plus four is equal to fifteen **h.** Eight times each number c minus nine is not greater than the number c plus seven **i.** Two times each number n plus six is less than three times the number n minus eight **j.** Each number b divided by five is greater than twenty **k.** Fifteen times each number x minus four is not greater than eighteen minus seven times the number x **l.** Each number h is greater than or equal to negative six **m.** Nineteen times each number w is less than or equal to thirty-eight **n.** Five times each number d plus seven is less than or equal to twelve **o.** Each number n minus sixteen is greater than or equal to four times the number n plus nine **p.** Each number n is greater than negative three and less than eight **q.** Each number x is less than eighteen and greater than zero **r.** Each number b is greater than or equal to four and less than or equal to twenty-seven **s.** Seven times each number r is less than or equal to twelve and greater than or equal to negative six **t.** Each number a is greater than two and less than or equal to forty-five

EXERCISE 4-16　Formulas

Practice Problems, Pages 480–481

1. **a.** $p = 4s$　**b.** $A = lw$　**c.** $c = 2\pi r$　**d.** $i = prt$　**e.** $r = d/2$　**f.** $A = p + i$　**g.** $A = bh$
h. $V = e^3$　**i.** $A = \pi r^2$　**j.** $A = \frac{1}{2}ab$　**k.** $A + B = 90°$　**l.** $n = l - d$　**m.** $V = \frac{1}{4}\pi d^2 h$
n. $p = b + 2e$　**o.** $r + c/s$　**p.** $p = 2(l + w)$　**q.** $A = \dfrac{h(b_1 + b_2)}{2}$　**r.** $A = 2lw + 2lh + 2wh$
s. $h = \sqrt{a^2 + b^2}$　**t.** $A = 2\pi r(h + r)$

EXERCISE 4-17　Evaluating Algebraic Expressions

Practice Problems, Pages 482–484

1. **a.** 18　**b.** 6　**c.** 72　**d.** 2　**e.** 144　**f.** 84　**g.** 24　**h.** 108　3. **a.** 192　**b.** 216　**c.** +60　**d.** −3
e. −28　**f.** 208　5. **a.** 620　**b.** 90　**c.** 82　**d.** 234　**e.** 1,592　7. **a.** 23　**b.** 375　**c.** 4　**d.** 1,224
e. 45　9. **a.** 2,376　**b.** 900　**c.** 2,374.625　**d.** 310,464　**e.** 398,654.4　11. **a.** 130　**b.** 880　**c.** 104
d. 54　**e.** 782　**f.** −25

EXERCISE 4-18　Evaluating Formulas

Practice Problems, Page 485

1. 170　3. 264　5. 202　7. 234　9. 22,608　11. 512　13. 1,640　15. 61　17. 300　19. 198

OPEN SENTENCES IN ONE VARIABLE

EXERCISE 4-19　Introduction to the Solution of Equations in One Variable

Practice Problems, Pages 487–488

1. $x = 4$　3. $r = 7$　5. $m = 0$　7. $d = 12$　9. $\frac{1}{2}$　11. 96　13. **a.** $7x + 9$　**b.** 42　15. $9x = 63$,
$x = 7, \dfrac{x}{7} = 1; 9x = 63, x = 7, \dfrac{x}{7} = 1; x = 7$

EXERCISE 4-20　Inverse Operations; Properties of Equality

Practice Problems, Page 491

1. **a.** 8　**b.** minus 22　**c.** minus 5　3. **a.** 5　**b.** plus 18　**c.** plus 12　5. **a.** 4　**b.** times 6
c. times 8　7. **a.** plus 10　**b.** divided by 16　**c.** minus 25　**d.** times 21　**e.** divided by 11
f. plus 37　9. **a.** 4; $w = 15$　**b.** 19; $s = 25$　**c.** 28; $h = 56$　**d.** 15; $x = 15$　11. **a.** 3; $t = 18$
b. 8; $m = 16$　**c.** 14; $x = 126$　**d.** 10; $a = 250$

EXERCISE 4-21　Solving Equations in One Variable

Practice Problems—Basic Type I, Page 493

1. $c = 8$　3. $i = 26$　5. $p = 77$　7. $a = 39$　9. $y = 26$　11. $d = 17$　13. $n = 0$　15. $i = \$7.50$
17. $x = 2\frac{1}{8}$　19. $a = \$1.50$　21. $b = -5$　23. $a = -8$　25. $g = -15$　27. $n = -3.2$　29. $s = 6.6$

Basic Type II, Pages 494–495

1. $c = 28$　3. $A = 166$　5. $b = 63$　7. $t = 9$　9. $c = 80$　11. $s = \$171$　13. $y = \$8.00$
15. $n = 16.6$　17. $c = -5$　19. $x = 8$　21. $x = 5$　23. $x = 75$　25. $t = 102$　27. $a = 105$
29. $x = \$1.10$

Basic Type III, Page 496

1. $a = 9$　3. $n = 3$　5. $x = 5$　7. $r = 9$　9. $w = 1$　11. $y = 0$　13. $c = 1\frac{3}{5}$　15. $y = \frac{1}{4}$
17. $n = 2\frac{5}{9}$　19. $r = 6\frac{3}{4}$　21. $t = .4$　23. $c = \$.09$　25. $c = 80$　27. $n = 36$　29. $p = \$4,000$

31. $x = -4$ **33.** $y = 8$ **35.** $c = 16$ **37.** $b = -2\frac{4}{9}$ **39.** $c = -5$ **41.** $m = 25$ **43.** $d = 4$
45. $h = 40$

Basic Type IV, Page 497

1. $d = 38$ **3.** $b = 21$ **5.** $t = 64$ **7.** $m = 225$ **9.** $x = 27$ **11.** $b = 7$ **13.** $c = \$.48$ **15.** $w = 150$
17. $n = -20$ **19.** $y = -36$ **21.** $d = 88$ **23.** $z = -35$

Solution by More Than One Axiom

Practice Problems, Page 498

1. $b = 14$ **3.** $y = 5$ **5.** $h = 10$ **7.** $t = 7$ **9.** $a = 6$ **11.** $n = 25$ **13.** $p = 650$ **15.** $a = 800$
17. $a = 5\frac{6}{7}$ **19.** $l = 300$ **21.** $d = \$1.20$ **23.** $n = 32$ **25.** $a = -4$ **27.** $b = -2$ **29.** $a = \$3.68$

EXERCISE 4-22   Solving Equations When the Replacements for the Variable Are Restricted

Practice Problems, Pages 499–500

1. a. $c = 7$ **b.** $n = 12$ **c.** no solution **d.** $t = 7$ **3. a.** no solution **b.** $c = -28$ **c.** $n = 8$
d. $a = -2$ **5. a.** $d = -2$ **b.** $x = 3\frac{1}{2}$ **c.** $p = 1,950$ **d.** $y = \frac{2}{3}$ **7. a.** no solution **b.** $h = 5$
c. no solution **d.** $c = 9$ **9. a.** $n = 6$ **b.** no solution **c.** no solution **d.** no solution
11. a. no solution **b.** no solution **c.** $x = -3$ **d.** $y = -7$

EXERCISE 4-23   Solving Equations with Absolute Values

Practice Problems, Page 501

1. $r = +7$ or -7 **3.** $x = +11$ or -11 **5.** $t = +12$ or -12 **7.** $a = +14$ or -14
9. $b = +20$ or -20

EXERCISE 4-24   Using Formulas

Practice Problems, Page 502

1. 6 **3.** 16 **5.** 7 **7.** 4 **9.** 300 **11.** 5 **13.** 33 **15.** 12,800 **17.** 7 **19.** 6 **21.** $12\frac{2}{3}$ **23.** 47.1
25. 16 **27.** -10 **29.** 400

EXERCISE 4-25   Graphing an Equation in One Variable on the Number Line

Practice Problems, Page 504

1.

a.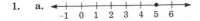

b.

c.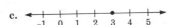

d.

e.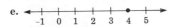

f.

g.

h.

i.

j.

k. no solution

l.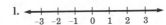

EXERCISE 4-26　Fraction, Decimal, and Percent Number Relationships—Problem Solving by Equation

Practice Problems, Page 506

1. a. 18　**b.** 27　**c.** 56　**d.** 215　**e.** 371　**3. a.** 4　**b.** 10.8　**c.** 98　**d.** 665　**e.** 245　**5. a.** .3　**b.** .68
c. .75　**d.** .875　**7. a.** 87　**b.** 70　**c.** 108　**d.** 48　**9. a.** 72　**b.** 650　**c.** 370　**d.** 45,000

EXERCISE 4-27　Ratio

Practice Problems, Pages 508–509

1. a. $\frac{5}{11}$　**b.** $\frac{9}{2}$　**c.** $\frac{16}{1}$　**d.** $n/12$　**3. a.** $\frac{5}{7}$　**b.** $\frac{8}{15}$　**c.** $\frac{1}{6}$　**d.** $\frac{4}{11}$　**5. a.** $\frac{4}{1}$ or 4　**b.** $\frac{4}{1}$ or 4　**c.** $\frac{5}{1}$ or 5
d. $\frac{3}{1}$ or 3　**7. a.** $\frac{5}{4}$　**b.** $\frac{8}{5}$　**c.** $\frac{5}{3}$　**d.** $\frac{10}{3}$　**9. a.** $\frac{2}{5}$　**b.** $\frac{1}{6}$　**c.** $\frac{3}{1}$ or 3　**d.** $\frac{2}{3}$　**e.** $\frac{5}{4}$　**f.** $\frac{9}{7}$　**g.** $\frac{1}{3}$　**h.** $\frac{3}{1}$ or 3
i. $\frac{4}{5}$　**j.** $\frac{5}{17}$　**k.** $\frac{10}{3}$　**l.** $\frac{7}{9}$　**m.** $\frac{40}{1}$ or 40　**n.** $\frac{7}{8}$　**o.** $\frac{3}{1}$ or 3　**p.** $\frac{2}{3}$　**11. a.** $\frac{88}{1}$ or 88　**b.** $\frac{5}{57}$　**c.** $\frac{25}{5}$
d. $\frac{180}{1}$ or 180　**e.** $\frac{16}{1}$ or 16　**f.** $\frac{8}{1}$ or 8　**g.** $\frac{1,800}{1}$ or 1,800　**h.** $\frac{4}{99}$　**i.** $\frac{35}{6}$　**j.** $\frac{50}{1}$ or 50　**13. a.** $\frac{2}{3}$　**b.** $\frac{3}{2}$
c. $\frac{3}{5}$　**d.** $\frac{2}{5}$　**15.** $\frac{48}{1}$ or 48

EXERCISE 4-28　Proportions

Practice Problems, Pages 511–512

1. a. $\frac{26}{13} = \frac{6}{3}$　**b.** $\frac{4}{12} = \frac{5}{15}$　**c.** $\frac{45}{80} = \frac{18}{32}$　**d.** $\frac{56}{14} = \frac{76}{19}$　**e.** $\frac{150}{100} = \frac{54}{36}$　**3.** c and d　**5.** $a = 5$; $y = 8$; $c = 65$;
$r = 24$; $t = .833$ or $\frac{5}{8}$　**7.** $a = 24$; $s = 63$; $n = 9$; $x = 20$; $13\frac{1}{8}$ or 13.125　**9.** $y = 4$; $x = 50$; $b = 60$;
$n = 45$; $x = 18$　**11.** 9　**13.** 285 km　**15.** $R_3 = 24$　**17.** $P = 19$

EXERCISE 4-29　Solving Inequalities in One Variable

Practice Problems—Type I, Page 515

1. a. $n > 3$, or all real numbers greater than 3　**b.** $a > 0$, or all real numbers greater than 0
c. $d > -11$, or all real numbers greater than -11　**3. a.** $b < 6$, or all real numbers less than 6
b. $g < -7$, or all real numbers less than -7　**c.** $m < 1.9$, or all real numbers less than 1.9
5. a. $m \geq 4$, or 4 and all real numbers greater than 4　**b.** $f \geq -17$, or -17 and all real numbers
greater than -17　**c.** $w \geq -16$, or -16 and all real numbers greater than -16　**7. a.** $y \leq 7$, or
7 and all real numbers less than 7　**b.** $e \leq \frac{1}{2}$, or $\frac{1}{2}$ and all real numbers less than $\frac{1}{2}$　**c.** $v \leq -3$,
or -3 and all real numbers less than -3　**9. a.** $s \neq 8$, or all real numbers except 8　**b.** $n \neq 1.3$,
or all real numbers except 1.3　**c.** $b \neq -11$, or all real numbers except -11　**11. a.** $x > 9$, or all
real numbers greater than 9　**b.** $y < -5$, or all real numbers less than -5　**c.** $n \geq -6$, or -6
and all real numbers greater than -6　**d.** $a \leq 20$, or 20 and all real numbers less than 20
e. $t \neq 0$, or all real numbers except 0　**f.** $b < -4$, or all real numbers less than -4　**g.** $r > -4$,
or all real numbers greater than -4　**h.** $z \geq 19$, or 19 and all real numbers greater than 19
i. $w \leq -10$, or -10 and all real numbers less than -10

Type II, Pages 515–516

1. a. $x > 2$, or all real numbers greater than 2　**b.** $y > -8$, or all real numbers greater than
-8　**c.** $a > \frac{1}{4}$, or all real numbers greater than $\frac{1}{4}$　**3. a.** $z < 5$, or all real numbers less than 5
b. $s < -8$, or all real numbers less than -8　**c.** $b < 1\frac{1}{2}$, or all real numbers less than $1\frac{1}{2}$
5. a. $n \geq 5$, or 5 and all real numbers greater than 5　**b.** $a \geq \frac{2}{3}$, or $\frac{2}{3}$ and all real numbers
greater than $\frac{2}{3}$　**c.** $x \geq 32$, or 32 and all real numbers greater than 32　**7. a.** $s \leq 9$, or 9 and all
real numbers less than 9　**b.** $h \leq 2$, or 2 and all real numbers less than 2　**c.** $n \leq 400$, or 400 and
all real numbers less than 400　**9. a.** $c \neq 16$, or all real numbers except 16　**b.** $x \neq 128$, or all
real numbers except 128　**c.** $f \neq 0$, or all real numbers except 0　**11. a.** $b < 4$, or all real
numbers less than 4　**b.** $a > -16$, or all real numbers greater than -16　**c.** $x \geq -\frac{3}{4}$, or $-\frac{3}{4}$ and

all real numbers greater than $-\frac{3}{4}$ **d.** $y \le 4\frac{1}{3}$, or $4\frac{1}{3}$ and all real numbers less than $4\frac{1}{3}$
e. $m \ne -8$, or all real numbers except -8 **f.** $t < 49$, or all real numbers less than 49 **g.** $n > 0$, or all real numbers greater than 0 **h.** $z \ge 75$, or 75 and all real numbers greater than 75
i. $x \le 128$, or 128 and all real numbers less than 128

Type III, Page 516

1. a. $d < -6$, or all real numbers less than -6 **b.** $c < -14$, or all real numbers less than -14
c. $h < -3$, or all real numbers less than -3 **3. a.** $x > -9$, or all real numbers greater than
-9 **b.** $r > -6$, or all real numbers greater than -6 **c.** $a > -4\frac{1}{2}$, or all real numbers greater
than $-4\frac{1}{2}$ **5. a.** $y \le -7$, -7 and all real numbers less than -7 **b.** $r \le 0$, or 0 and all real
numbers less than 0 **c.** $x \le -2$, or -2 and all real numbers less than -2 **7. a.** $s \ge -20$, or
-20 and all real numbers greater than -20 **b.** $y \ge 0$, or 0 and all real numbers greater than 0
c. $x \ge -5$, or -5 and all real numbers greater than -5 **9. a.** $n > -36$, or all real numbers
greater than -36 **b.** $b < 12$, or all real numbers less than 12 **c.** $y \ge -48$, or -48 and all real
numbers greater than -48 **d.** $\varsigma \le -90$, or -90 and all real numbers less than -90 **e.** $a > 21$,
or all real numbers greater than 21 **f.** $x < -40$, or all real numbers less than -40 **g.** $s \le 60$,
or 60 and all real numbers less than 60 **h.** $m < -36$ or all real numbers less than -36
i. $d \ge 4$, or 4 and all real numbers greater than 4

Miscellaneous, Page 517

1. a. $x < 4$, or all real numbers less than 4 **b.** $a > -5$, or all real numbers greater than -5
c. $m \le 2$, or 2 and all real numbers less than 2 **d.** $s \ge 2$, or 2 and all real numbers greater than
2 **e.** $n \ne -8$, or all real numbers except -8 **f.** $y < 9$, or all real numbers less than 9
g. $b > -\frac{1}{2}$, or all real numbers greater than $-\frac{1}{2}$ **h.** $r \ge 3$, or 3 and all real numbers greater
than 3 **i.** $t < -7$, or all real numbers less than -7 **j.** $z \le 5\frac{1}{2}$, or $5\frac{1}{2}$ and all real numbers less
than $5\frac{1}{2}$ **k.** $w > 14$, or all real numbers greater than 14 **l.** $x \ne \frac{1}{2}$, or all real numbers except $\frac{1}{2}$

Restricted Replacement, Page 517

1. 4, 5, 6, 7, 8, 9; 0, 1, 2, 3, 4, 5, 6; 0, 1, 2, 3; 0, 1, 2, 3, 4 **3.** 2, 3, 5; 2, 3, 5, 7, 11, 13, 17, 19, 23;
all prime numbers except 11; 2, 3 **5.** 0, 2, 4, 6; 8, 10, 12, 14, . . . ; 0, 2; 36, 38, 40, . . . ; **7.** no
replacements; no replacements; -2, -1, 0, 1, 2; -2, -1, 0, 1, 3 **9.** 16, 20, 24, 28; 16, 20, 24, 28;
no replacements; 12, 16, 20, 24

EXERCISE 4-30　Graphing on the Number Line an Inequality in One Variable

Practice Problems, Page 520

1.

a.

b.

c.

d.

e.

f.

g.

h.

i.

j.

k.

l.

m.
-1 0 1 2 3 4 5

o.
-4 -3 -2 -1 0 1 2

n. ◄——+——+——⊕——+——+——+——+——►
-6 -5 -4 -3 -2 -1 0

3. a. $a < 1$ **b.** $b > -4$ **c.** $-1 \leq c \leq 2$ **d.** $d \geq 5$ **e.** $-4 < e < 0$ **f.** $f \leq -1$

Review of Part 4, Pages 521–522

1. Integers: -48, $+\frac{20}{4}$ Rational Numbers: -48, $+6\frac{2}{5}$, $-.9$, $+\frac{20}{4}$, $-\frac{2}{3}$ Real Numbers: -48, $+6\frac{2}{5}$, $-.9$, $+\frac{20}{4}$, $-\sqrt{31}$, $-\frac{2}{3}$ **3. a.** True **b.** False **c.** True **d.** True **5. a.** -2 **b.** -9 **c.** -2 **d.** 4 **7. a.** -24 **b.** 63 **c.** 24 **d.** -243 **9.** $d = \frac{1}{2}(gt^2)$ **11.** $A = 2,112$ **13. a.** $n = -6$ **b.** $x = +5$ or -5

15. a. ◄——+——•——+——+——+——+——+——+——+——►
-5 -4 -3 -2 -1 0 1 2 3

c. ◄——+——+——+——+——+——+——+——•——+——+——►
-3 -2 -1 0 1 2 3 4 5

b. ◄——+——+——+——+——+——+——+——•——+——►
-3 -2 -1 0 1 2 3 4 5

17. a. $\frac{3}{4}$ **b.** $x = 24$ **19.** 2, 3, 5, 7

Competency Check Test, Pages 523–524

1. c **3.** c **5.** c **7.** b **9.** c **11.** b **13.** c **15.** d **17.** c **19.** b **21.** b **23.** b **25.** c

PART 5   GRAPHS, STATISTICS, AND PROBABILITY

Inventory Test, Page 526

1.

3.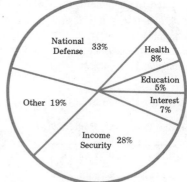

Average Annual Outlays of the Federal Government

5. 41 **7.** 5 **9.** 70th percentile **11.** $\frac{18}{30}$ or 60%

GRAPHS

EXERCISE 5-1   Bar Graph

Practice Problems, Pages 527–528

1. a. 400 miles; 100 miles **b.** Nile; Euphrates **c.** Congo 2,900 mi.; Nile 4,000 mi.; Volga 2,300 mi.; Euphrates 1,700 mi.; Amazon 3,900 mi.; Yangtze 3,100 mi.

3.

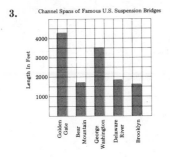

EXERCISE 5-2   Line Graph

Practice Problems, Page 529

1. a. 2 pupils; 1 pupil   **b.** Feb.; June   **c.** Sept., 6 pupils; Oct., 9 pupils; Nov., 13 pupils; Dec., 20 pupils; Jan., 21 pupils; Feb., 23 pupils; Mar., 15 pupils; Apr., 13 pupils; May, 7 pupils; June, 4 pupils
3.

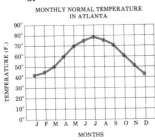

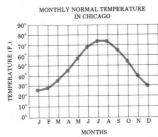

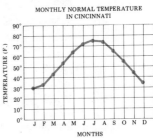

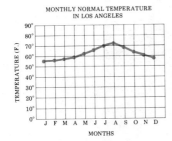

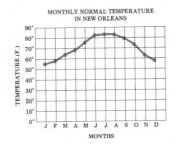

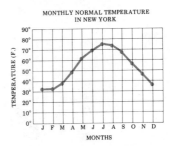

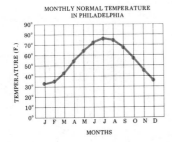

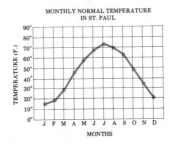

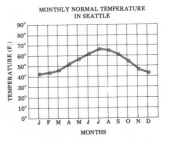

EXERCISE 5-3   Circle Graph

Practice Problems, Page 530

1. a. $\frac{1}{4}$; $\frac{1}{5}$ **b.** Twice as much   **c.** Food: $4,650; $387.50; $89.42 Savings: $1,860; $155; $35.77
Clothing: $2,232; $186; $42.92 Family Car: $1,860; $155; $35.77 Shelter: $3,720; $310; $71.54 Other
Expenses: $2,790; $232.50; $53.65 Operating Expenses: $1,488; $124; $28.62
3.

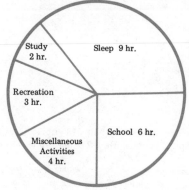

How Charles Spends His Day

STATISTICS

EXERCISE 5-4   Frequency Distribution; Histogram; Frequency Polygon

Practice Problems, Page 532

1. a.

Frequency Distribution Table		
Score	Tally	Frequency
100	1	1
95	11	2
90	1111	4
85	1111	4
80	111	3
75	1111	4
70	111	3
65	1	1
60	11	2
40	1	1

b. Histogram

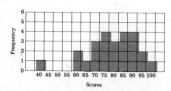

c. Frequency Polygon

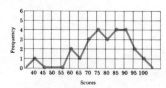

3. a.

Frequency Distribution Table		
Score	Tally	Frequency
52	11	2
51	111	3
50	1111	4
49	~~1111~~	5
48	111	3
47	111	3
46	11	2
45	~~1111~~ 11	7
44	111	3
43	111	3

b. Histogram

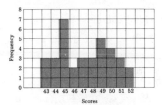

c. Frequency Polygon

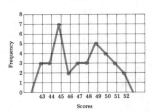

5. a.

Frequency Distribution Table		
Score	Tally	Frequency
10	111	3
9	~~1111~~ 1	6
8	~~1111~~ 1111	9
7	~~1111~~ 1	6
6	~~1111~~ 11	7
5	~~1111~~ 1	6
4	~~1111~~	5
3	111	3
2	1111	4
1	1	1

b. Histogram

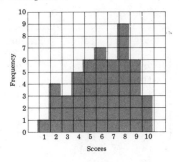

c. Frequency Polygon

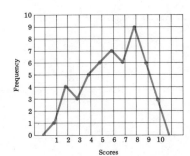

EXERCISE 5-5   Averages—Measures of Central Tendency

Practice Problems, Page 534

1. a. 84 **b.** 7 **c.** 46 **d.** 27   **3. a.** 15 **b.** 53 and 58 **c.** 7 **d.** 45

5. a.

Score	Tally	Frequency
9	1111 1	6
8	1111 1	6
7	1111	4
6	1111	4
5	11	2
4	11	2
3	1	1

Mean   7
Median 7
Mode   8 and 9

b.

Score	Tally	Frequency
50	1	1
49	11	2
48	111	3
47	1111	5
46	1	1
45	1	1
44	11	2
43	11	2
42	1	1
41		0
40	1	1

Mean   46
Median 47
Mode   47

c.

Score	Tally	Frequency
20	11	2
19	1111	4
18	1111	4
17	1111	4
16	1111	4
15	111	3
14	1111	5
13	11	2
12	11	2
11	1	1

Mean   16
Median 16
Mode   14

d.

Score	Tally	Frequency
64	1111	4
63	111	3
62	1111	4
61	1111	4
60	1111	4
59	1111 11	7
58	1111	4
57	111	3
56	11	2
55	1	1

Mean   60
Median 60
Mode   59

e.

Score	Tally	Frequency
10	11	2
9	~~1111~~ 111	8
8	~~1111~~ 1	6
7	~~1111~~	5
6	~~1111~~	5
5	~~1111~~ 11	7
4	~~1111~~	5
3	~~1111~~	5
2	111	3
1	1	1

Mean   6
Median   6
Mode   9

EXERCISE 5-6   Percentiles and Quartiles

Practice Problems, Pages 535–536

1. a. 70th percentile   **b.** 60th percentile   **c.** 20th percentile   **d.** 92nd percentile   **e.** $76\frac{2}{3}$ percentile   **3.** a, c and d would not be exempt   **5.** 6th; yes

PROBABILITY

EXERCISE 5-7   Probability; Odds

Practice Problems, Pages 537–539

1. a. $\frac{4}{9}$ **b.** $\frac{9}{15}$ or $\frac{3}{5}$ **c.** $\frac{8}{20}$ or $\frac{2}{5}$ **d.** $\frac{30}{48}$ or $\frac{5}{8}$ **e.** $\frac{36}{54}$ or $\frac{2}{3}$ **3. a.** $\frac{1}{5}$ **b.** $\frac{10}{6}$ or $\frac{5}{3}$ **c.** $\frac{9}{3}$ or 3 to 1   **d.** $\frac{30}{15}$ or 2 to 1   **e.** $\frac{9}{12}$ or $\frac{3}{4}$ **5. a.** (1) $\frac{26}{52}$ or $\frac{1}{2}$ (2) $\frac{12}{52}$ or $\frac{3}{13}$ (3) $\frac{4}{52}$ or $\frac{1}{13}$ (4) $\frac{1}{52}$ **b.** (1) $\frac{48}{52}$ or $\frac{12}{13}$ (2) $\frac{26}{52}$ or $\frac{1}{2}$ (3) $\frac{40}{52}$ or $\frac{10}{13}$ (4) $\frac{50}{52}$ or $\frac{25}{26}$ **c.** (1) $\frac{2}{50}$ or $\frac{1}{25}$ (2) $\frac{26}{26}$ or even (3) $\frac{1}{51}$ (4) $\frac{12}{40}$ or $\frac{3}{10}$ **d.** (1) $\frac{46}{6}$ or $\frac{23}{3}$ (2) $\frac{51}{1}$ or 51 to 1 (3) $\frac{48}{4}$ or 12 to 1 (4) $\frac{39}{13}$ or 3 to 1

Review of Part 5, Page 539

1.

Family Income Spending

3.

Score	Tally	Frequency
20	1	1
19	~~1111~~	5
18	111	3
17	~~1111~~	5
16	~~1111~~ 111	8
15	1	1
14	11	2

Mean   17   Mode   16
Median   17   Range   6

5. a. $\frac{8}{20}$ or $\frac{2}{5}$ **b.** $\frac{15}{5}$ or 3 to 1

Competency Check Test, Page 540

1. c   **3.** c   **5.** c   **7.** a   **9.** c

PART 6 EVERYDAY APPLICATIONS—PROBLEM SOLVING

INCOME, TAKE-HOME PAY, INCOME TAX

EXERCISE 6-1 Computing Income

Practice Problems, Pages 543-547

1. Answers will vary **3. a.** $193.20 **b.** $150.50 **c.** $361.15 **d.** $212.55 **e.** $229.14 **f.** $375.00
g. $234.60 **h.** $152.63 **5.** A, $247.65 B, $293.40 C, $304.85 D, $181.13 E, $247.00 **7.** Answers will
vary **9. a.** $4,420 **b.** $22,100 **c.** $5,434 **d.** $13,390 **e.** $18,837 **11. a.** $8,970 **b.** $11,580
c. $14,226 **d.** $11,648 **13. a.** $675 **b.** $265 **c.** $1,110 **d.** $520.50 **e.** $875.00 **15. a.** $300
b. $250 **c.** $525 **d.** $172.50 **e.** $425.75 **17. a.** $26 **b.** $43.60 **c.** $34 **d.** $48 **19.** $33.32
21. Charlotte, $105.70; Marilyn, $126.10; Leon, $236.70; Ronald, $208.24; Elaine, $127.79
23. $374.40 **25.** $176.22 **27.** $147.31 **29.** $243.17 **31.** $18,041 **33. a.** $11,940 **b.** $11,177
c. $14,783.10 **d.** $34,907.29 **e.** $153,301.65 **35.** $300 **37.** $560; $6,720; $129.23 **39.** $37.50; $150

EXERCISE 6-2 Withholding Tax

Practice Problems, Page 550

1. a. $4.38 **b.** $3.24 **c.** $45.14 **d.** $24.13 **e.** $91.69 **f.** $1.07 **g.** $15.34 **h.** $62.69 **i.** $24.47
j. $59.74 **3. a.** $3.60 **b.** $14.10 **c.** $32.30 **d.** $6.50 **e.** $29.80

EXERCISE 6-3 Social Security

Practice Problems, Page 551

1. a. $4.02 **b.** $7.71 **c.** $5.96 **d.** $15.21 **e.** $27.14 **f.** $41.21 **g.** $39.13 **h.** $18.09 **i.** $26.67
j. $48.58 **k.** $4.86 **l.** $6.82 **m.** $25.29 **n.** $35.68 **o.** $14.63 **p.** $3.61 **q.** $29.87 **r.** $14.05
s. $21.14 **t.** $40.75

EXERCISE 6-4 State or Local Income Tax—Wage Tax

Practice Problems, Page 551-552

1. a. $1.70 **b.** $2.32 **c.** $4.40 **d.** $3.49 **e.** $7.39 **3. a.** $4.55 **b.** $15.85 **c.** $14.25 **d.** $6.14
e. $10.17 **5. a.** $139.50 **b.** $186.75 **c.** $247.50 **d.** $334.13 **e.** $285.64

EXERCISE 6-5 Take-Home Pay

Practice Problems, Page 553

1. $13.30; $12.19; $157.54 **3.** $129.75 **5. a.** $84.90 **b.** $172.74 **c.** $165.77 **d.** $89.62 **e.** $254.34
f. $83.55 **g.** $172.74 **h.** $164.48 **i.** $258.24

EXERCISE 6-6 Federal Income Tax

Practice Problems, Pages 554-555

1. a. $357 **b.** $799 **c.** $748 **d.** $374 **e.** $450 **f.** $646 **g.** $808 **h.** $433 **i.** $842 **j.** $287 **3.** 4;
$12.50; $650; $459; overpaid by $191 **5. a.** overpaid $189.60 **b.** overpaid $230.00 **c.** overpaid
$243.00 **d.** overpaid $129.40 **e.** overpaid $212.20 **f.** overpaid $120.80 **g.** overpaid $211.60
h. overpaid $170.20 **i.** overpaid $243.00 **j.** overpaid $146.00

SPENDING MONEY

EXERCISE 6-7 Buying Food

Practice Problems, Pages 557–559

1. a. $.30 **b.** $.65 **c.** $.35 **d.** $2.90 **e.** $1.70 **f.** $6.75 **g.** $2.98 **h.** $.99 **i.** $.43 **j.** $.44 **k.** $3.70 **l.** $4.14 **m.** $.86 **n.** $1.20 **o.** $1.13 **p.** $2.18 **3. a.** $1.78 **b.** $1.12 **c.** $1.44 **d.** $5.28 **e.** $9.10 **f.** $1.32 **g.** $3.33 **h.** $4.03 **i.** $.96 **j.** $.66 **k.** $8.48 **l.** $13.65 **m.** $5.88 **n.** $1.34 **5. a.** $6.01 **b.** $11.59 **c.** $14.00 **d.** $23.26 **e.** $31.13 **f.** $16.24 **g.** $45.45

EXERCISE 6-8 Unit Pricing—Determining the Better Buy

Practice Problems, Pages 560–563

1. a. $.37 **b.** $.95 **c.** $.04 **d.** $.08 **e.** $.052 **f.** $.239 **g.** $.065 **h.** $.3075 **i.** $.291 **j.** $.4357 **3. a. (1)** Thirty-four and six tenths cents **(2)** Five and four tenths cents **(3)** Seventy-one and three tenths cents **(4)** Eight and two tenths cents **(5)** Sixty-nine and twenty-five hundredths cents **b. (1)** Two dollars and thirty-seven and four tenths cents **(2)** One dollar and thirty-two and nine tenths cents **(3)** Five dollars and six and seven tenths cents **(4)** Three dollars and ten and eight tenths cents **(5)** One dollar and forty-three and nine tenths cents **5. a.** 66¢ **b.** 24¢ **c.** 85.3¢ **d.** $1.60 **e.** 57.7¢ **7. a.** 90¢ **b.** $2.65 **c.** 83.3¢ **d.** $4.725 **e.** 21.3¢ **9. a.** $1.50 **b.** $1.67 **c.** $1.38; difference is $.29 **11. a.** 29¢ **b.** 9.5¢ **c.** $18.50 **d.** $41.25 **e.** $13\frac{1}{6}$¢ **13. a.** 10¢ **b.** 9¢ **c.** $17.01 **d.** 5¢ **e.** 20¢ **15. a.** 1 half-gallon at $1.01 **b.** 1 half-gallon at $1.99 **c.** 1 qt. at 53¢ **d.** 24-oz. bottle at $1.13 **e.** 24-oz. jar at 69¢ **f.** 1 pt. jar at $1.08 **g.** 5-lb 4 oz. pkg. at $2.57 **h.** 20-oz. jar at 89¢ **i.** 46-oz. can at $1.98 **j.** 1-lb. 10 oz. can at 59¢ **k.** 6 16-oz. bottles **l.** 8-oz. pkg. at 77¢ **m.** 1-lb. jar at $1.09 **n.** 13.4-oz. jar at $4.79 **o.** 7-oz. tube at $1.25 **17. a.** 23.8¢ **b.** 54.2¢ **c.** $2.438 **d.** 75.7¢ **19. a.** 2.26-kg bag at 75¢ **b.** 1.89-L bottle at 67¢ **c.** 454-g box at $1.05 **d.** 946-mL jar at 99¢

EXERCISE 6-9 Buying at a Sale

Practice Problems, Page 565

1. a. Tables, $55.30; lamps, $38.15; sofas, $297.50; bedroom sets, $629.30; bookcases, $50.93; mirrors, $26.59 **b.** Dresses, $26.00; coats, $60.00; shoes, $16.50; suits, $65.00; sweaters, $12.66; skirts, $14.30; ties, $2.34; shirts, $10.00 **c.** Refrigerators, $476; toaster-ovens, $31.96; hair dryers, $29.40; color TV, $356; blenders, $23.96; can openers, $11.75 **3.** $5,685.50 **5.** 20% **7.** $199.80 **9.** $37.41

EXERCISE 6-10 Making Change

Practice Problems, Page 566

1. a. (1) $.21 **(2)** $.52 **(3)** $.84 **(4)** $.91 **(5)** $.17 **b. (1)** $3.66 **(2)** $2.93 **(3)** $.42 **(4)** $4.28 **(5)** $1.04 **c. (1)** $3.59 **(2)** $6.85 **(3)** $.31 **(4)** $2.13 **(5)** $9.08 **d. (1)** $11.92 **(2)** $2.06 **(3)** $14.84 **(4)** $8.71 **(5)** $4.53

EXERCISE 6-11 Sales Tax

Practice Problems, Page 567

1. a. (1) $.10 **(2)** $.21 **(3)** $1.69 **(4)** $2.70 **b. (1)** $.96 **(2)** $.03 **(3)** $.54 **(4)** $6.46 **c. (1)** $.15 **(2)** $.48 **(3)** $2.87 **(4)** $10.63 **d. (1)** $7.28 **(2)** $.56 **(3)** $58.54 **(4)** $196.12 **e. (1)** $7.00 **(2)** $.21 **(3)** $1.11 **(4)** $71.34 **f. (1)** $.03 **(2)** $.02 **(3)** $.69 **(4)** $5.45 **g. (1)** $.42 **(2)** $.34 **(3)** $2.75 **(4)** $28.91 **3. a.** $.02 **b.** $.04 **c.** $.03 **d.** $.03 **e.** $.06 **f.** $.04 **g.** $.11 **h.** $.05 **i.** $.12 **j.** $.03 **k.** $.11 **l.** $.12

EXERCISE 6-12   Utilities and Telephone Services

Practice Problems, Pages 568–570

1. a. 8352; 8417 **b.** 6,500 cu. ft. **c.** $28.57 **3. a.** 6529; 6774 **b.** 245 kWh **c.** $21.67 **5. a.** 8.6¢
b. $34.4¢ **c.** .86¢ **d.** 12.9¢ **e.** 10.32¢ **f.** .344¢ **7.** 1,040 gallons; $1,312.88

Refresh Your Skills, Page 570

1. 70,293 **3.** 640,740 **5.** $8\frac{9}{20}$ **7.** $8\frac{4}{5}$ **9.** 7.06 **11.** 50,000 **13.** $513 **15.** $1,800 **17.** 2.53
19. 6,930 g

EXERCISE 6-13   Owning an Automobile

Practice Problems, Page 571

1. a. $1,631.25 **b.** 11,707 mi.; 11,919 mi.; 13,958 mi. **c.** (1) $3,147.83 (2) $262.32 (3) $60.54 (4) 33¢
3. 24 min. **5.** $287.45 **7.** $59\frac{1}{3}$ ft.; $111\frac{1}{2}$ ft.; $178\frac{2}{3}$ ft.; $260\frac{5}{6}$ ft.

EXERCISE 6-14   Buying Without Full Cash Payment

Practice Problems, Pages 573–576

1. $22 **3. a.** $45.00 **b.** $12.00 **c.** $31.80 **d.** $20.00 **e.** $9.30 **5.** $31.80 **7. a.** 88¢ **b.** 64¢
c. 96¢ **9. a.** 24% **b.** 9% **c.** 8% **d.** 10.5% **e.** 10% **11. a.** 14.25% **b.** 15.5% **c.** 18.0%
d. 17.0% **e.** 15.25%

EXERCISE 6-15   Fire Insurance

Practice Problems, Page 577

1. $50.40 **3.** $108.00 **5.** $30.60

EXERCISE 6-16   Life Insurance

Practice Problems, Pages 578–580

1. a. 78.3 yr. **b.** 71.8 yr. **c.** 7.3 yr. **3. a.** $0 **b.** $.98 **c.** $.97 **d.** $1.05 **e.** $.58 **5.** $163.40;
$10,000 **7.** $24,397.75

EXERCISE 6-17   Property Tax

Practice Problems, Pages 581–582

1. $.007; $.019; $.050; $.132; $.084; $.175; $.060; $.092 **3.** 40 mills; 80 mills; 5 mills; 26 mills; 61
mills; 105 mills **5.** 50 mills per $1; 33 mills per $1; 45 mills per $1; 190 mills per $1 **7.** $45 per
$1,000; $29 per $1,000; $42 per $1,000; $250 per $1,000 **9.** 2%; $2 per $100; $.02 per $1; 20 mills
per $1 **11.** $1,225.50 **13.** $779.10 **15.** increase in assessment; $9.50 greater

EXERCISE 6-18   Buying and Renting a House

Practice Problems, Page 583

1. a. $385.42 **b.** $187.50 **c.** $476.67 **3.** $19,000 **5.** The cost incurred during the first year when
buying the house is $303.25 more than the cost incurred when renting during the same year. Cost
of buying is $4,923.25; cost of renting is $4,620.

MANAGING MONEY

EXERCISE 6-20   Simple Interest

Practice Problems, Pages 587–588

1. a. $30 **b.** $60 **c.** $55 **d.** $287 **3. a.** $315 **b.** $22.50 **c.** $477 **d.** $493.02 **5. a.** $11.25
b. $64.75 **c.** $1,163.75 **d.** $3,720.94 **7. a.** $45 **b.** $215 **c.** $34.56 **9. a.** $4 **b.** $6.75 **c.** $60
d. $106.67 **e.** $.50 **f.** $25.00 **g.** $12 **h.** $21 **i.** $8.87 **j.** $10.80 **k.** $33.11 **l.** $34.38
11. a. $45 **b.** $193.75 **c.** $4.20 **d.** $10 **13.** $1,903.50 **15.** $64.75 **17.** 3% **19.** Neither, the
rates are the same.

EXERCISE 6-21   Borrowing Money

I. Discount Loan and Add-on-Interest Loan
Practice Problems, Pages 590–591

1. a. 12%; 3%; 18%; 7.2%; 9% **b.** 52%; 26%; 91%; 10.4%; 13% **c.** 365%; 36.5%; $273\frac{3}{4}$%; $547\frac{1}{2}$%;
$182\frac{1}{2}$% **3. a.** $564; $50 **b.** $1,350; $62.50 **c.** $7,650; $500 **d.** $2,362.50; $75 **e.** $1,656; $200
f. $3,450; $133.33 **g.** $6,300; $150 **h.** $8,550; $833.33 **i.** $13,781.25; $350 **j.** $2,952; $53.33
5. a. $75 **b.** $283.33 **c.** $20.70 **d.** $5.96 **e.** $478.08

II. Financing a Car
Practice Problems, Page 592

1. a. $45.00 **b.** $63.88 **c.** $68.75 **d.** $112.50 **3. a.** $349.84 **b.** $224.76 **c.** $250.00 **d.** $449.88
5. $87.30

III. Amortizing a Loan
Practice Problems, Pages 593–594

1. a. $322.10 **b.** $293.24 **c.** $393.90 **d.** $307.57 **e.** $347.13 **f.** $298.75 **3.** $192.23; $2,306.76;
$69,202.80; $44,202.80 **5.** $124,567.20; $158,352.00; 30 yr.; $33,784.80

EXERCISE 6-22   Compound Interest

Practice Problems, Pages 596–597

1. a. $219.50; $119.50 **b.** $1,708.47; $808.47 **c.** $2,185.25; $785.25 **d.** $20,789.28; $10,789.28
e. $14,567.47; $6,067.47 **f.** $18,500.57; $15,750.57 **3. a.** $443.37; $93.37 **b.** $2,345.16; $345.16
c. $1,623.65; $123.65 **d.** $26,532.80; $16,532.80 **e.** $7,989.85; $1,689.85 **f.** $93,648.50; $43,648.50
5. a. (1) $788.77 (2) $1,408.53 (3) $5,408.74 (4) $11,268.20 **b.** (1) $31.71 (2) $124.28 (3) $456.55
(4) $634.10 **c.** (1) $81.28 (2) $412.68 (3) $1,819.03 **d.** (1) $5.71 (2) $53.26 (3) $63.41

EXERCISE 6-23   Stocks, Bonds, Mutual Funds, Savings Certificates

Practice Problems, Pages 600–603

1. a. $4,000 **b.** $1,950 **c.** $4,175 **d.** $980 **e.** $2,081.25 **f.** $1,268.75 **g.** $770 **h.** $8,225
i. $1,887.50 **j.** $14,925 **3. a.** $4,097.04 **b.** $2,402.15 **c.** $2,449.50 **d.** $2,180.08 **e.** $4,158.98
f. $5,496.83 **g.** $1,494.11 **h.** $989.51 **i.** $2,764.14 **j.** $11,380.89 **5.** INA Cp, $.65; Idaho P, $.57;
Ill Powr, $.57; Inger R, $.79; Inlnd Stl, $.70; IBM, $.86; Int Harv, $.575; Int Min, $.65; Int
Paper, $.55; Int T&T, $.55; Iowa PS, $.60; Itek Cp, $— **7. a.** $696 **b.** $782.43 **c.** $232.56
d. $368.69 **e.** $190.34 **9. a.** (1) $1,506 (2) $1,226 (3) $524 (4) $972.75 **b.** (1) $1,508 (2) $610.80
(3) $1,784 (4) $1,289.45 **11.** (1) **a.** $890 **b.** $40 (2) **a.** $10,000 **b.** $462.50 (3) **a.** $44,100
b. $2,325 (4) **a.** $4,987.50 **b.** $230 (5) **a.** $18,850 **b.** $862.50 (6) **a.** $50,250 **b.** $1,312.50
13. a. $1,525 **b.** $4,222.50 **15.** $1,331.25

APPLICATIONS IN BUSINESS AND TRAVEL

EXERCISE 6-24   Discount

Practice Problems, Pages 605–606

1. $8.40; $47.60 **3.** $20.79; $109.16 **5.** $2.92; $94.48 **7.** $286.71 **9.** 24% **11.** 15% **13.** $320
15. $120.00

EXERCISE 6-25   Commission

Practice Problems, Pages 608–609

1. $3,045; $40,455 **3.** 12% **5.** 3%. **7.** $2,300; $520; $136.60; $390; $65.80 **9.** $225.28
11. $4,327.61 **13.** $8,567 **15.** $248.91

EXERCISE 6-26   Profit and Loss

Practice Problems, Pages 612–613

1. $796.50 **3.** $28.00; $29.85; $1.82; $55.40; $128.35 **5.** $90.00 **7.** 60%; 32%; 50%; 40%; 25%
9. $12\frac{1}{2}$% **11.** 21% **13. a.** $121.10 **b.** $337.50 **c.** $200.00 **d.** $36.00 **15.** $105 **17.** $93.31;
$16.51 **19.** $2.20

EXERCISE 6-27   Reading a Road Map

Practice Problems, Page 614

1. a. (52) to (74) ; Shelbyville, Rushville **b.** (65) all the way; Seymour, Franklin

c. (71) all the way; Carrollton **d.** Indianapolis to Cincinnati: (74) all the way;
Louisville to Indianapolis: (31) all the way; Cincinnati to Louisville: (42) all the way

EXERCISE 6-28   Reading a Timetable

Practice Problems, Pages 615–617

1. 7:00 P.M.; 10:00 P.M.; 12:17 A.M.; does not arrive in Wilmington **3.** 518 km; 6 hr. 59 min.;
74.2 km/h **5.** The Senator; 1:45 P.M.; 11:02 P.M.; 456 miles **7.** 10:45 A.M., PST; 4:00 P.M.,
PST; 9:50 P.M., PST; 7:45 A.M., MST **9.** 18 hr. 20 min. **11.** 8:20 A.M., CST; 10:15 A.M.,
CST; 1 hr. 55 min.; no **13.** 2 hr. 20 min. **15.** 3 hr. 40 min.; 55 min.

Competency Check Test, Pages 618–620

1. d **3.** a **5.** b **7.** c **9.** b **11.** b **13.** a **15.** d **17.** c **19.** c **21.** a **23.** c **25.** a